Lecture Notes in Computer Science 7116

Commenced Publication in 1973
Founding and Former Series Editors:
Gerhard Goos, Juris Hartmanis, and Jan van Leeuwen

Ivan Lirkov   Svetozar Margenov
Jerzy Waśniewski (Eds.)

Large-Scale Scientific Computing

8th International Conference, LSSC 2011
Sozopol, Bulgaria, June 6-10, 2011
Revised Selected Papers

 Springer

Volume Editors

Ivan Lirkov
Svetozar Margenov
Bulgarian Academy of Sciences
Institute of Information
and Communication Technologies
Acad. G. Bonchev, Block 25A
1113 Sofia, Bulgaria
E-mail:{ivan; margenov}@parallel.bas.bg

Jerzy Waśniewski
Technical University of Denmark
Department of Informatics
and Mathematical Modelling
Richard Petersens Plads
Building 321
2800 Kongens Lyngby, Denmark
E-mail: jw@imm.dtu.dk

ISSN 0302-9743   e-ISSN 1611-3349
ISBN 978-3-642-29842-4   e-ISBN 978-3-642-29843-1
DOI 10.1007/978-3-642-29843-1
Springer Heidelberg Dordrecht London New York

Library of Congress Control Number: 2012937730

CR Subject Classification (1998): G.1, D.1, F.1-2, C.2, I.6, J.2, J.6

LNCS Sublibrary: SL 1 – Theoretical Computer Science and General Issues

Typesetting: Camera-ready by author, data conversion by Scientific Publishing Services, Chennai, India

Printed on acid-free paper

Springer is part of Springer Science+Business Media (www.springer.com)

Preface

The 8th International Conference on Large-Scale Scientific Computations (LSSC 2011) was held in Sozopol, Bulgaria, June 6–10, 2011. The conference was organized and sponsored by the Institute of Information and Communication Technologies at the Bulgarian Academy of Sciences.

The plenary invited speakers and lectures were:

- E. Casas, Analysis of an Elliptic Control Problem with Non-differentiable Cost Functional
- O. Iliev, On Modeling and Simulation of Multiscale Filtration Processes
- U. Langer, A Non-standard Finite Element Method Based on Boundary Integral Operators
- P. Vassilevski, Construction and Properties of Coarse Finite Element Spaces by Algebraic Multigrid
- W. Wagner, Stochastic Particle Methods
- B. Wohlmuth, Do We Still Need Analysis in PDE-Based Simulation?

The success of the conference and the present volume in particular are outcome of the joint efforts of many partners from various institutions and organizations. First, thanks to all the members of the Scientific Committee for their valuable contribution forming the scientific face of the conference, as well as for their help in reviewing contributed papers. We especially thank the organizers of the special sessions. We are also grateful to the staff involved in the local organization.

Traditionally, the purpose of the conference is to bring together scientists working with large-scale computational models of environmental and industrial problems, and specialists in the field of numerical methods and algorithms for modern high-performance computers. The invited lectures reviewed some of the advanced achievements in the field of numerical methods and their efficient applications. The conference talks were presented by the university researchers and practical industry engineers including applied mathematicians, numerical analysts, and computer experts. The general theme for LSSC 2011 was "Large-Scale Scientific Computing" with a particular focus on the organized special sessions.

Special Sessions and Organizers:

- Robust Multigrid, Multilevel and Multiscale, Deterministic and Stochastic Methods for Modelling Highly Heterogeneous Media — R. Lazarov, P. Vassilevski, L. Zikatanov
- Advanced Methods for Transport — P. Bochev, D. Ridzal, M. Shashkov
- Control and Uncertain Systems — M. Krastanov, V. Veliov
- Applications of Metaheuristics to Large-Scale Problems — S. Fidanova, G. Luque

- Environmental Modeling — A. Ebel, K. Georgiev, Z. Zlatev
- Large-Scale Computing on Many-Core Architectures — G. Haase
- Multiscale Industrial, Environmental and Biomedical Problems — O. Iliev, P. Popov
- Efficient Algorithms of Computational Geometry — L. Dechevski, A. Lakså
- High-Performance Monte Carlo Simulations — A. Karaivanova, E. Atanassov, T. Gurov
- Voxel-Based Computations — P. Arbenz, R. Blaheta, M. Neytcheva

More than 150 participants from all over the world attended the conference representing some of the strongest research groups in the field of advanced large-scale scientific computing. This volume contains 77 papers submitted by authors from more than 20 countries.

The 9th International Conference LSSC 2013 will be held in June 2013.

January 2012

Ivan Lirkov
Svetozar Margenov
Jerzy Waśniewski

Table of Contents

Part IV: Control and Uncertain Systems

Part V: Applications of Metaheuristics to Large-Scale Problems

Part VI: Environmental Modeling

Part VII: Large-Scale Computing on Many-Core Architectures

Part XI: Voxel-Based Computations

Part XII: Contributed Papers

Part I

Plenary and Invited Papers

Smoothed Aggregation Spectral Element Agglomeration AMG: SA-ρAMGe*,**

Marian Brezina[1] and Panayot S. Vassilevski[2]

[1] Department of Applied Mathematics, Campus Box 526,
University of Colorado at Boulder, Boulder, CO 80309-0526, U.S.A.
marian.brezina@gmail.com
[2] Center for Applied Scientific Computing,
Lawrence Livermore National Laboratory,
P.O. Box 808, L-560, Livermore, CA 94550, U.S.A.
panayot@llnl.gov

Abstract. A two–level smoothed aggregation (or SA) scheme with tentative coarse space constructed by spectral element agglomeration method is shown to provide weak–approximation property in a weighted L_2-norm. The resulting method utilizing efficient (e.g., polynomial) smoothers is shown to have convergence factor independent of both the coarse and fine–grid mesh–sizes, as well as, to be independent of the contrast (i.e., possible large jumps in the PDE coefficient) for second order elliptic problems discretized on general unstructured meshes. The method allows for multilevel extensions. Presented numerical experiments exhibit behavior in agreement with the developed theory.

1 Introduction

This paper deals with the construction of two-grid iterative methods (or preconditioners) for solving elliptic problems with high contrast coefficient resolved only by a fine-grid discretization. The aim is to prove two–grid convergence bounds independent both of the fine and coarse mesh size as well as independent of the contrast. There are recent two-level results, [GE10], see also [SVZ], that deal with the construction of coarse spaces when used in combination with overlapping Schwarz method achieving this goal. The construction of the coarse spaces exploit local (element-based) procedures to construct local (overlapping) coarse spaces. These local spaces are patched using partition of unity. Similar construction goes back to the element–based algebraic multigrid (or AMGe) in [ρAMGi]. The latter method was modified in [ρAMGii], see also [LV08], to avoid the use of partition of unity and to allow for more straightforward multilevel extension

* This work was performed under the auspices of the U.S. Department of Energy by Lawrence Livermore National Laboratory under Contract DE-AC52-07NA27344.
** This work of the first author was sponsored by the Department of Energy under grant numbers DE-FG02-03ER25574 and DE-FC02-06ER25784, Lawrence Livermore National Laboratory under contract numbers B568677, and the National Science Foundation under grant numbers DMS-0621199, DMS-0749317, and DMS-0811275.

by applying recursion. So far this is the only known genuine multilevel AMGe method that does not lead to potentially highly ill-conditioned coarse bases (an issue arising from the partition of unity formed over overlapping domains used in a multilevel setting).

The present paper can be viewed as an extension of an early report [BHMV] from the point of view of handling problems with high contrast. That report contains also element-based construction of local coarse-spaces which are non–overlapping (or in terms of finite element functions they have minimal $\mathcal{O}(h)$ overlap). To improve the energy stability of the coarse basis, the polynomial smoothing idea from the multigrid smoothed aggregation (or SA) method is applied. In the present paper, we review this approach and also propose an alternative construction of the local (non–overlapping) spaces. We present a different analysis than [BHMV] following [BVV] to address both the high contrast coefficient and the aggressive coarsening. The two–level algorithm allows for multilevel extensions along the lines of [LV08] which provides an alternative to the formal multilevel extensions of the two–level methods exploiting partition of unity in [GE10] and [SVZ].

The remainder of the present paper is structured as follows. In Section 2 we provide two general conditions for the smoother and properties of the associated coarse space and prove that they ensure bounded two–level convergence factor. Section 3 contains the main results of the present paper. There, we review and extend the main results from [BHMV] paying attention to the high contrast coefficient and aggressive coarsening. Finally, Section 4 contains some numerical illustration of two- and multilevel results for problems with high–contrast.

2   A Two-Level Convergence Result

Consider the product iteration method involving "pre-smoothing" based on solving with M, coarse–grid solution involving restriction based on P^T, coarse-grid correction based on solving with $A_c = P^T A P$ and interpolation based on P, and finally using "post–smoothing" based on solving with M^T, represented by the following error-propagation matrix

$$E_{TL} \equiv (I - M^{-T}A)(I - PA_c^{-1}P^T A)(I - M^{-1}A). \tag{1}$$

Solving the system of linear equations, $A\mathbf{x} = \mathbf{b}$, where $\mathbf{b}$ is a given right-hand side and $\mathbf{x}_0$ an initial iterate, one iteration of the two-level process is described by the following algorithm:

Algorithm 1 (Two–level iteration algorithm)

 (i) *"Pre–smooth:" compute* $\mathbf{y}$ *from* $M(\mathbf{y} - \mathbf{x}_0) = \mathbf{b} - A\mathbf{x}_0$.
 (ii) *Restrict defect* $\mathbf{r}_c = P^T(\mathbf{b} - A\mathbf{y})$.
 (iii) *Solve for a coarse–grid correction* $\mathbf{x}_c$ *the coarse-grid problem* $(A_c = P^T A P)$,

$$A_c \mathbf{x}_c = \mathbf{r}_c.$$

(iii) Interpolate and update fine–grid iterate, $\mathbf{z} = \mathbf{y} + P\mathbf{x}_c$.
(iv) "Post-smooth", i.e., solve for $\mathbf{x}_{TL}$, $M^T(\mathbf{x}_{TL} - \mathbf{z}) = \mathbf{b} - A\mathbf{z}$.

The mapping $\mathbf{b} \mapsto \mathbf{x}_{TL} = B_{TL}^{-1}\mathbf{b}$ for $\mathbf{x}_0 = 0$ defines the inverse of the two–level operator B_{TL}. Alternatively, formula (1) defines an approximate inverse (or preconditioner) to A, B_{TL}^{-1}, from the identity $I - B_{TL}^{-1}A = E_{TL}$. If M is A–convergent, that is $\|I - A^{\frac{1}{2}}M^{-1}A^{\frac{1}{2}}\| < 1$ (or equivalently, $M + M^T - A$ is s.p.d.) we then have that B_{TL}^{-1} is s.p.d. and B_{TL} can be characterized by the following two-level version of the XZ-identity, cf. [Va08]:

$$\mathbf{v}^T B_{TL} \mathbf{v} = \min_{\mathbf{v} = \mathbf{v}_f + P\mathbf{v}_c} \left[\mathbf{v}_c^T A_c \mathbf{v}_c + \left(\mathbf{v}_f - M^{-T}AP\mathbf{v}_c \right)^T \overline{M} \left(\mathbf{v}_f - M^{-T}AP\mathbf{v}_c \right) \right].$$

Here $\overline{M} = M \left(M + M^T - A \right)^{-1} M^T$ is the symmetrized smoother. For any A–convergent smoother M and $A_c = P^T AP$, we have

$$\mathbf{v}^T A\mathbf{v} \leq \mathbf{v}^T B_{TL} \mathbf{v}.$$

In order to establish convergence properties of the two-level process, our goal is to estimate the constant, K_{TL}, in the upper bound,

$$\mathbf{v}^T B_{TL} \mathbf{v} \leq K_{TL} \mathbf{v}^T A\mathbf{v},$$

which is equivalent to $0 \leq \mathbf{v}^T A E_{TL} \mathbf{v} \leq \varrho_{TL} \|\mathbf{v}\|_A^2$, where $K_{TL} = \frac{1}{1 - \varrho_{TL}}$. For this we use the above two–level (XZ-) identity by selecting a proper coarse–grid approximation $P\mathbf{v}_c$ to any given $\mathbf{v}$ which gives

$$\mathbf{v}^T B_{TL} \mathbf{v} \leq \mathbf{v}_c^T A_c \mathbf{v}_c + 2\|\mathbf{v} - P\mathbf{v}_c\|_{\overline{M}}^2 + 2\mathbf{v}_c^T P^T A \left(M + M^T - A \right)^{-1} AP\mathbf{v}_c. \quad (2)$$

Consider the particular case when A is a s.p.d. sparse matrix coming from a finite element discretization of an elliptic PDE posed on a domain $\Omega \subset \mathbf{R}^d$, (plane polygon for $d = 2$, or polytope in the case of $d = 3$). We use quasi-uniform mesh $\mathcal{T}_h$ with mesh size h and respective finite element space V_h. Also, let $\mathcal{N}_h$ be the set of Lagrangian degrees of freedom, that in the case of piecewise linear finite element space V_h coincide with the values at the vertices $\{\mathbf{x}_i\}$ of the elements in $\mathcal{T}_h$. The corresponding elliptic bilinear form reads

$$a(u, \varphi) = \int_\Omega k(\mathbf{x})\nabla u \cdot \nabla \varphi \, d\mathbf{x}, \quad (3)$$

where $k = k(\mathbf{x})$ is a given positive coefficient. The coefficient may admit very large jumps in certain parts of Ω which are assumed resolved by the fine-grid $\mathcal{T}_h$, i.e., $k(\mathbf{x})$ varies smoothly within each fine–grid element $\tau \in \mathcal{T}_h$. Assume also that we have created a coarse finite element space $V_H \subset V_h$ where H reflects the characteristic diameter of the support of the coarse-basis functions. We also assume that the coarse space satisfies the following approximation property: For any $v \in V_h$ there is a $v_H \in V_H$ such that uniformly in h and H, we have

$$H^{-2}\|v - v_H\|_0^2 + a(v - v_H, v - v_H) \leq Ca(v, v). \quad (4)$$

Here, $\|.\|_0$ is the k–weighted L_2–norm, i.e., $\|\varphi\|_0^2 = \int_\Omega k(\mathbf{x}) \, \varphi^2(\mathbf{x}) \, d\mathbf{x}$. Without loss of generality, we may assume that k is piecewise constant with respect to the elements of $\mathcal{T}_h$. In that case, we have the following norm equivalence for any finite element function $v \in V_h$ and its nodal coefficient vector $\mathbf{v} = (v(\mathbf{x}_i))_{\mathbf{x}_i \in \mathcal{N}_h}$

$$\|v\|_0^2 = \int_\Omega k(\mathbf{x}) \, v^2(\mathbf{x}) \, d\mathbf{x} \simeq \sum_{\mathbf{x}_i \in \mathcal{N}_h} d_i \, v^2(\mathbf{x}_i) = \|\mathbf{v}\|_{D_G}^2. \tag{5}$$

Here, $D_G = \operatorname{diag}(d_i)$ is the diagonal of the k–weighted mass matrix G corresponding to the finite element space V_h and the weighted L_2 bilinear form $(k \cdot, \cdot)$. We have the following simple inverse inequality

$$a(v, \, v) \leq C_I h^{-2} \, \|v\|_0^2,$$

with a constant C_I independent of the coefficient $k = k(\mathbf{x})$ (which is assumed piecewise constant with respect to the elements $\tau \in \mathcal{T}_h$). Now, consider D_A– the diagonal of A and D_G–the diagonal of the weighted mass matrix G. The following norm equivalence

$$\mathbf{v}^T D_A \mathbf{v} \simeq h^{-2} \, \mathbf{v}^T D_G \mathbf{v}, \tag{6}$$

is easily seen to be uniform with respect to both $h \mapsto 0$ and the contrast (jumps of k). This can be proved directly based on an element-by-element local comparison of the two quadratic forms. Here, we use the assumption that $\mathcal{T}_h$ is quasi-uniform.

We consider smoothers M that possess the following "smoothing" property:

$$\mathbf{v}^T \overline{M} \mathbf{v} \leq \beta \left[\mathbf{v}^T A \mathbf{v} + \frac{b}{(H/h)^2} \|\mathbf{v}\|_\Lambda^2 \right]. \tag{7}$$

Here Λ is a given s.p.d. matrix. b satisfies $\|\Lambda^{-\frac{1}{2}} A \Lambda^{-\frac{1}{2}}\| \leq b$. In our case Λ will be spectrally equivalent to D_A. Consider the polynomial smoother M

$$M^{-1} = \left[I - p_\nu(b^{-1}\Lambda^{-1}A) \right] A^{-1},$$

where $p_\nu(t)$, $p_\nu(0) = 1$, is a properly chosen polynomial of degree ν, (cf., [BVV]), or Section 4 for details). The following estimate holds ([BVV])

$$\mathbf{x}^T \overline{M} \mathbf{x} \leq \beta \left(\mathbf{x}^T A \mathbf{x} + \frac{b}{(2\nu+1)^2} \, \|\mathbf{x}\|_\Lambda^2 \right). \tag{8}$$

Thus, if we choose the polynomial degree, ν, so that $2\nu + 1 \simeq \frac{H}{h}$, we obtain estimate (7).

Note that when $\Lambda = D_A$ (the diagonal of A) (or spectrally equivalent to D_A) this would make $b \simeq \|D_A^{-\frac{1}{2}} A D_A^{-\frac{1}{2}}\| = \mathcal{O}(1)$ (see Remark 1 below).

For the same polynomial smoother M, it has been shown in [BVV] that the following coercivity bound holds:

$$\mathbf{v}^T (M + M^T - A)\mathbf{v} \geq \alpha \, \mathbf{v}^T A \mathbf{v}. \tag{9}$$

The constant α in (9) is independent of both h and H.

Now, we are ready to prove the following two-level convergence result.

Theorem 1. *Assume that the A–convergent smoother M and its respective symmetrized version $\overline{M}$ satisfy "smoothing property" (7) where D_A is the diagonal of A, and in addition let the coercivity estimate (9) hold. Under the assumption of norm–equivalence of D_A and the diagonal of the weighted mass matrix G, $\|\mathbf{v}\|_{D_A} \simeq h^{-1} \|\mathbf{v}\|_{D_G}$, assuming that the coarse finite element space V_H ensures the approximation property (4), which rewritten in a matrix–vector form (letting $P\mathbf{v}_c$ be the coefficient vector of $v_H \in V_H$) reads,*

$$H^{-2}\|\mathbf{v} - P\mathbf{v}_c\|_G^2 + \|\mathbf{v} - P\mathbf{v}_c\|_A^2 \le C_a \|\mathbf{v}\|_A^2, \tag{10}$$

the following convergence estimate holds:

$$\frac{1}{1 - \varrho_{TL}} = K_{TL} \le Const \simeq 2\beta \ \max\{1,\, b\}C_a + (1 + \frac{2}{\alpha})\, (1 + \sqrt{C_a})^2.$$

That is, K_{TL} is bounded uniformly with respect to both h and H.

Proof. Using the approximation property (4) together with the assumed norm–equivalences shows

$$h^2\|\mathbf{v} - P\mathbf{v}_c\|_{D_A}^2 \simeq \|\mathbf{v} - P\mathbf{v}_c\|_{D_G}^2 \simeq \|\mathbf{v} - P\mathbf{v}_c\|_G^2 \le C_a H^2 \ \|\mathbf{v}\|_A^2.$$

By the triangle inequality, we also have that the coarse-grid interpolant $P\mathbf{v}_c$ is bounded in energy, i.e.,

$$\|\mathbf{v}_c\|_{A_c}^2 = \|P\mathbf{v}_c\|_A^2 \le (1 + \sqrt{C_a})^2 \ \|\mathbf{v}\|_A^2.$$

Now, from (2), using the smoothing property of M, (7), and the assumed coercivity (9), we obtain

$$\mathbf{v}^T B_{TL}\mathbf{v} \le \mathbf{v}_c^T A_c\mathbf{v}_c + 2\|\mathbf{v} - P\mathbf{v}_c\|_{\overline{M}}^2 + 2\mathbf{v}_c^T P^T A \left(M + M^T - A\right)^{-1} A P\mathbf{v}_c$$

$$\le \mathbf{v}_c^T A_c\mathbf{v}_c + \frac{2}{\alpha}\|P\mathbf{v}_c\|_A^2 + 2\beta \left(\|\mathbf{v} - P\mathbf{v}_c\|_A^2 + \frac{b}{(H/h)^2} \ \|\mathbf{v} - P\mathbf{v}_c\|_{D_A}^2\right)$$

$$\simeq (1 + \frac{2}{\alpha}) \ \|\mathbf{v}_c\|_{A_c}^2 + 2\beta \ (\|\mathbf{v} - P\mathbf{v}_c\|_A^2 + b \ H^{-2} \ \|\mathbf{v} - P\mathbf{v}_c\|_{D_G}^2)$$

$$\simeq (1 + \frac{2}{\alpha}) \ \|\mathbf{v}_c\|_{A_c}^2 + 2\beta \ (\|\mathbf{v} - P\mathbf{v}_c\|_A^2 + b \ H^{-2} \ \|\mathbf{v} - P\mathbf{v}_c\|_G^2)$$

$$\le (1 + \frac{2}{\alpha}) \ \|\mathbf{v}_c\|_{A_c}^2 + 2\beta \ \max\{1,\, b\} \ C_a \|\mathbf{v}\|_A^2$$

$$\le \left[2\beta \ \max\{1,\, b\}C_a + (1 + \frac{2}{\alpha})\, (1 + \sqrt{C_a})^2\right] \ \mathbf{v}^T A\mathbf{v}.$$

Remark 1. We remark that for any s.p.d. sparse matrix A, we have the simple estimate

$$\mathbf{v}^T A\mathbf{v} \le \kappa \ \mathbf{v}^T D_A\mathbf{v},$$

where $\kappa \ge 1$ stands for the maximal number of non-zero entries per row of A. This shows that $\|D_A^{-\frac{1}{2}} A D_A^{-\frac{1}{2}}\| \le \kappa$ and we can let $b = \kappa$. Therefore, under our assumption of quasi-uniform mesh, we may assume that in the estimates derived in Theorem 1, b is a constant that is both mesh-independent and independent of the coefficient $k(\mathbf{x})$.

Remark 2. Theorem 1 shows that K_{TL} will be bounded independently of possible large jumps in the PDE coefficient $k = k(\mathbf{x})$ as long as the approximation property (10) holds with a constant C_a independent of those jumps.

3 A Spectral, Element Agglomeration Based, Construction of Coarse Spaces with Weak Approximation Property

In the present section, we review an approach of constructing coarse spaces with weak approximation property (4), or its matrix–vector form (10), that requires knowledge of local element matrices. For details, we refer to the early papers [BHMV], [ρAMGi], [ρAMGii], and to some more recent ones that explicitly address the issue of high-contrast coefficients, [GE10], [SVZ]. Most of these methods are inherently two–level with the exception of [ρAMGii]. The methods, for example, in the more recent papers [GE10], [SVZ] can in principle be applied recursively, however issues related to the use of partition of unity have to be resolved since the resulting coarse matrices may get fairly ill-conditioned due to the (nearly) linear dependence of the respectively constructed coarse bases.

We consider the model second order elliptic bilinear form (3) and a quasi-uniform fine–grid triangulation $\mathcal{T}_h$ that is fine-enough to resolve the possible large jumps of the coefficient $k = k(\mathbf{x})$. Without loss of generality, we may assume that $k = k(\mathbf{x})$ is piecewise constant with respect to the elements of $\mathcal{T}_h$. We now present two spectral aggregation/element agglomeration based approaches to construct accurate enough coarse spaces. We assume that a set $\mathcal{T}_H$ of non–overlapping agglomerated elements $\{T\}$ has been constructed. This means that each T is a (connected) union of fine–grid elements $\{\tau\}$. Let the characteristic mesh size (diameter of $T \in \mathcal{T}_H$) be of order H. We do not assume that H is comparable to the fine–grid mesh–size h. For each T, we assemble the local stiffness matrix A_T and the local weighted mass matrix G_T. In addition to the set $\mathcal{T}_H$ of agglomerated elements T, we assume that we have respective aggregates $\{\mathcal{A}_i\}_{i=1}^{n_A}$ where each $\mathcal{A} = \mathcal{A}_i$ is contained in a unique T. The set $\{\mathcal{A}_i\}$ provides a non–overlapping partition of the set $\mathcal{N}_h$ of fine–degrees of freedom. The aggregates are easily constructed based on $\mathcal{T}_H$, by assigning the interface nodes, shared by multiple $\bar{T}$, $T \in \mathcal{T}_H$, to only one of the aggregates. We note that it is advantageous to assign the interface degrees of freedom to the aggregate with the largest value of the coefficient, k.

3.1 The Construction of Tentative Prolongator

In [BHMV] the following construction was used. Solve the generalized eigenvalue problem (we use the equivalent eigenvalue problem as proposed in [GE10])

$$A_T \mathbf{q}_k = \bar{\lambda}_k G_T \mathbf{q}_k, \; k = 1, \, \ldots, \, n_T, \tag{11}$$

where n_T is the number of fine–degrees of freedom in T. By choosing the first m_T eigenvectors in the lower part of the spectrum of $G_T^{-1} A_T$, we form the rectangular matrix $Q_T = [\mathbf{q}_1, \ldots, \mathbf{q}_{m_T}]$. Then, we extract the rows of Q_T with row–indices from the aggregate $\mathcal{A}$ (where $\mathcal{A} \subset T$) and form $Q_{\mathcal{A}}$. Finally, using for example SVD, we can form a linearly independent set from the columns of $Q_{\mathcal{A}}$. The resulting matrix $\widehat{P}_{\mathcal{A}}$ has orthogonal (hence linearly independent) columns. The global tentative prolongator is simply the block-diagonal matrix, with the blocking corresponding to individual aggregates,

$$\widehat{P} = \begin{bmatrix} \widehat{P}_{\mathcal{A}_1} & 0 & \cdots & 0 \\ 0 & \widehat{P}_{\mathcal{A}_2} & \cdots & 0 \\ 0 & 0 & \ddots & 0 \\ 0 & \cdots & 0 & \widehat{P}_{\mathcal{A}_{n_{\mathcal{A}}}} \end{bmatrix}. \tag{12}$$

Since for quasi-uniform mesh $\mathcal{T}_h$ it is easily seen that $h^2 D_T$, the scaled diagonal of A_T, and G_T, are uniformly spectrally equivalent, in practice we can instead solve (as in [BHMV]) the generalized eigenvalue problem equivalent to (11),

$$A_T \mathbf{q}_k = \lambda_k D_T \mathbf{q}_k, \ k = 1, \ldots, n_T. \tag{13}$$

Note that the respective eigenvalues are related as

$$\overline{\lambda}_k \simeq h^{-2} \lambda_k. \tag{14}$$

It is straightforward to see the following local estimate

$$\|\mathbf{v} - \widehat{P} \mathbf{v}_c\|_{D_T}^2 \leq \frac{1}{\lambda_{m_T+1}} \|\mathbf{v}\|_{A_T}^2. \tag{15}$$

By summing up the local estimates, using the equivalence between $h^2 D_T$ and G_T, we arrive at the weak approximation property (essentially proven in [BHMV])

$$H^{-2} \|\mathbf{v} - \widehat{P} \mathbf{v}_c\|_G^2 \leq \sigma \max_T \frac{h^2}{H^2 \lambda_{m_T+1}} \mathbf{v}^T A \mathbf{v}. \tag{16}$$

Here, σ is a constant independent of H, h and $k = k(\mathbf{x})$. Noticing now that $\frac{1}{\lambda_{m_T+1}}$ scales with $\left(\frac{H}{h}\right)^2$ (easily seen from (11), since $\overline{\lambda}_k$ scales as H^{-2} and $\lambda_k \simeq h^2 \overline{\lambda}_k$, see (14)), it follows that by choosing m_T appropriately (sufficiently large) a uniform bound with respect to both, H/h and the contrast (the size of jumps of $k = k(\mathbf{x})$ within each T) can be ensured. The size of m_T depends only on the number of subdomains of each T where k admits these large jumps. More detailed investigation on the optimal choice of m_T can be found in [GE10], see also [SVZ]. At any rate, an upper bound for m_T is the number of subdomains of T where k has substantially different values from the other subdomains.

3.2   Alternative Construction of Tentative Prolongator

With the purpose to allow for multilevel extension, we propose the following alternative procedure. Based on A_T, we compute its reduced Schur complement form S_A corresponding to the degrees of freedom of the aggregate $\mathcal{A}$, $\mathcal{A} \subset T$. Then, we solve the generalized eigenvalue problem

$$S_A \mathbf{p}_k = \lambda_k D_A \mathbf{p}_k, k = 1, \ldots, n_A,$$

where D_A is, for example, the diagonal of A_T restricted to $\mathcal{A}$. Here, n_A stands for the size of $\mathcal{A}$ (the number of degrees of freedom in $\mathcal{A}$). The tentative prolongator is simply the block–diagonal matrix $\widehat{P}$ with block-entries $\widehat{P}_A = [\mathbf{p}_1, \ldots, \mathbf{p}_{m_A}]$ for a sufficiently large $m_A \leq n_A$. In practice, we choose m_A such that for a given tolerance $\theta \in (0, 1]$,

$$\lambda_k \geq \theta \lambda_{\max} \text{ if } k > m_A \text{ (assuming } \lambda_s \leq \lambda_{s+1}). \tag{17}$$

It is straightforward to see the local estimates

$$\|\mathbf{v}_A - \widehat{P}_A \mathbf{v}_A^c\|_{D_A}^2 \leq \left(\frac{1}{\lambda_{m_A+1}}\right) \mathbf{v}_A^T S_A \mathbf{v}_A \leq \left(\frac{1}{\lambda_{m_A+1}}\right) \mathbf{v}_T^T A_T \mathbf{v}_T,$$

where $\mathbf{v}_T$ is any extension of $\mathbf{v}_A$ (defined on $\mathcal{A}$) to a vector $\mathbf{v}_T = [\mathbf{v}_A]$ defined on T. We use here the fact that S_A is a Schur complement of the symmetric positive semi–definite matrix A_T. By summing up the local estimates over $\mathcal{A}$, the following global one is obtained (D_A is the diagonal of A)

$$\|\mathbf{v} - \widehat{P}\mathbf{v}_c\|_{D_A}^2 \leq \left(\max_A \frac{1}{\lambda_{m_A+1}}\right) \mathbf{v}^T A \mathbf{v}. \tag{18}$$

Finally, using the uniform spectral equivalence between $h^2 D_A$ and D_G (and G), we obtain the desired weak approximation estimate of the form (16).

Above, D_A can be replaced with any other spectrally equivalent matrix D. For example, in Section 4, we consider a particular choice of D that makes $b = \|D^{-\frac{1}{2}} A D^{-\frac{1}{2}}\| = 1$.

3.3   The Smoothed Prolongator and Its Analysis

Based on the tentative prolongator $\widehat{P}$ and the matrix polynomial $S = s_\nu(b^{-1}D^{-1}A)$, where $b : \|D^{-\frac{1}{2}} A D^{-\frac{1}{2}}\| \leq b = \mathcal{O}(1)$ and

$$s_\nu(t) = (-1)^\nu \frac{1}{2\nu + 1} \frac{T_{2\nu+1}(\sqrt{t})}{\sqrt{t}}, \tag{19}$$

with $T_l(t)$ denoting the Chebyshev polynomial of the first kind and degree l over the interval $[-1, +1]$, we define the actual smoothed aggregation prolongation matrix P as

$$P = S\widehat{P}.$$

The effect of prolongation smoothing is that it makes the prolongation operator, P, stable in the energy norm. More specifically, letting Q be the D-orthogonal projection onto the space Range($\widehat{P}$), we get

$$\|SQ\mathbf{v}\|_A \leq \|S\mathbf{v}\|_A + \|S(\mathbf{v} - Q\mathbf{v})\|_A \leq \|\mathbf{v}\|_A + \frac{b^{\frac{1}{2}}}{2\nu + 1} \|\mathbf{v} - Q\mathbf{v}\|_D.$$

Here, we used the fact that $s_\nu^2(t) \in [0, 1]$ for $t \in (0, 1]$, and also its main property (cf., e.g., [Va08])

$$\sup_{t \in (0, 1]} |\sqrt{t} s_\nu(t)| \leq \frac{1}{2\nu + 1}.$$

Using now the proven weak approximation property of $Q\mathbf{v}$, i.e., estimate (18), we arrive at the following final energy norm bound of SQ,

$$\|SQ\mathbf{v}\|_A \leq \left[1 + \frac{b^{\frac{1}{2}}}{2\nu + 1} \frac{H}{h} \left(\max_{\mathcal{A}} \frac{h^2}{H^2 \lambda_{m_\mathcal{A}+1}} \right)^{\frac{1}{2}} \right] \|\mathbf{v}\|_A. \tag{20}$$

Choosing $\nu \simeq H/h$ ensures SQ being uniformly bounded in energy. Similarly,

$$\|\mathbf{v} - SQ\mathbf{v}\|_D \leq \|S(\mathbf{v} - Q\mathbf{v})\|_D + \|(I - S)\mathbf{v}\|_D \leq \|(I - Q)\mathbf{v}\|_D + \|(I - S)\mathbf{v}\|_D$$
$$= \|(I - Q)\mathbf{v}\|_D + \|D^{\frac{1}{2}} \left(I - s_\nu(b^{-1}D^{-1}A) \right) A^{-\frac{1}{2}}\| \|\mathbf{v}\|_A.$$

The last matrix norm is estimated as follows

$$\|D^{\frac{1}{2}} \left(I - s_\nu(b^{-1}D^{-1}A) \right) A^{-\frac{1}{2}}\| = \|X^{-\frac{1}{2}} \left(I - s_\nu(b^{-1}X) \right)\|, \quad X = A^{\frac{1}{2}} D^{-1} A^{\frac{1}{2}}.$$

Therefore,

$$\|\mathbf{v} - SQ\mathbf{v}\|_D \leq \|(I - Q)\mathbf{v}\|_D + b^{-\frac{1}{2}} \sup_{t \in (0, 1]} \frac{1 - s_\nu(t)}{\sqrt{t}} \|\mathbf{v}\|_A$$
$$\leq \left(\max_{\mathcal{A}} \frac{1}{\sqrt{\lambda_{m_\mathcal{A}+1}}} \right) \|\mathbf{v}\|_A + \frac{1}{\sqrt{b}} C_\nu \|\mathbf{v}\|_A.$$

Thus, using the equivalence $h^2 D \simeq G$, we have the final estimate

$$H^{-1} \|\mathbf{v} - SQ\mathbf{v}\|_G \simeq \frac{h}{H} \|\mathbf{v} - SQ\mathbf{v}\|_D$$
$$\leq \frac{h}{H} \left(\left(\max_{\mathcal{A}} \frac{1}{\sqrt{\lambda_{m_\mathcal{A}+1}}} \right) \|\mathbf{v}\|_A + \frac{1}{\sqrt{b}} C_\nu \|\mathbf{v}\|_A \right)$$
$$\leq \left(\left(\max_{\mathcal{A}} \frac{h}{H\sqrt{\lambda_{m_\mathcal{A}+1}}} \right) + b^{-\frac{1}{2}} \frac{C_\nu}{2\nu+1} \frac{2\nu+1}{\frac{H}{h}} \right) \|\mathbf{v}\|_A.$$

Since $\frac{C_\nu}{2\nu+1} = \mathcal{O}(1)$ (see [BVV]), by choosing $\nu \simeq \frac{H}{h}$, we obtain the desired uniform boundedness of $H^{-1} \|\mathbf{v} - SQ\mathbf{v}\|_G$ in terms of $\|\mathbf{v}\|_A$. We summarize:

Theorem 2. *The two–level spectral element agglomeration construction of tentative prolongator $\widehat{P}$ combined with the smoothed aggregation construction of the actual interpolation matrix $P = S\widehat{P}$ where $S = s_\nu(b^{-1}D^{-1}A)$ and s_ν is given in (19) for $\nu \simeq \frac{H}{h}$, ensures the weak approximation property with constant $\beta_1 = O(1)$:*

$$H^{-1} \|\mathbf{v} - SQ\mathbf{v}\|_G \leq \beta_1 \left(1 + \max_{\mathcal{A}} \left(\frac{h^2}{H^2 \lambda_{m_\mathcal{A}+1}} \right)^{\frac{1}{2}} \right) \|\mathbf{v}\|_A.$$

Also, the energy stability property holds with constant $\beta_2 = O(1)$:

$$\|SQ\mathbf{v}\|_A \leq \left(1 + \beta_2 \max_{\mathcal{A}} \left(\frac{h^2}{H^2 \lambda_{m_\mathcal{A}+1}} \right)^{\frac{1}{2}} \right) \|\mathbf{v}\|_A.$$

Based on Theorems 1–2, Remark 2 implies the following corollary.

Corollary 1. *The two–level spectral element agglomeration construction of tentative prolongator $\widehat{P}$ combined with the smoothed aggregation construction of the actual interpolation matrix as described in Theorem 2, exhibits convergence bound K_{TL} that is also contrast independent (in addition to the independence of both h and H).*

4   Numerical Experiments

We have implemented a multilevel version of the proposed SA spectral AMGe algorithm to test the performance of both the two-level (TL) method and its multilevel (ML) extension. We use agglomeration algorithm that exploits the fine–grid vertex coordinates, as described in [BVV]. In this way, the aggregates after the first coarse level have complexity similar to a geometric multigrid.

At coarse levels we use as multigrid relaxation a simple block Gauss–Seidel smoothing with blocks corresponding to the respective aggregates. At the finest level, we use as multigrid relaxation the polynomial smoother from [VBT],

$$p_\nu(t) = (1 - T_{2\nu+1}^2(\sqrt{t}))s_\nu(t).$$

Its smoothing and coercivity properties, (8)-(9), have been analyzed in [BVV].

To select the local eigenvectors (needed in the construction of the tentative prolongator, see (17)), we used fairly small tolerance $\theta = 0.01$. All experiments were run as a stationary iteration method with stopping criteria that halts the iteration once the preconditioned residual norm was reduced by a factor of 10^{-6} relative to the initial one.

In the experiments, we have chosen the weighted ℓ_1–smoother $\Lambda = \mathrm{diag}\,(d_i)_{i=1}^n$ where for the given s.p.d. matrix $A = (a_{ij})_{i,j=1}^n$ and any given positive weights $\{w_i\}$, we let $d_i = \sum_j |a_{ij}|\frac{w_i}{w_j}$. For any given positive weights $\{w_i\}$, we have $\mathbf{v}^T A \mathbf{v} \le \mathbf{v}^T \Lambda \mathbf{v}$. In the experiments, we have chosen $w_l = \sqrt{a_{ll}}$. In that case, we also have that Λ is spectrally equivalent to D_A (the diagonal of A), i.e., $\mathbf{v}^T D_A \mathbf{v} \le \mathbf{v}^T \Lambda \mathbf{v} \le \kappa\, \mathbf{v}^T D_A \mathbf{v}$, where κ is the maximum number of nonzero entries per row of A. We note that Theorem 1 applies with the above choice of Λ, since in its proof, we can replace D_A with any spectrally equivalent s.p.d. matrix Λ.

Our fine–grid problem is posed on the unit square domain Ω where the coefficient $k = k(\mathbf{x})$ takes two values: 1 and 10^c for various values of $c = -12, -9, -6, -3, 0, 1, 3, 6, 9, 12$. The region where $k = 10^c$ is shown in Figure 1 (right). The coefficient is resolved only on the finest mesh (see the middle of Figure 1) by using adaptive local refinement starting from an unstructured coarse mesh, shown in Figure 1 (left). The initial mesh does not resolve the coefficient. The distribution of the values 1 and 10^c is illustrated in Figure 1 (right).

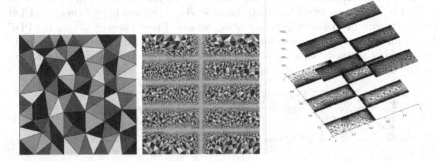

Fig. 1. Initial unstructured mesh (left) that does not resolve the discontinuous coefficient, its refinement (middle) that resolves the high-contrast discontinuous coefficient, and the actual distribution of the coefficient values 1 and 10^c, $c = 3$ (right)

Table 1. Convergence factor (ϱ) and number of iterations (n_{it}) for the two–level SA spectral AMGe with $\nu = 6$ (degree of polynomial smoother $3\nu + 1 = 19$). The fine (triangular) mesh is fixed with $465,712$ elements and $233,499$ dofs. The coarse mesh consists of $1,330$ elements. The jump in the PDE coefficient is 10^c.

c	# coarse grid dofs	n_{it}	ϱ
-12	2,729	53	0.74
-9	2,730	53	0.75
-6	2,730	51	0.74
-3	2,720	26	0.60
0	2,660	22	0.60
3	2,728	24	0.60
6	2,744	36	0.65
9	2,744	46	0.70
12	2,744	46	0.70

Table 2. Convergence factor (ϱ) and number of iterations (n_{it}) for both two-level (TL) and the multilevel (ML) SA spectral AMGe with $\nu = 6$ (degree of polynomial smoother $3\nu + 1 = 19$) at the finest level. At coarser level (in the ML case) block Gauss–Seidel smoother is used with blocks corresponding to the respective aggregates. The fine (triangular) mesh is fixed with $465,712$ elements and $233,499$ dofs. The coarse mesh consists of $5,250$ elements. The jump in the PDE coefficient is 10^c.

c	# coarse grid dofs	n_{it}^{TL}	ϱ_{TL}	n_{it}^{ML}	ϱ_{ML}
-12	11,191	57	0.73	77	0.79
-9	11,191	56	0.73	77	0.79
-6	11,191	46	0.70	67	0.78
-3	11,193	21	0.53	39	0.71
0	11,073	19	0.54	33	0.72
3	11,199	21	0.53	43	0.73
6	11,199	36	0.64	44	0.71
9	11,199	39	0.65	49	0.69
12	11,199	39	0.65	58	0.71

Form Table 1, we can see that the two-level (TL) SA spectral AMGe exhibits convergence factors that are fairly insensitive with respect to the contrast for a fixed fine-grid mesh size. Table 2 compares the behavior of the multilevel (ML) method versus the respective TL one. As we can see, the multilevel convergence

Table 3. Convergence factor (ϱ) for the two–level and multilevel SA spectral AMGe method with polynomial smoother at the finest level ($\nu = 6$) and block Gauss–Seidel at coarser levels with blocks corresponding to aggregates. Fixed jump of 10^{12} in the PDE coefficient and variable fine-grid mesh. The last column shows the operator complexity of the multilevel method; namely, the sum of the non-zero entries of all level matrices divided by the number of non–zero entries of the finest-level matrix.

# fine grid dofs	# fine grid elements	# dofs at 1st coarse level	# elements at 1st coarse level	n_{it}^{ML}	ϱ_{ML}	n_{it}^{TL}	ϱ_{TL}	operator complexity
40,377	80,192	1,237	570	17	0.36	13	0.26	1.13
100,613	200,416	3,410	1,596	29	0.55	19	0.49	1.14
233,499	465,712	11,199	5,250	58	0.71	39	0.65	1.19
578,017	1,153,792	43,026	20,860	84	0.80	38	0.63	1.29

factor seems also quite insensitive with respect to the contrast. Finally, Table 3 illustrates the performance of the TL and ML methods when we vary the fine–grid size and keep the contrast fixed. The TL convergence factors tend to stabilize when refining the mesh, whereas the ML ones exhibit some growth.

In conclusion, all TL-results are in good agreement with our theory; namely, the TL convergence factor stays bounded with both the contrast and is mesh-independent. Note that, all test were performed as stationary iterative method (not in a preconditioned CG iteration). If the method is used as a preconditioner in CG the convergence will be much better.

References

[ABHT] Adams, M., Brezina, M., Hu, J., Tuminaro, R.: Parallel multigrid smoothing: polynomial versus Gauss–Seidel. Journal of Computational Physics 188, 593–610 (2003)

[BHMV] Brezina, M., Heberton, C., Mandel, J., Vaněk, P.: An iterative method with convergence rate chosen a priori, UCD CCM Report 140 (1999),
http://www-math.cudenver.edu/~jmandel/papers/

[BVV] Brezina, M., Vaněk, P., Vassilevski, P.S.: An improved convergence analysis of smoothed aggregation algebraic multigrid. Numerical Linear Algebra with Applications (2011)

[ρAMGi] Chartier, T., Falgout, R., Henson, V.E., Jones, J., Manteuffel, T., McCormick, S., Ruge, J., Vassilevski, P.S.: Spectral AMGe (ρAMGe). SIAM J. Scientific Computing 25, 1–26 (2003)

[ρAMGii] Chartier, T., Falgout, R., Henson, V.E., Jones, J., Manteuffel, T., McCormick, S., Ruge, J., Vassilevski, P.S.: Spectral element agglomerate AMGe. In: Domain Decomposition Methods in Science and Engineering XVI. Lecture Notes in Computational Science and Engineering, vol. 55, pp. 515–524. Springer, Heidelberg (2007)

[GE10] Galvis, J., Efendiev, Y.: Domain decomposition preconditioners for multiscale flows in high-contrast media. Multiscale Modeling and Simulation 8, 1461–1483 (2010)

[LV08] Lashuk, I., Vassilevski, P.S.: On some versions of the element agglomeration AMGe method. Numerical Linear Algebra with Applications 15, 595–620 (2008)

[SVZ] Scheichl, R., Vassilevski, P.S., Zikatanov, L.T.: Weak approximation properties of elliptic projections with functional constraints, Available as LLNL Report
LLNL-JRNL-468811, February 3 (2011)

[VBT] Vaněk, P., Brezina, M., Tezaur, R.: Two-grid method for elasticity on unstructured meshes. SIAM Journal on Scientific Computing 21(3), 900–923 (1999)

[Va08] Vassilevski, P.S.: Multilevel Block Factorization Preconditioners, Matrix-based
Analysis and Algorithms for Solving Finite Element Equations, p. 514. Springer,
New York (2008)

Approximation of Sparse Controls in Semilinear Elliptic Equations

Eduardo Casas[1,*], Roland Herzog[2], and Gerd Wachsmuth[2]

[1] Departmento de Matemática Aplicada y Ciencias de la Computación
E.T.S.I. Industriales y de Telecomunicación
Universidad de Cantabria, Av. Los Castros s/n, 39005 Santander, Spain
eduardo.casas@unican.es
[2] Faculty of Mathematics, Chemnitz University of Technology
Reichenhainer Strasse 41, 09126 Chemnitz, Germany
{roland.herzog,gerd.wachsmuth}@mathematik.tu-chemnitz.de

Abstract. Semilinear elliptic optimal control problems involving the L^1 norm of the control in the objective are considered. Necessary and sufficient second-order optimality conditions are derived. A priori finite element error estimates for three different discretizations for the control problem are given. These discretizations differ in the use of piecewise constant, piecewise linear and continuous or non-discretized controls, respectively. Numerical results and implementation details are provided.

Keywords: optimal control of partial differential equations, sparse controls, non-differentiable objective, finite element discretization, a priori error estimates, semilinear equations.

1 Introduction

In this paper, we study the following control problem

$$\left.\begin{aligned} \text{minimize} \quad & J(u) \\ \text{such that} \quad & \alpha \le u(x) \le \beta \quad \text{for a.a. } x \in \Omega, \end{aligned}\right\} \tag{P}$$

where $J(u) = F(u) + \mu\, j(u)$, with $F : L^2(\Omega) \longrightarrow \mathbb{R}$ and $j : L^1(\Omega) \longrightarrow \mathbb{R}$ defined by

$$F(u) = \frac{1}{2}\|y(u) - y_d\|_{L^2(\Omega)}^2 + \frac{\nu}{2}\|u\|_{L^2(\Omega)}^2 \quad \text{and} \quad j(u) = \|u\|_{L^1(\Omega)},$$

$y(u)$ being the solution of the semilinear state equation

$$-\Delta y + a(\cdot, y) = u \quad \text{in } \Omega, \qquad \text{and} \qquad y = 0 \quad \text{on } \Gamma. \tag{1}$$

The most remarkable issue of the problem (**P**) is the presence of the non-differentiable function $j(u)$ in the cost functional. By regulating the parameter μ

* This author was partially supported by the Spanish Ministry of Science and Innovation under projects MTM2008-04206 and "Ingenio Mathematica (i-MATH)" CSD2006-00032 (Consolider Ingenio 2010).

I. Lirkov, S. Margenov, and J. Waśniewski (Eds.): LSSC 2011, LNCS 7116, pp. 16–27, 2012.

conveniently, we can obtain an optimal control with small support. In fact, for μ sufficiently large, the unique (local and global) solution of the control problem is $\bar{u} = 0$. For $\mu = 0$, the optimal control is usually non-zero at almost every point of the domain Ω. Taking some intermediate values of μ we get an optimal control which is active (non-zero) only in some subsets of Ω. In optimal control of distributed parameter systems, it is not possible or desirable to put the controllers at every point of the domain. Instead, it is required to localize the controllers in small regions. The main issue is, which are the most effective regions to set the controllers? The answer is obtained by solving the above problem.

The presence of the non-differentiable term $j(u)$ makes the analysis of (**P**) more complicated than usual. However, a complete analysis is possible. Indeed, we have got the first and second-order optimality conditions, we have proved the convergence of the discretizations and even we have obtained some error estimates. All these results will be presented in this paper along with some numerical computations and remarks on their practical realization.

Let us mention some previous papers dealing with the study of sparse controls for partial differential equations. As far as we know, the first paper devoted to this study for elliptic problems was published by Stadler [13]. The author studied mainly some numerical algorithms to solve the problem. In [14], a deep analysis of the problem was carried out. In both papers, the state equation was linear. The non-linear case has been analyzed in [3] and [4]. The theoretical results presented here are detailed in the last two references. The use of measures of small support to control a system has been considered in [7].

Let us finish this section by stating the assumptions on the elements involved in the problem (**P**).

Assumption 1.— Ω is an open bounded subset of $\mathbb{R}^n$, $n = 2$ or 3, with a $C^{1,1}$ boundary Γ. We also assume $-\infty < \alpha < 0 < \beta < +\infty$, $\mu > 0$, $\nu > 0$, and $y_d \in L^{\bar{p}}(\Omega)$ for some $\bar{p} > n$.

In the sequel, we will denote the set of feasible controls by

$$\mathbb{K} = \{u \in L^\infty(\Omega) : \alpha \leq u(x) \leq \beta \text{ for a.a. } x \in \Omega\}.$$

Assumption 2.— $a : \Omega \times \mathbb{R} \longrightarrow \mathbb{R}$ is a Carathéodory function of class C^2 with respect to the second variable, with $a(\cdot, 0) \in L^{\bar{p}}(\Omega)$ and satisfying

$$\frac{\partial a}{\partial y}(x, y) \geq 0 \ \forall y \in \mathbb{R} \text{ and for a.a. } x \in \Omega,$$

$$\forall M > 0 \ \exists C_M > 0 \text{ s.t. } \sum_{j=1}^{2} \left| \frac{\partial^j a}{\partial y^j}(x, y) \right| \leq C_M \ \forall |y| \leq M \text{ and for a.a. } x \in \Omega.$$

Note that these assumptions ensure in particular that the control-to-state map $L^p(\Omega) \longrightarrow W^{2,p}(\Omega)$ is well-defined and of class C^2 for $n/2 < p \leq \bar{p}$, see [3, Theorems 2.1 and 2.2].

2   First- and Second-Order Optimality Conditions

All the proofs of the results presented in this section can be found in [3]. The existence of a global solution to (**P**) is standard. Since the problem is not convex, we have to deal with local solutions.

Theorem 1 (First-order optimality conditions). *If $\bar{u}$ is a local minimum of (**P**), then there exist $\bar{\varphi} \in W^{2,\bar{p}}(\Omega)$ and a subgradient $\bar{\lambda} \in \partial j(\bar{u})$ such that*

$$-\Delta\bar{\varphi} + \frac{\partial a}{\partial y}(x, \bar{y})\,\bar{\varphi} = \bar{y} - y_d \quad in \ \Omega, \qquad \bar{\varphi} = 0 \quad on \ \Gamma, \qquad (2a)$$

$$\int_{\Omega} (\bar{\varphi} + \nu\,\bar{u} + \mu\,\bar{\lambda})(u - \bar{u})\,\mathrm{d}x \geq 0 \quad \forall u \in \mathbb{K}, \qquad (2b)$$

where $\bar{y} = y(\bar{u})$. Moreover, the following relations are satisfied by $\bar{u}$, $\bar{\varphi}$ and $\bar{\lambda}$:

$$\bar{u}(x) = \mathrm{Proj}_{[\alpha,\beta]}\left(-\frac{1}{\nu}\big(\bar{\varphi}(x) + \mu\,\bar{\lambda}(x)\big)\right), \qquad (3a)$$

$$\bar{u}(x) = 0 \quad \Leftrightarrow \quad |\bar{\varphi}(x)| \leq \mu, \qquad (3b)$$

$$\bar{\lambda}(x) = \mathrm{Proj}_{[-1,+1]}\left(-\frac{1}{\mu}\bar{\varphi}(x)\right). \qquad (3c)$$

When inserting (3c) into (3a), we obtain $\bar{u}$ as a function of the adjoint state $\bar{\varphi}$:

$$\bar{u}(x) = \mathcal{S}(\bar{\varphi}(x)). \qquad (4)$$

The function $\mathcal{S} : \mathbb{R} \to \mathbb{R}$ is known as the soft-thresholding operator which typically appears in sparsity-related optimization problems, see for instance [8, eq. (1.5)]. Here it must be supplemented with the truncation at the bounds α and β. Figure 1 shows a graph of $\mathcal{S}$ and in addition the dependence of $\bar{\lambda}$ on $\bar{\varphi}$.

It will be a crucial point in the numerical analysis and realization to establish relations parallel to (3) also for the discretized problem. We remark that (3a) and (3c) imply $\bar{u}, \bar{\lambda} \in C^{0,1}(\overline{\Omega}) = W^{1,\infty}(\Omega)$. Moreover, $\bar{\lambda}$ is unique.

In order to address the second-order optimality conditions we need to introduce the critical cone. Given a control $\bar{u} \in \mathbb{K}$ for which there exists $\bar{\lambda} \in \partial j(\bar{u})$ satisfying the first-order conditions (2), we define the closed convex cone

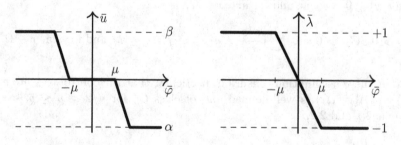

Fig. 1. Plot of the graph of $\mathcal{S}$ (4) and the projection formula (3c)

$$C_{\bar{u}} = \{v \in L^2(\Omega) \text{ satisfying } F'(\bar{u})\,v + \mu\,j'(\bar{u};v) = 0 \quad \text{and}$$
$$v(x) \geq 0 \text{ if } \bar{u}(x) = \alpha \quad \text{and} \quad v(x) \leq 0 \text{ if } \bar{u}(x) = \beta\}.$$

Theorem 2 (Second-order optimality conditions)

1. If $\bar{u}$ is a local minimum of (**P**), then $F''(\bar{u})\,v^2 \geq 0$ for every $v \in C_{\bar{u}}$.
2. Let $\bar{u} \in \mathbb{K}$ and $\bar{\lambda} \in \partial j(\bar{u})$ such that (2) hold. If $F''(\bar{u})\,v^2 > 0$ for all $v \in C_{\bar{u}} \setminus \{0\}$, then there exist $\delta > 0$ and $\varepsilon > 0$ such that

$$J(\bar{u}) + \frac{\delta}{2}\|u - \bar{u}\|_{L^2(\Omega)}^2 \leq J(u) \quad \forall u \in \mathbb{K} \cap B_\varepsilon(\bar{u}), \tag{5}$$

where $B_\varepsilon(\bar{u})$ denotes the $L^2(\Omega)$-ball of center $\bar{u}$ and radius ε.

3 Finite Element Approximation of (P)

The goal of this section is to study the approximation of problem (**P**) by finite elements. The state will be discretized by continuous piecewise linear functions. For the controls we analyze the three usual different approaches: piecewise constant, piecewise linear and continuous and non-discretization. The last case corresponds to the variational approach suggested by Hinze [11]. We prove the convergence of the discretization and derive error estimates for the optimal control in $L^2(\Omega)$ and $L^\infty(\Omega)$, as well as for the associated states and adjoint states.

Let us assume for simplicity that Ω is convex. We consider a family of triangulations $\{\mathcal{T}_h\}_{h>0}$ of $\overline{\Omega}$, defined in the standard way, e.g. in [1, Chapter 3.3]. The triangulations cover polygonal approximations $\Omega_h \subset \Omega$. We assume that they form a regular and uniform family, such that inverse inequalities hold, see [4, Section 3] for details.

We use piecewise linear approximations for the states, thus we set

$$\mathcal{Y}_h = \{y_h \in C(\overline{\Omega}) \mid y_{h|T} \in \mathcal{P}_1 \text{ for all } T \in \mathcal{T}_h, \text{ and } y_h = 0 \text{ in } \overline{\Omega} \setminus \Omega_h\},$$

where $\mathcal{P}_1$ is the space of polynomials of degree less than or equal to 1. We denote by $y_h(u) \in \mathcal{Y}_h$ the unique solution of the discrete state equation

$$\int_{\Omega_h} [\nabla y_h \cdot \nabla z_h + a(x,y_h)\,z_h]\,\mathrm{d}x = \int_{\Omega_h} u\,z_h\,\mathrm{d}x \quad \forall z_h \in \mathcal{Y}_h. \tag{6}$$

Thanks to the monotonicity of the nonlinear term of (6) and using Brouwer's fixed point theorem, it is easy to prove the existence and uniqueness of a solution $y_h(u)$ of (6) for any $u \in L^2(\Omega_h)$. Analogously, we denote by $\varphi_h(y)$ the unique solution of the discrete adjoint equation

$$\int_{\Omega_h} \left[\nabla z_h \cdot \nabla \varphi_h + \frac{\partial a}{\partial y}(x,y)\,\varphi_h\,z_h\right]\,\mathrm{d}x = \int_{\Omega_h} (y - y_d)\,z_h\,\mathrm{d}x \quad \forall z_h \in \mathcal{Y}_h. \tag{7}$$

Now, we analyze separately the discretization of the problem according to the different choices of the set $\mathbb{K}_h$.

3.1 Piecewise Constant Controls

We define the space of discrete controls by

$$\mathcal{U}_{h,0} = \{u_h \in L^2(\Omega_h) : u_{h|T} = \text{constant for all } T \in \mathcal{T}_h\}.$$

Every element $u_h \in \mathcal{U}_{h,0}$ can be written in the form $u_h = \sum_{T \in \mathcal{T}_h} u_T \chi_T$, where χ_T is the characteristic function of the cell T. The set of discrete feasible controls is given by

$$\mathbb{K}_{h,0} = \{u_h \in \mathcal{U}_{h,0} : \alpha \le u_T \le \beta \ \forall T \in \mathcal{T}_h\} = \mathbb{K} \cap \mathcal{U}_{h,0}.$$

Finally, the discrete control problem is formulated as follows

$$\left. \begin{array}{rl} \text{minimize} & J_h(u_h) = F_h(u_h) + \mu\, j_h(u_h) \\ \text{such that} & u_h \in \mathbb{K}_{h,0} \end{array} \right\} \qquad (\mathbf{P}_{h,0})$$

where $F_h : L^2(\Omega_h) \longrightarrow \mathbb{R}$ and $j_h : \mathcal{U}_h \longrightarrow \mathbb{R}$ are defined by

$$F_h(u_h) = \frac{1}{2}\int_{\Omega_h} (y_h(u_h) - y_d)^2 \, \mathrm{d}x + \frac{\nu}{2}\int_{\Omega_h} u_h^2 \, \mathrm{d}x \quad \text{and} \quad j_h(u_h) = \int_{\Omega_h} |u_h| \, \mathrm{d}x.$$

It is immediate that $(\mathbf{P}_{h,0})$ has at least one solution and we have the following first-order optimality conditions analogous to those of problem $(\mathbf{P})$, see Theorem 1.

Theorem 3. *If $\bar{u}_h$ is a local minimum of problem $(\mathbf{P}_{h,0})$, then there exists a subgradient $\bar{\lambda}_h \in \partial j_h(u_h)$ such that*

$$\int_{\Omega_h} (\bar{\varphi}_h + \nu\,\bar{u}_h + \mu\,\bar{\lambda}_h)\,(u_h - \bar{u}_h)\, \mathrm{d}x \ge 0 \quad \forall u_h \in \mathbb{K}_{h,0} \qquad (8)$$

holds, with the discrete state $\bar{y}_h = y_h(\bar{u}_h)$ and the discrete adjoint state $\bar{\varphi}_h = \varphi_h(\bar{y}_h)$. Moreover, we have

$$\bar{\lambda}_h = \sum_{T \in \mathcal{T}_h} \bar{\lambda}_T \chi_T \quad \text{with} \quad \begin{cases} \bar{\lambda}_T = +1 & \text{if } \bar{u}_T > 0, \\ \bar{\lambda}_T = -1 & \text{if } \bar{u}_T < 0, \\ \bar{\lambda}_T \in [-1, +1] & \text{if } \bar{u}_T = 0, \end{cases} \qquad (9)$$

where $\bar{u}_T$ are the coefficients of $\bar{u}_h = \sum_{T \in \mathcal{T}_h} \bar{u}_T \chi_T$.

Inequality (8) leads to the cellwise representation formula

$$\bar{u}_T = \text{Proj}_{[\alpha,\beta]}\left(-\frac{1}{\nu}\left[\frac{1}{|T|}\int_T \bar{\varphi}_h \, \mathrm{d}x + \mu\,\bar{\lambda}_T \right] \right). \qquad (10a)$$

Using (9) and (10a), we can prove for all $T \in \mathcal{T}_h$

$$\bar{u}_T = 0 \quad \Leftrightarrow \quad \frac{1}{|T|}\left| \int_T \bar{\varphi}_h \, \mathrm{d}x \right| \le \mu, \qquad (10b)$$

$$\bar{\lambda}_T = \text{Proj}_{[-1,+1]}\left(-\frac{1}{\mu\,|T|}\int_T \bar{\varphi}_h \, \mathrm{d}x \right). \qquad (10c)$$

In particular $\bar{\lambda}_h$ is unique. Note that (10) is the discrete analogue of (3). Combining (10a) and (10c) results in

$$\bar{u}_T = S\left(\frac{1}{|T|} \int_T \bar{\varphi}_h \, dx\right). \tag{11}$$

This replaces (4) in the case of piecewise constant control discretizations.

3.2   Piecewise Linear and Continuous Controls

Let us define the space of discrete controls

$$\mathcal{U}_{h,1} = \{u_h \in C(\overline{\Omega}_h) \mid u_{h|T} \in \mathcal{P}_1 \text{ for all } T \in \mathcal{T}_h\}$$

where $\mathcal{P}_1$ is the space of polynomials of degree less or equal than 1. We denote by $\{e_j\}_{j=1}^{N_v}$ the usual global nodal basis functions associated with the nodes of the triangulation x_j. Every element $u_h \in \mathcal{U}_{h,1}$ can be represented in the form

$$u_h = \sum_{j=1}^{N_v} u_j \, e_j = \sum_{j=1}^{N_v} u_h(x_j) \, e_j.$$

The set of discrete feasible controls is given by

$$\mathbb{K}_{h,1} = \{u_h \in \mathcal{U}_{h,1} : \alpha \le u_j \le \beta \quad \forall j = 1, \ldots, N_v\} = \mathbb{K} \cap \mathcal{U}_{h,1}.$$

Finally, the discrete control problem is formulated as follows

$$\left.\begin{array}{ll} \text{minimize} & J_h(u_h) = F_h(u_h) + \mu \, j_h(u_h) \\ \text{such that} & u_h \in \mathbb{K}_{h,1}. \end{array}\right\} \qquad (\mathbf{P}_{h,1})$$

The discrete functionals $F_h : \mathcal{U}_{h,1} \longrightarrow \mathbb{R}$ and $j_h : \mathcal{U}_{h,1} \longrightarrow \mathbb{R}$ are now defined in a non-standard way by

$$F_h(u_h) = \frac{1}{2} \int_{\Omega_h} (y_h(u_h) - y_d)^2 \, dx + \frac{\nu}{2} (u_h, u_h)_h \tag{12a}$$

$$j_h(u_h) = \sum_{j=1}^{N_v} |u_j| \int_{\Omega_h} e_j \, dx. \tag{12b}$$

Some words of explanation are in order. In the discretization of the smooth part (12a), we do not integrate $\|u_h\|_{L^2(\Omega_h)}^2$ exactly but rather discretize it by the nodal quadrature formula

$$(u_h, v_h)_h = \sum_{j=1}^{N_v} u_j v_j \int_{\Omega_h} e_j \, dx \quad \text{for } u_h, v_h \in \mathcal{U}_{h,1}. \tag{13}$$

Relation (13) defines a scalar product on $\mathcal{U}_{h,1}$ which is represented by a diagonally lumped mass matrix, see also Section 4 and [4, Remark 3.1]. Although this quadrature incurs an additional error, it will turn out to be crucial in order to obtain formulae parallel to (3) for the discretized problem $(\mathbf{P}_{h,1})$, see (17) below. These formulae are in turn essential in the derivation of error estimates. A similar quadrature formula is used for the discretization of the L^1 norm in j_h. This discretization of the L^1 norm was also used in [14, section 4.4].

It is immediate that $(\mathbf{P}_{h,1})$ has at least one solution. Now, we state the first-order optimality system for $(\mathbf{P}_{h,1})$.

Theorem 4 (Discrete first-order optimality conditions). *If $\bar{u}_h$ is a local minimum of problem $(\mathbf{P}_{h,1})$, then there exists a subgradient $\bar{\lambda}_h \in \partial j_h(\bar{u}_h)$ such that*

$$\int_{\Omega_h} \bar{\varphi}_h \, (u_h - \bar{u}_h) \, \mathrm{d}x + \big(\nu \, \bar{u}_h + \mu \, \bar{\lambda}_h, u_h - \bar{u}_h\big)_h \geq 0 \quad \forall u_h \in \mathbb{K}_{h,1} \qquad (14)$$

holds, with the discrete state $\bar{y}_h = y_h(\bar{u}_h)$ and the discrete adjoint state $\bar{\varphi}_h = \varphi_h(\bar{y}_h)$. Moreover, we have that the coefficients of the subgradient $\bar{\lambda}_h \in \mathcal{U}_{h,1}$ of j_h w.r.t. $(\cdot, \cdot)_h$ satisfy

$$\begin{cases} \bar{\lambda}_j = +1 & \text{if } \bar{u}_j > 0, \\ \bar{\lambda}_j = -1 & \text{if } \bar{u}_j < 0, \\ \bar{\lambda}_j \in [-1, 1] & \text{if } \bar{u}_j = 0. \end{cases} \qquad (15)$$

Analogous to (3) we wish to obtain representation formulae for $\bar{u}_h$ and $\bar{\lambda}_h$. It turns out that these are given naturally in terms of Carstensen's quasi-interpolation (see [2])

$$\Pi_h : L^1(\Omega_h) \longrightarrow \mathcal{U}_{h,1}, \qquad \Pi_h u = \sum_{j=1}^{N_v} \pi_j(u) \, e_j, \qquad \pi_j(u) = \frac{\int_{\Omega_h} u \, e_j \, \mathrm{d}x}{\int_{\Omega_h} e_j \, \mathrm{d}x}. \qquad (16)$$

Similarly to (10), we obtain

$$\bar{u}_j = \text{Proj}_{[\alpha,\beta]} \left(-\frac{1}{\nu} \big(\pi_j(\bar{\varphi}_h) + \mu \, \bar{\lambda}_j\big) \right), \qquad (17a)$$

$$\bar{u}_j = 0 \quad \Leftrightarrow \quad |\pi_j(\bar{\varphi}_h)| \leq \mu, \qquad (17b)$$

$$\bar{\lambda}_j = \text{Proj}_{[-1,+1]} \left(-\frac{1}{\mu}\pi_j(\bar{\varphi}_h) \right). \qquad (17c)$$

In particular $\bar{\lambda}_h$ is unique. Similar as Section 3.2, combining (17a) and (17c) results in

$$\bar{u}_j = \mathcal{S}\left(\pi_j(\bar{\varphi}_h) \right). \qquad (18)$$

This replaces (4) in the case of piecewise linear control discretizations.

3.3   Variational Discretization

In this section we consider a partial discretization of $(\mathbf{P})$. We do not discretize the controls and we set $\mathcal{U}_{h,3} = L^\infty(\Omega)$. Rather, the controls are implicitly discretized by the representation formula, see (19). This idea was introduced by

Hinze [11] and it was termed variational discretization of the control problem. This discretization is numerically implementable thanks to the fact that the optimal control $\bar{u}_h$ can be written in terms of the adjoint state $\bar{\varphi}_h$, see (19) below. This incomplete discretization leads to an improvement in the error estimate of $\bar{u} - \bar{u}_h$. The problem $(\mathbf{P}_{h,2})$ is defined as follows

$$\left.\begin{array}{ll} \text{minimize} & J_h(u_h) = F_h(u_h) + \mu\, j_h(u_h) \\ \text{such that} & u_h \in \mathbb{K} \end{array}\right\} \qquad (\mathbf{P}_{h,2})$$

where $F_h : L^2(\Omega_h) \longrightarrow \mathbb{R}$ and $j_h : L^1(\Omega_h) \longrightarrow \mathbb{R}$ are defined as in Section 3.1. The proof of the existence of a solution $\bar{u}_h$ of $(\mathbf{P}_{h,2})$ is standard. The optimality conditions satisfied by a local minimum of $(\mathbf{P}_{h,2})$ are given by Theorem 3 with $\mathbb{K}_{h,1}$ replaced by $\mathbb{K}$. This change leads to the same relations as given in Theorem 1, see formula (3), with $(\bar{u}, \bar{y}, \bar{\varphi}, \bar{\lambda})$ replaced by $(\bar{u}_h, \bar{y}_h, \bar{\varphi}_h, \bar{\lambda}_h)$. The same arguments that led to (4) now show

$$\bar{u}_h(x) = \mathcal{S}\left(\bar{\varphi}_h(x)\right). \qquad (19)$$

Similar as in (4), this relation holds for all $x \in \Omega_h$.

3.4 Analysis of the Convergence of the Discretizations

Let us denote by $(\mathbf{P}_h)$ any of the discrete problems $(\mathbf{P}_{h,i})$, $0 \le i \le 2$. Then, the following convergence results have been proved in [3] and [4].

Theorem 5. *For every $h > 0$ let $\bar{u}_h$ be a global solution of problem $(\mathbf{P}_h)$, then the sequence $\{\bar{u}_h\}_{h>0}$ is bounded in $L^\infty(\Omega)$ and there exist subsequences, denoted in the same way, converging to a point $\bar{u}$ in the weak-$\star$ topology on $L^\infty(\Omega)$. Any of these limit points is a solution of problem $(\mathbf{P})$. Moreover, we have*

$$\lim_{h\to 0}\left\{\|\bar{u} - \bar{u}_h\|_{L^\infty(\Omega_h)} + \|\bar{\lambda} - \bar{\lambda}_h\|_{L^\infty(\Omega_h)}\right\} = 0 \quad and \quad \lim_{h\to 0} J_h(\bar{u}_h) = J(\bar{u}). \ (20)$$

Theorem 6. *Let $\bar{u}$ be a strict local minimum of $(\mathbf{P})$, then there exists a sequence $\{\bar{u}_h\}_{h>0}$ of local minima of problems $(\mathbf{P}_h)$ such that (20) holds.*

Finally, we get the error estimates. To this end, let $\{\bar{u}_h\}_{h>0}$ denote a sequence of local minima of problems $(\mathbf{P}_h)$ such that $\|\bar{u} - \bar{u}_h\|_{L^\infty(\Omega_h)} \to 0$ when $h \to 0$, $\bar{u}$ being a local minimum of $(\mathbf{P})$; see the above theorems. Let us denote by $\bar{\lambda} \in \partial j(\bar{u})$ and $\bar{\lambda}_h \in \partial j_h(\bar{u}_h)$ the associated unique elements from the subdifferentials such that the first-order optimality conditions are satisfied. The goal is to obtain estimates of $\bar{u} - \bar{u}_h$ in the L^2 and L^∞ norms. We suppose that second-order sufficient conditions hold at $\bar{u}$. Then, the following estimate holds for problems $(\mathbf{P}_{h,0})$ and $(\mathbf{P}_{h,1})$

$$\|\bar{u} - \bar{u}_h\|_{L^\infty(\Omega_h)} + \|\bar{\lambda} - \bar{\lambda}_h\|_{L^\infty(\Omega_h)} + \|\bar{y} - \bar{y}_h\|_{L^\infty(\Omega_h)} + \|\bar{\varphi} - \bar{\varphi}_h\|_{L^\infty(\Omega_h)} \le C\,h.$$

The same estimate is obtained in the L^2 norm.

In the case of problem $(\mathbf{P}_{h,2})$, we have an error of order h^2 in the L^2 norm and of order $h^{2-n/p}\,|\log h|$ in the L^∞ norm.

4 Implementation and Numerical Examples

In this section we give some details concerning the implementation along with some numerical results. We focus on the particularities arising due to the presence of the $L^1(\Omega)$ norm in the objective. For simplicity of the presentation, we therefore restrict the discussion to a linear elliptic PDE with $a = 0$. The extension to semilinear problems will be obvious. Since the treatment of the variationally discretized problem differs in some aspects from the other two, we deal with the variational discretization separately.

4.1 Case of Discretized Control Space

We are going to present the discrete optimality systems of Sections 3.1 and 3.2 in terms of matrices and the coefficient-wise application of the soft-thresholding function $\mathcal{S}$, see (4) and Figure 1. In both cases, the state y and adjoint state φ are discretized by piecewise linear finite elements with standard nodal basis functions e_j, $j = 1, \dots, N_{v,i}$, where $N_{v,i}$ is the number of interior nodes. The control u is discretized either by the same basis functions, or by cell-wise constant basis functions f_j, $j = 1, \dots, N_T$, the number of cells in the mesh.

For simplicity of the presentation, we neglect that, in contrast to $\mathcal{Y}_h$, the discrete control space $\mathcal{U}_{h,1}$ of piecewise linears contains degrees of freedom located on the boundary Γ_h. In the continuous case, the condition (4) implies $\bar{u} = 0$ on Γ. The same holds for sufficiently fine discretizations, viz. whenever the coefficients on the boundary of the Carstensen quasi-interpolation of the adjoint state $\bar{\varphi}_h$ satisfy $|\pi_j(\bar{\varphi}_h)| \leq \mu$.

We define the following stiffness and mass matrices:

$$K = \left(\int_{\Omega_h} \nabla e_i \cdot \nabla e_j \, dx \right)_{i,j=1}^{N_{v,i}} \qquad M_{11} = \left(\int_{\Omega_h} e_i \, e_j \, dx \right)_{i,j=1}^{N_{v,i}}$$

$$M_{10} = \left(\int_{\Omega_h} e_i \, f_j \, dx \right)_{i,j=1}^{N_{v,i}, N_T} \qquad M_{00} = \left(\int_{\Omega_h} f_i \, f_j \, dx \right)_{i,j=1}^{N_T}$$

and $M_{01} = M_{10}^\top$.

Due to the quadrature formula (13) employed in the case of piecewise linear control discretizations, a diagonal mass matrix

$$M_{11,L} = \operatorname{diag} \left(\int_{\Omega_h} e_i \, dx \right)_{i=1}^{N_{v,i}}$$

appears in the objective (12) and thus in the discrete optimality system. This matrix can also be formed by taking the row sums (lumping) of the extended mass matrix (including boundary degrees of freedom), and then restricting it to the inner nodes. For convenience of notation, we also define $M_{00,L} = M_{00}$ since M_{00} is already diagonal.

Let us denote by $k \in \{0, 1\}$ the order of the control discretization. Denoting by $y_{d,h}$ the L^2-projection of y_d onto $\mathcal{Y}_h$, and identifying discrete functions with their coefficient vectors, we can write the discrete optimality system in the following way:

$$-M_{11}\,y_h \qquad\qquad\qquad +K\,\varphi_h \;=\; -M_{11}\,y_{d,h}, \qquad\qquad (21a)$$

$$u_h - \mathcal{S}\big(M_{kk,L}^{-1}M_{k1}\,\varphi_h\big) = 0, \qquad\qquad (21b)$$

$$K\,y_h - M_{1k}\,u_h \qquad\qquad\qquad = 0. \qquad\qquad (21c)$$

Equations (21a) and (21c) correspond to the discrete dual and primal state equations, respectively. Equation (21b) is simply an alternative formulation of the representation formulas (11) and (18), and it deserves an explanation.

We discuss the case $k = 0$ (piecewise constants) first. Then the i-th component of the argument of $\mathcal{S}$ is

$$\big[M_{00,L}^{-1}M_{01}\,\varphi_h\big]_i = \frac{1}{|T_i|}\int_{T_i}\varphi_h\,\mathrm{d}x, \quad i = 1,\ldots,N_T.$$

This shows the equivalence of (21b) and (11).

In the case $k = 1$ (piecewise linears), we first observe that

$$\big[M_{11}\,\varphi_h\big]_i = \int_{\Omega_h}\varphi_h\,e_i\,\mathrm{d}x,$$

giving in turn

$$\big[M_{11,L}^{-1}M_{11}\,\varphi_h\big]_i = \frac{1}{\int_{\Omega_h}e_i\,\mathrm{d}x}\int_{\Omega_h}\varphi_h\,e_i\,\mathrm{d}x = \pi_i(\varphi_h)$$

This shows the equivalence of (21b) and (18).

System (21) can be solved by a semi-smooth Newton method [6,10] acting on the coefficients of u_h. In a primal-dual active set formulation, one distinguishes three active sets, corresponding to $u \in \{\alpha, 0, \beta\}$, respectively. The implementation is quite simple since there are only two types of operations: matrix-vector multiplications and coefficient-wise evaluation of the nonlinearities $\mathcal{S}$ and $\mathcal{S}'$.

Remark 7. It is worth noting that the reasoning above shows that the Carstensen quasi-interpolation is represented by $M_L \Pi_h(v_h) = M\,v_h$ for all ansatz spaces with bases which sum up to one, where M is the mass matrix and M_L is the diagonally lumped mass matrix.

4.2   Case of Variational Discretization

In the case of variational control discretization, we use (19) to eliminate the control from the optimality system to obtain

$$-M_{11}\,y_h + K\,\varphi_h \;=\; -M_{11}\,y_{d,h}, \qquad\qquad (22a)$$

$$K\,y_h - b(\varphi_h) = 0, \qquad\qquad (22b)$$

where

$$b(\varphi_h) = \left(\int_{\Omega_h} \mathcal{S}(\varphi_h(x)) \, e_i(x) \, dx \right)_{i=1}^{N_{v,i}}.$$

System (22) can be solved again by a semi-smooth Newton method. Its implementation is more complicated since the evaluation of the residual and the generalized Jacobian involves nonlinearities which do not act in a coefficient-wise manner. In particular, one has to integrate over polyhedric subsets of cells since the integrand in $b(\varphi_h)$ and its derivative $b'(\varphi_h)$ has kinks and jumps, respectively.

4.3   Numerical Experiments

We conclude by presenting some numerical results. The FE library FEniCS [9,12] was used for the discretization related aspects of the implementation. In addition, the computational geometry toolbox CGAL [5] was used in the variationally

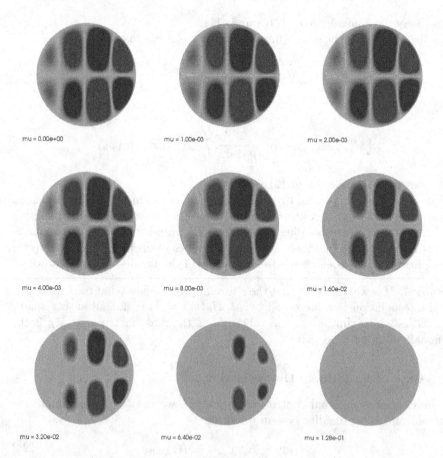

Fig. 2. Solution of an example problem for different values of $\mu = 0$ and $\mu = 2^i \cdot 10^{-3}, i = 0, \ldots, 7$ by piecewise linear controls

discrete case to determine the subsets for the integration of $b(\varphi_h)$. Figure 2 shows the dependence of the sparsity structure on the parameter μ weighting the L^1 norm. It confirms the expectation that the solution becomes sparser as μ grows. For sufficiently large μ, the unique (local and global) optimal control will be identically zero. The problem involves a semilinear equation and the complete data can be found in [3,4] along with more examples and a verification of the convergence orders.

References

1. Brenner, S.C., Scott, L.R.: The Mathematical Theory of Finite Element Methods. Springer, Heidelberg (1994)
2. Carstensen, C.: Quasi-interpolation and a posteriori error analysis in finite element methods. M2AN Math. Model. Numer. Anal. 33(6), 1187–1202 (1999)
3. Casas, E., Herzog, R., Wachsmuth, G.: Analysis of an elliptic control problem with non-differentiable cost functional. Technical report, TU Chemnitz (2010)
4. Casas, E., Herzog, R., Wachsmuth, G.: Approximation of sparse controls in semilinear equations by piecewise linear functions. Technical report, TU Chemnitz (2011)
5. CGAL. Computational Geometry Algorithms Library, http://www.cgal.org
6. Chen, X., Nashed, Z., Qi, L.: Smoothing methods and semismooth methods for nondifferentiable operator equations. SIAM Journal on Numerical Analysis 38, 1200–1216 (2000)
7. Clason, C., Kunisch, K.: A duality-based approach to elliptic control problems in non-reflexive Banach spaces. In: ESAIM: COCV (2010), doi:10.1051/cocv/2010003
8. Daubechies, I., Defrise, M., De Mol, C.: An iterative thresholding algorithm for linear inverse problems with a sparsity constraint. Communications on Pure and Applied Mathematics 57(11), 1413–1457 (2004)
9. FEniCS. FEniCS project (2007), http://www.fenicsproject.org/
10. Hintermüller, M., Ito, K., Kunisch, K.: The primal-dual active set strategy as a semismooth Newton method. SIAM Journal on Optimization 13(3), 865–888 (2002)
11. Hinze, M.: A variational discretization concept in control constrained optimization: The linear-quadratic case. Comp. Optim. Appls. 30, 45–61 (2005)
12. Logg, A., Wells, G.N.: Dolfin: Automated finite element computing. ACM Trans. Math. Softw. 37, 20:1–20:28 (2010)
13. Stadler, G.: Elliptic optimal control problems with L^1-control cost and applications for the placement of control devices. Comp. Optim. Appls. 44(2), 159–181 (2009)
14. Wachsmuth, G., Wachsmuth, D.: Convergence and regularization results for optimal control problems with sparsity functional. In: ESAIM: COCV (2010), doi:10.1051/cocv/2010027

A Non-standard Finite Element Method
Based on Boundary Integral Operators

Clemens Hofreither[1], Ulrich Langer[2], and Clemens Pechstein[2]

[1] Johannes Kepler University Linz, DK Computational Mathematics,
Altenberger Str. 69, 4040 Linz, Austria
`clemens.hofreither@dk-compmath.jku.at`
[2] Johannes Kepler University Linz, Institute of Computational Mathematics,
Altenberger Str. 69, 4040 Linz, Austria
`ulanger@numa.uni-linz.ac.at`,
`clemens.pechstein@numa.uni-linz.ac.at`

Abstract. This paper provides an overview over our results on the construction and analysis of a non-standard finite element method that is based on the use of boundary integral operators for constructing the element stiffness matrices. This approach permits polyhedral element shapes as well as meshes with hanging nodes. We consider the diffusion equation and convection-diffusion-reaction problems as our model problems, but the method can also be generalized to more general problems like systems of partial differential equations. We provide a rigorous H^1- and L_2-error analysis of the method for smooth and non-smooth solutions. This a priori discretization error analysis is only done for the diffusion equation. However, our numerical results also show good performance of our method for convection-dominated diffusion problems.

Keywords: non-standard FEM, boundary integral operators, Trefftz method, polyhedral meshes, convection-diffusion-reaction problems.

1 Introduction

We consider and analyze a non-standard finite element method (FEM) that was introduced by Copeland, Langer, and Pusch [4] and that is based on element-local boundary integral operators. This non-standard FEM permits polyhedral element shapes as well as meshes with hanging nodes. The method employs elementwise PDE-harmonic trial functions, and can thus be interpreted as a local Trefftz method. Indeed, E. Trefftz proposed to approximate the solution of the Dirichlet problem for the Laplace equation by means of a finite superposition of harmonic functions with unknown coefficients which are chosen in such a way that the Dirichlet boundary conditions are approximated in some weak sense [20]. In our approach, we do not explicitly use PDE-harmonic trial functions for computing the element stiffness matrices. Instead these local stiffness matrices are generated by a local boundary element method (BEM) since the intergrals over the polyhedral elements can be transformed to boundary integrals. This is the reason why we call this method BEM-based FEM. This construction

I. Lirkov, S. Margenov, and J. Waśniewski (Eds.): LSSC 2011, LNCS 7116, pp. 28–39, 2012.

principle requires the explicit knowledge of the fundamental solution of the partial differential operator, but only locally in every polyhedral element. This allows us to solve PDEs with elementwise constant coefficients. The BEM-based FEM has its historical root not only in the Trefftz method, but also in the symmetric boundary domain decomposition method proposed by Hsiao and Wendland [12]. Moreover, the idea of decomposing the computational domain into smaller subdomains can already be found in Trefftz' paper and is of course the central topic of Hsiao's and Wendland's paper on domain decomposition methods.

In this paper we consider the diffusion equation and convection-diffusion-reaction problems as our model problems, but the method can also be generalized to more general problems like the Helmholtz equation and systems of PDEs including the linear elasticity system and the time-harmonic Maxwell equations with elementwise constant or, at least, elementwise smooth coefficients [4,3]. We review the results of the papers [9] and [8] which provided a rigorous H^1- and L_2-error analysis, respectively. This analysis is not trivial since the geometric properties of the polyhedral elements and variational crimes arising from the approximation of the Steklov-Poincaré operator must be handled appropriately. We generalize these discretization error estimates to non-smooth solutions by means of space interpolation technique. We mention that non-smooth solutions are typical for PDE problems with jumping coefficients. This a priori discretization error analysis is only done for the diffusion equation. However, first experiments published in [10] have already shown very good numerical performance for convection-dominated diffusion problems as well.

2   The BEM-Based FEM for Convection-Diffusion-Reaction Problems

2.1   Skeletal Variational Formulation

We follow the approach taken in [9] for the Laplace equation in order to derive the so-called skeletal variational formulation for a convection-diffusion-reaction problem. We consider the Dirichlet boundary value problem

$$Lu = -\mathrm{div}(A\nabla u) + b \cdot \nabla u + cu = 0 \quad \text{in } \Omega,$$
$$u = g \quad \text{on } \partial\Omega \tag{1}$$

in a bounded Lipschitz domain $\Omega \subset \mathbb{R}^3$ with the boundary $\partial\Omega$ as our model problem. Here $A(x) \in \mathbb{R}^{3\times3}$, $b(x) \in \mathbb{R}^3$, and $c(x) \in \mathbb{R}$ are the coefficient functions of the partial differential operator L, and g is the given Dirichlet data in $H^{1/2}(\partial\Omega)$. We assume that $A(x)$ is symmetric and uniformly positive, and that $c(x) \geq 0$. While we restrict ourselves to the pure Dirichlet problem here, the generalization to mixed Dirichlet/Neumann problems is straightforward.

The corresponding variational formulation reads as follows: find $u \in H^1(\Omega)$ with $\gamma_\Omega^0 u = g$ such that

$$\int_\Omega (A\nabla u \cdot \nabla v + b \cdot \nabla u\, v + cuv)\, dx = 0 \quad \forall v \in H_0^1(\Omega), \tag{2}$$

where $\gamma_\Omega^0 : H^1(\Omega) \to H^{1/2}(\partial\Omega)$ refers to the Dirichlet trace operator from the domain Ω to its boundary and $H_0^1(\Omega) = \{v \in H^1(\Omega) : \gamma_\Omega^0 v = 0\}$. We require that the coefficients A, b, c are $L^\infty(\Omega)$ and that there exists a unique solution of (2).

Assume now that we are given a finite decomposition $\mathcal{T}$ of Ω into mutually disjoint Lipschitz polyhedra. As opposed to a standard finite element method, we do not require the existence of a reference element to which all elements $T \in \mathcal{T}$ can be mapped, but rather allow $\mathcal{T}$ to contain an arbitrary mixture of polyhedral element shapes. For reasons that will become clear later on, we require the coefficients $A(x)$, $b(x)$, and $c(x)$ to be piecewise constant with respect to the polyhedral mesh $\mathcal{T}$.

From the variational formulation and the density of $C_0^\infty(T)$ in $L_2(T)$, it follows that $\operatorname{div}(A\nabla u) = b \cdot \nabla u + cu \in L_2(T)$ for every element $T \in \mathcal{T}$, and we can conclude that the flux $A\nabla u$ is in $H(\operatorname{div}, T)$. Let n_T denote the outward unit normal vector on ∂T. Then the flux has a well-defined Neumann trace $\gamma_T^1 u :=$ $A\nabla u \cdot n_T \in H^{-1/2}(\partial T)$, also called the *conormal derivative* of u; cf. [6]. Moreover, we have the generalized Green's identity

$$\int_T A\nabla u \cdot \nabla v \, dx = -\int_T \operatorname{div}(A\nabla u)v \, dx + \langle \gamma_T^1 u, \gamma_T^0 v \rangle \qquad \forall v \in H^1(T). \quad (3)$$

Here and below, $\langle \cdot, \cdot \rangle$ denotes the duality pairing on $H^{-1/2}(\partial T) \times H^{1/2}(\partial T)$ (the particular element boundary ∂T will always be clear by context). Inserting (3) into (2) and remembering that $Lu = 0$ in $L_2(T)$, we obtain

$$0 = \sum_{T \in \mathcal{T}} \int_T (A\nabla u \cdot \nabla v + b \cdot \nabla u \, v + cuv) \, dx = \sum_{T \in \mathcal{T}} \Big(\underbrace{\int_T Lu \, v \, dx}_{=0} + \langle \gamma_T^1 u, \gamma_T^0 v \rangle \Big).$$

Fix now some element $T \in \mathcal{T}$ and observe that $u|_T$ is the unique solution of the local problem

$$L\varphi = 0, \qquad \gamma_T^0 \varphi = \gamma_T^0 u.$$

Here again we must require that these local problems do indeed have unique solutions. If we denote by $S_T : H^{1/2}(\partial T) \to H^{-1/2}(\partial T)$ the *Steklov-Poincaré operator* or *Dirichlet-to-Neumann map* for this local problem, we therefore have $\gamma_T^1 u = S_T \gamma_T^0 u$, yielding the equation

$$\sum_{T \in \mathcal{T}} \langle S_T \gamma_T^0 u, \gamma_T^0 v \rangle = 0 \qquad \forall v \in H_0^1(\Omega).$$

The above formulation operates not on the functions u and v themselves, but only on their Dirichlet traces on the element boundaries ∂T. The idea is therefore to work with function spaces which are defined only on these boundaries. We call $\Gamma_S := \bigcup_{T \in \mathcal{T}} \partial T$ the *skeleton* of the mesh $\mathcal{T}$ and introduce a skeletal function space $W = H^{1/2}(\Gamma_S)$ consisting of the traces of all functions from $H^1(\Omega)$ on Γ_S. We then search for a skeletal function $\hat{u} \in W$ which satisfies the Dirichlet boundary condition $\hat{u}|_{\partial\Omega} = g$ as well as the skeletal variational formulation

$$\sum_{T \in \mathcal{T}} \langle S_T \hat{u}|_{\partial T}, \, \hat{v}|_{\partial T} \rangle = 0 \qquad \forall \hat{v} \in W_0 = \{\hat{v} \in W : \hat{v}|_{\partial \Omega} = 0\}. \tag{4}$$

This skeletal variational formulation is equivalent to the standard variational formulation (2) in the sense that the traces $\hat{u}|_{\partial T} \in H^{1/2}(\partial T)$ obtained from (4) match the traces $\gamma_T^0 u$ of the function $u \in H^1(\Omega)$ obtained from (2). Conversely, $u|_T$ can be recovered from $\hat{u}|_{\partial T}$ by solving a local Dirichlet problem in T, i.e., $u|_T = \mathcal{H}_T \hat{u}|_{\partial T}$ with the element-local harmonic extension operator $\mathcal{H}_T : H^{1/2}(\partial T) \to H^1(T)$. This motivates using (4) as a starting point for discretization in order to obtain a numerical method for (1).

We remark that another interpretation of (4) is that of a weak enforcement of the continuity of conormal derivatives on inter-element boundaries.

2.2 Boundary Integral Operators

Evaluating the Dirichlet-to-Neumann map S_T used above essentially corresponds to solving a local problem $L\varphi = 0$ on T with the given Dirichlet data and then obtaining the conormal derivative $\gamma_T^1 \varphi$ of its solution. These local problems are in general not analytically solvable, and we approximate their solutions by the Galerkin discretization of element-local boundary integral equations. In the following, we outline some standard results on boundary integral operators. A more detailed treatment of these topics can be found in, e.g., [13,15,17,18].

We say that a *fundamental solution* of the partial differential operator L is a function $G(x, y)$ such that $L_x G(x, y) = \delta(y - x)$, where δ is the Dirac δ-distribution and $x, y \in \mathbb{R}^d$. Fundamental solutions for L from (1) with constant coefficients A, b, c are given in [17]; in particular, in $\mathbb{R}^3$ and under the assumption $c + \|b\|_{A^{-1}}^2 \geq 0$, we have

$$G(x - y) = G(z) = \frac{1}{4\pi \sqrt{\det A}} \frac{\exp\left(b^\top A^{-1} z - \lambda \|z\|_{A^{-1}}\right)}{\|z\|_{A^{-1}}},$$

where $\|x\|_{A^{-1}} = \sqrt{x^\top A^{-1} x}$ and $\lambda = \sqrt{c + \|b\|_{A^{-1}}^2}$.

More generally, we will assume that the coefficients A, b, c are constant only within each element. This leads to a potentially different fundamental solution in each element T, in the following denoted by $G_T(x, y)$, and allows us to treat PDEs with piecewise constant coefficients.

We now introduce the boundary integral operators

$$V_T : H^{-1/2}(\partial T) \to H^{1/2}(\partial T), \quad K_T : H^{1/2}(\partial T) \to H^{1/2}(\partial T),$$

$$K_T' : H^{-1/2}(\partial T) \to H^{-1/2}(\partial T), \quad D_T : H^{1/2}(\partial T) \to H^{-1/2}(\partial T),$$

called, in turn, the *single layer potential*, *double layer potential*, *adjoint double layer potential*, and *hypersingular* operators. For sufficiently regular arguments, they admit the integral representations

$$(V_T v)(y) = \int_{\partial T} G_T(x,y)v(x)\,ds_x,$$

$$(K_T u)(y) = \int_{\partial T} \widetilde{\gamma^1_{T,x}} G_T(x,y)u(x)\,ds_x,$$

$$(K'_T v)(y) = \int_{\partial T} \gamma^1_{T,y} G_T(x,y)v(x)\,ds_x,$$

$$(D_T u)(y) = -\gamma^1_{T,y} \int_{\partial T} \widetilde{\gamma^1_{T,x}} G_T(x,y)\big(u(x)-u(y)\big)\,ds_x,$$

where $\gamma^1_{T,y}$ refers to the conormal derivative γ^1_T with respect to the variable y, and $\widetilde{\gamma^1_{T,x}}$ refers to the modified conormal derivative

$$\widetilde{\gamma^1_T}u = \gamma^1_T u + (b \cdot n_T)\gamma^0_T u,$$

associated with the adjoint problem, with respect to the variable x.

In the literature, we find two representations of the Steklov-Poincaré operator in terms of the boundary integral operators, namely,

$$S_T = V_T^{-1}(\tfrac{1}{2}I + K_T) = D_T + (\tfrac{1}{2}I + K'_T)V_T^{-1}(\tfrac{1}{2}I + K_T). \qquad (5)$$

The first one is called the non-symmetric representation, and the latter is called the symmetric representation of the Steklov-Poincaré operator.

2.3   Discretization

Discretization of the Skeletal Function Space. We employ a Galerkin approach to the discretization of the skeletal variational formulation (4). To this end, we first discretize every element boundary ∂T by a conforming triangulation $\mathcal{F}_T$ composed of triangles. The number of triangular boundary elements per polyhedral element should be uniformly bounded. Furthermore, we assume that the element discretizations are *matching* in the sense that, for any two elements T_1 and T_2 having a common interface $\Gamma_{12} = \overline{T}_1 \cap \overline{T}_2 \neq \emptyset$, any triangle $\tau \in \mathcal{F}_{T_1}$ with $\tau \cap \Gamma_{12} \neq \emptyset$ should also belong to $\mathcal{F}_{T_2}$. In other words, inter-element boundaries must be triangulated identically in both elements. Finally, we require that the element meshes $\mathcal{F}_T$ are quasi-uniform and shape-regular in the usual sense, with constants which are uniform over all elements T.

This construction naturally gives us a triangulation $\mathcal{F} = \bigcup_{T \in \mathcal{T}} \mathcal{F}_T$ of the skeleton, on which we now construct a discrete trial space $W^h \subset W$ of piecewise (per boundary triangle $\tau \in \mathcal{F}$) linear and continuous functions. The space W^h is spanned by the nodal functions $\{\phi_i\}$, where i enumerates the vertices of the triangulation $\mathcal{F}$, and where each basis function ϕ_i has the value 1 in the i-th vertex, 0 in all other vertices, and is linear on each triangle $\tau \in \mathcal{F}$. The space $W_0^h = W^h \cap W_0$ is constructed from W^h by excluding those basis functions which belong to vertices on $\partial \Omega$.

Assuming that the given Dirichlet data g is piecewise linear, we thus arrive at the following Galerkin equations as the discrete version of (4): find $u^h \in W^h$ such that $u^h|_{\partial\Omega} = g$ and

$$\sum_{T\in\mathcal{T}} \langle S_T u^h|_{\partial T}, v^h|_{\partial T}\rangle = 0 \qquad \forall v^h \in W_0^h. \tag{6}$$

In the general case, the Dirichlet data g can be approximated piecewise linearly by interpolation (if continuous) or by L_2 projection onto the boundary triangles.

Discretization of the Dirichlet-to-Neumann Map. The boundary integral operator representation (5) of S_T contains the inverse of the single layer potential operator V_T, which is in general not computable exactly. Hence, we need to approximate the bilinear form $\langle S_T \cdot, \cdot\rangle$. To do this, we employ a mixed piecewise linear/piecewise constant scheme, where Dirichlet data is approximated piecewise linearly, while Neumann data is approximated by piecewise constants, as described in, e.g., [5,18,11].

Observe that, due to (5), the Neumann data corresponding to a function $u_T \in H^{1/2}(\partial T)$ can be written as

$$S_T u_T = D_T u_T + (\tfrac{1}{2}I + K_T')t_T \tag{7}$$

with $t_T = V_T^{-1}(\tfrac{1}{2}I + K_T)u_T \in H^{-1/2}(\partial T)$. Again referring to (5), we see that $t_T = S_T u_T$ is itself already the sought Neumann data. We will approximate S_T by first approximating t_T in a suitable space of piecewise constant functions and then substituting this approximation for t_T in (7).

Let $\{\phi_{T,i}\}$ denote the nodal piecewise linear functions restricted to the local mesh $\mathcal{F}_T$, where now i enumerates the vertices only of $\mathcal{F}_T$. Furthermore, introduce a space of piecewise constant boundary functions Z_T^h spanned by the basis $\{\psi_{T,k}\}$, where k enumerates the triangles $\tau \in \mathcal{F}_T$, such that $\psi_{T,k} \equiv 1$ on the k-th triangle and $\psi_{T,k} \equiv 0$ on all other triangles.

For any function $u_T \in H^{1/2}(\partial T)$, we compute the approximation $t_T^h \approx t_T$ of its Neumann data by the Galerkin projection of the equation $V_T t_T = (\tfrac{1}{2}I + K_T)u_T$ to the piecewise constant functions. In other words, we seek $t_T^h \in Z_T^h$ satisfying the variational equation

$$\langle \psi^h, V_T t_T^h\rangle = \langle \psi^h, (\tfrac{1}{2}I + K_T)u_T\rangle \qquad \forall \psi^h \in Z_T^h.$$

A computable approximation to $S_T u_T$ is then given by

$$\widetilde{S}_T u_T := D_T u_T + (\tfrac{1}{2}I + K_T')t_T^h,$$

and our fully discretized variational formulation takes the following form: find $u^h \in W^h$ such that $u^h|_{\partial\Omega} = g$ and

$$\sum_{T\in\mathcal{T}} \langle \widetilde{S}_T u^h|_{\partial T}, v^h|_{\partial T}\rangle = 0 \qquad \forall v^h \in W_0^h. \tag{8}$$

The corresponding stiffness matrix is assembled, as in the FEM, from the contributions from element stiffness matrices. The latter are computed using the fully numerical integration technique described by Sauter and Schwab [17].

3   Discretization Error Analysis for the Diffusion Equation

There is no complete error analysis of the presented method for general elliptic operators L of the form (1). However, for the special case of the Laplace operator, rigorous error estimates in the H^1 and L_2 norms are given in [9] and [8], respectively. Even in the Laplacian case, the analysis is nontrivial. Among the main technical hurdles are the general polyhedral element shapes, which means that no reference element is available. This rules out the standard technique widely used in FEM analysis of transforming to the reference element, estimating a quantity of interest there, and transforming back. In particular, this complicates the derivation of Dirichlet and Neumann approximation properties for our discrete skeletal function spaces. An auxiliary regular tetrahedral decomposition of each polyhedral element T is used as an analytical tool to overcome these difficulties. Furthermore, the approximation of S_T by $\widetilde{S}_T$ is a "variational crime" leading to a consistency error which has to be treated either by Strang's lemma [9] or by passing to an equivalent mixed formulation [8]. In both cases, estimating the consistency error requires novel results on explicit bounds for boundary integral operators [16] and a notion of regularity for polyhedral elements which relies on uniform bounds for Poincaré constants and the so-called Jones parameter. The regularity assumptions used in the previous works are outlined below.

Assumption 1. *We assume that the polyhedral mesh $\mathcal{T}$ satisfies the following conditions.*

- *There is a small, fixed integer uniformly bounding the number of boundary triangles of every element.*
- *Every element $T \in \mathcal{T}$ has an auxiliary conforming, quasi-regular, tetrahedral triangulation (cf. [2]) with regularity parameters which are uniform across all elements.*

Definition 1 (Uniform domain [14]). *A bounded and connected set $D \subset \mathbb{R}^d$ is called a* uniform domain *if there exists a constant C_U such that any pair of points $x_1 \in D$ and $x_2 \in D$ can be joined by a rectifiable curve $\gamma(t) : [0,1] \to D$ with $\gamma(0) = x_1$ and $\gamma(1) = x_2$, such that the arc length of γ is bounded by $C_U |x_1 - x_2|$ and*

$$\min_{i=1,2} |x_i - \gamma(t)| \le C_U \operatorname{dist}(\gamma(t), \partial D) \qquad \forall t \in [0,1].$$

If D is a uniform domain, we denote the smallest such constant C_U by $C_U(D)$ and call it the Jones parameter *of D.*

Any Lipschitz domain is a uniform domain. However, its Jones parameter may be arbitrarily large.

Definition 2 (Poincaré constant). *For a uniform domain D, let $C_P(D)$ be the smallest constant such that*

$$\inf_{c \in \mathbb{R}} \|v - c\|_{L_2(D)} \le C_P(D) \operatorname{diam}(D) |v|_{H^1(D)} \qquad \forall v \in H^1(D).$$

For convex domains D, one can show that $C_P(D) \leq \pi^{-1}$, cf. [1]. Estimates for star-shaped domains can be found in [19,21].

Assumption 2. *We assume that there are constants $C_U^* > 0$ and $C_P^* > 0$ such that, for all $T \in \mathcal{T}$,*

$$C_U(T) \leq C_U^*, \qquad C_U(B_T \setminus \overline{T}) \leq C_U^*,$$
$$C_P(T) \leq C_P^*, \qquad C_P(B_T \setminus \overline{T}) \leq C_P^*,$$

where B_T is a ball (or a suitable Lipschitz domain) enclosing T which satisfies $\operatorname{dist}(\partial B_T, \partial T) \geq \frac{1}{2} \operatorname{diam}(T)$.

Under the assumptions stated above, we may now formulate the main results on the discretization error. Note that the discrete solution $u^h \in W^h$ of (8) is only defined on the skeleton. In order to compare it with the exact solution $u_\Omega \in H^1(\Omega)$ of (2) with traces $u \in W$, we use the harmonic extension operator $\mathcal{H}_T$. Within any element T, the error to be estimated is thus

$$u_\Omega - \mathcal{H}_T(u^h|_{\partial T}) = \mathcal{H}_T\left((u - u^h)|_{\partial T}\right),$$

or globally $\mathcal{H}_S(u - u^h)$ with the piecewise harmonic extension operator

$$\mathcal{H}_S : H^{1/2}(\Gamma_S) \to H^1(\Omega), \qquad \forall T \in \mathcal{T} : (\mathcal{H}_S v)|_T = \mathcal{H}_T(v|_{\partial T}).$$

Theorem 1 (H^1 error estimate, [9]). *Let $L = -\Delta$, and let the mesh $\mathcal{T}$ satisfy Assumptions 1 and 2. Assume further that the given Dirichlet data g is piecewise linear and that the exact solution u_Ω of (2) lies in $H^2(\Omega)$. With $u \in W$ the solution of the skeletal variational formulation (4) and $u^h \in W^h$ the solution of the discretized skeletal formulation (8), we have the error estimate*

$$|\mathcal{H}_S(u - u^h)|_{H^1(\Omega)} \leq C \left(\sum_{T \in \mathcal{T}} h_T^2 |u_\Omega|_{H^2(T)}^2 \right)^{1/2} \leq C\, h\, |u_\Omega|_{H^2(\Omega)},$$

where the constant C depends only on the mesh regularity parameters, $h_T = \operatorname{diam} T$ denotes the element diameters, and $h = \max_T h_T$ denotes the mesh size.

Proof (Outline). The proof hinges on three results, namely, (i) a quasi-optimal bound for the discretization error in terms of the best approximation error for the exact Dirichlet and Neumann data, (ii) an approximation error estimate for the Dirichlet data, and (iii) an approximation error estimate for the Neumann data. All of these estimates need to be made explicit in terms of the mesh regularity parameters, and problem-adapted norms have to be used. The details can be found in [9].

Theorem 2 (L^2 error estimate, [8]). *Let the assumptions of Theorem 1 be satisfied. Assume further that the adjoint problem is H^2-coercive, i.e., that the solution $w \in H_0^1(\Omega)$ of*

$$\int_\Omega \nabla v \cdot \nabla w \, dx = \int_\Omega \mathcal{H}_S(u - u^h)\, v \, dx \qquad \forall v \in H_0^1(\Omega)$$

lies in $H^2(\Omega)$ and satisfies the estimate

$$|w|_{H^2(\Omega)} \leq C \, \|\mathcal{H}_S(u - u^h)\|_{L_2(\Omega)}.$$

Then we have the quasi-optimal L_2 discretization error estimate

$$\|\mathcal{H}_S(u - u^h)\|_{L_2(\Omega)} \leq C \, h^2 \, |u_\Omega|_{H^2(\Omega)},$$

where the constant C depends only on the mesh regularity parameters.

Proof (Outline). Due to the consistency error introduced by approximating the bilinear form $\langle S_{T\cdot}, \cdot \rangle$, Galerkin orthogonality is violated and the usual Aubin-Nitsche technique is not available. This may be remedied by passing to an equivalent mixed formulation, searching for the unknowns $(u, \bigotimes_{T \in \mathcal{T}} t_T)$, which makes the error by the approximation $\widetilde{S}_T \approx S_T$ explicit and thus restores Galerkin orthogonality. Many of the technical tools used to prove Theorem 1 can then be reused. The details can be found in [8].

In many practical applications, we have to deal with heterogeneous coefficients. If the coefficients in the PDE (1) have jumps across interfaces, or if the computational domain is non-convex and non-smooth, or if there are changes in the boundary conditions, then we cannot expect a smooth solution u_Ω, i.e., the assumption $u_\Omega \in H^2(\Omega)$ is too restrictive. The following theorem provides a convergence rate estimate for this case too.

Theorem 3 (H^1 error estimate under reduced regularity assumptions).
Let the assumptions of Theorem 1 hold with the exception of the regularity assumption imposed on the exact solution, $u_\Omega \in H^2(\Omega)$. Instead, we only assume $u_\Omega \in H^{1+s}(\Omega)$ with some $0 < s \leq 1$. Then we have the error estimate

$$\|\mathcal{H}_S(u - u^h)\|_{H^1(\Omega)} \leq C \, h^s \, \|u_\Omega\|_{H^{1+s}(\Omega)},$$

where the constant C again depends only on the mesh regularity parameters.

Proof (Outline). Theorem 1 together with Friedrichs' inequality implies

$$\|\mathcal{H}_S(u - u^h)\|_{H^1(\Omega)} \leq C \, h \, \|u_\Omega\|_{H^2(\Omega)}.$$

From the proof of Theorem 1, one easily obtains the stability estimate

$$\|\mathcal{H}_S(u - u^h)\|_{H^1(\Omega)} \leq C \, \|u_\Omega\|_{H^1(\Omega)}.$$

The statement then follows by a space interpolation argument (cf. [15]).

4   Numerical Results for a Convection-Diffusion Problem

The following example with strongly varying diffusion coefficients is adapted from [7]. We solve

$$-\operatorname{div}(A\nabla u) + (\beta, \beta, \beta)^\top \cdot \nabla u = 0 \quad \text{in } \Omega,$$

$$u = g \quad \text{on } \partial\Omega,$$

where $\Omega = (0,1)^3$ is the unit cube, $\beta = 50$, and

$$A(x,y,z) = \begin{cases} 10^4, & \frac{1}{3} < x,y,z < \frac{2}{3}, \\ 1, & \text{else,} \end{cases} \qquad g(x,y,z) = \begin{cases} 1, & z = 0, \\ 0, & \text{else.} \end{cases}$$

The non-symmetric linear systems resulting from either method are solved using GMRES, restarted every 500 iterations. Due to the strongly varying diffusion, GMRES takes a very high number of iterations to converge to a given accuracy, even on relatively small problems. A simple row scaling preconditioner can mitigate this problem ([7]), and we thus modify the linear system to be solved,

$$\underline{S}\,\underline{u}^h = \underline{f}^h \qquad \longrightarrow \qquad D\underline{S}\,\underline{u}^h = D\underline{f}^h,$$

where $\underline{S}$ is the stiffness matrix, $\underline{u}^h$ and $\underline{f}^h$ are the vectors corresponding to the discrete solution and right-hand side, respectively, and

$$D = \text{diag}(1/\|\underline{S}_1\|_p, \ldots, 1/\|\underline{S}_n\|_p)$$

is a diagonal matrix containing the reciprocal p-norms of the rows of the stiffness matrix $\underline{S}$. In our experiments, we chose $p = 1$.

Figure 1 shows a solution computed using the BEM-based FEM for a mesh with 456 769 vertices and mesh size $h \approx 0.0232924$. Table 1 displays GMRES iteration numbers for a standard FEM using piecewise linear trial functions and the BEM-based FEM, without and with the row scaling preconditioner.

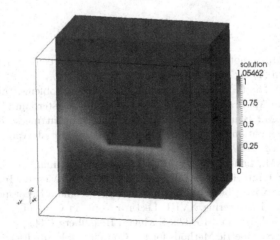

Fig. 1. Cross section through Ω at $x = 0.5$, computed by BEM-based FEM

Table 1. Iteration numbers using GMRES(500), without and with row scaling

		degrees of freedom				
		199	1153	7921	59041	456769
without row scaling	FEM	59	178	469	6042	19597
	BEM-based FEM	57	123	442	5356	15756
with row scaling	FEM	42	87	135	229	440
	BEM-based FEM	50	87	129	215	376

5 Conclusion and Outlook

We have summarized recent results on the so-called BEM-based FEM, giving a brief derivation of the method for a general elliptic partial differential equation and outlining the analysis in the Laplacian case for solutions of full as well as reduced regularity. We also have given new numerical results for a convection-diffusion benchmark problem.

The employed solution technique, namely, GMRES with a simple row scaling preconditioner, is clearly not optimal, and the development of optimal solvers will be a topic of future research. Furthermore, previous experiments in [10] have shown that, while stability of the method is superior to a standard piecewise linear FEM, additional stabilization is required for convection-dominated problems.

Acknowledgments. The support by the Austrian Science Fund (FWF) under grant DK W1214 is gratefully acknowledged.

References

1. Bebendorf, M.: A note on the Poincaré inequality for convex domains. Z. Anal. Anwendungen 22(4), 751–756 (2003)
2. Ciarlet, P.G.: The finite element method for elliptic problems. Studies in Mathematics and its Applications, vol. 4. North-Holland, Amsterdam (1987)
3. Copeland, D.M.: Boundary-element-based finite element methods for Helmholtz and Maxwell equations on general polyhedral meshes. Int. J. Appl. Math. Comput. Sci. 5(1), 60–73 (2009)
4. Copeland, D.M., Langer, U., Pusch, D.: From the Boundary Element Method to Local Trefftz Finite Element Methods on Polyhedral Meshes. In: Bercovier, M., Gander, M.J., Kornhuber, R., Widlund, O. (eds.) Domain Decomposition Methods in Science and Engineering XVIII. Lecture Notes in Computational Science and Engineering, vol. 70, pp. 315–322. Springer, Heidelberg (2009)
5. Costabel, M.: Symmetric Methods for the Coupling of Finite Elements and Boundary Elements. In: Brebbia, C., Wendland, W., Kuhn, G. (eds.) Boundary Elements IX, pp. 411–420. Springer, Heidelberg (1987)
6. Girault, V., Raviart, P.: Finite element methods for Navier-Stokes equations. Springer Series in Computational Mathematics, vol. 5. Springer, Berlin (1986)

7. Gordon, D., Gordon, R.: Row scaling as a preconditioner for some nonsymmetric linear systems with discontinuous coefficients. J. Computational Applied Mathematics 234(12), 3480–3495 (2010)
8. Hofreither, C.: L_2 error estimates for a nonstandard finite element method on polyhedral meshes. J. Numer. Math. 19(1), 27–39 (2011)
9. Hofreither, C., Langer, U., Pechstein, C.: Analysis of a non-standard finite element method based on boundary integral operators. Electronic Transactions on Numerical Analysis 37, 413–436 (2010),
http://etna.mcs.kent.edu/vol.37.2010/pp413-436.dir
10. Hofreither, C., Langer, U., Pechstein, C.: A non-standard finite element method for convection-diffusion-reaction problems on polyhedral meshes. In: Proceedings of the Third Conference of the Euro-American Consortium for Promoting the Application of Mathematics in Technical and Natural Sciences. American Institute of Physics (2011)
11. Hsiao, G.C., Steinbach, O., Wendland, W.L.: Domain decomposition methods via boundary integral equations. Journal of Computational and Applied Mathematics 125(1-2), 521–537 (2000)
12. Hsiao, G.C., Wendland, W.L.: Domain decomposition in boundary element methods. In: Glowinski, R., Kuznetsov, Y.A., Meurant, G., Périaux, J., Widlund, O.B. (eds.) Proceedings of the Fourth International Symposium on Domain Decomposition Methods for Partial Differential Equations, Moscow, May 21-25, 1990, pp. 41–49. SIAM, Philadelphia (1991)
13. Hsiao, G.C., Wendland, W.L.: Boundary Integral Equations. Springer, Heidelberg (2008)
14. Jones, P.W.: Quasiconformal mappings and extendability of functions in Sobolev spaces. Acta Math. 147, 71–88 (1981)
15. McLean, W.: Strongly Elliptic Systems and Boundary Integral Equations. Cambridge University Press, Cambridge (2000)
16. Pechstein, C.: Shape-explicit constants for some boundary integral operators. Appl. Anal. (December 2011) (published online), doi:10.1080/00036811.2011.643781
17. Sauter, S.A., Schwab, C.: Boundary Element Methods. Springer Series in Computational Mathematics, vol. 39. Springer, Heidelberg (2011)
18. Steinbach, O.: Numerical Approximation Methods for Elliptic Boundary Value Problems. Finite and Boundary Elements. Springer, New York (2008)
19. Thrun, A.: Über die Konstanten in Poincaréschen Ungleichungen. Master's thesis, Ruhr-Universität Bochum, Bochum (2003),
http://www.ruhr-uni-bochum.de/num1/files/theses/da_thrun.pdf
20. Trefftz, E.: Ein Gegenstück zum Ritzschen Verfahren. In: Verh. d. 2. Intern. Kongr. f. Techn. Mech., Zürich, pp. 131–137 (1926)
21. Veeser, A., Verfürth, R.: Poincaré constants of finite element stars. IMA J. Numer. Anal. (2011), first published online May 30, doi:10.1093/imanum/drr011

Part II

Robust Multigrid, Multilevel and Multiscale, Deterministic and Stochastic Methods for Modeling Highly Heterogeneous Media

Robust Solvers for Symmetric Positive Definite Operators and Weighted Poincaré Inequalities

Yalchin Efendiev[1], Juan Galvis[1], Raytcho Lazarov[1], and Joerg Willems[2]

[1] Dept. Mathematics, Texas A&M University, College Station, TX 77843, USA
[2] RICAM, Altenberger Strasse 69, 4040 Linz, Austria

Abstract. An abstract setting for robustly preconditioning symmetric positive definite (SPD) operators is presented. The term "robust" refers to the property of the condition numbers of the preconditioned systems being independent of mesh parameters and problem parameters. Important instances of such problem parameters are in particular (highly varying) coefficients. The method belongs to the class of additive Schwarz preconditioners. The paper gives an overview of the results obtained in a recent paper by the authors. It, furthermore, focuses on the importance of weighted Poincaré inequalities, whose notion is extended to general SPD operators, for the analysis of stable decompositions. To demonstrate the applicability of the abstract preconditioner the scalar elliptic equation and the stream function formulation of Brinkman's equations in two spatial dimensions are considered. Several numerical examples are presented.

Keywords: domain decomposition, robust additive Schwarz preconditioner, spectral coarse spaces, high contrast, Brinkman's problem, generalized weighted Poincaré inequalities.

1 Introduction

The robust preconditioning of symmetric positive definite (SPD) operators has been an important topic in the numerical analysis community. These operators correspond to symmetric coercive bilinear forms appearing in the weak formulation of various partial differential equations and systems modeling e.g. heat conduction or fluid flow in porous media. The condition numbers of the resulting linear systems typically depend on the mesh parameters of the underlying discretizations and variations in physical problem parameters, e.g. (highly) varying thermal conductivities in compound media. Thus, the convergence rates of iterative methods like conjugate gradients deteriorate as the mesh parameters decrease and the variations in problem parameters increase. One is, therefore, interested in designing preconditioners yielding preconditioned systems whose condition numbers are robust with respect to problem and mesh parameters.

Commonly used approaches include domain decomposition methods (cf. e.g. [8,12]) and multilevel/multigrid algorithms (cf. e.g. [14]). For certain classes of problems, including the scalar elliptic equation, $-\nabla \cdot (\kappa(\boldsymbol{x})\nabla\phi) = f$, these methods are successful in making the condition number of the preconditioned system

I. Lirkov, S. Margenov, and J. Waśniewski (Eds.): LSSC 2011, LNCS 7116, pp. 43–51, 2012.

independent of the mesh parameter. However, designing preconditioners that are robust with respect to variations in the physical parameters, e.g., the contrast in the conductivity $\max_{x \in \Omega} \kappa(x) / \min_{x \in \Omega} \kappa(x)$, where Ω is the domain, is more challenging. Some improvements in standard domain decomposition methods were made in the case of special arrangements of the highly conductive regions with respect to the coarse cells. The construction of preconditioners for these problems has been extensively studied in the last decades (see e.g. [6,8,12]). E.g., it was shown that nonoverlapping domain decomposition methods converge independently of the contrast (e.g. [9] and [12, Sections 6.4.4 and 10.2.4]) when conductivity variations within coarse regions are bounded.

Classical arguments to estimate the condition number of a two level overlapping domain decomposition method for the scalar elliptic case use weighted Poincaré inequalities of the form

$$\int_\omega \kappa(\psi - I_0^\omega \psi)^2 \, dx \leq C \int_\omega \kappa |\nabla \psi|^2 \, dx, \tag{1}$$

where $\omega \subset \Omega$ is a local subdomain and $\psi \in H^1(\omega)$. The operator $I_0^\omega \psi$ is a local representation of the function ψ in the coarse space. Since the constant C in (1) appears in the final bound for the condition number of the operator, it is desirable to obtain (1) with a constant independent of the contrast.

In particular two sets of coarse basis functions have been used in previous works for (1) to hold with a constant independent of the contrast in κ: (1) multiscale finite element functions with various boundary conditions (see e.g. [3,6,7]) and (2) energy minimizing or trace minimizing functions (see e.g. [13,16]). In these cases the coarse spaces have one coarse basis function per coarse node, and the corresponding overlapping domain decomposition methods are robust when the high-conductivity regions are isolated islands.

In [4,5] the construction of coarse basis functions is based on the solution of local generalized eigenvalue problems yielding (1) with contrast independent constants for general configurations of κ. Using multiscale partition of unity functions accounts for isolated local features within coarse blocks yielding a reduced dimension of the coarse space (see [5]). The idea of using local and global eigenvectors to construct coarse spaces within two-level and multi-level techniques has been used before (e.g. [1,11]). However, these authors did not study the convergence with respect to physical parameters and did not use generalized eigenvalue problems to achieve small dimensional coarse spaces.

In [2] the construction of coarse spaces based on generalized eigenfunctions has been generalized to abstract SPD bilinear forms. It is shown that the resulting coarse spaces yield robust additive Schwarz preconditioners, such that the resulting condition numbers are controlled by the maximal number of overlaps of subdomains and a predefined threshold determining which generalized eigenfunctions enter the coarse space construction. Also, to reduce the dimension of the coarse space multiscale partition of unity functions are considered, which as in [5] are shown to capture local features. The general framework of [2] was shown to be applicable to scalar elliptic equations and the stream function formulation of the corresponding mixed forms, Stokes' and Brinkman's equations. The latter models fluid flow in highly porous media and can be viewed as a generalization of Stokes' and Darcy's equations (see [15] and the references therein).

The robustness properties of the methods in [4,5] and [2] are similar. Nevertheless, the generalized eigenvalue problems of the general framework in [2] applied to the scalar elliptic case differ from those studied in [4,5]. In the paper at hand we investigate the relation between the two approaches. In particular we show that in the scalar elliptic case the validity of (1), with a coarse space as in [4], is equivalent to an estimate in [2] which only involves the bilinear form of the problem.

The paper is organized as follows. In Section 2 we introduce the problem setting and outline the construction of abstract robust preconditioners as discussed in [2]. Section 3 briefly addresses the application of this abstract framework to the scalar elliptic equation and Brinkman's problem. In Section 4, we discuss the relation between the abstract framework in [2] and weighted Poincaré inequalities. Section 5 is devoted to some numerical results showing the robustness of the abstract preconditioner when applied to the scalar elliptic and Brinkman's equations.

2   Coarse Spaces in Robust Stable Decompositions

Let $\Omega \subset \mathbb{R}^n$ be a bounded polyhedral domain, and let $\mathcal{T}_H$ be a quasiuniform quadrilateral ($n = 2$) or hexahedral ($n = 3$) triangulation of Ω with mesh-parameter H. For a node x_j of $\mathcal{T}_H$ let Ω_j be the union of cells $T \in \mathcal{T}_H$ adjacent to x_j. We define $I_j := \{i = 1, \ldots, n_x \,|\, \Omega_i \cap \Omega_j \neq \emptyset\}$ and set $n_I := \max_{j=1,\ldots,n_x} \#I_j$.

For a separable Hilbert space $\mathcal{V}_0 = \mathcal{V}_0(\Omega)$ of functions defined on Ω and for any subdomain $\omega \subset \Omega$ we set $\mathcal{V}(\omega) := \{\phi|_\omega \,|\, \phi \in \mathcal{V}_0\}$. Using this notation, we make the following assumptions:

(A1) $a_\omega(\cdot, \cdot) : (\mathcal{V}(\omega), \mathcal{V}(\omega)) \to \mathbb{R}$, is a symmetric positive semi-definite bounded bilinear form. Additionally, $a(\cdot, \cdot) := a_\Omega(\cdot, \cdot)$ is positive definite. For ease of notation we write $a_\omega(\phi, \psi)$ instead of $a_\omega(\phi|_\omega, \psi|_\omega)$ for all $\phi, \psi \in \mathcal{V}_0$.

(A2) For any $\phi \in \mathcal{V}_0$ and any family of pairwisely disjoint subdomains $\{\omega_j\}_{j=1}^{n_\omega}$ with $\cup_{j=1}^{n_\omega} \overline{\omega}_j = \overline{\Omega}$ we have $a(\phi, \phi) = \sum_{j=1}^{n_\omega} a_{\omega_j}(\phi, \phi)$.

(A3) For a suitable subspace $\mathcal{V}_0(\Omega_j)$ of $\mathcal{V}(\Omega_j)$ we have that $a_{\Omega_j}(\cdot, \cdot) : (\mathcal{V}_0(\Omega_j), \mathcal{V}_0(\Omega_j)) \to \mathbb{R}$ is positive definite for all $j = 1, \ldots, n_x$.

(A4) $\{\xi_j\}_{j=1}^{n_x} : \Omega \to [0, 1]$ is a family of functions such that: (a) $\sum_{j=1}^{n_x} \xi_j \equiv 1$ on Ω; (b) $\mathrm{supp}(\xi_j) = \overline{\Omega}_j$ for $j = 1, \ldots, n_x$; (c) For $\phi \in \mathcal{V}_0$ we have $\xi_j\phi \in \mathcal{V}_0$ and $(\xi_j\phi)|_{\Omega_j} \in \mathcal{V}_0(\Omega_j)$ for $j = 1, \ldots, n_x$.

We define the following symmetric bilinear form for $j = 1, \ldots, n_x$,

$$m_{\Omega_j}(\cdot, \cdot) : (\mathcal{V}(\Omega_j), \mathcal{V}(\Omega_j)) \to \mathbb{R}, \text{ with } m_{\Omega_j}(\phi, \psi) := \sum_{i \in I_j} a_{\Omega_j}(\xi_j\xi_i\phi, \xi_j\xi_i\psi) \quad (2)$$

Due to (A4) we have that (2) is well-defined. Also note, that since $\mathrm{supp}(\xi_j) = \overline{\Omega}_j$ we have $\xi_j\phi|_{\Omega_j} \equiv 0 \Leftrightarrow \phi|_{\Omega_j} \equiv 0$, which implies that $m_{\Omega_j}(\cdot, \cdot) : (\mathcal{V}(\Omega_j), \mathcal{V}(\Omega_j)) \to \mathbb{R}$ is positive definite.

Now for $j = 1, \ldots, n_x$ we consider the generalized eigenvalue problems: Find $(\lambda_i^j, \varphi_i^j) \in (\mathbb{R}, \mathscr{V}(\Omega_j))$ such that

$$a_{\Omega_j}\left(\psi, \varphi_i^j\right) = \lambda_i^j \, m_{\Omega_j}\left(\psi, \varphi_i^j\right), \quad \forall \psi \in \mathscr{V}(\Omega_j). \tag{3}$$

Without loss of generality we assume that the eigenvalues are ordered, i.e., $0 \le \lambda_1^j \le \lambda_2^j \le \ldots \le \lambda_i^j \le \lambda_{i+1}^j \le \ldots$. We now state our final assumption.

(A5) For a sufficiently small "threshold" $\tau_\lambda^{-1} > 0$ we may choose $L_j \in \mathbb{N}_0$ such that $\lambda_{L_j+1}^j \ge \tau_\lambda^{-1}$ for all $j = 1, \ldots, n_x$. For simplicity, we choose $\tau_\lambda^{-1} < 1$.

For $\phi \in \mathscr{V}_0$ let ϕ_0^j be the $m_{\Omega_j}(\cdot, \cdot)$-orthogonal projection of $\phi|_{\Omega_j}$ onto the first L_j eigenfunctions of (3). If $L_j = 0$, we set $\phi_0^j \equiv 0$. According to [2] we then have

$$m_{\Omega_j}\left(\phi - \phi_0^j, \phi - \phi_0^j\right) \le \tau_\lambda \, a_{\Omega_j}\left(\phi - \phi_0^j, \phi - \phi_0^j\right) \le \tau_\lambda \, a_{\Omega_j}(\phi, \phi). \tag{4}$$

Now, we specify the "coarse" space by

$$\mathscr{V}_H := \operatorname{span}\{\xi_j \varphi_i^j \,|\, j = 1, \ldots, n_x \text{ and } i = 1, \ldots, L_j\}. \tag{5}$$

Then, for any $\phi \in \mathscr{V}$ we have (here $j = 1, \ldots, n_x$)

$$\phi = \sum_{j=0}^{n_x} \phi_j, \text{ with } \phi_0 := \sum_{j=1}^{n_x} \xi_j \phi_0^j \in \mathscr{V}_H, \text{ and } \phi_j := (\xi_j(\phi - \phi_0))|_{\Omega_j} \in \mathscr{V}_0(\Omega_j). \tag{6}$$

Theorem 1. *Assuming that (A1)–(A5) hold, the decomposition defined in* (6) *satisfies* $\sum_{j=0}^{n_x} a(\phi_j, \phi_j) \le (2 + C\tau_\lambda) a(\phi, \phi)$, *where C only depends on n_I.*

Proof. See [2, Theorem 3.4]. ∎

By abstract domain decomposition theory (see e.g. [12, Section 2.3]) we thus know that the additive Schwarz preconditioner corresponding to the decomposition in (6) yields a condition number that only depends on n_I and τ_λ.

3   Applications

Scalar Elliptic Equation

The scalar elliptic equation is given by

$$-\nabla \cdot (\kappa \nabla \phi) = f, \quad x \in \Omega, \text{ and } \quad \phi = 0, \quad x \in \partial\Omega, \tag{7}$$

where $0 < \kappa \in L^\infty(\Omega)$, $\phi \in H_0^1(\Omega)$, and $f \in L^2(\Omega)$. With $\mathscr{V}_0 := H_0^1(\Omega)$, the corresponding variational formulation reads: Find $\phi \in \mathscr{V}_0$ such that for all $\psi \in \mathscr{V}_0$

$$a^{SE}(\phi, \psi) := \int_\Omega \kappa(x) \nabla \phi \cdot \nabla \psi \, dx = \int_\Omega f \psi \, dx.$$

It is easy to see that with $\mathscr{V}_0(\Omega_j) := H_0^1(\Omega_j) \subset \mathscr{V}_0|_{\Omega_j}$ and ξ_j the Lagrange finite element function of degree 1 corresponding to x_j, $j = 1, \ldots, n_x$, we have that (A1)–(A4) hold. Now consider (3) with $a(\cdot, \cdot)$ and $m(\cdot, \cdot)$ replaced by $a^{SE}(\cdot, \cdot)$ and $m^{SE}(\cdot, \cdot)$, where $m^{SE}(\cdot, \cdot)$ is given by (2) with $a^{SE}(\cdot, \cdot)$ instead of $a(\cdot, \cdot)$. It can then be shown (see [2, Section 4.1]), that for a binary medium, i.e., a medium with $\kappa(\boldsymbol{x}) = \kappa_{\min}$ ($\kappa_{\max}$) in Ω^s (Ω^p) with $\overline{\Omega}^s \cup \overline{\Omega}^p = \overline{\Omega}$ and $\kappa_{\max} \gg \kappa_{\min} > 0$ the $(L_j + 1)$-st eigenvalue of (3), i.e., λ_{L_j+1} is uniformly (with respect to H and $\kappa_{\max}/\kappa_{\min}$) bounded from below. Here L_j denotes the number of connected components of $\Omega^s \cap \Omega_j$. Thus, also (A5) is established with τ_λ independent of the contrast $\kappa_{\max}/\kappa_{\min}$. Note that by Theorem 1 this implies the robustness of the stable decomposition (6) and the corresponding additive Schwarz preconditioner.

Brinkman's Equations

Brinkman's equations modeling flows in highly porous media are given by

$$-\mu \Delta \boldsymbol{u} + \nabla p + \mu \kappa^{-1} \boldsymbol{u} = \boldsymbol{f} \text{ in } \Omega, \quad \nabla \cdot \boldsymbol{u} = 0 \text{ in } \Omega, \quad \text{and} \quad \boldsymbol{u} = \boldsymbol{0} \text{ on } \partial\Omega, \quad (8)$$

where $p \in L^2(\Omega)/\mathbb{R}$, $\boldsymbol{u} \in (H_0^1(\Omega))^2$, $\boldsymbol{f} \in (L^2(\Omega))^2$, $\mu \in \mathbb{R}^+$, and $\kappa \in L^\infty(\Omega)$, with $\kappa > 0$. Here we assume that $\Omega \subset \mathbb{R}^2$ is simply connected. The variational formulation of the Brinkman problem is: Find $(\boldsymbol{u}, p) \in ((H_0^1(\Omega))^2, L_0^2(\Omega))$ such that for all $(\boldsymbol{v}, q) \in ((H_0^1(\Omega))^2, L_0^2(\Omega))$ we have

$$\int_\Omega \mu \nabla \boldsymbol{u} : \nabla \boldsymbol{v} \, d\boldsymbol{x} + \int_\Omega \mu \kappa^{-1} \boldsymbol{u} \cdot \boldsymbol{v} \, d\boldsymbol{x} - \int_\Omega p \nabla \cdot \boldsymbol{v} \, d\boldsymbol{x} - \int_\Omega q \nabla \cdot \boldsymbol{u} \, d\boldsymbol{x} = \int_\Omega \boldsymbol{f} \cdot \boldsymbol{v} \, d\boldsymbol{x}. \quad (9)$$

As described in [2, Section 4] we may adopt the setting of stream functions. For $\mathscr{V}_0 := \left\{ \psi \in H^2(\Omega) \cap H_0^1(\Omega) \,|\, \frac{\partial \psi}{\partial \boldsymbol{n}}|_{\partial\Omega} = 0 \right\}$ the variational stream function formulation reads: Find $\phi \in \mathscr{V}_0$ such that for all $\psi \in \mathscr{V}_0$ we have

$$\int_\Omega \mu \left(\nabla(\nabla \times \phi) : \nabla(\nabla \times \psi) + \kappa^{-1} \nabla \times \phi \cdot \nabla \times \psi \right) d\boldsymbol{x} = \int_\Omega \boldsymbol{f} \cdot \nabla \times \psi \, d\boldsymbol{x}. \quad (10)$$

Here $\{\xi_j\}_{j=1}^{n_x}$ denotes a sufficiently regular partition of unity and $\mathscr{V}_0(\Omega_j) := \left\{ \psi \in H^2(\Omega_j) \cap H_0^1(\Omega_j) \,|\, \frac{\partial \psi}{\partial \boldsymbol{n}}|_{\partial\Omega_j} = 0 \right\}$. As shown in [2, Section 4.4] we can verify (A1)-(A5), where for the case of a binary medium, τ_λ in (A5) can again be chosen independently of the contrast $\kappa_{\max}/\kappa_{\min}$.

Remark 1. Instead of solving (10) for the stream function ϕ and then recovering $\boldsymbol{u} = \nabla \times \phi$, one may equivalently use the coarse space corresponding to (10) to construct a coarse space corresponding to the original formulation (9) by applying $\nabla \times$ to the coarse stream basis functions. By this, one obtains an equivalent robust additive Schwarz preconditioner for the formulation of Brinkman's problem in the primal variables $\boldsymbol{u}$ and p (for details see [8, Section 10.4.2]).

4   Connection to Weighted Poincaré Inequalities

Poincaré type inequalities play a crucial role in the analysis of domain decomposition methods. In the scalar elliptic case, taking $\mathscr{V}_H$ to be the space of Lagrange finite elements of degree 1 corresponding to the coarse mesh $\mathcal{T}_H$, one can show by using the standard Poincaré inequality that the constant in the corresponding decomposition is independent of the mesh parameter H (see [12, Chapter 3] and the references therein).

Recently, for scalar elliptic problems with varying $\kappa(\boldsymbol{x})$, so called "weighted" Poincaré inequalities have received increasing attention (see [4,5,10]). According to [4,5,10] the validity of such an appropriate estimate with a constant independent of the contrast $\kappa_{max}/\kappa_{min}$ yields a corresponding decomposition with a constant independent of the contrast.

For giving the precise form of the inequality used in [4] we consider the following generalized eigenvalue problem for each Ω_j, $j = 1, \ldots, n_{\boldsymbol{x}}$: Find $(\widehat{\lambda}_i^j, \widehat{\varphi}_i^j) \in (\mathbb{R}, H^1(\Omega_j))$ such that

$$a_{\Omega_j}^{SE}\left(\psi, \widehat{\varphi}_i^j\right) = \widehat{\lambda}_i^j \widehat{m}_{\Omega_j}^{SE}\left(\psi, \widehat{\varphi}_i^j\right), \quad \forall \psi \in H^1(\Omega_j), \tag{11}$$

where $\widehat{m}_{\Omega_j}^{SE}(\phi, \psi) := H^{-2} \int_{\Omega_j} \kappa(\boldsymbol{x}) \phi \psi \, d\boldsymbol{x}$ for $\phi, \psi \in H^1(\Omega_j)$ is the weighted mass matrix scaled by H^{-2}. This scaling is not considered in [4], however, we introduce it here to simplify the exposition.

The weighted Poincaré inequality derived in [4] then reads as follows:

$$\widehat{m}_{\Omega_j}^{SE}\left(\phi - \widehat{\phi}_0^j, \phi - \widehat{\phi}_0^j\right) \le \tau_{\widehat{\lambda}} a_{\Omega_j}^{SE}(\phi, \phi), \tag{12}$$

where similarly to above $\widehat{\phi}_0^j$ is the $\widehat{m}_{\Omega_j}^{SE}(\cdot, \cdot)$-orthogonal projection of $\phi|_{\Omega_j}$ onto those eigenfunctions of (11) whose eigenvalues are below a threshold $\tau_{\widehat{\lambda}}^{-1}$. Due to the scaling by H^{-2} the eigenvalues in (11) are independent of H.

Before pointing out a connection between inequality (12) and our estimate (4), we introduce a triangular (for $n = 2$) or tetrahedral (for $n = 3$) mesh corresponding to $\mathcal{T}_H$. It is well-known that we can obtain a quasiuniform triangular/tetrahedral mesh, denoted by $\widetilde{\mathcal{T}}_H$, by subdividing cells in $\mathcal{T}_H$ without introducing new nodes.

Proposition 1. *Let $\{\xi_j\}_{j=1}^{n_{\boldsymbol{x}}}$ be the piecewise linear Lagrange finite element functions corresponding to $\widetilde{\mathcal{T}}_H$. Then we have that inequalities (4) (with $a(\cdot, \cdot) = a^{SE}(\cdot, \cdot)$ and $m(\cdot, \cdot) = m^{SE}(\cdot, \cdot)$) and (12) are up to constants equivalent in the following sense:*

For $\widehat{\phi}_0^j$, $\tau_{\widehat{\lambda}}$ as in (12) we have $m_{\Omega_j}^{SE}\left(\phi - \widehat{\phi}_0^j, \phi - \widehat{\phi}_0^j\right) \le C\tau_{\widehat{\lambda}} a_{\Omega_j}^{SE}(\phi, \phi)$. $\tag{13}$

For ϕ_0^j, τ_{λ} as in (4) we have $\widehat{m}_{\Omega_j}^{SE}\left(\phi - \phi_0^j, \phi - \phi_0^j\right) \le C\tau_{\lambda} a_{\Omega_j}^{SE}(\phi, \phi)$. $\tag{14}$

Here the constants C are independent of H, $\tau_{\widehat{\lambda}}$, and τ_{λ} but may depend on n_I and the geometry of Ω_j.

Proof. Since ξ_j is piecewise linear we note that

$$\frac{1}{CH} \le \min\{|\nabla \xi_j(\boldsymbol{x})| \mid \boldsymbol{x} \in \mathring{T}, \text{ for } T \in \widetilde{\mathcal{T}}_H, \mathring{T} \subset \Omega_j\} \text{ and } \|\nabla \xi_i\|_{L^\infty(\Omega_j)} \le \frac{C}{H} \text{ for } i \in I_j, \tag{15}$$

where $|\cdot|$ denotes some norm on $\mathbb{R}^n$ and C only depends on the geometry of Ω_j.

For proving (13) we note that by (2) we have

$$m_{\Omega_j}^{SE}\left(\phi - \widehat{\phi}_0^j, \phi - \widehat{\phi}_0^j\right) = \sum_{i \in I_j} a_{\Omega_j}^{SE}\left(\xi_j \xi_i (\phi - \widehat{\phi}_0^j), \xi_j \xi_i (\phi - \widehat{\phi}_0^j)\right)$$

$$= \sum_{i \in I_j} \int_{\Omega_j} \kappa \left| \nabla \left(\xi_j \xi_i (\phi - \widehat{\phi}_0^j)\right) \right|^2 d\boldsymbol{x} \leq 2 \sum_{i \in I_j} \int_{\Omega_j} \kappa \left(\left| \nabla (\phi - \widehat{\phi}_0^j) \right|^2 + \left| \nabla (\xi_j \xi_i)(\phi - \widehat{\phi}_0^j) \right|^2 \right) d\boldsymbol{x},$$

where we have used Schwarz' inequality and the fact that $|\xi_i| < 1$ for $i \in I_j$. Thus by (15),

$$m_{\Omega_j}^{SE}\left(\phi - \widehat{\phi}_0^j, \phi - \widehat{\phi}_0^j\right) \leq C a_{\Omega_j}^{SE}\left(\phi - \widehat{\phi}_0^j, \phi - \widehat{\phi}_0^j\right) + \frac{C}{H^2} \sum_{i \in I_j} \int_{\Omega_j} \kappa (\phi - \widehat{\phi}_0^j)^2 d\boldsymbol{x}$$

$$\leq C\, a_{\Omega_j}^{SE}(\phi, \phi) + C\, \widehat{m}_{\Omega_j}^{SE}\left(\phi - \widehat{\phi}_0^j, \phi - \widehat{\phi}_0^j\right) \leq C \tau_{\widehat{\lambda}}\, a_{\Omega_j}^{SE}(\phi, \phi),$$

where we have used that $\widehat{\phi}_0^j$ satisfies (12), $\tau_{\widehat{\lambda}} > 1$, and $a_{\Omega_j}^{SE}\left(\phi - \widehat{\phi}_0^j, \phi - \widehat{\phi}_0^j\right) \leq a_{\Omega_j}^{SE}(\phi, \phi)$. This establishes (13). For verifying (14) we first note that by the definition of $\widehat{m}_{\Omega_j}^{SE}(\cdot, \cdot)$ and (15) we have

$$\widehat{m}_{\Omega_j}^{SE}\left(\phi - \phi_0^j, \phi - \phi_0^j\right) = H^{-2} \int_{\Omega_j} \kappa (\phi - \phi_0^j)^2 d\boldsymbol{x}$$

$$\leq C \int_{\Omega_j} \kappa |\nabla \xi_j|^2 (\phi - \phi_0^j)^2 d\boldsymbol{x} = C \int_{\Omega_j} \kappa \left| \sum_{i \in I_j} \nabla (\xi_i \xi_j)(\phi - \phi_0^j) \right|^2 d\boldsymbol{x},$$

where we have used that $\sum_{i \in I_j} \xi_i \equiv 1$ in Ω_j. Since $\nabla(\xi_i \xi_j (\phi - \phi_0^j)) = \nabla(\xi_i \xi_j)(\phi - \phi_0^j) + \xi_i \xi_j \nabla(\phi - \phi_0^j)$, Schwarz' inequality yields

$$\frac{1}{C}\, \widehat{m}_{\Omega_j}^{SE}\left(\phi - \phi_0^j, \phi - \phi_0^j\right) \leq \sum_{i \in I_j} \int_{\Omega_j} \kappa \left| \nabla(\xi_i \xi_j (\phi - \phi_0^j)) \right|^2 + \kappa \underbrace{(\xi_i \xi_j)^2}_{\leq 1} \left| \nabla(\phi - \phi_0^j) \right|^2 d\boldsymbol{x}$$

$$\leq \sum_{i \in I_j} a_{\Omega_j}^{SE}\left(\xi_i \xi_j (\phi - \phi_0^j), \xi_i \xi_j (\phi - \phi_0^j)\right) + C a_{\Omega_j}^{SE}\left(\phi - \phi_0^j, \phi - \phi_0^j\right)$$

$$\leq m_{\Omega_j}^{SE}\left(\phi - \phi_0^j, \phi - \phi_0^j\right) + C a_{\Omega_j}^{SE}(\phi, \phi),$$

where we have used the second inequality in (4). Thus, $\widehat{m}_{\Omega_j}^{SE}\left(\phi - \phi_0^j, \phi - \phi_0^j\right) \leq C \tau_\lambda\, a_{\Omega_j}^{SE}(\phi, \phi)$, by our assumption that ϕ_0^j satisfies (4) and $\tau_\lambda > 1$.

Remark 2 (Generalized Weighted Poincaré Inequality). Proposition 1 allows the interpretation of (4) as a *generalized* weighted Poincaré inequality. For the scalar elliptic case, i.e., for $a(\cdot, \cdot) = a^{SE}(\cdot, \cdot)$ and $m(\cdot, \cdot) = m^{SE}(\cdot, \cdot)$, Proposition 1 shows that in terms of the robustness of the stable decomposition it does not matter whether the coarse space is based on eigenmodes corresponding to (3) or (11). For more complicated bilinear forms, such as the one appearing in (10), it may, however, be rather difficult to formulate a suitable analogue to the weighted Poincaré inequality (12). Using our abstract approach this is straightforward, since the bilinear form $m_{\Omega_j}(\cdot, \cdot)$ is entirely based on the partition of unity $\{\xi_i\}_{i=1}^{n_x}$ and the bilinear form $a(\cdot, \cdot)$ (see (2)).

5   Numerical Results

We give some numerical examples showing the robustness of the additive Schwarz preconditioner corresponding to the decomposition (6). We consider applications to the equations in Section 3 considering the geometry shown on the left (coarse triangulation: 8×8, fine triangulation: 64×64) with varying contrast $\kappa_{\max}/\kappa_{\min}$. Here the black parts denote regions of high conductivity (scalar elliptic case) and low permeability (Brinkman case), respectively. Tables 1(a) and 1(b) show the results, i.e., coarse space dimensions and condition number estimates.

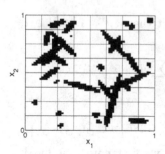

In the scalar elliptic case we use a conforming finite element discretization with bilinear Lagrange elements yielding a fine space of dimension 4225. The right hand side f in (7) is chosen to compensate for the boundary condition given by $\phi(\boldsymbol{x}) = 1 - x_1$ on $\partial\Omega$. As partition of unity $\{\xi_j\}_{j=1}^{n_x}$, we use bilinear Lagrange finite element functions corresponding to the coarse mesh $\mathcal{T}_H$.

Brinkman's equations are discretized with an $H(div)$-conforming Discontinuous Galerkin discretization (cf. [15] and the references therein) using Raviart-Thomas finite elements of degree 1 yielding a fine space of dimension 49408. It is well-known that in two spatial dimensions the corresponding stream function space is given by Lagrange biquadratic finite elements. The coarse (divergence free) velocity space is constructed as outlined in Remark 1. The right hand side f is chosen to compensate for the boundary condition given by $\boldsymbol{u} = \boldsymbol{e}_1$ on $\partial\Omega$. As partition of unity we choose piecewise polynomials of degree 3 with sufficient global regularity.

For comparison reasons we also provide numerical results for additive Schwarz preconditioners using standard coarse spaces, denoted by $\mathcal{V}_H^{st}$. For the scalar elliptic case this is given by coarse bilinear Lagrange finite element functions. For the Brinkman problem $\mathcal{V}_H^{st}$ is the span of the curl of the partition of unity functions corresponding to interior coarse mesh nodes. The numerical results show the robustness of the preconditioners using spectral coarse spaces, whereas the preconditioners using standard coarse spaces deteriorate with increasing contrasts.

Table 1. Results for standard and spectral coarse spaces. Threshold: $\tau_\lambda^{-1} = 0.5$.

(a) Scalar elliptic case.

$\kappa_{\max}/\kappa_{\min}$	dim $\mathcal{V}_H^{st}$	cond. #	dim. $\mathcal{V}_H$	cond. #
1e2	49	3.50e1	70	15.5
1e3	49	3.11e2	124	26.7
1e4	49	3.06e3	145	7.92
1e5	49	3.06e4	148	7.90
1e6	49	2.71e5	148	7.90

(b) Brinkman case.

$\kappa_{\max}/\kappa_{\min}$	dim $\mathcal{V}_H^{st}$	cond. #	dim. $\mathcal{V}_H$	cond. #
1e2	49	1.88e1	66	12.6
1e3	49	2.68e1	75	19.2
1e4	49	1.32e2	99	21.9
1e5	49	1.00e3	125	17.7
1e6	49	9.31e3	145	26.0

Acknowledgments. The research of Y. Efendiev was partially supported by the DOE and NSF (DMS 0934837, DMS 0724704, and DMS 0811180). The research of Y. Efendiev, J. Galvis, and R. Lazarov was supported in parts by award KUS-C1-016-04, made by King Abdullah University of Science and Technology (KAUST). The research of R. Lazarov and J. Willems was supported in parts by NSF Grant DMS-1016525.

References

1. Chartier, T., Falgout, R.D., Henson, V.E., Jones, J., Manteuffel, T., McCormick, S., Ruge, J., Vassilevski, P.S.: Spectral AMGe (ρAMGe). SIAM J. Sci. Comput. 25(1), 1–26 (2003)
2. Efendiev, Y., Galvis, J., Lazarov, R., Willems, J.: Robust domain decomposition preconditioners for abstract symmetric positive definite bilinear forms. Technical Report 2011-05, RICAM (2011), submitted to Math. Model. Numer. Anal.
3. Efendiev, Y., Hou, T.Y.: Multiscale finite element methods. Surveys and Tutorials in the Applied Mathematical Sciences, vol. 4. Springer, New York (2009); Theory and applications
4. Galvis, J., Efendiev, Y.: Domain decomposition preconditioners for multiscale flows in high-contrast media. Multiscale Model. Simul. 8(4), 1461–1483 (2010)
5. Galvis, J., Efendiev, Y.: Domain decomposition preconditioners for multiscale flows in high contrast media: reduced dimension coarse spaces. Multiscale Model. Simul. 8(5), 1621–1644 (2010)
6. Graham, I.G., Lechner, P.O., Scheichl, R.: Domain decomposition for multiscale PDEs. Numer. Math. 106(4), 589–626 (2007)
7. Hou, T.Y., Wu, X.-H., Cai, Z.: Convergence of a multiscale finite element method for elliptic problems with rapidly oscillating coefficients. Math. Comp. 68(227), 913–943 (1999)
8. Mathew, T.P.A.: Domain Decomposition Methods for the Numerical Solution of Partial Differential Equations. Lecture Notes in Computational Science and Engineering. Springer, Berlin (2008)
9. Nepomnyaschikh, S.V.: Mesh theorems on traces, normalizations of function traces and their inversion. Sov. J. Numer. Anal. Math. Modelling 6(2), 151–168 (1991)
10. Pechstein, C., Scheichl, R.: Weighted Poincaré inequalities. Technical Report 2010-10, Inst. of Comp. Math., Johannes Kepler University (2010)
11. Sarkis, M.: Nonstandard coarse spaces and Schwarz methods for elliptic problems with discontinuous coefficients using non-conforming elements. Numer. Math. 77(3), 383–406 (1997)
12. Toselli, A., Widlund, O.: Domain Decomposition Methods – Algorithms and Theory. Springer Series in Computational Mathematics. Springer, Heidelberg (2005)
13. Van Lent, J., Scheichl, R., Graham, I.G.: Energy-minimizing coarse spaces for two-level Schwarz methods for multiscale PDEs. Numer. Linear Algebra Appl. 16(10), 775–799 (2009)
14. Vassilevski, P.S.: Multilevel block-factorization preconditioners. Matrix-based analysis and algorithms for solving finite element equations. Springer, New York (2008)
15. Willems, J.: Numerical Upscaling for Multiscale Flow Problems. PhD thesis, University of Kaiserslautern (2009)
16. Xu, J., Zikatanov, L.T.: On an energy minimizing basis for algebraic multigrid methods. Comput. Vis. Sci. 7(3-4), 121–127 (2004)

Additive Schur Complement Approximation for Elliptic Problems with Oscillatory Coefficients

J. Kraus

Johann Radon Institute for Computational and Applied Mathematics,
Austrian Academy of Sciences, Altenberger Straße 69, A-4040 Linz, Austria
johannes.kraus@oeaw.ac.at

Abstract. We introduce an algorithm for Additive Schur Complement Approximation (ASCA) that can be used in various iterative methods for solving systems of linear algebraic equations arising from finite element discretization of Partial Differential Equations (PDE). Here we consider a model problem of a scalar elliptic PDE with highly oscillatory (piecewise constant) diffusion coefficient. The main ideas are illustrated by three different examples that reveal the key point of constructing a robust sparse ASCA. We also demonstrate how the quality of the ASCA can be improved and how its sparsity is controlled.

1 Introduction

Sparse Schur complement approximations play an important role in many algorithms for the numerical solution of partial differential equations (PDE). The exact Schur complements arising in classical multilevel block factorization preconditioners are typically dense matrices. In many cases a spectrally equivalent sparse approximation of the exact Schur complement is a crucial step in the design of optimal or nearly optimal order iterative solution methods, see, e.g., [6] and the references therein. The focus of the present paper is on constructing an Additive Schur Complement Approximation (ASCA) that can be employed recursively in various multilevel methods.

Throughout this paper we want to assume that the system of linear algebraic equations, whose Schur complement we want to approximate, stems from a finite element discretization of a given PDE, and that we have access to the individual element (stiffness) matrices.

The approximation technique presented in this paper can be viewed as a generalization of the method that has been introduced in [2] and reconsidered and further investigated in [1,3,4] for different classes of problems/matrices. An important point in the conception of the ASCA is that it allows (and in general also makes use of) coverings of the domain by overlapping subdomains whereas similar additive Schur complement approximations that have been studied earlier are based on non-overlapping coverings. This feature not only improves the quality of the approximation significantly but also yields the desired robustness when dealing with problems arising in the modeling of highly heterogeneous media.

I. Lirkov, S. Margenov, and J. Waśniewski (Eds.): LSSC 2011, LNCS 7116, pp. 52–59, 2012.

Preliminary numerical tests have indicated that the proposed ASCA technique can be applied successfully to various problem classes, including systems of PDE and certain indefinite problems. In the present paper, we introduce the basic algorithm only and try to give a flavor of its potential. For this purpose we consider a model problem of a scalar elliptic PDE with highly oscillatory (piecewise constant) coefficient. Similar problems have been addressed recently by different authors, see, e.g., [5] and the references therein.

2   Problem Formulation

Let us consider the following second-order elliptic boundary value problem

$$- \nabla \cdot (a(x)\nabla u(x)) = f(x) \quad in \ \ \Omega, \tag{1a}$$
$$u = \quad 0 \quad on \ \Gamma_D, \tag{1b}$$
$$(\mathbf{a}(x)\nabla u(x)) \cdot \mathbf{n} = \quad 0 \quad on \ \Gamma_N, \tag{1c}$$

where Ω is a convex polygonal (polyhedral) domain in $\mathbb{R}^d$, $d = 2, 3$, $f(x)$ is a given source term in $L_2(\Omega)$, $\mathbf{n}$ is the outward unit vector normal to the boundary $\Gamma = \partial\Omega$, $\Gamma = \Gamma_D \cup \Gamma_N$, and the coefficient matrix $\mathbf{a}(x) = (a_{ij}(x))_{i,j=1}^d$ is symmetric positive definite (SPD), and uniformly bounded in Ω.

In particular, we will consider Problem (1) with diffusion tensor

$$\mathbf{a}(x) = \alpha(x)I = \alpha_e I \quad \forall e \in \mathcal{T}_h \tag{2}$$

where $\alpha_e > 0$ is a scalar quantity that may vary over several orders of magnitude across element interfaces. Without loss of generality, (after rescaling) we may assume that $\alpha_e \in (0, 1]$ for all $e \in \mathcal{T}_h$. The weak formulation of (1) reads: Find $u \in H_D^1(\Omega) := \{v \in H^1(\Omega) : v = 0 \text{ on } \Gamma_D\}$ such that

$$\int_\Omega \alpha(x)\nabla u \cdot \nabla v = \int_\Omega fv \quad \forall v \in H_D^1(\Omega). \tag{3}$$

In this paper we want to consider a conforming finite element method for the solution of (3) in the subspace of continuous piecewise bilinear functions. This finally leads to the problem of solving a system of linear algebraic equations $A_h u_h = \mathbf{f}_h$ where $A_h \in \mathbb{R}^{n \times n}$ is sparse and symmetric positive definite (SPD).

3   Additive Schur Complement Approximation

3.1   Preliminaries and Notation

Let $\mathcal{T}_h = \cup_e\{e\}$ be a non-overlapping partition of Ω_h into elements e, which we will also refer to as the *(fine) mesh.*

Then by F we denote any union of elements e from $\mathcal{T}_h$ and we shall call F a *structure*; Further let $\mathcal{F} = \mathcal{F}_h = \{F = F_i : i = 1, 2, \ldots, n_F\}$ denote a set of structures that covers $\mathcal{T}_h$, i.e., for all $e \in \mathcal{T}_h$ there exists a structure $F \in \mathcal{F}_h$ such that $e \subset F$.

Next, by G we denote any union of structures F from $\mathcal{F}_h$ and we shall call G a *macro structure*; Further let $\mathcal{G} = \mathcal{G}_h = \{G = G_i : i = 1, 2, \ldots, n_G\}$ denote a set of macro structures that covers $\mathcal{F}_h$, i.e., for all $F \in \mathcal{F}_h$ there exists a macro structure $G \in \mathcal{G}_h$ such that $F \subset G$.

Depending on whether the intersection of any two mutually distinct (macro) structures is empty or there exists at least one pair of (macro) structures with nonempty intersection, we will refer to $\mathcal{F}_h$ (or $\mathcal{G}_h$) either as a *non-overlapping* or as an *overlapping covering* of $\mathcal{T}_h$ (or $\mathcal{F}_h$), respectively.

Throughout this paper we will assume that we have access to the individual element (stiffness) matrices, which we will denote by A_e. In general, by A_X, or B_X, or S_X, etc., where $X \in \{e, F, G\}$, we will denote a small-sized matrix that is associated with either an element e or a structure F or a macro structure G, respectively. The corresponding restriction operator we will denote by R_X (or $R_{Y \mapsto X}$), where $X \in \{e, F, G\}$ (and $Y \in \{F, G\}$). Note that the transpose of this operator, R_X^T defines the natural inclusion, which maps any vector $\mathbf{v}_X$ that is defined on X to a global vector $\mathbf{v}$ by extending $\mathbf{v}_X$ with zeros outside of X. Hence, the assembly of the stiffness matrix $A = A_h$ can be written in the form $A = \sum_{e \in \mathcal{T}_h} R_e^T A_e R_e$. Alternatively, we can also represent A in terms of local matrices A_F associated with the set of structures $\mathcal{F}$, i.e.,

$$A = \sum_{F \in \mathcal{F}} R_F^T A_F R_F, \tag{4}$$

or, we can assemble A from local matrices A_G associated with the covering $\mathcal{G}$ by macro structures, i.e.,

$$A = \sum_{G \in \mathcal{G}} R_G^T A_G R_G, \tag{5}$$

where

$$A_F = \sum_{e \subset F} \sigma_{e,F} R_{F \mapsto e}^T A_e R_{F \mapsto e}, \quad A_G = \sum_{F \subset G} \sigma_{F,G} R_{G \mapsto F}^T A_F R_{G \mapsto F}. \tag{6}$$

Here A_F and A_G denote the *structure* and *macro structure matrices*, respectively. The non-negative scaling factors $\sigma_{e,F}$ and $\sigma_{F,G}$ in (6) have to be chosen in such a way that the assembling properties (4) and (5) are satisfied, which implies

$$\sum_{F \supset e} \sigma_{e,F} = 1 \quad \forall e \in \mathcal{T}, \tag{7a}$$

$$\sum_{G \supset F} \sigma_{F,G} = 1 \quad \forall F \in \mathcal{F}. \tag{7b}$$

Moreover, any *macro structure matrix* A_G can also be assembled directly, i.e., $A_G = \sum_{e \subset G} \sigma_{e,G} R_{G \mapsto e}^T A_e R_{G \mapsto e}$, which requires the weights for the element matrices to satisfy the condition

$$\sum_{G \supset e} \sigma_{e,G} = 1 \quad \forall e \in \mathcal{T}. \tag{8}$$

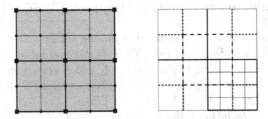

Fig. 1. Macro structure $G_{2\times 2}$ composed of 2×2 non-overlapping structures $F_{2\times 2}$ (left); Covering $\mathcal{G}_{1/2}$ with an overlap of $1/2$ of the width of one macro structure (right picture)

One possibility to satisfy (7)–(8) is to choose the scaling factors according to

$$\sigma_{e,F} = \frac{1}{\sum_{F\supset e} 1}, \quad \sigma_{F,G} = \frac{1}{\sum_{G\supset F} 1}, \quad \sigma_{e,G} = \frac{1}{\sum_{G\supset e} 1}. \tag{9}$$

Note that if $\mathcal{A}_{\mathcal{F}} := \{A_F : F \in \mathcal{F}\}$ has the assembling property (4) then (7b) guarantees that $\mathcal{A}_{\mathcal{G}} := \{A_G : G \in \mathcal{G}\}$ has the assembling property (5).

Figure 1 illustrates the composition of one macro structure $G = G_{2\times 2}$ from 2×2 non-overlapping structures $F = F_{2\times 2}$ of size 2×2 elements (left picture), and an overlapping covering $\mathcal{G} = \mathcal{G}_{1/2}$ of a set $\mathcal{F}$ of $4 \times 4 = 16$ structures F by $3 \times 3 = 9$ macro structures G (right picture).

3.2   Algorithm

For the algorithmic representation of the method of additive Schur complement approximation (ASCA) assume that we are given:

- a set $\mathcal{D}$ of degrees of freedom (DOF) which is the union of a set $\mathcal{D}_c$ of coarse degrees of freedom (CDOF) and its complement $\mathcal{D}_f := \mathcal{D} \setminus \mathcal{D}_c$ in $\mathcal{D}$;
- a non-overlapping or an overlapping covering $\mathcal{F}$ of $\mathcal{T}$;
- a set of structure matrices $\mathcal{A}_{\mathcal{F}} := \{A_F : F \in \mathcal{F}\}$ that satisfy the assembling property (4);

Then we permute the rows and columns of A according to a two-level partitioning of $\mathcal{D}$, which means that the rows and columns that correspond to $\mathcal{D}_f := \mathcal{D} \setminus \mathcal{D}_c$ are numbered first and those corresponding to $\mathcal{D}_c$ are numbered last, i.e.,

$$A = A_h = \begin{bmatrix} A_{ff} & A_{fc} \\ A_{cf} & A_{cc} \end{bmatrix} \begin{matrix} \} \mathcal{D}_f \\ \} \mathcal{D}_c \end{matrix} \tag{10}$$

The corresponding Schur complement we denote by

$$S = A_{cc} - A_{cf} A_{ff}^{-1} A_{fc}. \tag{11}$$

Then the following algorithm can be used in order to compute an additive Schur complement approximation Q for the exact Schur complement S.

Algorithm 1 (Add. Schur complement approx.). $ASCA(\mathcal{F}, \mathcal{A}_{\mathcal{F}}, \mathcal{D}_f, \mathcal{D}_c)$

1. Determine a global two-level numbering for the set $\mathcal{D} = \mathcal{D}_f \cup \mathcal{D}_c$ of all DOF with consecutive numbering of the CDOF in the second block and apply the corresponding permutation to the matrix A, which results in (10).
2. Determine a (non-overlapping or overlapping) covering $\mathcal{G}$ of $\mathcal{F}$ and a set of scaling factors $\{\sigma_{F,G} : F \in \mathcal{F}\}$ satisfying (7b).
3. For all $G \in \mathcal{G}$ perform the following steps:
   (a) Determine a "local" two-level numbering of the DOF of G and assemble the corresponding macro structure matrix A_G according to (6). Then

$$A_G = \begin{bmatrix} A_{G:\text{ff}} & A_{G:\text{fc}} \\ A_{G:\text{cf}} & A_{G:\text{cc}} \end{bmatrix} \begin{matrix} \} \mathcal{D}_{G:f} \\ \} \mathcal{D}_{G:c} \end{matrix} \qquad (12)$$

   where $\mathcal{D}_{G:f}$ and $\mathcal{D}_{G:c}$ denote the "local" sets of FDOF and CDOF.
   (b) Compute the "local" Schur complement

$$S_G = A_{G:\text{cc}} - A_{G:\text{cf}} A_{G:\text{ff}}^{-1} A_{G:\text{fc}}. \qquad (13)$$

   (c) Determine the "local-to-global" mapping for the CDOF in $\mathcal{D}_{G:c}$, i.e., determine the global numbers of the "local" CDOF and define $R_{G:c}$.
4. Assemble a global Schur complement approximation Q from the "local" contributions S_G:

$$Q = \sum_{G \in \mathcal{G}} R_{G:c}^T S_G R_{G:c} \qquad (14)$$

In the following we will describe three different examples for constructing an ASCA that will also illustrate the generality of this concept.

3.3   Examples

As already pointed out any ASCA is determined by its corresponding covering $\mathcal{G}$ and by the scaling factors $\sigma_{F,G}$ used in the assembly (6) of macro structure matrices from structure matrices.

In all three examples given below we consider a conforming finite element discretization of Problem (1) based on bilinear elements where $\Omega = \Omega_h = (0,1)^2$, and $\mathcal{T}_h$ is a uniform mesh composed of square elements with mesh size $h = h_x = h_y = 1/(N+1)$, i.e., $n = (N+1)^2$, where N is of the form $N = 2k$ and $k \in \mathbb{N}$. The sets $\mathcal{D}_f$ and $\mathcal{D}_c$ of FDOF and CDOF can be associated with the vertices of the elements (nodes) of $\mathcal{T}_h$ and $\mathcal{T}_H$, respectively, where $\mathcal{T}_h$ has been obtained from a uniform coarse mesh $\mathcal{T}_H$ by subdividing each coarse element of size H into four similar elements of size $h = H/2$.

Example 1. Let $\mathcal{F} = \mathcal{F}_0 = \{F_{2 \times 2}\}$ be the set of 2×2 structures that provides a non-overlapping covering of Ω, see left picture of Figure 1. Next, we consider a non-overlapping covering $\mathcal{G} = \mathcal{G}_0$ of $\mathcal{F} = \mathcal{F}_0$ by macro structures $G_{2 \times 2}$. The index 0 of $\mathcal{F}_0$ and $\mathcal{G}_0$ indicates that neither structures nor macro structures overlap in this example.

Example 2. Consider the overlapping covering $\mathcal{G} = \mathcal{G}_{1/2}$ of $\mathcal{F} = \mathcal{F}_0 = \{F_{2\times2}\}$ by macro structures $G_{2\times2}$, as illustrated by Figure 1. The set of structures provides a non-overlapping partition as in Example 1. The index $1/2$ of $\mathcal{G}_{1/2}$ indicates that in this example macro structures overlap with $1/2$ of their width, see right picture of Figure 1.

With this example we want to show that introducing an overlap in the covering $\mathcal{G}$ of $\mathcal{F}$ is the key to the robustness of the ASCA with respect to inter-element jumps of the PDE coefficient.

The proof of the following result will be included in an upcoming paper.

Theorem 2. *Consider the discretization of Problem (3) based on conforming bilinear elements on a uniform mesh with mesh size h. Let Q denote the ASCA based on the overlapping covering $\mathcal{G} = \mathcal{G}_{1/2}$ of Example 2. Then, independently of the jumps of the coefficient α_e (on the fine mesh) the following estimate holds*

$$\frac{1}{4}\mathbf{v}^T S \mathbf{v} \le \mathbf{v}^T Q \mathbf{v} \le \mathbf{v}^T S \mathbf{v} \quad \forall \mathbf{v}.$$

Example 3. In this example let us start with a covering $\mathcal{F} = \mathcal{F}_{1/2} = \{F_{3\times3}\}$ of Ω by overlapping structures (with an "overlap of width $1/2$"). Then we consider the overlapping covering $\mathcal{G} = \mathcal{G}_{1/2}$ of $\mathcal{F} = \mathcal{F}_{1/2}$ by macro structures $G_{3\times3}$ of size 3×3 (structures), as shown in Figure 2. That is, in this example neighboring structures $F \in \mathcal{F}_{1/2}$ overlap each other with half of their width, and neighboring macro structures $G \in \mathcal{G}_{1/2}$ overlap each other with half of their width as well.

With this last example we want to demonstrate two things. First, in general, both, structures and macro structures may overlap, and second, by increasing the size of the structures (and that of the macro structures and their overlap) one can improve the quality of the ASCA at the price of a gradual loss of sparsity.

4 Numerical Results

The numerical results presented below are for the model problem (3) where the coefficient is defined by (2), $\alpha_e = 10^{-p}$, and $p \in \{0, 1, \ldots, q\}$ is a uniformly distributed random integer exponent. We study Examples 1 to 3.

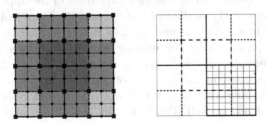

Fig. 2. Macro structure $G_{3\times3}$ composed of 3×3 overlapping structures $F_{3\times3}$ (left); Covering $\mathcal{G}_{1/2}$ with an overlap of $1/2$ of the width of one macro structure (right picture)

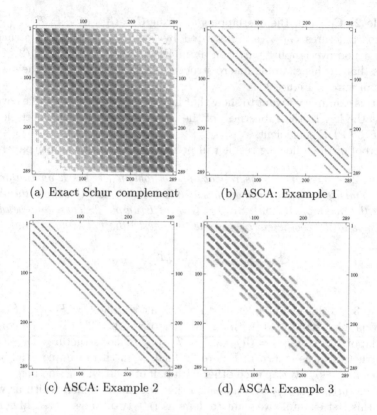

(a) Exact Schur complement   (b) ASCA: Example 1

(c) ASCA: Example 2   (d) ASCA: Example 3

Fig. 3. Nonzero pattern

We compute (effective) relative condition numbers (since no essential boundary conditions are imposed) as a measure of the quality of the approximations.

As it can be seen from Figure 3(a) the exact Schur complement is a dense matrix, whereas the additive Schur complement approximations defined in Examples 1 to 3 are sparse matrices with a gradual loss of sparsity as the (macro) structures and their overlap in the covering $\mathcal{G}$ increase, see Figure 3(b)–3(d).

On the other hand, the results in Table 1 show that the overlap of the macro structures provides the key to the desired robustness of the approximation with respect to the jumps of the coefficients. We report the largest observed relative condition number in a sample of one hundred different random distributions of $\alpha_e = 10^{-p}$, $p \in \{0, 1, \ldots, q\}$. The number of nonzero entries $\mathrm{nnz}(Q_i)$ in the different approximations, as considered in Example i, $i \in \{1, 2, 3\}$, is denoted by $\mathrm{nnz}(Q_i)$, and is listed in the last row of Table 1. As Table 2 confirms, the (effective) relative condition number $\kappa = \kappa(Q_i, S)$ is uniformly bounded with respect to the mesh size as well. The results in Table 2 are for Example 2.

Table 1. $\kappa(Q_i, S)$ and nnz(Q_i) for a uniform mesh of 32×32 elements

κ	Example 1	Example 2	Example 3
$q = 0$	1.3	1.3	1.1
$q = 1$	2.0	1.6	1.3
$q = 2$	4.8	2.0	1.4
$q = 4$	54.1	2.5	1.5
$q = 6$	204.7	2.9	1.6
$q = 8$	2697.3	3.1	1.6
nnz(Q_i)	4225	6241	14641

Table 2. $\kappa(Q_2, S)$ for varying mesh size h: Example 2

κ	$h = 1/8$	$h = 1/16$	$h = 1/32$	$h = 1/64$
$q = 0$	1.2	1.3	1.3	1.3
$q = 1$	1.6	1.7	1.7	1.7
$q = 2$	1.9	1.9	2.0	2.0
$q = 4$	2.4	2.6	2.6	2.6
$q = 6$	2.6	2.7	2.9	2.9
$q = 8$	2.7	2.8	3.1	3.0

Acknowledgment. The author would like to thank Maria Lymbery for the fruitful discussions and for producing the figures in this paper. The support by the Austrian Science Fund (FWF) grant P22989-N18 is highly appreciated.

References

1. Axelsson, O., Blaheta, R., Neytcheva, M.: Preconditioning of boundary value problems using elementwise Schur complements. SIAM J. Matrix Anal. Appl. 31, 767–789 (2009)
2. Kraus, J.: Algebraic multilevel preconditioning of finite element matrices using local Schur complements. Numer. Linear Algebra Appl. 13, 49–70 (2006)
3. Neytcheva, M., Do-Quang, M., Xin, H.: Element-by-Element Schur Complement Approximations for General Nonsymmetric Matrices of Two-by-Two Block Form. In: Lirkov, I., Margenov, S., Waśniewski, J. (eds.) LSSC 2009. LNCS, vol. 5910, pp. 108–115. Springer, Heidelberg (2010)
4. Neytcheva, M.: On element-by-element Schur complement approximations. Linear Algebra and Its Applications 434, 2308–2324 (2011)
5. Scheichl, R., Vassilevski, P., Zikatanov, L.: Weak approximation properties of elliptic projections with functional constraints, Tech. report, LLNL (February 2011) (submitted to Multiscale Modeling & Simulation)
6. Vassilevski, P.: Multilevel Block Factorization Preconditioners. Springer, New York (2008)

Part III

Advanced Methods
for Transport

Optimization–Based Modeling with Applications to Transport: Part 1. Abstract Formulation

Pavel Bochev[1], Denis Ridzal[2], and Joseph Young[1]

[1] Numerical Analysis and Applications
[2] Optimization and Uncertainty Quantification,
Sandia National Laboratories*, Albuquerque, NM 87185-1320, USA
{pbboche,dridzal,josyoun}@sandia.gov

Abstract. This paper is the first of three related articles, which develop and demonstrate a new, optimization–based framework for computational modeling. The framework uses optimization and control ideas to assemble and decompose multiphysics operators and to preserve their fundamental physical properties in the discretization process. An optimization–based monotone, linearity preserving algorithm for transport (OBT) demonstrates the scope of the framework. The second and the third parts of this work focus on the formulation of efficient optimization algorithms for the solution of the OBT problem, and computational studies of its accuracy and efficacy.

1 Introduction

In this, and two companion papers [10,8], we formulate, apply and study computationally a new optimization-based framework for computational modeling. The framework uses optimization and control ideas to (i) assemble and decompose multiphysics operators and (ii) preserve their fundamental physical properties in the discretization process. It further develops the approach in [3,2], which demonstrates an optimization-based synthesis of fast solvers. Here we focus on application of the framework for the preservation of physical properties. We develop an optimization-based algorithm for transport (OBT) of a positive scalar function (density), which is monotone and preserves local bounds and linear functions on arbitrary unstructured grids.

The OBT algorithm combines the incremental remap (constrained interpolation) strategy for transport in [5] with the reformulation of the remap step as an inequality constrained quadratic program (QP) [4]. The objective in this QP is to minimize the discrepancy between target high-order mass fluxes and the approximate mass fluxes subject to inequality constraints derived from physically motivated bounds on the primitive variable (density). The merger of these

---

* Sandia National Laboratories is a multi-program laboratory managed and operated by Sandia Corporation, a wholly owned subsidiary of Lockheed Martin Corporation, for the U.S. Department of Energy's National Nuclear Security Administration under contract DE-AC04-94AL85000.

I. Lirkov, S. Margenov, and J. Waśniewski (Eds.): LSSC 2011, LNCS 7116, pp. 63–71, 2012.

ideas yields a new type of transport algorithms that can be applied to arbitrary unstructured grids and extended to higher than second-order accuracy by using suitably defined target fluxes.

Our approach differs substantially from the dominant methods for transport, which preserve the physical properties directly in the discretization process through monotonic reconstruction of the fields. The slope and flux limiters used for this purpose tie together preservation of physical properties with restrictions on the mesh geometry and/or the accuracy. As a result, many of them do not preserve linear functions on irregular grids [1], which impacts accuracy and robustness. An alternative is to use sophisticated "repair" procedures [6] or error compensation algorithms [7], which fix the out-of-bound values and maintain positivity on arbitrary unstructured grids. However, limiters and "repair" procedures obscure the sources of discretization errors, which complicates the analysis of the transport schemes, and their higher-order extensions on unstructured grids are very complex.

2   An Abstract Framework for Optimization–Based Modeling

The abstract framework for optimization–based modeling in this section merges at an abstract level many of the ideas explored in [3,2,4]. To explain the basic concepts it suffices to consider a coupled problem with two "physics" components, $\mathcal{L}_1 : X_1 \times X_2 \times D_1 \mapsto Y$ and $\mathcal{L}_2 : X_2 \times X_1 \times D_2 \mapsto Y$ respectively, where X_i, $i = 1, 2$, and Y are Banach spaces for the solution and the data, respectively. The spaces D_i, $i = 1, 2$, supply the model parameters. Given a set of parameters $d_i \in D_i$ the multiphysics problem is to seek $\{u_1, u_2\} \in X_1 \times X_2$ such that

$$\mathcal{L}_1(u_1, u_2; d_1) + \mathcal{L}_2(u_2, u_1; d_2) = 0. \tag{1}$$

Regarding (1) we make the following assumptions

A.1 There exist subsets $\mathcal{U}_i \subseteq X_i$ such that $u_i \in \mathcal{U}_i$ for any solution of (1).

A.2 $\mathcal{L}_i$ are well-posed: given $\tilde{u}_2 \in \mathcal{U}_2$ the equation $\mathcal{L}_1(u_1, \tilde{u}_2; d_1) = 0$ has a unique solution $u_1 \in \mathcal{U}_1$. Conversely, if $\tilde{u}_1 \in \mathcal{U}_1$, then $\mathcal{L}_2(u_2, \tilde{u}_1; d_2) = 0$ has a unique solution $u_2 \in \mathcal{U}_2$.

A.3 There are discrete spaces X_{ij}^h, $i, j = 1, 2$, and Y^h as well as discrete operators $\mathcal{L}_1^h : X_{11}^h \times X_{12}^h \times D_1 \mapsto Y^h$ and $\mathcal{L}_2^h : X_{22}^h \times X_{21}^h \times D_2 \mapsto Y^h$ such that for $\tilde{u}_2^h \in X_{12}^h$, $\tilde{u}_1^h \in X_{21}^h$ the problems $\mathcal{L}_1^h(u_1^h, \tilde{u}_2^h; d_1) = 0$ and $\mathcal{L}_2^h(u_2^h, \tilde{u}_1^h; d_2) = 0$, have unique solutions, and $\lim_{h \to 0} \|u_i - u_i^h\|_{X_{ii}^h} = 0$ for $i = 1, 2$.

A.4 Robust and efficient solvers $(\mathcal{L}_i^h)^{-1}$ exist for each discrete constituent component operator $\mathcal{L}_i^h$.

Remark 1. A.1 is a formal statement of intrinsic physical properties (maximum principle, positivity, monotonicity, and etc.) of the exact solution of (1). A.2 states that the component physics operators $\mathcal{L}_i$ are well-posed and preserve these properties so long as the inputs are admissible. A.3–A.4 imply that each $\mathcal{L}_i$ can be discretized and solved in a stable, accurate and efficient manner. However, in A.3–A.4 it is not assumed that $\mathcal{L}_i^h$ preserve the physical properties in A.1.

Assumptions A.1–A.4 can be satisfied for most problems of interest by breaking down the multiphysics model into sufficiently small constituent components. However, in general, $X_{11}^h \neq X_{21}^h$ and $X_{22}^h \neq X_{12}^h$, which is a formal way of saying that stable discretizations of $\mathcal{L}_1$ and $\mathcal{L}_2$ may require *mutually exclusive* field representations. Consequently, $\mathcal{L}_1^h + \mathcal{L}_2^h$ is not guaranteed to be a meaningful, let alone a well-posed discrete operator. This means that even with A.1–A.4 holding, we are not assured that (i) a stable and accurate *monolithic* discretization of the multiphysics problem (1) is readily available, and (ii) if such a discretization exists, the resulting problem can be solved in a robust and efficient manner.

Our strategy for dealing with these two fundamental issues arising in the discretization of complex multiphysics problems is based on non-standard application of optimization and control ideas. Specifically, we reformulate (1) into an equivalent *multi-objective constrained optimization problem*. The cost functional in this problem minimizes discrepancies between multiple versions of the exact solution subject to constraints derived from the component physics operators and the condition that physical properties are preserved in the optimal solution. In so doing our approach exposes the component physics operators and separates the preservation of physical properties from mesh geometry and field representations. It is an example of a "divide-and-conquer" strategy, which *decomposes the operator space and relieves the discretization process of tasks that it is not well-equipped to handle in a robust and efficient manner*. Due to the limited space, demonstration of the framework will be restricted to preservation of physical properties in a single physics setting.

2.1 Optimization-Based Reformulation of Multiphysics Operators

We reformulate the multiphysics model (1) into a multi-objective constrained optimization problem in three stages. For simplicity, we assume that all necessary spaces are Hilbertian and postpone the discussion of data assimilation to the end of the section. In the first stage we modify (1) into the equivalent problem

$$\mathcal{L}_1(u_1, u_2; d_1) + \mathcal{R}(\theta) + \mathcal{L}_2(u_2, u_1; d_2) - \mathcal{R}(\theta) = 0, \tag{2}$$

where X is a Hilbert space, $\theta \in X$ is a control function, and $\mathcal{R} : X \mapsto Y$ is a suitable operator. At the second stage we fix $\theta \in X$, $u_{12} \in \mathcal{U}_2$, and $u_{21} \in \mathcal{U}_1$ and split (2) into two independent problems: seek $u_{11} \in X_1$ and $u_{22} \in X_2$ such that

$$\mathcal{L}_1(u_{11}, u_{12}; d_1) + \mathcal{R}(\theta) = 0 \quad \text{and} \quad \mathcal{L}_2(u_{22}, u_{21}; d_2) - \mathcal{R}(\theta) = 0 \tag{3}$$

The third stage, reconnects these problems using the multi-objective functional

$$J_{\vec{\alpha}}(u_{11}, u_{12}; u_{22}, u_{21}; \theta) = \frac{1}{2}\left(\alpha_1 \|u_{11} - u_{21}\|_W^2 + \alpha_2 \|u_{22} - u_{12}\|_W^2 + \alpha_3 \|\theta\|_X^2\right), \tag{4}$$

where $\vec{\alpha} = \{\alpha_1, \alpha_2, \alpha_3\}$, and W is such that $X_i \subseteq W$. In this stage we replace (1) by the constrained optimization problem[1]:

$$\min \ J_{\vec{\alpha}}(u_{11}, u_{12}; u_{22}, u_{21}; \theta) \quad \text{subject to} \quad \begin{cases} \mathcal{L}_1(u_{11}, u_{12}; d_1) + \mathcal{R}(\theta) = 0 \\ \mathcal{L}_2(u_{22}, u_{21}; d_2) - \mathcal{R}(\theta) = 0 \\ u_{i1} \in \mathcal{U}_1 \,; u_{i2} \in \mathcal{U}_2 \,, i = 1, 2 \end{cases} . \quad (5)$$

The first two constraints in (5) are defined by the physics operators, and the third enforces the physical properties on the optimal solution. Its structure depends on these properties and can include both equality and inequality constraints.

2.2 Discretization of the Reformulated Problem

The gist of our strategy is to use the optimization problem (5) as an instrument to fuse stable and accurate discretizations $\mathcal{L}_i^h$ of the constituent physics operators into feature-preserving discretizations of the multiphysics problem (1). We assume that X_{ij}^h and $\mathcal{L}_i^h$ are as in A.1-A.4. Stable and accurate discretizations of the component operators may require mutually exclusive discrete spaces $X_{11}^h \neq X_{21}^h$ and $X_{22}^h \neq X_{12}^h$. This is a serious problem for monolithic discretizations of (1) but is *completely benign* for the discretization of (5) where physics operators are decoupled and can be approximated independently.

To separate the accuracy considerations from the preservation of physical properties we treat the discrete solutions u_{ij}^h of $\mathcal{L}_i^h$ as *targets* that provide the best possible accuracy and impose the constraints on a separate set of variables $\widehat{u}_{ij}^h$. To this end, we modify the multi-objective functional by adding terms which force these new variables $\widehat{u}_{ij}^h$ close to the optimally accurate targets u_{ij}^h:

$$\widehat{J}_{\vec{\alpha}}(\widehat{u}_{ij}^h, u_{ij}^h; \theta^h) = \frac{1}{2}\left(\sum_{i,j=1}^{2} \widehat{\alpha}_{ij} \|\widehat{u}_{ij}^h - u_{ij}^h\|_W^2 + \sum_{j=1}^{2} \alpha_j \|u_{1j}^h - u_{2j}^h\|_W^2 + \alpha_3 \|\theta^h\|_X^2 \right), \quad (6)$$

where $\vec{\alpha} = \{\{\widehat{\alpha}_{ij}\}, \alpha_1, \alpha_2, \alpha_3\}$. The discrete optimization problem then reads

$$\min \ \widehat{J}_{\vec{\alpha}}(\widehat{u}_{ij}^h, u_{ij}^h; \theta^h) \quad \text{subject to} \quad \begin{cases} \mathcal{L}_1^h(u_{11}^h, u_{12}^h; d_1) + \mathcal{R}^h(\theta^h) = 0 \\ \mathcal{L}_2^h(u_{22}^h, u_{21}^h; d_2) - \mathcal{R}^h(\theta^h) = 0 \\ \widehat{u}_{i1}^h, \in \mathcal{U}_1 \,; \ \widehat{u}_{i2}^h \in \mathcal{U}_2 \,; \ i = 1, 2 \end{cases} . \quad (7)$$

The discrete optimization problem (7) retains the key features of (5) and thus, it can be used both for the **synthesis** of approximate multiphysics operators from discretizations of their constituent physics components, and for the **decomposition** of such operators into simpler parts.

---

[1] Problem (5) differs from those encountered in conventional PDE-constrained optimization. For instance, because θ is a virtual control, it is not subject to constraints, and does not reduce solution regularity relative to the original problem (1).

3   Application to Transport Problems

We use the optimization framework in Sections 2.1–2.2 to develop a new class of conservative, monotone and bounds preserving methods for the scalar transport equation

$$\partial_t \rho + \nabla \cdot \rho \mathbf{v} = 0 \quad \text{on } \Omega \times [0, T] \quad \text{and} \quad \rho(\mathbf{x}, 0) = \rho^0(\mathbf{x}), \tag{8}$$

where $T > 0$ is the final time, $\rho(\mathbf{x}, t)$ is a positive density function (the primitive variable) on $\Omega \times [0, T]$ with initial distribution $\rho^0(\mathbf{x})$, and $\mathbf{v}$ is a velocity field. For simplicity, we assume that $\rho(\mathbf{x}, t) = 0$ on $\partial\Omega \times [0, T]$. Let $K_h(\Omega)$ denote a partition of Ω into cells κ_i, $i = 1, \ldots, K$. We solve (8) using a cell-centered discretization of the density. The degrees of freedom ρ_i^n approximate the mean cell density at time $t = t_n$:

$$\rho_i(t_n) = \frac{\int_{\kappa_i} \rho(\mathbf{x}, t_n) dV}{\int_{\kappa_i} dV} = \frac{\int_{\kappa_i} \rho(\mathbf{x}, t_n) dV}{\text{vol}(\kappa_i)}.$$

The approximate mass in cell κ_i at time t_n is $m_i^n = \rho_i^n \text{vol}(\kappa_i)$.

To solve (8) we proceed as follows. Numerical integration of $\rho^0(\mathbf{x})$ on each grid cell κ_i yields the initial cell masses $\vec{m}^0 = (m_1^0, \ldots, m_K^0)$ and the initial density distribution $\vec{\rho}^0 = (\rho_1^0, \ldots, \rho_K^0)$ on $K_h(\Omega)$, where $\rho_i^0 = m_i^0/\text{vol}(\kappa_i)$. Suppose that the approximate solution $\vec{\rho}^n = (\rho_1^n, \ldots, \rho_K^n)$ is known on $K_h(\Omega)$ at time $0 \le t_n < T$ and Δt_n is an admissible *explicit* time step. To find the approximate density distribution $\vec{\rho}^{n+1} = (\rho_1^{n+1}, \ldots, \rho_K^{n+1})$ on $K_h(\Omega)$ at the new time step $t_{n+1} = t_n + \Delta t_n$, we apply the forward *incremental remapping* algorithm [5]. This algorithm advances the solution of (8) to the next time step using that the mass of a Lagrangian volume $V_L(t)$ is conserved along the trajectories $d\mathbf{x}/dt = \mathbf{v}$:

$$\int_{V_L(t_{n+1})} \rho dV = \int_{V_L(t_n)} \rho dV. \tag{9}$$

In particular, if $V_L(t_n) = \Omega$, and $\widetilde{\Omega} = V_L(t_{n+1})$ is the deformed region, the total mass is conserved:

$$M(\Omega) = \int_\Omega \rho(\mathbf{x}, t) dV = \int_{\widetilde{\Omega}} \rho(\mathbf{x}, t) dV = M(\widetilde{\Omega}).$$

The idea of the incremental remap approach is to evolve the computational grid $K_h(\Omega)$ into a grid $\widetilde{K}_h(\widetilde{\Omega})$ on the deformed region $\widetilde{\Omega}$ at t_{n+1}, compute the mean density on this grid and interpolate it back to $K_h(\Omega)$. Specifically, if we set $V_L(t_n) = \kappa_i$ then, according to (9) the mass $\widetilde{m}_i$ in $V_L(t_{n+1})$ equals the mass m_i^n in $V_L(t_n)$ and the mean density on $V_L(t_{n+1})$ is $\widetilde{\rho}_i = m_i^n/\text{vol}(V_L(t_{n+1}))$.

In practice, $V_L(t_{n+1})$ is not known exactly and must be approximated. A simple strategy is to evolve the vertices $\{\mathbf{x}_p\}$ of κ_i along the trajectories using, e.g., an explicit Euler method. This yields a cell $\widetilde{\kappa}_i$ with vertices $\widetilde{\mathbf{x}}_p = \mathbf{x}_p + \Delta t_n \mathbf{v}$, which approximates $V_L(t_{n+1})$. The mass $\widetilde{m}_i$ and the mean density $\widetilde{\rho}_i$ on $\widetilde{\kappa}_i$ are

$$\widetilde{m}_i = m_i^n \quad \text{and} \quad \widetilde{\rho}_i = \frac{m_i^n}{\text{vol}(\widetilde{\kappa}_i)}; \quad i = 1, \ldots, K.$$

Algorithm 1. One forward step of incremental remapping

input : Density approximation $\vec{\rho}^n = (\rho_1^n, \ldots, \rho_K^n)$ at time t_n, time step Δt_n
output: Density approximation $\vec{\rho}^{n+1} = (\rho_1^{n+1}, \ldots, \rho_K^{n+1})$ at time t_{n+1}

1 Project grid: $K_h(\Omega) \ni \mathbf{x}_p \mapsto \mathbf{x}_p + \Delta t_n \mathbf{v} = \tilde{\mathbf{x}}_p \in \tilde{K}_h(\tilde{\Omega})$

2 Transport m and ρ: $\forall \tilde{\kappa}_i \in \tilde{K}_h(\tilde{\Omega})$ set $\tilde{m}_i = m_i^n$ and $\tilde{\rho}_i = \tilde{m}_i / \mathrm{vol}(\tilde{\kappa}_i)$

3 Remap density: $\vec{\rho}^{n+1} = \mathcal{R}(\{\tilde{\rho}_1, \ldots, \tilde{\rho}_K\})$

Conservative interpolation (remap) of the mean density values $\tilde{\rho}_i$ from the deformed mesh $\tilde{K}_h(\tilde{\Omega})$ onto the original mesh $K_h(\Omega)$ gives the approximate mean cell density $\vec{\rho}^{n+1} = (\rho_1^{n+1}, \ldots, \rho_K^{n+1})$ at the next time level; see Algorithm 1.

The conservative interpolation (remap) operator $\mathcal{R}$ is the key ingredient of Algorithm 1. To state the requirements on $\mathcal{R}$ without going into unnecessary technical details, it is convenient to assume that $\mathbf{v} \cdot \mathbf{n} = 0$. In this case the original and deformed regions coincide: $\Omega = \tilde{\Omega}$, and the mass is conserved at all times. Let $\tilde{N}_i$, and N_i denote the neighborhoods of $\tilde{\kappa}_i \in \tilde{K}_h(\tilde{\Omega})$, and $\kappa_i \in K_h(\Omega)$, resp., i.e., all cells that share vertex or an edge or a face with $\tilde{\kappa}_i$ or κ_i. Define

$$\tilde{\rho}_i^{\min} = \min_{j \in \tilde{N}_i} \tilde{\rho}_j ; \qquad \tilde{\rho}_i^{\max} = \max_{j \in \tilde{N}_i} \tilde{\rho}_j ;$$

Under the assumptions stated above, $\mathcal{R}$ must satisfy the following requirements:

R.1 local bounds are preserved: $\tilde{\rho}_i^{\min} \leq \rho_i^{n+1} \leq \tilde{\rho}_i^{\max}$;

R.2 total mass is conserved: $\sum_{i=0}^K m_i^{n+1} = \sum_{i=0}^K \tilde{m}_i = \sum_{i=0}^K m_i^n$;

R.3 linearity is preserved: $m_i^{n+1} = \int_{\kappa_i} \rho(\mathbf{x}, t_{n+1}) dV$ if $\rho(\mathbf{x}, t)$ is linear in $\mathbf{x}$.

We use the optimization framework in Section 2.2 to define a constrained interpolation operator that satisfies R.1–R.3. The starting point is the flux-form formula[2] for the cell masses on $K_h(\Omega)$ corresponding to the new time level:

$$m_i^{n+1} = \tilde{m}_i + \sum_{\tilde{\kappa}_j \in \tilde{N}_i} \tilde{F}_{ij}^h ; \quad i = 1, \ldots, K . \tag{10}$$

The mass fluxes $\tilde{F}_{ij}^h$ approximate the mass exchanges between the cells in the neighborhood $\tilde{N}_i$ of $\tilde{\kappa}_i$. We specialize the abstract optimization problem (7) as follows. Clearly, (8) is a "single-physics" equation and it suffices to consider a single "physics" operator with a single target field u^h and a single approximation field $\hat{u}^h$. We identify $\hat{u}^h$ with the mass fluxes in (10) and u^h – with a target flux $\tilde{F}_{ij}^T$ which is *exact for linear density functions*. In this context, the single "physics" operator $\mathcal{L}^h(u^h)$ is the target flux reconstruction

$$\mathcal{L}^h(\tilde{F}_{ij}^T) = \tilde{F}_{ij}^T = \int_{\kappa_i \cap \tilde{\kappa}_j} \tilde{\rho}_j^\ell(\mathbf{x}) dV - \int_{\tilde{\kappa}_i \cap \kappa_j} \tilde{\rho}_i^\ell(\mathbf{x}) dV ; \quad \tilde{\kappa}_j \in \tilde{N}_i ; \kappa_j \in N_i$$

---

[2] The conditions under which this formula holds [7] can be always satisfied for the settings of interest to us.

where $\widetilde{\rho}_i^\ell(\mathbf{x})$ is density reconstruction on cell $\widetilde{\kappa}_i$, which is exact for linear functions. Finally, from R.1 we obtain bounds for the mass on the new time level:

$$\widetilde{\rho}_i^{\min}\mathrm{vol}(\kappa_i) = m_i^{\min} \leq m_i^{n+1} \leq m_i^{\max} = \widetilde{\rho}_i^{\max}\mathrm{vol}(\kappa_i)$$

Thus, the abstract discrete problem (7) specializes to the following QP:

$$\underset{\widetilde{F}_{ij}^h}{\text{minimize}} \sum_{i=1}^{K} \sum_{\widetilde{\kappa}_j \in \widetilde{N}_i} (\widetilde{F}_{ij}^h - \widetilde{F}_{ij}^T)^2 \quad \text{subject to}$$

$$\begin{cases} \widetilde{F}_{ij}^T = \displaystyle\int_{\kappa_i \cap \widetilde{\kappa}_j} \widetilde{\rho}_j^\ell(\mathbf{x})dV - \int_{\widetilde{\kappa}_i \cap \kappa_j} \widetilde{\rho}_i^\ell(\mathbf{x})dV & \leftarrow \text{``physics'' operator} \\ \widetilde{F}_{ij}^h = -\widetilde{F}_{ji}^h & \leftarrow \text{mass conservation} \\ m_i^{\min} \leq \widetilde{m}_i + \displaystyle\sum_{\widetilde{\kappa}_j \in \widetilde{N}_i} \widetilde{F}_{ij}^h \leq m_i^{\max} & \leftarrow \text{local bounds}. \end{cases} \quad (11)$$

The optimization-based formulation (11) for constrained interpolation is of independent interest for Arbitrary Lagrangian-Eulerian methods [4]. It separates enforcement of the physical properties R.1 and R.2, which is done through the constraints, from the enforcement of the accuracy R.3, which is achieved through the objective functional. As a result, (11) is impervious to cell shapes and can be used on arbitrary grids. We conclude Part 1 with a proof that (11) has optimal solution. Part 2 [10] develops efficient algorithms for (11) and Part 3 [8] presents implementation of Algorithm 1 and computational studies.

Theorem 1. *Assume that $K_h(\Omega)$ and $\widetilde{K}_h(\widetilde{\Omega})$ are such that every cell $\kappa_i \in K_h(\Omega)$ is contained in the neighborhood $\widetilde{N}_i$ of its image $\widetilde{\kappa}_i \in \widetilde{K}_h(\widetilde{\Omega})$. For any given set of masses $\widetilde{m}_i$ and associated densities $\widetilde{\rho}_i = \widetilde{m}_i/\mathrm{vol}(\widetilde{\kappa}_i)$ on $\widetilde{K}_h(\widetilde{\Omega})$ there exist antisymmetric fluxes $\{\widetilde{F}_{ij}\}$ which satisfy the inequality constraints in (11).*

Proof. We need to show that there are antisymmetric fluxes $\widetilde{F}_{ij}$ such that

$$\widetilde{\rho}_i^{\min}\mathrm{vol}(\kappa_i) \leq \widetilde{\rho}_i\mathrm{vol}(\widetilde{\kappa}_i) + \sum_{\widetilde{\kappa}_j \in \widetilde{N}_i} \widetilde{F}_{ij} \leq \widetilde{\rho}_i^{\max}\mathrm{vol}(\kappa_i)$$

Fix a cell index $1 \leq i \leq K$, and choose $\widehat{\rho}_j$, for $\widetilde{\kappa}_j \in \widetilde{N}_j$ according to

$$\widetilde{\rho}_i^{\min} \leq \widehat{\rho}_j \leq \widetilde{\rho}_i^{\max} \quad \text{for } j \neq i \quad \text{and} \quad \widehat{\rho}_i = \widetilde{\rho}_i. \quad (12)$$

Define the fluxes

$$\widetilde{F}_{ij} = \widehat{\rho}_j\mathrm{vol}(\kappa_i \cap \widetilde{\kappa}_j) - \widehat{\rho}_i\mathrm{vol}(\widetilde{\kappa}_i \cap \kappa_j). \quad (13)$$

Clearly, $\widetilde{F}_{ij} = -\widetilde{F}_{ji}$. Using the definition (13)

$$\widetilde{\rho}_i\mathrm{vol}(\widetilde{\kappa}_i) + \sum_{\widetilde{\kappa}_j \in \widetilde{N}_i} \widetilde{F}_{ij} = \widetilde{\rho}_i\Big[\mathrm{vol}(\widetilde{\kappa}_i) - \sum_{j \neq i}\mathrm{vol}(\widetilde{\kappa}_i \cap \kappa_j)\Big] + \sum_{j \neq i}\widehat{\rho}_j\mathrm{vol}(\kappa_i \cap \widetilde{\kappa}_j)$$

$$= \widehat{\rho}_i\mathrm{vol}(\widetilde{\kappa}_i \cap \kappa_i) + \sum_{j \neq i}\widehat{\rho}_j\mathrm{vol}(\kappa_i \cap \widetilde{\kappa}_j) = \sum_{\widetilde{\kappa}_j \in \widetilde{N}_i}\widehat{\rho}_j\mathrm{vol}(\kappa_i \cap \widetilde{\kappa}_j).$$

From $\kappa_i = \cup_k(\kappa_i \cap \tilde{\kappa}_j)$ and the bounds in (12) it follows that

$$\sum_{\tilde{\kappa}_j \in \tilde{N}_i} \hat{\rho}_j \mathrm{vol}(\kappa_i \cap \tilde{\kappa}_j) \leq \tilde{\rho}_i^{\max} \sum_{\tilde{\kappa}_j \in \tilde{N}_i} \mathrm{vol}(\kappa_i \cap \tilde{\kappa}_j) = \tilde{\rho}_i^{\max}\mathrm{vol}(\kappa_i) \, ;$$

$$\sum_{\tilde{\kappa}_j \in \tilde{N}_i} \hat{\rho}_j \mathrm{vol}(\kappa_i \cap \tilde{\kappa}_j) \geq \tilde{\rho}_i^{\min} \sum_{\tilde{\kappa}_j \in \tilde{N}_i} \mathrm{vol}(\kappa_i \cap \tilde{\kappa}_j) = \tilde{\rho}_i^{\min}\mathrm{vol}(\kappa_i) \, . \qquad \square$$

In [4] we prove that (11) preserves linear densities if the barycenter of κ_i remains in the convex hull of the barycenters of the cells in $\tilde{N}_i$ for all $1 \leq i \leq K$. This condition is less restrictive than the one required for linearity preservation by Van Leer limiting [9] and is valid for any unstructured grid. In summary, using Algorithm 1 in conjunction with an operator $\mathcal{R}$ defined by the QP (11), yields a conservative and monotone transport algorithm that is applicable to *arbitrary cell shapes*, including polygons and polyhedra.

References

1. Berger, M., Murman, S.M., Aftosmis, M.J.: Analysis of slope limiters on irregular grids. In: Proceedings of the 43rd AIAA Aerospace Sciences Meeting. No. AIAA2005-0490, AIAA, Reno, NV, January 10-13 (2005)
2. Bochev, P., Ridzal, D.: Additive Operator Decomposition and Optimization–Based Reconnection with Applications. In: Lirkov, I., Margenov, S., Waśniewski, J. (eds.) LSSC 2009. LNCS, vol. 5910, pp. 645–652. Springer, Heidelberg (2010)
3. Bochev, P., Ridzal, D.: An optimization-based approach for the design of PDE solution algorithms. SIAM Journal on Numerical Analysis 47(5), 3938–3955 (2009), http://link.aip.org/link/?SNA/47/3938/1
4. Bochev, P., Ridzal, D., Scovazzi, G., Shashkov, M.: Formulation, analysis and numerical study of an optimization-based conservative interpolation (remap) of scalar fields for arbitrary lagrangian-eulerian methods. Journal of Computational Physics 230(13), 5199–5225 (2011), http://www.sciencedirect.com/science/article/B6WHY-52F895B-2/2/5e30ada70a5c6053464dfe9ceb74cf26
5. Dukowicz, J.K., Baumgardner, J.R.: Incremental remapping as a transport/advection algorithm. Journal of Computational Physics 160(1), 318–335 (2000), http://www.sciencedirect.com/science/article/B6WHY-45FC8N8-6F/2/179cbfc9634bb79579b68754cebd5525
6. Kucharik, M., Shashkov, M., Wendroff, B.: An efficient linearity-and-bound-preserving remapping method. Journal of Computational Physics 188(2), 462–471 (2003), http://www.sciencedirect.com/science/article/B6WHY-48CWYJW-2/2/d264d65dcfa253e387aea5bdebfd433f
7. Margolin, L.G., Shashkov, M.: Second-order sign-preserving conservative interpolation (remapping) on general grids. Journal of Computational Physics 184(1), 266–298 (2003), http://www.sciencedirect.com/science/article/B6WHY-47HS5PX-4/2/9acf255c80d91bf5873398d5b929303e

8. Ridzal, D., Bochev, P., Young, J., Peterson, K.: Optimization–Based Modeling with Applications to Transport. Part 3. Implementation and Computational Studies. In: Lirkov, I., Margenov, S., Wańsiewski, J. (eds.) LSSC 2011. LNCS, vol. 7116, pp. 81–88. Springer, Heidelberg (2012)

9. Swartz, B.: Good neighborhoods for multidimensional Van Leer limiting. Journal of Computational Physics 154(1), 237–241 (1999), http://www.sciencedirect.com/science/article/B6WHY-45GMW6B-25/2/5ba96d929cffd2519d4a04719509a5e7

10. Young, J., Ridzal, D., Bochev, P.: Optimization–Based Modeling with Applications to Transport. Part 2. Optimization Algorithm. In: Lirkov, I., Margenov, S., Wańsiewski, J. (eds.) LSSC 2011. LNCS, vol. 7116, pp. 72–80. Springer, Heidelberg (2012)

Optimization-Based Modeling
with Applications to Transport:
Part 2. The Optimization Algorithm

Joseph Young[1], Denis Ridzal[2], and Pavel Bochev[1]

[1] Numerical Analysis and Applications,
[2] Optimization and Uncertainty Quantification,
Sandia National Laboratories*, Albuquerque, NM 87185-1320, USA
{josyoun,dridzal,pbboche}@sandia.gov

Abstract. This paper is the second of three related articles that develop and demonstrate a new optimization-based framework for computational modeling. The framework uses optimization and control ideas to assemble and decompose multiphysics operators and to preserve their fundamental physical properties in the discretization process. One application of the framework is in the formulation of robust algorithms for optimization-based transport (OBT). Based on the theoretical foundations established in Part 1, this paper focuses on the development of an efficient optimization algorithm for the solution of the *remap subproblem* that is at the heart of OBT.

1 Introduction

In this and two companion papers [1,5] we formulate and study a new optimization-based framework for computational modeling. One application of the framework, introduced in Part 1 [1], is in the formulation of a new class of optimization-based transport (OBT) schemes. OBT schemes combine *incremental remap* [4] with the reformulation of the *remap subproblem* as an inequality-constrained quadratic program (QP) [2]. In this paper we develop and analyze an efficient optimization algorithm for the solution of the remap subproblem.

Our algorithm is based on the *dual* formulation of the remap subproblem. Our previous work [2] uses the reflective Newton method by Coleman and Li [3] for the solution of the dual remap subproblem. The Coleman-Li approach handles general bound-constrained QPs and ensures convergence from remote starting points using a trust-region globalization. In this paper we focus solely on the derivation and solution of a first-order optimality system that is specific to the remap subproblem; in other words, we disregard globalization. In practice, see Part 3 [5], the resulting Newton method proves sufficiently accurate and robust

* Sandia National Laboratories is a multi-program laboratory managed and operated by Sandia Corporation, a wholly owned subsidiary of Lockheed Martin Corporation, for the U.S. Department of Energy's National Nuclear Security Administration under contract DE-AC04-94AL85000.

I. Lirkov, S. Margenov, and J. Waśniewski (Eds.): LSSC 2011, LNCS 7116, pp. 72–80, 2012.
© Springer-Verlag Berlin Heidelberg 2012

in the context of incremental remapping where a nearly feasible and optimal initial guess for the remap subproblem is typically available.

2   The Remap Subproblem: Optimization Theory

Optimization-based transport, see Part 1 [1], requires the solution of the *remap subproblem*,

$$\min_{\widetilde{F}_{ij}^h} \sum_{i=1}^{K} \sum_{\widetilde{\kappa}_j \in \widetilde{N}_i} (\widetilde{F}_{ij}^h - \widetilde{F}_{ij}^T)^2 \quad \text{subject to}$$

$$\begin{cases} \widetilde{F}_{ij}^h = -\widetilde{F}_{ji}^h \\ m_i^{\min} \leq \widetilde{m}_i + \sum_{\widetilde{\kappa}_j \in \widetilde{N}_i} \widetilde{F}_{ij}^h \leq m_i^{\max} \,, \end{cases} \tag{1}$$

where $\widetilde{\kappa}_j \in \widetilde{N}_i$ denotes $j \in \{j : \widetilde{\kappa}_j \in \widetilde{N}_i\}$, $\widetilde{F}_{ij}^h$ are the unknown mass fluxes, $\widetilde{F}_{ij}^T$ are the given target fluxes, $\widetilde{\kappa}_i$ are the cells of the deformed mesh, $\widetilde{N}_i$ are the corresponding cell neighborhoods, $m_i^{\min}$ and $m_i^{\max}$ are the given local mass extrema, and $\widetilde{m}_i$ are the given masses on the deformed mesh. We enforce the antisymmetry constraint $\widetilde{F}_{ij}^h = -\widetilde{F}_{ji}^h$ explicitly by using only the fluxes $\widetilde{F}_{pq}^h$ for which $p < q$. This results in the simplified remap subproblem

$$\min_{\widetilde{F}_{ij}^h} \sum_{i=1}^{K} \sum_{\substack{\widetilde{\kappa}_j \in \widetilde{N}_i \\ i<j}} (\widetilde{F}_{ij}^h - \widetilde{F}_{ij}^T)^2 \quad \text{subject to}$$

$$m_i^{\min} - \widetilde{m}_i \leq \sum_{\substack{\widetilde{\kappa}_j \in \widetilde{N}_i \\ i<j}} \widetilde{F}_{ij}^h - \sum_{\substack{\widetilde{\kappa}_j \in \widetilde{N}_i \\ i>j}} \widetilde{F}_{ji}^h \leq m_i^{\max} - \widetilde{m}_i \,. \tag{2}$$

In compact matrix / vector notation problem (2) has the form

$$\min_{\vec{F} \in \mathbb{R}^M} \frac{1}{2}(\vec{F} - \vec{F}^H)^{\mathsf{T}}(\vec{F} - \vec{F}^H) \quad \text{subject to}$$

$$\vec{b}_{\min} \leq \mathbf{A}\vec{F} \leq \vec{b}_{\max} \,, \tag{3}$$

where M denotes the number of unique flux variables, $\widetilde{F}_{ij}^h$. We also define $\vec{F} \in \mathbb{R}^M$, $\vec{F}^H \in \mathbb{R}^M$, $\vec{b}_{\min} \in \mathbb{R}^K$ and $\vec{b}_{\max} \in \mathbb{R}^K$ such that $\vec{F}_{\iota(i,j)} = \widetilde{F}_{ij}^h$, $\vec{F}_{\iota(i,j)}^H = \widetilde{F}_{ij}^T$, $(\vec{b}_{\min})_i = m_i^{\min} - \widetilde{m}_i$ and $(\vec{b}_{\max})_i = m_i^{\max} - \widetilde{m}_i$, respectively, where ι is an indexing function. Finally we let $\mathbf{A} \in \mathbb{R}^{K \times M}$ be a matrix with entries -1, 0 and 1 defining the inequality constraints in (2) or a related proxy (see swept-region approximation, [2, Sec. 4.1,4.2]). The matrix $\mathbf{A}$ is typically very sparse, with $M > K$ in 2D and 3D.

In what follows we use two conventions. First, we define the *Euclidean inner product*, $\langle \cdot, \cdot \rangle : \mathbb{R}^{2m} \to \mathbb{R}$, as $\langle \vec{x}, \vec{y} \rangle = \vec{x}^{\mathsf{T}}\vec{y}$, and the *Euclidian norm* $\|x\|_2^2 =$

$\langle \vec{x}, \vec{x} \rangle = \vec{x}^{\mathsf{T}} \vec{x}$. Second, we abbreviate the *nonnegative orthant* as $\mathbb{R}_+^m = \{ \vec{x} \in \mathbb{R}^m : \vec{x} \geq 0 \}$. To be precise, $\vec{x} \geq 0$ denotes componentwise inequality.

Rather than solving (3) directly, we focus on its dual formulation. This allows us to reformulate the problem into a simpler, *bound-constrained* optimization problem.

Theorem 1. *Given the definitions of $\vec{F}^H \in \mathbb{R}^M$, $\vec{b}_{\min} \in \mathbb{R}^K$, $\vec{b}_{\max} \in \mathbb{R}^K$, and $\mathbf{A} \in \mathbb{R}^{K \times M}$ from above, let us define $J_p : \mathbb{R}^M \to \mathbb{R}$ and $J_d : \mathbb{R}^{2K} \to \mathbb{R}$ as*

$$J_p(\vec{F}) = \frac{1}{2} \| \vec{F} - \vec{F}^H \|_2^2$$

and

$$J_d(\vec{\lambda}, \vec{\mu}) = \frac{1}{2} \| \mathbf{A}^{\mathsf{T}} \vec{\lambda} - \mathbf{A}^{\mathsf{T}} \vec{\mu} \|_2^2 - \langle \vec{\lambda}, \vec{b}_{\min} - \mathbf{A} \vec{F}^H \rangle - \langle \vec{\mu}, -\vec{b}_{\max} + \mathbf{A} \vec{F}^H \rangle.$$

Then, we have that

$$\min_{\vec{F} \in \mathbb{R}^M} \left\{ J_p(\vec{F}) : \vec{b}_{\min} \leq \mathbf{A} \vec{F} \leq \vec{b}_{\max} \right\} = \min_{(\vec{\lambda}, \vec{\mu}) \in \mathbb{R}_+^{2K}} \left\{ J_d(\vec{\lambda}, \vec{\mu}) \right\}$$

where we call the first problem the primal and the second problem the dual. Furthermore,

$$\left\{ \vec{F}^H + \mathbf{A}^{\mathsf{T}} (\vec{\lambda}^* - \vec{\mu}^*) \right\} = \arg \min_{\vec{F} \in \mathbb{R}^M} \left\{ J_p(\vec{F}) : \vec{b}_{\min} \leq \mathbf{A} \vec{F} \leq \vec{b}_{\max} \right\}$$

whenever

$$(\vec{\lambda}^*, \vec{\mu}^*) \in \arg \min_{(\vec{\lambda}, \vec{\mu}) \in \mathbb{R}_+^{2K}} \left\{ J_d(\vec{\lambda}, \vec{\mu}) \right\}.$$

Proof. We begin with the observation that J_p denotes a strictly convex, continuous function and that $\{ \vec{F} \in \mathbb{R}^M : \vec{b}_{\min} \leq \mathbf{A} \vec{F} \leq \vec{b}_{\max} \}$ denotes a bounded, closed, convex set. Therefore, a unique minimum exists and is attained. Furthermore, since there exists an $\vec{F}$ such that $\vec{b}_{\min} < \mathbf{A} \vec{F} < \vec{b}_{\max}$ [1], we satisfy Slater's constraint qualification. This tells us that strong duality holds, which implies that the Lagrangian dual exists and possesses the same optimal value as the original problem.

Based on this knowledge, we notice that

$$\min_{\vec{F} \in \mathbb{R}^M} \left\{ J_p(\vec{F}) : \vec{b}_{\min} \leq \mathbf{A} \vec{F} \leq \vec{b}_{\max} \right\}$$

$$= \min_{\vec{F} \in \mathbb{R}^M} \max_{(\vec{\lambda}, \vec{\mu}) \in \mathbb{R}_+^{2K}} \left\{ J_p(\vec{F}) - \langle \mathbf{A} \vec{F} - \vec{b}_{\min}, \vec{\lambda} \rangle - \langle \vec{b}_{\max} - \mathbf{A} F, \vec{\mu} \rangle \right\}$$

$$= \max_{(\vec{\lambda}, \vec{\mu}) \in \mathbb{R}_+^{2K}} \min_{\vec{F} \in \mathbb{R}^M} \left\{ J_p(\vec{F}) - \langle \vec{F}, \mathbf{A}^{\mathsf{T}} (\vec{\lambda} - \vec{\mu}) \rangle + \langle \vec{b}_{\min}, \vec{\lambda} \rangle - \langle \vec{b}_{\max}, \vec{\mu} \rangle \right\}.$$

Next, we consider the function $J : \mathbb{R}^M \to \mathbb{R}$ where

$$J(\vec{F}) = J_p(\vec{F}) - \langle \vec{F}, \mathbf{A}^{\mathsf{T}} (\vec{\lambda} - \vec{\mu}) \rangle$$

and $(\vec{\lambda}, \vec{\mu}) \in \mathbb{R}^{2K}$ are fixed. We see that J is strictly convex. Therefore, it attains its unique minimum when $\nabla J = 0$. Specifically, when

$$\vec{F} - \vec{F}^H - \mathbf{A}^\mathsf{T}(\vec{\lambda} - \vec{\mu}) = 0,$$

which occurs if and only if

$$\vec{F} = \vec{F}^H + \mathbf{A}^\mathsf{T}(\vec{\lambda} - \vec{\mu}).$$

Therefore, we may find the optimal solution to our original problem with this equation when $(\vec{\lambda}, \vec{\mu})$ are optimal. In addition, we may use this knowledge to simplify our derivation of the dual. Let $\omega = \mathbf{A}^\mathsf{T}(\vec{\lambda} - \vec{\mu})$ and notice that

$$\max_{(\vec{\lambda},\vec{\mu})\in\mathbb{R}_+^{2K}} \min_{\vec{F}\in\mathbb{R}^M} \left\{ J_p(\vec{F}) - \langle \vec{F}, \mathbf{A}^\mathsf{T}(\vec{\lambda} - \vec{\mu})\rangle + \langle b_{\min}, \vec{\lambda}\rangle - \langle b_{\max}, \vec{\mu}\rangle \right\}$$

$$= \max_{(\vec{\lambda},\vec{\mu})\in\mathbb{R}_+^{2K}} \left\{ J_p(\vec{F}^H + \omega) - \langle \vec{F}^H + \omega, \omega\rangle + \langle b_{\min}, \vec{\lambda}\rangle - \langle b_{\max}, \vec{\mu}\rangle \right\}$$

$$= \max_{(\vec{\lambda},\vec{\mu})\in\mathbb{R}_+^{2K}} \left\{ \frac{1}{2}\|\omega\|_2^2 - \langle \vec{F}^H, \omega\rangle - \|\omega\|_2^2 + \langle b_{\min}, \vec{\lambda}\rangle - \langle b_{\max}, \vec{\mu}\rangle \right\}$$

$$= \max_{(\vec{\lambda},\vec{\mu})\in\mathbb{R}_+^{2K}} \left\{ -\frac{1}{2}\|\mathbf{A}^\mathsf{T}(\vec{\lambda} - \vec{\mu})\|_2^2 - \langle \mathbf{A}\vec{F}^H, \vec{\lambda} - \vec{\mu}\rangle + \langle b_{\min}, \vec{\lambda}\rangle - \langle b_{\max}, \vec{\mu}\rangle \right\}$$

$$= \min_{(\vec{\lambda},\vec{\mu})\in\mathbb{R}_+^{2K}} \left\{ \frac{1}{2}\|\mathbf{A}^\mathsf{T}(\vec{\lambda} - \vec{\mu})\|_2^2 + \langle \mathbf{A}\vec{F}^H, \vec{\lambda} - \vec{\mu}\rangle - \langle b_{\min}, \vec{\lambda}\rangle + \langle b_{\max}, \vec{\mu}\rangle \right\}$$

$$= \min_{(\vec{\lambda},\vec{\mu})\in\mathbb{R}_+^{2K}} \left\{ \frac{1}{2}\|\mathbf{A}^\mathsf{T}\vec{\lambda} - \mathbf{A}^\mathsf{T}\vec{\mu}\|_2^2 - \langle \vec{\lambda}, \vec{b}_{\min} - \mathbf{A}\vec{F}^H\rangle - \langle \vec{\mu}, -\vec{b}_{\max} + \mathbf{A}\vec{F}^H\rangle \right\}$$

$$= \min_{(\vec{\lambda},\vec{\mu})\in\mathbb{R}_+^{2K}} \left\{ J_d(\vec{\lambda}, \vec{\mu}) \right\}.$$

Hence, we see the equivalence between our two optimization problems and note that the equation $\vec{F} = \vec{F}^H + \mathbf{A}^\mathsf{T}(\vec{\lambda} - \vec{\mu})$ allows us to find an optimal primal solution given an optimal solution to the dual. $\qquad\square$

Although the primal problem is strictly convex and possesses a unique optimal solution, the dual formulation does not. Rather, the dual problem is convex, but not strictly convex, so multiple minima may exist. Second, our formula for reconstructing the primal solution from the dual depends on an optimal dual solution. If the solution to the dual is not optimal, the reconstruction formula may generate infeasible solutions. With these points in mind, we require two additional definitions before we may proceed to our optimization algorithm.

Definition 1. *We define the diagonal operator,* $\mathrm{Diag} : \mathbb{R}^m \to \mathbb{R}^{m \times m}$, *as*

$$[\mathrm{Diag}(\vec{x})]_{ij} = \begin{cases} \vec{x}_i & \text{when} \quad i = j \\ 0 & \text{''} \quad i \neq j \end{cases}.$$

Definition 2. *For some symmetric, positive semidefinite* $\mathbf{H} \in \mathbb{R}^{m \times m}$ *and some* $\vec{b} \in \mathbb{R}^m$, *we define the operator* $v_{\mathbf{H},\vec{b}} : \mathbb{R}^m \to \mathbb{R}^m$ *as*

$$v_{\mathbf{H},\vec{b}}(\vec{x}) = \begin{cases} \vec{x}_i & \text{when} & [\mathbf{H}\vec{x} + \vec{b}]_i \geq 0 \\ 1 & \text{"} & [\mathbf{H}\vec{x} + \vec{b}]_i < 0 \end{cases}.$$

When both $\mathbf{H}$ *and* $\vec{b}$ *are clear from the context, we abbreviate this function as* v.

In order to solve the dual optimization problem, we use a simplified version of the locally convergent Coleman-Li algorithm [3]. The key to this algorithm follows from the following lemma.

Lemma 1. *Let* $\mathbf{H} \in \mathbb{R}^{m \times m}$ *be symmetric, positive semidefinite and let* $\vec{b} \in \mathbb{R}^m$. *Then, for some* $\vec{x}^* \geq 0$, *we have that*

$$\vec{x}^* \in \arg \min_{x \in \mathbb{R}^m_+} \left\{ \frac{1}{2} \langle \mathbf{H}\vec{x}, \vec{x} \rangle + \langle \vec{b}, \vec{x} \rangle \right\} \iff \mathrm{Diag}(v(\vec{x}^*))(\mathbf{H}\vec{x}^* + \vec{b}) = 0.$$

Proof. We begin with the observation that since $\mathbf{H}$ is symmetric, positive semidefinite, the problem

$$\min_{x \in \mathbb{R}^m_+} \left\{ \frac{1}{2} \langle \mathbf{H}\vec{x}, \vec{x} \rangle + \langle \vec{b}, x \rangle \right\}$$

represents a convex optimization problem with a coercive objective and a closed, convex set of constraints. Therefore, a minimum exists and the first order optimality conditions become sufficient for optimality.

In the forward direction, we assume that we have an optimal pair $(\vec{x}^*, \vec{\lambda}^*)$ that satisfy the first order optimality conditions,

$$\mathbf{H}\vec{x}^* + \vec{b} - \vec{\lambda}^* = 0$$

$$\vec{x}^* \geq 0, \vec{\lambda}^* \geq 0$$

$$\mathrm{Diag}(\vec{x}^*)\vec{\lambda}^* = 0.$$

According to these equations, $\vec{\lambda}^* = \mathbf{H}\vec{x}^* + \vec{b}$ and $\vec{\lambda}^* \geq 0$. This implies that $\mathbf{H}\vec{x}^* + \vec{b} \geq 0$. Therefore, according to the definition of v, $[\mathrm{Diag}(v(\vec{x}^*))]_{ii} = \vec{x}^*_i$ for all i. This tells us that

$$[\mathrm{Diag}(v(\vec{x}^*))(\mathbf{H}\vec{x}^* + \vec{b})]_i = \vec{x}^*_i[\mathbf{H}\vec{x}^* + \vec{b}]_i = \vec{x}^*_i \vec{\lambda}^*_i = 0$$

where the final equality follows from our fourth optimality condition, complementary slackness.

In the reverse direction, we assume that $\mathrm{Diag}(v(\vec{x}^*))(\mathbf{H}\vec{x}^* + \vec{b}) = 0$ for some $\vec{x}^* \in \mathbb{R}^m_+$. Since the problem

$$\min_{x \in \mathbb{R}^m_+} \left\{ \frac{1}{2} \langle \mathbf{H}\vec{x}, \vec{x} \rangle + \langle \vec{b}, \vec{x} \rangle \right\}$$

represents a convex optimization problem, it is sufficient to show that the first order optimality conditions hold for $\vec{x}^*$ and some $\vec{\lambda}^*$. Of course, we immediately see that we satisfy primal feasibility since $\vec{x}^* \geq 0$ by assumption.

Due to the definition of v, our initial assumption implies that $\mathbf{H}\vec{x}^* + \vec{b} \geq 0$. If this was not the case, then there would exist an i such that $[\mathbf{H}\vec{x}^* + \vec{b}]_i < 0$. In this case, we see that $[v(\vec{x}^*)]_i = 1$ and that $[\mathrm{Diag}(v(\vec{x}^*))(\mathbf{H}\vec{x}^* + \vec{b})]_i = [\mathbf{H}\vec{x}^* + \vec{b}]_i < 0$, which contradicts our initial assumption. Therefore, $\mathbf{H}\vec{x}^* + \vec{b} \geq 0$. As a result, let us set $\vec{\lambda}^* = \mathbf{H}\vec{x}^* + \vec{b}$. This allows us to satisfy our first optimality condition, $\mathbf{H}\vec{x}^* + \vec{b} - \vec{\lambda}^* = 0$ as well as our third, $\vec{\lambda}^* \geq 0$.

In order to show that we satisfy complementary slackness, we combine our initial assumption as well as our knowledge that $\mathbf{H}\vec{x}^* + \vec{b} \geq 0$ to see that

$$0 = \mathrm{Diag}(v(\vec{x}^*))(\mathbf{H}\vec{x}^* + \vec{b})$$
$$= \mathrm{Diag}(\vec{x}^*)(\mathbf{H}\vec{x}^* + \vec{b})$$
$$= \mathrm{Diag}(\vec{x}^*)\vec{\lambda}^*.$$

Therefore, we satisfy our final optimality condition and, hence, $\vec{x}^*$ denotes an optimal solution to the optimization problem. □

The above lemma allows us to recast a bound-constrained, convex quadratic optimization problem into a piecewise differentiable system of equations. In order to solve this system of equations, we apply Newton's method. Before we do so, we require one additional definition and a lemma.

Definition 3. *For some symmetric, positive semidefinite* $\mathbf{H} \in \mathbb{R}^{m \times m}$ *and some* $\vec{b} \in \mathbb{R}^m$, *we define the operator* $K_{\mathbf{H},\vec{b}} : \mathbb{R}^m \to \mathbb{R}^{m \times m}$ *as*

$$[K_{\mathbf{H},\vec{b}}(\vec{x})]_{ij} = \begin{cases} 1 & \text{when} & [\mathbf{H}\vec{x} + \vec{b}]_i \geq 0 \\ 0 & \text{"} & [\mathbf{H}\vec{x} + \vec{b}]_i < 0 \end{cases}.$$

When both $\mathbf{H}$ *and* $\vec{b}$ *are clear from the context, we abbreviate this operator as* K.

Lemma 2. *Let* $\mathbf{H} \in \mathbb{R}^{m \times m}$ *be symmetric, positive definite,* $\vec{b} \in \mathbb{R}^m$, *and define the function* $J : \mathbb{R}^m \to \mathbb{R}$ *as*

$$J(\vec{x}) = \mathrm{Diag}(v(\vec{x}))(\mathbf{H}\vec{x} + \vec{b}).$$

Then, we have that

$$J'(\vec{x}) = K(\vec{x})\mathrm{Diag}(\mathbf{H}\vec{x} + \vec{b}) + \mathrm{Diag}(v(x))H.$$

Proof. Let us begin by assessing the derivative of v. We notice that

$$[v(\vec{x} + t\vec{\eta})]_i = \begin{cases} \vec{x}_i + t\vec{\eta}_i & \text{when} & [\mathbf{H}\vec{x} + b]_i \geq 0 \\ 1 & \text{"} & [\mathbf{H}\vec{x} + b]_i < 0 \end{cases}.$$

Therefore, from a piecewise application of Taylor's theorem, we see that

$$[v'(\vec{x})\vec{\eta}]_i = \begin{cases} \vec{\eta}_i & \text{when} & [\mathbf{H}\vec{x} + b]_i \geq 0 \\ 0 & \text{"} & [\mathbf{H}\vec{x} + b]_i < 0 \end{cases}.$$

Next, we apply a similar technique to J. Let us define $g : \mathbb{R}^m \to \mathbb{R}$ so that $g(\vec{x}) = \mathbf{H}\vec{x} + \vec{b}$. Then, we see that

$$
\begin{aligned}
J(\vec{x} + t\vec{\eta}) &= \mathrm{Diag}(v(\vec{x} + t\vec{\eta}))(\mathbf{H}(\vec{x} + t\vec{\eta}) + \vec{b}) \\
&= \mathrm{Diag}(v(\vec{x}) + tv'(\vec{x})\vec{\eta} + o(|t|))(\mathbf{H}\vec{x} + \vec{b} + t\vec{\eta}) \\
&= \mathrm{Diag}(v(\vec{x}))g(\bar{x}) + t\left(\mathrm{Diag}(v(\vec{x}))\mathbf{H}\vec{\eta} + \mathrm{Diag}(v'(\vec{x})\vec{\eta})g(\bar{x})\right) + o(|t|).
\end{aligned}
$$

Hence, from a piecewise application of Taylor's theorem, we have that

$$
\begin{aligned}
J'(\vec{x})\vec{\eta} &= \mathrm{Diag}(v(\vec{x}))\mathbf{H}\vec{\eta} + \mathrm{Diag}(v'(\vec{x})\vec{\eta})(\mathbf{H}\vec{x} + \vec{b}) \\
&= \mathrm{Diag}(v(\vec{x}))\mathbf{H}\vec{\eta} + K(\vec{x})\mathrm{Diag}(\mathbf{H}\vec{x} + \vec{b})\vec{\eta}.
\end{aligned}
$$

Therefore, $J'(\vec{x}) = K(\vec{x})\mathrm{Diag}(\mathbf{H}\vec{x} + \vec{b}) + \mathrm{Diag}(v(\vec{x}))\mathbf{H}$. □

The preceding lemma allows us to formulate Newton's method where we seek a step $\vec{p} \in \mathbb{R}^m$ such that $J'(\vec{x})\vec{p} = -J(\vec{x})$. Although the operator $J'(\vec{x})$ is well structured, it is nonsymmetric. We symmetrize the system as follows.

Definition 4. *For some symmetric, positive semidefinite* $\mathbf{H} \in \mathbb{R}^{m \times m}$ *and some* $\vec{b} \in \mathbb{R}^m$, *we define the operator* $D_{\mathbf{H},\vec{b}} : \mathbb{R}_+^m \to \mathbb{R}^{m \times m}$ *as*

$$
D_{\mathbf{H},\vec{b}}(\vec{x}) = \mathrm{Diag}(v_{\mathbf{H},\vec{b}}(\vec{x}))^{1/2}.
$$

When both $\mathbf{H}$ *and* $\vec{b}$ *are clear from the context, we abbreviate this operator as* D.

Lemma 3. *Let* $\mathbf{H} \in \mathbb{R}^{m \times m}$ *be symmetric, positive semidefinite and let* $\vec{b} \in \mathbb{R}^m$. *Then, we have that*

$$
\begin{aligned}
&(K(\vec{x})\mathrm{Diag}(\mathbf{H}\vec{x} + \vec{b}) + \mathrm{Diag}(v(\vec{x}))\mathbf{H})\vec{p} = -\mathrm{Diag}(v(\vec{x}))(\mathbf{H}\vec{x} + \vec{b}) \\
\Longleftrightarrow\; &(K(\vec{x})\mathrm{Diag}(\mathbf{H}\vec{x} + \vec{b}) + D(\vec{x})\mathbf{H}D(\vec{x}))\vec{q} = -D(\vec{x})(\mathbf{H}\vec{x} + \vec{b})
\end{aligned}
$$

where $\vec{p} = D(x)\vec{q}$.

Proof. Notice that

$$
\begin{aligned}
0 &= (K(\vec{x})\mathrm{Diag}(\mathbf{H}\vec{x} + \vec{b}) + \mathrm{Diag}(v(\vec{x}))\mathbf{H})\vec{p} + \mathrm{Diag}(v(\vec{x}))(\mathbf{H}\vec{x} + \vec{b}) \\
&= (K(\vec{x})\mathrm{Diag}(\mathbf{H}\vec{x} + \vec{b}) + D(\vec{x})^2\mathbf{H})\vec{p} + D(\vec{x})^2(\mathbf{H}\vec{x} + \vec{b}) \\
&= D(\vec{x})((D(\vec{x})^{-1}K(\vec{x})\mathrm{Diag}(\mathbf{H}\vec{x} + \vec{b}) + D(\vec{x})\mathbf{H})\vec{p} + D(\vec{x})(\mathbf{H}\vec{x} + \vec{b})) \\
&= D(\vec{x})((D(\vec{x})^{-1}K(\vec{x})\mathrm{Diag}(\mathbf{H}\vec{x} + \vec{b}) + D(\vec{x})\mathbf{H})D(\vec{x})\vec{q} + D(\vec{x})(\mathbf{H}\vec{x} + \vec{b})) \\
&= D(\vec{x})((K(\vec{x})\mathrm{Diag}(\mathbf{H}\vec{x} + \vec{b}) + D(\vec{x})\mathbf{H}D(\vec{x}))\vec{q} + D(\vec{x})(\mathbf{H}\vec{x} + \vec{b})),
\end{aligned}
$$

which occurs if and only if

$$
0 = (K(\vec{x})\mathrm{Diag}(\mathbf{H}\vec{x} + \vec{b}) + D(\vec{x})\mathbf{H}D(\vec{x}))\vec{q} + D(\vec{x})(\mathbf{H}\vec{x} + \vec{b})
$$

since $D(\vec{x})$ is nonsingular. □

Properly, we require a line search to ensure feasible iterates. However, we can be far more aggressive in practice. In order to initialize the algorithm, we use the starting iterate of $(\vec{\lambda}, \vec{\mu}) = (\vec{0}, \vec{0})$. This corresponds to a primal solution where $\vec{F} = \vec{F}^H$. Since the optimal solution to the primal problem is close to the target $\vec{F}^H$, we expect the optimal solution to the dual problem to reside in a neighborhood close to zero. As a result, Newton's method should converge quadratically to the solution with a step size equal to one. Therefore, we ignore the feasibility constraint and always use a unit step size. Sometimes, this allows the dual solution to become slightly infeasible, but the amount of infeasibility tends to be small. In practice, the corresponding primal solution is always feasible and produces good results. In order to allow infeasible solutions, we must use the original formulation of Newton's method rather than the symmetric reformulation. Namely, the operator D becomes ill-defined for infeasible points.

When we combine the above pieces, we arrive at the final algorithm.

Algorithm 1. Dual algorithm for the solution of the remap subproblem

---

1. Define $H \in \mathbb{R}^{2K \times 2K}$ and $b \in \mathbb{R}^{2K}$ as

$$\mathbf{H} = \begin{bmatrix} \mathbf{AA}^\mathsf{T} & -\mathbf{AA}^\mathsf{T} \\ -\mathbf{AA}^\mathsf{T} & \mathbf{AA}^\mathsf{T} \end{bmatrix} \qquad \vec{b} = \begin{bmatrix} \mathbf{A}\vec{F}^H - \vec{b}_{\min} \\ -\mathbf{A}\vec{F}^H + \vec{b}_{\max} \end{bmatrix}.$$

2. Initialize $\vec{x} = \vec{0}$.
3. Until $\|\mathrm{Diag}(v(\vec{x}))(\mathbf{H}\vec{x} + \vec{b})\|$ becomes small or we exceed a fixed number of iterations.
   (a) When feasible, solve

   $$(K(\vec{x})\mathrm{Diag}(\mathbf{H}\vec{x} + \vec{b}) + D(\vec{x})\mathbf{H}D(\vec{x}))\vec{q} = -D(\vec{x})(\mathbf{H}\vec{x} + \vec{b})$$

   and set $\vec{p} = D(x)\vec{q}$. Otherwise, solve

   $$(K(\vec{x})\mathrm{Diag}(\mathbf{H}\vec{x} + \vec{b}) + \mathrm{Diag}(v(\vec{x}))\mathbf{H})\vec{p} = -\mathrm{Diag}(v(\vec{x}))(\mathbf{H}\vec{x} + \vec{b}).$$

   (b) Set $\vec{x} = \vec{x} + \vec{p}$.

References

1. Bochev, P., Ridzal, D., Young, J.: Optimization–Based Modeling with Applications to Transport. Part 1. Abstract Formulation. In: Lirkov, I., Margenov, S., Waśniewski, J. (eds.) LSSC 2011. LNCS, vol. 7116, pp. 63–71. Springer, Heidelberg (2012)
2. Bochev, P., Ridzal, D., Scovazzi, G., Shashkov, M.: Formulation, analysis and numerical study of an optimization-based conservative interpolation (remap) of scalar fields for arbitrary lagrangian-eulerian methods. Journal of Computational Physics 230(13), 5199–5225 (2011),
   http://www.sciencedirect.com/science/article/B6WHY-52F895B-2/2/5e30ada70a5c6053464dfe9ceb74cf26

3. Coleman, T.F., Li, Y.: A reflective newton method for minimizing a quadratic function subject to bounds on some of the variables. SIAM Journal on Optimization 6(4), 1040–1058 (1996), http://link.aip.org/link/?SJE/6/1040/1
4. Dukowicz, J.K., Baumgardner, J.R.: Incremental remapping as a transport/advection algorithm. Journal of Computational Physics 160(1), 318–335 (2000), http://www.sciencedirect.com/science/article/B6WHY-45FC8N8-6F/2/179cbfc9634bb79579b68754cebd5525
5. Ridzal, D., Bochev, P., Young, J., Peterson, K.: Optimization–Based Modeling with Applications to Transport. Part 3. Implementation and Computational Studies. In: Lirkov, I., Margenov, S., Waśniewski, J. (eds.) LSSC 2011. LNCS, vol. 7116, pp. 81–88. Springer, Heidelberg (2012)

Optimization-Based Modeling with Applications to Transport: Part 3. Computational Studies

Denis Ridzal[2], Joseph Young[1], Pavel Bochev[1], and Kara Peterson[1]

[1] Numerical Analysis and Applications
[2] Optimization and Uncertainty Quantification,
Sandia National Laboratories*, Albuquerque, NM 87185-1320, USA
{dridzal,josyoun,pbboche,kjpeter}@sandia.gov

Abstract. This paper is the final of three related articles that develop and demonstrate a new optimization-based framework for computational modeling. The framework uses optimization and control ideas to assemble and decompose multiphysics operators and to preserve their fundamental physical properties in the discretization process. One application of the framework is in the formulation of robust algorithms for optimization-based transport (OBT). Based on the theoretical foundations established in Part 1 and the optimization algorithm for the solution of the remap subproblem, derived in Part 2, this paper focuses on the application of OBT to a set of benchmark transport problems. Numerical comparisons with two other transport schemes based on incremental remapping, featuring flux-corrected remap and the linear reconstruction with van Leer limiting, respectively, demonstrate that OBT is a competitive transport algorithm.

1   Introduction

In this and two companion papers [1,7] we formulate and study a new optimization-based framework for computational modeling. One application of the framework, introduced in Part 1 [1], is in the formulation of a new class of optimization-based transport (OBT) schemes, which combine *incremental remap* [3] with the reformulation of the *remap subproblem* as an inequality-constrained quadratic program (QP) [2]. An efficient algorithm for the solution of the remap subproblem is presented in Part 2 [7]. In this paper we apply the OBT framework to a series of benchmark transport problems cited in [4].

Numerical comparisons with two other transport schemes based on incremental remapping are presented. The first scheme solves the remap subproblem using flux-corrected remap (FCR); for an FCR reference see [5]. We denote this scheme by FCRT (FCR based Transport). The second scheme solves the remap

---

* Sandia National Laboratories is a multi-program laboratory managed and operated by Sandia Corporation, a wholly owned subsidiary of Lockheed Martin Corporation, for the U.S. Department of Energy's National Nuclear Security Administration under contract DE-AC04-94AL85000.

I. Lirkov, S. Margenov, and J. Waśniewski (Eds.): LSSC 2011, LNCS 7116, pp. 81–88, 2012.

subproblem via a linear flux reconstruction with van Leer limiting, see [3] and references therein. We denote the latter transport scheme by LVLT (Linear Van Leer based Transport). In comparisons with FCRT and LVLT, we demonstrate that OBT, while computationally more expensive, can be more accurate and significantly more robust.

2   Implementation

The OBT framework is developed in Part 1 [1]. To summarize, OBT for mass density relies on an incremental remap procedure with the following steps: (1) move an original computational grid in the direction of the advection and obtain a new grid; (2) compute mass density updates on the new grid; and (3) remap mass density onto the original grid. The remap subproblem in step (3) is formulated as an inequality-constrained quadratic program (QP) and solved using Newton's method for piecewise differentiable systems, derived in Part 2 [7].

The QP describing the remap subproblem has the form

$$
\min_{\vec{F} \in \mathbb{R}^M} \frac{1}{2}(\vec{F} - \vec{F}^H)^\mathsf{T}(\vec{F} - \vec{F}^H) \quad \text{subject to}
$$
$$
\vec{b}_{\min} \leq \mathbf{A}\vec{F} \leq \vec{b}_{\max}
$$

(1)

where $\vec{F}^H \in \mathbb{R}^M$ are the given discrete high-order fluxes, $\vec{b}_{\min} \in \mathbb{R}^K$ and $\vec{b}_{\max} \in \mathbb{R}^K$ are lower and upper bounds obtained from local mass density bounds on the new grid, and $\mathbf{A} \in \mathbb{R}^{K \times M}$ is an inequality-constraint matrix. Below we define the dimensions K and M for a concrete implementation of OBT. We also elaborate on the computation of $\vec{F}^H$ and $\mathbf{A}$.

For the implementation of OBT, FCRT and LVLT algorithms we use structured quadrilateral grids. If N_x and N_y are the numbers of intervals in x and y directions, respectively, then $K = N_x N_y$. The high-order flux vector $\vec{F}^H$ can be computed via integration over exact cell intersections, following the theory in Part 1 [1]. We avoid this potentially costly computation by using the concept of *swept regions*, see [6,2], where mass exchanges are allowed only between cells that share a side. This simplifies the computation of high-order fluxes used in OBT, as well as the computation of low and high-order fluxes used in FCRT and LVLT. Following the swept-region approximation, the dimension M is given by $M = (N_x + 1)N_y + (N_y + 1)N_x$. Assuming a dimensional partitioning of flux variables, Figure 1 gives the inequality-constraint matrix $\mathbf{A}$ for a structured grid with $N_x = 3$ and $N_y = 4$.

We implement OBT, FCRT and LVLT in Matlab™ and rely on vectorized arithmetic and efficient data structures for the storage of mesh data. We remark that such implementation can rival the computational performance of mathematically equivalent Fortran code, see [2, Sec. 6.4]. The global linear systems involving the matrices $\mathbf{A}\mathbf{A}^\mathsf{T}$, see Part 2 [7], are solved using sparse Cholesky and/or LU factorizations.

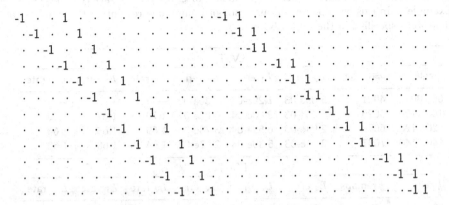

Fig. 1. The inequality-constraint matrix **A** for a structured grid with $N_x = 3$ and $N_y = 4$. For compactness, zeros have been replaced by dots.

3 Results

In this section, we study OBT, FCRT and LVLT on a number of numerical examples similar to those given by LeVeque [4]. We do not give a direct comparison to all methods implemented by LeVeque, such as the transport with superbee limiting, however, we provide analogous figures so that an interested reader can make a qualitative comparison.

In Example 1, we compute a solid body rotation of a smooth hump in a circular flow. Specifically, we define the initial hump to be

$$q(x, y, 0) = \frac{1}{4}(1 + \cos(\pi r(x, y))),$$

where

$$r(x, y) = \min\{\sqrt{(x - x_0)^2 + (y - y_0)^2}, r_0\}/r_0,$$

and define the rotating flow as

$$u = -(y - 1/2), \qquad v = (x - 1/2).$$

In order to test the accuracy of the methods, we set $r_0 = 0.15$, $x_0 = 0.25$, and $y_0 = 0.5$ and compute the solution over the domain $[0, 1] \times [0, 1]$. In order to insure that we rotate the hump one complete revolution, we use $\lfloor 8(2\pi n) \rfloor$ time steps, where n denotes the size of the computational grid in both the x and y directions, resulting in a maximum CFL number of about $1/8$. We summarize the amount of error and the convergence rates for each method in Table 1.

As we can see from Table 1, the three methods give similar numerical results. OBT exhibits slightly better asymptotic convergence than FCRT or LVLT. The lack of perfect second order convergence may be attributed to the time discretization.

Table 1. Errors and convergence rate estimates after applying a variety of methods to Example 1. In this example, we rotate a smooth hump one revolution for a number of time steps specified in the table above.

			LVLT				
#cells	#remaps	L_2 err	L_1 err	L_∞ err	L_2 rate	L_1 rate	L_∞ rate
80×80	4021	5.85e-03	1.29e-03	7.36e-02	—	—	—
100×100	5026	4.00e-03	8.88e-04	5.08e-02	1.70	1.67	1.66
120×120	6031	2.94e-03	6.59e-04	3.78e-02	1.69	1.65	1.64
140×140	7037	2.35e-03	5.30e-04	2.97e-02	1.64	1.60	1.62

			FCRT				
#cells	#remaps	L_2 err	L_1 err	L_∞ err	L_2 rate	L_1 rate	L_∞ rate
80×80	4021	5.66e-03	1.24e-03	5.42e-02	—	—	—
100×100	5026	3.89e-03	8.63e-04	3.66e-02	1.68	1.62	1.76
120×120	6031	2.85e-03	6.45e-04	2.57e-02	1.69	1.61	1.84
140×140	7037	2.29e-03	5.21e-04	1.98e-02	1.63	1.56	1.82

			OBT				
#cells	#remaps	L_2 err	L_1 err	L_∞ err	L_2 rate	L_1 rate	L_∞ rate
80×80	4021	6.15e-03	1.40e-03	5.71e-02	—	—	—
100×100	5026	4.11e-03	9.38e-04	3.76e-02	1.81	1.81	1.88
120×120	6031	2.95e-03	6.83e-04	2.60e-02	1.82	1.78	1.94
140×140	7037	2.33e-03	5.46e-04	1.98e-02	1.75	1.70	1.91

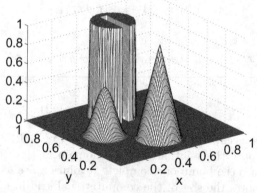

Fig. 2. Initial data for the solid body rotation tests

In Example 2, we use the same rotating flow from above, but rotate a combination of a smooth hump, cone, and slotted disk shown in Figure 2. In each test, we use the same smooth hump as above and use a cone and slotted disk with radii 0.15 and heights 1 centered at $(0.5, 0.25)$ and $(0.5, 0.75)$, respectively. In the

Table 2. The amount of time required to compute Example 2 using OBR, FCR, and linear reconstruction with van Leer limiting in seconds. In each test, we compute one revolution of the smooth hump, cone, and slotted disk on a variety of different grid sizes. For the number of time steps, we use $\lfloor 2\pi n \rfloor$ where n denotes the grid size.

Grid Size	40×40	80×80	160×160	320×320
OBT	4.00	34.21	422.85	4108.27
FCRT	0.83	5.48	45.27	375.90
LVLT	0.89	5.84	45.38	362.65

latter, we define the slot by $[0.475, 0.525] \times [0.6, 0.85]$. As before, we rotate the objects one full revolution, but use $\lfloor 2\pi n \rfloor$ time steps, where n denotes the size of the computational grid. In our first test, we summarize a comparison of the runtime required for each method in Table 2. In addition, we give a qualitative comparison between OBT and FCRT in Figures 3 and 4, respectively.

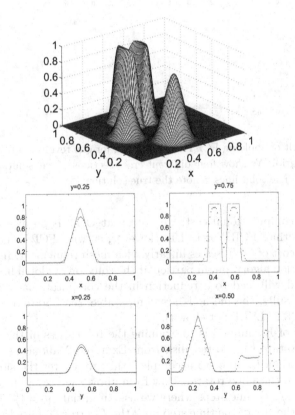

Fig. 3. The result of applying OBT to Example 2 for one revolution (628 time steps) on a 100×100 grid. We show four different cross sections of the solution along with the surface plot. The solid lines denote the true solution.

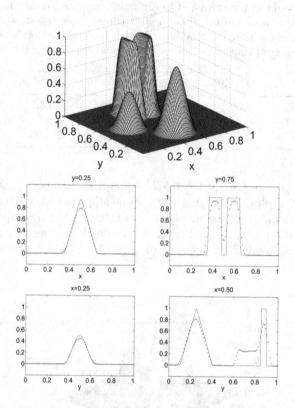

Fig. 4. The result of applying FCRT to Example 2 for one revolution (628 time steps) on a 100×100 grid. We show four different cross sections of the solution along with the surface plot. The solid lines denote the true solution.

In terms of computational cost, we note that OBT is on average 10 times more expensive than FCRT or LVLT. However, as with FCRT and LVLT, the computational cost of OBT scales linearly with mesh refinement. Improvements in the optimization algorithm, in particular the solution of global linear systems, are possible and will lead to a reduction in the computational cost. In terms of qualitative results, the methods are comparable, see Figure 3 for OBT and Figure 4 for FCRT (LVLT not shown).

The purpose of Example 3 is to examine the robustness of the methods. In particular, we rotate the slotted disk from Example 2 about its axis for one complete revolution. In order to accomplish this, we center the slotted disk at $(0.5, 0.5)$ and use the same rotating flow from above. We compute the result of rotation for $2\pi(1/\Delta t)$ time steps where we use an initial $\Delta t = 1/100$. Then, we slowly increase the size of the time step until the L_1 error doubles from its initial value computed at $\Delta t = 1/100$. We give the result of this test on a 100×100 grid in Table 3 and a qualitative depiction of OBT in Figure 5.

Table 3. L_1 errors in OBR, FCR, and linear reconstruction with van Leer limiting on Example 3. In this example, we rotate a slotted disk centered at the point (0.5,0.5) one revolution ($\lfloor 2\pi\Delta t \rfloor$ time steps). Our goal is to determine the largest time step in which the error measured in the L_1 norm doubles given a baseline where $1/\Delta t=100$. Results that are better than this error bound are given in bold.

	$1/\Delta t=100$ CFL=1.00	$1/\Delta t=62$ CFL=1.60	$1/\Delta t=61$ CFL=1.62	$1/\Delta t=45$ CFL=2.20	$1/\Delta t=44$ CFL=2.25	$1/\Delta t=19$ CFL=5.50	$1/\Delta t=18$ CFL=5.21
OBT	**2.14e-02**	**2.37e-02**	**2.38e-02**	**2.60e-02**	**2.62e-02**	**4.02e-02**	4.36e-02
FCRT	**1.97e-02**	**2.19e-02**	**2.21e-02**	3.00e-02	6.00e+06	9.45e+38	1.83e+40
LVLT	**2.14e-02**	**2.36e-02**	8.15e-01	3.47e+54	2.85e+56	2.83e+79	6.23e+77

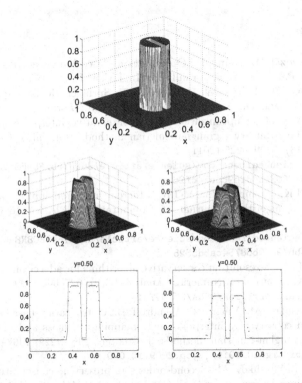

Fig. 5. The result of applying OBT to Example 3. In this example we rotate a slotted disk centered at the point (0.5,0.5) one revolution. In the above plots, we show the original disk on top. On the bottom, we show surface and cross section plots of the disk after one revolution where $\Delta t = .01$ (CFL=1.00) on the left and $\Delta t = 0.047$ (CFL=4.7) on the right.

As we can see, OBT exhibits a level of robustness that far exceeds FCRT and LVLT. In fact, OBT produces qualitatively reasonable results for CFL numbers of about 5, while both FCRT and LVLT break down numerically at CFL numbers beyond 2.25.

4   Conclusion

In this paper we applied the framework for optimization-based transport, formalized in [1,7], to benchmark transport problems presented in [4].

Numerical comparisons with transport schemes based on incremental flux-corrected remap and incremental linear flux reconstruction with van Leer limiting demonstrate that optimization-based transport is a competitive alternative. In particular, while computationally more expensive due to the solution of a globally constrained optimization problem, optimization-based transport is shown to be more accurate asymptotically and significantly more robust.

Future work includes performance optimizations and applications to the transport of systems.

References

1. Bochev, P., Ridzal, D., Young, D.: Optimization–Based Modeling with Applications to Transport. Part 1. Abstract Formulation. In: Lirkov, I., Margenov, S., Waśniewski, J. (eds.) LSSC 2011. LNCS, vol. 7116, pp. 63–71. Springer, Heidelberg (2012)
2. Bochev, P., Ridzal, D., Scovazzi, G., Shashkov, M.: Formulation, analysis and numerical study of an optimization-based conservative interpolation (remap) of scalar fields for arbitrary lagrangian-eulerian methods. Journal of Computational Physics 230(13), 5199–5225 (2011),
   http://www.sciencedirect.com/science/article/B6WHY-52F895B-2/2/
   5e30ada70a5c6053464dfe9ceb74cf26
3. Dukowicz, J.K., Baumgardner, J.R.: Incremental remapping as a transport/advection algorithm. Journal of Computational Physics 160(1), 318–335 (2000),
   http://www.sciencedirect.com/science/article/B6WHY-45FC8N8-6F/2/
   179cbfc9634bb79579b68754cebd5525
4. LeVeque, R.J.: High-resolution conservative algorithms for advection in incompressible flow. SIAM Journal on Numerical Analysis 33(2), 627–665 (1996),
   http://link.aip.org/link/?SNA/33/627/1
5. Liska, R., Shashkov, M., Váchal, P., Wendroff, B.: Optimization-based synchronized flux-corrected conservative interpolation (remapping) of mass and momentum for arbitrary Lagrangian-Eulerian methods. J. Comput. Phys. 229, 1467–1497 (2010),
   http://dx.doi.org/10.1016/j.jcp.2009.10.039
6. Margolin, L.G., Shashkov, M.: Second-order sign-preserving conservative interpolation (remapping) on general grids. J. Comput. Phys. 184(1), 266–298 (2003)
7. Young, J., Ridzal, D., Bochev, P.: Optimization–Based Modeling with Applications to Transport. Part 2. Optimization Algorithm. In: Lirkov, I., Margenov, S., Waśniewski, J. (eds.) LSSC 2011. LNCS, vol. 7116, pp. 72–80. Springer, Heidelberg (2012)

Part IV

Control and Uncertain Systems

Newton's Method and Secant Method for Set-Valued Mappings

Robert Baier[1] and Mirko Hessel-von Molo[2]

[1] University of Bayreuth, Chair of Applied Math., 95440 Bayreuth, Germany
robert.baier@uni-bayreuth.de
[2] University of Paderborn, Chair of Applied Math., 33098 Paderborn, Germany
mirkoh@math.uni-paderborn.de

Abstract. For finding zeros or fixed points of set-valued maps, the fact that the space of convex, compact, nonempty sets of $\mathbb{R}^n$ is not a vector space presents a major disadvantage. Therefore, fixed point iterations or variants of Newton's method, in which the derivative is applied only to a smooth single-valued part of the set-valued map, are often applied for calculations. We will embed the set-valued map with convex, compact images (i.e. by embedding its images) and shift the problem to the Banach space of directed sets. This Banach space extends the arithmetic operations of convex sets and allows to consider the Fréchet-derivative or divided differences of maps that have embedded convex images. For the transformed problem, Newton's method and the secant method in Banach spaces are applied via directed sets. The results can be visualized as usual nonconvex sets in $\mathbb{R}^n$.

Keywords: set-valued Newton's method, set-valued secant method, Gauß-Newton method, directed sets, embedding of convex compact sets.

1 Introduction

In this article we illustrate how Newton's method and the similar secant method can be applied to solve two standard problems for set-valued mappings with convex, compact images. Our key method is to embed these problems into a Banach space setting, which allows to reformulate them as zero-finding problems. For references on uses of Newton's method in Banach spaces see e.g. [12,6]. We would like to point out that the ideas presented here differ from the pioneering works (e.g. [10,1]) based on homotopy methods or mutational equations. We start this article with a brief introduction to the Banach space of directed sets. In Sections 2 and 3, we apply this to introduce the Fréchet-derivative as well as Newton's method and the secant method for set-valued maps. We conclude with several numerical examples in Section 4.

1.1 Directed Sets

Directed sets provide a way to interpret compact, convex, nonempty sets of $\mathbb{R}^n$, i.e. sets in $\mathcal{C}(\mathbb{R}^n)$, as elements of a Banach space. They were introduced in [3]

I. Lirkov, S. Margenov, and J. Waśniewski (Eds.): LSSC 2011, LNCS 7116, pp. 91–98, 2012.

(see also references therein), their visualization is studied in [3, Part II]. For $C, D \in \mathcal{C}(\mathbb{R}^n)$ and $M \in \mathbb{R}^{m \times n}$, we write $C + D$ for the *Minkowski sum*, $\lambda \cdot C$ for the *multiplication with non-negative scalars* and $M \cdot C$ for the *linear image*.

Definition 1. Directed intervals resp. sets $\overrightarrow{A} \in \mathcal{D}(\mathbb{R}^n)$ *are defined recursively.*

For $n = 1$, *a directed interval is a pair* $\overrightarrow{A} = (a_1(l))_{l=\pm 1}$ *of two real numbers.* *(We will also use the notation* $\overrightarrow{A} = \overrightarrow{[c,d]}$, *where* $a_1(-1) = -c$ *and* $a_1(1) = d$.) *In this case, the norm of* $\overrightarrow{A}$ *is given by* $\|\overrightarrow{A}\|_1 = \max_{l=\pm 1} |a_1(l)|$.

For $n \geq 2$, *a directed set has two components* $\overrightarrow{A} = \left(\overrightarrow{A_{n-1}(l)}, a_n(l)\right)_{l \in S_{n-1}}$, *given by a map* $\overrightarrow{A_{n-1}} : S_{n-1} \to \mathcal{D}(\mathbb{R}^{n-1})$ *that is uniformly bounded with respect to* $\| \cdot \|_{n-1}$, *and a continuous real-valued function* $a_n(\cdot)$. *The norm is defined by*

$$\|\overrightarrow{A}\| := \|\overrightarrow{A}\|_n := \max\{ \sup_{l \in S_{n-1}} \|\overrightarrow{A_{n-1}(l)}\|_{n-1}, \max_{l \in S_{n-1}} |a_n(l)| \}.$$

As a space of functions on S_{n-1} with values in a linear space, with the usual pointwise definitions of addition and scalar multiplication, $\mathcal{D}(\mathbb{R}^n)$ is a linear space. With the norm defined above, $\mathcal{D}(\mathbb{R}^n)$ is a Banach space that can be equipped with a partial ordering that represents inclusion of sets (see [3, Part I]). Sets in $\mathcal{C}(\mathbb{R}^n)$ can be *embedded* in this Banach space via $J_n : \mathcal{C}(\mathbb{R}^n) \to \mathcal{D}(\mathbb{R}^n)$ as

$$J_1([c,d]) := \overrightarrow{[c,d]} = (-c, d),$$
$$J_n(C) := \left(J_{n-1}(\Pi_{n-1,l} \cdot Y(l,C)), \delta^*(l,C) \right)_{l \in S_{n-1}} \quad (n \geq 2).$$

Here, $\delta^*(l, C)$ denotes the *support function* of C in direction $l \in S_{n-1}$, $Y(l,C)$ the corresponding *supporting face* (i.e. the maximizers of $\langle l, \cdot \rangle$ on the set C), and $\Pi_{n-1,l}$ is a linear *projection* which maps the orthogonal complement of $\mathrm{span}\{l\}$ into $\mathbb{R}^{n-1}$ and l to the origin in $\mathbb{R}^{n-1}$, see [3, Part I] for details.

Proposition 1 ([3]). *The embedding* $J_n : \mathcal{C}(\mathbb{R}^n) \to \mathcal{D}(\mathbb{R}^n)$ *is positively linear:*

$$J_n(\lambda \cdot C + \mu \cdot D) = \lambda \cdot J_n(C) + \mu \cdot J_n(D) \quad (C, D \in \mathcal{C}(\mathbb{R}^n), \ \lambda, \mu \geq 0)$$

In contrast to formulations using pairs of sets (see [3] for references), a directed set (and thus a difference of embedded convex sets) can be visualized as a subset in $\mathbb{R}^n$ which is usually nonconvex. The visualization of the directed set is plotted with arrows (outer normals for embedded convex sets, inner ones for their inverses) on the boundaries (see [3, Part II] for details).

2   Fréchet-Derivative for Set-Valued Maps

As $\mathcal{D}(\mathbb{R}^m)$ forms a Banach space, we use the usual definition of the Fréchet-derivative of a function $\overrightarrow{F} : \mathbb{R}^n \to \mathcal{D}(\mathbb{R}^m)$. To lift a set-valued map $F : \mathbb{R}^n \rightrightarrows \mathbb{R}^m$ with images in $\mathcal{C}(\mathbb{R}^m)$ to the space $\mathcal{D}(\mathbb{R}^m)$, we will use the notation $\overrightarrow{F}(x) = J_m(F(x))$. If $\overrightarrow{F} : \mathbb{R}^n \to \mathcal{D}(\mathbb{R}^m)$ is Fréchet-differentiable, F is called *directed differentiable*. This notion of the derivative generalizes the one of set-valued maps for one variable studied in [8,4].

Lemma 1 (see [2]). *Let* $F : \mathbb{R}^n \to \mathcal{C}(\mathbb{R}^m)$ *and* $w \in \mathbb{R}^n$.

(i) *If* $F(x) = U$, *then* $\vec{F}'(x)(w) = J_m(\{0\})$.

(ii) *If* $F(x) = r(x) \cdot G(x) + h(x)$ *with* $r : \mathbb{R}^n \to [0, \infty)$, $h : \mathbb{R}^n \to \mathbb{R}^m$, $G : \mathbb{R}^n \to \mathcal{C}(\mathbb{R}^m)$ *which are (directed) Fréchet-differentiable, then* $\vec{F}'(x)(w) = r'(x) \cdot w \cdot \vec{G}(x) + r(x) \cdot \vec{G}'(x)(w) + J_m(\{h'(x) \cdot w\})$.

Especially, if $F(x) = r(x) \cdot U$ *with* $U \in \mathcal{C}(\mathbb{R}^m)$, *then* $\vec{F}'(x)(w) = r'(x) \cdot w \cdot \vec{U}$.

(iii) *If* $\vec{F}(x) = r(x) \cdot \vec{G}(x) + \vec{H}(x)$ *and* $r : \mathbb{R}^n \to \mathbb{R}$, $\vec{G}, \vec{H} : \mathbb{R}^n \to \mathcal{D}(\mathbb{R}^m)$ *are Fréchet-differentiable, then* $\vec{F}'(x)(w) = r'(x) \cdot w \cdot \vec{G}(x) + r(x) \cdot \vec{G}'(x)(w) + \vec{H}'(x)(w)$.

Proof. (i) A constant map in a Banach space has 0 as Fréchet-differential.
(ii) follows immediately from (iii) with $\vec{G}(x) = J_m(G(x))$, $J_m(r(x) \cdot G(x)) = r(x) \cdot J_m(G(x))$ and $\vec{H}(x) = J_m(\{h(x)\})$. Since $-J_m(\{h(x)\}) = J_m(\{-h(x)\})$ we can directly show from the definition that $\vec{H}'(x)(w) = J_m(\{h'(x) \cdot w\})$.
(iii) Clearly, the sum and the product of two functions are Fréchet-differentiable, if the operands have this property. □

We now study explicit formulas for the Fréchet-derivative.

Proposition 2. *Let* $F : \mathbb{R}^n \to \mathcal{C}(\mathbb{R}^m)$ *and* $w, x \in \mathbb{R}^n$.
(i) *If* $m = 1$ *and* $F(x) = [g(x), h(x)]$ *with two differentiable functions* $g, h : \mathbb{R}^n \to \mathbb{R}$ *and* $g(x) \leq h(x)$ *for all* $x \in \mathbb{R}^n$, *then* $\vec{F}$ *is differentiable with*

$$\vec{F}'(x)(w) = \overrightarrow{[g'(x) \cdot w, h'(x) \cdot w]}.$$

(ii) *If* $m \geq 2$, *the Fréchet-differential of* $\vec{F}(x) = (\vec{F}_{m-1}(l; x), f_m(l; x))_{l \in S_{m-1}}$ *is*

$$\vec{F}'(x)(w) = (\vec{F}'_{m-1}(l; x)(w), f'_m(l; x) \cdot w)_{l \in S_{m-1}}.$$

Proof. (i) apply Lemma 1(ii) to $F(x) = g(x) + (h(x) - g(x)) \cdot [-1, 1]$
(ii) for a proof see [2, Proposition 3.6] □

3   Newton's Method and Secant Method

For the sake of brevity, we refer to the literature for convergence results about iterative methods in Banach spaces. For Newton's method see e.g. [12,5,6], for the secant method e.g. [9] and for Gauß-Newton method e.g. [7].

3.1   Newton's Method for Directed Sets

Problem 1. Let $F : \mathbb{R}^n \to \mathcal{C}(\mathbb{R}^m)$ be given. We are looking for a solution $\hat{x} \in \mathbb{R}^n$ of $0 \in F(\hat{x})$. Using the partial order "$\leq$" on $\mathcal{D}(\mathbb{R}^m)$ and the embedded map $\vec{F}$, this problem can be transformed into $J_m(\{0\}) \leq \vec{F}(\hat{x})$.

To apply Newton's iteration for directed sets to Problem 1 we choose a starting point $x^0 \in \mathbb{R}^n$ and iteratively obtain $x^{k+1} = x^k + d^k$, where $d^k \in \mathbb{R}^n$ has to solve the linear inequality

$$-\overrightarrow{F}(x^k) \leq \overrightarrow{F}'(x^k)(d^k). \tag{1}$$

Checking this inequality essentially means checking the positivity of a function, which is easily realized algorithmically.

Remark 1. In many publications, e.g. in the study of necessary optimality conditions in non-smooth optimization or optimal control problems, the set-valued map F is given in the form $F(x) = g(x) + H(x)$ with a smooth function $g : \mathbb{R}^n \to \mathbb{R}^m$ and an u.s.c. map $H : \mathbb{R}^n \Rightarrow \mathbb{R}^m$ (representing subdifferentials or normal cones in the mentioned applications), see e.g. [5,6]. The set-valued derivative of F is avoided and an inclusion step of the form

$$0 \in g(x^k) + g'(x^k) \cdot d^k + H(x^k + d^k)$$

is studied. While in this case the computation of d^k typically requires a nonlinear inclusion to be solved, for this splitting of F (with Fréchet-differentiable $\overrightarrow{H}$) Newton's method as suggested above leads to an inclusion problem linearly in d^k, since Newton's method in (1) demands a solution of

$$-J_m(\{g(x^k)\}) - \overrightarrow{H}(x^k) \leq J_m(\{g'(x^k) \cdot d^k\}) + \overrightarrow{H}'(x^k)(d^k).$$

Problem 2. Let $G, H : \mathbb{R}^n \to \mathcal{C}(\mathbb{R}^m)$ be given. We are looking for a solution $\hat{x} \in \mathbb{R}^n$ of $G(\hat{x}) = H(\hat{x})$. Transformed to $\mathcal{D}(\mathbb{R}^m)$, this reads $\overrightarrow{G}(\hat{x}) = \overrightarrow{H}(\hat{x})$.

Using $\overrightarrow{F} = \overrightarrow{G} - \overrightarrow{H}$, Newton's iteration with directed sets for this problem proceeds similarly to the one for Problem 1, with the difference that the updates d^k have to solve the linear equation

$$\overrightarrow{F}'(x^k)(d^k) = -\overrightarrow{F}(x^k). \tag{2}$$

3.2   Secant Method Based on Directed Sets

To apply the secant method as studied e.g. in [11,9], we need a notion of divided differences. Set-valued divided differences in one variable are studied e.g. in [8,4]. Following [11], we introduce multivariate divided differences for directed sets.

Definition 2. *Let $\overrightarrow{F} : \mathbb{R}^n \to \mathcal{D}(\mathbb{R}^m)$ and $u, v, w \in \mathbb{R}^n$ be given such that $u_j \neq v_j$ for $j = 1, \ldots n$. We define*

$$\overrightarrow{F}[u,v]^j := \frac{1}{u_j - v_j} \cdot \Big(\overrightarrow{F}(u_1, \ldots, u_{j-1}, u_j, v_{j+1}, \ldots, v_n)$$

$$- \overrightarrow{F}(u_1, \ldots, u_{j-1}, v_j, v_{j+1}, \ldots, v_n) \Big) \in \mathcal{D}(\mathbb{R}^m),$$

$$\overrightarrow{F}[u,v](w) := \sum_{j=1}^n w_j \cdot \overrightarrow{F}[u,v]^j \in \mathcal{D}(\mathbb{R}^m).$$

In the limit for $u \to v$, this definition yields the Fréchet-differential $\overrightarrow{F}'(u)$.

This approach generalizes the divided differences in one variable and satisfies the characteristic equation

$$\overrightarrow{F}[u,v](u-v) = \overrightarrow{F}(u) - \overrightarrow{F}(v) \quad (u,v \in \mathbb{R}^n). \tag{3}$$

To apply the *secant method* to Problem 1, one chooses two starting points $x^{-1}, x^0 \in \mathbb{R}^n$ and iteratively computes $x^{k+1} = x^k + d^k$, where d^k solves

$$-\overrightarrow{F}(x^k) \leq \overrightarrow{F}[x^k, x^{k-1}](d^k). \tag{4}$$

To treat Problem 2 with the secant method, we again set $\overrightarrow{F} = \overrightarrow{G} - \overrightarrow{H}$, choose initial points x^{-1} and x^0 and compute $x^{k+1} = x^k + d^k$, where d^k solves

$$\overrightarrow{F}[x^k, x^{k-1}](d^k) = -\overrightarrow{F}(x^k). \tag{5}$$

An advantage of (5) is that the derivative $\overrightarrow{F}'$ is avoided and replaced by differences of function values. It is obvious that the iterations (2) and (5) generalize their pointwise analogues.

4 Examples

In this section we present numerical examples for the procedures proposed above.

Example 1. Let $U_1, \ldots, U_M \in \mathcal{C}(\mathbb{R}^n)$ be such that the vectors $J_n(U_i)$ are linearly independent in $\mathcal{D}(\mathbb{R}^n)$. Consider an unknown convex combination $U_0 = \sum_{i=1}^M \lambda_i \cdot U_i$ with coefficients $\lambda_i \geq 0$, $i = 1, \ldots, M$ and $\sum_{i=1}^M \lambda_i = 1$. In order to apply Newton's method for Problem 2 to compute the vector $\lambda \in \mathbb{R}^M$, we define $G(x) = \sum_{i=1}^M x_i \cdot U_i$ for $x \in \mathbb{R}^M$ and $H(x) = U_0$, and set $\overrightarrow{F}(x) = \sum_{i=1}^M x_i \cdot J_n(U_i) - J_n(U_0)$. By Proposition 1

$$\overrightarrow{F}(x) = \sum_{i=1}^M (x_i - \lambda_i) \cdot J_n(U_i) \quad \text{and} \quad \overrightarrow{F}'(x)(w) = \sum_{i=1}^M w_i \cdot J_n(U_i),$$

so that the Newton step results in

$$\sum_{i=1}^M d_i^k \cdot J_n(U_i) = -\sum_{i=1}^M (x_i^k - \lambda_i) \cdot J_n(U_i) \quad \text{and} \quad \sum_{i=1}^M (x_i^k - \lambda_i + d_i^k) \cdot J_n(U_i) = 0.$$

From the assumption of linear independence we obtain $d_i^k - \lambda_i + x_i^k = 0$ for $i = 1, \ldots, M$, and hence $d^k = -x^k + \lambda$. This means that already the first iteration step leads to the solution $x^1 = \hat{x} = \lambda$. We obtain the same picture for the secant method. Since

$$\overrightarrow{F}[x^k, x^{k-1}]^i = \frac{1}{x_i^k - x_i^{k-1}} \cdot \left(x_i^k \cdot J_n(U_i) - x_i^{k-1} \cdot J_n(U_i) \right) = J_n(U_i),$$

$$\overrightarrow{F}[x^k, x^{k-1}](d^k) = \sum_{i=1}^M d_i^k \cdot \overrightarrow{F}[x^k, x^{k-1}]^i = \sum_{i=1}^M d_i^k \cdot J_n(U_i),$$

the left-hand side of (5) equals the one of the Newton step (2).

From now on, we consider Problem 2.

Example 2. Let $\alpha, \beta \in \mathbb{R}$ and consider the functions $g(v) = \alpha\|v\|_2^2 v + \beta v$, $h(v) = \|v\|_2 v$ for $v \in \mathbb{R}^2$. We consider Problem 2 with $G, H : \mathbb{R} \to \mathcal{C}(\mathbb{R}^2)$, $G(x) = g(x \cdot B_1(0))$, $H(x) = h(x \cdot B_1(0))$. A simple calculation shows that $g(x \cdot B_1(0)) = (\alpha x^3 + \beta x) \cdot B_1(0)$ and $h(x \cdot B_1(0)) = x^2 \cdot B_1(0)$. The solutions for $\alpha = -1$, $\beta = \frac{3}{4}$ are $\hat{x} \in \{0, -\frac{3}{2}, \frac{1}{2}\}$. To eliminate the trival solution $\hat{x} = 0$, we divide the leading term by x and obtain $\overrightarrow{F}(x) = (\alpha x^2 + \beta - x) \cdot J_2(B_1(0))$. For this, one has

$$\overrightarrow{F}'(x) = (2\alpha x - 1) \cdot \overrightarrow{B} \quad \text{with } \overrightarrow{B} = J_2(B_1(0)) \text{ and}$$

$$\overrightarrow{F}[u, v] = \frac{1}{u - v} \cdot \left((\alpha u^2 - u + \beta) - (\alpha v^2 - v + \beta)\right) \cdot \overrightarrow{B} = (\alpha(u + v) - 1) \cdot \overrightarrow{B}.$$

For the Newton step we thus get $d^k = -\frac{\alpha(x^k)^2 - x^k + \beta}{2\alpha x^k - 1}$, while the iteration for the secant method results in $d^k = -\frac{\alpha(x^k)^2 - x^k + \beta}{\alpha(x^k + x^{k-1}) - 1}$. In Fig. 1(a) the iterates $\overrightarrow{X}^k := x^k \cdot J_2(B_1(0))$ of the secant method are depicted for the starting values $x^{-1} = 0$, $x^0 = -\frac{3}{4}$. Here, the iterates $k = 0, 2, 3$ are inverses of embedded sets (inner normals), whereas the other iterates are embedded convex sets (outer normals). For $k = -1$, the embedded origin is visualized in the plot.

In Example 3, we solve Problem 2 using the Gauß-Newton method in [7] and the Gauß-secant method for two-dimensional directed sets. Similar to the approximation of convex sets with finitely many supporting hyperplanes, we approximate a set-valued map $\overrightarrow{F} : \mathbb{R}^n \to \mathcal{D}(\mathbb{R}^2)$ by choosing a finite number of unit vectors $l^\mu \in S_1$, $\mu = 1, \ldots, M$ and evaluating $\overrightarrow{F}(x) = (f_1(x; \eta; l))_{\eta = \pm 1}, f_2(x; l))_{l \in S_1}$ in these directions, leading to a discretized map $\overrightarrow{F}_M : \mathbb{R}^n \to \mathbb{R}^{3M}$ defined by

$$\left(\overrightarrow{F}_M(x)\right)_i = \begin{cases} f_1(x; -1; l^i) & \text{if } i = 1, \ldots, M, \\ f_1(x; 1; l^{i-M}) & \text{if } i = M+1, \ldots, 2M, \\ f_2(x; l^{i-2M}) & \text{if } i = 2M+1, \ldots, 3M. \end{cases}$$

Recalling the ideas behind the Gauß-Newton method, we replace $\overrightarrow{F}_M(x) = 0$ by the minimization problem $\min_{x \in \mathbb{R}^n} \|\overrightarrow{F}_M(x)\|_2^2$ which is solved iteratively by minimizing $\|\overrightarrow{F}_M(x^k) + \overrightarrow{F}'_M(x^k)(d^k)\|_2^2$ or $\|\overrightarrow{F}_M(x^k) + \overrightarrow{F}_M[x^k, x^{k-1}](d^k)\|_2^2$, respectively. This leads to the normal equations

$$\overrightarrow{F}'_M(x^k)^\top \overrightarrow{F}'_M(x^k)(d^k) = -\overrightarrow{F}'_M(x^k)^\top \overrightarrow{F}_M(x^k) \quad \text{and} \tag{6}$$

$$\overrightarrow{F}[x^k, x^{k-1}]^\top \overrightarrow{F}[x^k, x^{k-1}](d^k) = -\overrightarrow{F}[x^k, x^{k-1}]^\top \overrightarrow{F}_M(x^k) \tag{7}$$

as iterative steps. The (i, j)-th entry of the matrix of divided differences is given for $j = 1, 2$ by the expressions

$$\frac{f_1(x_1^k, x_2^{k+j-2}; -1; l^i) - f_1(x_1^{k+j-2}, x_2^{k-1}; -1; l^i)}{x_j^k - x_j^{k-1}} \quad \text{for } i = 1, \ldots, M,$$

$$\frac{f_1(x_1^k, x_2^{k+j-2}; 1; l^i) - f_1(x_1^{k+j-2}, x_2^{k-1}; 1; l^i)}{x_j^k - x_j^{k-1}} \quad \text{for } i = M+1, \ldots, 2M,$$

$$\frac{f_2(x_1^k, x_2^{k+j-2}; l^{i-2M}) - f_2(x_1^{k+j-2}, x_2^{k-1}; l^{i-2M})}{x_j^k - x_j^{k-1}} \quad \text{for } i = 2M+1, \ldots, 3M,$$

which are much simpler to compute than the corresponding partial derivatives.

In the next nonlinear test example we are looking for parameters in the determining matrix of an ellipsoid centered in the origin which coincides with the unit ball.

Example 3. For $x \in \mathbb{R}^2$ we consider $G(x) = A(x) \cdot B_1(0)$ with $A(x) = \begin{pmatrix} x_1 & 0 \\ 0 & x_2 \end{pmatrix}$, $H(x) = B_1(0)$ in Problem 2. In this example, it is rather complicated to evaluate the derivative of $\overrightarrow{F}(x) = \overrightarrow{G}(x) - \overrightarrow{H}(x)$ analytically. Clearly, $\hat{x} = (1,1)^\top$ is one solution and $\overrightarrow{F}'(x) = (\overrightarrow{G}'(x; l), g_2(x; l))_{l \in S_1}$ with

$$f_2(x; l) = g_2(x; l) - 1 = \delta^*(l, G(x)) - 1 = \sqrt{l_1^2 x_1^2 + l_2^2 x_2^2} - 1,$$

$$\frac{\partial f_2}{\partial x}(x; l) = \frac{1}{\sqrt{l_1^2 x_1^2 + l_2^2 x_2^2}} \begin{pmatrix} l_1^2 x_1 \\ l_2^2 x_2 \end{pmatrix}^\top.$$

The derivative of $\overrightarrow{F}'_1(x; l)$ within the Gauß-Newton method is complicated, since it involves the projection to $\mathbb{R}$, whereas the calculation of $\overrightarrow{F}[u, v](w)$ in the Gauß-secant method only requires the weighted sum of differences of function values. Table 1 demonstrates that the Gauß-secant method converges rather rapidly, but slower than the Gauß-Newton method. In Fig. 1(b) the iterates $\overrightarrow{X}^k := J_2(A(x^k) \cdot B_1(0))$ of the secant method are depicted for the starting values $x^{-1} = (4,3)^\top$, $x^0 = (3,2)^\top$. Depending on the starting values, convergence to one of the solutions $(\pm 1, \pm 1)^\top$ can be observed.

To sketch the outline of further research, we mention [2]. In this publication, Newton's method for set-valued maps is used to approximate invariant sets, i.e. to find sets P for which $g(P) = P$. This task is similar to Problem 2, if the set is parametrized e.g. as $P(x) = x \cdot B_1(0)$ (see Example 2) and $G(x) = g(P(x))$, $H(x) = h(P(x))$, $h(x) = x$, but considerably more complicated, as the map $\overrightarrow{G}$ depends on a directed set. This leads to a zero-finding problems of the form $0 = \overrightarrow{F}(\overrightarrow{X}) = J_n(g(V_n(\overrightarrow{X}))) - \overrightarrow{X}$, the solution of which requires much more advanced Banach space techniques than in the present paper.

Table 1. Iterates x^k of the Gauß-Newton (values on the left) and the secant method (values on the right) for Example 3

k	x_1^k	x_2^k	$\frac{1}{2}\|\overrightarrow{F}_M(x^k)\|_2^2$	x_1^k	x_2^k	$\frac{1}{2}\|\overrightarrow{F}_M(x^k)\|_2^2$
-1				4.00000	3.00000	149.2847951
0	3.00000000	2.00000000	68.997425636149	3.00000	2.00000	68.9974256
1	1.01002370	1.00647268	0.001644629887	0.98748	0.98710	0.0032343
2	1.00000078	1.00000078	0.000000000012	0.99942	0.99933	0.0000081

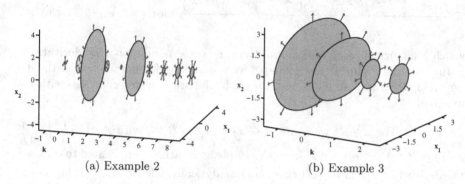

(a) Example 2   (b) Example 3

Fig. 1. Iterates of the secant method for two examples

References

1. Aubin, J.-P.: Mutational and Morphological Analysis. Tools for Shape Evolution and Morphogenesis. Systems & Control: Foundations & Applications. Birkhäuser Boston Inc., Boston (1999)
2. Baier, R., Dellnitz, M., Hessel-von Molo, M., Kevrekidis, I.G., Sertl, S.: The computation of invariant sets via Newton's method, 21 pages (May 2010) (submitted)
3. Baier, R., Farkhi, E.: Differences of Convex Compact Sets in the Space of Directed Sets. Part I: The Space of Directed Sets. Set-Valued Anal. 9(3), 217–245 (2001); Part II: Visualization of Directed Sets 9(3), 247–252 (2001)
4. Baier, R., Perria, G.: Set-valued Hermite interpolation. J. Approx. Theory 163(10), 1349–1372 (2011)
5. Dontchev, A.L., Hager, W.W., Veliov, V.M.: Uniform convergence and mesh independence of Newton's method for discretized variational problems. SIAM J. Control Optim. 39(3), 961–980 (2000) (electronic)
6. Dontchev, A.L., Veliov, V.M.: Metric regularity under approximations. Control Cybernet. 38(4B), 1283–1303 (2009)
7. Li, C., Zhang, W.H., Jin, X.Q.: Convergence and uniqueness properties of Gauss-Newton's method. Comput. Math. Appl. 47(6-7), 1057–1067 (2004)
8. Perria, G.: Set-valued interpolation. Bayreuth. Math. Schr. 79, 154 pages (2007)
9. Potra, F.-A.: An error analysis for the secant method. Numer. Math. 38(3), 427–445 (1981/1982)
10. Saint-Pierre, P.: Newton and other continuation methods for multivalued inclusions. Set-Valued Anal. 3(2), 143–156 (1995)
11. Schmidt, J.W.: Eine Übertragung der Regula Falsi auf Gleichungen in Banachräumen. II. Nichtlineare Gleichungssysteme. Z. Angew. Math. Mech. 43(3), 97–110 (1963)
12. Yamamoto, T.: A method for finding sharp error bounds for Newton's method under the Kantorovich assumptions. Numer. Math. 49(2-3), 203–220 (1986)

Optimal Control of Multibody Systems in Resistive Media

F.L. Chernousko

Institute for Problems in Mechanics,
Russian Academy of Sciences,
Moscow, Russia
chern@ipmnet.ru

Abstract. Locomotion of a mechanical system consisting of a main body and one or two links attached to it by cylindrical joints is considered. The system moves in a resistive medium and is controlled by periodic angular oscillations of the links relative to the main body. The resistance force acting upon each body is a quadratic function of its velocity. Under certain assumptions, a nonlinear equation of motion is derived and simplified. The optimal control of oscillations is found that corresponds to the maximal average locomotion speed.

Keywords: optimal control, nonlinear dynamics, robotics, locomotion.

1 Introduction

It is well-known that a multibody mechanical system, whose links perform specific oscillations relative to each other, can move progressively in a resistive medium. This locomotion principle is used by fish, snakes, insects, and some animals [1–3]. In robotics, the same principle is applied to locomotion of snake-like robots along a surface [4].

Dynamics and optimization of snake-like multilink mechanisms that move along a plane in the presence of Coulomb's dry friction forces acting between the mechanism and the plane, have been studied in [5–7].

Various aspects of fish-like locomotion in a fluid are considered in many papers, and a number of swimming robotic systems have been developed [8–11].

In this paper, we consider a progressive motion of a multilink system in a fluid in the presence of resistance forces proportional to the squared velocity of the moving body. The mechanical model is described and simplified. The optimal control problem for the motion of the links is formulated, and its exact solution is presented.

2 Mechanical Model

Consider a mechanical system consisting of the main body and two symmetric links OA and $O'A'$ attached to it by cylindrical joints (Fig. 1). The length of

I. Lirkov, S. Margenov, and J. Waśniewski (Eds.): LSSC 2011, LNCS 7116, pp. 99–105, 2012.

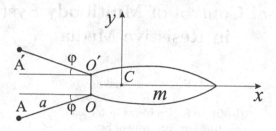

Fig. 1. System with two links

the links is denoted by a, and their mass is negligible compared to the mass m of the main body.

Let us introduce the Cartesian coordinate frame Cxy connected with the main body and denote by $\mathbf{i}$ and $\mathbf{j}$ the unit vectors directed along the axes Cx and Cy, respectively.

The links perform symmetric periodic oscillations of period T about the joints O and O' so that the angle φ between the links and the axis Cx satisfies the condition

$$\varphi(t + T) = \varphi(t) \tag{1}$$

for each time instant t. During the period $[0, T]$, the angle φ first increases from 0 to φ_0 and then decreases from φ_0 to 0.

Denote by v the velocity of the progressive motion of the main body along the axis Cx. We will consider only forward motions, so that $v \geq 0$. The value of the resistance force acting upon the body is denoted by $c_0 v^2$, where c_0 is a positive constant.

Suppose for simplicity that the resistance forces acting upon each link can be reduced to a force applied at the end points A and A'. Denote by $\mathbf{V}$ the velocity of point A and by ω the angular velocity of the link OA. We have (see Fig. 1):

$$\mathbf{V} = v\mathbf{i} + a\omega \sin \varphi \mathbf{i} - a\omega \cos \varphi \mathbf{j}, \quad \omega = \dot{\varphi}. \tag{2}$$

Here and below, dots denote derivatives with respect to time t.

The quadratic resistance force applied to the point A is given by

$$\mathbf{F} = -k_0 V \mathbf{V}, \tag{3}$$

where $k_0 > 0$ is a constant coefficient.

Under the assumptions made, the equation of the progressive motion of the main body can be written as follows:

$$m\dot{v} = -c_0 v^2 + 2F_x , \tag{4}$$

where F_x is the projection of the vector $\mathbf{F}$ from (3) onto the axis Cx. Note that the projection of the resistance force acting upon the link $O'A'$ is also equal to F_x. The added mass of the main body can be included into m. Introducing the denotations

$$c_0/m = c, \qquad 2k_0/m = k$$

and using equations (2) and (3) to determine F_x, we convert equation (4) to the form:

$$\dot{v} = -cv^2$$
$$-k(v + a\omega \sin\varphi)\sqrt{v^2 + a^2\omega^2 + 2va\omega \sin\varphi}\,. \qquad (5)$$

Similarly, the system with one link attached to the main body can be considered [12]. The system with one link imitates a fish with a tail, whereas the system with two symmetric links is a model of a swimming animal with two extremities. Under certain symmetry conditions, the system with one link can be described by the same equation (5).

We assume that the resistance of the links is much less than the resistance of the main body: $\mu = k/c \ll 1$ and suppose that the oscillations of the links have a short period and small amplitude.

Let us introduce new variables:

$$\varphi = \mu\psi, \quad T = \mu T_0,$$
$$t = T\tau = \mu T_0\tau, \quad v = \mu(a/T_0)u\,. \qquad (6)$$

Here, the new variable ψ and constant T_0 are of order $O(1)$, τ is a new (fast) time, and u is the new non-dimensional velocity.

Let us transform equation (5) using (6) and omitting terms of higher order of μ. After simplifications, we obtain the equation

$$\frac{du}{d\tau} = -\varepsilon\left[u^2 + \left(u + \psi\frac{d\psi}{d\tau}\right)\left|\frac{d\psi}{d\tau}\right|\right], \quad \varepsilon = ca\mu^2\,. \qquad (7)$$

Here, ε is a small parameter, and $\psi(\tau)$ is a periodic function of τ with a period equal to 1. Applying the asymptotic methods of averaging [13] to equation (7), we obtain the equation of the first approximation as follows:

$$\frac{du}{d\tau} = -\varepsilon(u^2 + I_1 u + I_0)\,, \qquad (8)$$

$$I_1 = \int_0^1 \left|\frac{d\psi}{d\tau}\right| d\tau, \quad I_0 = \int_0^1 \psi\frac{d\psi}{d\tau}\left|\frac{d\psi}{d\tau}\right| d\tau \qquad (9)$$

The solution $u(\tau)$ of the averaged equation (8) differs from the solution of equation (7), under the same initial conditions, by terms of order of ε for the large time interval of order of ε^{-1}.

According to (9), we have $I_1 > 0$.

If $I_0 > 0$, then the right-hand side of equation (8) is positive for all $u > 0$. Hence, $du/d\tau < -\varepsilon I_0 < 0$, the velocity decreases and reaches zero in finite time. In this case, the forward motion of the system is impossible.

We will consider below a more interesting case, where $I_0 < 0$. Then equation (8) has a unique positive stationary solution

$$u^* = [-I_0 + (I_1^2/4)]^{1/2} - I_1/2 > 0 \qquad (10)$$

which is globally asymptotically stable. Thus, for any initial condition $u(\tau_0) = u_0 \geq 0$, we have $u(\tau) \to u^*$ as $\tau \to \infty$.

To check the inequality $I_0 < 0$ and evaluate the velocity u^*, we are to specify the periodic function $\psi(\tau)$ subject to condition (1) and calculate the integrals I_1 and I_0 from (9).

3   Optimal Control

Let us consider the optimal control problem for the angular motion of the links. We will regard the dimensionless angular velocity Ω as the control subject to the constraints

$$-\Omega_- \leq \Omega = d\psi/d\tau \leq \Omega_+, \qquad (11)$$

where Ω_- and Ω_+ are given positive constants.

Suppose that the normalized angle ψ changes over the interval $\tau \in [0,1]$ as follows: it grows from $\psi(0) = 0$ to $\psi(\theta) = \psi_0 > 0$ and then decreases from ψ_0 to $\psi(1) = 0$. Here, $\theta \in (0,1)$ and $\psi_0 > 0$ are constant parameters.

The problem is to find functions $\Omega(\tau)$ and $\psi(\tau)$ that satisfy (11), the boundary conditions imposed above and maximize the average velocity u^* defined by (10).

The solution of this problem is obtained by means of Pontryagin's maximum principle [14]. After that, the parameter $\theta \in (0,1)$ is chosen in order to maximize u^*. Omitting this rather lengthy analysis, we present below the final results.

The optimal control $\Omega(\tau)$ and the corresponding optimal time history of the normalized angle $\psi(\tau)$ are given by equations

$$\Omega = \Omega_+, \quad \psi = \Omega_+ \tau \quad \text{for} \quad \tau \in (0, \tau_*),$$

$$\Omega = \Omega_+ \left[1 + \frac{3(\tau - \tau_*)}{2\tau_*} \right]^{-1/3},$$

$$\psi = \Omega_+ \tau_* \left[1 + \frac{3(\tau - \tau_*)}{2\tau_*} \right]^{2/3} \qquad (12)$$

$$\text{for} \quad \tau \in (\tau_*, \theta),$$

$$\Omega = -\Omega_-, \ \psi = \Omega_-(1 - \tau) \ \text{for } \tau \in (\theta, 1),$$

$$\tau_* = s\theta, \quad \theta = 1 - \varphi_0/\Omega_-.$$

Here, s is the only root of the cubic equation

$$s(3 - s)^2 = 4(\psi_0/\Omega_+)^3(1 - \psi_0/\Omega_-)^{-3} \qquad (13)$$

lying in the interval $s \in (0,1)$. Equations (12) and (13) define the functions $\Omega(\tau)$ and $\psi(\tau)$ for the interval $(0,1)$.

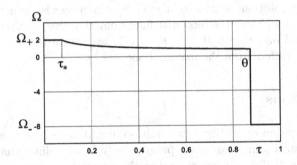

Fig. 2. Optimal control $\Omega(\tau)$

The maximum value of the average stationary velocity u^* that corresponds to the solution presented above is given by

$$u^* = [\psi_0^2 + \psi_0^2 \Omega_- /2 - \\ -2(\psi_0^3/\theta)(2-s)(3-s)^{-2}]^{1/2} - \psi_0, \qquad (14)$$

where $s \in (0,1)$ is the root of equation (13).

Thus, the optimal control is completely determined in terms of normalized variables. To return to the original dimensional ones, one is to use equations (6).

4   Example

Let us consider a numerical example. We assume that $\psi_0 = 1$, $\Omega_+ = 2$, $\Omega_- = 8$ and obtain from the optimal solution (12)–(14):

$$\theta = 0.875, \ s = 0.088, \ \tau_* = 0.077, \ u^* = 1.118 \qquad (15)$$

The time histories of functions $\Omega(\tau)$ and $\psi(\tau)$ from equation (12) are shown in Figs. 2 and 3, respectively.

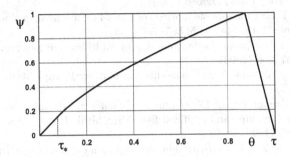

Fig. 3. Optimal trajectory $\psi(\tau)$

This optimal solution is close to the case of a piecewise constant angular velocity $\omega(\tau)$. If we choose the piecewise linear function $\psi(\tau)$ so that it coincides with the optimal one at $\tau = 0$, $\tau = \theta$, $\tau = 1$, we obtain the average velocity $u_* = 1.104$ that differs from the maximal one less by 1.5%.

5 Conclusions

A mechanical system consisting of a main body and one or two links attached to it by cylindrical joints can move progressively in a medium that acts upon moving bodies with forces proportional to the squared velocities of the bodies. Under assumptions made, the equation of motion is simplified, and the average velocity of the progressive motion is evaluated.

The optimal time history of the angular oscillations of the links is obtained that corresponds to the maximal, under the conditions imposed, average speed of the progressive motion.

The optimal solution has the following property: the deflection of the links from the axis of locomotion is carried out with a decreasing angular velocity, whereas their retrieval is always performed with the maximal admissible angular velocity.

The obtained results correlate well with observations of the process of swimming.

Acknowledgements. The work was supported by the Russian Foundation for Basic Research (Project 11-01-00513) and by the Program of Support for Leading Scientific Schools (NSh–64817.2010.1).

References

1. Gray, J.: Animal Locomotion. Norton, New York (1968)
2. Lighthill, J.: Mathematical Biofluiddynamics. SIAM, Philadelphia (1975)
3. Blake, R.W.: Fish Locomotion. Cambridge University Press, Cambridge (1983)
4. Hirose, S.: Biologically Inspired Robots: Snake-like Locomotors and Manipulators. Oxford University Press, Oxford (1993)
5. Chernousko, F.L.: Controllable motions of a two-link mechanism along a horizontal plane. J. Appl. Math. Mech. 65, 565–577 (2001)
6. Chernousko, F.L.: Snake-like locomotions of multilink mechanisms. J. Vibration Control 9, 237–256 (2003)
7. Chernousko, F.L.: Modelling of snake-like locomotion. J. Appl. Math. Comput. 164, 415–434 (2005)
8. Terada, Y., Yamamoto, I.: Development of oscillating fin propulsion system and its application to ships and artificial fish. Mitsubishi Heavy Industries Tech. Review 36, 84–88 (1999)
9. Mason, R., Burdick, J.: Construction and modelling of a carangiform robotic fish. In: Korcke, P., Trevelyan, J. (eds.) Experimental Robotics VI. LNCIS, vol. 250, pp. 235–242. Springer, Heidelberg (2000)

10. Colgate, J.E., Lynch, K.M.: Mechanics and control of swimming: a review. IEEE J. Oceanic Engng. 29, 660–673 (2004)
11. http://en.wikipedia.org/wiki/RoboTuna#References
12. Chernousko, F.L.: Optimal motion of a two-body system in a resistive medium. Journal of Optimization Theory and Applications 147, 278–297 (2010)
13. Bogoliubov, N.N., Mitropolsky, Y.A.: Asymptotic Methods in the Theory of Non-linear Oscillations. Gordon and Breach, New York (1961)
14. Pontryagin, L.S., Boltyanskii, V.G., Gamkrelidze, R.V., Mishchenko, E.F.: The Mathematical Theory of Optimal Processes. Gordon and Breach, New York (1986)

Classical and Relaxed Progressively Refining Discretization-Optimization Methods for Optimal Control Problems Defined by Ordinary Differential Equations

I. Chryssoverghi, J. Coletsos, and B. Kokkinis

Department of Mathematics, School of Applied Mathematics and Physics,
National Technical University of Athens
Zografou Campus, 15780 Athens, Greece
ichriso@math.ntua.gr

Abstract. An optimal control problem is considered, for systems defined by nonlinear ordinary differential equations, with control and pointwise state constraints. Since the problem may have no classical solutions, it is also formulated in the relaxed form. Various necessary/sufficient conditions for optimality are first given for both formulations. In order to solve these problems numerically, we then propose a discrete penalized gradient projection method generating classical controls, and a discrete penalised conditional descent method generating relaxed controls. In both methods, the discretization procedure is progressively refining in order to achieve efficiency with reduced computational cost. Results are given concerning the behaviour in the limit of these methods. Finally, numerical examples are provided.

1 Classical and Relaxed Optimal Control Problems

Consider the following optimal control problem. The state equation is given by
$$y'(t) = f(t, y(t), w(t)), \quad t \in I := [0, T], \quad y(0) = y^0,$$
where $y(t) \in \mathbb{R}^d$, the constraints on the control w are $w(t) \in U$, for $t \in I$, where U is a compact subset of $\mathbb{R}^{d'}$, the pointwise constraints on the state $y := y_w$ are
$$G_1(w)(s) := g_1(s, y(s)) \leqslant 0, \text{ for } s \in I,$$
where g_1 takes values in $\mathbb{R}^p$, and the cost functional is
$$G_0(w) := \tilde{g}_0(y(T)) + \int_I g_0(t, y(t), w(t))dt.$$
Additional equality/inequality terminal/integral control/state constraints could also be included, but they are omitted for compactness of exposition.

Define first the set of *classical controls*
$$W := \{w : I \to U \mid w \text{ measurable}\} \subset L^2(I, \mathbb{R}^{d'}).$$
The *classical optimal control problem* is to minimize $G_0(w)$ subject to $w \in W$ and to the above state constraints.

It is well known that, even if U is convex, the classical problem may have no solutions. The existence of such a solution is usually proved under strong, unrealistic for

I. Lirkov, S. Margenov, and J. Waśniewski (Eds.): LSSC 2011, LNCS 7116, pp. 106–114, 2012.
© Springer-Verlag Berlin Heidelberg 2012

nonlinear systems, convexity assumptions, such as the Cesari property. Reformulated in the relaxed form, the problem is convexified and partially linearized in some sense and has thus a solution in a larger space under weaker assumptions.

Next, we define the set of *relaxed controls* (see [9], [8]) by

$$R := \{r : I \to M_1(U) \mid r \text{ weakly measurable}\} \subset L_w^\infty(I, M(U)) \equiv L^1(I, C(U))^*,$$

where $M(U)$ (resp. $M_1(U)$) is the set of Radon (resp. probability) measures on U. The set W (resp. R) is endowed with the relative strong (resp. weak star) topology, and R is convex, metrizable and compact. If each classical control $w(\cdot)$ is identified with its associated Dirac relaxed control $r(\cdot) := \delta_{w(\cdot)}$, then W *may also* be considered as a subset of R, and W is thus dense in R. For a given $\phi \in L^1(I; C(U; \mathbb{R}^n))$ (or equivalently $\phi \in B(I, U; \mathbb{R}^n)$, where $B(I, U; \mathbb{R}^n)$ is the set of Caratheodory vector functions bounded by an integrable function), and for $r \in R$ we shall use for simplicity the notation

$$\varphi(t, r(t)) := \int_U \varphi(t, u) r(t)(du).$$

The *relaxed optimal control problem* is then defined by *replacing w by r*, with the above notation, and W by R in the classical problem.

We suppose that the function f is defined on $I \times \mathbb{R}^d \times U$, measurable for y, u fixed, continuous for t fixed, and satisfies

$\|f(t, y, u)\| \leq \psi(t)(1 + \|y\|)$, for every $(t, y, u) \in I \times \mathbb{R}^d \times U$, with $\psi \in L^1(I)$,
$\|f(t, y_1, u) - f(t, y_2, u)\| \leq L \|y_1 - y_2\|$, for every $(t, y_1, y_2, u) \in I \times \mathbb{R}^{2d} \times U$.

The theorems of Sections 1 and 2 are proved by using the techniques of [9] and [7].

Theorem 1. *For every relaxed (or classical, as $W \subset R$) control $r \in R$, the state equation has a unique absolutely continuous solution $y := y_r$. Moreover, there exists a constant b such that $\|y_r\|_\infty \leq b$ for every $r \in R$.*

Let B denote the closed ball in $\mathbb{R}^d$ with center 0 and radius b (defined in Theorem 1). We suppose now in addition that the function g_0 is defined on $I \times B \times U$, measurable for fixed y, u, continuous for fixed t, and satisfies

$$|g_0(t, y, u)| \leq \zeta_0(t), \text{ for every } (t, y, u) \in I \times B \times U,$$

with $\zeta_0 \in L^1(I)$, that the function g_1 is continuous on $I \times B$, and that the function $\tilde{g}_0$ is continuous on B.

Theorem 2. *If the relaxed problem has an admissible control, then it has a solution.*

2 Optimality Conditions

In order to state optimality conditions, we suppose in addition that the functions f, g_0, f_y, f_u, g_{0y}, g_{0u}, are defined on $I \times B' \times U'$, where B' (resp. U') is an open set containing B (resp. U), measurable on I for fixed $(y, u) \in B \times U$, continuous on $B \times U$ for fixed $t \in I$, and satisfy

$$\|f_y(t, y, u)\| \leq \xi(t), \quad \|f_u(t, y, u)\| \leq \eta(t),$$
$$\|g_{0y}(t, y, u)\| \leq \zeta_{01}(t), \quad \|g_{0u}(t, y, u)\| \leq \zeta_{02}(t),$$

for every $(t, y, u) \in I \times B \times U$, with $\xi, \eta, \zeta_{01}, \zeta_{02} \in L^1(I)$, that the function $\tilde{g}_{0y}$ is defined on B' and continuous on B, and that the function g_1 is defined on $I \times B'$ and continuous on $I \times B$.

Theorem 3. *(i) If U is convex, the directional derivative of $G_0 : W \to \mathbb{R}$ is given by*

$$DG_0(w, w' - w) := \lim_{\alpha \to 0^+} \frac{G_0(w+\alpha(w'-w))-G_0(w)}{\alpha}$$

$$= \int_0^T [z(t)f_u(t, y(t), w(t)) + g_{0u}(t, y(t), w(t))][w'(t) - w(t)]dt, \text{ for } w, w' \in W,$$

where $y := y_w$, and the adjoint state $z := z_w$, a row n-vector, is the solution of the classical linear adjoint equation

$$z'(t) = -z(t)f_y(t, y(t), w(t)) - g_{0y}(t, y(t), w(t)), \ t \in I,$$
$$z(T) = \tilde{g}_{0y}(y(T)), \text{with } y := y_w,$$

The directional derivative of $G_1 : W \to C(I)^p$ is given by the $p \times n$-matrix function

$$DG_1(w, w' - w)(s)$$
$$= g_{1y}(s, y(s))Z(s)^{-1} \int_0^s Z(t)f(t, y(t), w(t))[w'(t) - w(t)]dt, \ s \in I,$$

where the $n \times n$-matrix function $Z := Z_w$ satisfies the fundamental matrix equation

$$Z'(t) = -Z(t)f_y(t, y(t), w(t)), \quad t \in I, \ Z(T) = E \ (E \text{ identity matrix}).$$

(ii) The directional derivative of $G_0 : R \to \mathbb{R}$ is given by

$$DG_0(r, r' - r) := \lim_{\alpha \to 0^+} \frac{G_0(r+\alpha(r'-r))-G_0(r)}{\alpha}$$

$$= \int_0^T [z(t)f(t, y(t), r'(t) - r(t)) + g_0(t, y(t), r'(t) - r(t))]dt, \text{ for } r, r' \in R,$$

where $y = y_r$, and the relaxed adjoint $z := z_r$ is the solution of the relaxed linear adjoint equation

$$z'(t) = -z(t)f_y(t, y(t), r(t)) - g_{0y}(t, y(t), r(t)), \ t \in I, \ z(T) = \tilde{g}_{0y}(y(T)), y := y_r.$$

The directional derivative of $G_1 : R \to C(I)^p$ is given by

$$DG_1(r, r' - r)(s) = g_{1y}(s, y(s))Z(s)^{-1} \int_0^s Z(t)f(t, y(t), r'(t) - r(t))dt, \ s \in I,$$

where $Z := Z_r$ is defined as in (i), but with w replaced by r.

The following theorem addresses various optimality conditions.

Theorem 4. *(i) We suppose that U is convex. If $w \in W$ is optimal for the classical problem, then w is weakly extremal classical, i.e. there exist multipliers*

$$\lambda_0 \in \mathbb{R}, \quad \lambda_1 \in [C(I)^p]^* \equiv M(I)^p,$$

with $\lambda_0 \geqslant 0$, $\lambda_1 \geqslant 0$, $|\lambda_0| + \|\lambda_1\|_* = 1$, where $\|\lambda_1\|_* := \sum_{j=1}^p \|\lambda_{1j}\|_M$,

such that

$$\lambda_0 DG_0(w, w' - w) + \int_0^T \lambda_1(ds)DG_1(w, w' - w)(s)$$
$$= \int_0^T \{\lambda_0[z(t)f_u(t, y(t), w(t)) + g_{0u}(t, y(t), w(t))]$$
$$+ \left(\int_t^T \lambda_1(ds)g_{1y}(s, y(s))Z(s)^{-1}\right) Z(t)f_u(t, y(t), w(t))\}[w'(t) - w(t)]dt \geqslant 0,$$

for every $w' \in W$, and
$$\int_0^T \lambda_1(ds)G_1(w)(s) = 0 \text{ (classical transversality conditions)}.$$

(ii) With U not necessarily convex, if $r \in R$ is optimal for either the relaxed or the classical problem, then r is strongly extremal relaxed, i.e. there exist multipliers as in (i), such that

(1) $\lambda_0 DG_0(r, r' - r) + \int_0^T \lambda_1(ds) DG_1(r, r' - r)(s)$

$\quad = \int_0^T \{\lambda_0[z(t)f(t, y(t), r'(t) - r(t)) + g_0(t, y(t), r'(t) - r(t))]$

$\quad + \left(\int_t^T \lambda_1(ds) g_{1y}(s, y(s)) Z(s)^{-1}\right) Z(t)f(t, y(t), r'(t) - r(t))\}dt \geqslant 0,$

$\quad$ *for every* $r' \in R$,

and

(2) $\int_0^T \lambda_1(ds) G_1(r)(s) = 0$ *(relaxed transversality conditions).*

If U is convex, then condition (1) implies the global weak relaxed condition

(3) $\int_0^T \{\lambda_0[z(t)f_u(t, y(t), r(t)) + g_{0u}(t, y(t), r(t))]$

$\quad + \left(\int_t^T \lambda_1(ds) g_{1y}(s, y(s)) Z(s)^{-1}\right) Z(t) f_u(t, y(t), r(t))\}[\phi(t, r(t)) - r(t)]dt \geqslant 0,$

$\quad$ *for every* $\phi \in B(I, U; U)$.

A control r satisfying conditions (2) and (3) is called weakly extremal relaxed.

(iii) Suppose that the derivatives of f and g_0 in u are excluded (resp. included) in the last assumptions, and that the data are such that G_0 and G_1 are convex. If $r \in R$ (resp. $w \in W$, with U convex) is admissible and strongly extremal relaxed (resp. weakly extremal classical) for the relaxed (resp. classical) problem, with some $\lambda_0 > 0$, then r (resp. $w \in W$) is optimal for this problem.

3   Discretizations

For each integer $n = 1, 2, ...,$ let $t_{n0} := 0, ..., t_{nN} := T$ $(N := N_n)$ be a subdivision of I into subintervals of equal length $h := h_n$, such that either $h_{n+1} := h_n$ or $h_{n+1} := h_n/2$, and $h_n \to 0$. The discrete classical state equation (which has a unique solution for h_n sufficiently small) is defined by applying the implicit midpoint scheme to the continuous equation

$$y_{n,i+1} = y_{ni} + hf(\bar{t}_{ni}, \bar{y}_{ni}, w_{ni}), \ i = 0, ..., N - 1, \ y_{n0} = y^0,$$

where $\bar{t}_{ni} := (t_{ni} + t_{n,i+1})/2$, $\bar{y}_{ni} := (y_{ni} + y_{n,i+1})/2$, and $(w_{ni})_{i=0,...,N-1} \in W_n$ is the discrete piecewise constant control (we have $W_n \subset W_{n+1}$). The discrete state constraints are

$$G_{1nj}(w_n) := g_1(\bar{t}_{nj}, \bar{y}_{nj}) \leqslant 0, \ j = 0, ..., N - 1,$$

and the discrete cost functional

$$G_{0n}(w_n) := \tilde{g}_0(y_{nN}) + h \sum_{i=0}^{N-1} g_0(\bar{t}_{ni}, \bar{y}_{ni}, w_{ni}).$$

The directional derivative of G_{0n} is given by

$$DG_{0n}(w_n, w'_n - w_n) = h \sum_{i=0}^{N-1} [\bar{z}_{0ni} f_u(\bar{t}_{ni}, \bar{y}_{ni}, w_{ni}) + g_{0u}(\bar{t}_{ni}, \bar{y}_{ni}, w_{ni})](w'_{ni} - w_{ni}),$$

where the corresponding discrete adjoint state z_{0n} is defined by

$$z_{0ni} = z_{0n,i+1} + h[\bar{z}_{0ni} f_y(\bar{t}_{ni}, \bar{y}_{ni}, w_{ni}) + g_{0y}(\bar{t}_{ni}, \bar{y}_{ni}, w_{ni})], \ i = N-1, ..., 0,$$

$$z_{0nN} = \tilde{g}_{0y}(y_{nN}),$$

where $\bar{z}_{0ni} := (z_{0ni} + z_{0n,i+1})/2$.

The directional derivative of G_{1nj}, $j = 0, ..., N-1$, is given by

$$DG_{1nj}(w_n, w'_n - w_n) = h \sum_{i=0}^{N-1} \bar{z}_{1nji} f_u(\bar{t}_{ni}, \bar{y}_{ni}, w_{ni})(w'_{ni} - w_{ni}),$$

where $\bar{z}_{1nji} := (z_{1nji} + z_{1nj,i+1})/2$ and the corresponding adjoints z_{1nj} are defined by (δ_{ij} Kronecker symbol)

$$z_{1nji} = z_{1nj,i+1} + h\bar{z}_{1nji} f_y(\bar{t}_{ni}, \bar{y}_{ni}, w_{ni}) + \delta_{ij} g_{1y}(\bar{t}_{ni}, \bar{y}_{ni}), \ i = N-1, ..., 0, \ z_{1njN} = 0.$$

The corresponding definitions in the relaxed case are similar.

4 Discretization-Optimization Methods

Let (M^m) be a positive increasing sequence such that $M^m \to \infty$ as $m \to \infty$, $\gamma > 0$, $b, c \in (0,1)$, and $(\beta^m), (\zeta_k)$ positive sequences, with (β^m) decreasing such that $\beta^m \to 0$, and $\zeta_k \leqslant 1$. Note that the parameters $h_n, M^m, \beta^m, \zeta_k$ are given in advance in the below algorithms 1 and 2. Define first the *penalized functionals* on W_n

$$G_n^m(w_n) := G_{0n}(w_n) + \tfrac{1}{2} M^m h \sum_{i=0}^{N-1} [\max(0, G_{1ni}(w_n))]^2,$$

where the max is taken component-wise, the *progressively refining classical penalized gradient projection method* is described by the following Algorithm, where U is assumed to be convex. The progressively refining procedure has the advantage of reducing computational cost and practically keeping the efficiency of the original method with the fixed finest discretization. This is explained by the fact that a finer discretization becomes more crucial as the iterate gets closer to the limit control.

Algorithm 1

Step 1. Set $k := 0$, $m := 1$. Choose a value of n and an initial control $w_{10}^1 \in W_n$.
Step 2. Find $v_{nk}^m \in W_n$ such that

$$e_k := DG_n^m(w_{nk}^m, v_{nk}^m - w_{nk}^m) + (\gamma/2)h \sum_{i=0}^{N-1} \|v_{nik}^m - w_{nik}^m\|^2$$

$$= \min_{v'_n \in W_n} [DG_n^m(w_{nk}^m, v'_n - w_{nk}^m) + (\gamma/2)h \sum_{i=0}^{N-1} \|v'_{ni} - w_{nik}^m\|^2] \text{ (Projection)},$$

and set $d_k := DG_n^m(w_{nk}^m, v_{nk}^m - w_{nk}^m)$.
Step 3. If $|d_k| \leqslant \beta^m$, set $w_n^m := w_{nk}^m$, $w_{n+1,k}^{m+1} := w_{nk}^m$, $v_n^m := v_{nk}^m$, $e^m := e_k$, $d^m := d_k$, $m := m+1$, $n := n+1$. Go to Step 2.
Step 4. (*Armijo Step Search*) Find the lowest integer $s \in \mathbb{Z}$, say $\bar{s}$, such that $\alpha(s) = c^s \zeta_k \in (0, 1]$ and $\alpha(s)$ satisfy the inequality
$$G_n^m(w_{nk}^m + \alpha(s)(v_{nk}^m - w_{nk}^m)) - G_n^m(w_{nk}^m) \leqslant \alpha(s) b d_k.$$
Set $\alpha_k := \alpha(\bar{s})$.
Step 5. Set $w_{n,k+1}^m := w_{nk}^m + \alpha_k(v_{nk}^m - w_{nk}^m)$, $k := k+1$. Go to Step 2.

A (strongly or weakly, classical or relaxed) extremal control is called *abnormal* if there exist multipliers as in the corresponding optimality conditions, with $\lambda_0 = 0$. A control is admissible *and* abnormal extremal in rather exceptional situations (see [9]). With $w^m_{n(m)}$ as defined in Step 3, define the sequence of discrete $\mathbb{R}^p$-valued function *multipliers*

$$\lambda^m_i := M^m \max(0, G_{1ni}(w^m_n)) = M^m \max(0, g_1(\bar{t}_{ni}, \bar{y}^m_{ni})), \quad i = 0, ..., N-1.$$

The proof of the following theorem is similar to that of theorem 5.1 in [6].

Theorem 5. *(i) In the presence of state constraints, if the whole sequence $(w^{m(k)}_{n(k),k})$ generated by Algorithm 1 converges to some $w \in W$ in L^2 strongly and the sequence of piecewise constant functions $((\lambda^m_i)_{i=0,...,N-1})$ in $L^1(I)^p$ is bounded, then w is admissible and weakly extremal classical for the classical problem. In the absence of state constraints, if a subsequence $(w^{m(k)}_{n(k),k})_{k \in K}$ converges to some $w \in W$ in L^2 strongly, then w is weakly extremal classical for the classical problem.*

(ii) In the presence of state constraints, if a subsequence $(w^m_{n(m)})_{m \in M}$ (in R) of the sequence generated by Algorithm 1 in Step 3 converges to some r in R and the sequence $((\lambda^m_i)_{i=0,...,N-1})$ in $L^1(I)^p$ is bounded, then r is admissible and weakly extremal relaxed for the relaxed problem. In the absence of state constraints, if a subsequence $(w^{m(k)}_{n(k),k})_{k \in K}$ converges to some r in R, then r is weakly extremal relaxed for the relaxed problem.

(iii) In any of the convergence cases (i) or (ii) with state constraints, suppose that the classical, or relaxed, problem has no admissible, abnormal extremal, controls. If the limit control is admissible, then the sequence of multipliers is bounded, and this control is extremal as above.

Next, define the *penalized functionals* on R_n

$$G^m_n(r_n) := G_{0n}(r_n) + \tfrac{1}{2} M^m h \sum_{i=0}^{N-1} [\max(0, G_{1ni}(r_n))]^2.$$

The *progressively refining relaxed penalized conditional descent method* is described by the following Algorithm, where U is *not necessarily convex*.

Algorithm 2

Step 1. Set $k := 0$, $m := 1$. Choose a value of n and an initial control $r^1_{n0} \in R_n$.

Step 2. Find $\bar{r}^m_{nk} \in R_n$ such that
$$d_k := DG^m_n(r^m_{nk}, \bar{r}^m_{nk} - r^m_{nk}) = \min_{r'_n \in R_n} DG^m_n(r^m_{nk}, r'_n - r^m_{nk}) \text{ (Conditional gradient).}$$

Step 3. If $|d_k| \leqslant \beta^m$, set $r^m_n := r^m_{nk}, r^{m+1}_{n+1,k} := r^m_{nk}, \bar{r}^m := \bar{r}^m_{nk}, d^m := d_k, m := m+1$, $n := n+1$. Go to Step 2.

Step 4. *(Armijo Step Search)* Find the lowest integer $s \in \mathbb{Z}$, say $\bar{s}$, such that $\alpha(s) = c^s \zeta_k \in (0, 1]$ and $\alpha(s)$ satisfy the inequality
$$G^m_n(r^m_{nk} + \alpha(s)(\bar{r}^m_{nk} - r^m_{nk})) - G^m_n(r^m_{nk}) \leqslant \alpha(s) b d_k.$$
Set $\alpha_k := \alpha(\bar{s})$.

Step 5. Choose any $r^m_{n,k+1} \in R_n$ such that
$$G^m_n(r^m_{n,k+1}) \leqslant G^m_n(r^m_{nk} + \alpha_k(\bar{r}^m_{nk} - r^m_{nk})).$$
Set $k := k + 1$. Go to Step 2.

If the chosen initial control r^1_{n0} is classical, using Caratheodory's theorem it can be shown by induction that the control $r^m_{n,k+1}$ in Step 5 can be chosen, for each k, to be a *Gamkrelidze relaxed control*, i.e. a convex combination of a *fixed* number of classical (Dirac) controls. The Gamkrelidze relaxed controls thus computed can then be simulated by classical ones using a standard procedure (see [3]).

With r_n^m as defined in Step 3, define the sequence of *multipliers*

$$\lambda_i^m := M^m \max(0, G_{1n}(r_n^m)) = M^m \max(0, g_1(\bar{t}_{ni}, \bar{y}_{ni}^m)), \quad i = 0, ..., N-1.$$

The proof of the following theorem is similar to that of theorem 5.1 in [5].

Theorem 6. *We suppose that the derivatives in u are excluded in the last assumptions of Section 2.*
(i) In the presence of state constraints, if a subsequence $(r_{n(m)}^m)_{m \in L}$ of the sequence generated by Algorithm 2 in Step 3 converges to some r in R and the sequence of piecewise constant functions $((\lambda_i^m)_{i=0,...,N-1})$ in $L^1(I)^p$ is bounded, then r is admissible and strongly extremal relaxed for the relaxed problem. In the absence of state constraints, if a subsequence $(w_{n(k),k}^{m(k)})_{k \in K}$ converges to some r in R, then r is strongly extremal relaxed for the relaxed problem.
(ii) In case (i) with state constraints, suppose that the relaxed problem has no admissible, abnormal extremal, controls. If r is admissible, then the sequence of multipliers is bounded and r is also strongly extremal relaxed for the relaxed problem.

If the additional assumptions of Theorem 4 (iii) are also satisfied, then Algorithms 1 and 2 compute *optimal* controls (in case (i), for Algorithm 1).

5   Numerical Examples

Example 1. Set $I := [0, 1]$. Define the reference state and control

$$\bar{y}_1(t) := e^{-t}, \ \bar{y}_2(t) := e^{-2t}, \ \bar{y}_3(t) := e^{-3t}, \ \bar{w}(t) := \min(1, -1 + 2.5t),$$

and consider the following optimal control problem, with state equations

$$y_1' = -y_1 + y_3 - e^{-3t} + \sin y_1 - \sin \bar{y}_1 + w_1 - \bar{w}, \ t \in I,$$
$$y_2' = y_1 - 2y_2 - e^{-t} + w_2 - \bar{w}, \ t \in I,$$
$$y_3' = y_2 - 3y_3 - e^{-2t} + w_3 - \bar{w}, \ t \in I,$$
$$y_1(0) = y_2(0) = y_3(0) = 1,$$

control constraint set $U = [-1, 1]$, and cost functional

$$G_0(w) := 0.5 \int_0^1 \left\{ \sum_{i=1}^3 [(y_i - \bar{y}_i)^2 + (w_i - \bar{w})^2] \right\} dt.$$

The optimal control and state are clearly $w^* := (\bar{w}, \bar{w}, \bar{w})$ and $y^* := (\bar{y}_1, \bar{y}_2, \bar{y}_3)$. In all these examples, the midpoint scheme was used with smallest step size $h = 1/500$. Algorithm 1, without penalty, was applied here with $\gamma = 0.5$ and initial control zero. After 15 iterations in k, we obtained

$$G_0(w_{nk}) = 1.366 \cdot 10^{-12}, \ d_k = -4.979 \cdot 10^{-15}, \ \varepsilon_k = 4.316 \cdot 10^{-7}, \ \zeta_k = 4.789 \cdot 10^{-7},$$

where d_k was defined in Step 2 of Algorithm 1, ε_k is the max-error for the state at the nodes, and ζ_k the max-error for the control at the midpoints of the intervals.

Example 2. With the same state equations, cost, and parameters as in Example 1, but with constraint set $U = [-0.75, 1]$, additional pointwise state constraints

$$G_1(w)(t) := 0.8 - 0.4t - y_1(t) \leqslant 0, \quad t \in I,$$

and applying the penalized Algorithm 1, we obtained after 85 iterations in k

$$G_0(w_{nk}^m) = 3.765340220 \cdot 10^{-3}, \quad d_k = -4.502 \cdot 10^{-6},$$

and the maximum state constraint violation

$$\theta_k := \max_i [\max(0, \ 0.8 - 0.4t_{ni} - y_{1nik}^m)] = 1.444 \cdot 10^{-5}.$$

Example 3. Consider the following problem, with state equations

$$y_1' = -y_2 + w_1, \quad y_2' = -y_1 + w_2, \ t \in [0, 0.5),$$
$$y_1' = -y_2 + w_1 - t + 0.5, \quad y_2' = -y_1 + w_2 - t + 0.5, \ t \in [0.5, 1],$$
$$y_1(0) = y_2(0) = 1,$$

nonconvex control constraint set

$$U := \{(u_1, 0) \in \mathbb{R}^2 \,|\, 0 \leqslant u_1 \leqslant 1\} \cup \{(0, u_2) \in \mathbb{R}^2 \,|\, 0 \leqslant u_2 \leqslant 1\},$$

and cost functional

$$G_0(w) := 0.5 \int_0^1 [(y_1 - \bar{y})^2 + (y_2 - \bar{y})^2] dt,$$

where $\bar{y} = e^{-t}$. Clearly, the unique optimal relaxed control is

$$r^*(t) := \delta_{(0,0)}, \ t \in [0, 0.5) \text{ (classical)}$$
$$r^*(t) := 2(1-t)\delta_{(0,0)} + (t-0.5)\delta_{(1,0)} + (t-0.5)\delta_{(0,1)}, \ t \in [0.5, 1] \text{ (non-classical)},$$

where δ denotes Dirac measures, which yields the optimal state $(y_1^*, y_2^*) := (\bar{y}, \bar{y})$, and cost $G_0(r^*) := 0$. The application of Algorithm 2, without penalty and with initial control zero, yielded after 120 iterations in k the results

$$G_0(r_{nk}) = 2.552 \cdot 10^{-6}, \quad d_k = -3.393 \cdot 10^{-5}.$$

Example 4. Consider the following problem, with state equation

$$y' = -y + w, \quad t \in I, \quad y(0) = 1,$$

control constraint set $U = [-1, 1]$, pointwise and terminal state constraints

$$G_1(w)(s) := 0.3 - y(s) \leqslant 0, \quad s \in I, \quad G_2(w) := y(1) - 0.5 = 0,$$

and *nonconvex* cost functional

$$G_0(w) := \int_0^1 (0.5y^2 - w^2) dt.$$

Writing the solution of the state equation in closed form via the fundamental solution, first forward with initial condition $y(0) = 1$, and then backward with terminal condition $y(1) = 0.5$, we can see that the unique optimal relaxed control and state are

$$r^*(t) = \begin{cases} \delta_{-1}, \ t \in [0, \rho) \text{ (classical)} \\ 0.35\delta_{-1} + 0.65\delta_1, \ t \in [\rho, \sigma) \text{ (non - classical)} \\ \delta_1, \ t \in [\sigma, 1] \text{ (classical)} \end{cases}$$

$$y^*(t) = \begin{cases} -1 + 2e^{-t}, \ t \in [0, \rho) \\ 0.3, \ t \in [\rho, \sigma) \\ 1 - 0.5e^{1-t}, \ t \in [\sigma, 1] \end{cases}$$

where $\rho := -\ln 0.65 \approx 0.43$ is such that $-1 + 2e^{-\rho} = 0.3$, and $\sigma = 1 - \ln 1.4 \approx 0.66$ such that $1 - 0.5e^{1-\sigma} = 0.3$. The optimal cost is $G_0(r^*) \approx -0.868398913624$. The application of the penalized Algorithm 2 and with initial control -1 yielded after 200 iterations in k the results

$$G_0(r_{nk}^m) = -0.868200567558, \quad \max_i[\max(0, 0.3 - y_{1nik}^m)] = 2.962 \cdot 10^{-4},$$
$$G_2(r_{nk}^m) = -9.732 \cdot 10^{-6}, \quad d_k = -1.236 \cdot 10^{-3}.$$

References

1. Azhmyakov, V., Schmidt, W.: Approximations of Relaxed Optimal Control Problems. J. of Optimization Theory and Applications 130(1), 61–77 (2006)
2. Chryssoverghi, I., Bacopoulos, A.: Discrete approximation of relaxed optimal control problems. J. of Optimization Theory and Applications 65(3), 395–407 (1990)
3. Chryssoverghi, I., Coletsos, J., Kokkinis, B.: Approximate relaxed descent method for optimal control problems. Control and Cybernetics 30(4), 385–404 (2001)
4. Chryssoverghi, I.: Discrete gradient projection method with general Runge-Kutta schemes and control parameterizations for optimal control problems. Control and Cybernetics 34, 425–451 (2005)
5. Chryssoverghi, I., Coletsos, J., Kokkinis, B.: Discretization methods for nonconvex optimal control problems with state constraints. Numer. Funct. Anal. Optim. 26, 321–348 (2005)
6. Chryssoverghi, I., Coletsos, J., Kokkinis, B.: Discretization methods for optimal control problems with state constraints. J. of Computional and Applied Mathematics 191, 1–31 (2006)
7. Chryssoverghi, I.: Discretization methods for semilinear parabolic optimal control problems. Intern. J. of Numerical Analysis and Modeling 3, 437–458 (2006)
8. Roubíček, T.: Relaxation in Optimization Theory and Variational Calculus. Walter de Gruyter, Berlin (1997)
9. Warga, J.: Optimal control of Differential and Functional Equations. Academic Press, New York (1972)
10. Warga, J.: Steepest descent with relaxed controls. SIAM J. Control Optim., 674–682 (1977)

On the Asymptotic Stabilization
of an Uncertain Bioprocess Model*

Neli S. Dimitrova and Mikhail I. Krastanov

Institute of Mathematics and Informatics
Bulgarian Academy of Sciences
Acad. G. Bonchev Str. Bl. 8, 1113 Sofia, Bulgaria
nelid@bio.bas.bg, krast@math.bas.bg

Abstract. We study a nonlinear model of a biological digestion process, involving two microbial populations and two substrates and producing biogas (methane). A feedback control law for asymptotic stabilization of the closed-loop system is proposed. An extremum seeking algorithm is applied to stabilize the system towards the maximum methane flow rate.

1 Introduction

We consider a model of a continuously stirred tank bioreactor presented by the following nonlinear system of ordinary differential equations [4], [7], [8], [10], [11]:

$$\frac{ds_1}{dt} = u(s_1^i - s_1) - k_1 \mu_1(s_1)x_1 \tag{1}$$

$$\frac{dx_1}{dt} = (\mu_1(s_1) - \alpha u)x_1 \tag{2}$$

$$\frac{ds_2}{dt} = u(s_2^i - s_2) + k_2 \mu_1(s_1)x_1 - k_3 \mu_2(s_2)x_2 \tag{3}$$

$$\frac{dx_2}{dt} = (\mu_2(s_2) - \alpha u)x_2 \tag{4}$$

with output

$$Q = k_4 \mu_2(s_2)x_2. \tag{5}$$

The state variables s_1, s_2 and x_1, x_2 denote substrate and biomass concentrations, respectively: s_1 represents the organic substrate, characterized by its chemical oxygen demand (COD), s_2 denotes the volatile fatty acids (VFA), x_1 and x_2 are the acidogenic and methanogenic bacteria respectively. The parameter $\alpha \in (0, 1]$ represents the proportion of bacteria that are affected by the dilution.

---

* This research has been partially supported by the Bulgarian Science Fund under grant Nr. DO 02–359/2008 and by the bilateral project "Variational Analysis and Applications" between the Bulgarian Academy of Sciences and the Izrael Academy of Sciences.

I. Lirkov, S. Margenov, and J. Waśniewski (Eds.): LSSC 2011, LNCS 7116, pp. 115–122, 2012.

The constants k_1, k_2 and k_3 are yield coefficients related to COD degradation, VFA production and VFA consumption respectively; k_4 is a coefficient.

It is assumed that the input substrate concentrations s_1^i and s_2^i are constant and the methane flow rate Q is a measurable output. The dilution rate u is considered as a control input.

The functions $\mu_1(s_1)$ and $\mu_2(s_2)$ model the specific growth rates of the microorganisms. Following [11] we impose the following assumption on μ_1 and μ_2:

Assumption A1: $\mu_j(s_j)$ is defined for $s_j \in [0, +\infty)$, $\mu_1(s_1^i) \geq \mu_2(s_2^i)$, $\mu_j(0) = 0$, $\mu_j(s_j) > 0$ for $s_j > 0$; $\mu_j(s_j)$ is continuously differentiable and bounded for all $s_j \in [0, +\infty)$, $j = 1, 2$.

This model has been investigated in [10], [11], where a controller for regulating the effluent COD is proposed and its robustness is illustrated by a simulation study.

The main goal of the paper is to construct a feedback control law based on online measurements for asymptotic stabilization of the system (1)–(4). Then, by means of a numerical extremum seeking algorithm, the closed loop system is steered to that equilibrium point where the maximum methane output is achieved among all other equilibrium points.

2 Asymptotic Stabilization

Define $s^i := \dfrac{k_2}{k_1} s_1^i + s_2^i$ and let the following assumption be satisfied:

Assumption A2: Lower bounds s^{i-} and k_4^- for the values of s^i and k_4, as well as an upper bound k_3^+ for the value of k_3 are known.

Consider the control system (1)–(4) in the state space $p = (s_1, x_1, s_2, x_2)$ and define the following feedback control law:

$$k(p) := \beta \, k_4 \, \mu_2(s_2) \, x_2 \quad \text{with} \quad \beta \in \left(\frac{k_3^+}{s^{i-} \cdot k_4^-}, +\infty \right). \tag{6}$$

The feedback k depends only on β and Q, i. e. $k = k(\beta, Q) = \beta \cdot Q$.

Obviously, the number $\bar{s} := s^i - \dfrac{k_3}{\beta k_4}$ belongs to the interval $(0, s^i)$.

Denote by Σ the closed-loop system obtained from (1)–(4) by substituting the control variable u by the feedback $k(p)$.

Assumption A3. There exists a point $\bar{s}_1$ such that

$$\mu_1(\bar{s}_1) = \mu_2 \left(\bar{s} - \frac{k_2}{k_1} \bar{s}_1 \right), \quad \bar{s}_1 \in (0, s_1^i).$$

The above Assumption A3 is called in [7] regulability of the system.

Define

$$\bar{s}_2 = \bar{s} - \frac{k_2}{k_1} \bar{s}_1, \quad \bar{x}_1 = \frac{s_1^i - \bar{s}_1}{\alpha k_1}, \quad \bar{x}_2 = \frac{1}{\alpha \beta k_4}. \tag{7}$$

It is straightforward to see that the point

$$\bar{p} := (\bar{s}_1, \bar{s}_2, \bar{x}_1, \bar{x}_2)$$

is an equilibrium point for the system (1)–(4). We shall prove below that the feedback law (6) asymptotically stabilizes the closed-loop system Σ to $\bar{p}$.
  Denote

$$s := \frac{k_2}{k_1} s_1 + s_2$$

and define the following sets

$$\Omega_0 = \{(s_1, x_1, s_2, x_2)| \ s_1 > 0, \ x_1 > 0, \ s_2 > 0, \ x_2 > 0\},$$

$$\Omega_1 = \left\{(s_1, x_1, s_2, x_2)| \ s_1 + k_1 x_1 \leq \frac{s_1^i}{\alpha}, \ s + k_3 x_2 \leq \frac{s^i}{\alpha}\right\},$$

$$\Omega_2 = \left\{\left(s_1, x_1, \bar{s} - \frac{k_2}{k_1} s_1, \bar{x}_2\right)| \ 0 < s_1 < \frac{k_1}{k_2}\bar{s}, \ x_1 > 0\right\},$$

$$\Omega = \Omega_0 \cap \Omega_1.$$

Assumption A4. Let $\mu_1'(s_1) + \dfrac{k_2}{k_1}\mu_2'\left(\bar{s} - \dfrac{k_2}{k_1} s_1\right) > 0$ be satisfied on $\Omega \cap \Omega_2$.

  Assumption A4 is technical and is used in the proof of the main result. It will be discussed in more details later in Section 4, where the growth rates μ_1 and μ_2 are specified as the Monod and the Haldane laws and numerical values for the model coefficient are introduced.

Theorem 1. *Let Assumptions A1, A2, A3 and A4 be satisfied. Let us fix an arbitrary number $\beta \in \left(\dfrac{k_3^+}{s^{i-} \cdot k_4^-}, +\infty\right)$ and let $\bar{p} = (\bar{s}_1, \bar{x}_1, \bar{s}_2, \bar{x}_2)$ be the corresponding equilibrium point. Then the feedback control law $k(\cdot)$ defined by (6) stabilizes asymptotically the control system (1)–(4) to the point $\bar{p}$ for each starting point p_0 from the set Ω_0.*

Proof. Let us fix an arbitrary point $p_0 \in \Omega_0$ and a positive value $u_0 > 0$ for the control. According to Lemma 1 from [7] there exists $T > 0$ such that the value of the corresponding trajectory of (1)–(4) for $t = T$ belongs to the set Ω. Hence the corresponding trajectory of (1)–(4) starting from the point p_0 enters the set Ω after a finite time. Moreover one can directly check that each trajectory staring from a point from the set Ω remains in Ω. For that reason we shall consider the control system (1)–(4) only on the set Ω.
  Let us remind that by Σ we have denoted the closed-loop system obtained from (1)–(4) by substituting the control variable u by the feedback $k(p)$. Then one can directly check that the following ordinary differential equations

$$\begin{aligned}
\frac{ds}{dt} &= -\beta k_4 \mu_2(s_2) x_2 (s - \bar{s})\\
\frac{dx_2}{dt} &= -\alpha\beta k_4 \mu_2(s_2) x_2 (x_2 - \bar{x}_2)
\end{aligned} \tag{8}$$

are satisfied. Let us define the function

$$V(p) = (s - \bar{s})^2 + (x_2 - \bar{x}_2)^2.$$

Clearly, the values of this function are nonnegative. If we denote by $\dot{V}(p)$ the Lie derivative of V with respect to the right-hand side of (8) then for each point p of Ω,

$$\dot{V}(p) = -\beta k_4 \mu_2(s_2) x_2 (s - \bar{s})^2 - \alpha \beta k_4 \mu_2(s_2) x_2 (x_2 - \bar{x}_2)^2 \leq 0.$$

Applying LaSalle's invariance principle (cf. [9]), it follows that every solution of Σ starting from a point of Ω is defined on the interval $[0, +\infty)$ and approaches the largest invariant set with respect to Σ, which is contained in the set Ω_∞, where Ω_∞ is the closure of the set $\Omega \cap \Omega_2$. It can be directly checked that the dynamics of Σ on Ω_∞ is described by the system

$$\frac{ds_1}{dt} = \frac{1}{\alpha} \chi(s_1)(s_1^i - s_1) - k_1 \mu_1(s_1) x_1$$

$$\frac{dx_1}{dt} = (\mu_1(s_1) - \chi(s_1)) x_1,$$

where $\chi(s_1) := \mu_2 \left(\bar{s} - \dfrac{k_2}{k_1} s_1 \right)$ (remind that $\bar{s} := s^i - \dfrac{k_3}{\beta k_4}$). According to (7) we have that $\bar{s} = \dfrac{k_2}{k_1} \bar{s}_1 + \bar{s}_2$ and $s_1^i = \bar{s}_1 + \alpha k_1 \bar{x}_1$. Then the dynamics of Σ on the set Ω_∞ can be written as follows:

$$\frac{ds_1}{dt} = -\frac{1}{\alpha} \chi(s_1) \cdot (s_1 - \bar{s}_1 + \alpha k_1 (x_1 - \bar{x}_1)) - k_1 (\mu_1(s_1) - \chi(s_1)) \cdot x_1$$

$$\frac{dx_1}{dt} = (\mu_1(s_1) - \chi(s_1)) \cdot x_1.$$

Consider the function

$$W(s_1, x_1) = (s_1 - \bar{s}_1 + \alpha k_1 (x_1 - \bar{x}_1))^2 + \alpha(1 - \alpha) k_1^2 (x_1 - \bar{x}_1)^2. \qquad (9)$$

This function takes nonnegative values. It can be directly checked that for each point $\left(s_1, x_1, \bar{s} - \frac{k_2}{k_1} s_1, \bar{x}_2 \right)$ of the set Ω_∞,

$$\dot{W}(s_1, x_1) = -\frac{2}{\alpha} \chi(s_1)(s_1 - \bar{s}_1 + \alpha k_1 (x_1 - \bar{x}_1))^2$$
$$- 2(1 - \alpha) k_1 x_1 (s_1 - \bar{s}_1)(\mu_1(s_1) - \chi(s_1)).$$

Assumptions A3 and A4 imply

$$\mu_1(s_1) - \chi(s_1) = \mu_1(s_1) - \mu_2 \left(\bar{s} - \frac{k_2}{k_1} s_1 \right) = \mu_1(s_1) - \mu_2 \left(\bar{s}_2 - (s_1 - \bar{s}_1) \frac{k_2}{k_1} \right)$$

$$= \mu_1(\bar{s}_1) + \int_{\bar{s}_1}^{s_1} \mu_1'(\theta) \, d\theta - \mu_2(\bar{s}_2) + \frac{k_2}{k_1} \int_{\bar{s}_1}^{s_1} \mu_2' \left(\bar{s}_2 - (\theta - \bar{s}_1) \frac{k_2}{k_1} \right) d\theta$$

$$= \int_{\bar{s}_1}^{s_1} \left(\mu_1'(\theta) + \frac{k_2}{k_1} \mu_2' \left(\bar{s}_2 - (\theta - \bar{s}_1) \frac{k_2}{k_1} \right) \right) d\theta,$$

and therefore
$$\dot{W}(s_1, x_1) \le 0 \qquad (10)$$

for each point $\left(s_1, x_1, \bar{s} - \frac{k_2}{k_1} s_1, \bar{x}_2\right)$ from the set Ω_∞.

To complete the proof we use a refinement of LaSalle's invariance principle recently obtained in [2] (cf. also [6] and [12], where similar stabilizability problems are studied). Denote by $\phi(t, p)$ the value of the trajectory of the closed-loop system Σ at time t starting from the point $p \in \Omega$. The positive limit set (or ω-limit set) of the solution $\phi(t, p)$ of the closed-loop system Σ is defined as

$$L^+(p) = \{\tilde{p} : \text{ there exists a sequence } \{t_n\} \to \infty \text{ with } \tilde{p} = \lim_{t_n \to +\infty} \phi(t_n, p)\}.$$

Let us fix an arbitrary point $p_0 \in \Omega$. The invariance of the set Ω with respect to the trajectories of Σ and the LaSalle invariance principle imply that the ω-limit set $L^+(p_0)$ is a nonempty connected invariant subset of Ω_∞.

Now consider again the function $W(\cdot, \cdot)$ defined by (9). The restriction of the Lie derivative $W(\cdot, \cdot)$ on Ω_∞ is semidefinite, meaning $\dot{W}(s_1, x_1) \le 0$ for each point $(s_1, x_1, \bar{s} - \frac{k_2}{k_1} s_1, \bar{x}_2) \in \Omega_\infty$. The proof of Theorem 6 from [2] implies that $L^+(p_0)$ is contained in one connected component of the set $L^\infty := \{(s_1, x_1, \bar{s} - \frac{k_2}{k_1} s_1, \bar{x}_2) \in \Omega_\infty : \dot{W}(s_1, x_1) = 0\}$. Taking into account (10) and Assumption A1, one can obtain that $L^\infty = \{\bar{p}\}$, and hence $L^+(p_0) = \{\bar{p}\}$. Moreover, one can verify that $\bar{p}$ is a Lyapunov stable equilibrium point for the closed-loop system Σ. This completes the proof.

3   Extremum Seeking

Let the assumptions A1, A2, A3 and A4 hold true. Denote by $\beta \in \left(\frac{k_3^+}{s^{i-} \cdot k_4^-}, +\infty\right)$ some constant. Consider $\bar{p}_\beta = (\bar{s}_1, \bar{x}_1, \bar{s}_2, \bar{x}_2)$ where $\bar{s}_1$, $\bar{x}_1$, $\bar{s}_2$ and $\bar{x}_2$ are computed according to (7). Assume further that the static characteristic

$$Q(\bar{p}_\beta) = k_4\, \mu_2(\bar{s}_2)\, \bar{x}_2,$$

which is defined on the set of all steady states $\bar{p}_\beta$ has a maximum at a unique steady state point

$$p_{\beta_*}^{\max} = (s_1^*, x_1^*, s_2^*, x_2^*),$$

that is $Q_{\max} := Q(p_{\beta_*}^{\max})$.

Our goal now is to stabilize the dynamic system towards the (unknown) maximum methane flow rate $Q_{\max}$. We apply the feedback control law

$$(Q, \beta) \longmapsto k(Q, \beta) = \beta \cdot Q. \qquad (11)$$

According to Theorem 1, this feedback will asymptotically stabilize the control system (1)–(4) to the point $\bar{p}_\beta$.

To stabilize the dynamics (1)–(4) towards $Q_{\max}$ by means of the feedback (11), we use an iterative extremum seeking algorithm. This algorithm is presented in details in [5] and applied to a two-dimensional bioreactor model with adaptive feedback. The algorithm can easily be adapted for the model considered here. The main idea of the algorithm is based on the fact that Theorem 1 is valid for any value of $\beta > \dfrac{k_3^+}{s^{i-} \cdot k_4^-}$. Thus we can construct a sequence of points $\beta^1, \beta^2, \ldots, \beta^n, \ldots$, converging to β_*, and generate in a proper way a sequence of values $Q^1, Q^2, \ldots, Q^n, \ldots$ which converges to $Q_{\max}$. The algorithm is carried out in two stages: on Stage I, an interval $[\beta] = [\beta^-, \beta^+]$ is found such that $[\beta] > \dfrac{k_3^+}{s^{i-} \cdot k_4^-}$ and $\beta_* \in [\beta]$; on Stage II, the interval $[\beta]$ is refined using an elimination procedure based on a Fibonacci search technique. Stage II produces the final interval $[\bar{\beta}] = [\bar{\beta}^-, \bar{\beta}^+]$ such that $\beta_* \in [\bar{\beta}]$ and $\bar{\beta}^+ - \bar{\beta}^- \leq \varepsilon$, where the tolerance $\varepsilon > 0$ is specified by the user.

4   Numerical Simulation

In the computer simulation, we consider for $\mu_1(s_1)$ and $\mu_2(s_2)$ the Monod and the Haldane model functions for the specific growth rates, which are used in the original model [1], [3], [4], [7], [8]:

$$\mu_1(s_1) = \frac{\mu_m s_1}{k_{s_1} + s_1}, \qquad \mu_2(s_2) = \frac{\mu_0 s_2}{k_{s_2} + s_2 + \left(\dfrac{s_2}{k_I}\right)^2}. \qquad (12)$$

Here μ_m, k_{s_1}, μ_0, k_{s_2} and k_I are kinetic coefficients. Obviously, $\mu_1(s_1)$ and $\mu_2(s_2)$ satisfy Assumption A1: $\mu_1(s_1)$ is monotone increasing and bounded by μ_m; there is a point $\tilde{s}_2$ such that $\mu_2(s_2)$ achieves its maximum at $\tilde{s}_2 = k_I \sqrt{k_{s_2}}$. Simple derivative calculations imply that if $\bar{s}$ is chosen such that $0 < \bar{s} \leq \tilde{s}_2$ then $\mu_2'\left(\bar{s} - \frac{k_2}{k_1} s_1\right) \geq 0$ holds true thus Assumption A4 is satisfied. Moreover, if the point $\bar{s}$ is sufficiently small, then Assumptions A3 and A4 are simultaneously satisfied.

Usually the formulation of the growth rates is based on experimental results, and therefore it is not possible to have an exact analytic form of these functions, but only some quantitative bounds. Assume that we know bounds for $\mu_1(s_1)$ and $\mu_2(s_2)$, i. e.

$$\mu_j(s_j) \in [\mu_j(s_j)] = [\mu_j^-(s_j), \mu_j^+(s_j)] \text{ for all } s_j \geq 0, \quad j = 1, 2.$$

This uncertainty can be simulated by assuming in (12) that instead of exact values for the coefficients μ_m, k_{s_1}, μ_0, k_{s_2} and k_I we have compact intervals for them: $\mu_m \in [\mu_m]$, $k_{s_1} \in [k_{s_1}]$, $\mu_0 \in [\mu_0]$, $k_{s_2} \in [k_{s_2}]$, $k_I \in [k_I]$. Then any $\mu_j(s_j) \in [\mu_j(s_j)]$, $j = 1, 2$, satisfies Assumption 1; it follows also that there exist intervals for the kinetic coefficients, such that Assumption 4 is satisfied for any $\mu_j(s_j) \in [\mu_j(s_j)]$, $j = 1, 2$. Such intervals are for example the following:

$$[\mu_m] = [1.2,\ 1.4], \qquad [k_{s_1}] = [6.5,\ 7.2],$$
$$[\mu_0] = [0.64,\ 0.84], \qquad [k_{s_2}] = [9,\ 10.28], \quad [k_I] = [15,\ 17].$$

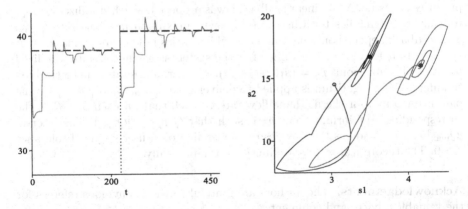

Fig. 1. Time evolution of $Q(t)$ (left) and a trajectory in the (s_1, s_2)-phase plane (right)

To simulate Assumption A2 we assume intervals for the coefficients k_j to be given, i. e. $k_j \in [k_j] = [k_j^-, k_j^+]$, $j = 1, 2, 3, 4$. Such numerical intervals are

$$[k_1] = [9.5, 11.5], \quad [k_2] = [27.6, 29.6], \quad [k_3] = [1064, 1084], \quad [k_4] = [650, 700].$$

All above intervals are chosen to enclose the numerical coefficients values derived by experimental measurements [1]; the values $\alpha = 0.5$, $s_1^i = 7.5$, $s_2^i = 75$ are also taken from [1]. In this case the inequality $\mu_1^-(s_1^i) > \mu_2^+(s_2^i)$ is fulfilled, which means that $\mu_1(s_1^i) > \mu_2(s_2^i)$ for any $\mu_j \in [\mu_j]$, $j = 1, 2$, thus Assumption A1 is satisfied.

In the simulation process we proceed in the following way. At the initial time $t_0 = 0$ we take random values for the coefficients from the corresponding intervals. We apply the extremum seeking algorithm to stabilize the system towards $Q_{\max}$. Then, at some time $t_1 > t_0$, we choose another set of random coefficient values and repeat the process; thereby the last computed values for the phase variables (s_1, x_1, s_2, x_2) are taken as initial conditions.

The left plot in Figure 1 shows the time profile of $Q(t)$; there the vertical dot-line segment marks the time moment t_1, when the new coefficients values are taken in a random way from the corresponding intervals. The horizontal dash-line segments go through $Q_{\max}$. The "jumps" in the graph correspond to the different choices of β by executing the algorithmic steps. The right plot in Figure 1 shows a projection of the trajectory in the phase plane (s_1, s_2); the empty circle denotes the initial point $(s_1(0), s_2(0))$. It is well seen how the trajectory consecutively approaches the two steady states (solid circles), corresponding to the different choice of the coefficient values.

5   Conclusion

The paper is devoted to the stabilization of a four-dimensional nonlinear dynamic system, which models anaerobic degradation of organic wastes and produces methane. A nonlinear feedback law is proposed, which stabilizes asymptotically the dynamics towards the (unknown) maximum methane rate $Q_{\max}$. The feedback depends on a parameter β, which varies in known bounds. First it is shown that for any chosen value of β, the system is asymptotically stabilized to an equilibrium point $\bar{p}_\beta = (\bar{s}_1, \bar{x}_1, \bar{s}_2, \bar{x}_2)$. Further, an iterative numerical extremum seeking algorithm is applied to deliver bounds $[\beta_m]$ and $[Q^{\max}]$ for the parameter β and for the methane flow rate Q, such that for each $\beta \in [\beta_m]$, the corresponding equilibrium point $\bar{p}_\beta$ is such that $Q(\bar{p}_\beta) \in [Q_{\max}]$. The interval $[Q_{\max}]$ can be made as tight as desired depending on a user specified tolerance $\varepsilon > 0$. The theoretical results are illustrated numerically.

Acknowledgements. The authors are grateful to the anonymous referees for the valuable advices and comments.

References

1. Alcaraz-González, V., Harmand, J., Rapaport, A., Steyer, J., González–Alvarez, V., Pelayo-Ortiz, C.: Software sensors for highly uncertain WWTPs: a new apprach based on interval observers. Water Research 36, 2515–2524 (2002)
2. Arsie, A., Ebenbauer, C.: Locating omega-limit sets using height functions. Journal of Differential Equations 248(10), 2458–2469 (2010)
3. Bernard, O., Hadj-Sadok, Z., Dochain, D.: Advanced monitoring and control of anaerobic wastewate treatment plants: dynamic model development and identification. In: Proceedings of Fifth IWA Inter. Symp. WATERMATEX, Gent, Belgium, pp. 3.57–3.64 (2000)
4. Bernard, O., Hadj-Sadok, Z., Dochain, D., Genovesi, A., Steyer, J.-P.: Dynamical model development and parameter identification for an anaerobic wastewater treatment process. Biotechnology and Bioengineering 75, 424–438 (2001)
5. Dimitrova, N., Krastanov, M.I.: Nonlinear Adaptive Control of a Model of an Uncertain Fermentation Process. Int. Journ. of Robust and Nonlinear Control 20, 1001–1009 (2010)
6. El-Hawwary, M.I., Maggiore, M.: A Reduction Principles and the Stabilization of Closed Sets for Passive Systems. arXiv:0907.0686vl [mathOC], 1–25 (2009)
7. Grognard, F., Bernard, O.: Stability analysis of a wastewater treatment plant with saturated control. Water Science Technology 53, 149–157 (2006)
8. Hess, J., Bernard, O.: Design and study of a risk management criterion for an unstable anaerobic wastewater treatment process. Journal of Process Control 18(1), 71–79 (2008)
9. Khalil, H.: Nonlinear Systems. Macmillan Publishing Company, New York (1992)
10. Maillert, L., Bernard, O.: A simple robust controller to stabilize an anaerobic digestion processes. In: Proc. of the 8th Conference on Computer Applications in Biotechnology, Quebec, pp. 213–218 (2001)
11. Maillert, L., Bernard, O., Steyer, J.-P.: Robust regulation of anaerobic digestion processes. Water Science and Technology 48(6), 87–94 (2003)
12. Mischaikow, K., Smith, H., Tieme, H.: Asymptotically autonomous semiflows: chain recurence and Lyapunov functions. TAMS 347(5), 1669–1685 (1995)

Reachable Sets of Impulsive Control System with Cone Constraint on the Control and Their Estimates

Tatiana F. Filippova and Oksana G. Matviychuk

Department of Optimal Control, Institute of Mathematics and Mechanics,
Russian Academy of Sciences,
16 S. Kovalevskaya str., Ekaterinburg 620990, Russia
{ftf,vog}@imm.uran.ru
http://www.imm.uran.ru

Abstract. The problem of estimating reachable sets of linear measure driven (impulsive) dynamical control system with uncertainty in initial data is considered. It is assumed that the impulsive controls in the dynamical system belong to the intersection of a special cone with a generalized ellipsoid both taken in the space of functions of bounded variation. The algorithms for constructing the external ellipsoidal estimates of reachable sets for such control systems are given. Numerical simulation results relating to the proposed procedures are also discussed.

1 Introduction

One of the key issues in the optimal control theory [2, 7, 9–11, 17] and in the theory of differential games [8] is the construction and the study of reachable (attainable) sets for a dynamical system, i.e., the sets of all the states of the space to which the point can be taken from the initial state (or from the set of initial states) in a prescribed time with the help of admissible controls. A complete theory based on the technique of ellipsoidal calculus [2, 12] is presently developed for the construction of external and internal estimates of reachable sets for control systems described by linear differential equations with classical (measurable) controls. In the framework of this approach, the principal problem consists in finding in the state space an ellipsoid (or a family of ellipsoids) that provides outer or inner (with respect to the operation of inclusion of sets) estimates for the required reachable set.

In [4, 5, 13, 20], schemes of constructing set-valued estimates of reachable sets of impulsive control systems [1, 3, 14, 15, 18, 21] based on ideas and methods of ellipsoidal calculus were considered. In this case, the constraint on control impulsive functions was given in the form of a special generalized "ellipsoid" in the space of functions of bounded variation. Another approach based on polyhedral estimation techniques for impulsive control systems was presented in [6] where internal and external polyhedral approximations of reachable sets were given.

In the present paper, we assume that impulsive controls in the linear dynamical system have to belong to the intersection of a special cone with a generalized

I. Lirkov, S. Margenov, and J. Waśniewski (Eds.): LSSC 2011, LNCS 7116, pp. 123–130, 2012.

ellipsoid both taken in the space of functions of bounded variation. In particular, under such restriction vectors of impulsive jumps of admissible controls are taken from a given finite-dimensional ellipsoid intersected with the cone. We study here the special case of such restricting cone when only nonnegative (in each coordinate) control measures are allowed in the system [1, 5, 9].

The properties and the special structure of solutions of differential systems with such impulsive controls are studied. New state estimation algorithms which allow to construct the ellipsoidal estimates for a convex envelope of the union of related ellipsoids of a finite dimensional space are introduced. An example of external ellipsoidal estimates of reachable sets of a linear impulsive control system is also given.

2   Problem Formulation

Let us start by introducing the following basic notations. Let R^n denote the n-dimensional Euclidean space. We denote by (x, y) the scalar product of vectors $x, y \in R^n$ and by $\|x\| = (x, x)^{1/2}$ the Euclidean norm of the vector x, the symbol $'$ stands for transposition, and $B(a, r) = \{x \in R^n : \|x - a\| \leq r\}$. Let $E(y, Y) = \{x \in R^n : (x - y)'Y^{-1}(x - y) \leq 1\}$ be the ellipsoid in R^n with center y and a symmetric positive definite $n \times n$-matrix Y, and let I be the identity $n \times n$-matrix.

Consider a dynamic control system described by a differential equation with impulsive control (measure) $u(\cdot)$:

$$dx(t) = A(t)x(t)dt + du(t), \quad x(t_0 - 0) = x_0, \quad t_0 \leq t \leq T, \tag{1}$$

or in the integral form,

$$x(t) = x(t; u(\cdot), x_0) = K(t, t_0)x_0 + \int_{t_0}^{t} K(t, \tau)du(\tau), \quad t_0 \leq t \leq T. \tag{2}$$

Here we assume that $A(t)$ is continuous $n \times n$ - matrix function, $\Phi(t)$ is the Cauchy matrix solution $\dot{\Phi} = A(t)\Phi$ ($\Phi(t_0) = I$), $K(t, \tau) = \Phi(t)\Phi^{-1}(\tau)$. We assume that $u(\cdot) \in V_p^n$ ($1 \leq p < \infty$) where V_p^n means the space of n-vector functions $u(\cdot)$ such that $u(t)$ is continuous from the right on $[t_0, T)$ with $u(t_0 - 0) = 0$ and

$$V_p[u(\cdot)] = \sup_{\{t_i | 0 = t_0 < \ldots < t_k = T\}} \sum_{i=1}^{k} \|u(t_i) - u(t_{i-1})\|_p < \infty$$

$$\|u\|_p = \left(\sum_{i=1}^{n} |u_i|^p \right)^{\frac{1}{p}}, \quad u = (u_1, \ldots, u_n).$$

Denote C_q^n the space of continuous n-vector functions $y(\cdot)$ with the norm

$$\|y(\cdot)\|_{\infty, q} = \max_{t_0 \leq t \leq T} \|y(t)\|_q.$$

It is well known that the space V_p^n is the dual of C_q^n ($V_p^n = C_q^{n*}$) where $p = 1$ if $q = \infty$, $p = \infty$ if $q = 1$ and $1 < p < \infty$ if $q = (1 - p^{-1})^{-1}$.

Let $E_0 = E(0, Q_0^{-1})$ be an ellipsoid in R^n with a center at the origin and with Q_0 being a given symmetric positive definite $n \times n$ matrix. Consider the so-called generalized "ellipsoid" [4, 19, 20] E in C_q^n :

$$E = \{y(\cdot) \in C_q^n \mid y(t) \in E_0 \ \forall \, t \in [t_0, T]\} \tag{3}$$

and its conjugate ellipsoid [16] $E^* \subset V_p^n$,

$$E^* = \{u(\cdot) \in V_p^m \mid \int_{t_0}^T (y(t))' du(t) \le 1 \ \forall y(\cdot) \in E\}. \tag{4}$$

Admissible impulsive controls $u(\cdot)$ in (1) are assumed to be restricted by a set U which is defined as follows. Denote

$$K_0 = R_+^n = \{u \in R^n \mid u = (u_1, \ldots, u_n), \ u_i \ge 0, \ i = 1, \ldots, n\},$$

$$K = \{y(\cdot) \in C_q^n \mid y(t) \in K_0 \ \forall \, t \in [t_0, T]\}$$

and take the conjugate cone [16]

$$K^* = \{u(\cdot) \in V_p^n \mid \int_{t_0}^T (y(t))' du(t) \ge 0 \ \forall y(\cdot) \in K\}. \tag{5}$$

Definition 1. *The function $u(\cdot) \in V_p^n$ will be called admissible control if $u(\cdot) \in U = E^* \cap K^*$.*

Remark 1. Let $u(\cdot)$ be a piecewise constant function on $[t_0, T]$ with discontinuity instants $\{t_i\}$ and with jump vectors

$$\Delta u(t_i) = u(t_{i+1}) - u(t_i) \in E(0, Q_0) \cap K_0$$

$$= \{z \in R^n \mid z' Q_0^{-1} z \le 1, \ z = (z_1, \ldots, z_n), \ z_i \ge 0, \ i = 1, \ldots, n\}. \tag{6}$$

Then $u(\cdot)$ is admissible.

We will assume also that the initial state x_0 to the system (1) is unknown but bounded with a given ellipsoidal bound,

$$x_0 \in X_0 = E(r, R) \tag{7}$$

where R is a symmetric positive definite $n \times n$ matrix and $r \in R^n$.

Denote

$$X(t; U, X_0) = \bigcup_{x_0 \in X_0} \bigcup_{u \in U} \{x(t; u(\cdot), x_0)\}.$$

Definition 2. *The set $X(T; U, X_0)$ is called reachable set at the instant T of the impulsive differential system (1) from the initial set X_0 under controls $u(\cdot) \in U$.*

So the main problem of the paper is to find the external estimates of ellipsoidal type for the reachable set $X(T; U, X_0)$ basing on the special ellipsoidal structure of sets X_0 and U.

3 Preliminaries

We mention here some auxiliary results describing the properties of $X(T; U, X_0)$.

Theorem 1. *The reachable set $X(T; U, X_0)$ is convex and compact in R^n. Every state vector $x \in X(T; U, X_0)$ may be generated by a solution $x(\cdot; u(\cdot), x_0)$ of (1) with $x(T; u(\cdot), x_0) = x$ where $x_0 \in X_0$ and a piecewise constant control $u(\cdot)$ has no more than $(n+1)$ jumps $\Delta u(t_i) \in E_0^* \cap K_0^*$. If the vector x belongs to the boundary of $X(T; U, X_0)$ then the number of related control jumps may be reduced to n.*

Proof. The proof follows from classical theorems of impulsive control theory [7, 9] and taking into account details related to cone and ellipsoidal constraints as it was done in [1, 4, 13, 19, 20]. □

For a set $A \subset R^n$ we denote its closed convex hull [16] as $\bar{\mathrm{co}}\, A$.

Theorem 2. *The following equalities hold*

$$X(T; U, E(r, R)) = E(r^*, R^*) + X(T; U, 0),$$

$$X(T; U, 0) = \bar{\mathrm{co}}\left(\bigcup_{\tau \in [t_0, T]} K(T, \tau)(E(0, Q_0) \cap K_0) \right), \tag{8}$$

where $r^* = K(T, t_0)r$, $R^* = K(T, t_0)R(K(T, t_0))'$.

Proof. The equalities (8) follow from the linearity of the system (2) and from the properties of solutions of impulsive systems formulated in Theorem 1. □

Remark 2. Assume that we find an ellipsoid E^+ such that $X(T; U, 0) \subseteq E^+$. Then applying well-known formulas for calculating the external ellipsoidal estimate for the sum of two ellipsoids $E(r^*, R^*)$ and E^+ [2, 12] we can find the resulting external ellipsoid for $X(T; U, X_0))$ in (8). So the main difficulty is in constructing the outer ellipsoidal estimate for $X(T; U, 0)$.

4 Main Results

In this section we apply the techniques of the ellipsoidal calculus for finding the external estimates of $X(T; U, 0)$ for which we will use the shorter notation $X(T) = X(T; U, 0)$.

The idea of constructing the external estimates for $X(T)$ is based on general results of ellipsoidal calculus [2, 12] and on the new procedures of external approximation of a closed convex hull of the union of a variety of some ellipsoids [4, 13, 19, 20].

The solution of the above problem will be found as a result of several sequential steps.

Theorem 3. *For any $i = 1, \ldots, n$ the following inclusion holds*

$$\{z \in R^n \mid z'Q_0^{-1}z \leq 1, \; z = (z_1, \ldots, z_n), \; z_i \geq 0\} \subseteq E(a_i, Q_i) \qquad (9)$$

with

$$a_i = (n+1)^{-1}((Q_0)_{ii})^{-1/2}Q_0^{(i)},$$
$$Q_i = (n^{-2}(n^2-1)Q_0^{-1} + 2(n+1)n^{-2}((Q_0)_{ii})^{-1}V^{(i)})^{-1}, \qquad (10)$$

where $Q_0^{(i)}$ and $(Q_0)_{ii}$ denote, respectively, the i-column and (i,i)-element of $n \times n$-matrix Q_0, $V^{(i)}$ is the $n \times n$-matrix with elements $V_{sk}^{(i)}$ $(1 \leq s, k \leq n)$ such that $V_{sk}^{(i)} = 1$ if $s = k = i$, otherwise $V_{sk}^{(i)} = 0$.

Proof. The proof of this theorem follows from the properties of ellipsoids and the result ([2], pp.224-226). □

Directly from Theorem 3 we come to the estimate.

Corollary 1. *The following external estimate is valid*

$$(E(0, Q_0) \cap K_0) \subseteq \bigcap_{1 \leq i \leq n} E(a_i, Q_i), \qquad (11)$$

where a_i, Q_i are defined by formulas (10).

Theorem 4. *For any parameter vector $\alpha = (\alpha_1, \ldots, \alpha_n)$ with $\alpha_i \geq 0$ and $\sum_{i=1}^n \alpha_i = 1$ the following inclusion holds*

$$(E(0, Q_0) \cap K_0) \subseteq E(a_\alpha, Q_\alpha) \qquad (12)$$

where

$$a_\alpha = \Big(\sum_{i=1}^n \alpha_i Q_i^{-1}\Big)^{-1} \cdot \sum_{i=1}^n \alpha_i Q_i^{-1}a_i, \; Q_\alpha = (1 - h^2(\alpha))\Big(\sum_{i=1}^n \alpha_i Q_i^{-1}\Big)^{-1}, \qquad (13)$$

$$h^2(\alpha) = \sum_{i=1}^n \alpha_i a_i' Q_i^{-1}a_i - \Big(\sum_{i=1}^n \alpha_i Q_i^{-1}a_i\Big)'\Big(\sum_{i=1}^n \alpha_i Q_i^{-1}\Big)^{-1}\Big(\sum_{i=1}^n \alpha_i Q_i^{-1}a_i\Big), \qquad (14)$$

where a_i, Q_i are defined in (10).

Proof. The inclusion (12) is derived here using the idea [2, 12, 17] for construction the outer estimate of the intersection of finite number of ellipsoids. □

From Theorems 2 and 4 we come to the upper estimate.

Corollary 2. *The following inclusion holds*

$$X(T) \subseteq \bar{co}\Big(\bigcup_{t_0 \leq \tau \leq T} E(a_\alpha^\tau, Q_\alpha^\tau)\Big),$$
$$a_\alpha^\tau = K(T, \tau)a_\alpha, \; Q_\alpha^\tau = K(T, \tau)Q_\alpha(K(T, \tau))'. \qquad (15)$$

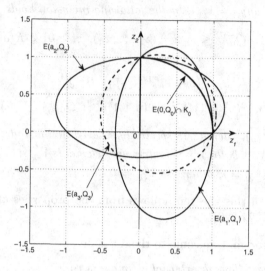

Fig. 1. Ellipsoidal estimate of the intersection $E(0, Q_0) \cap K_0$

Remark 3. The estimate (12) is valid for any convex combination $\{\alpha_i\}$ ($\alpha_i \geq 0$, $i = 1, \ldots, n$, $\sum \alpha_i = 1$). To simplify the calculations we may take $\alpha_i = 1/n$ for all $i = 1, \ldots, n$, see also the example in Section 5.

The last result (15) allows us to apply numerical algorithms developed in [5, 13, 19, 20] for finding the external ellipsoidal estimate E^+ of the closed convex hull of a union of ellipsoids $\bar{co}(\bigcup_{t_0 \leq \tau \leq T} E(a_\alpha^\tau, Q_\alpha^\tau))$. Therefore by Theorems 1–4 this resulting ellipsoid E^+ will present the external estimate of the reachable $X(T; U, X_0)$. Details related to the construction of the ellipsoid E^+ may be found in [19].

5   Numerical Simulation: Example

Consider the following control system (see also Example 1 in [20], note however that in [20] we did not claim the conic constraints on admissible controls):

$$\begin{cases} dx_1(t) = x_2(t)dt + du_1(t), \\ dx_2(t) = du_2(t), \end{cases} \quad 0 \leq t \leq 1. \tag{16}$$

We assume that $x_0 = 0$, $E(0, Q_0) = E(0, I) = B(0, 1)$ and $K_0 = \{u \in R^2 | u = (u_1, u_2), u_1 \geq 0, u_2 \geq 0\}$.

The ellipsoid $E(a_3, Q_3) = E(a_\alpha, Q_\alpha)$ shown on Fig. 1 is found by formulas (13)–(14) of Theorem 4 with $\alpha = (1/2, 1/2)$ and we have indeed $E(0, Q_0) \cap K_0 \subseteq E(a_1, Q_1) \cap E(a_2, Q_2) \subseteq E(a_3, Q_3)$.

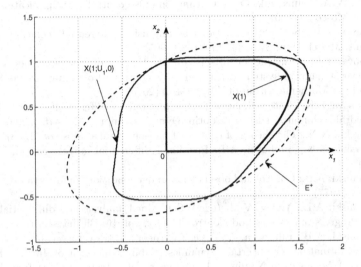

Fig. 2. Ellipsoidal estimate of the reachable set $X(1)$

From (10) in Theorem 3 we have here

$$h^2(\alpha) = 1/16, \quad a_1 = (1/3, 0), \quad a_2 = (0, 1/3),$$

$$Q_1 = \frac{4}{9} \cdot \begin{pmatrix} 1 & 0 \\ 0 & 3 \end{pmatrix}, \quad Q_2 = \frac{4}{9} \cdot \begin{pmatrix} 3 & 0 \\ 0 & 1 \end{pmatrix},$$

and therefore $E(a_3, Q_3) = B(a_3, (5/8)^{1/2})$ with $a_3 = (1/4, 1/4)$.

Denote

$$U_1 = \text{cl}^* \big(\text{co} \, \{ \, u(\cdot) \in V_p^n \mid \Delta u(t) = u(t) - u(t-0) \in E(a_\alpha, Q_\alpha), \, \forall t \in [t_0, T] \} \big),$$

where cl^* means $*$-weak closure in V_p^n.

Fig. 2 shows three sets: (a) the exact reachable set $X(1)$ of the system (16), (b) the reachable set $X(1; U_1, 0)$ of the system (16) but with controls $u(\cdot) \in U_1$ and (c) the resulting ellipsoid E^+ drawn a dotted line. The ellipsoid E^+ is found here basing on Theorems 1–4 and using the numerical algorithms described in [5, 13, 19, 20]. So Fig. 2 confirms the validity of the proposed estimates $X(1) \subseteq X(1; U_1, 0) \subseteq E^+$.

Acknowledgments. The research was supported by the RFBR Project No. 09-01-00223 and by Fundamental Research Program of the Presidium of RAS "Mathematical Control Theory" (Project No. 09-P-1-1014).

References

1. Anan'ina, T.F.: On optimization of a controlled system in the class of generalized controls. Differ. Equ. 11(4), 595–603 (1975)
2. Chernousko, F.L.: State estimation for dynamic systems. Nauka, Moscow (1988)

3. Dykhta, V.A., Sumsonuk, O.N.: Optimal impulse control with applications. Fizmatgiz, Moscow (2000)
4. Filippova, T.F.: Set-valued solutions to impulsive differential inclusions. Math. Comput. Model. Dyn. Syst. 11, 149–158 (2005)
5. Filippova, T.F.: Construction of set-valued estimates of reachable sets for some nonlinear dynamical systems with impulsive control. Proceedings of the Steklov Institute of Mathematics suppl. (2), 95–102 (2010)
6. Kostousova, E.K.: State estimation for linear impulsive differential systems through polyhedral techniques. Discrete Contin. Dyn. Syst. suppl., 466–475 (2009)
7. Krasovskii, N.N.: The theory of control of motion. Nauka, Moscow (1968)
8. Krasovskii, N.N., Subbotin, A.I.: Positional differential games. Nauka, Moscow (1974)
9. Kurzhanski, A.B.: Control and observation under conditions of uncertainty. Nauka, Moscow (1977)
10. Kurzhanski, A.B., Veliov, V.M. (eds.): Set-valued analysis and differential inclusions. Progress in Systems and Control Theory, vol. 16. Birkhäuser, Basel (1990)
11. Kurzhanski, A.B., Filippova, T.F.: On the theory of trajectory tubes — a mathematical formalism for uncertain dynamics, viability and control. In: Kurzhanski, A.B. (ed.) Advances in Nonlinear Dynamics and Control: a Report from Russia. Progress in Systems and Control Theory, vol. 17, pp. 122–188. Birkhäuser, Boston (1993)
12. Kurzhanski, A.B., Valyi, I.: Ellipsoidal calculus for estimation and control. Birkhäuser, Basel (1997)
13. Matviychuk, O.G.: Estimation problem for linear impulsive control systems under uncertainty. In: Fortuna, L., Fradkov, A., Frasca, M. (eds.) From Physics To Control Through An Emergent View, Singapore. World Scientific Series on Nonlinear Science, Series B, vol. 15, pp. 271–276 (2010)
14. Miller, B.M.: The generalized solutions of nonlinear optimization problems with impulse control. SIAM J. Control Optim. 34(4), 1420–1440 (1996)
15. Pereira, F.L., Filippova, T.F.: On a solution concept to impulsive differential systems. In: Proc. of the 4th MathTools Conference, pp. 350–355. S.-Petersburg, Russia (2003)
16. Rockafellar, R.T.: Convex analysis. Princeton University Press, Princeton (1970)
17. Schweppe, F.C.: Uncertain dynamical systems. Prentice-Hall, Englewood Cliffs (1973)
18. Vinter, R.B., Pereira, F.M.F.L.: A maximum principle for optimal processes with discontinuous trajectories. SIAM J. Control Optim. 26, 155–167 (1988)
19. Vzdornova, O.G., Filippova, T.F.: External ellipsoidal estimates of the attainability sets of differential impulse systems. J. Comput. System Sci. 45(1), 34–43 (2006)
20. Vzdornova, O.G., Filippova, T.F.: Pulse control problems under ellipsoidal constraints: Constraint parameters sensitivity. Autom. Remote Control 68(11), 2015–2028 (2007)
21. Zavalishchin, S.T., Sesekin, A.N.: Impulsive processes. Models and applications. Nauka, Moscow (1991)

Optimal Mass Transportation-Based Models for Neuronal Fibers

Antonio Marigonda and Giandomenico Orlandi

Department of Computer Sciences, University of Verona
Strada Le Grazie 15 - I-37134 Verona, Italy
{antonio.marigonda,giandomenico.orlandi}@univr.it

Abstract. Diffusion Magnetic Resonance Imaging (MRI) is used to (non-invasively) study neuronal fibers in the brain white matter. Reconstructing fiber paths from such data (tractography problem) is relevant in particular to study the connectivity between two given cerebral regions. By considering the fiber paths between two given areas as geodesics of a suitable well-posed optimal control problem (related to optimal mass transportation), we are able to provide a quantitative criterion to estimate the connectivity between two given cerebral regions, and to recover the actual distribution of neuronal fibers between them.

1   Introduction

Diffusion Magnetic Resonance Imaging (MRI) is a powerful non-invasive method producing images of biological tissues exploiting the water molecules diffusion into the living tissues under a magnetic field. This technique enhances the highly non-homogenous character of the diffusion medium, revealing underlying microstructure. Recently, this method has been widely applied to the study of the neuronal fibers in the brain white matter, and several methods have been proposed to reconstruct the fiber paths from such data (tractography). Results rely on the model chosen to represent water molecules diffusion into an MRI voxel: we recall e.g. the Diffusion Spectrum Imaging (DSI) model, which describes the diffusion inside each voxel as a probability density function defined on a set of predefined directions inside the voxel. DSI is able to successfully describe more complex tissue configurations than other models as, for example, Diffusion Tensor Imaging (DTI), but lacks to consider the density of fibers going to make up a bundle trajectory among adjacent voxels, preventing any evaluation of the real physical dimension of neuronal fiber bundles. We will describe a new approach, based on ideas from mass transportation theory and optimal control, that takes into account the whole information given by DSI in order to reconstruct the underlying water diffusion process, and recover the actual distribution of neuronal fibers. We will focus on the problem of finding the bundle of fibers connecting two given cerebral regions, by considering it as the superposition of paths that fulfill the following requirements: they lie in the higher water density regions, and the distribution of their tangents fits as much as possible with the velocity

I. Lirkov, S. Margenov, and J. Waśniewski (Eds.): LSSC 2011, LNCS 7116, pp. 131–138, 2012.

distribution given by the displacement density function (DDF). We may thus assign a cost to each diffusive process satisfying the continuity equation (granting mass conservation) and steering the initial mass configuration to the final one along a family of paths. This cost will be higher for velocities that are far from the set of averaged observed velocities given by the DDF. The processes which minimize this cost give a reasonable picture of the connecting neuronal fiber bundle.

2   Mathematical Foundation

The DDF function assigns to each voxel a probability distribution on the space of velocities. A voxel is represented by a set $Q(\alpha)$, defined as the unit cube centered at $\alpha \in \mathbb{Z}^3$. For each $(\alpha, v) \in \mathbb{Z}^3 \times \mathbb{R}^3$, let $N(\alpha)$ be the total number of molecules contained in $Q(\alpha)$, and $N(\alpha, v)$ be the number of molecules in $Q(\alpha)$ displaced of $v\delta t$ in the time δt. Define the *(averaged) displacement density function* (DDF) $f_D : \mathbb{Z}^3 \to [0, +\infty[$ by setting $f(\alpha, v) := \dfrac{N(\alpha, v)}{N(\alpha)}$. Due to the highly anisotropic character of the medium where diffusion occurs, the fraction of molecules displaced in the direction of a neuronal fiber will be significatively different from the fraction of molecules displaced in the orthogonal directions. Moreover, many fibers aligned in the same direction will result in an higher value of the DDF corresponding to that direction than a single fiber.

The main properties of f_D are the following:

1. if we fix α and a direction $w \in \mathbb{S}^2$, we have that the probability of a displacement in direction w follows a normal distribution according to its magnitude;
2. for α fixed, we have that if two directions are sufficiently close, the respective distributions are close, i.e. $w \mapsto f_D(\alpha, w)$ is continuous;

Of particular significance for the construction of our model is the set $M(\alpha)$ of displacements with cumulative probability less or equal $1/2$. Remark that this set may be different from the graph of the Oriented Distribution Function usually considered in current models.

The set $M(\alpha)$ contains the zero vector and its boundary is given by those vectors v whose modulus is the median of the magnitudes of observed displacements in the direction $v/|v|$. This set is in general nonconvex, however it is centrally symmetric. We can think about $\partial M(\alpha)$ as a set of averaged observed velocities.

We associate the set $M(\alpha)$ to the center of the voxel. We consider a suitable continuous extension $f(x, v)$ of the DDF to the whole voxel in such a way that

$$f_D(\alpha, v) = \int_{Q(\alpha)} f(x, v)\, dx,$$

and, for any direction $w \in \mathbb{S}^2$ and $\lambda \in \mathbb{R}$, the profile of $f(x, \lambda w)$ is similar to $f_D(\alpha, \lambda w)$. We set for instance $f(x, \lambda w) = a(x, w)e^{-\frac{\lambda^2}{2r(x,w)}}$, where $a, r : \mathbb{R}^3 \times$

$\mathbb{S}^2 \to]0, +\infty[$ are continuous strictly positive functions bounded away from 0 such that

$$\int_{w \in \mathbb{S}^2} \int_0^{+\infty} f(x, \lambda w) \lambda^2 \, d\lambda \, d\Sigma(w) = 1,$$

with $d\Sigma$ the area element of $\mathbb{S}^2$. We assume that, for fixed x, $w \mapsto r(x, w)$ and $w \mapsto a(x, w)$ are Lipschitz continuous functions.

We may hence define the following set of displacements $M(x)$ for any $x \in M(\alpha)$:

$$M(x) := \left\{ v \in \mathbb{R}^3 \setminus \{0\} : \int_0^{|v|} f\left(x, s \frac{v}{|v|}\right) ds \le \int_{|v|}^{+\infty} f\left(x, s \frac{v}{|v|}\right) ds \right\} \cup \{0\}$$

Boundary points of this set are precisely those displacements with cumulative probability function $1/2$. The sets $M(x)$ inherit from f some regularity properties: $x \mapsto M(x)$ is continuous with respect to the Hausdorff distance, moreover $M(x)$ is compact, star-shaped and centrally symmetric. These properties of $M(x)$ will be crucial in the regularity issues for the optimal mass transportation model, defined in term of the sets $M(x)$, that we discuss in the next sections.

2.1 Relation with Optimal Transportation and Hamilton-Jacobi Theory

In [6] it was proposed a variational model for diffusion MR, settled in a continuous framework, based on ideas from control theory.

In this model, it is imagined that an infinite number of particles starts at time 0 from the boundary ∂S of a given *seed region* S evolving along the streamline given by the gradient ∇C_S^* of a suitable *cost function* C_S^* which assigns to each point x a cost to reach the seed region S related to an optimal control problem. At time t the evolution front (i.e. the level curves of C_S^*) will represent the propagation along the medium. If the seed region is sufficiently small, for short time $0 < t < \tau$ we have an approximate picture of the neural net near the seed region. This method turns out to be very useful to establish the connectivity of a *single point p* with the region S, since it is sufficient to consider the optimal trajectory (minimizer of the cost) steering p on S, but gives few information about the connectivity between two *regions*. Experimental data show that neuronal fibers are not uniformly distributed in the brain, hence it is natural to associate also a *neuronal density* to the seed region and let this density evolve along the streamlines. This approach is not covered by [6], since they do not consider the possibility to associate a density to the seed region. Indeed, this will provide more robust information about the real path of neuronal fibers, moreover the problem can be set in the more general framework of optimal transportation theory.

Assume to have two compactly supported mass distributions with densities ρ_0 and ρ_1, such that each total mass is 1, representing respectively the initial and the final configuration of a distribution of particles transported trough the neuronal fiber paths. We consider the problem of transporting ρ_0 in ρ_1, granting mass conservation, minimizing a suitable cost which depends on the path followed by the mass particles and which reflects the behavior of the extended DDF. This problem is referred to as a Monge-Kantorovich problem, and it admits an equivalent fluid dynamics interpretation due to Benamou and Brenier (see Chapter 8 in [8]). It is well known that Monge's problem requires some regularity assumptions to admit a solution. But due to lack of compactness the problem cannot be restricted in general to the class of absolutely continuous measures. A relaxed formulation on the space of probability measures, introduced by Kantorovich in 1942, involves the notion of *transport plans* rather than transport paths or optimal transport vector fields, leading to generalized solutions. Consider the set of *transport plans* from μ_0 to μ_1 defined by

$$\Pi(\mu_0, \mu_1) = \left\{ \pi \in \mathscr{P}(\mathbb{R}^3 \times \mathbb{R}^3) : \begin{array}{l} \pi(A \times \mathbb{R}^3) = \mu_0(A), \ \forall \ \mu_0\text{-measurable } A \\ \pi(\mathbb{R}^3 \times B) = \mu_1(B), \ \forall \ \mu_1\text{-measurable } B \end{array} \right\}.$$

The Kantorovich's problem is to find

$$\inf_{\pi \in \Pi(\mu_0, \mu_1)} \int_{\mathbb{R}^3 \times \mathbb{R}^3} c(x, y) \, d\pi(x, y). \tag{1}$$

This is of course a relaxation of the Monge's problem: indeed if $\gamma_x(1)$ is optimal in the Monge's problem, then $(\text{id} \times \gamma_x(1)) \sharp \mu_0$ is optimal in the Kantorovich's problem ($\sharp$ denotes the push-forward of measures). Under very mild hypothesis on the cost function c (see Theorem 1.3 in Ref. [8]) it can be stated existence of a solution for Kantorovich's problem.

The model case is given by $c(x, y) = |y - x|^p$, i.e. a power of the Euclidean distance between x and y. In this case we can give the following:

Definition 1. *Let* $X = Y = \mathbb{R}^3$, $1 \leq p < +\infty$, *and* μ_0, μ_1 *be probability measures on* $\mathbb{R}^3$. *We define the p-Wasserstein distance by setting*

$$W_p(\mu_0, \mu_1) := \inf_{\pi \in \Pi(\mu_0, \mu_1)} \left(\int_{\mathbb{R}^3 \times \mathbb{R}^3} |x - y|^p \, d\pi(x, y) \right)^{1/p},$$

this turns out to be a metric on the space of probability measures with compact support (see [8]).

To have more insight on optimal transport plans, we recall the following (cfr. Definition 2.33 in [8])

Definition 2. *Let* $c : \mathbb{R}^3 \times \mathbb{R}^3 \to [0, +\infty]$ *be lower semicontinuos. A function* $g : \mathbb{R}^3 \to \mathbb{R}$ *is c-concave if there exists a function* ψ *such that* $g(y) = \psi^c(y) := \inf\{c(x, y) - \psi(x) : x \in \mathbb{R}^3\}$.

c-concave functions possess a PDE characterization that turns out to be extremely useful in our setting (we refer to [2] for a comprehensive introduction to the theory of viscosity solutions of Hamilton-Jacobi equations):

Proposition 1. *Let f be a Lipschitz continuous function, and $L : \mathbb{R}^3 \times \mathbb{R}^3 \to \mathbb{R}$ be continuous in x-variables and such that $v \mapsto L(x, v)$ is convex for every $x \in \mathbb{R}^3$. Consider the function*

$$\varphi(t, x) = \inf \left\{ f(y) + \int_0^t L(\xi(t), \dot{\xi}(t)) \, dt : \xi(0) = y, \, \xi(t) = x \right\} \tag{2}$$

Then we have that $\varphi(0, x) = f(x)$, $\varphi(1, x)$ is c-concave, and $\varphi(t, x)$ is the viscosity solution of the following Hamilton-Jacobi equation:

$$\begin{cases} \partial_t \varphi(t, x) + H(x, D_x \varphi(t, x)) = 0, \\ \varphi(0, x) = f(x), \end{cases} \tag{3}$$

where H is the Legendre-Fenchel transform of L (see Corollary 3.6 p.151 in [2]).

We recall next the following characterization of optimal transport plans, referring to Section 2.4 in Ref. [8] and [3] for the proof and further references:

Theorem 1. *Assume that there exists a μ_0-measurable function $c_1 : \mathbb{R}^3 \to \mathbb{R}$ and a μ_1-measurable function $c_2 : \mathbb{R}^3 \to \mathbb{R}$ such that $c(x, y) \leq c_1(x) + c_2(y)$ for all $(x, y) \in \mathbb{R}^3 \times \mathbb{R}^3$ and that $\inf_{\pi \in \Pi(\mu_0, \mu_1)} \int_{\mathbb{R}^3 \times \mathbb{R}^3} c(x, y) \, d\pi(x, y) < +\infty$. Then:*

1. *Kantorovich's problem (1) has a solution, and a plan $\pi \in \Pi(\mu_0, \mu_1)$ is optimal iff π is concentrated on the set $\{(x, y) \in \mathbb{R}^3 \times \mathbb{R}^3 : f(x) + f^c(y) = c(x, y)\}$, where $f \in L^1(\mathbb{R}^3, \mu_0)$ is a c-concave function such that $f^c \in L^1(\mathbb{R}^3, \mu_1)$,*
2. *If $c(x, y)$ is continuous, bounded below, and μ_0, μ_1 are compactly supported, then f, f^c are upper semicontinuous. If $x \mapsto c(x, y)$ is locally Lipschitz on a set U and the Lipschitz constant is locally independent of y, then f can be chosen to be locally Lipschitz on U.*

We notice that Proposition 1 applied taking as f the c-concave function given by Theorem 1 produces a function φ, solution of (3) (for further details on this, we recall Subsection 2.5.2, Subsection 5.4.6 in Ref. [8]). If φ is sufficiently smooth, problem (1) admits a solution in the original Monge's formulation.

More precisely we can state the main result of the paper:

Theorem 2. *Assume that*

1. *$L : \mathbb{R}^3 \times \mathbb{R}^3 \to \mathbb{R}$ is continuous in the x-variables and convex in the v-variables;*
2. *L is symmetric in the v-variables, i.e. $L(x, v) = L(x, -v)$ for every $x, v \in \mathbb{R}^3$;*
3. *the Legendre transform H of L belongs to $\mathrm{Lip}_{\mathrm{loc}}(\mathbb{R}^3 \times \mathbb{R}^3)$ and it is C^1 in the p-variables;*
4. *the cost $c(x, y)$ satisfies hypothesis of Theorem 1.*

Then problem (1) admits a solution with a $C^{1,1}$ displacement interpolation, and the corresponding optimal interpolating velocity vector field v_t in the characteristic system $\dot{T}_t(x) = v_t(T_t(x))$, $T_0(x) = x$, is given by $v_t(x) = \nabla_p H(x, D_x \varphi(t, x))$.

Proof. The proof is inspired by the remark at the beginning of p.181 in [8]. We recall that a function $h(x)$ is called *semiconcave* if there exist $C > 0$ such that the function $x \mapsto h(x) - C|x|^2$ is concave. According to Theorem 1, the initial datum f of Equation (2) can be chosen to be locally Lipschitz continuous. By Theorem 5.3.8 in Ref. [5], we have that viscosity solutions of Equation (2) are locally semiconcave in $]0, T[\times\mathbb{R}^3$. Observe that since $L(x, q)$, and hence $H(x, p)$ are even respectively in q and p, we have $c(x, y) = c(y, x)$. Let us now consider the optimal transport from μ_1 to μ_0. By the strict convexity assumption on L, it is unique and it is simply obtained by reversing the velocity vector field v_t into $-v_{T-t}$, where v_t is the optimal interpolating vector field related to optimal transportation from μ_0 to μ_1 for $0 \leq t \leq T$. By Equation 13.5 in Ref. [9], we have $\partial_v L(x, v_t) = D_x\varphi(t, x)$ and $\partial_v L(x, -v_{T-t}) = D_x\psi(t, x)$ a.e., where ψ solves the Hamilton Jacobi equation $\partial_t\psi + H(x, D_x\psi) = 0$, $\psi(0, x) = f^c(x)$ corresponding to the optimal transport from μ_1 to μ_0. This yields $D_x\varphi(t, x) = -D_x\psi(T-t, x)$. Since both φ and ψ are semiconcave in $]0, T[\times\mathbb{R}^3$, we have that there exists $K > 0$ such that for fixed $s, t \in]0, T[$ we have $D_x^2\varphi(t, x) \leq K$ and $D_x^2\psi(s, x) \leq K$ in the sense of distributions of $\mathbb{R}^3$, moreover also $-D_x^2\varphi(t, x) = D_x^2\psi(T - t, x) \leq K$. This implies that $x \mapsto \varphi(t, x)$ is semiconcave and semiconvex in $\mathbb{R}^3$ and hence it is $C^{1,1}(\mathbb{R}^3)$ (see [5] for details on properties of semiconcave functions). According to Chapter 13, Equation 13.5 in Ref. [9], the two relations $\nabla_v L(x, v_t(x)) = D_x\varphi(t, x)$ and $v_t(x) = \nabla_p H(x, D_x\varphi(t, x))$ define a vector field v_t which, through the solution $T_t(x)$ of the characteristic system $\dot{T}_t(x) = v_t(T(x))$, $T_0(x) = x$, realizes the minimum in the transport problem between ρ_0 and $T_t\sharp\rho_0$ for every $t \in]0, T[$, i.e. v_t is the optimal interpolating velocity vector field.

2.2   Minimum Time and Intrinsic Metric Models

The first model we propose corresponds essentially to choose $L(x, v) = 1$ if $v \in M(x)$, $+\infty$ elsewhere. The transport cost associated to this choice is given by:

$$c_1^\varepsilon(x, y) := \inf \left\{ \int_0^T \left(1 + \varepsilon\frac{|\dot{z}(t)|^2}{2}\right) dt : z : [0, T] \to \mathbb{R}^3, \right.$$
$$\left. z(0) = x, \ z(T) = y, \dot{z}(t) \in M(z(t)) \text{ for a.e. } t \in [0, T]\right\}.$$

where we add a quadratic perturbation depending on a small parameter $\varepsilon > 0$ in order to obtain smooth approximate solutions. The case $\varepsilon = 0$ corresponds to the minimum time needed to steer x to y with curves satisfying $\dot{\gamma} \in M(\gamma(t))$. The Lagrangian function associated is $L_1^\varepsilon(x, v) := 1 + \varepsilon|v|^2/2$. We notice that Theorem 2 applies for L_1^ε, yielding a sufficiently smooth solution φ_ε of (2) and the optimal interpolating vector field $v_t^\varepsilon(x) = \nabla_p H_1^\varepsilon(x, D_x\varphi_\varepsilon(t, x))$ for a.e. x. We associate to this cost a *generalized* Wasserstein distance, defined, for two given probability measures μ_0, μ_1, by $W_2^{c_1^\varepsilon}(\mu_0, \mu_1) = \inf \left\{ \int_{\mathbb{R}^3 \times \mathbb{R}^3} c_1^\varepsilon(x, y) \, d\pi : \pi \in \Pi(\mu_0, \mu_1)\right\}^{1/2}$.

In this second model, we modify the infinitesimal metric on $\mathbb{R}^3$ in order to take into account the anisotropic character of the diffusion as expressed by the

DDF, by using the gauge function $\gamma_{M(x)}(v) = \inf\{\lambda > 0 : v/\lambda \in M(x)\}$ to penalize $v_t(x)$ according to its distance to $M(x)$. This equips $\mathbb{R}^3$ with the structure of a geodesic space, where the geodesics correspond to the optimal transport trajectories.

Due to technical reasons, we relax the problem replacing the set $M(x)$ with its convex hull and set $L_2(x, v) = \gamma^2_{\overline{co}M(x)}(v)$. We define the following cost function $c_2(x, y) := \inf\left\{\int_0^1 \gamma^2_{\overline{co}M(z(t))}(\dot{z}(t))\, dt : z(0) = x, z(1) = y\right\}$, and accordingly we define the distance $W_2^{c_2}(\mu_0, \mu_1) = \inf\left\{\int_{\mathbb{R}^3 \times \mathbb{R}^3} c_2(x, y)\, d\pi : \pi \in \Pi(\mu_0, \mu_1)\right\}^{1/2}$. Also in this case, Theorem 2 applies, due to strict convexity of L_2, so we obtain a solution of the transport problem and the optimal interpolating vector field v_t is given by $v_t(x) = \nabla_p H_2(x, D_x\varphi(t, x))$ for a.e. x.

2.3   A Quantitative Brain Connectivity Criterion

From a medical point of view, it turns out to be important to know not only whether or not two regions are connected by a fiber bundle, but also the quantity of fibers forming that connecting bundle (*connectivity problem*), e.g. in situations where, due to medical pathologies, we have two cerebral areas connected by fibers that pass through a damaged region.

We suggest the following criterion in order to quantitatively estimate the connectivity of two regions D_0 and D_1, by means of the transport distance $W_2^{c_i}$: let μ_0 and μ_1 be two absolutely continuous measures with the same mass, constant densities, and support coinciding respectively with D_0 and D_1. Define the ratio

$$Q(D_0, D_1) = \frac{W_2^{c_i}(\mu_0, \mu_1)}{d_H(D_0, D_1)}, \qquad i = 1, 2, \qquad (4)$$

of the transport distance $W_2^{c_i}$ (related to the transport cost c_i between μ_0 and μ_1, as defined in Subsection 2.2) and the Hausdorff distance d_H (related to the Euclidean metric) between D_0 and D_1. A small quotient in (4) implies a high fiber density connection between D_0 and D_1. This kind of information is apparently not provided by other tractography models that ignore information on fiber density.

3   Conclusions

It is an experimental fact that the measurements in each voxel enjoy very poor robustness properties, and sometimes we can have a voxel where no information are available, surrounded by regular voxels. This *missing cell problem* is not adequately faced by classical models, since they assume that no fibers can cross such voxels. We propose to assign to each empty voxel Q_α a constant profile $M(\alpha)$ given by a ball, whose radius should be determined by experimental data. In this way, the missing information in the voxel will be in some sense *reconstructed* by the information contained in the neighboring ones: fibers will be propagated in the missing voxel according to the behavior in the nearby ones.

From a mathematical point of view, the main issue here is to provide from an optimal transport plan, solution of the Monge-Kantorovic problem, a unique optimal interpolating velocity vector field v_t describing the evolutive process. The obstruction is given by the *lack of smoothness* of v_t, due to the presence in a voxel of multiple fibers crossing along different directions. Explicit representation formulas for the solutions, regularization procedures and, in the case of nonuniqueness, also selection principles can be used to face the problem. We refer to [8] for a complete analysis of these topics, just recalling that the use of selection principles is extensively used also in other tractography models.

In the second model, the problem was relaxed by making a *convexification* of the sets $M(x)$, similar to the (implicit) convexification made in [6] by using the Legendre transform. This relaxation has no physical meaning, since the sets $M(x)$, giving the average displacement probabilities, are characterized by strong anisotropy, hence are nonconvex. However, in the nonconvex case, it is not possible in general to define a optimal interpolating velocity vector field v_t based on characteristic curves. For details and counterexamples on this delicate point, we refer to [7].

Reconstructing a reliable picture of the whole net of neuronal fibers will actually require a large number of implementations of our model. The algorithmic implementation and the numerical validation of our model is currently still under investigation. Also a *mean field games* reformulation seems promising from the viewpoints of the associated numerical methods.

References

1. Agrachev, A., Lee, P.: Optimal transportation under nonholonomic constraints. Trans. Amer. Math. Soc. 361(11), 6019–6047 (2009)
2. Bardi, M., Capuzzo-Dolcetta, I.: Optimal control and viscosity solutions of Hamilton-Jacobi-Bellman equations. Birkhäuser Boston, Boston (1997)
3. Bernard, P., Buffoni, B.: Optimal mass transportation and Mather theory. J. Eur. Math. Soc. (JEMS) 9(1), 85–121 (2007)
4. Clarke, F.H., Ledyaev, Y.S., Stern, R.J., Wolenski, P.R.: Nonsmooth analysis and control theory. Graduate Texts in Mathematics, vol. 178. Springer, New York (1998)
5. Cannarsa, P., Sinestrari, C.: Semiconcave functions, Hamilton-Jacobi equations, and optimal control. Birkhäuser Boston, Boston (2004)
6. Pichon, E., Westin, C.-F., Tannenbaum, A.: A Hamilton-Jacobi-Bellman Approach to High Angular Resolution Diffusion Tractography. In: Duncan, J.S., Gerig, G. (eds.) MICCAI 2005. LNCS, vol. 3749, pp. 180–187. Springer, Heidelberg (2005)
7. Siconolfi, A.: Metric character of Hamilton-Jacobi equations. Trans. Amer. Math. Soc. 355(5), 1987–2009 (2003)
8. Villani, C.: Topics in optimal transportation. Graduate Studies in Mathematics, vol. 58. American Mathematical Society, Providence (2003)
9. Villani, C.: Optimal transport old and new. Grundlehren der Mathematischen Wissenschaften, vol. 338. Springer, Berlin (2009)

Optimal Controls in Models
of Economic Growth and the Environment

Elke Moser[1,2], Alexia Prskawetz[1,2], and Gernot Tragler[1]

[1] Vienna University of Technology (TU Wien),
Institute for Mathematical Methods in Economics (IWM), A-1040 Wien, Austria
[2] Vienna Institute of Demography (VID),
Austrian Academy of Sciences (OeAW), A-1040 Wien, Austria
moser_elke@hotmail.com, afp@econ.tuwien.ac.at,
tragler@eos.tuwien.ac.at

Abstract. We investigate the impact of environmental quality standards on the accumulation of two types of capital ("brown" vs. "green") and the corresponding R&D investments in an endogenous growth model. We show that environmental regulation as economic policy instrument rather represses economic growth in the long run but fosters green R&D and the accumulation of green capital in the short run. In addition we show that subsidies may support a shift to a greener production and can be used to counteract against repressed accumulation of green capital.

1 Introduction

Climate change and the current environmental situation have increasingly become central topics in the world wide media, politics, and science. Scientific evidence on global warming (e.g. [5]) underlines the importance of climate mitigation and the need to act. Among other possible mitigation options, the usage of environmental policy instruments seems to be promising. However, the question arises how they can be utilized in the most effective way and whether strict environmental regulation has a supporting or rather repressing impact on innovation and economic growth. To answer these questions we build on a recent paper by Rauscher [7] who already addressed this issue by constructing a simple dynamic environmental-economic model in an endogenous growth framework to investigate whether tighter environmental standards will induce a shift from end-of-pipe emission abatement to a process-integrated one. While in [7] the problem formulation is kept rather general without assuming specific model functions, in this paper we introduce two types of production functions (Cobb Douglas vs. CES) and apply optimal control theory to solve for the dynamic paths of this two-state two-control environmental-economic system. We also present a model extension, which additionally considers subsidies as economic policy instrument.[1]

---

[1] As a general reference on the global environment, natural resources, and economic growth we highly recommend [3].

I. Lirkov, S. Margenov, and J. Waśniewski (Eds.): LSSC 2011, LNCS 7116, pp. 139–146, 2012.

2   The Base Model

Consider a competitive market economy where a continuum of identical firms using identical technology produce a homogenous GDP good. In this economy, two types of capital can be accumulated: first, there is *conventional capital*, also called *brown capital*, which is pollutive, and secondly, non-polluting *green capital* can be chosen. Additionally, the government sets environmental standards which the entrepreneurs are obligated to meet, while the necessary abatement effort as well as the according costs depend on the stringency of these regulations. The benefit of less abatement effort due to the adoption of cleaner technology comes at the cost of less profitability of green R&D.

The representative agent, who is assumed to be a capital-owning entrepreneur doing his/her R&D in-house and who invests and consumes at the same time, maximizes his/her present value of future utility, given as

$$\int_0^\infty e^{-rt}(U(C(t)) + u(\epsilon))dt , \tag{1}$$

where $C(t) > 0$ is the consumption or dividend income, $U(C(t))$ describes the utility one can get out of it, and r is the discount rate. Further on, ϵ specifies the exogenously given environmental quality determined by the government, which is an index between 0 and 1, with $\epsilon = 0$ denoting the *laissez-faire* scenario (no environmental regulation exists and therefore environmental quality is low) and $\epsilon = 1$ standing for the maximal attainable environmental quality. The private sector's utility of environmental quality is denoted as $u(\epsilon)$ and for the analysis will be set as $u(\epsilon) = c\epsilon^\gamma$ with $c > 0$ and $0 < \gamma < 1$.

Conventional capital $K(t)$ and/or green capital $G(t)$ are used to produce an output $F(K(t), G(t))$. One of the central assumptions in this model is that the output is used completely for consumption, for the coverage of opportunity costs due to R&D investments of either type, and for end-of-pipe emission abatement, which is summarized in the following budget constraint:

$$F(K(t), G(t)) - C(t) - w(R_K(t) + R_G(t)) - \chi(\epsilon)K(t) = 0. \tag{2}$$

Note that as of here, we will often omit the time argument t for the ease of exposition. R_K and R_G denote the investments for R&D to generate new capital of types K and G, respectively. The parameter $w \in [0, 1]$ represents the exogenous opportunity costs. The abatement costs for achieving the binding environmental constraints set by the government are proportional to the installed conventional capital K. The costs per unit capital are given as $\chi(\epsilon)$, which is increasing and convex in the stringency of environmental regulation, i.e. $\chi' > 0, \chi'' > 0$, and will be set for this analysis as $\chi(\epsilon) = a\epsilon^\beta$ with $a > 0$ and $\beta > 1$.

Regarding the state dynamics for the two types of capital, we use a Cobb Douglas production function with decreasing returns to scale (see (3a) and (3b)).The existing capital stock itself has a positive feedback on the accumulation. Assuming that this positive feedback is weaker than the contribution of new technology due to R&D, the partial elasticity of production of the capital stock is supposed to be less than the one of the R&D investments. Additionally, it is more likely that conventional capital is more established in the economy than green one and therefore accumulation of the former is easier. Hence, the partial elasticities of green capital G are supposed to be less or equal to those of conventional capital K (i.e. $\sigma_1 \leq \delta_1$ and $\sigma_2 \leq \delta_2$). Further on, one has to consider that capital of either type is subject to depreciation over time, which is taken into account by including depreciation rates ϕ and ψ.

Solving equation (2) for consumption C together with (1) leads to an optimal control problem with R_K and R_G as control variables and the two available types of capital as states, which is given as

$$\max_{R_K, R_G} \int_0^\infty e^{-rt} \Big(\ln \big(\tau + F(K,G) - w(R_K + R_G) - a\epsilon^\beta K \big) + c\epsilon^\gamma \Big) dt \quad (3)$$

$$\text{s.t.:} \quad \dot{K} = dK^{\delta_1} R_K^{\delta_2} - \phi K \tag{3a}$$

$$\dot{G} = eG^{\sigma_1} R_G^{\sigma_2} - \psi G \tag{3b}$$

$$0 \leq R_K(t) \quad \forall t \geq 0 \tag{3c}$$

$$0 \leq R_G(t) \quad \forall t \geq 0 \tag{3d}$$

$$0 \leq F(K,G) - w(R_K + R_G) - a\epsilon^\beta K = C \tag{3e}$$

$$0 < \alpha_1, \alpha_2, \gamma, w < 1 \quad \text{and} \quad \alpha_1 + \alpha_2 \leq 1 \tag{3f}$$

$$0 < \delta_1, \delta_2 < 1 \quad \text{and} \quad \delta_1 + \delta_2 < 1 \tag{3g}$$

$$0 < \sigma_1, \sigma_2 < 1 \quad \text{and} \quad \sigma_1 + \sigma_2 < 1 \tag{3h}$$

$$0 < a, c, d, e, r, \tau, \phi, \psi, \ 1 < \beta, \ 0 \leq \epsilon \leq 1, \tag{3i}$$

where the mixed path constraint (3e) ensures that consumption remains non-negative and τ avoids infinite slope (and hence problems in the numerical analysis) in case this constraint is active. Table 1 gives an overview of the parameters and the corresponding values used for the following analyses. The parameters in the production function $F(K,G)$ as well as the environmental quality index ϵ will be a matter of variation to investigate different scenarios and therefore are not listed yet.

3   Analysis of the Base Model

We analyze our optimal control model (3)-(3i) by applying Pontryagin's maximum principle (see, e.g., [2]). As for this paper we face a tight page limit, we omit displaying the resulting optimality conditions, which can be found in [6]. For the computation of optimal trajectories, we use the toolbox OCMat with the parameter values from Tab. 1.[2]

---

[2] OCMat is a toolbox using the MATLAB language to analyze specifically discounted infinite-time optimal control models and is freely accessible at http://orcos.tuwien.ac.at/research/ocmat_software/.

Table 1. Description and values of the parameters

Parameter	Value	Description
a	1	Constant of proportionality of abatement costs
c	5	Scale parameter describing the utility from environmental quality
d	1	Scale parameter of brown capital accumulation
e	1	Scale parameter of green capital accumulation
r	0.05	Discount rate
w	0.1	Opportunity cost of R&D
β	2	Exponent of abatement costs
γ	0.4	Exponent of the utility from environmental quality
δ_1	0.3	Production elasticity of K in $\dot{K}$
δ_2	0.5	Production elasticity of R_K in $\dot{K}$
σ_1	0.3	Production elasticity of G in $\dot{G}$
σ_2	0.4	Production elasticity of R_G in $\dot{G}$
τ	1	Additive constant in the utility from consumption
ϕ	0.05	Depreciation rate of brown capital
ψ	0.05	Depreciation rate of green capital

For the production function $F(K, G)$ we consider both a Cobb Douglas $(K^{\alpha_1} G^{\alpha_2})$ and a CES function $((\alpha_1 K^{\alpha_2} + (1 - \alpha_1) G^{\alpha_2})^{\frac{1}{\alpha_2}})$. As the respective results are very similar (see [6]), for reasons of brevity we here restrict ourselves to the Cobb Douglas function.

In Fig. 1 we find the shares of the equilibrium levels of K and G in the total capital stock for varying ϵ.[3] As one can see, the ratio of G follows a convex-concave shape. At the beginning, the usage of G is quite low and does not change much with increasing ϵ. In this area, the abatement costs are still too low to change the advantage of conventional capital. The inflexion point is at $\epsilon = 0.362$ where green capital starts to dominate conventional capital. From here on the ratio of G grows quite quickly until it converges to almost 100%. Accordingly, the ratio of K follows a concave-convex decrease. The dependence on ϵ of the equilibrium ratios of the corresponding R&D investments is qualitatively the same, while the position of the inflexion point is slightly lower at $\epsilon = 0.263$.

Figure 2 provides a phase portrait in the (K, G)-space for $\alpha_1 = 0.6$, $\alpha_2 = 0.2$, and $\epsilon = 0.4$. The trajectories, some of which are divided into a black and a gray part, show the paths converging to the equilibrium. For very high levels of brown capital, abatement costs are so high that initially nothing is invested in brown R&D, so $R_K = 0$ (black parts of the trajectories). At the dashed line, the control constraint (3c) becomes inactive, so R_K starts being positive (gray parts of the trajectories in the interior of the admissible region). The gray shaded area on the right hand side marks the inadmissible region where consumption is negative. Within this area, abatement costs already exceed the total production

---

[3] With "equilibrium" we denote a steady state of the canonical system, where all dynamics are equal to zero.

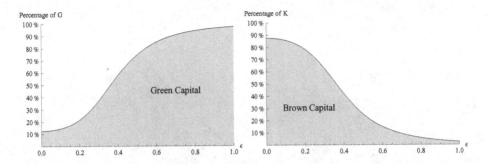

Fig. 1. Equilibrium ratios of K and G for varying ϵ with $\alpha_1 = 0.6$, $\alpha_2 = 0.2$

output, so no admissible control ($C \geq 0$) exists to satisfy the budget contraint. Hence, firms starting inside this area are already in debt and because we do not allow disinvestments (negative R_K and/or R_G), there exists no admissible path leading to the equilibrium.

4   Subsidization for Environmental-Friendly Production

We distinguish between subsidization of *green R&D* and subsidization of *green capital*. In the former case, the objective function (3) is replaced by

$$\max_{R_K, R_G} \int_0^\infty e^{-rt} \Big(\ln \big(\underbrace{\tau + K^{\alpha_1} G^{\alpha_2} - w(R_K + R_G) - a\epsilon^\beta K + swR_G}_{=C \geq 0} \big) + c\epsilon^\gamma \Big) dt \;,$$

and in the latter case the objective function transforms to

$$\max_{R_K, R_G} \int_0^\infty e^{-rt} \Big(\ln \big(\underbrace{\tau + K^{\alpha_1} G^{\alpha_2} - w(R_K + R_G) - a\epsilon^\beta K + sG}_{=C \geq 0} \big) + c\epsilon^\gamma \Big) dt \;,$$

where the (exogenous) parameter $s \in [0, 1]$ represents the subsidy rate the government is willing to pay for green R&D investments or green capital stock, respectively.

The phase portraits look very similar to those in Fig. 2, but there is one main difference to be observed *only* for the case of subsidization of green *capital*. While in both, the base model without subsidization and the case of subsidized green R&D, the equilibrium levels of K and G tend to zero as $\epsilon \to 1$ (left panel in Fig. 3), the subsidization of green capital succeeds in maintaining a clearly positive equilibrium level of G for all values of ϵ. In Fig. 4 we see exemplarily how an initially brown capital-dominated production is quickly transformed into a green one under the subsidization of G.

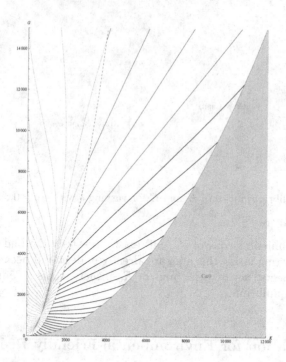

Fig. 2. Phase portrait in (K, G)-space for $\alpha_1 = 0.6$, $\alpha_2 = 0.2$, and $\epsilon = 0.4$

5   Conclusions and Possible Model Extensions

Our analysis underlines Rauscher's [7] conclusions that increasing stringency of environmental regulation causes a decline in both types of capital and consequently also in production output, so it rather suppresses than supports economic growth. However, increasing environmental regulation indeed has a positive impact on the accumulation of green capital and on the increase of green R&D investments (at least relative to their brown counterparts), leading to an overall greener production for increasing environmental quality standards. We further introduced subsidization as additional policy instrument and showed that subsidization of green capital has potential to be a promising environmental policy instrument. Summing up, we conclude that environmental regulation can cause a shift to greener production but only at the cost of repressed economic growth. However, if environmental economic instruments are used wisely by considering a trade-off between losses in growth and achievements towards a greener technology, a compromise may well be found yielding more satisfying results.

We want to close this paper with discussing a few model extensions we want to deal with in the future. In the presented model, environmental quality is determined *exogenously* through the required standards set by the government. Therefore, the condition of the environment and the limits the agents are obligated to meet are considered in a one-to-one relation, which may be unrealistic. It is rather the case that the desired environmental aims of the government and

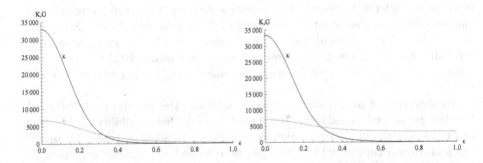

Fig. 3. Equilibrium levels of K and G for varying ϵ under the subsidization of R_G (left panel, $s = 0.5$) and G (right panel, $s = 0.1$) for $\alpha_1 = 0.6, \alpha_2 = 0.2$

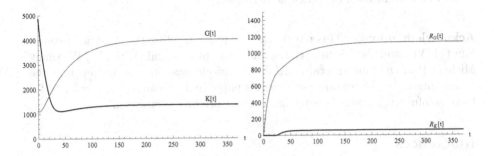

Fig. 4. Optimal time paths of states and controls starting from a definitely brown capital-dominated production under the subsidization of G for $\alpha_1 = 0.6, \alpha_2 = 0.2$, $\epsilon = 0.4$, and $s = 0.1$

the real environmental conditions do not coincide. Additionally, it is assumed in our model that the abatement effort exactly equals the necessary level needed to satisfy the required standards and therefore is not object of the agent's decision. Consequently, we initially exclude the possibility that an agent who is motivated by some incentives like utility out of cleaner environment, taxes, or subsidies, puts his/her abatement effort beyond the necessary level. Taking these points into account, a possible modification of the present model would be to introduce an emission function $E(K, G)$ describing the emissions of brown (and possibly also green) capital during the production process. Such a function is considered in [8], where the emission generating inputs are capital and labor used for production. As capital and labor input for the abatement process are also used as further controls in that model, the decision about the abatement effort turns out to be independent of the obligatory emission limits. Accordingly, using the abatement share a of the generated emissions $E(K, G)$ as third control in our model would probably lead to similar conditions. In addition to the emission *flow*, an emission *stock* directly related to the current emissions $E(K, G)$ and inversely related to the current abatement effort as well as to a natural self-cleaning rate could be considered to reflect the environmental quality. Formulating such an emission function also has the advantage that the emission standards are not further defined in a general way but now can be specified in more detail, e.g. according to one of the five different types of standards given in [4] such as standards as a set level of total output, per unit output, or per unit of specified input.

Another lack of our model we have to admit is the missing explanation of how the introduced subsidies in Sect. 4 are financed. One possibility to close this model in an economically coherent way would be to simultaneously introduce taxes on brown capital, which can then be refunded as subsidies according to a corresponding governmental budget restriction. Following [1], we alternatively think about changing the model into a micro-economic problem considering the entrepreneurs' profit as objective function while the government sets its environmental policies in form of tradeable permits.

Acknowledgements. This paper was partly financed by the Austrian Science Fund (FWF) and the Vienna Science and Technology Fund (WWTF). We thank Michael Rauscher for several fruitful discussions enabling us to step into the highly interesting issues addressed in this paper and two anonymous referees for their helpful suggestions how to improve our original submission.

References

1. Chevallier, J., Etner, J., Jouvet, P.A.: Bankable pollution permits under uncertainty and optimal risk management rules – theory and empirical evidence. EconomiX Working Paper 25 (2008)
2. Grass, D., Caulkins, J.P., Feichtinger, G., Tragler, G., Behrens, D.A.: Optimal Control of Nonlinear Processes – With Applications in Drugs, Corruption, and Terror. Springer, Heidelberg (2008)
3. Greiner, A., Semmler, W.: The Global Environment, Natural Resources, and Economic Growth. Oxford University Press, New York (2008)
4. Helfand, G.E.: Standards versus standards – the effects of different pollution restrictions. American Economic Review 81, 622–634 (1992)
5. Intergovernmental Panel on Climate Change: IPCC Fourth Assessment Report – Climate Change, AR4 (2007)
6. Moser, E.: Optimal Controls in Models of Economic Growth and the Environment. Master Thesis, Vienna University of Technology (2010)
7. Rauscher, M.: Green R&D versus end-of-pipe emission abatement – a model of directed technical change. Thuenen-Series of Applied Economic Theory 106 (2009)
8. Xepapadeas, A.P.: Environmental policy, adjustment costs and behavior of the firm. Journal of Environmental Economics and Management 23, 258–275 (1992)

On the Minimum Time Problem for Dodgem Car–Like Bang–Singular Extremals

L. Poggiolini and G. Stefani

Dipartimento di Sistemi e Informatica – Università degli Studi di Firenze, Italy

Abstract. In this paper we analyse second order conditions for a bang–singular extremal in a class of problems including the Dodgem–car one. Moreover we state some result on optimality and structural stability whose proof will appear elsewhere.

1 Introduction

This paper aims to be a preliminary result in the study of optimality, structural stability and sensitivity analysis for a class of problems which includes the Dodgem car problem for appropriate initial conditions.

In many classical problems Pontryagin Maximum Principle (PMP), together with the coercivity of the second variation, yields the optimality of a reference trajectory. These conditions for the nominal problem together with the uniqueness and normality of the adjoint covector yield structural stability and possibly allow to perform sensitivity analysis.

Here we focus on second order conditions for the following problem:

$$(\mathbf{P_0}) \qquad \text{minimise } T \text{ subject to} \tag{1}$$

$$\dot{\xi}(t) = f_0(\xi(t)) + u(t)f_1(\xi(t)) \tag{2}$$

$$\xi(0) = \widehat{x}_0, \quad \xi(T) \in N_f, \quad u(t) \in [-1,1] \tag{3}$$

where N_f is the integral line of f_1 emanating from a given point y, i.e. $N_f := \{\exp s f_1(y): s \in \mathbb{R}\}$. The state space is $\mathbb{R}^n$ and f_0, $f_1: \mathbb{R}^n \to \mathbb{R}^n$ are smooth (C^∞) vector fields.

We assume we are given a reference triple $(\widehat{T}, \widehat{\xi}, \widehat{u})$ and we consider strong local optimality according to the following definitions:

Definition 1. *The trajectory $\widehat{\xi}$ is a* (time, state)–local minimiser *of* $(\mathbf{P_0})$ *if there exist $\varepsilon > 0$ and a neighbourhood $\widetilde{\mathcal{U}}$ of its graph in $\mathbb{R} \times \mathbb{R}^n$ such that no admissible trajectory ξ whose graph is in $\widetilde{\mathcal{U}}$ reaches the target in time $T \in (\widehat{T} - \varepsilon, \widehat{T})$.*

We point out that this kind of optimality is independent of the values of the associated control and it is local both with respect to final time and space. A *stronger version* of strong local optimality is the following:

Definition 2. *The trajectory $\widehat{\xi}$ is a* state–local minimiser *of* $(\mathbf{P_0})$ *if there is a neighbourhood $\mathcal{U}$ of its range in $\mathbb{R}^n$ such that $\widehat{\xi}$ is a minimiser among the admissible trajectories whose range is in $\mathcal{U}$.*

I. Lirkov, S. Margenov, and J. Waśniewski (Eds.): LSSC 2011, LNCS 7116, pp. 147–154, 2012.

Motivated by the Dodgem car problem as in [3], we assume that the reference triple $(\widehat{T}, \widehat{\xi}, \widehat{u})$ is a normal bang–singular Pontryagin extremal for problem $(\mathbf{P_0})$. By bang–singular trajectory (or triple, or control) we mean that $\widehat{u}$ has the following structure

$$\widehat{u}(t) \equiv u_1 \in \{-1, 1\} \ \forall t \in [0, \widehat{\tau}), \qquad \widehat{u}(t) \in (-1, 1) \ \forall t \in (\widehat{\tau}, \widehat{T}]. \tag{4}$$

We denote by $\widehat{f}_t := f_0 + \widehat{u}(t)f_1$ the time–dependent vector field associated to the reference control. $h_1 := f_0 + u_1 f_1$ is its restriction to the bang interval $[0, \widehat{\tau})$.

By normal Pontryagin extremal we mean that the reference triplet satisfies Pontryagin Maximum Principle (PMP) in normal form.

Assumption 1.1 (PMP) *There exists a solution* $\widehat{\lambda} : t \in [0, \widehat{T}] \to \widehat{\lambda}(t) \in (\mathbb{R}^n)^*$ *of the adjoint equation* $\dot{\lambda}(t) = -\langle \lambda(t), D\widehat{f}_t(\widehat{\xi}(t)) \rangle$ *such that*

$$\langle \widehat{\lambda}(\widehat{T}), f_1(\widehat{\xi}(\widehat{T})) \rangle = 0 \tag{5}$$

$$\langle \widehat{\lambda}(t), \widehat{f}_t(\widehat{\xi}(t)) \rangle = \max_{|u| \leq 1} \{ \langle \widehat{\lambda}(t), (f_0 + uf_1)(\widehat{\xi}(t)) \rangle \} = 1 \qquad a.e. \ t \in [0, \widehat{T}]. \tag{6}$$

$\widehat{\lambda}$ is called *adjoint covector*. We denote its junction point between the bang and the singular arc and its final point as $\ell_1 := \widehat{\lambda}(\widehat{\tau})$, $\ell_f := \widehat{\lambda}(\widehat{T})$, respectively.

Notice that the *transversality condition* in (5) gives no new piece of information but it is a consequence of (6).

In [5] we proved sufficient conditions for (time, state)–local optimality of a bang–singular trajectory in the case when the singular control is state–feedback. In that paper we required second order conditions for the problem with fixed end–points and we proved the result for the constraints (3), thanks to the special form of the final constraint N_f.

In [6] second order conditions were given for both kinds of local optimality for a general singular control and fixed end–points. We point out that the paper considers also bang–singular–bang extremals.

Here we consider a general bang–singular control and we focus on sufficient second order conditions. Problem $(\mathbf{P_0})$ is linear with respect to the control, therefore the standard second variation associated to the singular arc is completely degenerate. We use a coordinate free version of *Goh's transformation* in order to obtain suitable sufficient conditions on the singular arc and we end up with a quadratic form on variations $(\varepsilon_0, \varepsilon_1, w) \in \mathbb{R}^2 \times L^2([\widehat{\tau}, \widehat{T}], \mathbb{R})$, see Assumption 2.1. The couple (ε_1, w) comes from Goh's transformation and ε_0 takes into account the variation of the initial point of the singular arc due to the variation of the bang time interval. Due to the particular form of the final constraint we cannot require coercivity with respect to the variations of the final point, see Lemma 1.

This feature is observed in the Dodgem car problem and justifies the results in [4] where a lack of stability has been proven when the final constraint of the perturbed problem is not an integral line of the controlled vector field.

We point out that, as in [6], we require the coercivity of the quadratic form for the minimum time problem between the end–points of the singular arc.

Finally we announce results on optimality and structural stability which will appear elsewhere.

2   The Extended Second Variation

In what follows we shall refer to $\widehat{\tau}$ as the *switching time* of the reference control $\widehat{u}$. The flow of $\widehat{f}_t$ starting at time $\widehat{\tau}$ is defined in a neighbourhood of $\widehat{x}_1 := \widehat{\xi}(\widehat{\tau})$ and we denote it as $\widehat{S}_t \colon x \in \mathbb{R}^n \mapsto \widehat{S}_t(x) \in \mathbb{R}^n$; finally let $\widehat{x}_f := \widehat{\xi}(\widehat{T})$.

We denote the derivative of a smooth function $\beta \colon \mathbb{R}^n \to \mathbb{R}$ in the direction of a smooth vector field f as $f \cdot \beta(x) := \langle D\beta(x), f(x) \rangle$. The Lie bracket between the vector fields f and g is denoted by $[f, g](x) := (Dg\, f - Df\, g)(x)$.

In order to obtain second order conditions we transform problem ($\mathbf{P_0}$) into an equivalent Mayer problem on the fixed time interval $[0, \widehat{T}]$ and state space $\mathbb{R} \times \mathbb{R}^n$, see also [1] and [7]. We consider a new control $u_0 \in L^\infty([0, \widehat{T}], (0, \infty))$, which allows any perturbation of the length both of the bang and the singular time interval, reparametrising the time by $t = u_0$. We obtain that problem ($\mathbf{P_0}$) is equivalent to the Mayer problem on $\mathbb{R} \times \mathbb{R}^n$ described below.

For $\boldsymbol{x} = (x^0, x) \in \mathbb{R} \times \mathbb{R}^n$, set $\boldsymbol{f}_0(\boldsymbol{x}) := (1, f_0(x))$ and $\boldsymbol{f}_1(\boldsymbol{x}) := (0, f_1(x))$. Problem ($\mathbf{P_0}$) becomes:

$$\text{minimise} \quad \xi^0(\widehat{T}) \quad \text{subject to}$$

$$\dot{\boldsymbol{\xi}}(t) = u_0(t)\, \boldsymbol{f}_0(\boldsymbol{\xi}(t)) + u_0(t) u(t)\, \boldsymbol{f}_1(\boldsymbol{\xi}(t)) \quad t \in [0, \widehat{T}] \tag{7}$$

$$\boldsymbol{\xi}(0) = (0, \widehat{x}_0), \quad \boldsymbol{\xi}(\widehat{T}) \in \mathbb{R} \times N_f, \quad (u_0(t), u(t)) \in (0, +\infty) \times [-1, 1],$$

where $\boldsymbol{\xi} = (\xi^0, \xi)$. The trajectory $\widehat{\boldsymbol{\xi}}(t) := (t, \widehat{\xi}(t))$, associated to $(u_0, u) = (1, \widehat{u})$, is an extremal with adjoint covector $\widehat{\boldsymbol{\lambda}} \colon t \mapsto (-1, \widehat{\lambda}(t)) \in \mathbb{R} \times (\mathbb{R}^n)^*$.

In order to get the coordinate–free second variation as defined in [2], we consider the dragged system, at time $\widehat{\tau}$, along the reference flow, by setting $\boldsymbol{\eta}(t) := (\eta^0, \eta)(t) = (\xi^0(t) + (\widehat{\tau} - t), \widehat{S}_t^{-1} \circ \xi(t))$. We define the dragged vector fields at time $\widehat{\tau}$, along the reference flow as

$$g_{i,t} := D\widehat{S}_t^{-1} f_i \widehat{S}_t, \ i = 0, 1, \qquad \widehat{g}_t := D\widehat{S}_t^{-1} \widehat{f}_t \widehat{S}_t = g_{0,t} + \widehat{u}(t) g_{1,t}. \tag{8}$$

Problem (7) becomes

$$\text{minimise} \quad \eta^0(\widehat{T}) \quad \text{subject to}$$

$$\begin{aligned}
&\dot{\eta}^0(t) = u_0(t) - 1 \\
&\dot{\eta}(t) = (u_0(t) - 1)\, g_{0,t}(\eta(t)) + (u_0(t) u(t) - \widehat{u}(t))\, g_{1,t}(\eta(t)) \\
&\eta^0(0) = \widehat{\tau}, \quad \eta(0) = \widehat{x}_1 \quad \eta^0(\widehat{T}) \in \mathbb{R}, \quad \eta(\widehat{T}) \in \widehat{S}_{\widehat{T}}^{-1}(N_f).
\end{aligned} \tag{9}$$

The reference trajectory is transformed into the constant trajectory $t \to (\widehat{\tau}, \widehat{x}_1)$ and the reference flow is the constant one. Moreover $N_f^1 := \widehat{S}_{\widehat{T}}^{-1}(N_f)$ is the integral line of $g_{1,\widehat{T}}$ through $\widehat{x}_1$ and $\widehat{g}_t \equiv h_1$ for any $t \in [0, \widehat{\tau})$.

The second variation for problem (9), as defined in [2], is a quadratic form defined through the solutions of the linearization of system (9) given by

$$\dot{\delta\eta}^0(t) = \delta u_0(t)$$
$$\dot{\delta\eta}(t) = \delta u_0(t)\widehat{g}_t(\widehat{x}_1) + \delta u(t)g_{1,t}(\widehat{x}_1)$$
$$\delta\eta^0(0) = 0, \quad \delta\eta^0(\widehat{T}) \in \mathbb{R}$$
$$\delta\eta(0) = 0, \quad \delta\eta(\widehat{T}) \in \text{span}\{g_{1,\widehat{T}}(\widehat{x}_1)\}$$
$$(\delta u_0, \delta u) \in L^2((0,\widehat{T}), \mathbb{R}^2), \quad \delta u = 0 \text{ on } [0,\widehat{\tau}].$$

Integrating on the bang arc and setting $\varepsilon := \int_0^{\widehat{\tau}} \delta u_0(t)\,dt$ we obtain the equivalent system on $[\widehat{\tau}, \widehat{T}]$

$$\begin{aligned}
&\dot{\delta\eta}^0(t) = \delta u_0(t) &&\dot{\delta\eta}(t) = \delta u_0(t)\widehat{g}_t(\widehat{x}_1) + \delta u(t)g_{1,t}(\widehat{x}_1)\\
&\delta\eta^0(\widehat{\tau}) = \varepsilon, &&\delta\eta(\widehat{\tau}) = \varepsilon h_1(\widehat{x}_1)\\
&\delta\eta^0(\widehat{T}) \in \mathbb{R}, &&\delta\eta(\widehat{T}) = \varepsilon_f g_{1,\widehat{T}}(\widehat{x}_1)\\
&(\varepsilon, \varepsilon_f) \in \mathbb{R}^2, &&(\delta u_0, \delta u) \in L^2((\widehat{\tau}, \widehat{T}), \mathbb{R}^2).
\end{aligned} \tag{10}$$

Given any function β_1 such that $\beta_1|_{N_f^1} = 0$ and $d\beta_1(\widehat{x}_1) = -\widehat{\ell}_1 \in (\mathbb{R}^n)^*$, the quadratic form is given by

$$J''[\varepsilon, \varepsilon_f, \delta u_0, \delta u]^2 = \frac{\varepsilon^2}{2}h_1^2 \cdot \beta_1(\widehat{x}_1) +$$
$$+ \int_{\widehat{\tau}}^{\widehat{T}} \delta\eta(s) \cdot (\delta u_0(s)\,\widehat{g}_s + \delta u(s)\,g_{1,s}) \cdot \beta_1(\widehat{x}_1)\,ds \tag{11}$$

where $\varepsilon, \varepsilon_f, \delta u_0, \delta u, \delta\eta$ satisfy system (10). Define $v_0, v, v_1, w: [\widehat{\tau}, \widehat{T}] \to \mathbb{R}$ and $\varepsilon_0, \varepsilon_1$ by

$$\begin{aligned}
&\dot{v}_0(t) = -\delta u_0(t), &&\dot{v}(t) = -\delta u(t), &&\dot{v}_1(t) = -v_0(t)\dot{\widehat{u}}(t),\\
&v_0(\widehat{T}) = 0, &&v(\widehat{T}) = 0, &&v_1(\widehat{T}) = 0,\\
&w := v_1 + v, &&\varepsilon_0 := v_0(\widehat{\tau}), &&\varepsilon_1 := v_0(\widehat{\tau})\widehat{u}(\widehat{\tau}+) + w(\widehat{\tau}).
\end{aligned}$$

Integrating by parts as in [6], taking into account that on $[\widehat{\tau}, \widehat{T}]$

1. $[\widehat{g}_s, g_{1,s}] \cdot \beta_1(\widehat{x}_1) = \langle\lambda(s), [f_0, f_1](\widehat{\xi}(s))\rangle \equiv 0$
2. $[\widehat{g}_s, \dot{g}_{1,s}] \cdot \beta_1(\widehat{x}_1) = \langle\widehat{\lambda}(s), ([f_0, [f_0, f_1]] + \widehat{u}(s)[f_1, [f_0, f_1]])(\widehat{\xi}(s))\rangle \equiv 0$

and that $g_{1,\widehat{T}} \cdot g_{1,\widehat{T}} \cdot \beta_1(\widehat{x}_1) = 0$, we get that (11) can be written as

$$J''[\varepsilon, \varepsilon_f, \varepsilon_0, \varepsilon_1, w]^2 = \frac{1}{2}(\varepsilon h_1 + \varepsilon_0 f_0 + \varepsilon_1 f_1) \cdot (\varepsilon h_1 + \varepsilon_0 f_0 + \varepsilon_1 f_1) \cdot \beta_1(\widehat{x}_1) +$$
$$+ \frac{1}{2}\int_{\widehat{\tau}}^{\widehat{T}} (w^2(s)[\dot{g}_{1,s}, g_{1,s}] \cdot \beta_1(\widehat{x}_1) + 2w(s)\zeta(s) \cdot \dot{g}_{1,s} \cdot \beta_1(\widehat{x}_1))\,ds \tag{12}$$

where $(\varepsilon, \varepsilon_f, \varepsilon_0, \varepsilon_1, w) \in \mathbb{R}^4 \times L^2([\widehat{\tau}, \widehat{T}], \mathbb{R})$ satisfy the linear system

$$\zeta(s) = w(s)\dot{g}_{1,s}(\widehat{x}_1),$$ (13)

$$\zeta(\widehat{\tau}) = \varepsilon h_1(\widehat{x}_1) + \varepsilon_0 f_0(\widehat{x}_1) + \varepsilon_1 f_1(\widehat{x}_1), \qquad \zeta(\widehat{T}) = \varepsilon_f g_{1,\widehat{T}}(\widehat{x}_1).$$ (14)

Lemma 1. *The maximal sub–space on which we can require coercivity is the space $\mathcal{W}$ of $\mathbb{R}^4 \times L^2([\widehat{\tau}, \widehat{T}], \mathbb{R})$ of the quintuples $(\varepsilon, \varepsilon_f, \varepsilon_0, \varepsilon_1, w)$ such that $\varepsilon = \varepsilon_f = 0$ and the following linear system admits a solution ζ:*

$$\dot{\zeta}(t) = w(t)\dot{g}_{1,t}(\widehat{x}_1), \quad \zeta(\widehat{\tau}) = \varepsilon_0 f_0(\widehat{x}_1) + \varepsilon_1 f_1(\widehat{x}_1), \quad \zeta(\widehat{T}) = 0.$$ (15)

Namely $\mathcal{W}$ is the space one obtains considering the minimum time problem with fixed end–points $\xi(\widehat{\tau}) = \widehat{x}_1$ $\xi(\widehat{T}) = \widehat{x}_f$, see [6].

Proof. Since the vector fields h_1, f_0, f_1 are linearly dependent at $\widehat{x}_1$, it is clear that the quadratic form (12)–(14) cannot be coercive. Thus since $\zeta(\widehat{\tau})$ is constrained to $\mathrm{span}\{f_0, f_1\}(\widehat{x}_1)$, without loss of generality we restrict ourselves to $\varepsilon = 0$ obtaining a first reduction.

Choosing $\varepsilon_0 = 0$, $\varepsilon_1 = 1$, $\varepsilon_f = 1$ and $w \equiv 1$, we get $\zeta(t) = g_{1,t}(\widehat{x}_1) \, \forall t \in [\widehat{\tau}, \widehat{T}]$. Substituting in (12) and recalling that $g_{1,\widehat{T}} \cdot g_{1,\widehat{T}} \cdot \beta_1(\widehat{x}_1) = 0$ we get $J'' = 0$, so that the claim is proven. □

We shall call *extended second variation* J''_{ext} the quadratic form defined by the right–hand side of (12) on the space $\mathcal{W}$.

In the following lemma we state some easy consequences of the coercivity of J''_{ext} which allow to write the quadratic form more conveniently.

Lemma 2. 1. *If J''_{ext} is coercive then f_0 and f_1 are linearly independent along the singular arc of the reference trajectory.*

2. *J''_{ext} is independent of the choice of β_1 with property $\mathrm{d}\beta_1(\widehat{x}_1) = -\widehat{\ell}_1$.*

Proof. 1. Choose $\varepsilon_0 = 0$, $\varepsilon_1 = 1$ and $w(s) = 1$, $s \in [\widehat{\tau}, t]$, $w(s) = 0$, $s \in (t, \widehat{T}]$. An easy computation shows that coercivity implies that $g_{1,t}(\widehat{x}_1) \neq 0$ so that f_1 is not zero along the reference trajectory. Normality of the adjoint covector completes the proof.

2. Consider two functions β_1 and γ_1 such that $\mathrm{d}\beta_1(\widehat{x}_1) = \mathrm{d}\gamma_1(\widehat{x}_1) = -\widehat{\ell}_1$. Taking into account that we consider only solutions of system (15) with $\zeta(\widehat{T}) = 0$, an integration by parts shows that the right–hand side of (12) is the same if we use either β_1 or γ_1.

Thanks to the previous lemma we choose special coordinates around $\widehat{x}_1$ in such a way that $f_1 \equiv \frac{\partial}{\partial x_1}$, $f_0 = \frac{\partial}{\partial x_2} - x_1 ([f_0, f_1](\widehat{x}_1) + O(x))$.

In such coordinates choose $\beta_1(x) := -\sum_{i=2}^{n} \lambda_i x_i$, where $(0, \lambda_2, \ldots, \lambda_n)$ are the coordinates of $\widehat{\ell}_1$. We get $\lambda_2 = 1$, $f_1 \cdot \beta_1 \equiv 0$, and $f_0 \cdot f_0 \cdot \beta_1(\widehat{x}_1) = 0$. Thus the quadratic form in (12) can be written as

$$J''_{ext}[\varepsilon_0, \varepsilon_1, w]^2 = \frac{1}{2} \int_{\widehat{\tau}}^{\widehat{T}} (w^2(t)[\dot{g}_{1,t}, g_{1,t}] \cdot \beta_1(\widehat{x}_1) +$$

$$+ 2w(t)\, \zeta(t) \cdot \dot{g}_{1,t} \cdot \beta_1(\widehat{x}_1))\, dt$$ (16)

Assumption 2.1 (Coercivity) *The quadratic form defined in (16) is coercive on the subspace* W *of* $\mathbb{R}^2 \times L^2([\widehat{\tau}, \widehat{T}], \mathbb{R})$ *of the variations* $\delta e = (\varepsilon_0, \varepsilon_1, w)$ *such that system (15) admits solution.*

Remark 1. 1. J''_{ext} *is a quadratic form defined on the space* $\mathbb{R}^2 \times L^2([\widehat{\tau}, \widehat{T}], \mathbb{R})$, *but only its restriction to* W *is coordinate free.*

2. $\dot{g}_{1,t}(x) = D\widehat{S}_t^{-1}[f_0, f_1]\widehat{S}_t(x)$

3. *Since* $[\dot{g}_{1,t}, g_{1,t}] \cdot \beta_1(\widehat{x}_1) = \langle \widehat{\lambda}(t), [f_1, [f_0, f_1]](\widehat{\xi}(t)) \rangle$, *the coercivity of the extended second variation implies the so called* **strong generalised Legendre condition (SGLC)** *i.e.*

$$\langle \widehat{\lambda}(t), [f_1, [f_0, f_1]](\widehat{\xi}(t)) \rangle \geq k > 0 \ \forall t \in [\widehat{\tau}, \widehat{T}]. \tag{SGLC}$$

This specifies the singular arc as a *singular extremal of the first kind.*

Using the results of [8] we can state necessary and sufficient conditions for the coercivity of the extended second variation through the Hamiltonian and the Lagrangian subspace associated to J''_{ext}. Namely, let $H''_t: (\mathbb{R}^n)^* \times \mathbb{R}^n \to \mathbb{R}$ and L'' be defined by

$$H''_t(\omega, \delta x) = \frac{-1}{2R(t)} \left(\langle \omega, \dot{g}_{1,t}(\widehat{x}_1) \rangle + \delta x \cdot \dot{g}_{1,t} \cdot \beta_1(\widehat{x}_1) \right)^2,$$

$$L'' = \mathbb{R}(0, f_1(\widehat{x}_1)) \oplus \mathbb{R}(0, f_0(\widehat{x}_1)) \oplus \{f_1(\widehat{x}_1), f_0(\widehat{x}_1)\}^{\perp}. \tag{17}$$

An easy application of Theorem 2.6 in [2] proves the following lemma.

Lemma 3. *Let* $\mathcal{H}''_t$ *be the flow of the Hamiltonian system associated to* H''_t. *The followings are equivalent*

1. J''_{ext} *is coercive on* W
2. *If* $\pi_* \mathcal{H}''_t(\omega, \delta x) = 0$ *for some* $t \in [\widehat{\tau}, \widehat{T}]$ *and some* $(\omega, \delta x) \in L''$ *then* $\mathcal{H}''_s(\omega, \delta x) = (\omega, 0)$ *for any* $s \in [\widehat{\tau}, t]$.

3 Optimality and Stability Results

The results on optimality and structural stability can be obtained under some regularity assumptions which are known to be the streghtening of necessary conditions derived from PMP, namely:

Assumption 3.1 $u_1 \langle \widehat{\lambda}(t), f_1(\widehat{\xi}(t)) \rangle > 0 \ \ \forall t \in [0, \widehat{\tau})$.

Assumption 3.2 $\langle \widehat{\lambda}(\widehat{\tau}), (u_1[f_0, [f_0, f_1]] + [f_1, [f_0, f_1]]) (\widehat{\xi}(\widehat{\tau})) \rangle > 0$. *Equivalently, under (SGLC),* $\widehat{u}$ *is discontinuous at time* $\widehat{\tau}$, *see e.g. [6].*

The following theorem gives the strong local optimality result for the reference trajectory $\widehat{\xi}$. The proof will appear elsewhere, although it works as in [6].

Theorem 1. *Suppose that $\widehat{\xi}$ is a bang–singular trajectory satisfying Assumptions 1.1–3.1–3.2.*

1. *If the quadratic form J''_{ext} is coercive on the subspace of $\mathcal{W}$ defined by $\varepsilon_0 = 0$, then $\widehat{\xi}$ is a strict (time, state)–local minimiser for the minimum time problem between $\widehat{x}_0$ and N_f.*
2. *If the quadratic form J''_{ext} is coercive on $\mathcal{W}$, and $\widehat{\xi}$ has no self–intersection, then it is a strict state–local minimiser for the minimum time problem between $\widehat{x}_0$ and N_f.*

Concerning structural stability issues, the work is in progress. We announce here some preliminary result. We consider problem (**P₀**) as the nominal problem for the following parametrised family of minimum time problems

$$(\mathbf{P_r}) \qquad \dot{\xi}(t) = f_0^r(\xi(t)) + u(t)f_1^r(\xi(t)) \tag{18}$$

$$\xi(0) = a^r, \quad \xi(T) \in N_f^r, \quad u(t) \in [-1, 1] \tag{19}$$

where $r \in \mathbb{R}^k$ and N_f^r is a given integral line of f_1^r, i.e. $N_f^r = \{\exp s f_1^r(y^r): s \in \mathbb{R}\}$. We assume that all the data depend smoothly on $r \in \mathbb{R}^m$ and we point out that the special form of the final constraint is maintained in (**P_r**).

We add the following assumption on (**P₀**) which can be proven to be equivalent to controllability of system (15),

Assumption 3.3 $\widehat{\lambda}\big|_{[\widehat{\tau},\widehat{T}]}$ is the unique adjoint covector associated to $\widehat{\xi}\big|_{[\widehat{\tau},\widehat{T}]}$ for the minimum time problem between $\widehat{x}_1$ and $\widehat{x}_f$.

The preliminary result that we are able to prove using Hamiltonian methods is the following lemma.

Lemma 4. *Under Assumptions 1.1, 2.1, 3.1, 3.2 and 3.3, there exist $\delta > 0$ such that for any $r \in \mathbb{R}^k$ with $\|r\| < \delta$, there exists an unique normal bang–singular extremal of (**P_r**), say (λ^r, ξ^r) with switching time τ^r and final time T^r such that*

$$\|\lambda^r(0) - \widehat{\lambda}(0)\| + \|\xi^r(T^r) - \widehat{\xi}(\widehat{T})\| + |\tau^r - \widehat{\tau}| + \left|T^r - \widehat{T}\right| < \delta.$$

*Moreover such extremal is a strict (time, state)–minimiser of (**P_r**) and, if it has no self–intersection, it is a strict state–minimiser.*

4   The Dodgem Car Problem

Consider the Dodgem car problem as defined in [3], i.e. the minimum time problem for the following control system in $\mathbb{R}^3$, see also [4] where some stability results are proven.

$$
\begin{aligned}
\dot{x}_1(t) &= \cos(x_3) & x_1(0) &= h & x_1(T) &= 0 \\
\dot{x}_2(t) &= \sin(x_3) & x_2(0) &= 0 & x_2(T) &= 0 & u \in [-1, 1]. \\
\dot{x}_3(t) &= u & x_3(0) &= \frac{\pi}{2} & x_3(T) &\in \mathbb{R}
\end{aligned} \tag{20}
$$

It is known that for $h > 2$ there is a bang-singular optimiser and that any singular control is null. In [5] we showed that any bang-singular extremal of this problem satisfies Assumptions 3.1 - 3.2 and the coercivity assumption on the subspace of $\mathcal{W}$, defined by $\varepsilon_0 = 0$.

It turns out that any bang-singular extremal of (20) satisfies also Assumption 2 of Theorem 1 and Assumption 3.3. Indeed in this case an easy computation shows that the extremal trajectory has no self–intersection. Moreover $\dot{g}_{1,t} = [f_0, f_1]$ so that the dynamics of system (15) reduces to $\dot{\zeta}(t) = w(t)[f_0, f_1](\widehat{x}_1)$.

Since $f_0, f_1, [f_0, f_1]$ are linearly independent on $\mathbb{R}^3$, system (15) can be solved if and only if $\varepsilon_0 = \varepsilon_1 = 0$ and $\int_{\widehat{\tau}}^{\widehat{T}} w(t)\,\mathrm{d}t = 0$ so that the quadratic form (16) is coercive.

Concerning Assumption 3.3, again the linear independence of $f_0, f_1, [f_0, f_1]$ determines $\widehat{\lambda}(\widehat{\tau})$ and hence the adjoint covector.

Therefore any bang-singular extremal is a strict state-local optimiser and this property is stable (in the sense of Lemma 4) under small perturbations of the data which preserve the final constraint as an integral manifold of the controlled vector field.

Notice that our result does not contradict the results in [4], where under a perturbation which does not preserve the special form of the final constraint, a new bang arc appears after the singular one. It is the opinion of the authors that this occurs because the *natural* second variation for the nominal problem is not coercive, see Lemma 3.

References

1. Agrachev, A.A., Sachkov, Y.L.: Control Theory from the Geometric Viewpoint. Springer, Heidelberg (2004)
2. Agrachev, A.A., Stefani, G., Zezza, P.: An invariant second variation in optimal control. Internat. J. Control 71(5), 689–715 (1998)
3. Craven, B.D.: Control and optimization. Chapman & Hall (1995)
4. Felgenhauer, U.: Structural Stability Investigation of Bang-Singular-Bang Optimal Controls. Journal of Optimization Theory and Applications 152, 605–631 (2012)
5. Poggiolini, L., Stefani, G.: Sufficient optimality conditions for a bang–singular extremal in the minimum time problem. Control and Cybernetics 37(2), 469–490 (2008)
6. Poggiolini, L., Stefani, G.: Bang-singular-bang extremals: sufficient optimality conditions. Journal of Dynamical and Control Systems 17(4), 469–514 (2011)
7. Pontryagin, L.S., Boltyanskii, V.G., Gamkrelidze, R.V., Mishenko, E.F.: The Mathematical Theory of Optimal Processes. Wiley (1962)
8. Stefani, G., Zezza, P.: Constrained regular LQ-control problems. SIAM J. Control Optim. 35(3), 876–900 (1997)

Perturbation Bounds for the Nonlinear Matrix Equation $X + A^H X^{-1} A + B^H X^{-1} B = I$

Ivan Popchev[1], Petko Petkov[2], Mihail Konstantinov[3], and Vera Angelova[1]

[1] Institute of Information and Communication Technologies,
Bulgarian Academy of Sciences, 1113 Sofia, Bulgaria
{ipopchev,vangelova}@iit.bas.bg
[2] Department of Automatics, Technical University of Sofia
1756 Sofia, Bulgaria
php@tu-sofia.acad.bg
[3] University of Architecture, Civil Engineering and Geodesy
1046 Sofia, Bulgaria
mmk_fte@uacg.bg

Abstract. In this paper we make a complete perturbation analysis of the nonlinear matrix equation $X + A^H X^{-1} A + B^H X^{-1} B = I$, where A and B are square complex matrices, A^H denotes the complex conjugate transpose of the matrix A and I is the identity matrix. We obtain local (first order) perturbation bounds and a non-local perturbation bound for the solution to the equation. The perturbation bounds allow to derive condition and accuracy estimates for the computed solution, when using a stable numerical algorithm to solve the equation.

Keywords: perturbation analysis, nonlinear matrix equations, condition numbers, perturbation bounds, Lyapunov majorants.

1 Introduction

The efficient and reliable numerical computations require knowledge on the convergence and way of bounding errors in the computed results.

In this paper we consider the sensitivity of the nonlinear matrix equation

$$X + A^H X^{-1} A + B^H X^{-1} B = I \,, \tag{1}$$

with A and B being square complex matrices, and I being the identity matrix. The notation A^H stands for the complex conjugate transpose of the matrix A. The problem of solving the matrix equation (1) is related to the problem of solving the system of linear equations $Px = f$. When in solving the system of linear equations, it is transformed to two linear systems with lower and upper triangular block coefficient matrix, respectively, the existence of the matrix decomposition is related to the solution of equation (1) (see [3] for examples). It is proved in [3] that equation (1) has a positive definite solution and two iterative algorithms for finding the positive definite solution of equation (1) are proposed.

I. Lirkov, S. Margenov, and J. Waśniewski (Eds.): LSSC 2011, LNCS 7116, pp. 155–162, 2012.

Many authors have studied the sensitivity of the particular cases of (1) - the matrix equations $X \pm A^* X^{-1} A = Q$ and $X \pm A^* X^{-1} A = I$, where A^* denotes the complex conjugate transpose of A in the complex case and the transpose of A in the real case. Several perturbation bounds are proposed. In [7], a comparative analysis of the effectiveness and the reliability of these bounds on the base of several non-trivial numerical examples is made. The results show that our bounds from [5], based on the Fréchet derivatives, are reliable, accurate and give the sharpest estimates. The bounds from [2] do not reach the estimated value and due to constructive reasons, have closer field of application. The other considered bounds are more pessimistic. In [8] linear sensitivity analysis of the matrix inequality $(A + B_k)^\top P(A + B_k) - P < C^\top C$, $P > 0$ is considered. The analysis is done in a similar way as for matrix equality after introducing a suitable right hand side, which is slightly perturbed.

In this paper we make a complete perturbation analysis of equation (1). To the best of our knowledge, this is the first study on the sensitivity of the non-linear matrix equation (1). Local perturbation bounds are derived using the Fréchet derivatives. To derive the non-local perturbation bound, the perturbation analysis problem is written in equivalent form as a matrix equation for the perturbation in the solution, and the technique of Lyapunov majorants and the fixed point principle of Schauder are used. The local bounds are obtained neglecting second and higher order terms and are only asymptotically valid. For the calculation of the condition numbers and the local bounds, the technique [4], based on the theory of additive operators, must be applied because the function $A \to A^H$ is not homogeneous. The non-local bound gives a domain of admissible perturbations and a non-linear function, which estimates the perturbations in the solution to equation (1). The inclusion of the perturbations to the domain of admissibility guaranties that the perturbed equation has a unique solution in a neighbourhood of the unperturbed solution. The non-local bound is rigorous, but may not exist or may be pessimistic in some cases.

Throughout the paper, we denote by $\mathbb{C}^{n \times n}$ and $\mathbb{R}^{n \times n}$ the sets of $n \times n$ complex and real matrices, respectively; $A^\top$ denotes the transpose of A; $A \otimes B = (a_{ij} B)$ denotes the Kronecker product of A and B; $\text{vec}(A) = (a_1^\top, a_2^\top, \ldots, a_n^\top)^\top$ is the vector representation of the matrix A, where $A = (a_{ij}), a_1, \ldots, a_n \in \mathbb{C}^n$ are the columns of A; $\|.\|_2$ and $\|.\|_F$ are the Euclidean and the Frobenius matrix norm, respectively. The notation ':=' stands for 'equal by definition'.

The paper is organised as follows. The problem is briefly stated in section 2. In section 3, local perturbation bounds are derived. The non-local perturbation bound is obtained in section 4. In section 5, the effectiveness of the bounds proposed is illustrated by three numerical examples. Section 6 contains our conclusions.

2   Statement of the Problem

Rewrite equation (1) in the equivalent form as an operator equation

$$F(X, S) := X + A^H X^{-1} A + B^H X^{-1} B - I = 0, \tag{2}$$

where $S := (A, B)$ denotes the collection of data matrices A and B. Consider the perturbed operator equation

$$F(X + \delta X, S + \delta S) := X + \delta X + (A + \delta A)^{\mathrm{H}}(X + \delta X)^{-1}(A + \delta A) \qquad (3)$$
$$+ (B + \delta B)^{\mathrm{H}}(X + \delta X)^{-1}(B + \delta B) - I = 0 ,$$

where the data matrices A and B from (2) are perturbed with small perturbations δA, δB, respectively. The perturbed data result in perturbation δX in the solution $X + \delta X$ to the perturbed equation (3). We assume that the unperturbed equation (2) and the perturbed equation (3) have positive definite solutions X and $X + \delta X$, respectively. Our goal is to estimate the norm of the perturbation δX in the solution by the norm of the perturbations δA and δB in the data matrices A and B, respectively.

3 Local Perturbation Bounds

Based on the identity

$$(X + \delta X)^{-1} = X^{-1} - X^{-1}\delta X X^{-1} + (X^{-1}\delta X)^2(I_n + X^{-1}\delta X)^{-1}X^{-1} ,$$

we rewrite the perturbed equation (3) in terms of the Fréchet derivatives

$$F(X + \delta X, S + \delta S) = F(X, S) + F_X(X, S)(\delta X) \qquad (4)$$
$$+ \sum_{Z \in S} F_Z(X, S)(\delta Z) + G(X, S)(\delta X, \delta S) ,$$

where

$$F_X(X, S)(\delta X) := \delta X - A^{\mathrm{H}}X^{-1}\delta X X^{-1}A - B^{\mathrm{H}}X^{-1}\delta X X^{-1}B : \mathbb{C}^{n \times n} \to \mathbb{C}^{n \times n}$$

is the partial Fréchet derivative of $F(X, S)$ in X. The terms $F_Z(X, S)(\delta Z)$ are the partial Fréchet pseudo-derivatives of $F(X, S)$ in Z, and Z stands for A and B

$$F_A(X, S)(Y) := Y^{\mathrm{H}}X^{-1}A + A^{\mathrm{H}}X^{-1}Y : \mathbb{C}^{n \times n} \to \mathbb{C}^{n \times n} ,$$
$$F_B(X, S)(Y) := Y^{\mathrm{H}}X^{-1}B + B^{\mathrm{H}}X^{-1}Y : \mathbb{C}^{n \times n} \to \mathbb{C}^{n \times n} .$$

The operators $F_A(X, S)$ and $F_B(X, S)$ are additive but not homogeneous. The term $G(X, S)(\delta X, \delta S)$ contains second and higher order terms in δX, δS.

We suppose that the operator $F_X(X, S)(\delta X)$ is invertible. Then, from equation (4) it follows that

$$\delta X = -F_X^{-1}(\delta A^{\mathrm{H}}X^{-1}A + A^{\mathrm{H}}X^{-1}\delta A)$$
$$- F_X^{-1}(\delta B^{\mathrm{H}}X^{-1}B + B^{\mathrm{H}}X^{-1}\delta B) - F_X^{-1}(G(X, S)(\delta X, \delta S)) ,$$

or in vector form

$$\mathrm{vec}(\delta X) = - \sum_{Z \in S} L^{-1}\left(((X^{-1}Z)^{\top} \otimes I_n)\mathrm{vec}(\delta Z^{\mathrm{H}}) + (I_n \otimes Z^{\mathrm{H}}X^{-1})\mathrm{vec}(\delta Z)\right) (5)$$
$$+ \mathrm{O}(\delta X, \delta S)^2 .$$

Here $L := \mathrm{Mat}(F_X(X,S)) = I_{n^2} - (X^{-1}A)^\top \otimes (A^H X^{-1}) - (X^{-1}B)^\top \otimes (B^H X^{-1})$ is the matrix of the operator $F_X(X,S)$.

From equation (5) and applying the technique [4] for additive complex operators, we obtain

$$\mathrm{vec}(\delta X) = \sum_{Z \in S} M_Z \mathrm{vec}(\delta Z) + \mathrm{O}(\delta X, \delta S)^2 \,, \tag{6}$$

where $M_Z := \begin{bmatrix} L_{Zr} + L_{\bar{Z}r} & L_{Zi} - L_{\bar{Z}i} \\ L_{Zi} + L_{\bar{Z}i} & L_{Zr} - L_{\bar{Z}r} \end{bmatrix}$ with

$$L_Z = L_{Zr} + iL_{Zi} := -L^{-1}\left[I_n \otimes (Z^H X^{-1}) \right],$$
$$L_{\bar{Z}} = L_{\bar{Z}r} + iL_{\bar{Z}i} := -L^{-1}\left[((X^{-1}Z)^\top \otimes I_n)\mathcal{P}_{n^2} \right], \quad Z \in S\,.$$

Here, $\mathcal{P}_{n^2} \in \mathbb{C}^{n^2 \times n^2}$ is the so called vec-permutation matrix such that for each $Y \in \mathbb{C}^{n \times n}$ it is fulfilled $\mathrm{vec}(Y^\top) = \mathcal{P}_{n^2}\mathrm{vec}(Y)$.

Set $\delta_X := \|\delta X\|_F$, $\delta_A := \|\delta A\|_F$, $\delta_B := \|\delta B\|_F$, $\delta := \begin{bmatrix} \delta_A & \delta_B \end{bmatrix}^\top \in \mathbb{R}_+^2$, On the base of equation (6), we formulate

- absolute condition numbers K_Z, $Z \in S$ for the solution of equation (1), relative to the matrix coefficients A and B

$$K_Z := \|M_Z\|_2 \,,$$

- relative condition numbers k_Z, $Z \in S$ for the solution of equation (1), relative to the matrix coefficients A and B

$$k_Z := \frac{K_Z \|Z\|_2}{\|X\|_2} \,,$$

- first local estimate, based on the condition numbers

$$\delta_X \le \mathrm{est}_1(\delta) + \mathrm{O}(\|\delta\|_2^2), \quad \delta \to 0 \,,$$
$$\mathrm{est}_1(\delta) := \sum_{Z \in S} K_Z \delta_Z \,,$$

- second local estimate

$$\delta_X \le \mathrm{est}_2(\delta) + \mathrm{O}(\|\delta\|_2^2), \quad \delta \to 0 \,, \tag{7}$$
$$\mathrm{est}_2(\delta) := \left\| \begin{bmatrix} M_A & M_B \end{bmatrix} \right\|_2 \|\delta\|_2 \,,$$

- third local estimate

$$\delta_X \le \mathrm{est}_3(\delta) + \mathrm{O}(\|\delta\|_2^2), \quad \delta \to 0 \,, \tag{8}$$
$$\mathrm{est}_3(\delta) := \sqrt{\delta^\top \begin{bmatrix} \|M_A^\top M_A\|_2 & \|M_A^\top M_B\|_2 \\ \|M_B^\top M_A\|_2 & \|M_B^\top M_B\|_2 \end{bmatrix} \delta} \,.$$

The bound $est_3(\delta)$ is always superior to the bound $est_1(\delta)$ for all δ. The bounds $est_2(\delta)$ and $est_3(\delta)$ are alternative for some non-negative vector δ. Hence, we formulate the following local perturbation bound for the solution to equation (1)

Theorem 1. *For small perturbations in the data, the Frobenius norm δ_X of the perturbation δX in the solution X of equation (1) satisfies the local perturbation bound*

$$\delta_X \leq g(\delta) + O(\|\delta\|_2^2), \quad \delta \to 0 , \tag{9}$$
$$g(\delta) := \min\{est_2(\delta), est_3(\delta)\} ,$$

where $est_2(\delta)$ and $est_3(\delta)$ are given by (7), (8), respectively.

4   Non-local Perturbation Bound

Consider the perturbed equation written in terms of the Fréchet derivatives (4) under the assumption of invertibility of the operator $F_X(X, S)(\delta X)$. In the local bound, we neglected the terms of second and higher order in δX and δS. Here, we consider equation (4) in its entirety and for this purpose we write out the term $G(X, S)(\delta X, \delta S)$. We rewrite equation (4) in the equivalent form

$$F_X(X, S)(\delta X) = \Phi_0(\delta S) + \Phi_1(\delta X, \delta S) + \Phi_2(\delta X, \delta S) ,$$

where

$$\Phi_0(\delta S) := - \sum_{Z \in S} (Z^H X^{-1} \delta Z + \delta Z^H X^{-1} Z + \delta Z^H X^{-1} \delta Z) ,$$

$$\Phi_1(\delta X, \delta S) := \sum_{Z \in S} (Z^H X^{-1} \delta X X^{-1} \delta Z + \delta Z^H X^{-1} \delta X X^{-1} Z$$
$$+ \delta Z^H X^{-1} \delta X X^{-1} \delta Z) ,$$

$$\Phi_2(\delta X, \delta S) := - \sum_{Z \in S} (Z + \delta Z)^H E(Z + \delta Z) .$$

Here $E := (X^{-1} \delta X)^2 (I_n + X^{-1} \delta X)^{-1} X^{-1}$.

As the operator $F_X(X, S)(\delta X)$ is invertible by assumption, we get the operator equation

$$\delta X = \Phi(\delta X, \delta S) := F_X^{-1}(\Phi_0) + F_X^{-1}(\Phi_1) + F_X^{-1}(\Phi_2) . \tag{10}$$

Set $\mu := \|X^{-1}\|_2$, $l := \|L^{-1}\|_2$ and suppose that $\delta_X \leq \rho$, where $\rho < 1/\mu$ is a positive quantity. The following inequality holds true

$$\|F(\delta X, \delta S)\|_F \leq h(\rho, \delta) := a_0(\delta) + a_1(\delta)\rho + \frac{a_2(\delta)\rho^2}{1 - \mu\rho} .$$

Here

$$a_0(\delta) := g(\delta) + l\mu(\delta_A^2 + \delta_B^2) ,$$

$$a_1(\delta) := \sum_{Z \in S} \left[\mu(\|L^{-1}(I_n \otimes Z^H X^{-1})\|_2 \right.$$

$$\left. + \|L^{-1}((X^{-1}Z)^T \otimes I_n)\mathcal{P}_{n^2}\|_2)\delta_Z + l\mu^2\delta_Z^2 \right] ,$$

$$a_2(\delta) := \sum_{Z \in S} \mu^3 \left[\|L^{-1}(Z^T \otimes Z^H)\|_2 \right.$$

$$\left. + \|L^{-1}(Z^T \otimes I_n)\mathcal{P}_{n^2} + L^{-1}(I_n \otimes Z^H)\|_2\delta_Z + l\delta_Z^2 \right] .$$

The function $h(\rho, \delta)$ is a Lyapunov majorant for the operator $\Phi(\delta X, \delta S)$, see [1,4,6]. The corresponding majorant equation is $\rho = h(\rho, \delta)$. For $\rho < 1/\mu$ it is equivalent to the quadratic equation

$$(a_2(\delta) + \mu(1 - a_1(\delta)))\rho^2 - (1 - a_1(\delta) + \mu a_0(\delta))\rho + a_0(\delta) = 0 .$$

The majorant equation $\rho = h(\rho, \delta)$ has a root

$$\rho(\delta) := f(\delta) = \frac{2a_0(\delta)}{1 - a_1(\delta) + \mu a_0(\delta) + \sqrt{d(\delta)}} , \quad (11)$$

$$d(\delta) := (1 - a_1(\delta) + \mu a_0(\delta))^2 - 4a_0(\delta)(a_2(\delta) + \mu(1 - a_1(\delta)))$$

for

$$\delta \in \Omega , \quad (12)$$

$$\Omega := \left\{ \delta \in \mathbb{R}_+^2 : a_1(\delta) - \mu a_0(\delta) + 2\sqrt{a_0(\delta)(a_2(\delta) + \mu(1 - a_1(\delta)))} \le 1 \right\} .$$

Hence, for $\delta \in \Omega$ the operator $\Phi(., \delta S)$ maps the closed central ball

$$\mathcal{B}_{f(\delta)} := \left\{ \text{vec}(\delta X) \in \mathbb{C}^{n^2} : \|\text{vec}(\delta X)\|_2 \le f(\delta) \right\}$$

of radius $f(\delta)$ into itself, where $f(\delta)$ is continuous and $f(0) = 0$. Then, according to the Schauder fixed point principle, there exists a solution $\delta X \in \mathcal{B}_{f(\delta)}$ of the operator equation (10). In what follows we can formulate the theorem

Theorem 2. *For $\delta \in \Omega$ (12) the non-local perturbation bound $\|\delta X\|_F \le f(\delta)$ is valid for equation (1). The function $f(\delta)$ is determined by (11).*

5   Numerical Examples

In this section, we illustrate the effectiveness of the local bound $g(\delta)$ (9) and the non-local bound $f(\delta)$, (11), Ω (12), proposed in Theorems 1 and 2. We design three numerical problems for estimation of the perturbation in the solution to equation (1) - a real and a complex third order equations and a real fifth order equation. For the perturbed equation the perturbations in the data coefficients are chosen as

$$\delta A = V \delta A_0 V, \quad \delta B = V \delta B_0 V, \text{ with } V = I_3 - 2vv^\top/3, \quad v = \begin{bmatrix} 1 & 1 & 1 \end{bmatrix},$$
$$\delta A_0 = 10^{-j}\text{diag}(0.1 \ 0.2 \ -0.1) \text{ and } \delta B_0 = 10^{-j}\text{diag}(0 \ -0.1 \ 0.1),$$
$$j = 10, 9, \ldots, 1$$

for the third order equation, and

$$\delta A_0 = 10^{-j}\text{diag}(0.1 \ 0.2 \ -0.1 \ -0.2 \ 0.1), \quad \delta B_0 = 10^{-j}\text{diag}(0 \ 0.1 \ 0 \ -0.1 \ 0)$$

for the fifth order equation. These problems are designed so as to have closed form solutions X and δX to equation (1) and to the error equation, respectively.

Example 5.1

Consider the non-linear third order real matrix equation (1) with matrix coefficients $A = V A_0 V$, $B = V B_0 V$, where

$$A_0 = 10^{-1}\text{diag}(9.3058 \ 9.3462 \ 9.4312),$$
$$B_0 = 10^{-1}\text{diag}(2.0514 \ 2.0778 \ 3.0835).$$

In Table 1 the ratios of the perturbation bound to the estimated value for $j = 10, 8, 6, 4, 2$ are listed. The case, when the condition of existence (12) of the non-local bound is violated, is denoted by asterisk.

Example 5.2

Consider the non-linear fifth order real matrix equation (1) with matrix coefficients $A = V A_0 V$, $B = V B_0 V$,

$$A_0 = 10^{-1}\text{diag}(7.9369 \ 5.5019 \ 6.5012 \ 7.2362 \ 5.1788)$$
$$B_0 = 10^{-1}\text{diag}(9.6412 \ 8.1557 \ 9.1476 \ 6.6802 \ 7.8844).$$

Table 1. Numerical examples

		$j = 2$	$j = 4$	$j = 6$	$j = 8$	$j = 10$
		Example 5.1				
$\frac{g(\delta)}{\|\delta X\|_F}$	(9)	1.0856	1.0860	1.0860	1.0860	1.0860
$\frac{f(\delta)}{\|\delta X\|_F}$	(11, 12)	*	1.2134	1.0870	1.0860	1.0860
		Example 5.2				
$\frac{g(\delta)}{\|\delta X\|_F}$	(9)	3.1944	3.1949	3.1949	3.1943	3.1949
$\frac{f(\delta)}{\|\delta X\|_F}$	(11, 12)	*	*	*	3.2138	3.1951
		Example 5.3				
$\frac{g(\delta)}{\|\delta X\|_F}$	(9)	6.7684	6.8247	6.8252	6.8252	6.8252
$\frac{f(\delta)}{\|\delta X\|_F}$	(11, 12)	*	6.8986	6.8260	6.8252	6.8252

The ratios of the perturbation bound to the Frobenius norm of the perturbation in the solution are listed in Table 1. The cases, when the condition of existence (12) of the non-local bound is violated, are denoted by asterisk.

Example 5.3

Consider equation (1) of third order with the complex matrix

$$A = V A_0 V, \quad A_0 = 10^{-2} \text{diag}(5.4462 + i8.2163 \ \ 5.4462 - i8.2163 \ \ 25.1076)$$

and the real matrix

$$B = V B_0 V, \quad B_0 = 10^{-1} \text{diag}(1.6180 \ -2.9470 \ -1.1510) .$$

For the ratios of the perturbation bound to δ_X we obtain the results, listed in Table 1.

6   Conclusions

The results of the numerical examples show that for the considered class of numerical problems, the perturbation bounds, proposed in this paper, give satisfactory accurate estimates of the perturbation in the solution to equation (1).

Acknowledgments. The research work was partially supported by grant 210182.

References

1. Grebenikov, E.A., Ryabov, Y.A.: Constructive Methods for Analysis of Nonlinear Systems, Nauka, Moscow (1979) (in Russian)
2. Hasanov, V.I.: Notes on Two Perturbation Estimates of the Extreme Solutions to the Equations $X \pm A^* X^{-1} A = Q$. Appl. Math. Comput. 216, 1355–1362 (2010)
3. Long, J.-H., Hu, X.-Y., Zhang, L.: On the Hermitian Positive Definite Solution of the Nonlinear Matrix Equation $X + A^* X^{-1} A + B^* X^{-1} B = I$. Bull. Braz. Math. Soc., New Series. Sociedade Brasileira de Matemática, vol. 39(3), pp. 371–386 (2008)
4. Konstantinov, M.M., Gu, D.W., Mehrmann, V., Petkov, P.H.: Perturbation Theory for Matrix Equations. North-Holland, Amsterdam (2003) ISBN 0-444-51315-9
5. Konstantinov, M.M., Petkov, P.H., Angelova, V.A., Popchev, I.P.: Sensitivity of a Complex Fractional–Affine Matrix Equation. In: Jub. Sci. Conf. Univ. Arch. Civil Eng. Geod., Sofia, vol. 8, pp. 495–504 (2002)
6. Konstantinov, M.M., Petkov, P.H.: The Method of Splitting Operators and Lyapunov Majorants in Perturbation Linear Algebra and Control. Numer. Func. Anal. Optim. 23, 529–572 (2002)
7. Popchev, I.P., Angelova, V.A.: On the Sensitivity of the Matrix Equations $X \pm A^* X^{-1} A = Q$. Cybernetics and Information Technologies 10(4), 36–61 (2010)
8. Yonchev, A.: Full Perturbation Analysis of the Discrete-Time LMI Based Bounded Output Energy Problem. In: Conf. of Automatics and Informatics, Sofia, vol. 1, pp. 5–8 (2010)

Part V

Applications of Metaheuristics to Large-Scale Problems

Sensitivity Analysis for the Purposes of Parameter Identification of a *S. cerevisiae* Fed-Batch Cultivation

Maria Angelova and Tania Pencheva

Institute of Biophysics and Biomedical Engineering
- Bulgarian Academy of Sciences
105 Acad. G. Bonchev Str., 1113 Sofia, Bulgaria
{maria.angelova,tania.pencheva}@clbme.bas.bg

Abstract. The application of the sensitivity analysis to parameter identification of a *S. cerevisiae* fed-batch cultivation is presented. Parameter identification of a cultivation described with a fifth order non linear mathematical model is a difficult task because of the high number of parameters to be estimated. The aim of the study is the sensitivity analysis to be applied to determine the most significant parameters and to answer the question which parameters are most easily estimated. For that purpose a sensitivity model of a *S. cerevisiae* fed-batch cultivation is developed and as a result a stepwise parameter identification procedure is proposed.

1 Introduction

Biotechnological and particularly fermentation processes (FP) are widely used in different branches of industry. FP as a fundamental mode takes an important part in the production of pharmaceuticals, chemicals and enzymes, yeast, foods and beverages. Live microorganisms play a crucial role in such processes so their peculiarities predetermine some special features of FP as modeling and control objects. As complex, nonlinear, dynamic systems with interdependence and time-varying process variables, FP are a serious challenge for modelling and further high-quality control. Many parameters involved in FP models are often unidentifiable and meanwhile high accuracy in model parameters estimation is essential for developed models adequacy. The theory of sensitivity analysis (SA) extensively employed for chemical reactors is of recent origin for biological reactions [3,5]. SA applied for biotechnological processes including FP [2,7,8] allows the influence of the models' parameters to be investigated. SA also can help in solving the parameter estimation problem. Obtained knowledge is a powerful tool for elucidation a system's behavior due to variations in the parameters that affect the system dynamics.

The present study focuses on the sensitivity analysis of the fifth order non-linear model determined the interdependence between main state variables of *S. cerevisiae* fed-batch cultivation, namely biomass, substrate, ethanol and dissolved oxygen concentrations. For that purpose, some remarks on sensitivity

I. Lirkov, S. Margenov, and J. Waśniewski (Eds.): LSSC 2011, LNCS 7116, pp. 165–172, 2012.

analysis are firstly presented, followed by the description of the fermentation process model. In order to provide a precisely and accurate model parameter identification a sensitivity analysis using sensitivity functions is applied that results in a stepwise parameter identification procedure, demonstrated on the *S. cerevisiae* fed-batch cultivation.

2 Sensitivity Analysis

As an important tool to clarify the model characteristic, the sensitivity analysis can be used to express the influence of a parameter change on the state variables. Therefore, sensitivity analysis can provide strong indications as to which parameters are the most difficult to identify either because of their limited influence on the total system behaviour, or due to the fact that some parameters compensate for the effects of others. High sensitivity to a parameter suggests that the system's performance can drastically change with small variations in the parameter. Vice versa, low sensitivity suggests little change in the performance. Accurate and efficient calculation of parametric sensitivity coefficients (i.e., values of first and second partial derivatives of the state variables with respect to the model parameters) is essential for many aspects of parameter estimation in nonlinear models of processes in chemical engineering, especially determination of gradients for optimization and evaluation of parameter precision.

A general mathematical model of fed-batch cultivation can be given as [7,8]:

$$\frac{dx_j}{dt} = f_j(x_1, \ ..., \ x_m, \ t, p_1, \ ..., p_n) \tag{1}$$

where x_j are state variables and p_i are model parameters.

By definition sensitivity functions are the partial derivatives of the state variables with respect to the parameters. According to [4] sensitivity functions are defined as follows:

$$s_{ji} = \frac{\partial x_j\,(p_i,\ t)}{\partial p_i}\,|_{p=p_0} \ , \ i = 1 \div n, \ j = 1 \div m, \tag{2}$$

where s_{ji} are the sensitivity function of i-th parameter according j-th variable.

Differentiation of (1) regarding p leads to

$$\frac{\partial f_j\,(p,\ t)}{\partial p_i} = \sum_{j=1}^{m} \frac{\partial f_j\,(x,\ p,\ t)}{\partial x_j}\frac{\partial x_j}{\partial p_i} + \frac{\partial f_j\,(x,\ p,\ t)}{\partial p_i} \tag{3}$$

The derivatives $\frac{\partial x_j}{\partial p_i}$ are obtained by solving the sensitivity equation:

$$\frac{ds_{ji}}{dt} = \sum_{j=1}^{m} \frac{\partial f_j}{dx_j} s_{ji}\,(t,p) + \frac{\partial f_j}{dp_i} \tag{4}$$

Thus sensitivity model of considered system is formed from the mathematical model (1) and the sensitivity equations (4).

3   Mathematical and Sensitivity Model of *S. cerevisiae* Fed-Batch Cultivation

Sensitivity analysis of local models parameters are based on experimental data set from *S. cerevisiae* fed-batch cultivation obtained in the *Institute of Technical Chemistry - University of Hannover*, Germany [6]. The cultivation of the yeast is performed in a 1.5 l reactor, using a Schatzmann medium. Glucose in feeding solution is 50 g/l. The temperature was controlled at 30°C, the pH at 5.7. The stirrer speed was set to 500 rpm. The aeration rate was kept at 300 l/h. Biomass and ethanol were measured off-line while substrate (glucose) and dissolved oxygen were measured on-line. The considered here fed-batch cultivation of *S. cerevisiae* is characterized by keeping glucose concentration equal or below to its critical level ($S_{crit} = 0.05$ g/l), sufficient dissolved oxygen in the broth $O_2 \geq O_{2crit}$ ($O_{2crit} = 18\%$), as well as presence of ethanol in the broth.

Mathematical model of *S. cerevisiae* fed-batch cultivation is commonly described as follows, according to the mass balance:

$$\frac{dX}{dt} = \mu X - \frac{F}{V} X \tag{5}$$

$$\frac{dS}{dt} = -q_S X + \frac{F}{V} (S_{in} - S) \tag{6}$$

$$\frac{dE}{dt} = q_E X - \frac{F}{V} E \tag{7}$$

$$\frac{dO_2}{dt} = -q_{O_2} X + k_L a (O_2^* - O_2) \tag{8}$$

$$\frac{dV}{dt} = F \tag{9}$$

where X is the concentration of biomass, [g/l]; S - concentration of substrate (glucose), [g/l]; E - concentration of ethanol, [g/l]; O_2 - concentration of oxygen, [%]; O_2^* - dissolved oxygen saturation concentration, [%]; F - feeding rate, [l/h]; V - volume of bioreactor, [l]; $k_L a$ - volumetric oxygen transfer coefficient, [1/h]; S_{in} - glucose concentration in the feeding solution, [g/l]; μ, q_S, q_E, q_{O_2} - respectively specific rates of growth, substrate utilization, ethanol production and dissolved oxygen consumption, [1/h].

In general, Monod kinetics is used when a global model of fermentation processes is being developed. Thus the following functions described Monod kinetics should be implemented into model (5)-(9):

$$\mu = \mu_{\max} \frac{S}{S + k_S}, q_S = \frac{1}{Y_{S/X}} \frac{\mu_{\max} S}{S + k_S},$$

$$q_E = \frac{1}{Y_{E/S}} \frac{\mu_{\max} S}{S + k_S}, q_{O_2} = \frac{1}{Y_{O/X}} \frac{\mu_{\max} S}{S + k_S}, \tag{10}$$

where $\mu_{\max}$ is the maximum growth rates, [1/h]; k_S - saturation constant, [g/l]; Y_{ij} - yield coefficients, [g/g].

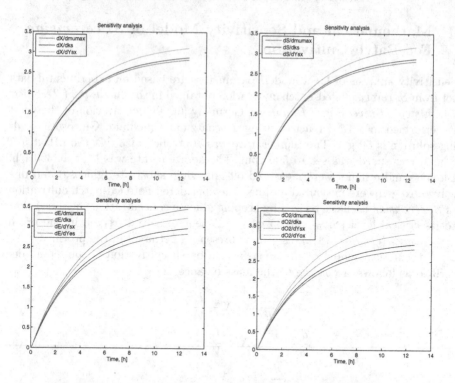

Fig. 1. Simulation curves from sensitivity analysis of Monod model

For the purposes of sensitivity analysis application to considered model (5)-(9) the state variable vector is presented as $x = [X, S, E, O_2]$, while the model parameters set consists of $p = [\mu_{max} \; k_S \; Y_{S/X} \; Y_{E/S} \; Y_{O/X}]$. Solving the sensitivity model of the system, the following parameter values are used: $p_0 = [0.5 \; 0.5 \; 0.5 \; 0.5 \; 0.5]$. Sensitivity models are analytically worked out and sensitivity functions are calculated. Fig. 1 presents simulation curves from sensitivity analysis of considered model.

The results from the sensitivity analysis for considered model (5)-(9) with Monod kinetics can be summarized as follows: the highest sensitivity is featured to yield coefficients $Y_{S/X}$, $Y_{E/S}$, $Y_{O/X}$, followed by maximum growth rate μ_{max} and saturation constant k_S.

Presented investigation based on the sensitivity analysis can result in the following 3-step parameter identification procedure:

- Step 1. estimation of parameters μ_{max}, k_S, $Y_{S/X}$ based on experimental data set for dynamics of S and X. The system (5), (6) and (9) with vector parameter $p = [\mu_{max}, k_S, Y_{S/X}]$ is considered.
- Step 2. estimation of parameter $Y_{E/S}$ based on an experimental data set for dynamics of E.
- Step 3. estimation of parameter $Y_{O/X}$ based on an experimental data set for dynamics of O_2.

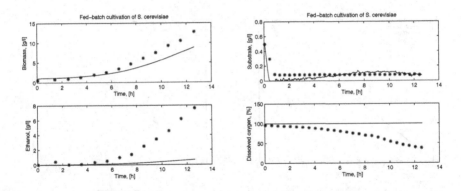

Fig. 2. Experimental data and model prediction of *S. cerevisiae* fed-batch cultivation applying Monod kinetics

Matlab environment is used for applying proposed parameter identification procedure. Under *Matlab* environment *Simulink* model of the investigated process has been developed. *Simulink* model describes concrete differential equations, taking into account process parameters and initial variable values. In order to identify kinetic parameters in each *Simulink* model, a script contained the necessary instructions for the corresponding identification procedure has been also developed. Different conventional optimization methods, such as *simplex*, *minimax* etc. have been applied.

Fig. 2 presents the best results obtained from the parameter identification when the proposed here 3-step parameter identification procedure has been applied to considered here *S. cerevisiae* fed-batch cultivation, described with model (5)-(9) applying Monod kinetics (10). In all presented figures * denotes the experimental data, while - denotes the model prediction.

As it could be seen by figures, Monod kinetics could not describe dynamics of considered here yeast growth process - neither for biomass growth and substrate utilization, moreover for ethanol production and dissolved oxygen consumption. Even the fact that proposed stepwise procedure for parameter identification leads to reducing the number of simultaneous estimated parameters, it is obvious that this is not always a precondition for obtaining of an adequate model.

These investigations provoke the idea some other specific rates to be explored in order to be obtained an adequate mathematical description of considered yeast cultivation. It could be applied the models, presented by Zhang et al. [9]. Applying the rules for functional states recognition [6], first ethanol production state is identified for considered here *S. cerevisiae* fed-batch cultivation. As such, local models in (5)-(9) have structures originally presented by Zhang et al. [9]:

$$\mu = \mu_{\max}, q_S = \frac{\mu_{\max}}{Y_{S/X}} \frac{S}{S + k_s},$$

$$q_E = (q_s - q_{Scrit}) Y_{E/S}, q_{O_2} = \frac{\mu_{max}}{Y_{O/X}}, \tag{11}$$

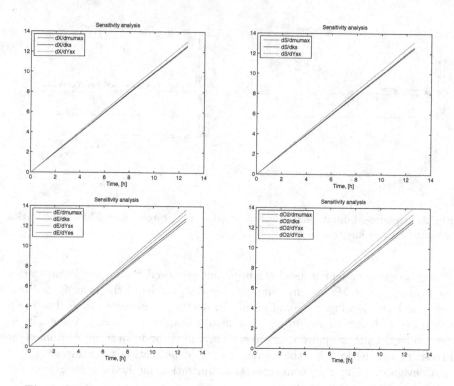

Fig. 3. Simulation curves from sensitivity analysis according to Zhang et al.

where the previously mentioned symbols keep their meaning, while q_{Scrit} is the value of q_S at the critical value of S_{crit}.

For the purposes of sensitivity analysis application to considered model (5)-(9) but when specific growth rates according to Zhang et al. have been considered the same state variable vector as above is used, as well as the same model parameters set with equal initial conditions. Sensitivity models are again analytically worked out and sensitivity functions are newly calculated. Fig. 3 presents simulation curves from sensitivity analysis of considered model (5)-(9) with specific growth rates according to Zhang et al.

In comparison to results obtained when Monod kinetics has been applied, results from the sensitivity analysis in this case are more closely grouped with no prominent parameter. Moreover, attempts to be applied the proposed stepwise parameter identification procedure using mentioned above conventional optimization methods, do not lead to satisfactory results. That is why the idea of using genetic algorithms (GA) as an alternative stochastic technique, intuitively appeared. Applying GA [1] for a parameter identification of considered here *S. cerevisiae* fed-batch cultivation, described with model (5)-(9) applying local

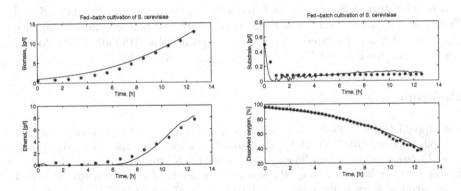

Fig. 4. Experimental data and model prediction of *S. cerevisiae* fed-batch cultivation according to Zhang et al.

models of Zhang et al. (11), the following parameter values have been obtained [6]: $p = [0.24\,0.28\,0.09\,1.48\,1600]$. Fig. 4 presents results obtained from the parameter identification when GA has been applied to considered here *S. cerevisiae* fed-batch cultivation, described with model (5)-(9) local models of Zhang et al. (11).

4   Analysis and Conclusions

In this investigation the sensitivity analysis of a fifth order non linear mathematical model of a *S. cerevisiae* fed-batch cultivation process is studied. Two different structures of functions described specific rates are examined. Sensitivity analysis is firstly applied in respect to state variables and model parameters in case when Monod kinetics is generally used in a global model of considered cultivation process. As a result 3-step parameter identification procedure is proposed but despite the fact that such procedure leads to reducing the number of simultaneous estimated parameters, this is not always a precondition for obtaining of an adequate model. In considered here cultivation process Monod kinetics could not describe adequately the process dynamics. Thus sensitivity analysis has been further analytically worked out to state variables and model parameters when specific growth rates have been described according to Zhang et al. In this case model parameters are more closely grouped with no prominent parameter. Moreover, when proposed stepwise parameter identification procedure has been applied using conventional optimization methods, no satisfactory results have been obtained. That provokes using of genetic algorithms as an alternative stochastic technique. Presented here methodology could be implemented also for another fermentation process, that may result in lower or higher order of non linear mathematical model. Corresponding to the parameter identification purposes, sensitivity analysis can be applied to other specific growth rates, as well as to different optimization criteria.

Acknowledgements. This work is partially supported by the European Social Fund and Bulgarian Ministry of Education, Youth and Science under Operative Program "Human Resources Development", grant BG051PO001-3.3.04/40 and National Science Fund of Bulgaria, grant number DID 02-29.

References

1. Chipperfield, A.J., Fleming, P., Pohlheim, H., Fonseca, C.M.: Genetic algorithm toolbox for use with MATLAB. User's guide, version 1.2. Dept. of Automatic Control and System Engineering, University of Sheffield, UK (1994)
2. Gorry, B., Ireland, A., King, P.: Sensitivity analysis of real-time systems. World Academy of Science, Engineering and Technology 37, 12–19 (2008)
3. King, J.M.P., Titchener-Hooker, N.J., Zhou, Y.: Ranking bioprocess variables using global sensitivity analysis: a case study in centrifugation. Bioprocess and Biosystems Engineering 30(2), 123–134 (2007)
4. Noycova, N.A., Gyllenberg, M.: Sensitivity analysis and parameter estimation in a model of anaerobic waste water treatment processes with substrate inhibition. Bioprocess Engineering 23, 343–349 (2000)
5. Patnaik, P.R.: Transient sensitivity analysis of a cybernetic model of microbial growth on two substrates. Bioprocess Engineering 21, 135–140 (1999)
6. Pencheva, T., Roeva, O., Hristozov, I.: Functional state approach to fermentation processes modelling. In: Tzonkov, S., Hitzmann, B. (eds.) Prof. Marin Drinov. Academic Publishing House, Sofia (2006)
7. Roeva, O.: Parameter Estimation of a Monod-Type Model Based on Genetic Algorithms and Sensitivity Analysis. In: Lirkov, I., Margenov, S., Waśniewski, J. (eds.) LSSC 2007. LNCS, vol. 4818, pp. 601–608. Springer, Heidelberg (2008)
8. Roeva, O.: Sensitivity analysis of E. coli fed-batch cultivation local models. Mathematica Balkanica 4 (2011)
9. Zhang, X.-C., Visala, A., Halme, A., Linco, P.: Functional state modelling approach for bioprosesses: local models for aerobic yeast growth processes. J. Proc. Contr. 4(3), 127–134 (1994)

A Matheuristic Algorithm for a Large-Scale Energy Management Problem

D. Anghinolfi[1], L.M. Gambardella[2], R. Montemanni[2],
C. Nattero[1], M. Paolucci[1], and N.E. Toklu[2],*

[1] Dipartimento di Informatica, Sistemistica e Telematica, University of Genoa, Genoa, Italy
[2] Istituto Dalle Molle di Studi sull'Intelligenza Artificiale, Lugano, Switzerland
engin@idsia.ch

Abstract. The demand for electrical energy is globally growing very quickly. For this reason, the optimization of power plant productions and power plant maintenance scheduling have become important research topics. A Large Scale Energy Management (LSEM) problem is studied in this paper. Two types of power plants are considered: power plants of type 1 can be refueled while still operating. Power plants of type 2 need to be shut down from time to time, for refueling and ordinary maintenance (these are typically nuclear plants). Considering these two types of power plants, LSEM is the problem of optimizing production plans and scheduling of maintenances of type 2 plants, with the objective of keeping the production cost as low as possible, while fulfilling the customers demand. Uncertainty about the customers demand is taken into account in the model considered. In this article, a matheuristic optimization approach based on problem decomposition is proposed. The approach involves mixed integer linear programming and simulated annealing optimization methods. Computational results on some realistic instances are presented.

1 Introduction

As the requirement for electrical energy is growing rapidly, optimizing and coordinating the operations of power plants in an economical way to satisfy energy demand has become a crucial practical issue. Maintenance scheduling and production planning of power plants are important for making sure that available human resources and material resources for maintenance are used in reasonable amounts and the equipment involved are kept in perfect efficiency, while satisfying the customers demand [9].

Studies have been made on power plant maintenance scheduling, by using mixed integer programming [1, 3] and also by using metaheuristic methods, including genetic algorithms [5], and ant colony optimization [4].

In this paper a Large Scale Energy Management (LSEM) problem of practical importance is considered. In this problem, two types of power plants, based on different power generation technologies, are used to satisfy the power demands. Power plants of type 1 are those that can be refueled while still operating. Power plants of type 2 are those that need to be shut down from time to time to be refueled and maintained (these

* Corresponding author. Supported by SNF through project 200021-119720.

I. Lirkov, S. Margenov, and J. Waśniewski (Eds.): LSSC 2011, LNCS 7116, pp. 173–181, 2012.

are typically nuclear power plants). Overall, the problem considered is to optimize both the maintenance scheduling for the nuclear power plants (taking into account different resources and technical constraints) and the production planning for all the power plants (again, taking into account technical constraints), with the aim of minimizing overall costs while fulfilling energy demands. Demands are affected by uncertainty in the model considered.

Since the LSEM problem considered is a very large and complex problem, the approach proposed here involves the decomposition of the problem into smaller sub-problems, giving the opportunity of working separately on the maintenance scheduling for nuclear plants and on the production planning of all the plants. The method used for this decomposition approach is a combination of mixed integer linear programming for production planning, and of a simulated annealing metaheuristics [6] (embedding again a mixed integer linear programming core) for maintenance schedule.

The combination of mathematical programming methods with metaheuristic approaches leads to the research framework of matheuristic algorithms. Recent examples of matheuristic approaches can be found in [7] and [8].

In section 2, a formal definition of the problem can be found. Section 3 describes the matheuristic approach we developed for the problem. In section 4 some experimental results of the new approach on real instances are presented and discussed. Conclusions are finally drawn in section 5.

2   Problem Definition

The problem studied in this paper is that used within the ROADEF/EURO 2010 Challenge [9]. According to the problem definition, there are two types of power plants. The plants of the first type can be refueled while production. The plants of the second type must be shut down for refueling. The data to decide on this problem are the refueling amounts and production plans of all power plants, and also the scheduling of the maintenance shutdown (*outage*) weeks for the power plants of the second type. Multiple scenarios, modeling alternative customer demands, are also considered in the problem: a feasible solution is able to satisfy all possible scenarios.

More in details, the problem includes constraints (numbered as 1 to 21), which effect the production plans of power plants and scheduling the outages of type 2 plants. These constraints will be formally defined in the reminder of this section, together with the other elements used to describe each problem instance. We refer the reader to [9] for the official description of the problem.

Main Concepts. The definition of the LSEM problem is based on the basic concepts defined in the reminder of this section.

A *timestep* is the most elementary time unit of the problem. The index variable for timestep is t. The first timestep is $t = 0$ and the last timestep is $t = T - 1$.

Some events depend on *weeks*, where each week contains an instance-dependent number of timesteps. The index variable for weeks is h. The first week is $h = 0$ and the last week is $h = H - 1$.

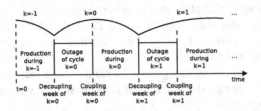

Fig. 1. Sequence of cycles within a T2 power plant

Type 1 (T1) power plants can be refueled during production. The index variable for these plants is j. The first T1 plant is $j = 0$ and the last is $j = J - 1$.

Type 2 (T2) power plants cannot be refueled during production. The index variable for these plants is i. The first T2 plant is $i = 0$ and the last is $i = I - 1$.

Each T2 plant has $K + 1$ *cycles* (indexed as k, ranging between $k = -1$ – starting cycle – and $k = K - 1$), which are based on weeks and represent the production/outage alternation. More precisely, within a cycle of T2 plant i, the following elements are defined: the *outage weeks* are those weeks in which a T2 plant is shut off to be refueled; the first outage week is referred to as the *decoupling week*; a set of consecutive weeks in which a T2 plant is active and producing is called *production campaign*. the *coupling week* is finally the first week of a production campaign. The sequence of cycles of a T2 plant is summarized in figure 1. Note that outage weeks are to be decided during the optimization.

Each instance has multiple *scenarios*. Each scenario provide alternative values for customer demands and different production limits for T1 plants. The index variable for the scenarios is s. The first scenario is $s = 0$ and the last is $s = S - 1$.

Decision Variables. The LSEM problem has the following decision variables.

$ha(i,k)$: decoupling week of T2 plant i at cycle k.
$p(j,t,s)$: production of T1 plant j at timestep t in scenario s.
$p(i,t,s)$: production of T2 plant i at timestep t in scenario s.
$r(i,k)$: amount of refueling during the outage k of T2 plant i.

The following variable can be inferred by the previous ones, but it is useful to define them explicitly.

$x(i,t,s)$: stock level of T2 plant i at timestep t in scenario s.
$ec(i,k)$: timesteps of the production campaign of cycle k for T2 plant i.
$ea(i,k)$: outage weeks of cycle k for T2 plant i.

Constraints. The LSEM problem considers the following 21 constraints.

[CT1]: For each scenario s the total production at each timestep t must be equal to the given demand $DEM^{t,s}$.
[CT2]: The production of each T1 plant must be between predefined time and scenario dependent given boundaries $PMIN_j^{t,s}$ and $PMAX_j^{t,s}$.
[CT3]: Production of a T2 plant must be zero, when it is on outage.

[CT4]: The production of a T2 plant cannot be negative.

[CT5]: During each production campaign k of a T2 plant i, if the fuel level is greater than or equal to a given threshold $BO_{i,k}$, the production of that plant cannot be more than a given threshold $PMAX_i^t$.

[CT6]: During the production campaign of a T2 plant i, if the fuel level is less than the given threshold $BO_{i,k}$, the production of that plant must follow a given profile $PB_{i,k}$, with a given tolerance ε.

[CT7]: The stock refilled during outage k of T2 plant i must be between given boundaries $RMIN_{i,k}$ and $RMAX_{i,k}$.

[CT8]: Fuel stock at $t = 0$ must be equal to a given initial fuel stock XI_i.

[CT9]: The fuel stock $x(i,t,s)$ of T2 plant i evolves according to production $p(i,t,s)$ and to the given length D^t of timestep t.

[CT10]: The fuel stock of T2 plant i right after outage k is defined according to some given refueling coefficients $Q_{i,k}$.

[CT11]: The fuel stock levels before and after the refueling are bounded by the given threshold $AMAX_{i,k}$ and $SMAX_{i,k}$.

[CT12]: Modulation (oscillation) in the power produced by T2 plants causes a wear of the equipment. Modulation is therefore constrained by two given parameters $PMAX_i^t$ and $MMAX_{i,k}$.

The remaining constraints are specific to T2 plants outage scheduling.

[CT13]: Outage k of T2 plant i must start between given boundaries $TO_{i,k}$ and $TA_{i,k}$. Moreover, outages cannot overlap: $DA(i,k)$ is the given length of outage k for plant i.

[CT14]: The entire outage periods of T2 plants in a given set A_m must be scheduled at least Se_m weeks apart from each other.

[CT15]: This constraint is same with [CT14], except that a specific time interval $[ID_m; IF_m]$ is given, and the constraint only applies within the given time interval.

[CT16]: All decoupling weeks of all T2 plants within the given set A_m must be scheduled at least Se_m weeks apart from each other.

[CT17]: All coupling weeks of all T2 plants within the given set A_m must be scheduled at least Se_m weeks apart from each other.

[CT18]: All coupling weeks of all T2 plants within the given set A_m must be scheduled at least Se_m weeks apart from all decoupling weeks of the plants in A_m.

[CT19]: The outages of T2 plants require a number of regional-limited shared human resources. The number of simultaneous outages in a same region is therefore constrained: A_m is a set of T2 plants; $L_{i,k,m} \in [0, DA_{i,k}]$ is a time interval (in weeks) indicating how long after the beginning of the outage k the shared resources is required; $TU_{i,k,m}$ specifies the number of weeks the shared resource is required during outage k; Q_m finally defines the available quantity for the regional-limited shared resource.

[CT20]: It is forbidden to have too many simultaneous T2 plants' outages. The following instance-dependent parameters are required: $A_m(h)$ is the set of T2 plants. The set depends on the week index h; $N_m(h)$ is the maximum number of simultaneous outages allowed in week h.

[CT21]: The power capacity of the T2 plants on outage in a same region, in a given time period, has to be below a given threshold. The following instance-dependent

Fig. 2. The overall architecture of the MATHDEC approach. In this figure, the rectangles represent modules and the arrows represent the data passed between these modules.

parameters are required: C_m is a set of T2 plants; IT_m is a time period in weeks; $IMAX_m$ is the threshold for total offline power capacity in a week.

Objective Function. The objective of the problem is to minimize the sum of:
(i) the cost of production for T1 plants averaged over all scenarios
(ii) the total cost of refueling for T2 plants, reduced by the scenario-averaged cost of the remaining fuel at the end of the time horizon.

3 The MATHDEC Approach

LSEM is a difficult problem which requires both discrete decisions like outage dates and continuous decisions like production plans and refueling/stock values.

Considering the complex nature of the problem and the huge values of the parameters encountered in real instances, the problem can be classified, as a matter of facts, as a very difficult problem to solve as a whole. The method we propose, MATHDEC, decomposes the problem into smaller and easier-to-solve subproblems. They cover, namely, the scheduling of the outages, a week-based production planning and a more detailed timestep-based production planning.

To deal with the different subproblems, different modules have been developed. The overall architecture of the MATHDEC approach is depicted in figure 2, where the interactions among the different modules are shown. The remainder of this section is devoted to the description of these modules.

3.1 The Heuristic Outage Schedule Generator Module

The Heuristic Outage Schedule Generator (HOSG) module is an iterative heuristic approach to generate outage scheduling for T2 plants. It does not consider timesteps, but only weeks and does not take car of production/refueling issues. The main idea of HOSG is to have as many active cycles as possible (i.e. make $ha(i,k) \neq -1$ for as many (i,k) possible), so that the work load on T1 plants will be less: it has been experimentally found that high quality solutions have this characteristic.

HOSG iteratively produces outage schedules and calls the *MIP outage scheduler* submodule, which is in charge of checking if a scheduling is feasible or not according to constraints [CT13] to [CT21]. It is based on a mixed integer programming model. A detailed definition of the model can be found in Appendix A, in the online addendum [2].

In its first stage, HOSG executes MILP Outage Scheduler with all week intervals containing -1, which means that it gives the disabling option for all cycles of each T2 plant. After receiving a feasible schedule, HOSG initiates its second stage.

In the second stage, HOSG initializes $k' = 1$, which represents a cycle. Iteratively increasing k', HOSG executes MILP Outage Scheduler, with intervals up to k' excluding -1 and the other intervals including -1. This means, cycles up to k' must be executed, they do not have the disabling option (i.e. $ha(i,k) \neq -1 \mid k \leq k'$). When a k' is reached where a failure signal is received from MILP Outage Scheduler, HOSG initiates its final stage.

The final stage is another iterative process which takes the last feasible schedule received as a base. At each iteration, HOSG randomly picks a small subset of the T2 plants and increases their number of mandatory cycles (i.e. number of cycles k with $ha(i,k) \neq -1$). Then, MILP Outage Scheduler is executed. If a feasible schedule is received, that schedule becomes the new base schedule for the next iteration. This loop is exited after a given computation time (between 20 and 30 minutes in our implementation) is elapsed.

The output of this module is a feasible outage schedule given by $ha(i,k)$ values, typically with the number of mandatory cycles increased with respect to the initial one.

3.2 The Local Search Module

The Local Search (LS) module starts from the solution provided by HOSG, and tries to improve it. LS still works on weeks only, but now production and refueling are taken into account, although in a relaxed way: they are aggregated per weeks (instead of timesteps). A simulated annealing algorithm [6] is implemented where random variation to outage weeks are introduced at each iteration. A solution vector, for each T2 plant i, stores K decoupling week values ($ha(i,k)$). Each $ha(i,k)$ is extended into a time interval $[ha(i,k) - \delta, ha(i,k) + \delta]$, where δ is a radius value monotonically decreased during the computation. In our implementation, the temperature T is in $[0.1;1.0]$. The search starts with $T = 1.0$ and at each iteration T is reduced by multiplying it by a cooling factor of 0.999. When $T \leq 0.1$ the search is restarted with $T = 1.0$. The set of time intervals obtained is sent to the MIP outage scheduler submodule (see 3.1 and Appendix A in the online addendum [2] for more details) that return a feasible schedule fulfilling the given time intervals if it exists, or a negative answer otherwise. In case a feasible schedule is found, it is passed to the *Weekly Production Planner* submodule (WPP), which returns weekly production plans together with an estimation of the objective function of the problem obtained by aggregating timesteps into weeks. More in details, the WPP submodule receives an outage schedule $ha(i,k)$ and determines a rough approximation of the plant production, refueling and consequent costs. In particular, WPP solves a linear programming model, which explicitly considers all the scenarios but takes planning decisions still on a weekly base, and generates feasible refueling values. A detailed description of the linear program is available in Appendix B, in the online addendum [2].

3.3   The Timestep Production Planner Module

The Timestep Production Planner (TSPP) module implements the final stage for solving the problem. The input of this module is an outage schedule values $ha(i, k)$ generated by HOSG and improved by LS, and modulation reference values mod_{iks} and weekly refueling values generated by WPP.

The main idea of TSPP is to generate all production and refueling values expanded into timesteps, with deviations from the reference modulation values minimized.

At first, the weekly refueling plan is expanded into timesteps and adjusted to satisfy [CT7] and [CT11]. Then, for each scenario, a linear programming model is executed iteratively to generate the final solution. A complete description of the linear program can be found in Appendix C, in the online addendum [2].

The linear programming model considers [CT1] to [CT5], as other constraints were already handled by other modules. The objective here is to minimize the total deviations from the reference modulation given by WPP, and also the average refueling cost over all scenarios.

For each scenario, a loop is executed, in which this linear programming formulation is iteratively used. At the end of each iteration, the program adjusts the mod_{iks} values if the deviation values are greater than 0. The program also checks [CT7] and [CT11]. If they are not satisfied, it adjusts the fuel stock values. When the deviation values are 0 and [CT7] and [CT11] are satisfied, the loop reaches its end.

Finally, the module gives the timestep-based and scenario-aware production, fuel stock and refueling values, which, together with the outage schedule generated previously, form a solution for the problem instance.

The module is started 12 minutes before the end of the available computation time, in order to safely have a feasible solution before the deadline.

4   Experimental Results

The algorithm MATHDEC, described in section 3, has been implemented in C++, and ILOG CPLEX 11.0 [10] has been used to solve mixed integer linear programs.

In Table 1 the results obtained by MATHDEC on the 10 realistic benchmark instances [11] used within the ROADEF/EURO 2010 Challenge [9] are compared with those obtained by the 18 teams that took part into the competition [12]. Among the instances considered, the following maximum values were encountered: number of timesteps 5817; number of weeks 277; number of T1 plants 27; number of type T2 plants 56; number of campaigns 6 and number of scenarios 121. For each instance, the best and the average solution costs obtained within the competition (1 hour on a Intel Xeon 5420 2.5GHz/8GB computer) are reported together with those obtained by the MATHDEC approach (1 hour on a less performing but comparable Intel Core 2 Duo 2.4GHz/3GB computer). The averages for each column are reported in the last line of the table.

Table 1 suggests that the method MATHDEC obtains results comparable with those of the methods presented at the ROADEF/EURO 2010 Challenge: MATHDEC is never able to improve best challenge results, but finds better results than the average ones found during the challenge for all but 2 instances. MATHDEC is also able to retrieve

Table 1. Results on the ROADEF/EURO 2010 Challenge instances

Instance	Best [12]	Average [12]	MATHDEC
B6	8.34E+10	8.79E+10	9.11E+10
B7	8.12E+10	9.23E+10	8.63E+10
B8	8.19E+10	5.85E+11	3.19E+11
B9	8.18E+10	6.84E+11	2.50E+11
B10	7.78E+10	9.45E+10	8.52E+10
X11	7.91E+10	8.67E+10	8.43E+10
X12	7.76E+10	8.45E+10	8.33E+10
X13	7.64E+10	8.16E+10	8.37E+10
X14	7.62E+10	8.97E+10	8.49E+10
X15	7.51E+10	9.73E+10	8.02E+10
Averages	*7.90E+10*	*1.98E+11*	*1.25E+11*

results of the same order of magnitude of the best challenge results for 8 instances out of the 10 considered (interestingly enough, the 2 problematic instances do not coincide with those for which MATHDEC is worse than the average results). The method presented is therefore fairly robust: the best solutions of the competition were found by different methods, each one typically "specialized" on some of the instances. Notice that the method MATHDEC depends on heuristic choices, and multiple runs on the same instance provide results deviating from those reported in Table 1 by relatively factors in the order of 1.00E+8. This does not affect our general considerations.

5   Conclusions

A very large-scale real energy management problem has been described, and a new matheuristic optimization approach to solve it has been discussed. The method decomposes the initial problem into smaller subproblem, that are solved by mathematical programming and metaheuristic tools. The benefit brought by the decomposition have been shown empirically, through experimental results on realistic instances.

References

1. Ahmad, A., Kothari, D.P.: A practical model for generator maintenance scheduling with transmission constraints. Electric Power Components and Systems 28, 501–513 (2000)
2. Anghinolfi, D., Gambardella, L.M., Montemanni, R., Nattero, C., Paolucci, M., Toklu, N.E.: Addendum to: A matheuristic algorithm for a large-scale energy management problem (February 2011), http://www.idsia.ch/~roberto/roadef_addendum.pdf
3. Dopazo, J.F., Merrill, H.M.: Optimal generator maintenance scheduling using integer programming. IEEE Transactions on Power Apparatus and Systems PAS-94(5), 1537–1545 (1975)
4. Foong, W.K., Maier, H.R., Simpson, A.R., Stolp, S.: Ant colony optimization for power plant maintenance scheduling optimization. Annals of Operations Research 159(1), 433–450 (2008)
5. Gil, E., Bustos, J., Rudnick, H.: Short-term hydrothermal generation scheduling model using a genetic algorithm. IEEE Transactions on Power Systems 18(4), 1256–1264 (2003)
6. Kirkpatrick, S., Gelatt, C.G., Vecchi, M.P.: Optimization by simulated annealing. Science 220(4598), 671–680 (1983)

7. Maniezzo, V., Stützle, T., Voß, S. (eds.): Hybridizing Metaheuristics and Mathematical Programming. Annals of Information Systems, vol. 10. Springer, Heidelberg (2009)
8. Montemanni, R., Mahdabi, P.: A linear programming-based evolutionary algorithm for the minimum power broadcast problem in wireless sensor networks. Journal of Mathematical Modelling and Algorithms 10(2), 145–162 (2011),
   http://dx.doi.org/10.1007/s10852-010-9146-9
9. Porcheron, M., Gorge, A., Juan, O., Simovic, T., Dereu, G.: Challenge ROADEF/EURO 2010: A large-scale energy management problem with varied constraints. Tech. rep., Électricité de France (EDF) R&D (2010)
10. IBM ILOG CPLEX Solver (December 2010), http://www.cplex.com
11. Data instances of ROADEF/EURO 2010 challenge (July 2010),
    http://challenge.roadef.org/2010/instances.en.htm
12. Results of the ROADEF/EURO 2010 challenge (December 2010),
    http://challenge.roadef.org/2010/result.en.htm

On a Game-Method for Modelling with Intuitionistic Fuzzy Estimations: Part 1

Lilija Atanassova[1] and Krassimir Atanassov[2]

[1] Institute of Information and Communication Technologies,
Bulgarian Academy of Sciences, Acad. G. Bonchev Str., bl. 2, Sofia 1113, Bulgaria
l.c.atanassova@gmail.com,
[2] Bioinformatics and Mathematical Modelling Dept.,
Institute of Biophysics and Biomedical Engineering, Bulgarian Academy of Sciences,
Acad. G. Bonchev Str., bl. 105, Sofia 1113, Bulgaria
krat@bas.bg

Abstract. A new extension of Conway's Game of Life is introduced. It is based on a previous Conway's game extension, given by the authors. Now we use elements of intuitionistic fuzziness that give more detailed estimations of the degrees of existence and of the non-existence of the objects occuring the cells of the game plane. Rules for the motions and rules for the interactions among the objects are dicsussed.

1 Introduction

In a series of papers (see [7,11]), the authors extended the standard Conway's Game of Life (CGL) (see, e.g., [17]), adding in it intuitionistic fuzzy estimations (for intuitionistic fuzziness see [5]). On the other hand, more than 30 years ago the authors introduced the idea for another extension of CGL, called "game-method for modelling" (GMM), and its application in astronomy and combinatorics (see [1,2,3,8,16]). This idea found particular application in growth and dynamics of forest stands (see, [12,13]) and forest fires (see [14]).

Here we will introduce a new extension of the standard CGL on the basis of both modifications, discussed by us.

2 Description of the Game-Method for Modelling from Crisp Point of View

The standard CGL has a "universe" which is an infinite two-dimensional orthogonal grid of square cells, each of which is in one of two possible states, alive or dead, or (as an equivalent definition) in the square there is an asterisk, or not. The first situation corresponds to the case when the cell is alive and the second – to the case when the cell is dead.

Now, following the ideas from [1,2,8], we will extend the CGL.

Let us have a set of symbols S and an n-dimensional simplex (in the sense of [15]) comprising of n-dimensional cubes (at $n = 2$, a two-dimensional net of

I. Lirkov, S. Margenov, and J. Waśniewski (Eds.): LSSC 2011, LNCS 7116, pp. 182–189, 2012.
© Springer-Verlag Berlin Heidelberg 2012

squares). Let material points (or, for brief, objects) be found in some of the vertices of the simplex and let a set of rules A be given, containing:

1) rules for the motions of the objects along the vertices of the simplex;
2) rules for the interactions among the objects.

Let the rules from the i-th type be marked as i-rules, where $i = 1, 2$.

When $S = \{*\}$, we obtain the standard CGL.

To each object its number is associated, n-tuple of coordinates characterizing its location in the simplex and a symbol from S reflecting the peculiarity of the object (e.g. in physical applications – mass, charge, concentration, etc.). We shall call an *initial configuration* every ordered set of $(n + 2)$-tuples with an initial component being the number of the object; the second, third, etc. until the $(n + 1)$-st – its coordinates; and the $(n + 2)$-nd – its symbol from S. We shall call a *final configuration* the ordered set of $(n + 2)$-tuples having the above form and being a result of a (fixed) initial configuration, modified during a given number of times when the rules from A have been applied.

The single application of a rule from A over a given configuration K will be called an elementary step in the transformation of the model and will be denoted by $A_1(K)$. In this sense, if K is an initial configuration, and L is a final configuration derived from K through multiple application the rules from A, then configurations $K_0, K_1, ..., K_m$ will exist, for which $K_0 = K, K_{i+1} = A_1(K_i)$ for $0 \le i \le m - 1, K_m = L$, (the equality "=" is used in the sense of coincidence in the configurations) and this will be denoted by

$$L = A(K) \equiv A_1(A_1(...A_1(K)...)).$$

Let a rule P be given, which juxtaposes to a combination of configurations M a single configuration $P(M)$ being the mean of the given ones. We shall call this rule a *concentrate rule*. The concentration can be made either over the values of the symbols from S for the objects, or over their coordinates, (not over both of them simultaneously).

For example, if k-th element of M ($1 \le k \le s$, where s is the number of elements of M) is a rectangular with $p \times q$ squares and if the square staying on (i, j)-th place ($1 \le i \le p, 1 \le j \le q$) contains number $d_{i,j}^k \in \{0, 1, ..., 9\}$, then on the (i, j)-th place of $P(M)$ stays:

– minimal number

$$d_{i,j} = \left\lfloor \frac{1}{s} \sum_{k=1}^{s} d_{i,j}^k \right\rfloor,$$

– maximal number

$$d_{i,j} = \left\lceil \frac{1}{s} \sum_{k=1}^{s} d_{i,j}^k \right\rceil,$$

– average number

$$d_{i,j} = \left\lfloor \frac{1}{s} \sum_{k=1}^{s} d_{i,j}^k + \frac{1}{2} \right\rfloor,$$

where for real number $x = a + \alpha$, where a is an integer number and $\alpha \in [0, 1)$:
$[x] = a$ and

$$[x] = \begin{cases} a, & \text{if } \alpha = 0 \\ a+1, & \text{if } \alpha > 0 \end{cases}$$

Let B be a criterion derived from physical or mathematical considerations. For two given configurations K_1 and K_2, it answers the question whether they are close to each other or not. For example, for two configurations K_1 and K_2 having the form from the above example,

$$B(K_1, K_2) = \frac{1}{p.q} \sum_{i=1}^{p} \sum_{j=1}^{q} |d_{i,j}^1 - d_{i,j}^2| < C_1$$

or

$$B(K_1, K_2) = \left(\frac{1}{p.q} \sum_{i=1}^{p} \sum_{j=1}^{q} \left(d_{i,j}^1 - d_{i,j}^2 \right)^2 \right)^{\frac{1}{2}} < C_2,$$

where C_1 and C_2 are given constants.

For the set of configurations M and the set of rules A we shall define the set of configurations

$$A(M) = \{L | (\exists K \in M)(L = A(K))\}.$$

The rules A will be called statistically correct, if for a great enough (from a statistical point of view) natural number N:

$$(\forall m > N)(\forall M = \{K_1, K_2, ..., K_m\})$$

$$(B(A(P(M)), P(\{L_i | L_i = A(K_i), 1 \leq i \leq m\})) = 1). \qquad (*)$$

The essence of the method is in the following: the set of rules A, the proximity criterion B and the concentrate rule P are fixed preliminarily. A set of initial configurations M is chosen and the set of the corresponding final configurations is constructed. If the equation $(*)$ is valid we may assume that the rules from the set A are correct in the frames of the model, i.e. they are logically consistent. Otherwise, we replace a part (or all) of them with others. If the rules become correct, then we can add to the set some new ones or transform some of the existing and check permanently the correctness of the newly constructed system of rules. Thus, in consequtive steps, extending and complicating the rules in set A and checking their correctness, we construct the model of the given process. Afterwards we may check the temporal development (as regards the final system of rules A) of a particular initial configuration.

We initially check the correctness of the modelling rules and only then we proceed to the actual modelling. To a great deal this is due to the fact that we work over discrete objects with rules that are convenient for computer implementation. Thus, a series of checks of the equation $(*)$ can be performed only to construct the configuration $A(K)$ for a given configuration K and a set of rules A.

For example, if we would like to model (in a plane) the Solar System, we can mark the Sun by "9", Jupiter and Saturn – by "8", the Earth and Venus – by "7", the rest of the planets – by "6", the Moon and the rest bigger satelits – by "5", the smaller satelits – by "4", the bigger asteroids – by "3", the smaller asteroids – by "2" and the cosmic dust – by "1".

If we would like to model the forest dynamics, then the digits will correspond to the number of the territory trees in a square unit of the forest teritory. If a river flows through the forest, then we can mark its cells, e.g., by letter "R"; if there are stones without trees, we can mark them by cells with letter "S".

3   On the Game-Method for Modelling with Intuitionistic Fuzzy Estimations

The intuitionistic fuzzy propositional calculus has been introduced more than 20 years ago (see, e.g., [4,5]). In it, if x is a variable, then its truth-value is represented by the ordered couple

$$V(x) = \langle a, b \rangle,$$

so that $a, b, a + b \in [0, 1]$, where a and b are the degrees of validity (existence, membership, etc., and of non-validity, non-existence, etc.) of x and there the following definitions are given.

Below, we shall assume that for the two variables x and y the equalities: $V(x) = \langle a, b \rangle, V(y) = \langle c, d \rangle$ $(a, b, c, d, a + b, c + d \in [0, 1])$ hold.

For two variables x and y the operations "conjunction" (&), "disjunction" ($\vee$), "implication" ($\rightarrow$), and "(standard) negation" ($\neg$) are defined by:

$$V(x \& y) = \langle \min(a, c), \max(b, d) \rangle,$$

$$V(x \vee y) = \langle \max(a, c), \min(b, d) \rangle,$$

$$V(x \rightarrow y) = \langle \max(b, c), \min(a, d) \rangle,$$

$$V(\neg x) = \langle b, a \rangle.$$

In [6,9,10] the following two operations, which are analogues to operation "disjunction", are defined

$$V(x + y) = \langle a, b \rangle + \langle c, d \rangle = \langle a + c - ac, bd \rangle,$$

$$V(x \vee_L y) = \langle a, b \rangle \vee_L \langle c, d \rangle = \langle \min(1, a + c), \max(0, b + d - 1) \rangle,$$

$$V(x \vee_Z y) = \langle a, b \rangle \vee_Z \langle c, d \rangle = \langle \max(a, \min(b, c)), \min(b, d) \rangle.$$

The following relation of partial order is defined in IF logic (see, e.g., [5]) for every two variables x and y:

$$x \leq y \text{ if and only if } a \leq c \text{ and } b \geq d.$$

It can be easily seen that

$$V(x \vee_Z y) \leq V(x \vee y) \leq V(x + y) \leq V(x \vee_L y).$$

As we saw above, each asterisk from CGL corresponds to a symbol in the GMM. Now, we will perform the next step of extension. For this step, we have two possibilities that are equivalent in general case.

First, we can change the symbols from S, with which the objects in GMM are marked, to IF-couples, determining the degree of existence and degree of non-existence of this object.

Second, we can keep the symbols from S and attach the same IF-couple to them.

Let us discuss the second case, because, obviously, we can obtain the first one directly.

As a first step of the research in the present direction, it is convenient to assume that the objects will not change their IF-parameters as a result of movement from one cell to another, that has a common side with the previous cell. In a future research, we will discuss the more complex situation, when in the result of the movement the IF-parameters changes (e.g., will decrease). Therefore, the criteria for existence of an object before and after its movement will be the same and we can use the conditions from [11], but here we will extend its list of criteria.

Let us assume that the square $\langle i, j \rangle$ is assigned a pair of real numbers $\langle \mu_{i,j}, \nu_{i,j} \rangle$, so that $\mu_{i,j} + \nu_{i,j} \leq 1$. We can call the numbers $\mu_{i,j}$ and $\nu_{i,j}$ degree of existence and degree of non-existence of an object, or (in CGL and its IF-extension), of a symbol "*" in square $\langle i, j \rangle$. Therefore, $\pi(i, j) = 1 - \mu_{i,j} - \nu_{i,j} \leq 1$ will correspond to the degree of uncertainty, e.g., lack of information about existence of an asterisk in the respective cell.

Below, we will formulate seven criteria for existence of an object in a cell, that will include as a particular case the standard game. We note that the list below is longer than the list from [11], where six criteria are given.

We will suppose that there exists an object in square $\langle i, j \rangle$ if:

(1) $\mu_{i,j} = 1.0$. Therefore, $\nu_{i,j} = 0.0$. This is the situation in the standard CGL, when an asterisk exists in square $\langle i, j \rangle$.

(2) $\mu_{i,j} > 0.5$. Therefore, $\nu_{i,j} < 0.5$. In the partial case, when $\mu_{i,j} = 1 > 0.5$, we obtain $\nu_{i,j} = 0 < 0.5$, i.e., the standard existence of the object – case 1.

(3) $\mu_{i,j} \geq 0.5$. Therefore, $\nu_{i,j} \leq 0.5$. Obviously, if case 2 is valid, then case 3 also will be valid.

(4) $\mu_{i,j} > \nu_{i,j}$. Obviously, cases (1) and (2) are partial cases of the present one, but case (3) is not included in the currently discussed case for $\mu_{i,j} = 0.5 = \nu_{i,j}$.

(5) $\mu_{i,j} \geq \nu_{i,j}$. Obviously, all previous cases are partial cases of the present one.

(6) $\mu_{i,j} > 0$. Obviously, cases (1), (2), (3), and (4) are partial cases of the present one, but case (5) is not included in the currently discussed case for $\mu_{i,j} = 0.0 = \nu_{i,j}$.

(7) $\nu_{i,j} < 1$. Obviously, cases (1), (2), (3), and (4) are partial cases of the present one.

From these criteria it follows that if one is valid – let it be s-th criterion ($1 \leq s \leq 7$) then we can assert that the object exists with respect to the s-th criterion and, therefore, it will exist with respect to all other criteria the validity of which follows from the validity of the s-th criterion.

On the other hand, if s-th criterion is not valid, then we will say that the object does not exist with respect to s-th criterion. It is very important that in this case the square may not be totally empty. We may tell that the square is "s-full" if it contains an object with respect to s-th criterion, or, that it is "s-empty" if it is empty or contains an object, that does not satisfy the s-th criterion.

For the aims of the GMM, it will be suitable to use (with respect to the type of the concrete model) one of the first four criteria for existence of an object. Let us say for each fixed square $\langle i, j \rangle$ that therein is an object by s-th criterion ($1 \leq s \leq 4$), whether this criterion confirms the existence of the object.

We must mention that in CGL there are no rules for object (asterisk) transfer. On the other hand, in GMM a new situation arises, which does not exist in CGL, but is an extension of the CGL-situations related to bourning and death of asterisks. In the present case, two or more objects can enter one cell and now we have to use the rules for interaction. Let us discuss the simpler case when only two objects can enter one cell, although the formulas below will be valid for $n \geq 2$ objects, as well because each of the discussed below operations is associative:

Let two objects have IF-estimations $\langle a, b \rangle$ and $\langle c, d \rangle$. Then, in the cell a new object with IF-estimations will be generated. The values of these estimations can be different in respect of the user preferences:

1 - strongly high: $\langle \min(1, a + c), \max(0, b + d - 1) \rangle$,
2 - high: $\langle a + c - ac, bd \rangle$,
3 - standard: $\langle \max(a, c), \min(b, d) \rangle$.
4 - weak: $\langle \max(a, \min(b, c)), \min(b, d) \rangle$.

For example, if we would like to model a situation of entering some people in a lift cage, then we can use the first formula, if a, b, c, d are the people's parameters of weight, normalized by the lift capacity. If modelling forest dynamics we would like to describe a case when trees from two regions (represented by squares) enter one and the same cell, then we can use the second formula. The fact that a planet in the Solar System swallows cosmic dust, of course, must be interpreted by the third formula.

4   Conclusion

In the next research in this direction of extension of CGL, we will discuss the possibility for constructing different aggregation procedures for obtaining a single configuration $P(M)$, that is juxtaposed to a combination of configurations M. Therefore, we plan to construct an intuitionistic fuzzy concentrate rule.

Also, we will introduce different intuitionistic fuzzy criteria for proximity between two configurations. The concept of the statistical correctness of some rules
also will obtain an intuitionistic fuzzy interpretation.

Acknowledgement. The second author is grateful for the support provided by
the projects DID-02-29 "Modelling processes with fixed development rules" and
BIn-2/09 "Design and development of intuitionistic fuzzy logic tools in information technologies" funded by the National Science Fund, Bulgarian Ministry of
Education, Youth, and Science.

References

1. Atanassov, K.: On a combinatorial game-method for modelling. Advances in Modelling & Analysis 19(2), 41–47 (1994)
2. Atanassov, K.: Application of a combinatorial game-method in combinatorial geometry. Part 1: Combinatorial algorithms for solving variants of the Steiner-
Rosenbaum's problem. Advances in Modelling & Analysis 2(1-2), 23–29 (1998)
3. Atanassov, K.: Application of a combinatorial game-method in combinatorial geometry. Part 2: Algorithm for grouping and transferring of points and a general
algorithm. Advances in Modelling & Analysis 2(1-2), 31–36 (1998)
4. Atanassov, K.: Two variants of intuitonistic fuzzy propositional calculus. Preprint
IM-MFAIS-5-88, Sofia (1988)
5. Atanassov, K.: Intuitionistic Fuzzy Sets. Springer Physica-Verlag, Heidelberg
(1999)
6. Atanassov, K.: Remarks on the conjunctions, disjunctions and implications of the
intuitionistic fuzzy logic. Int. J. of Uncertainty, Fuzziness and Knowledge-Based
Systems 9(1), 55–65 (2001)
7. Atanassov, K., Atanassova, L.: A game-method for modelling. In: Proc. of the
3rd International School Automation and Scientific Instrumentation, Varna, pp.
229–232 (October 1984)
8. Atanassov, K., Atanassova, L., Sasselov, D.: On the combinatorial game-metod
for modelling in astronomy. Comptes Rendus de l'Academie bulgare des Sciences,
Tome 47(9), 5–7 (1994)
9. Atanassov, K., Tcvetkov, R.: On Zadeh's intuitionistic fuzzy disjusnction and conjunction. Notes on Intuitionistic Fuzzy Sets 17(1), 1–4 (2011)
10. Atanassov, K., Tcvetkov, R.: On Lukasiewicz's intuitionistic fuzzy disjusnction
and conjunction. Annual of "Informatics" Section Union of Scientists in Bulgaria 3
(2010)
11. Atanassova, V., Atanassov, K.: Ant Colony Optimization Approach to Tokens'
Movement within Generalized Nets. In: Dimov, I., Dimova, S., Kolkovska, N. (eds.)
NMA 2010. LNCS, vol. 6046, pp. 240–247. Springer, Heidelberg (2011)
12. Dimitrov, D.: Modelling the growth and dynamics of forest stands by game-method.
Advances in Modelling & Analysis 2(1-2), 11–22 (1998)
13. Dimitrov, D.: Modelling the growth and dynamics of forest stands by extended
game-method. Advances in Modelling & Analysis 4(1-2), 7–21 (1999)
14. Dobrinkova, N., Fidanova, S., Atanassov, K.: Game-Method Model for Field Fires.
In: Lirkov, I., Margenov, S., Waśniewski, J. (eds.) LSSC 2009. LNCS, vol. 5910,
pp. 173–179. Springer, Heidelberg (2010)

15. Kuratovski, K.: Topology. Academic Press, New York (1966)
16. Sasselov, D., Atanassov, K.: On the generalized nets realization of a combinatorial game-method for modelling in astronomy. Advances in Modelling & Analysis 23(4), 59–64 (1995)
17. Wikipedia contributors: Conway's Game of Life. Wikipedia, The Free Encyclopedia, http://en.wikipedia.org/w/index.php?title=Conway's_Game_of_Life (accessed March 20, 2011)

A Generalized Net with an ACO-Algorithm Optimization Component

Vassia Atanassova[1], Stefka Fidanova[1],
Panagiotis Chountas[2], and Krassimir Atanassov[3]

[1] Institute of Information and Communication Technologies,
Bulgarian Academy of Sciences, Acad. G. Bonchev Str., bl. 2, Sofia 1113, Bulgaria
vassia.atanassova@gmail.com, stefka@parallel.bas.bg
[2] University of Westminster, Dept. of Business Information Systems,
School of Electronics & Computer Science,
115 New Cavendish Street, London W1W 6UW, United Kingdom
p.i.chountas@westminster.ac.uk
[3] Bioinformatics and Mathematical Modelling Dept.,
Institute of Biophysics and Biomedical Engineering,
Bulgarian Academy of Sciences,
Acad. G. Bonchev Str., bl. 105, Sofia 1113, Bulgaria
krat@bas.bg

Abstract. In the paper we describe a generalized net G_{ACOA} realizing an arbitrary algorithms for ant colony optimization. In this sense, this net is universal for all standard algorithms for ant colony optimization, since it describes the way of functioning and results of their work. Then, we discuss the way of constructing a GN that includes the G_{ACOA} as a subnet. In this way, we ensure the generalized net tokens' optimal transfer with regard to the results of G_{ACOA}. Thus, we construct a generalized net, featuring an optimization component and thus optimally functioning.

1 Introduction

Generalized nets (GNs; see [1,4,6]) are an apparatus for modelling of parallel and concurrent processes, developed as an extension of the concept of Petri nets and some of their modifications. During the last 25 years it was shown that the GNs can be used for constructing of universal tools, describing the functioning and the result of the work of the other types of Petri nets, of the finite automata and Turing machine, of expert systems and machine learning processes, data bases and data warehouses, etc. (see, e.g. [3,5,7,12]).

In a series of papers by some of the authors, it was shown that the GNs can represent the functioning and the result of the work of different Ant Colony Optimization (ACO) algorithms (see, e.g. [8,9,10]). On the other hand, in [1] it was shown that we can construct special types of GNs, featuring an optimization component. By optimization component we will understand a subnet of a given generalized net, which describes a particular optimization problem and whose

I. Lirkov, S. Margenov, and J. Waśniewski (Eds.): LSSC 2011, LNCS 7116, pp. 190–197, 2012.

results from functioning can be used in the main net with the purpose of optimizing its behaviour. So far, such optimization components have been defined for the Transportational Problem and the Travelling Salesman Problem [1,2].

Here, for the first time it will be proposed an optimization problem, realizing the ACO-algorithm. This component guarantees that the GN functions optimally with respect of the results from the ACO-algorithm performance.

2 Short Remarks on GNs

Formally, every GN-transition is described by a seven-tuple:

$$Z = \langle L', L'', t_1, t_2, r, M, \square \rangle,$$

where:

(a) L' and L'' are finite, non-empty sets of places (the transition's input places (*inputs*) and output places (*outputs*), respectively).
(b) t_1 is the current time-moment of the transition's firing;
(c) t_2 is the current value of the duration of its active state;
(d) r is the transition's *condition* determining which tokens will pass (or *transfer*) from the transition's inputs to its outputs; it has the form of an *Index Matrix* (IM; see [1,4]):

$$r = \begin{array}{c|ccccc} & l''_1 & \dots & l''_j & \dots & l''_n \\ \hline l'_1 & & & & & \\ \vdots & & & r_{i,j} & & \\ l'_i & & (r_{i,j} & - & \text{predicate}) & \\ \vdots & & (1 \le i \le m, 1 \le j \le n) & & \\ l'_m & & & & & \end{array} \quad ;$$

$r_{i,j}$ is the predicate which corresponds to the i-th input and j-th output places. When its truth value is *"true"*, a token from i-th input place can be transferred to j-th output place; otherwise, this is not possible;
(e) M is an IM of the capacities of transition's arcs, its elements being natural numbers corresponding to the number of tokens that may transfer through the transition at a time (see [6]);
(f) $\square$ is an object having a form similar to a Boolean expression. It may contain as variables the symbols which serve as labels for transition's input places, and is an expression built up from variables and the Boolean connectives $\wedge$ and $\vee$. When the value of a type (calculated as a Boolean expression) is *"true"*, the transition can become active, otherwise it cannot.

The ordered four-tuple

$$E = \langle \langle A, \pi_A, \pi_L, c, f, \theta_1, \theta_2 \rangle, \langle K, \pi_K, \theta_K \rangle, \langle T, t^o, t^* \rangle, \langle X, \Phi, b \rangle \rangle$$

is called a *Generalized Net* (GN) if:

(a) A is a set of transitions;

(b) π_A is a function giving the priorities of the transitions, i.e., $\pi_A : A \rightarrow N$, where $N = \{0, 1, 2, \ldots\} \cup \{\infty\}$;

(c) π_L is a function giving the priorities of the places, i.e., $\pi_L : L \rightarrow N$, where $L = pr_1 A \cup pr_2 A$, and $pr_i X$ is the i-th projection of the n-dimensional set, where $n \in N, n \geq 1$ and $1 \leq k \leq n$ (obviously, L is the set of all GN-places);

(d) c is a function giving the capacities of the places, i.e., $c : L \rightarrow N$;

(e) f is a function which calculates the truth values of the predicates of the transition's conditions (for the GN described here let the function f have the value "$false$" or "$true$", i.e., a value from the set $\{0, 1\}$;

(f) θ_1 is a function giving the next time-moment when a given transition Z can be activated, i.e., $\theta_1(t) = t'$, where $pr_3 Z = t, t' \in [T, T+t^*]$ and $t \leq t'$. The value of this function is calculated at the moment when the transition terminates its functioning;

(g) θ_2 is a function giving the duration of the active state of a given transition Z, i. e., $\theta_2(t) = t'$, where $pr_4 Z = t \in [T, T + t^*]$ and $t' \geq 0$. The value of this function is calculated at the moment when the transition starts functioning;

(h) K is the set of the GN's tokens;

(i) π_K is a function giving the priorities of the tokens, i.e., $\pi_K : K \rightarrow N$;

(j) θ_K is a function giving the time-moment when a given token can enter the net, i.e., $\theta_K(\alpha) = t$, where $\alpha \in K$ and $t \in [T, T + t^*]$;

(k) T is the time-moment when the GN starts functioning. This moment is determined with respect to a fixed (global) time-scale;

(l) t^o is an elementary time-step, related to the fixed (global) time-scale;

(m) t^* is the duration of the GN functioning;

(n) X is the set of all initial characteristics the tokens can receive when they enter the net;

(o) Φ is a characteristic function which assigns new characteristics to every token when it makes a transfer from an input to an output place of a given transition;

(p) b is a function giving the maximum number of characteristics a particular token can receive, i.e., $b : K \rightarrow N$.

When the GN has only a part of the above components, it is called reduced GN. Below we shall use a reduced GN without the temporal components characterizing the net or the transition, without tokens', places' and transitions' priorities, as well as without arcs' and places' capacities.

In [1,4] different operations, relations and operators are defined over GNs. One of them, namely $\cup$, can merge two given GN-models.

3   A GN Universal for the ACO-Algorithms

Following [8], we shall construct the GN that is universal for the ACO-algorithms. This GN will describe how a set of n artificial ants move along the ACO-algorithm graph and let this graph consist of m vertices and k arcs. Let us assume that any quantity of pheromone left in the graph is known in advance.

Let us denote the so described GN by G_{ACOA}. It has 3 transitions, 16 places (see Fig. 1) and four types' $(\alpha, \beta, \gamma$ and $\varepsilon)$ of tokens. The α token is designed to contain information about the ants' location within the graph. The β token obtains the values of the number of vertices (m) and the number of arcs (k). The γ token contains the structure of the graph, i.e. a matrix of incidence, while the ϵ token contains the information about the vertices/arcs where pheromone has been laid by the ants, and the quantities of this pheromone. All these tokens enter the input places of the GN with certain initial characteristics, that are explained below.

Token α enters place l_1 with the initial characteristic *"The n-dimensional vector with elements being the ants' locations within the graph"*.

The β token enters place l_2 with the initial characteristic

"⟨m-dimensional vector with elements – the graph vertices

or l-dimensional vector with elements – the graph arcs;

objective function⟩",

where m is the number of the nodes of the graph of the problem and l is the number of the arcs of the graph; l_{11} – with the initial characteristic

"the graph structure with m vertices and l arcs";

l_{12} – with the initial characteristic

"initial data for the places and quantities of the pheromones".

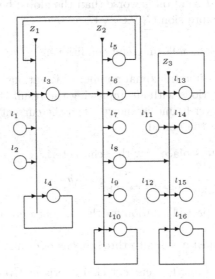

Fig. 1. GN net model for ACO

$$Z_1 = < \{l_1, l_2, l_4, l_5, l_6\}, \{l_3, l_4\},$$

	l_3	l_4
l_1	true	false
l_2	false	true
l_4	false	true
l_5	true	false
l_6	true	false

$>$.

Token α, occupying either place l_1, or place l_5, or place l_6 enters place l_3 with a characteristic *"Vector of current transition function results $\langle \varphi_{1,cu}, \varphi_{2,cu}, ..., \varphi_{n,cu} \rangle$"*, while token ε remains looping in place l_4 obtaining the characteristic

"new m-dimensional vector with elements – the graph vertices,

or new l-dimensional vector with elements – the graph arcs".

$$Z_2 = < \{l_3, l_{10}\}, \{l_5, l_6, l_7, l_8, l_9, l_{10}\},$$

	l_5	l_6	l_7	l_8	l_9	l_{10}
l_3	$W_{3,5}$	$W_{3,6}$	$W_{3,7}$	true	$W_{3,9}$	$W_{3,10}$
l_{10}	false	false	true	true	$W_{10,9}$	$W_{10,10}$

$>$,

where

$W_{3,5} = $ "The current iteration has not finished",
$W_{3,6} = W_{3,10} = \neg W_{3,5} \vee \neg W_{10,9}$,
$W_{10,7} = $ "The current best solution is worse than the global best solution",
$W_{10,9} = $ "Truth-value of expression $C_1 \vee C_2 \vee C_3$ is *true*",
$W_{10,10} = \neg W_{10,9}$,
where C_1, C_2, and C_3 are the following end-conditions:

- C_1 – "Computational time (maximal number of iterations) is achieved",
- C_2 – "Number of iterations without improving the result is achieved",
- C_3 – "If the upper/lower bound is known, then the current results are close (e.g., less than 5%) to the bound".

Token α from place l_3 enters place l_5 with a characteristic

$$"\langle S_{1,cu}, S_{2,cu}, ..., S_{n,cu} \rangle",$$

where $S_{i,cu}$ is the current partial solution for the current iteration, made by the i-th ant $(1 \leq i \leq n)$.

If $W_{3,6} = true$, then token α splits to three tokens α, α' and α''.

- Token α enters place l_6 with a current characteristic *"New n-dimensional vector with elements being the new ants' locations"*.
- Token α' enters place l_8 with the last α's characteristic (before the splitting).

- Token α'' enters place l_{10} with a current characteristic being the pair "$\langle$ *The best solution for the current iteration; its number* $\rangle$".

When $W_{10,9} = true$ token α'' can enter place l_9 where it obtains the characteristic "*The best achieved result*".

In place l_7 one of both tokens from place l_{10} enters, which has worse; the worse values as a current characteristic, while in place l_{10} the token, containing the best values as a current characteristic, is preserved to keep looping.

$$Z_3 = < \{l_8, l_{11}, l_{12}, l_{13}, l_{16}\}, \{l_{13}, l_{14}, l_{15}, l_{16}\},$$

	l_{13}	l_{14}	l_{15}	l_{16}
l_8	false	false	false	true
l_{11}	true	false	false	false
l_{12}	false	false	false	true
l_{13}	$W_{13,13}$	$W_{13,14}$	false	false
l_{16}	false	false	$W_{16,15}$	$W_{16,16}$

$>,$

where

$W_{13,14} = W_{16,15} =$ "truth-value of expression $C_1 \vee C_2 \vee C_3$ is *true*",
$W_{13,13} = W_{16,16} = \neg W_{13,14}.$

Tokens γ from place l_{11} and β from place l_{12} with above mentioned characteristics enter, respectively, places l_{13} and l_{16} without any characteristic.

Token α from place l_8 enters place l_{16} and unites with token β (the new token is again β) with characteristic "*Value of the pheromone updating function with respect to the values of the objective function*".

Tokens β and γ enter, respectively, places l_{14} and l_{15} without any characteristics.

4   GNs with Optimization Component G_{ACOA}

Based on the definition of the concept of a GN, we shall construct a new type of GNs – GNs with optimization components (GNOC).

The tokens in GN-transitions transfer in the GN by the algorithms are discussed in [1,4]. These algorithms are based on the checking of the truth-values of the transition condition predicates.

Let E be an arbitrary GN and let us desire to control its tokens transfer optimally on the base of solutions obtained by the ACO-algorithm. Now, we can unite (by operation $\cup$ between two GNs, mentioned above) the GNs E and G_{ACOA} and in the new GN we can organize the optimal way for tokens transfer.

According to the algorithm for tokens transfer from [4], this transfer is realized at every time-moment of the functioning of the given GN in the frames of one active transition - the so called "abstract transition", which is a union of all active GN-transitions at this moment.

By this reason, we can describe the functioning of only one GN-transition, and, in particular, the abstract transition.

Let the transition Z from E be given. Each of its input places (in the terms of the ACO-algorithm) will correspond to an input vertex of the graph. Let us determine all GN-tokens in the inputs of Z. Their number will be the initial characteristic of token α from G_{ACOA} (in the form of input vertices of the given graph). The information about the places that contain these tokens will be given as an initial characteristic of token β of the same GN. The information about the graph can be inserted in the GN G_{ACOA} before the beginning of the process and it will be kept during the whole simulation.

Having in mind the characteristic functions of the output places of Z, we can determine those of the vertices of the same graph that exhibit sufficiently high weights.

Each token from the Z-inputs will correspond to an ant that has to search for a path to the vertex with the maximal weight.

The GN-transition components t_1, t_2, and M are not necessary (for example, we can assume that the values of t_1 are sequential natural numbers, $t_2 = 1$ and the elements of M are equal to ∞). The transition type has the form of disjunction.

On the basis of the above determined characteristics for the tokens of G_{ACOA}, these tokens go through the GN and in a result determine the trajectories of the separate ants (and therefore tokens from Z) from input vertices of the graph to its vertices with highest values of the weights (that correspond respective to the input and output places of Z). So, using the solutions from G_{ACOA} we can determine the way of transfer of the tokens from input to output places of Z.

This procedure can be used for control of each one of the transitions of GN E. In this case, in G_{ACOA} we will put respective number of α- and β-tokens, so that their number will correspond to the number of the transitions in E.

5   Conclusion

The described procedure is universal in nature. It is applicable to each one GN. In [1] a GN was described and this net has an optimization component solving a transportational problem. The present model is the second one, but in future similar models can be constructed for other optimization procedures, e.g., traveling salesman problems, knapsack-problem and others.

The herewith proposed procedure can find various applications, for instance in emergency medicine. Following the ideas proposed in [11,13] it is possible to construct generalized net models of the organization and management of emergency medicine, and using ant colony optimization component to search for optimality in the decision making processes.

Acknowledgments. The authors would like to acknowledge the support provided by the Bulgarian National Science Fund under Grants Ref. No. DID-02-29 "Modelling Processes with Fixed Development Rules" and DTK-02-44 "Effective Monte Carlo Methods for Large-Scale Scientific Problems". This work has also been partially supported by Royal Society, Grant Ref. No. JP100372 "Generalised net models and intuitionistic fuzzy sets in intelligent systems".

References

1. Atanassov, K.: Generalized Nets. World Scientific, Singapore (1991)
2. Atanassov, K.: Generalized nets and some travelling salesman problems. In: Atanassov, K. (ed.) Applications of Generalized Nets, pp. 68–81. World Scientific, Singapore (1993)
3. Atanassov, K.: Generalized Nets in Artificial Intelligence. Generalized nets and Expert Systems, vol. 1. "Prof. M. Drinov". Academic Publishing House, Sofia (1998)
4. Atanassov, K.: On Generalized Nets Theory. "Prof. M. Drinov". Academic Publishing House, Sofia (2007)
5. Atanassov, K., Aladjov, H.: Generalized Nets in Artificial Intelligence. Generalized nets and Machine Learning, vol. 2. "Prof. M. Drinov". Academic Publishing House, Sofia (2001)
6. Atanassova, V.: Generalized nets, Transition, Ifigenia, the wiki for intuitionistic fuzzy sets and generalized nets (2009),
   http://ifigenia.org/wiki/Generalized_nets,
   http://ifigenia.org/wiki/Transition
7. Chountas, P., Kolev, B., Rogova, E., Tasseva, V., Atanassov, K.: Generalized Nets in Artificial Intelligence. Generalized nets, Uncertain Data and Knowledge Engineering, vol. 4. "Prof. M. Drinov". Academic Publishing House, Sofia (2007)
8. Fidanova, S., Atanassov, K.: Generalized net models of the process of ant colony optimization with intuitionistic fuzzy estimations. In: Atanassov, K., Shannon, A. (eds.) Proceedings of the Ninth International Workshop on Generalized Nets, Sofia, July 4, vol. 1, pp. 41–48 (2008)
9. Fidanova, S., Atanassov, K.: Generalized net models for the process of hybrid ant colony optimization. Comptes Rendus de l'Academie bulgare des Sciences, Tome 61(12), 1535–1540 (2008)
10. Fidanova, S., Atanassov, K.: Generalized Nets as Tools for Modeling of the Ant Colony Optimization Algorithms. In: Lirkov, I., Margenov, S., Waśniewski, J. (eds.) LSSC 2009. LNCS, vol. 5910, pp. 326–333. Springer, Heidelberg (2010)
11. Ibri, S., Drias, H., Nourelfath, M.: Integrated Emergency Vehicle Dispatching and Covering: A Parallel Ant-Tabu Approach. In: 8th International Conference of Modeling and Simulation (MOSIM 2010) Hammamet, Tunisia, Lavoisier, May 10-12, pp. 1039–1045 (2010)
12. Kolev, B., El-Darzi, E., Sotirova, E., Petronias, I., Atanassov, K., Chountas, P., Kodogiannis, V.: Generalized Nets in Artificial Intelligence. Generalized nets, Relational Data Bases and Expert Systems, vol. 3. "Prof. M. Drinov". Academic Publishing House, Sofia (2006)
13. Liu, N., Huang, B., Pan, X.: Using the Ant Algorithm to Derive Pareto Fronts for Multiobjective Siting of Emergency Service Facilities. Transportation Research Record: Journal of the Transportation Research Board, No. 1935, Transportation Research Board of the National Academies, Washington, D.C., 120–129 (2005)

Time Series Prediction by Artificial Neural Networks and Differential Evolution in Distributed Environment

Todor Balabanov[1], Iliyan Zankinski[2], and Nina Dobrinkova[1]

[1] Institute of Information and Communication Technologies,
Bulgarian Academy of Science,
Acad. G. Bonchev St., Block 2, 1113 Sofia, Bulgaria
todorb@iinf.bas.bg
http://www.iict.bas.bg/
[2] Faculty of Computer Systems and Management,
Technical University of Sofia,
8 Kliment Ohridski Bld, 1756 Sofia, Bulgaria

Abstract. Current work will present a model for time series prediction by the usage of Artificial Neural Networks (ANN) trained with Differential Evolution (DE) in distributed computational environment. Time series prediction is a complex work and demand development of more effective and faster algorithms. ANN is used as a base and it is trained with historical data. One of the main problems is how to select accurate ANN training algorithm. There are two general possibilities — exact numeric optimization methods and heuristic optimization methods. When the right heuristic is applied the training can be done in distributed computational environment. In this case there is much faster and realistic output, which helps to achieve better prediction.

Keywords: forecasting, ANN, DE, distributed computing.

1 Introduction

Time series prediction is a process of predicting future values in a series of data based on their known values in the past moments. Decision makers are in the position of taking very responsible steps during formation of investment strategy. To invest it means to take acceptable risk with expectation of certain amount of profit. The most important aspect of investment is the balance between risk taken and expected profit. On the currency market (FOREX) the main trading is done by exchanging currencies. Currency is the most volatile in price changing object of trading. During the process of trading on such market as FOREX decision makers needs to take three important decisions: 1. Price will go up or down; 2. What volume to buy or sell; 3. How long to keep the opened position. Even it sounds simple in fact it is very difficult to estimate price changing direction, because the huge amount of factors influence on it. The order volume is directly

I. Lirkov, S. Margenov, and J. Waśniewski (Eds.): LSSC 2011, LNCS 7116, pp. 198–205, 2012.

related to amount of risk taken. High volume order can lead to high profit if price changing direction is well estimated, but it can lead to high loss in other case. How long to keep the opened position is related to make the profit even bigger or to make the loss as smaller as possible. Financial forecasting is most important for the traders on the currency market, because of the high price dynamics.

Development of effective and reliable financial predictions is demanding and complex task. Therefore, the field is very relevant for the development and utilization of self-organizing and self-learning systems for prediction. The application of ANN for predicting time series in the field of economy have been considered by various authors, e.g. Dunis [1], Giles [2], and Moody [3]. A common method employs so-called Feed Forward Neural Networks (FFNN) (Haykin [4]). Networks of the FFNN type are very effective, but suffer from a fundamental flaw, namely they lack short-term memory. This problem could be avoided by using so-called Recurrent Neural Networks (RNN). However, RNN on other hand are yielding difficulties due to their difficulty to employ precise gradient methods during the learning phase (Werbos [5]). Possible solution is combined approach for training ANN by evolutionary algorithms. It has been proposed by Yao [6] and several other authors. Evolutionary algorithms show significantly better results for optimum search in complex multidimensional spaces with the presence of many local optima, in which case gradient-based methods are likely to get stuck (Holland [7]). In this paper model of self-learning system for predicting time series is presented. The system is based on ANN and DE in distributed environment.

This paper is organized as follows: Section 2 defines problem of time series prediction. Section 3 introduces ANN and DE based methodology of problem solving. Section 4 presents proposed model to be used for solving of the problem. Section 5 marks the aspects of future work to be done. Finally, Section 6 presents the conclusions.

2   Problem Definition

In the field of time series prediction there are few specific aspects. There are series of values strongly related to the axis of time (time-value pairs). Values in time series are not independent values. Those values are related to each other in a way that newer values are consequence of older values, for example — the rate between EUR and USD. It is extremely unusual this rate to differ dramatically in two sequent trading days. The rate is changing smoothly but constantly. This rate in common illustrates trading process between Europe and USA. Problem in financial time series prediction is by knowing past price values to develop prediction model and by using this model to estimate future price values. If the model is accurate predicted values can be used by decision makers to decrease investment risk.

3 Methodology

3.1 Artificial Neural Networks

ANN is a mathematical model of natural neural networks. It consists of artificial neurons with set of connections between them. Information is processed by entering the network and feed-forwarded into internal layers (neurons) to the output. ANNs are designed to be self-adaptive systems capable to change their internal structure based on external or internal information that flows in during learning process. ANNs are used to model complex relation between inputs and outputs (functional relation) or fitting a pattern in set of data (data mining).

The idea of ANN was inspired from biological central nervous systems and their neurons, axons, dendrites, and synapses. Generally ANN consists of network of simple processing elements that represent complex global behavior realized by connections between processing units and unit parameters. From the practical point of view ANN use comes with algorithms implemented to alter the strength (weights) of the connections to achieve a desired signals flow.

ANNs are pretty similar to natural neural networks in a way that network functions are done collectively and in parallel by the processing elements. In modern computing ANN approach is combined with non-adaptive methods for achieving better practical results. The main advantage of using ANN is its ability to infer a function from examples observation. This is particularly useful in problems where the complexity of investigated process (or its data) makes implementation of such function by hand very difficult.

3.2 Differential Evolution

DE is population based evolutionary metaheuristic optimization method. DE is developed by Kenneth Price and Rainer Storn [8] and it is based on classical genetic algorithms (GA). Methaheuristics like DE do not guarantee that optimal solution will be found. DE is used for complex (high dimensional spaces) search spaces exploration. It can be applied mainly in continues problems (real-values) and it is less useful for discrete problems even that it is not required the problem to be differentiable (as it is in classical optimization methods such as gradient descent).

Like classical GA, DE uses population of candidate solutions and creates new candidate solutions by combining existing ones according to rules of crossover, mutation and selection (best candidates are kept according objective function calculation). By estimating objective function values and selecting the best candidates there is no need to use gradient.

In comparison between DE and GA the main difference is in mutation operator. DE's mutation is based on calculation of difference vector which makes mutation much more effective than GA's mutation by changing each value of mutated chromosome. Disadvantage of using difference vector is related to difficulties of applying small differences on discrete values.

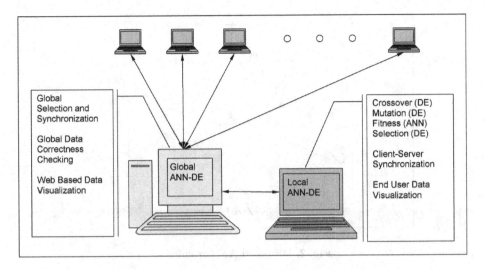

Fig. 1. System — conceptual model

4   Model Description

Proposed mathematical model is based on standard ANN and combined train-ing with DE and back-propagation (BP). ANN topology itself is a matter of research interest therefore topology is parameterized on the remote server. Con-nections between neurons can be — only forward, forward-backward and even full-connected (connected to all other neurons and itself). As properties of each neuron linear summation function and sigmoid activation function is used.

In DE, as variance of genetic algorithms (GA), each chromosome presents one set of weights for a particular topology of ANN. As shown in Fig. 1 global DE population is presented on the remote server. Each computational unit (client machines) is loading subset of global DE population (step 1 in Fig. 3) and performing local DE-BP mixed ANN training (step 2 in Fig. 3). On regular basis local machines are connecting to remote server. During this connection local population is updated and local computational results are reported (step 3 in Fig. 3). Because of DE's high level of parallelism there is no theoretical limit in number of computational units. As technical limitation bottle neck of the system is the remote server. Technical limit of the simultaneous connected clients is the limit of the server. It is not needed each client to have constant connection with the remote server. By this way technical limitation is easily overcome. Each computational unit can calculate during weeks and even months before remote server connection to be done.

As described in Fig. 2, DE based training process consists of five major steps: 1. Loading set of weights (a chromosome) from DE to ANN; 2. Loading training examples into ANN; 3. Calculate prediction values; 4. Calculate total prediction error (older data have less impact on the calculated error); 5. Estimate chromo-some fitness. Each set of weights (chromosome) is loaded into ANN structure

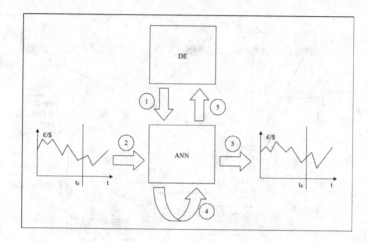

Fig. 2. DE based ANN training

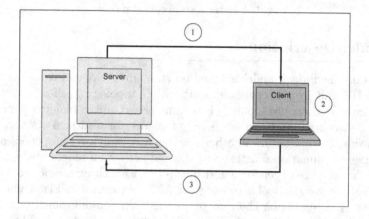

Fig. 3. Local machine computations

after that each input value is feed-forwarded. Each output value is compared with expected value and this difference is used to form total error achieved by this particular set of weights. Calculated total error is provided as fitness value into DE population. The main goal of DE optimization is to minimize ANN's total error. Because training process is real time learning (new data are rising up with the time going) total error minimization is continuous process.

All computations are done locally as shown in Fig. 3. Local computations consist of DE training loop — crossover and mutation. Both operations are done as described in Fig. 4 and Fig. 5. Parent chromosomes are selected for crossover by GA selection rule. There are different selection rules, but in current model two parents are selected randomly and by survival percent it is decided which

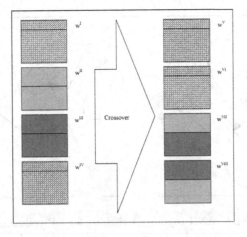

Fig. 4. DE crossover

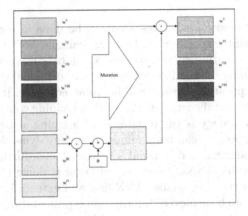

Fig. 5. DE mutation

one of both to survive. Crossover operation is considered as destructive during ANN training process, because of that it can be regulated as parameter. After that mutation is performed. Classical DE mutation is a sum of mutated chromosome and weighted difference vector (difference between other two randomly selected chromosomes multiplied by weight coefficient). Mutation operator is the advantage of DE, compared with GA.

For parallel training local ANN copies are sent to each computational node. At local machine DE and BP are applied and constant synchronization with remote server is done. Problem of slow learning rate, when ANN is trained by DE, can be addressed by switching to BP training. This process is done by removal of all recurrent connections, as shown in Fig. 6.

Learning convergence rate can be used as indicator for switching between DE and BP and vice versa. BP can be represented as special case of GA mutation.

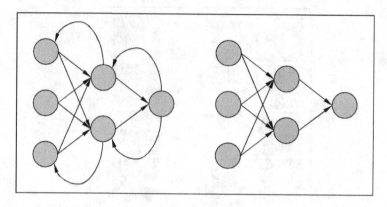

Fig. 6. ANN topology switching for BP training

5   Future Work

With the proposed model further research can be done in the following directions: 1. Parallel training speed up analysis; 2. Dynamic ANN topology training;
3. Input-output information representation; 4. Global DE population selection
for faster ANN training. Increasing number of computational units is not always
guarantee that computational result will be obtained faster. It is subject of additional research how parallelism will affect ANN training convergence. Training
ANNs with static topology is the usual case of learning algorithm. It is interesting how training process can be modified to employ dynamic ANN topology.
Rules for neurons addition and removal are base of such further research. It is
well known that smaller ANNs are subject of faster training than larger ANNs.
If there are rules for training smaller ANN first and extend ANN topology then
better convergence can be achieved. Information representation in the input level
of ANN is one of the strongest arguments of ANN topology size. Better input information representation can lead to smaller ANN topology and better training
convergence. In the case of DE it is interesting how global chromosome population will be controlled. Very important parameter for investigation is how often
local DE populations will be synchronized with the global population. For better
search space exploration global population can be manipulated with less effective
(as fitness value) chromosomes. Development of proper algorithms (scheduling
algorithms) for global population subset (local DE population is always subset of global DE population) distribution is common task for better usage of
distributed computing model.

6   Conclusions

DE, as training algorithm, prevents ANN from overtraining. Also DE allows
training of ANN with recurrent connections. There is no need for training examples to be shuffled. There is no limitation of parallel training processes number.

Malicious computation nodes are not problem they are even advantage by making DE population even richer as genotype. Main disadvantage of ANN used for time series prediction is slow learning rate, even with state of the art learning algorithms. Using distributed computing calculations can be done very cost efficient. By providing only one well organized centralized server computational power of supercomputer can be achieved. Main goal in further research is implementation of algorithms for better and faster ANN training algorithms.

Acknowledgements. This work is supported by the European Social Fund and Bulgarian Ministry of Education, Youth, and Science under Operative Program 'Human Resources Development', Grant BG051PO001-3.3.04/40.

References

1. Dunis, C.L., Williams, M.: Modelling and trading the eur/usd exchange rate: Do neural network models perform better? Derivatives Use, Trading and Regulation 8(3), 211–239 (2002)
2. Giles, C.L., Lawrence, S., Tsoi, A.C.: Noisy time series prediction using a recurrent neural network and grammatical inference. Machine Learning 44(1/2), 161–183 (2001)
3. Moody, J.E.: Economic forecasting: Challenges and neural network solutions. In: Proceedings of the International Symposium on Artificial Neural Networks, Hsinchu, Taiwan (1995)
4. Haykin, S.: Neural Networks, A Comprehensive Foundation, 2nd edn. Prentice-Hall, Inc. (1999)
5. Werbos, P.: Backpropagation through time: what it does and how to do it. Proceedings of the IEEE 78(10), 1550–1560 (1990)
6. Yao, X.: Evolving artificial neural networks. Proc. of the IEEE 87(9), 1423–1447 (1999)
7. Holland, J.: Adaptation in Natural and Artificial Systems. The University of Michigan Press (1975)
8. Storn, R., Price, K.: Differential evolution — a simple and efficient heuristic for global optimization over continuous spaces. Journal of Global Optimization 11, 341–359 (1997)

Differential Evolution Applied to Large Scale Parametric Interval Linear Systems

Jerzy Duda and Iwona Skalna

AGH University of Science and Technology, Krakow, Poland
jduda@zarz.agh.edu.pl,
skalna@agh.edu.pl

Abstract. Differential evolution (DE) is regarded to be a very effective optimisation method for continuous problems in terms of both good optimal solution approximation and short computation time. The authors applied DE method to the problem of solving large scale interval linear systems. Different variants of DE were compared and different strategies were used to ensure that candidate solutions generated in the process of recombination mechanism were always feasible. For the large scale problems the method occurred to be very sensitive to the constraint handling strategy used, so finding an appropriate strategy was very important to achieve good solutions in a reasonable time. Real world large optimisation problems coming from structural engineering were used as the test problems. Additionally DE performance was compared with evolutionary optimisation method presented in [10].

1 Introduction

Imprecision, approximation, or uncertainty, which are often involved in practical problems, can be modelled conveniently by intervals. In interval analysis, an unknown or imprecise parameter $\tilde{p}$ is replaced by an interval number $\boldsymbol{p}$. Then, the radius $r(\boldsymbol{p})$ is a measure for the absolute accuracy of the midpoint $\check{p}$ considered as an approximation of an unknown value $\tilde{p}$ contained in $\boldsymbol{p}$.

Now, consider parametric linear systems with coefficient that are nonlinear functions of parameters

$$A(p)x(p) = b(p), \tag{1}$$

where $A(p)$ is an $n \times n$ matrix, $b(p)$ is n-dimensional vector, and $p = (p_1, \ldots, p_k)^T$ is a k-dimensional parameter vector. Solving such systems is an important part of many scientific and engineering problems.

Assuming that some of the parameters are unknown but bounded, varying within prescribed intervals $p_i \in \boldsymbol{p}_i$ $(i = 1, \ldots, k)$, the following family of parametric linear system is obtained

$$A(p)x(p) = b(p), \ p \in \boldsymbol{p}. \tag{2}$$

I. Lirkov, S. Margenov, and J. Waśniewski (Eds.): LSSC 2011, LNCS 7116, pp. 206–213, 2012.

The set of all solutions to the point linear systems from the family (2) is called a *parametric solution set* and is defined as

$$S(\boldsymbol{p}) = \{x \in \mathbb{R} \mid \exists p \in \boldsymbol{p} \ \ A(p)x = b(p)\}. \tag{3}$$

It is generally of a complicated nonconvex structure [1]. In practice, therefore, an interval vector $\boldsymbol{x}^*$, called an *outer solution*, satisfying $\Box S(\boldsymbol{p}) \subseteq \boldsymbol{x}^*$ is computed. It is closed and bounded if $A(p)$ is nonsingular for each $p \in \boldsymbol{p}$. An interval vector $\boldsymbol{x}^{**}$, called *inner solution*, satisfying $\boldsymbol{x}^{**} \subseteq \Box S(\boldsymbol{p})$ is often calculated as well. If inner and outer solutions are close, this gives a demonstration of the quality of both of them. However, if the outer solution is much wider than the inner one, one or both of the solutions are poor.

The tightest outer solution is called a *hull solution* and is denoted by $\Box S(\boldsymbol{p})$. The problem of computing the hull solution can be written as a problem of solving $2n$ constrained optimisation problems. The following theorem holds true.

Theorem 1. *Let* $A(p)x(p) = b(p)$, $p \in \boldsymbol{p}$, *be a family of parametric linear systems, and let*

$$\underline{x}_i = \min \{x_i \mid x \in S(\boldsymbol{p})\}, \tag{4a}$$

$$\overline{x}_i = \max \{x_i \mid x \in S(\boldsymbol{p})\}. \tag{4b}$$

Then $\Box S(\boldsymbol{p}) = [\underline{x}, \ \overline{x}]$.

Computing the hull solution is, in general case, NP-hard [9]. Some algorithms for its computation were suggested in the literature, e.g. the combinatorial approach, the monotonicity method of Rao & Berke [8], and the vertex sensitivity method of Pownuk [6]. Although, they are often exact at small uncertainty, they do not always give the hull and may underestimate, even at small uncertainties [4].

In this paper, various variants of differential evolution algorithms are applied to approximate the hull solution of parametric linear systems of a large scale. The algorithm is described in Section 2 along with the discussion about constraints handling. Section 3 provides the results of the computational experiments. The paper ends with conclusions giving also the suggestions for the future work.

2   Differential Evolution Method for Parametric Interval Linear Systems

2.1   Method Basics

Differential evolution (DE) is regarded as a very effective optimisation method especially for continuous problems [2]. The algorithm itself can be treated as a variation of evolutionary algorithm, as the method is founded on the principles of selection, crossover and mutation. However in this case the main process of the algorithm is focused on the way new individuals are created. A new solution is created by combining the weighted difference between some individuals and

---

Initialise P of pop_{size} at random
while $(i < gen_{max})$ **do**
   while $(j < pop_{size})$ **do**
      Choose n individuals at random $p_1, p_2,..., p_n \in P$
      Generate mutant p_m from $p_1, p_2,..., p_n \cup P_{best}$
      $p'_j \longleftarrow$ Crossover$(p_j, p_m); j \longleftarrow j+1$
   end while
   $i \longleftarrow i+1$
end while

Fig. 1. Differential evolution method

another one (best or randomly chosen) [2]. The outline of DE method is shown in Fig. 1. A crucial role in differential evolution is played by a mutation operator. Many strategies with different variant of this operator were defined in literature ([7,3]). In basic strategies described as $/rand/1/bin$ and $/rand/1/exp$ a mutated individual p_m called a *trial vector* is created as follows:

$$p_m = p_1 + s(p_2 - p_3) \ , \tag{5}$$

where s is a scale parameter called also an *amplification factor*.

These two strategies, however, use different approach to the crossover operation. In the binomial variant of the crossover (denoted as $*/*/bin$) the mutated individual p_m is mixed with the original p_j individual with a probability C_R in the following way:

$$p'_{jk} = \begin{cases} p_{mk}, & \text{if } r \leqslant C_R \text{ or } k = r_n \\ p_{jk}, & \text{if } r > C_R \text{ and } k \neq r_n \end{cases} , \tag{6}$$

where $r \in [0, 1]$ is a uniformly distributed random number and r_n is a randomly chosen index from $\{1, 2, ..., d\}$, while d is the problem dimension ensuring that at least one element of p'_j is exchanged with p_m.

In exponential variant of the crossover (denoted as $*/*/exp$) the starting position of the crossover r_n is chosen randomly from $\{1, 2, ..., d\}$ and next $L \leq d$ subsequent elements are taken from the mutated individual p_m in a circular manner. This is shown in Fig. 2. Strategies $/rand/2/bin$ and $/rand/2/exp$ are similar to the strategies presented earlier, but the mutant individual p_m is created from four randomly chosen points instead of two.

$$p_m = p_1 + s(p_2 + p_3 - p_4 - p_5) \ , \tag{7}$$

$p' \longleftarrow p; k \longleftarrow r_n$
while $r \leq CR$ and $L < d$ **do**
   $p'_{jk} \longleftarrow p_{mk}$
   $k \longleftarrow k+1; L \longleftarrow L+1$
end while

Fig. 2. Exponential crossover in DE

The only difference between strategies denoted as /best/ * /* and /rand/ * /* is that in the first ones the p_1 individual is not chosen randomly, but is set to the best solution found so far p_{best}. In two remaining strategies, which were tested in our experiments denoted as /randToBest/1/*, the p_m is created as follows:

$$p_m = s(p_{best} - p_j) + s(p_1 - p_2) \ . \tag{8}$$

2.2 Constraints Handling

The way the mutated individuals p_m are created, what has been presented in the previous subsection, can lead to infeasible solutions. While the solution must be in a given range $[p_{min}, p_{max}]$, if for example randomly chosen individual $p_1 > p_{max}/2$ and other randomly chosen individuals meet the condition $s(p_2 - p_3) \geq p_{max}$, then the mutated solution p_m will be out of upper bound.

The simplest way to deal with this problem is to repeat the stage of choosing random individuals until calculated p_m fits in its bounds. Such tactics works well for relatively small problems, when the computational cost of fitness function is relatively negligible. For the problems of parametric interval linear systems it worked well for the number of parameters up to 50.

For larger problems it was necessary to find out another way of ensuring that newly created individuals fit their bounds. Instead of banning an infeasible solution, a simple repair mechanism was used. When an element p_{mk} of the solution was out of its bounds it was replaced with the element from the best solution found so far. Two other strategies were also tested. In the first one a not-fitted element was replaced by the one of its bounds (with the probability of 0.5). In another one, that element was replaced with a randomly chosen value from uniform distribution $u([p_{min}, p_{max}])$.

The results of the experiments with various constraints handling strategies conducted for a small size problem (with 15 nodes and 38 elements) and a larger one (with 36 nodes and 110 elements) are given in Table 1. Values in the cells show the average percentage increase over the worst performing strategy, which was the strategy of replacing the element out of its range with a randomly generated value from the correct range.

It is worth to notice that the strategy of repeating the process of random individuals choice until the solution will be feasible, was not the best strategy for the smaller instance. This strategy completely failed for the problem with

Table 1. Difference between various strategies of treating infeasible solutions

strategy	number of nodes/number of elements	
	15/38	36/110
repeat choice	2.1%	time exceeded
replace with best	4.8%	14.2%
replace with bound	2.3%	6.6%
replace with random	0.0%	0.0%

110 elements and 36 nodes, as its computational cost exceeded ten times computational cost for other strategies. Thus for the remaining experiments provided in the next section, the strategy of replacement the not fitted element with the element coming from the best solution found so far, was chosen.

3   Computational Experiments

In order to find out which variant of differential evolution will perform the best for large size parametric interval linear systems and how it compares with another metaheuristic method - evolutionary optimisation method [10], three real size test problems with increasing number of parameters, where formulated and evaluated.

3.1   First Test Problem

All the problems presented in the paper come from structural engineering and describe parameters for calculating the displacement of certain points in bays under some downward forces.

First test problem, which has been introduced in the previous section, consists of 36 nodes and 110 elements. All the parameters as well as the load have uncertainty level of 10%.

To maintain the same computation time restriction for differential evolution and evolutionary algorithm (EOM), and after a series of experiments, the following parameters were taken: $pop_{size}=10$, $gen_{max}=50$, $s=0.8$ and $C_R=0.9$ for DE and $pop_{size}=6$ and $max_{gen}=5$ for EOM.

The results achieved by the algorithms were compared using the standard measure of overestimation. For two intervals, such that $[a] \subseteq [b]$, the measure is defined as [5]:

$$O_\omega([a], [b]) := 100\%(1 - \omega([a])/\omega([b])) , \qquad (9)$$

where $\omega([x])$ gives the length of interval $[x]$.

As EOM algorithm gave the narrowest estimation of intervals for all the test problems, different strategies of DE were compared related to the interval length generated by EOM. In our experiments a wider interval means better solution, so the standard measure O_ω indicates by what percentage one solution better estimates the hull solution than another (in our case given by EOM algorithm). We also performed pairwise t-tests for equal means between the results achieved by different strategies. Last column indicates which of the differences are significant assuming $\alpha = 0.05$.

Results achieved for the first test problem are gathered in Table 2.

The best performing DE strategies were *best/2/bin* and *randToBest/1/bin*, while the less efficient was rand/1/exp strategy. Generally strategies with binomial crossover outperformed the strategies with exponential crossover. However, the results generated by them were on average at least 15% better than the results achieved by evolutionary method.

Table 2. DE comparison for the first problem

no.	strategy	average	significant difference
0	eom	0.0%	1,2,3,4,5,6,7,8,9,10
1	best/1/exp	17.8%	0,5,6,7,8,9
2	rand/1/exp	15.7%	0,5,6,7,8,9
3	randToBest/1/exp	17.3%	0,5,6,7,8,9
4	best/2/exp	19.2%	0,5,6,7,8,9
5	rand/2/exp	16.5%	0,5,6,7,8,9
6	best/1/bin	32.8%	0,1,2,3,4,9
7	rand/1/bin	30.0%	0,1,2,3,4,7,8
8	randToBest/1/bin	34.2%	0,1,2,3,4,6,9
9	best/2/bin	34.3%	0,1,2,3,4,6,9
10	rand/2/bin	28.3%	0,1,2,3,4,5,7,8

3.2 Second Test Problem

The second test problem is much larger than the first one. It defines 120 nodes and 420 elements. In this case parameters and the load have uncertainty level of 6%.

Evolutionary method could not deal with the problem of such a large size in the given time restriction (20 times longer than for the first test problem), despite that the parameters for it were set for minimal values of pop_{size}=4 and max_{gen}=2. For differential evolution the parameters were also set to small values pop_{size}=5, max_{gen}=4, but in spite of that, the average computational time was almost 6 times longer than for the first test problem.

The results are shown in Table 2. This time the results are compared relatively to the worst performing /$rand/1/exp$ DE strategy. Strategies $rand/2/exp$ and $rand/2/bin$ were omitted as the pop_{size} was to small for them. Also this time we calculated t-Student tests for equal means. This showed that differences between the tested strategies are rarely significant at α =0.05, but the best performing strategy /$best/2/bin$ significantly differs from 4 other strategies.

Table 3. DE comparison for the second problem

no.	strategy	average	significant difference
0	best/1/exp	21.6%	1,3
1	rand/1/exp	0.0%	0,2,7
2	randToBest/1/exp	20.1%	1,3
3	best/2/exp	0.1%	0,2,7
4	best/1/bin	24.5%	1,3
5	rand/1/bin	9.1%	7
6	randToBest/1/bin	9.0%	7
7	best/2/bin	31.0%	1,3,5,6

Table 4. DE comparison for the third problem

no.	strategy	average	significant difference
0	best/1/exp	0.5%	3,4,5
1	rand/1/exp	0.1%	3,4,5
2	randToBest/1/exp	0.0%	3,4,5
3	best/1/bin	35.4%	0,1,2,5
4	rand/1/bin	34.7%	0,1,2,5
5	randToBest/1/bin	5.1%	0,1,2,3,4

For such a large problem the results were different than for the first test problem. This time strategies with mutations, which exploit the best solution found so far were the winning ones. This, however, may be a side effect of the relatively small generations number calculated.

3.3   Third Test Problem

The third test problem consists of 225 nodes and 812 elements. 25% of case parameters and all the load have uncertainty level of 6%.

This time only three generations were calculated with four-individuals population size. Average computational time for DE was about 5 times longer than for the second test problem, which means 30 times longer than for the first one.

The comparison results and t-Student tests for equal means are given in Table 4. This time *rand/2/best* strategy occurred to be the worst performer. Due to pop_{size} limit strategies *rand/2/exp*, *rand/2/bin*, *best/2/exp*, and *best/2/bin* could not be applied.

The results were similar to the results achieved for the first test problem in this sense, that the strategies with binomial crossover significantly outperformed strategies with exponential crossover.

4   Conclusions

An attempt of applying general metaheurisitcs for solving parametric interval linear system with coefficients that are arbitrary functions of parameters is described in the paper. Differential evolution occurred to be more effective than evolutionary algorithm, which could not deal with the problems of larger size in a reasonable time.

Computational experiments showed that choosing an appropriate strategy of differential evolution method allowed for performance gain up to 35%. Different strategies of treating infeasible solutions were also tested, and the conclusions coming from those experiments could be applied not only to the presented class of problems, but for a much wider area of constrained problems.

Still, further improvements have to be introduced into DE algorithm allowing for providing good solutions for large scale parametric interval linear systems in shorter time. For the third test problem presented in the paper a single pass of DE algorithm took over 100 minutes.

To achieve this it is necessary to speed up the process of the fitness function calculation. This can be done for example by involving a graphical processor unit (GPU) in this process. Additionally some calculations in the DE algorithm itself can be done in parallel, which can further reduce overall computation time.

References

1. Alefeld, G., Kreinovich, V., Mayer, G.: The Shape of the Solution Set for Systems of Interval Linear Equations with Dependent Coefficients. Mathematische Nachrichten 192(1), 23–36 (2006)
2. Dréo, J., Pétrowski, A., Siarry, P., Taillard, E.: Metaheuristics for Hard Optimization. Springer, Heidelberg (2006)
3. Feoktistov, V.: Differential Evolution in Search of Solutions. Springer, Heidelberg (2006)
4. Neumaier, A., Pownuk, A.: Linear Systems with Large Uncertainties with Applications to Truss Structures. Reliable Computing 13(2), 149–172 (2007)
5. Popova, E., Iankov, R., Bonev, Z.: Bounding the Response of Mechanical Structures with Uncertainties in all the Parameters. In: Muhannah, R.L., Mullen, R.L. (eds.) Proceedings of the NSF Workshop on Reliable Engineering Computing (REC), Savannah, Georgia, USA, pp. 245–265 (2006)
6. Pownuk, A.: Efficient Method of Solution of Large Scale Engineering Problems with Interval Parameters Based on Sensitivity Analysis. In: Muhanna, R.L., Mullen, R.L. (eds.) Proc. NSF Workshop Reliable Engineering Computing, Savannah, Georgia, USA, pp. 305–316 (2004)
7. Price, K., Storn, R.M., Lampinen, J.A.: Differential Evolution. A Practical Approach to Global Optimization. Springer, Berlin (2005)
8. Rao, S.S., Berke, L.: Analysis of uncertain structural systems using interval analysis. AIAA Journal 35, 727–735 (1997)
9. Rohn, J., Kreinovich, V.: Computing exact componentwise bounds on solutions of linear systems with interval data is NP-hard. SIAM Journal on Matrix Analysis and Applications (SIMAX) 16, 415–420 (1995)
10. Skalna, I.: Evolutionary Optimization Method for Approximating the Solution Set Hull of Parametric Linear Systems. In: Boyanov, T., Dimova, S., Georgiev, K., Nikolov, G. (eds.) NMA 2006. LNCS, vol. 4310, pp. 361–368. Springer, Heidelberg (2007)

User-Centric Optimization
with Evolutionary and Memetic Systems

Javier Espinar, Carlos Cotta, and Antonio J. Fernández-Leiva

Dept. Lenguajes y Ciencias de la Computación, ETSI Informática,
Campus de Teatinos, Universidad de Málaga,
29071 Málaga – Spain
{ccottap,afdez}@lcc.uma.es

Abstract. One of the lessons learned in the last years in the meta-heuristics community, and most prominently in the area of evolutionary computation (EC), is the need of exploiting problem knowledge in order to come up with effective optimization tools. This problem-knowledge can be provided in a variety of ways, but there are situations in which endowing the optimization algorithm with this knowledge is a very elusive task. This may be the case when this problem-awareness is hard to encapsulate within a specific algorithmic description, e.g., they belong more to the space of human-expert's intuition than elsewhere. An extreme case of this situation can take place when the evaluation itself of solutions is not algorithmic, but needs the introduction of a human to critically assess the quality of solutions. The above use of a combined human-user/evolutionary-algorithm approach is commonly termed interactive EC. The term user-centric EC is however more appropriate since it hints possibilities for the system to be proactive rather than merely interactive, i.e., to anticipate some of the user behavior and/or exhibit some degree of creativity. Such features constitute ambitious goals that require a good grasp of the basic underlying issues surrounding interactive optimization. An overview of these is presented in this paper, along with some hints on what the future may bring to this area. An application example is provided in the context of the search for Optimal Golomb Rulers, a very hard combinatorial problem.

1 Introduction

Evolutionary algorithms (EA), and metaheuristics in general, have been shown to be adequate tools for combinatorial optimization in many different areas. However, in spite of their proved efficiency as optimization methods, recently it has been evident the need of exploiting problem knowledge in order to both obtain solutions of better quality and accelerate the optimization process [7,12]. This has been done for instance via the hybridization with another techniques [11,9], by designing specific genetic operators or by defining intelligent representations with inherent information on them. Other forms of adding problem-knowledge are possible though.

I. Lirkov, S. Margenov, and J. Waśniewski (Eds.): LSSC 2011, LNCS 7116, pp. 214–221, 2012.
© Springer-Verlag Berlin Heidelberg 2012

However, the incorporation of problem knowledge is not always enough to optimize those problems in which the search has to be conducted on spaces comprising candidate solutions which are not easy to evaluate mathematically (e.g., a psychological space). A solution within the framework of the metaheuristics –and more specifically within evolutionary computing (EC)– is the so-called interactive EC (IEC) that basically means that the user can influence the evolutionary process when this is being executed. In the most traditional form of IEC one can find a reactive search-based mechanism in which the user provides some feedback to the demands of the running evolutionary algorithm. Advanced IEC techniques are devoted to reduce the fatigue of the human user, produced by the continuous feedback that the subjacent EC technique demands from him. To this end, it can also employ proactive algorithms that are able of guessing the further user interactions and thus reduce the requirement of user interventions.

This paper provides a general overview on user-centric evolutionary computation mainly focused in hybrid evolutionary algorithms (i.e., memetic algorithms), providing principles for their design and identifying the places where human can interact with the subjacent algorithm. A study case for the optimization of Golomb rulers is also described to show the adequacy of human-guided hybrid EA-based optimization.

2   Human Guided Memetic Algorithm

The most common implementation of memetic algorithm (MA) consists of combining an EA with a procedure to perform a local search (LS) that is usually done after evaluation, although it must be noted however that MAs do not simply reduce themselves to this particular scheme. In fact, the purpose of using LS inside an EA is to provide specific knowledge that can lead to a better optimization process. Ideally, this combination should be synergistic, providing better results that either the EA or the LS by themselves.

Generally speaking, IEC (also termed in the following indistinctly as usercentric EC or human-guided EC, although admittedly some nuances exist in this terminology) is an optimization paradigm that promotes the communication between a human user and an automated evolutionary algorithm (AEA). The intervention of the human is usually required by the AEA for instance to provide subjective fitness evaluation of candidate solutions. Recently, [5] presents a survey on interactive evolutionary computation; now, this paper complements and extends that paper, and deals with human-guided memetic algorithms (HGMA). The aim is to help the reader to understand the mechanisms of HGMAs and provide some indications for their design.

No general approach for the design of effective interactive memetic algorithms (IMA) exists in a well-defined sense, and hence this design phase must be addressed from an intuitive point of view as well. Figure 1 shows a possible schema for human guided memetic algorithms (i.e., interactive hybrid EAs); note that this schema might also be valid for a human-guided EA (i.e., IEC) by simply replacing the MA component by its corresponding EA. The part enclosed by a

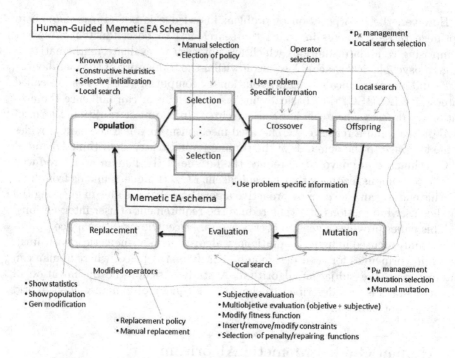

Fig. 1. A possible schema of human guided memetic algorithm

line corresponds with a classical memetic algorithm; the places to incorporate problem knowledge (in form of local search) within an evolutionary algorithm, according to [6] are indicated. In a more general sense, the most typical schema of a MA consists of applying a LS technique to improve the offspring generated in each generation, however there are many possibilities to incorporate knowledge information inside a MA (also different forms of MA are also possible though) and in the following we list some of the places where this can be done:

- During the initialization of the population some heuristic method may be used to generate high quality initial solutions.
- Also, it is possible to use a heuristic method as a recombination or mutation operator. Such operators can be intelligently designed so that specific problem knowledge is used in order to improve the offspring.
- Problem knowledge can also be incorporated in the genotype to phenotype mapping, e.g., when repairing an infeasible solution.
- Another place where an optimization method (e.g., LS) can be particularly useful when used inside a MA is in the optimization of problems where different representations are considered. In these cases, knowledge-based methods can be specifically useful in the encoding-decoding phase.

Distinct places and forms where human might interact with the MA are also indicated in Figure 1; in general the user can influence the optimization process in several ways that basically can be summarized as follows:

- The user may order (or classify) the set of candidates according to some (possibly psychological) criteria provided specifically by her. The user might also affect the evolution process by selecting the individuals to be maintained or removed in further generations (in case of elitist evolutionary algorithms). The user might even influence the individual representation at the genetic or memetic level. Of course, the user might also indicate the type of initialization for the initial population (e.g., randomly, heuristically, etc.).
- The user might elect the policy of both replacement and parent selection.
- The user might determine the choice of the genetic operators (playing thus the role of a hyper-heuristics selector that works inside the subjacent EA mechanism) and also change dynamically some parameters of the algorithm (e.g., the operator probabilities $p_X, p_M, \ldots$). In a deeper form of interaction, the user might also execute manually the corresponding genetic operation (e.g., crossover or mutation).
- The user can control the application of LS (if we consider the most classical form of an MA) in several levels. For instance, it is well known that applying always LS in each generation of the MA (or initially on each individual in the initial population) is not always the best option. Partial Lamarckianism [8], namely applying LS only to a fraction of individuals, can result in better performance. The user might then decide, not only initially but also during the evolving process, the probability of applying LS. Also the individuals to which local search will be applied might be chosen by the (possibly heuristics based) criteria imposed by the user (or at a low level, even manually).
- The user might evaluate –even in eventual ways– the candidate solutions that are generated by the evolutionary algorithm during the evolving procedure. Moreover, the user might add the subjective evaluation as an additional component to the objective evaluation (for instance as an addend with some associated weight) or use the subjective value and the objective values as two different objectives to optimize (transforming thus the model in an interactive multiobjective memetic algorithm).
- The user might reformulate the objective function. For instance the user interaction might consist of refining this function by adjusting the weighing of specific terms.
- The user might also modify the set of constraints attached originally to the definition of the problem. More specifically, the user might provide new (or remove existing) constraints dynamically; this means that the user can impose additional constraints in the form of hard constraints (i.e., their satisfaction is mandatory) or soft constraints (i.e., they can be violated but in this case the violation generates an extra penalty value to be added to the fitness value of the solution).
- The user might incorporate additional mechanisms for improving the optimization process. This is for instance the case when the user decide dynamically to add a restarting mechanism during the optimization process in case of stagnation (or premature convergence). Moreover, the user might decide if the phases of restarting/local-improvement should be performed only on a reduced subset of the population.

Which of the above possibilities is most appropriate is a problem-specific issue, although those strategies leading to involve the user in the way solutions are evaluated are possibly the simplest approach. From a global perspective, the basic idea is to let the user affect the search dynamics with the objective of driving (resp. deviating) the search towards (resp. from) specific regions of the solution space. In the case of human-guided multi-objective MAs the aim is to direct the exploration toward particular regions of the Pareto front.

3   A Human-Guided Hybrid Evolutionary Approach for Optimal Golomb Rulers Search

This section presents a study case to show how human interaction can affect the optimization process in memetic algorithms. The problem under consideration is the search of optimal Golomb rulers. Golomb Rulers [1], are a class of undirected graphs that, unlike usual rulers, measure more discrete lengths than the number of marks they carry. This is due to the fact that on any given Golomb ruler, all differences between pairs of marks are unique. Thus, an *n-mark Golomb ruler* is an ordered set of n distinct non-negative integers, called *marks*, $a_1 < ... < a_n$, such that all the differences $a_i - a_j$ $(i > j)$ are distinct. Clearly we may assume $a_1 = 0$. By convention, a_n is the *length of the Golomb ruler*. An *Optimal Golomb Ruler* (OGR) is defined as the shortest Golomb ruler for a number of marks.

The task of obtaining a Golomb ruler is relatively easy, but finding the optimal for a specified number of marks is an extremely difficult task, and to date, the highest Golomb ruler whose shortest length is known is that with 26 marks. Proof of this difficulty is that the OGRs with 24, 25 and 26 marks were obtained via distributed massively parallel searches (taking several months - even years - to find them). In general, complete methods are inherently limited by the complexity to obtain OGRs and thus metaheuristics (and hybrids thereof) have already been applied to the search of OGRs; to the best of our knowledge, the 16-marks Golomb ruler is the biggest instance of the problem found via heuristics methods (in particular via a memetic algorithms designed as scatter search hybridized with a GRASP-like procedure and local search [4]).

3.1   The Hybrid Algorithm and Experimental Analysis

Our basic model consists of a EA that incorporates ideas from greedy randomized adaptive search procedures (GRASP) in order to perform the genotype-to-phenotype mapping. This hybrid EA was described in [3]. As regards interaction, our human-guided MA (HGMA) enables the user to dynamically (i.e., in running time) change the constraints associated to the problem; this is done by giving the user the capability to change the range of values each gene can take. The user can also request to display statistical information of the population (frequencies of each allele, correlation of an allele with fitness, quartile associated to each gene, etc.), as well as individual solutions, and take decisions such as

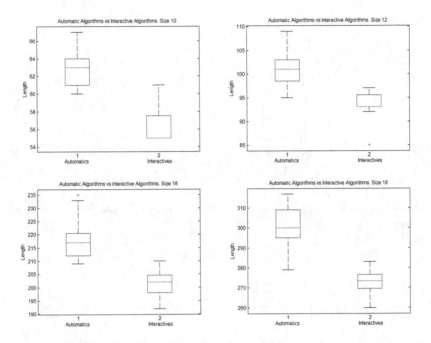

Fig. 2. Results obtained by automated and human-guided algorithms on several instances of the problem

manually selecting solutions for breeding or mutating a specific candidate, and change operators and parameters, just to cite a few.

To alleviate the fatigue of the user (that usually appears in the context of human-guided search when the user is continuously requested by the subjacent algorithm to supply feedback), our approach conceives the interaction as a purely asynchronous feature, i.e., our HGMA behaves as an automated EA (AEA) and hence is capable of working without user intervention. However, it is also capable of being interrupted by the user who can interact with the algorithm and resume the search subsequently[1].

The experiments have been done with an (μ, λ)-EA ($popsize = 100$, $\mu = 10$, $\lambda = 10$, binary tournament selection, $p_X = 1.0$, $p_M = 0.06$, one-point crossover, and mutation by random value change). The objective is to search for an n-mark Golomb ruler and values of n in $\{10, 12, 14, 16, 18, 20\}$ were considered. 20 runs per value of n were executed without user intervention and other 20 runs when the user intervenes 1, 2, or 4 times (5 runs per number of interactions). The maximum number of evaluations is set to $maxevals = 1000n^2$ and the initial range of values for each gene is $[0, n)$. The main goal of these experiments is

---

[1] Note however that there are some ways to alleviate this fatigue as for instance by reducing the accuracy of the judgements required to the user [10] or by replacing the reactive collaboration by a proactive reaction in which the intervention of the user is optional and the algorithm runs autonomously [2].

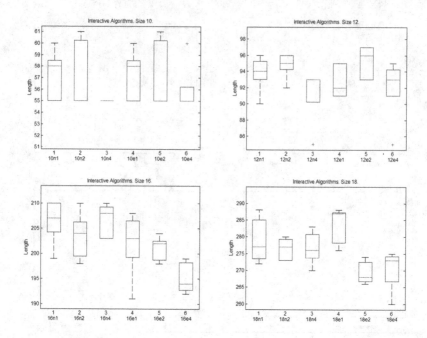

Fig. 3. Results obtained by distinct versions, denoted as nUi, of our hgEA on several problems instances (i.e., $n \in \{10, 12, 16, 18\}$) considering two types of user (i.e., $U \in \{n=\text{non-expert}, e=\text{expert}\}$) and i user interventions (i.e., $i \in \{1, 2, 4\}$)

determining if human-guidance results in a benefit, rather than the absolute quality of the results. Observe that our human-guide (or interactive) memetic algorithm (i.e., HGMA) provides the best solutions compared to those obtained by the automatic algorithms (i.e., those with no user intervention) as shown in Fig. 2; in addition we also proved that the average differences among these algorithms were statistically significant in favor of the HGMA.

Subsequently we have evaluated how the number of user interventions as well as the 'nature' (i.e., expert or non-expert) of the user affect the performance of the HGMA. Figure 3 show that the intervention of an expert is specially manifested in the harder instances. Also, if the user is non-expert, then increasing the number of interventions might be even harmful.

4   Conclusions

This work has presented some principles for the design of human-guided hybrid evolutionary algorithms that should help the interested reader to understand the basic mechanisms of this kind of algorithms. The paper has also described a user-centric memetic algorithm that was implemented for the search of optimal Golomb rulers, a very hard to solve combinatorial problem. An experimental analysis has shown the usefulness of the human guide. In any case, as already indicated, these results should be taken carefully as this paper represents a preliminary study on this issue. Further analysis on a wider set of problem instances as well as distinct problems should also be conducted.

Acknowledgements. This work is supported by project NEMESIS (TIN-2008-05941) of the Spanish Ministerio de Ciencia e Innovación, and project TIC-6083 of Junta de Andalucía.

References

1. Babcock, W.C.: Intermodulation interference in radio systems. Bell Systems Technical Journal, 63–73 (1953)
2. Breukelaar, R., Emmerich, M., Bäck, T.: On Interactive Evolution Strategies. In: Rothlauf, F., Branke, J., Cagnoni, S., Costa, E., Cotta, C., Drechsler, R., Lutton, E., Machado, P., Moore, J.H., Romero, J., Smith, G.D., Squillero, G., Takagi, H. (eds.) EvoWorkshops 2006. LNCS, vol. 3907, pp. 530–541. Springer, Heidelberg (2006)
3. Cotta, C., Fernández, A.: A Hybrid GRASP – Evolutionary Algorithm Approach to Golomb Ruler Search. In: Yao, X., Burke, E.K., Lozano, J.A., Smith, J., Merelo-Guervós, J.J., Bullinaria, J.A., Rowe, J.E., Tiňo, P., Kabán, A., Schwefel, H.-P. (eds.) PPSN 2004. LNCS, vol. 3242, pp. 481–490. Springer, Heidelberg (2004)
4. Cotta, C., Dotú, I., Fernández, A.J., Hentenryck, P.V.: Local search-based hybrid algorithms for finding Golomb rulers. Constraints 12(3), 263–291 (2007)
5. Cotta, C., Fernández-Leiva, A.J.: Bio-inspired Combinatorial Optimization: Notes on Reactive and Proactive Interaction. In: Cabestany, J., Rojas, I., Joya, G. (eds.) IWANN 2011, Part II. LNCS, vol. 6692, pp. 348–355. Springer, Heidelberg (2011)
6. Eiben, A.E., Smith, J.E.: Introduction to evolutionary computation. Springer, Heidelberg (2003)
7. Hart, W.E., Belew, R.K.: Optimizing an arbitrary function is hard for the genetic algorithm. In: Belew, R.K., Booker, L.B. (eds.) 4th International Conference on Genetic Algorithms, pp. 190–195. Morgan Kaufmann, San Mateo CA (1991)
8. Houck, C., Joines, J., Kay, M., Wilson, J.: Empirical investigation of the benefits of partial Lamarckianism. Evolutionary Computation 5(1), 31–60 (1997)
9. Moscato, P., Cotta, C.: A modern introduction to memetic algorithms. In: Gendreau, M., Potvin, J.-Y. (eds.) Handbook of Metaheuristics, 2nd edn. International Series in Operations Research and Management Science, vol. 146, pp. 141–183. Springer, Heidelberg (2010)
10. Ohsaki, M., Takagi, H., Ohya, K.: An input method using discrete fitness values for interactive GA. Journal of Intelligent and Fuzzy Systems 6(1), 131–145 (1998)
11. Puchinger, J., Raidl, G.R.: Combining Metaheuristics and Exact Algorithms in Combinatorial Optimization: A Survey and Classification. In: Mira, J., Álvarez, J.R. (eds.) IWINAC 2005, Part II. LNCS, vol. 3562, pp. 41–53. Springer, Heidelberg (2005)
12. Wolpert, D.H., Macready, W.G.: No free lunch theorems for optimization. IEEE Transactions on Evolutionary Computation 1(1), 67–82 (1997)

Intuitionistic Fuzzy Estimation of the Ant Colony Optimization Starting Points

Stefka Fidanova[1], Krassimir Atanassov[2], and Pencho Marinov[1]

[1] IICT – Bulgarian Academy of Sciences,
Acad. G. Bonchev str. bl.25A, 1113 Sofia, Bulgaria
stefka,pencho@parallel.bas.bg
[2] CLBME – Bulgarian Academy of Science,
Acad. G. Bonchev str, bl 105, 1113 Sofia, Bulgaria
krat@bas.bg

Abstract. The ability of ant colonies to form paths for carrying food is rather fascinating. The problem is solved collectively by the whole colony. This ability is explained by the fact that ants communicate in an indirect way by laying trails of pheromone. The higher the pheromone trail within a particular direction, the higher the probability of choosing this direction. The collective problem solving mechanism has given rise to a metaheuristic referred to as Ant Colony Optimization. On this work we use intoitionistic fuzzy estimation of start nodes with respect to the quality of the solution. Various start strategies are offered. Sensitivity analysis of the algorithm behavior according to estimation parameters is made. As a test problem Multidimensional (Multiple) Knapsack Problem is used.

1 Introduction

A large number of real-life optimization problems in science, engineering, economics, and business are complex and difficult to solve. They can not be solved in an exact manner within a reasonable amount of computational resources. Using approximate algorithms is the main alternative to solve this class of problems. The approximate algorithms are specific heuristics, which are problem dependent, and metaheuristics, which are more general approximate algorithms applicable to a large variety of optimization problems. One of the most successful metaheuristic is Ant Colony Optimization (ACO) [2,4,13].

ACO algorithms have been inspired by the real ants behavior. In nature, ants usually wander randomly, and upon finding food return to their nest while laying down pheromone trails. If other ants find such a path, they are likely not to keep traveling at random, but to instead follow the trail, returning and reinforcing it if they eventually find food. However, as time passes, the pheromone evaporate. The more time it takes for an ant to travel down the path and back again, the more time the pheromone has to evaporate and the path to become less prominent. A shorter path will be visited by more ants and thus the pheromone density remains high for a longer time. ACO is implemented as a team of intelligent agents which simulate the ants behavior, walking around the graph

I. Lirkov, S. Margenov, and J. Waśniewski (Eds.): LSSC 2011, LNCS 7116, pp. 222–229, 2012.

representing the problem to solve using mechanisms of cooperation and adaptation. Examples of optimization problems are Traveling Salesman Problem [12], Vehicle Routing [14], Minimum Spanning Tree [10], Multiple Knapsack Problem [6], etc.

The transition probability $p_{i,j}$, to choose the node j from the graph of the problem, when the current node is i, is based on the heuristic information $\eta_{i,j}$ and the pheromone trail level $\tau_{i,j}$ of the move, where $i, j = 1, \ldots, n$.

$$p_{i,j} = \frac{\tau_{i,j}^a \eta_{i,j}^b}{\sum\limits_{k \in Unused} \tau_{i,k}^a \eta_{i,k}^b}, \tag{1}$$

where $Unused$ is the set of unused nodes of the graph.

The higher value of the pheromone and the heuristic information, the more profitable it is to select this move and resume the search. In the beginning, the initial pheromone level is set to a small positive constant value τ_0; later, the ants update this value after completing the construction stage. ACO algorithms adopt different criteria to update the pheromone level. The pheromone trail update rule is given by:

$$\tau_{i,j} \leftarrow \rho \tau_{i,j} + \Delta \tau_{i,j}, \tag{2}$$

where ρ models evaporation in the nature and $\Delta \tau_{i,j}$ is new added pheromone which is proportional to the quality of the solution.

The novelty in this paper is the use of intuitionistic fuzzy estimations of start nodes with respect to the quality of the solution and thus to better manage the search process. Intuitionistic logic or constructive logic is a symbolic logic. In this logic any statement can be truth, false, or uncertainty. Thus truth plus false can not be equal to one. On the basis of the estimations we offer several start strategies and their combinations. Like a benchmark problem Multiple Knapsack Problem (MKP) is used, which is a representative of the class of subset problems. A lot of real world problems can be represented by it. Moreover MKP arises like a subproblem in many optimization problems.

The rest of the paper is organized as follows: in section 2 intuitionistic fuzzy estimation of start node is introduced and several start strategies are proposed. In section 3 the strategies are applied on MKP and sensitivity analysis of the algorithm according strategy parameters is made. At the end some conclusions and directions for future work are done.

2   Intuitionistic Fuzzy Estimation

The known ACO algorithms create a solution starting from random node. But for some problems, especially subset problems, it is important from which node the search process starts. For example if an ant starts from node which does not belong to the optimal solution, probability to construct it is zero. In this paper it is offered intuitionistic fuzzy estimation of start nodes and after that several start strategies are proposed. The aim is to use the experience of the ants from previous iteration to choose the better starting node. Other authors use this experience only by the pheromone, when the ants construct the solutions.

2.1   Intuitionistic Fuzzy Sets

At the beginning we will define the intuitionistic fuzzy sets [1]. Let X, Y, and Z be ordinary finite non-empty sets. An intuitionistic fuzzy set in X is an expression A given by:

$$A = \{< x, \mu_A(x), \nu_A(x) > | x \in X\}, \quad \text{where} \tag{3}$$

$$\mu_A : X \rightarrow [0, 1] \quad \nu_A : X \rightarrow [0, 1]$$

with the condition $0 \le \mu_A(x) + \nu_A(x) \le 1$ for all $x \in A$. The numbers $\mu_A(x)$ and $\nu_A(X)$ denote respectively the degree of membership and the degree of non-membership of the element x in the set A. When $\nu_A(x) = 1 - \mu_A(x)$, set A is a fuzzy set. When $\nu_A(x) \le 1 - \mu_A(x)$, set A is an intuitionistic fuzzy.

Let the graph of the problem has m nodes. The set of nodes is divided on N subsets. There are different ways for dividing. Normally, the nodes of the graph are randomly enumerated. An example for creating of the subsets, without lost of generality, is: the node number one is in the first subset, the node number two - in the second subset, etc., the node number N is in the N-th subset, the node number $N + 1$ is in the first subset, etc. Thus the number of nodes in the separate subsets are almost equal. After the first iteration the estimations $D_j(i)$ and $E_j(i)$ are introduced of the node subsets, where $i \ge 2$ is the number of the current iteration and $D_j(i)$ and $E_j(i)$ are weight coefficients of $j-th$ node subset $(1 \le j \le N)$, which are calculated by the following formulas:

$$D_j(i) = \varphi.D_j(i - 1) + (\psi - \varphi).F_j(i), \tag{4}$$

$$E_j(i) = \varphi.E_j(i - 1) + (\psi - \varphi).G_j(i), \tag{5}$$

where $i \ge 2$ is the current process iteration and for each j $(1 \le j \le N)$:

$$F_j(i) = \begin{cases} \dfrac{f_{j,A}}{n_j} & \text{if } n_j \ne 0 \\ F_j(i - 1) & \text{otherwise} \end{cases}, \tag{6}$$

$$G_j(i) = \begin{cases} \dfrac{g_{j,B}}{n_j} & \text{if } n_j \ne 0 \\ G_j(i - 1) & \text{otherwise} \end{cases}, \tag{7}$$

$f_{j,A}$ is the number of the solutions among the best $A\%$, $g_{j,B}$ is the number of the solutions among the worst $B\%$, where $A + B \le 100$, $i \ge 2$ and

$$\sum_{j=1}^{N} n_j = n, \tag{8}$$

where n_j $(1 \le j \le N)$ is the number of solutions obtained by ants starting from nodes subset j, n is the number of ants. Initial values of the weight coefficients are: $D_j(1) = 1$ and $E_j(1) = 0$. The parameter φ, $0 \leq \varphi \leq 1$, shows the weight of the information from the previous iterations and from the last iteration. When $\varphi = 0$ only the information from the last iteration is taken in to account. If $\varphi = 0.5 * \psi$ the influence of the previous iterations versus the last is equal. When $\varphi = \psi$ only the information from the previous iterations is taken in to account. The balance between the weights of the previous iterations and the last is important. At the beginning when the current best solution is far from the optimal one, some of the node subsets can be estimated as good. Therefore, if the value of the parameter φ is too high the estimation can be distorted. If the weight of the last iteration is too high then information for good and bad solutions from previous iterations is ignored, which can distort estimation too.

We try to use the experience of the ants from previous iteration to choose the better starting node. Other authors use this experience only by the pheromone, when the ants construct the solutions [5].

2.2   Start Strategies

Let us fix threshold E for $E_j(i)$ and D for $D_j(i)$, then we construct several strategies to choose start node for every ant, the threshold E increases every iteration with $1/i$ where i is the number of the current iteration:

1. If $E_j(i)/D_j(i) > E$ then the subset j is forbidden for current iteration and we choose the starting node randomly from $\{j \mid j$ is not forbidden$\}$;
2. If $E_j(i)/D_j(i) > E$ then the subset j is forbidden for current simulation and we choose the starting node randomly from $\{j \mid j$ is not forbidden$\}$;
3. If $E_j(i)/D_j(i) > E$ then the subset j is forbidden for K_1 consecutive iterations and we choose the starting node randomly from $\{j \mid j$ is not forbidden$\}$;
4. Let $r_1 \in [R, 1)$ be a random number. Let $r_2 \in [0, 1]$ be a random number. If $r_2 > r_1$ we randomly choose node from subset $\{j \mid D_j(i) > D\}$, otherwise we randomly chose a node from the not forbidden subsets, r_1 is chosen and fixed at the beginning.
5. Let $r_1 \in [R, 1)$ be a random number. Let $r_2 \in [0, 1]$ be a random number. If $r_2 > r_1$ we randomly choose node from subset $\{j \mid D_j(i) > D\}$, otherwise we randomly chose a node from the not forbidden subsets, r_1 is chosen at the beginning and increase with r_3 every iteration.

Where $0 \leq K_1 \leq$ "number of iterations" is a parameter. If $K_1 = 0$, then strategy 3 is equal to the random choose of the start node. If $K_1 = 1$, then strategy 3 is equal to the strategy 1. If $K_1 =$ "maximal number of iterations", then strategy 3 is equal to the strategy 2.

We can use more than one strategy for choosing the start node, but there are strategies which can not be combined. We distribute the strategies into two sets: $St1 = \{strategy1, strategy2, strategy3\}$ and $St2 = \{strategy4, strategy5\}$. The strategies from same set can not be used at once. When we combine strategies from $St1$ and $St2$, first we apply the strategy from $St1$ and according it some of the regions (node subsets) become forbidden, and after that we choose the starting node from not forbidden subsets according the strategy from $St2$.

3   Experimental Results

The intuitionistic fuzzy estimation and start strategies performance are analyzed in this section. Like test Multiple Knapsack Problem is used because it is subset problem. The Multiple Knapsack Problem has numerous applications in theory as well as in practice. It also arises as a subproblem in several algorithms for more complex problems and these algorithms will benefit from any improvement in the field of MKP. The following major applications can be mentioned: problems in cargo loading, cutting stock, bin-packing, budget control, and financial management may be formulated as MKP. In [11] it is proposed to use the MKP in fault tolerance problem and in [3] it is designed a public cryptography scheme whose security realize on the difficulty of solving the MKP. In [9] it is mentioned that two-processor scheduling problems can be solved as a MKP. Other applications are industrial management, naval, aerospace, computational complexity theory.

The MKP can be thought as a resource allocation problem, where there are m resources (the knapsacks) and n objects and every object j has a profit p_j. Each resource has its own budget c_j (knapsack capacity) and consumption r_{ij} of resource i by object j. The aim is maximizing the sum of the profits, while working with a limited budget.

The MKP can be formulated as follows:

$$\max \sum_{j=1}^{n} p_j x_j$$

$$\text{subject to } \sum_{j=1}^{n} r_{ij} x_j \le c_i \quad i = 1, \dots, m \tag{9}$$

$$x_j \in \{0, 1\} \quad j = 1, \dots, n$$

x_j is 1 if the object j is chosen and 0 otherwise.

There are m constraints in this problem, so MKP is also called m-dimensional knapsack problem. Let $I = \{1, \dots, m\}$ and $J = \{1, \dots, n\}$, with $c_i \ge 0$ for all $i \in I$. A well-stated MKP assumes that $p_j > 0$ and $r_{ij} \le c_i \le \sum_{j=1}^{n} r_{ij}$ for all $i \in I$ and $j \in J$. Note that the $[r_{ij}]_{m \times n}$ matrix and $[c_i]_m$ vector are both non-negative.

In the MKP one is not interested in solutions giving a particular order. Therefore a partial solution is represented by $S = \{i_1, i_2, \dots, i_j\}$ and the most recent elements incorporated to S, i_j need not to be involved in the process for selecting the next element. Moreover, solutions for ordering problems have a fixed length as one search for a permutation of a known number of elements. Solutions for MKP, however, do not have a fixed length. The graph of the problem is defined as follows: the nodes correspond to the items, the arcs fully connect nodes. Fully connected graph means that after the object i one can choose the object j for every i and j if there are enough resources and object j is not chosen yet.

The computational experience of the ACO algorithm is shown using 10 MKP instances from "OR-Library" available within WWW access at http://people. brunel.ac.uk/~mastjjb/jeb/orlib, with 100 objects and 10 constraints. To provide a fair comparison for the above implemented ACO algorithm, a pre-defined number of iterations, $k = 100$, is fixed for all the runs. The developed technique has been coded in C++ language and implemented on a Pentium 4 (2.8 GHz). The parameters are fixed as follows: $\rho = 0.5$, $a = 1$, $b = 1$, number of used ants is 20, $A = 30$, $B = 30$, $D = 1.5$, $E = 0.5$, $K_1 = 5$, $R = 0.5$, $r_3 = 0.01$. The values of ACO parameters (ρ, a, b) are from [7] and experimentally is found that they are best for MKP. The tests are run with 1, 2, 4, 5, and 10 nodes within the nodes subsets. The following combinations for parameters φ and ψ are used: $(0.125, 0.25)$, $(0.125, 0.5)$, $(0.125, 0.825)$, $(0.25, 0.5)$, $(0.25, 0.75)$, $(0.25, 0.825)$, $(0.5, 0.75)$, $(0.5, 0.825)$, $(0.75, 0.825)$, $(0.5, 1)$. For every experiment, the results are obtained by performing 30 independent runs, then averaging the fitness values obtained in order to ensure statistical confidence of the observed difference. The computational time, which start strategies takes, is negligible with respect to running time of the algorithm. Tests with all combinations of strategies and with random start (12 combinations) are run. Thus we perform 180 000 tests.

Average achieved result by some strategy, is better than without any strategy, for every test problem. For fair comparison, the difference d between the worst and best average result for every problem is divided by 10. If the average result for some strategy is between the worst average result and worst average plus $d/10$ it is appreciated with 1. If it is between the worst average plus $d/10$ and worst average plus $2d/10$ it is appreciated with 2 and so on. If it is between the best average minus $d/10$ and the best average, it is appreciated with 10. Thus for a test problem the achieved results for every strategy and every nodes division is appreciated from 1 to 10. After that the rate of all test problems for every strategy and every nodes division is summed. So theirs rate becomes between 10 and 100 (see Table 1). It is like percentage of successes. The best results are achieved when there is only one node in the node subset, for all combinations of the values of φ and ψ. When the node subsets consist one node then from them preferably bad or preferably good solutions start. Therefore on Table 1 we show only this case.

With respect to the strategies (rows), the best rates are achieved with strategy 3 and combinations between strategy 1 and strategies 4 and 5. With respect to the values of the parameters φ and ψ the best rate is when $\varphi = 0.5$ and $\psi = 0.75$. In this case there is good balance between previous and current iterations and the rate of the intuitionistic fuzziness is low. On Table 1, we observe that when the rate of the intuitionistic fuzziness is high (the value of ψ is small), the ACO algorithm achieves worse solutions. If there is not intuitionistic fuzziness ($\psi = 1$), then the rate of the solution is bad too [8]. The algorithm performance is similar when the difference between ψ and φ is big, which is bad balance between current and previous iterations.

Table 1. Estimation of strategies and rate of fuzziness

φ ψ	0.125 0.25	0.125 0.5	0.125 0.75	0.125 0.825	0.25 0.5	0.25 0.75	0.25 0.825	0.5 0.75	0.5 0.825	0.75 0.825	0.5 1
random	32	32	32	32	32	32	32	32	32	32	32
strat. 1	95	93	93	93	93	92	92	94	93	93	96
strat. 2	82	79	79	79	79	85	85	82	77	83	78
strat. 3	93	92	92	92	92	93	94	**99**	94	93	97
strat. 4	83	83	83	83	83	83	83	83	83	83	83
strat. 5	83	83	83	83	83	83	83	83	83	83	83
strat. 1-4	96	96	96	96	96	96	96	92	94	95	97
strat. 1-5	96	96	96	96	96	96	96	92	94	95	97
strat. 2-4	84	83	83	83	83	83	83	82	81	86	83
strat. 2-5	84	83	83	83	83	83	83	82	81	86	83
strat. 3-4	94	93	93	93	93	94	94	93	93	97	96
strat. 3-5	94	93	93	93	93	94	94	93	93	97	96

4   Conclusion

This paper is addressed to ant colony optimization algorithm with controlled start using intuitionistic fuzzy estimation. So, the start node of each ant depends of the goodness of the respective region. We have analyzed the behavior of the ACO algorithm. The rate of the achieved solutions is higher when the rate of the fuzziness is not very high and when the difference between the coefficients ψ and φ is not very big. In this case there is good balance between results achieved in previous iterations and current iteration. The future work will be focused on parameter settings which manage the starting procedure. It will be investigated on influence of the parameters to algorithm performance. The aim is to study in detail the relationships between the start nodes and the quality of the achieved solutions.

Acknowledgments. This work has been partially supported by the Bulgarian National Scientific Fund under the grants ID-Modeling Processes with fixed development rules DID 02/29 and TK-Effective Monte Carlo Methods for large-scale scientific problems DTK 02/44.

References

1. Atanassov, K.: Intuitionistic Fuzzy Sets. Springer, Heidelberg (1999)
2. Bonabeau, E., Dorigo, M., Theraulaz, G.: Swarm Intelligence: From Natural to Artificial Systems. Oxford University Press, New York (1999)
3. Diffe, W., Hellman, M.E.: New direction in cryptography. IEEE Trans Inf. Theory IT-36, 644–654 (1976)
4. Dorigo, M., Gambardella, L.M.: Ant Colony System: A Cooperative Learning Approach to the Traveling Salesman Problem. IEEE Transactions on Evolutionary Computation 1, 53–66 (1997)
5. Dorigo, M., Stutzle, T.: Ant Colony Optimization. MIT Press (2004)
6. Fidanova, S.: Evolutionary Algorithm for Multiple Knapsack Problem. In: Int. Conference Parallel Problems Solving from Nature, Real World Optimization Using Evolutionary Computing, Granada, Spain (2002) ISBN No 0-9543481-0-9
7. Fidanova, S.: Ant colony optimization and multiple knapsack problem. In: Renard, J.P. (ed.) Handbook of Research on Nature Inspired Computing for Economics and Management, pp. 498–509. Idea Grup. Inc. (2006) ISBN 1-59140-984-5
8. Fidanova, S., Atanassov, K., Marinov, P., Parvathi, R.: Ant Colony Optimization for Multiple Knapsack Problems with Controlled Starts. Int. J. Bioautomation 13(4), 271–280 (2008)
9. Martello, S., Toth, P.: A mixtures of dynamic programming and branch-and-bound for the subset-sum problem. Management Science 30, 756–771 (1984)
10. Reiman, M., Laumanns, M.: A Hybrid ACO algorithm for the Capacitated Minimum Spanning Tree Problem. In: Proc. of First Int. Workshop on Hybrid Metahuristics, Valencia, Spain, pp. 1–10 (2004)
11. Sinha, A., Zoltner, A.A.: The multiple-choice knapsack problem. J. Operational Research 27, 503–515 (1979)
12. Stutzle, T., Dorigo, M.: ACO Algorithm for the Traveling Salesman Problem. In: Miettinen, K., Makela, M., Neittaanmaki, P., Periaux, J. (eds.) Evolutionary Algorithms in Engineering and Computer Science, pp. 163–183. Wiley (1999)
13. Stutzle, T., Hoos, H.H.: MAX-MIN Ant System. In: Dorigo, M., Stutzle, T., Di Caro, G. (eds.) Future Generation Computer Systems, vol. 16, pp. 889–914 (2000)
14. Zhang, T., Wang, S., Tian, W., Zhang, Y.: ACO-VRPTWRV: A New Algorithm for the Vehicle Routing Problems with Time Windows and Re-used Vehicles based on Ant Colony Optimization. In: Sixth International Conference on Intelligent Systems Design and Applications, pp. 390–395. IEEE Press (2006)

Variable Neighborhood Search for Robust Optimization and Applications to Aerodynamics

A. Mucherino[1], M. Fuchs[1], X. Vasseur[1], and S. Gratton[1,2]

[1] CERFACS, Toulouse, France
{mucherino,martin.fuchs,xavier.vasseur}@cerfacs.fr
[2] INPT-IRIT, University of Toulouse, Toulouse, France
serge.gratton@enseeiht.fr

Abstract. Many real-life applications lead to the definition of robust optimization problems where the objective function is a black box. This may be due, for example, to the fact that the objective function is evaluated through computer simulations, and that some parameters are uncertain. When this is the case, existing algorithms for optimization are not able to provide good-quality solutions in general. We propose a heuristic algorithm for solving black box robust optimization problems, which is based on a bilevel Variable Neighborhood Search to solve the minimax formulation of the problem. We also apply this algorithm for the solution of a wing shape optimization where the objective function is a computationally expensive black box. Preliminary computational experiments are reported.

1 Introduction

Global optimization aims at finding solutions to a problem for which the objective function value is optimized and all constraints, if any, are satisfied. Many optimization problems arising in real-life applications have been proved to be NP-hard. Moreover, there are applications in which the complexity of the problem is increased by the necessity of considering some uncertain parameters. Uncertainty may concern, for example, prices, demands, traveling conditions, working time, etc. [15].

Robust optimization is a way to manage uncertainty in optimization problems [1]. The main difficulty is that the set defined by the uncertain parameters may be quite large in size and therefore lead to the definition of a large-scale optimization problem. In recent years, the scientific community has been giving a lot of attention to algorithms and methods for solving this class of problems.

Different approaches have been proposed over time for dealing with uncertainty [4,10,15]. In this work, we consider the approach in which the uncertain parameters are transformed in decision variables. Naturally, the newly introduced variables (that, in the following, we will refer to as *uncertain variables*) cannot play the same role of the original decision variables. In other words, two different sets of variables are considered in the optimization problem, and they need to be optimized in two different ways.

I. Lirkov, S. Margenov, and J. Waśniewski (Eds.): LSSC 2011, LNCS 7116, pp. 230–237, 2012.

When considering this approach, the uncertain variables are optimized so that they correspond to the *worst-case scenario* they are able to describe. In this way, the original variables are optimized under the situation in which the worst values for the uncertain variables are considered. If the found solution is optimal for the worst-case scenario, then it *should* still be acceptable when the uncertain variables have values different from the nominal ones.

Let us suppose that the following global optimization problem needs to be solved:

$$\min_{x \in B} f(x, y),$$

where x is a vector of decision variables, $B \subseteq \Re^{d_x}$, y is a vector of uncertain parameters, $y \in C \subseteq \Re^{d_y}$, and $f : B \times C \to \Re$. In order to solve this optimization problem by considering the worst-case scenario that the uncertain variables are able to give, we can rewrite it as the following bilevel program:

$$\begin{aligned}
\min_{x \in B} & \ f(x, t) \\
\text{s.t.} & \ t = \arg\max_{y \in C} f(x, y).
\end{aligned} \tag{1}$$

Note that the inner and the outer problem of this bilevel program have both the same objective function $f(x, y)$, but the actual variables are the x's in the outer problem, and the y's in the inner problem. Since the original problem is a minimization problem, the worst-case scenario given by the uncertain variables y can be found by maximizing $f(x, y)$. It is important to note that there is a strong dependence between the two problems, so that they cannot simply be solved one after another (e.g. first solve the inner problem and then solve the outer one), but rather simultaneously. This makes the solution of bilevel programs quite difficult. This kind of problem is also known in the scientific community as the *minimax* optimization problem. We remark that minimax problems with convex-concave payoff and convex constraints can be reduced to single optimization problems [2], but problems having these properties are quite rare in real-life applications.

Optimization problems formulated like (1) are usually quite difficult to solve, and, in some cases, the NP-hardness of such problems has been formally proved. Moreover, an additional aspect to consider is that, in real-life applications, $f(x, y)$ may be represented by a *black box*. In other words, the mathematical expression of $f(x, y)$ is not known, and it is rather evaluated through computer simulations, which might be expensive. For this reason, heuristics are usually employed for solving this class of problems. Some examples are given in [9,14,16].

In this paper, we propose a heuristic specifically designed for solving bilevel programs and we apply this algorithm for the solution of minimax problems arising in the context of robust optimization. The Variable Neighborhood Search (VNS) [7,12] is a meta-heuristic search based on the idea of exploring neighbors of current problem solutions in order to discover better ones. Our heuristic implements two VNS's, closely combined to each other. This idea was already mentioned in the paper that introduced the VNS [12], and it has been already exploited in different forms (see for example [11]). Our idea is to make the two VNS's work simultaneously on the outer and on the inner problem of the bilevel program. While one VNS works for optimizing the decision variables x, the other one works for finding the worst-case scenario given by the uncertain variables y. We will refer to this heuristic algorithm as the *bilevel* VNS.

The paper is organized as follows. In Section 2 we give a description of the bilevel VNS algorithm for solving minimax optimization problems. Two examples of robust optimization problems are considered in Section 3. The first one is an academic problem that we consider for testing our heuristic on a problem for which the solution can be estimated (Section 3.1). The second problem is the one of finding the optimal shape of aircraft's wings in the worst-case scenario given by the flight operating conditions (Section 3.2). In the latter example, the objective function can be evaluated through a computationally expensive computer simulation. Final remarks and conclusions are given in Section 4.

2   The Bilevel VNS

The meta-heuristic Variable Neighborhood Search (VNS) [7,12] is one of the most successful heuristics for global optimization. It is based on the idea of exploring small neighbors of currently known solutions, which may increase in size only if no better solutions can be found in the current neighbors. VNS makes usually use of local search algorithms for reducing the search to local minima only. In this way, a path of local optima is defined, that may lead to the global optimum of the considered problem.

The VNS is currently widely employed for solving problems arising in various applications. In some very recent works, the VNS has been used, for example, for solving the p-median clustering problem [6], feature selection problems in data mining [13], and routing [8] and task scheduling [17] problems.

When the objective function of the considered optimization problem is a black box (or not easily differentiable), at each iteration of the VNS, the local search can be replaced by another execution of the VNS. In this case, the algorithm may be referred to as *double* VNS, where two VNS's are combined. This is done in order to improve the quality of the found solutions. In practice, even if the second VNS does not represent a valid alternative to a (deterministic) local search, it can help in finding solutions that are closer to the optimal ones.

In this work, we rather consider two VNS's for solving simultaneously the inner and the outer problem of a bilevel program. The basic idea is to make only one VNS work on the inner problem, while the second VNS works on the outer problem. No local searches (neither deterministic ones nor heuristic ones) are employed for reducing the search to local minima only. This is done for avoiding that the execution of the heuristic becomes too expensive: we suppose that our objective function can only be evaluated by employing an expensive computer simulation, and therefore the total number of possible function evaluations is very limited.

Algorithm 1 is a sketch of our bilevel VNS heuristic for bilevel programs (1).

At the beginning, random values for both variables x and y can be chosen so that the corresponding constraints, if any, are satisfied. Thereafter, the first

Algorithm 1. A bilevel VNS heuristic for bilevel programs.

1: let $iter_x = 0$; let $iter_y = 0$;
2: let $\varepsilon_x = \varepsilon_{min}$; let $\varepsilon_y = \varepsilon_{min}$;
3: randomly generate $x \in B$;
4: randomly generate $y \in C$;
5: let $x^{best} = x$; let $y^{best} = y$; let $f^{best} = f(x,y)$;
6: **while** $(\varepsilon_y \le \varepsilon_{max})$ **do**
7:    let $iter_y = iter_y + 1$;
8:    randomly pick $y \in C$ from a neighborhood of y^{best} with radius ε_y;
9:    **if** $(f(x^{best}, y) > f^{best})$ **then**
10:       let $\varepsilon_y = \varepsilon_{min}$;
11:       let $y^{best} = y$; let $f^{best} = f(x^{best}, y)$;
12:       **while** $(\varepsilon_x \le \varepsilon_{max})$ **do**
13:          let $iter_x = iter_x + 1$;
14:          randomly pick $x \in B$ from a neighborhood of x^{best} with radius ε_x;
15:          **if** $(f(x, y^{best}) < f^{best})$ **then**
16:             let $\varepsilon_x = \varepsilon_{min}$;
17:             let $x^{best} = x$; let $f^{best} = f(x, y^{best})$;
18:          **else**
19:             modify ε_x;
20:          **end if**
21:       **end while**
22:    **else**
23:       modify ε_y;
24:    **end if**
25: **end while**

VNS starts. Its execution is controlled by a while loop which stops when ε_y gets larger than ε_{max}. ε_y represents the radius of the current neighbor of y_{best} where solutions to the problem can be picked randomly. It can only range in the interval $[\varepsilon_{min}, \varepsilon_{max}]$. When a better solution is found, ε_y is set back to ε_{min}, otherwise it is modified and it may be increased in value after a certain number of iterations without any improvement.

The first VNS considers the uncertain variables y. Therefore, for each improved solution of the inner problem of the bilevel program, a full execution of the second VNS, on the decision variables x, is performed. The two VNS's behave exactly in the same way. The only difference is that, while the first VNS works on the variables y and tries to maximize the objective function, the second VNS works on the variables x and tries to minimize the function value.

By borrowing some notions from game theory, we can say that the bilevel VNS implements a two-person game, where each person uses one of the two VNS's to optimize his strategies, i.e. the objective function (minimization against maximization) [3]. In case there is a solution which is optimal for both players (the so-called *saddle point*), then the two VNS's do not really play one against the other and the bilevel VNS should converge towards that solution. Otherwise, one of the two players may overpower the other. This is a typical situation which is hard to manage by deterministic methods, because there is the risk of cycling. Since our bilevel VNS is instead a heuristic with randomized parts, it will mostly like not cycle.

3   Development and Testing

We implemented our bilevel VNS heuristic for bilevel programs in Python. We chose this programming language because part of the code for evaluating the objective function of the problem described in Section 3.2 was already written in this language. The experiments have been performed on one core of an Intel Xeon 8 CPU E5630 @ 2.53 GHz with 6GB RAM, running Linux. For the experiments in Section 3.2, each function evaluation has actually been performed on a different machine, where the necessary software was installed. More details are given in Section 3.2.

3.1   CPU Scheduling

This is the problem of efficiently scheduling CPUs of a parallel machine [9]. We consider this problem with the aim of testing the performances of our bilevel VNS algorithm. Deterministic or hybrid methods may also be used for its solution.

Let us suppose there is a parallel machine composed by m CPUs having exactly the same properties. Let us suppose we have a set of n jobs to be assigned to the CPUs of this parallel computer. For each job $j \in J$, where $J = \{1, 2, \ldots, n\}$, and for each CPU $k \in K$, where $K = \{1, 2, \ldots, m\}$, we can define the binary decision variable x_{jk}, which is 1 if the jth job is assigned to the kth CPU, and it is 0 otherwise. For each job, there is uncertainty on the total processing time, and therefore a lower bound p_j and an upper bound q_j are available for each job j. The objective function for our minimax problem is therefore:

$$f(x, y) = \max_{k \in K} \left\{ \sum_{j=1}^{n} x_{jk} y_j \right\},$$

where each uncertain variable y_j is constrained between p_j and q_j, and $\forall j \in J$, $\sum_{k \in K} x_{jk} = 1$, because every job needs to be assigned to exactly one CPU.

Table 1 shows some computational experiments.

Instances of the problem have been created for different choices of n and m. For all instances, $p_j = 1$ and $q_j = 2$ for each job $j \in J$, and therefore the worst-case scenario is given by the situation $y_j = 2$ for all $j \in J$. In the table, nsg and nna are algorithm parameters: nsg is the maximum number of solutions that are randomly generated in our bilevel VNS before increasing in value ε_x or ε_y; nna is the total number of neighbors which are allowed during the execution of the algorithm. As a consequence, only nna values in $[\varepsilon_{min}, \varepsilon_{max}]$ can be chosen in both VNS's. Finally, for each experiment, $iter_x$ and $iter_y$ represent the final number of performed iterations in the two VNS's. Therefore, the total number of function evaluations is $iter_x + iter_y$.

Table 1. Experiments on the CPU scheduling problem

instance name	m	n	nsg	nna	$iter_x$	$iter_y$	$f(x,y)$
test1	2	5	2	4	16	8	5.43
test1	2	5	4	6	70	25	6.28
test1	2	5	6	10	2137	749	6.12
test2	4	9	4	6	105	42	4.81
test2	4	9	6	10	319	116	6.65
test2	4	9	8	12	437	263	6.20

For both instances test1 and test2, the lower bound lb on the objective function value is 6 when the variables y represent the worst-case scenario. For the instance test1, the found solutions are relatively close to lb in all experiments. In the first experiment, the algorithm was not able to find the worst-case scenario, because $f(x,y)$ is smaller than lb. In the last two experiments, instead, *worse* scenarios were identified, and solutions with $f(x,y)$ close to lb were also found. Similar results are obtained for the instance test2. Each experiment took no more than 10 seconds of CPU time. However, they could have been much more expensive if the objective function were a black box. The most expensive experiment in Table 1 is the one in which the total number of function evaluations is almost 3000. Therefore, if we suppose that the objective function requires a 5 minute computer simulation, the same result could have been obtained in about 10 days of CPU time.

3.2 Wing Shape Optimization

We consider in this section an important problem arising in aerodynamics, which is the one of finding a robust shape for aircraft's wings. Naturally, many parameters are uncertain in this application, because flight conditions can change during time, such as the flying speed. Moreover, the objective function can only be evaluated by a computer simulation, which is able to evaluate the drag and the lift of the wing depending on its shape via Computational Fluid Dynamics (CFD) simulations [4]. The output of the black box is actually the pressure drag coefficient of the wing shape plus a penalty term for the case that the lift of the wing is not sufficiently large. The black box is currently installed on a machine belonging to the company *Airbus*. Therefore, solver and black box are in practice executed by different computers.

In this application, each black box call includes the generation of a finite volume mesh associated with the wing shape chosen, and the CFD calculations of the drag and lift properties of the wing, based on either Euler or Navier-Stokes flow. The input of the black box is the specification of the shape geometry and a set of parameters subject to uncertainty (which are usually 5). The geometry is specified by the Free-Form Deformation (FFD) model [4], where a certain number of *bumps* are employed for representing the wing. Each bump is represented by 3 continuous variables.

Table 2. Experiments on the wing shape optimization problem

instance name	bumps	nsg	nna	$iter_x$	$iter_y$	$f(x,y)$
Euler1	1	2	2	6	16	54.35
Euler1	1	3	3	4	20	60.91
Euler1	1	4	4	5	29	59.50

Table 2 shows some preliminary experiments. We consider an instance where the wing shape is represented by using one bump, and the CFD calculations are based on the Euler flow. In the second experiment, the obtained value for $f(x,y)$ is worse with respect to the one found in the first experiment, probably because the algorithm was able to find a *worse* scenario. In the third experiment, since more neighbors are considered in the two VNS's, a better solution has been found.

4   Conclusions

We proposed a bilevel VNS for robust optimization. This heuristic algorithm is able to manage minimax optimization problems where the objective function is represented by a computationally expensive black box. The algorithm is based on the well-known meta-heuristic algorithm Variable Neighborhood Search (VNS), and on the idea of combining two VNS's, in order to solve simultaneously the inner and the outer problem of the bilevel program.

This paper represents the basis of our work in this field. In the near future, it is our intention to consider and compare other approaches to robust optimization (the minimax formulation may be indeed inadequate if we prioritize optimizing the expected case instead of the worst-case) and to conceive suitable heuristics specifically designed for problems with black boxes. Another path we follow is to consider cloud models, such as the one described in [5]. Finally, developed algorithms, including the bilevel VNS proposed in this paper, should be compared to existing algorithms, in order to study and improve their performances.

Acknowledgments. The authors gratefully acknowledge partial support of the *Fondation de Recherche pour l'Aéronautique et l'Espace* (FRAE) in the framework of the MEMORIA project. We also would like to thank Matthieu Meaux, Marc Montagnac, Melodie Mouffe, and Anke Troeltzsch (CERFACS) for the fruitful discussions.

References

1. Ben-Tal, A., Ghaoui, L.E., Nemirovski, A.: Robust Optimization. Princeton University Press, Princeton (2009)
2. Canon, M.D.: Monoextremal representations of a class of minimax problems. Management Science 15(5), 228–238 (1969)
3. van Damme, E.: Stability and Perfection of Nash Equilibria, 2nd edn. Springer, Heidelberg (2002)

4. Duvigneau, R.: Aerodynamic shape optimization with uncertain operating conditions using metamodels. Tech. Rep. INRIA/RR–6143–FR+ENG, INRIA (2007)
5. Fuchs, M., Neumaier, A.: Potential based clouds in robust design optimization. Journal of Statistical Theory and Practice 3(1), 225–238 (2008)
6. Hansen, P., Brimberg, J., Urosevic, D., Mladenovic, N.: Solving large p-median clustering problems by primal–dual variable neighborhood search. Data Mining and Knowledge Discovery 19(3), 351–375 (2009)
7. Hansen, P., Mladenovic, N.: Variable neighborhood search: Principles and applications. European Journal of Operational Research 130(3), 449–467 (2001)
8. Hemmelmayr, V.C., Doerner, K.F., Hartl, R.F.: A variable neighborhood search heuristic for periodic routing problems. European Journal of Operational Research 195(3), 791–802 (2009)
9. Herrmann, J.: A genetic algorithm for minimax optimization problems. In: IEEE Conference Proceedings Congress on Evolutionary Computation (CEC 1999), Washington, DC, USA, pp. 1–5 (1999)
10. Huyse, L., Lewis, R.: Aerodynamic shape optimization of twodimensional airfoils under uncertain operating conditions. Tech. Rep. 2001-1, NASA Langley Research Center Hampton, VA, USA (2001)
11. Liberti, L., Lavor, C., Maculan, N.: Double VNS for the molecular distance geometry problem. In: Proceedings of the Mini Euro Conference on Variable Neighbourhood Search, Tenerife, Spain (2005)
12. Mladenovic, M., Hansen, P.: Variable neighborhood search. Computers and Operations Research 24, 1097–1100 (1997)
13. Mucherino, A., Urtubia, A.: Consistent biclustering and applications to agriculture. In: Proceedings of the Industrial Conference on Data Mining (ICDM 2010), IbaI Conference Proceedings Workshop on Data Mining and Agriculture (DMA 2010), pp. 105–113. Springer, Berlin (2010)
14. Rogalsky, T., Derksen, R.W., Kocabiyik, S.: Differential evolution in aerodynamic optimization. In: Proceedings of the 46th Annual Conference of the Canadian Aeronautics and Space Institute, pp. 29–36. Canadian Aeronautics and Space Institute (1999)
15. Sahinidis, N.V.: Optimization under uncertainty: state-of-the-art and opportunities. Computers and Chemical Engineering 28, 971–983 (2004)
16. Vicini, A., Quagliarella, D.: Airfoil and wing design through hybrid optimization strategies. American Institute of Aerodynamics and Astronautics 37(5) (1999)
17. Wen, Y., Xu, H., Yang, J.: A heuristic-based hybrid genetic-variable neighborhood search algorithm for task scheduling in heterogeneous multiprocessor system. Information Sciences 181(3), 567–581 (2011)

Processor Array Design
with the Use of Genetic Algorithm

Piotr Ratuszniak

Koszalin University of Technology, ul. Śniadeckich 2, 75-453 Koszalin, Poland
ratusz@ie.tu.koszalin.pl

Abstract. In this paper a method for processors arrays design dedicated
to realization of specimen linear algebra algorithms in FPGA devices is
presented. Within an allocation mapping process a genetic algorithm
for information dependency graph projection is used and the runtime
of the given algorithm is optimized. For larger input matrices, graph
decomposition is used which allows the projection results to be obtained.
The obtained projection results, with and without graph decomposition,
for a specimen linear algebra algorithm are compared. Additionally, a
parallel realization of the evolutionary algorithm for multicore processors
is presented, which allows projection results to be obtained for larger
input matrix sizes.

1   Introduction

Nowadays, an increasing interest is observed in making computations using parallel architecture. At present, the most popular platforms for parallel computation include: supercomputers, workstations with multicore processors, graphical
processor units and FPGA devices. Only the use of FPGA platforms allows
adaptation of the platform architecture to the algorithm being considered [1].
In comparison to alternative platforms, FPGA devices are not expensive and
consume less power [2,3]. Because of this, FPGA devices are used in green computing, although knowledge of one of the hardware description languages (HDL
such as VHDL or Verilog) and design experience are needed for the design process to be effective. At present, there are several translators available from the
C language to one of the HDL languages, such as ImpuleC. Unfortunately, the
architectures obtained from these translators are not optimal [1]. In order to
increase the effectiveness of the design process, complete functional block generators dedicated to an implementation in the FPGA device are used [4]. These
generators are called intellectual property cores (IPcore). The author has designed his own IPcore generator of computational accelerators for linear algebra
algorithms dedicated to implementation in popular FPGA devices. Parallel architectures in FPGA devices are implemented by two approaches: the first is
global in sequence and locally parallel and the second is globally parallel and local in sequence [5]. For the first approach, the author created a program module
for designing a parallel architecture by using a genetic algorithm and constraint
programming. This parallel architecture is dedicated to an implementation in

I. Lirkov, S. Margenov, and J. Waśniewski (Eds.): LSSC 2011, LNCS 7116, pp. 238–246, 2012.

multi-context FPGA devices. The results of the design process are presented in [6]. In the current paper the author describes the design process results for the second approach. Parallel architectures obtained using this approach are called processor arrays [5,7]. The design process consists of two steps. In the first step the operations from the given algorithm are mapped to the processor element from a processor array. The first step is called allocation mapping or information dependencies graph space projection and it has one constraint in the form of a connection locality. The connection locality constraint provides the low signal latency and high maximum work frequency for an FPGA structure. In the second step (known as time projection or schedule mapping) the execution time (the tact number) for each operation is calculated. The author proposes the use of a genetic algorithm for allocation mapping and optimization of the runtime (in tacts or macrotacts) of the complete linear algebra algorithm as given.

At present, there are several described methods for linear and nonlinear space and time projection of the information dependencies graph into processor arrays [7,8,9]. Only the proposed method allows for exact definition of the structure of a parallel architecture, and the realization runtime needed is often smaller in comparison with similar architectures (with similar hardware complexity) obtained with the use of linear and nonlinear methods. In preliminary experiments, the allocation mapping was computed for an algorithm with small input matrices (an information dependencies graph with nodes numbering fewer than 300). For bigger information dependencies graphs the author could not obtain permissible space projection solutions during runtime of the designed algorithm of less than one hour. For this reason the author decided to use information dependency graph decomposition in the allocation mapping process by using a genetic algorithm. In the proposed method the solution "grows up" during the runtime of the designed method. The genetic algorithm operators worked at the next assumed limits of a position in a chromosome. The proposed approach allowed them to obtain the allocation mapping for the processor arrays for larger input matrices. In this paper, the author present the allocation mapping for processor array results with and without information dependencies graph decomposition. We also designed a parallel realization of the genetic algorithm which allowed us to obtain permissible results for larger graphs with higher efficiency and in a short assumed design runtime.

2   Information Dependency Graph Projection in a Processor Array

Fig. 1 presents a specimen information dependency graph of the Gauss LU decomposition algorithm for a sample banded matrix. In information dependencies graphs each node represents an operation and each edge represents the transfer of data. This graph contains two types of arithmetical operation: division (2) and subtraction with multiplication (1).

In the allocation mapping for each node one processor was assigned from an array. The graph projection process, as presented, has one main limitation.

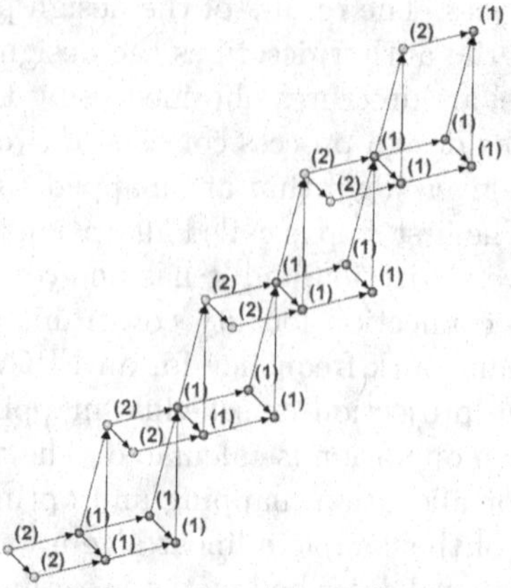

Fig. 1. An example of an information dependency graph for the Gauss LU algorithm (band width=5, matrix size=7)

A condition is necessary for a local connection between the processors in a complete processor array. The connection locality means that each graph edge can connect two nodes only from a neighbour processors. After the allocation mapping process, the tact number for each operation in a whole processor element is estimated. In an optimization process, the program minimized the maximum tact number needed for realization of the complete given algorithm.

3   The Genetic Algorithm for Information Dependencies Graphs Projection in a Processor Array

There are several popular methods for regular information dependencies graphs projection in processor array [7,8,9]. These methods are not automatic and additionally, in the case of the use of linear or non-linear functions of allocation mapping, it is difficult to achieve a set structure or parallel architecture. Some methods use graph transformations specific to the given algorithm [5], but these transformations cannot be generalized for different algorithms. For these reasons, the author decided to use a genetic algorithm, which allows for exact definition of the structure of the designed processor array and also allows for full automation of the design and the optimization process because the definition of projection functions is not required. This allows implementation of this method in the IP-Core generator for different linear algebra algorithms. The main problem with the use of a genetic algorithm for information dependency graph projection is the graph size for larger input matrices. The author decided to split the information dependency graph into subgraphs. This allowed a permissible solution of the graphs projection to be obtained in a short assumed runtime. The number of subgraphs was one of the program parameters defined by the user. In the genetic algorithm a division of groups coding by using numbers in a chromosome is used. Each number in the chromosome represents a single node from the information dependencies graph and the values represent the number of processors in the complete processor array. The genetic algorithm population contains 100 chromosomes and the initial population was generated randomly. In this case the complete graph nodes were randomly assigned to a processor element inside the processor array. This method usually generates a non-permissible solution. The authors of [6] described the use of constraints programming for generating initial permissible solutions. This allows correct solutions to be obtained during short program runtimes, but sometimes the genetic algorithm converges into the local minimum. Therefore, for a short program runtime, the algorithm remains in the local minimum and the randomly generated population often allows better results to be obtained. In the current paper the author describes how a projection method with graph decomposition, then a genetic algorithm working with smaller subgraphs and a random initial population, was sufficient, because for small graphs the algorithm can find permissible solutions for a short runtime. In the first step of the program operation, the range of graph nodes was calculated for recombination operators and periods for each subgraph were assigned. Subgraphs grew up according to the program runtime, but recombination operators were applied to a position in a chromosome within the calculated boundaries in the final parts of the chromosomes for the current subgraph. A standard one-point crossover operator with a fixed probability was used. After many experiments with several kinds and dimensions of input graph, the value of probability was experimentally chosen at the 0.2 level and a mutation operator with a variable probability was used. For the first stage of the algorithm operation, the value of the mutation probability was calculated, then exactly one position in a chromosome was mutated. This caused a constant number of mutations for different graph sizes. Another change was the introduction of variable probability for the mutation in accordance with the runtime of the algorithm. After assumed periods of time without improvement of the best solution, the probability of the mutation increased. The value of the mutation probability was incremented according to the periods that followed. Usually, within 20 minutes of running the genetic algorithm, the modified positions number in the chromosome was smaller than 10 for each subgraph. The described genetic algorithm was operated in two stages for each subgraph. In the first stage, before finding a permissible solution, the objective function depended on space projection errors such as a non-local connection between nodes from other processor elements in the array. The objective function F1 in the first stage of the algorithm operation is presented in equation (1)

$$F1 = 1 + EN * (EN - SE) \tag{1}$$

where:
EN - edges number in a whole graph, SE - space projection errors (non local connections between the processors)

After finding a permissible solution, the value of the objective function was calculated using two methods. For the permissible solutions the value of the objective function F2 was additionally dependent on the number of clock cycles needed to realize the whole graph of the given linear algebra algorithm. For other solutions, which do not fulfill the locality constraint, the objective function F3 gives much lower values, similar to the penalty function [4]. Detailed formulas for the objective function in the second stage of the algorithm work are presented in equation (2) and (3).

$$F2 = 1 + EN^2 + NN - T \tag{2}$$

$$F3 = 1 + EN - SE \tag{3}$$

where:
NN - nodes number in a whole graph, T - number of clock cycles,
SE - realization time in tacts

In the proposed genetic algorithm, the standard elitist selection model was used. With this method of selection, the best solution is always copied to a new population. This selection model was used because, in the mutation process, a chromosome could easily be changed into a non-permissible solution. Strict time limits for the runtime of the proposed genetic algorithm are assumed, because the algorithm is used in the developed IPCore generator dedicated to parallel architectures for linear algebra algorithms. The maximum runtime of the algorithm was limited to 20 or 30 minutes. The genetic algorithm is excellent for parallel processing, because the objective function value for each chromosome can be calculated separately. Nowadays, computers with multicore processors are very popular but a standard serial coding program can use only one core at any moment. The whole program was created using the Microsoft .NET platform and all the calculations presented were obtained on a computer with an Intel Core 2 Quad 2.2 GHz processor which contains four processor cores. In tandem with the parallel processing of the genetic algorithm the Microsoft ParallelFX extension is used. In this case, the program can use all four cores in the genetic calculations. Figure 2 presents the processor cores loaded for serial genetic computations and Figure 3 presents the same cores loaded for the parallel implementation.

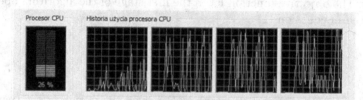

Fig. 2. An example of CPU processor cores loads for serial realization described genetic algorithm

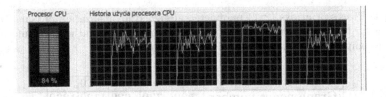

Fig. 3. CPU processor cores loads for parallel realization genetic algorithm obtained with the use of Microsoft ParallelFX extension

For serial genetic calculations, the average value of total CPU load was near to 25% and for the parallel processing this value was usually above 80%. This parallel implementation provides a higher value for calculated generations over the same computation time. The exact number of generations for a speed-up analysis is not used, because the algorithm works in two stages, using other objective functions with different computation complexity. The generations number in whole program runtimes depends on the period of time within the program finds a permissible solution. The advantage of parallel program processing is presented in final part of section 4.

4 Experiments. Designed Processors Arrays Parameters

The main processor arrays parameter as designed is the minimum tact number needed for execution of the complete given linear algebra algorithm. The minimum tact number obtained depends on the input matrix size and the number of assumed subgraphs. Table 1 presents the minimum tact numbers needed to perform the given Gauss LU algorithm realized in four processor array(2x2). For the graph projection, a few variants of the graph decomposition for 50, 100, and 150 subgraphs were compared with a method without decomposition. The table contains the best and average results, the percentage values equivalent to the permissible solutions found and the average processor loads. Table 2 presents similar values obtained within 30 minutes runtime.

Based on the results presented in Table 1, one can conclude that the graphs decomposition for the processors array design allowed a solution to be obtained for a greater input matrix size. For a higher number of subgraphs, the program can design processor arrays for information dependencies graphs with higher efficiency. On the other hand, for smaller graphs a higher number of subgraphs provides worse projection parameters. The results presented in Table 2 allow one to infer that longer program runtimes provide similar solutions and higher program efficiency. Table 3 presents the genetic algorithm efficiency for standard serial and parallel realizations for multicore processors. These results also prove the higher program efficiency obtainable for bigger input matrix sizes, similar to longer program runtimes.

Table 1. Processor arrays (2x2) parameters designed for realization of Gauss LU algorithms for banded matrices (band width=9) obtained with and without information graph decomposition for 20 minutes of program runtime

Matrix size		30	40	50	60	70
Graph nodes number		520	720	920	1120	1320
Graph edges number		1213	1683	2153	2623	3093
whole graph	find	0%	0%	0%	0%	0%
50 subgraphs	average	207	272	355	-	-
	best	200	275	355	-	-
	find	80%	80%	20%	0%	0%
	proc. load	63%	64%	65%	-	-
100 subgraphs	average	223	320	385	-	-
	best	209	307	368	-	-
	find	100%	100%	100%	0%	0%
	proc. load	58%	56%	60%	-	-
150 subgraphs	average	258	305	384	498	576
	best	245	294	375	478	576
	find	100%	100%	60%	80%	20%
	proc. load	50%	59%	60%	56%	57%

Table 2. Processor arrays (2x2) parameters designed for realization of Gauss LU algorithms for banded matrices (band width=9) obtained with and without information graph decomposition for 30 minutes of program runtime

Matrix size		30	40	50	60	70
Graph nodes number		520	720	920	1120	1320
Graph edges number		1213	1683	2153	2623	3093
whole graph	find	0%	0%	0%	0%	0%
50 subgraphs	average	210	274		-	-
	best	200	264	-	-	-
	find	100%	100%	0%	0%	0%
	proc. load	62%	66%	-	-	-
100 subgraphs	average	229	313	378	483	-
	best	215	297	370	483	-
	find	100%	80%	100%	20%	0%
	proc. load	57%	57%	61%	58%	-
150 subgraphs	average	264	297	393	494	557
	best	256	269	380	492	550
	find	100%	100%	100%	40%	60%
	proc. load	49%	61%	59%	57%	59%

Table 3. Processor arrays (2x2) parameters obtained by using serial and parallel genetic algorithms for 50 subgraphs

Matrix size		30	40	50	60	70
serial realization	best	200	264	-	-	-
	find	100%	100%	0%	0%	0%
parallel realization	best	207	259	335	404	-
	find	100%	100%	80%	20%	0%

5 Conclusions

In this paper, the author has presented design process results for an exemplary processor array dedicated to the realization of selected linear algebra algorithm. In this process a genetic algorithm was used and for given algorithm realization tact number was minimized. In comparison with other popular methods, the advantage of the genetic algorithm is that it allows exact definition of the processor array structure. Moreover, this method does not require a space and time projection definition from the program user and, because of this, it is useful for an implementation in the form of the IPcore generator for users not having knowledge of the processor array domain. The obtained results prove that graph decomposition allows one to obtain solutions for a larger size of matrices of given input linear algebra algorithms. The parallel processing of the proposed genetic algorithm allows one additionally to increment input matrix sizes or improve the percentage of permissible solutions.

References

1. Peterson, M.: FPGA Acceleration for outstanding performance. Challenges and Opportunities. In: Parallel Processing and Applied Mathematics, Wroclaw, Poland (2009)
2. Kestur, S., Davis, J., Williams, O.: BLAS Comparision on FPGA, CPU and GPU. In: IEEE Computer Society Symposium on VLSI (2010)
3. Williams, J., George, A.D., Richardson, J., Gosrani, K., Suresh, S.: Computational Density of Fixed and Reconfigurable Multi-Core Devices for Application Acceleration. In: Proc. 4th Reconf. Sys. Inst., Nat'l Center for Supercomp. App., Illinois (2008)
4. Chen, Y.K., Kung, S.Y.: Trend and Challenge on System-on-a-Chip Designs. Journal of Signal Processing Systems 53, 217–229 (2008)
5. Maslennikow, O.: Podstawy teorii zautomatyzowanego projektowania reprogramowalnych równoległych jednostek przetwarzajacych dla jednoukładowych systemów czasu rzeczywistego. Wyd. Uczelniane Politechniki Koszalińskiej, stron 273 (2004)
6. Ratuszniak, P., Maslennikow, O.: New Conception and Algorithm of Allocation Mapping for Processor Arrays Implemented into Multi-Context FPGA Devices. Mathematica Balkanica 23 (2009)

7. Kung, S.Y.: VLSI Array Processors. Prentice Hall, Englewood Cliffs (1988)
8. Quinton, P., Robert, Y.: Systolic algorithms and architectures. Prentice Hall (1991)
9. Sergyienko, A., Kaniewski, J., Maslennikow, O., Wyrzykowski, R.: A metod for mapping DSP algorithm into application specific processor. In: Proc. 24th Euromicro Conference on Parallel and Distributed Processing, vol. 1. IEEE Comp. Soc. Press, Vasteras (1998)

A Hybrid Genetic Algorithm for Parameter Identification of Bioprocess Models

Olympia Roeva

Institute of Biophysics and Biomedical Engineering, BAS
105 Acad. G. Bonchev Str., 1113 Sofia, Bulgaria
olympia@clbme.bas.bg

Abstract. In this paper a hybrid scheme using GA and SQP method is introduced. In the hybrid GA-SQP the role of the GA is to explore the search place in order to either isolate the most promising region of the search space. The role of the SQP is to exploit the information gathered by the GA. To demonstrate the usefulness of the presented approach, two cases for parameter identification of different complexity are considered. The hybrid scheme is applied for modeling of *E. coli MC4110* fed-batch cultivation process. The results show that the GA-SQP takes the advantages of both GA's global search ability and SQP's local search ability, hence enhances the overall search ability and computational efficiency.

1 Introduction

Robust and efficient methods for parameter identification are of key importance in system biology and related areas. Nowadays the most common direct methods used for global optimization are evolutionary algorithms such as genetic algorithms (GA). The principal advantages of GA are domain independence, non-linearity and robustness. The GA effectiveness has been already demonstrated for identification of fed-batch cultivation processes [2,11]. The same qualities that make the GA so robust also can make it more computationally intensive and slower than other methods [8]. On the contrary, local search methods have faster convergence due to the use of local information for determination of the most promising search direction by creating logical movements. One of the leading methods for solving constrained non-linear optimization problems is sequential quadratic programming (SQP) [3,5]. Algorithms in this class guarantee global convergence and typically require few iterations to locate a solution point. However, local search methods can easily be entrapped in local minima. An approach that overcomes the above disadvantages is to combine GA with local search methods, to design more efficient methods with relatively faster convergence than the pure GA. Hybrid GA have received significant interest in recent years and are being increasingly used to solve real-world problems [1]. Different local search methods have got attention in such combinations [7,9,13].

In this paper a parameter identification of an *E. coli MC4110* fed-batch fermentation process using hybrid GA is proposed. To improve the performance of

I. Lirkov, S. Margenov, and J. Waśniewski (Eds.): LSSC 2011, LNCS 7116, pp. 247–255, 2012.
© Springer-Verlag Berlin Heidelberg 2012

the conventional GA, a combine scheme using the GA and SQP method is introduced. Thus, optimizers work jointly to locate efficiently quality design points better than either could alone.

This paper is organized as follows. Outline of the introduced hybrid algorithm is described in Section 2. In Section 3 a discussion of the obtained numerical results of E. coli cultivation process model parameter identification is presented. Conclusion remarks are done in Section 4.

2   Outline of the Hybrid GA-SQP

GA is very effective at finding optimal solutions to a variety of complex optimization problems because it does not impose many of the limitations of the traditional techniques. The same characteristics that make the GA so robust can make it more computationally intensive and hence slower than other methods. To improve the performance of the conventional GA, a hybrid scheme using GA and SQP method is proposed.

Background of the GA. GA was developed to model adaptation processes mainly operating on binary strings and using a recombination operator with mutation as a background operator. The GA maintains a population of individuals, $P(t) = x_1^t, ..., x_n^t$ for generation t. Each individual represents a potential solution to the problem and is implemented as some data structure S. Each solution is evaluated to give some measure of its "fitness". Fitness of an individual is assigned proportionally to the value of the objective function of the individuals. Then, a new population (generation $t + 1$) is formed by selecting more fit individuals (selected step). Some members of the new population undergo transformations by means of "genetic" operators to form new solution. There are unary transformations m_i (mutation type), which create new individuals by a small change in a single individual ($m_i : S \rightarrow S$), and higher order transformations c_j (crossover type), which create new individuals by combining parts from several individuals ($c_j : S \times ... \times S \rightarrow S$). After some number of generations the algorithm converges - it is expected that the best individual represents a near-optimum (reasonable) solution. The combined effect of selection, crossover and mutation gives so-called reproductive scheme growth equation [4]:

$$\xi\left(S, t+1\right) \geq \xi\left(S, t\right) \cdot eval\left(S, t\right) / \bar{F}\left(t\right)\left[1 - p_c \cdot \frac{\delta\left(S\right)}{m - 1} - o\left(S\right) \cdot p_m\right].$$

A pseudo code of a GA is presented as:

1 Set generation number to zero ($t = 0$)
2 Initialise usually random population of individuals ($P(0)$)
3 Evaluate fitness of all initial individuals of population
4 Begin major generation loop in k:
    4.1 Test for termination criterion
    4.2 Increase the generation number
    4.3 Select a sub-population (select $P(i)$ from $P(i - 1)$)

4.4 Recombine the genes of selected parents (recombine $P(i)$)

4.5 Perturb the mated population stochastically (mutate $P(i)$)

4.6 Evaluate the new fitness (evaluate $P(i)$)

5 End major generation loop

Background of the SQP algorithm. SQP is one of the most popular and robust algorithms for nonlinear continuous optimization. The general optimization problem to minimize an objective function f under nonlinear equality and inequality constraints is [5]:

$$\min_{x \in \mathcal{R}^n} f(x), \; c(x) = 0, \; b(x) \geq 0, \; x_l \leq x \leq x_u,$$

where x is an n-dimensional parameter vector. It is assumed that all problem functions $f(x)$, $c(x)$, and $b(x)$ are continuously differentiable on the whole $\mathcal{R}^n$.

At an iteration x_k (for the equality constrains), a basic SQP algorithm defines an appropriate search direction d_k as a solution to the QP subproblem

$$\min_{d \in \mathcal{R}^n} f(x_k) + g(x_k)^T d + \tfrac{1}{2} d^T \nabla^2_{xx} f(x) + \sum_{i=1}^{t} \lambda^i \nabla^2_{xx} c^i(x)$$

$$\text{s.t. } c(x_k) + A(x_k) d = 0$$

is equal to, or is a symmetric approximation for, the Hessian of the Lagrangian. A pseudo code of SQP algorithm could be presented as:

1 Set the initial point $x = x_0$
2 Set the Hessian matrix $(H_0 = I)$
3 Evaluate f_0, g_0, c_0 and A_0
4 Solve the QP subproblem to find search direction d_k
5 Update $x_{k+1} = x_k + \alpha_k d_k$
6 Evaluate $f_{k+1}, g_{k+1}, c_{k+1}$ and A_{k+1}
7 Convergence check
 If yes, go to exit
 If no, obtain H_{k+1} by updating H_k and go back to Step 4

3   Numerical Results and Discussion

E. coli MC4110 fed-batch cultivation model. The mathematical model of the considered process can be represented by [11]:

$$\frac{dX}{dt} = \mu_{max} \frac{S}{k_S + S} X - \frac{F_{in}}{V} X \tag{1}$$

$$\frac{dS}{dt} = -\frac{1}{Y_{S/X}} \mu_{max} \frac{S}{k_S + S} X + \frac{F_{in}}{V}(S_{in} - S) \tag{2}$$

$$\frac{dA}{dt} = \frac{1}{Y_{A/X}} \mu_{max} \frac{A}{k_A + A} X - \frac{F_{in}}{V} A \tag{3}$$

$$\frac{dpO_2}{dt} = -\frac{1}{Y_{pO_2/X}} \mu_{max} \frac{pO_2}{k_{pO_2} + pO_2} X + k_L a(pO_2^* - pO_2) - \frac{F_{in}}{V} pO_2 \tag{4}$$

$$\frac{dV}{dt} = F_{in} \tag{5}$$

where: X is biomass concentration, [g/l]; S - substrate concentration, [g/l]; A - acetate concentration, [g/l]; pO_2 - dissolved oxygen concentration, [%]; pO_2^* - saturation concentration of dissolved oxygen, [%]; F_{in} - feeding rate, [l/h]; V - bioreactor volume, [l]; S_{in} - substrate concentration in the feeding solution, [g/l]; μ_{max} - maximum value of the specific growth rate, $[h^{-1}]$; k_i - saturation constants; k_La - volumetric oxygen transfer coefficient, $[h^{-1}]$; $Y_{i/X}$ - yield coefficients, [-]. For the parameter estimation problem real experimental data of the *E. coli MC4110* fed-batch cultivation process are used. The cultivation condition and the experimental data have been presented in [12].

For comparison of the performance of the presented here hybrid GA-SQP with pure GA and SQP two cases are examined. In the first case (*Case 1*) the system (1)-(2) and (5) is considered. The estimated parameters are: μ_{max}, k_S, and $Y_{S/X}$. In the second case (*Case 2*) the full system (1)-(5) is considered with unknown parameters: μ_{max}, k_S, k_A, k_{pO_2}, $Y_{S/X}$, $Y_{A/X}$, $Y_{pO_2/X}$, and pO_2^*. The objective function is presented as a minimization of a distance measure J between experimental and model predicted values, represented by the vector $\mathbf{y}$:

$$J = \sum_{i=1}^{n} \sum_{j=1}^{m} \{[\mathbf{y}_{exp}(i) - \mathbf{y}_{mod}(i)]_j\}^2 \to min \tag{6}$$

where n is the number of data for each state variable m; $\mathbf{y}_{exp}$ - the experimental data; $\mathbf{y}_{mod}$ - model predictions with a given set of the parameters.

Algorithm parameters. Based on results in [11,10], genetic algorithm operators and parameters for considered here parameter identification of fermentation process are as follows: A binary 20 bit representation is considered. The selection method used here is the roulette wheel selection. A double point crossover and a bit inversion mutation are applied. Crossover rate should generally be high - here it is set to 70%. Mutation is randomly applied with low probability - 0.01. A value of 97% for the rate of the selected individuals (generation gap) is accepted. Particularly important parameters of GA are the population size and number of generations. If there are too low number of chromosomes, GA has a few possibilities to perform crossover and only a small part of search space is explored. On the other hand, if there are too many chromosomes, GA slows down. The number of individuals is set to 100. The number of generations is 200 (pure GA) and 10 (hybrid GA-SQP). The division of the hybrid's time between the two methods influences the efficiency and the effectiveness of the search process. Numerous tests are performed to find the optimal division of the algorithm's time. The GA is run for 5, 10, 15, 20, 25, 30 generations before the SQP algorithm is started. The obtained results show that the optimal number of generations is 10. For 10 generations the GA reaches near optimum solution, which is a good initial point for the SQP algorithm. The use of 20 or 30 generations reflects mainly on the computational cost and has negligible improvement on the initial point.

Results from parameter identification. All computations are performed using a PC/Intel Core 2 Quad CPU Q8200 @2.34GHz platform running Windows

Table 1. Search parameters utilized in the different algorithms

Search parameter	GA	SQP	GA-SQP
μ_{max}	0.4780	0.4742	0.4741
k_S	0.0145	0.0148	0.0148
$1/Y_{S/X}$	2.0313	2.0137	2.0137
$1/Y_{A/X}$	8.6169	12.3012	9.3365
$1/Y_{pO_2/X}$	0.0340	0.0388	0.0368
k_A	54.3274	67.1266	50.9262
k_{pO_2}	0.0017	0.001	0.001
$k_L a$	282.4246	300.0062	282.9080
pO_2^*	21.2696	21.2988	21.2988

XP, Matlab 7.5 environment. Initially the algorithms (GA, SQP, and GA-SQP) were tested for parameter estimation of model (1)-(5) using generated data. The results were explicit: (i) The estimates of SQP were very sensitive to the initial points. The algorithm reached the value of J between 0.0063 and 0.013 according to the considered initial point. The computational time was 40-60 s. (ii) The best result of GA is J=0.0137 (for 5 runs). The computational time was 117.04 s. (iii) The result from GA-SQP was J=0.0063 for each run of the algorithm. The GA is run for 10 generations. The computational time varied between 55-75 s. In the second step the algorithms are used for parameter estimation of two considered models (*Case 1* and *Case 2*) using real experimental data. The experimental data were used without filtration or any processing. The idea was to test the algorithms in such hard real conditions. The numerical results (*Case 2*) from the parameter identification are presented in Table 1.

In Table 1, considering GA and GA-SQP the average values of 30 runs are presented. One-factor ANOVA analysis is performed to see if the means of the obtained 30 groups are equal. The results are displayed in Fig. 1. The estimated parameter values of the algorithms are in admissible ranges [6,14]. The parameters estimates obtained by the three algorithms are very close. The only exceptions are the parameters $Y_{A/X}$ and k_A. The acetate concentration during the cultivation process is very small in comparison with the concentrations of the other state variables. So, the influence of acetate error on the objective function is smaller than the other errors. In contrast, the influence of biomass error is tolerable and the result is almost equal estimates of μ_{max}, k_S, and $Y_{S/X}$.

```
                              ANOVA Table
Source        SS          df      MS          F             Prob>F
------------------------------------------------------------------
Columns    1678385.6916    8    209798.2    4.91484e+006      0
Error           11.1412  261         0
Total      1678396.8328  269
```

Fig. 1. One-factor ANOVA results

Table 2. Results of the search methods in Case 2

Criterion	GA	SQP	GA-SQP
J	6.5617	$6.5483^1/6.6822^2$	6.5470
CPU time, s	208.7500	$67.8594^1/66.3281^2$	72.17198

[1] "good" initial point, [2] "bad" initial point

For the simple task (*Case 1*) the SQP and the hybrid GA-SQP algorithms obtain same values of the objective function. Moreover, the pure SQP has better convergence time. SQP enjoy global convergence guarantees and requires few function evaluations to locate a solution point. The pure GA has obtained almost the same J but for greater computational time - 88.1875 s. GA reaches the area near an optimum point relatively quickly but it took many function evaluations to achieve convergence. In *Case 2* when 9 parameters were estimated simultaneously the effectiveness of the hybrid GA-SQP is more evident. The result values of the objective function (Eq. (6)) and computation time are given in Table 2. The best result ($J = 6.5470$) is obtained using the hybrid GA-SQP after a total computation time of 72.1719 s. The GA-SQP hybrid technique merges a lot of features of the GA and the SQP optimality criterion. A combination of a genetic algorithm and a local search method speed up the search to locate the exact global optimum. It exhibits the robust global search capability of the GA while preserving the efficient local search capability afforded by SQP. Very close result was obtained by SQP with a "good" initial point. If the initial point is "bad" the algorithm falls in another local extrema with $J = 6.6822$. The solution depends on the choice of the start points as the pure SQP usually seeks a solution in the neighborhood of the start point. The obtained result from the pure GA is $J = 6.5617$ for longer computational time of 208.75 s. Since pure GAs consider a group of points in each search space in each generation, they are best suited for global search. But their main operations (i.e. reproduction, crossover, and mutation) are not very efficient for local search. The obtained through generations objective function values are presented in Fig. 2. As it can be seen the best performance show hybrid GA-SQP.

A quantitative measure of the differences between modelled and measured values is another important criterion for the adequacy of a model. The graphical results of the comparison between the model predictions of state variables, based on hybrid GA-SQP algorithm estimations, and the experimental data points of the real *E. coli* cultivation are presented in Fig. 3.

The presented graphics show a very good correlation between the experimental and predicted data.

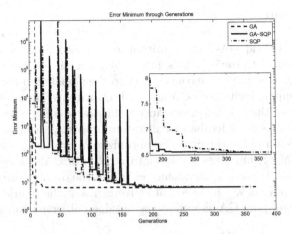

Fig. 2. Objective function values trough algorithms generations

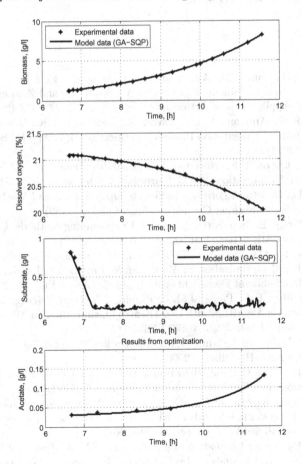

Fig. 3. Comparison between the model predictions and the real process variables

4 Conclusion

In this paper a hybrid GA-SQP algorithm is proposed. In such a hybrid, applying a local search to the solutions guided by a genetic algorithm in the most promising region can accelerate convergence to the global optimum. The hybrid algorithm is compared with pure GA and SQP algorithms for parameter identification procedure. Algorithms performance is illustrated using a set of non-linear models of *E. coli MC4110* fed-batch cultivation process. As evident from graphical and numerical results, the proposed optimization hybrid algorithm performs very well. The algorithm takes the advantages of both GA's global search ability and SQP's local search ability, hence enhances the overall search ability and computational efficiency. The speed of convergence of the hybrid algorithm is superior to that of pure GA as well as the obtained objective function.

Acknowledgements. This work is supported by the National Science Fund Grants DMU 02/4 and DID 02-29.

References

1. Akpınar, S., Bayhan, G.M.: A Hybrid Genetic Algorithm for Mixed Model Assembly Line Balancing Problem with Parallel Workstations and Zoning Constraints. Engineering Applications of Artificial Intelligence 24(3), 449–457 (2011)
2. Benjamin, K.K., Ammanuel, A.N., David, A., Benjamin, Y.K.: Genetic Algorithm using for a Batch Fermentation Process Identification. J. of Applied Sciences 8(12), 2272–2278 (2008)
3. Byrd, R.H., Curtis, F.E., Nocedal, J.: Infeasibility Detection and SQP Methods for Nonlinear Optimization. SIAM Journal on Optimization 20, 2281–2299 (2010)
4. Goldberg, D.: Genetic Algorithms in Search, Optimization and Machine Learning. Addison-Wesley Publishing Company, Massachusetts (1989)
5. Gill, P.E., Wong, E.: Sequential Quadratic Programming Methods, UCSD Department of Mathematics, Technical Report NA-10-03 (2010)
6. Levisauskas, D., Galvanauskas, V., Henrich, S., Wilhelm, K., Volk, N., Lubbert, A.: Model-based Optimization of Viral Capsid Protein Production in Fed-batch Culture of Recombinant Escherichia coli. Biopr. & Biosys. Eng. 25, 255–262 (2003)
7. Mateus da Silva, F.J., Pérez, J.M.S., Pulido, J.A.G., Rodríguez, M.A.V.: AlineaGA - A Genetic Algorithm with Local Search Optimization for Multiple Sequence Alignment. Appl. Intell. 32, 164–172 (2010)
8. Nocedal, J., Wright, S.J.: Numerical Optimization. Springer Series in Operations Research. Springer, Heidelberg (2006)
9. Paplinski, J.P.: The Genetic Algorithm with Simplex Crossover for Identification of Time Delays. Intelligent Information Systems, 337–346 (2010)
10. Roeva, O.: Improvement of Genetic Algorithm Performance for Identification of Cultivation Process Models, Advanced Topics on Evolutionary Computing. Artificial Intelligence Series-WSEAS, pp. 34–39 (2008)
11. Roeva, O.: Parameter Estimation of a Monod-Type Model Based on Genetic Algorithms and Sensitivity Analysis. In: Lirkov, I., Margenov, S., Waśniewski, J. (eds.) LSSC 2007. LNCS, vol. 4818, pp. 601–608. Springer, Heidelberg (2008)

12. Roeva, O., Pencheva, T., Hitzmann, B., Tzonkov, S.: A Genetic Algorithms Based Approach for Identification of Escherichia coli Fed-batch Fermentation. Int. J. Bioautomation 1, 30–41 (2004)
13. Tseng, L.-Y., Lin, Y.-T.: A Hybrid Genetic Local Search Algorithm for the Permutation Flowshop Scheduling Problem. European J. of Operational Res. 198(1), 84–92 (2009)
14. Zelic, B., Vasic-Racki, D., Wandrey, C., Takors, R.: Modeling of the Pyruvate Production with Escherichia coli in a Fed-batch Bioreactor. Biopr. & Biosys. Eng. 26, 249–258 (2004)

A General Frame
for Building Optimal Multiple SVM Kernels

Dana Simian and Florin Stoica

"Lucian Blaga" University of Sibiu, Romania

Abstract. The aim of this paper is to define a general frame for building optimal multiple SVM kernels. Our scheme follows 5 steps: formal
representation of the multiple kernels, structural representation, choice
of genetic algorithm, SVM algorithm, and model evaluation. The computation of the optimal parameter values of SVM kernels is performed
using an evolutionary method based on the SVM algorithm for evaluation of the quality of chromosomes. After the multiple kernel is found by
the genetic algorithm we apply cross validation method for estimating
the performance of our predictive model. We implemented and compared
many hybrid methods derived from this scheme. Improved co-mutation
operators are used and a comparative study about their effect on the
predictive model performances is made. We tested our multiple kernels
for classification tasks but they can be also used for other types of tasks.

1 Introduction

Classification task can be found in many fields of activity (medicine, biology,
bibliomining, webmining, etc.). The classifier is strong dependent on data type. A
good classifier for a specific class of problems or data might have a bad behaviour
for other types of data sets. It is necessary to train the classifier on specific sets
of data. Support Vector Machines (SVMs) represents an important and popular
tool for machine learning tasks, especially for classification and regression. SVMs
are supervised learning methods introduced by Vapnik [16]. Being given a set
of training and testing data, defined by their label (target value) and a set of
features (attributes), the goal of SVMs is to produce, using the training set of
data, a model which predicts target value of data instances from a testing set
which are given only the features. The accuracy of the model for a specific test
set is defined as the percentage of test set items that are correctly classified
by the model. If the accuracy is acceptable, the model can be used to classify
data for which the class label is unknown. If the data is linear separable, an
optimal separating hyperplane with maximal margin is obtained [16,3]. In the
case of non separable data, the kernel method is mapping the original data
into a higher dimensional features space where they become separable despite
being non-separable by a hyperplane in the original input space. The kernel
substitution method, known as "kernel trick" was first published by Aizerman
et al. in [1]. A kernel function must satisfy Mercer's theorem (see [8]), which
states that for any function $K : X \times X \to R$, continuous, symmetric, positive,

I. Lirkov, S. Margenov, and J. Waśniewski (Eds.): LSSC 2011, LNCS 7116, pp. 256–263, 2012.

semi-definite, there exists a function ϕ defined on an inner product space of possibly high dimension, such that $K(x_i, x_j) = \langle \phi(x_i), \phi(x_j) \rangle$. The form of the feature map ϕ does not need to be known, it is implicitly defined by the choice of kernel function K. Different kind of kernels can be found in literature [3,5,6]. The most important are:

$$\text{Polynomial: } K_{pol}^{d,r}(x_1, x_2) = (x_1 \cdot x_2 + r)^d, \; r, d \in Z_+ \qquad (1)$$

$$\text{RBF: } K_{RBF}^{\gamma}(x_1, x_2) = \exp\left(\frac{-1}{2\gamma^2}|x_1 - x_2|^2\right) \qquad (2)$$

$$\text{Sigmoidal: } K_{sig}^{\gamma}(x_1, x_2) = \tanh(\gamma \cdot x_1 \cdot x_2 + 1) \qquad (3)$$

In the following we take into account only the binary classification tasks and SVMs.

Standard SVMs use one kernel from (1)-(3) and the prediction supposes the choice of kernel parameters. Usually the choice of the kernel is made in an empirical way. The real problems require more complex kernels. The aim of this paper is to obtain optimal multiple kernels for given sets of data. We present a general hybrid scheme for obtaining optimal multiple kernels for given type of data. Many particular methods derived from this frame are implemented. We evaluate the performances of these multiple SVM kernels using cross-validation. The paper is organized as follows. In section 2 we present our main theoretical result, the frame of a hybrid method for building SVM kernels. Different solutions for constructing SVM kernels using evolutionary and hybrid approaches [7,10,11,13,14,15] can be integrated in our frame. Implementation details are presented in section 3. Results of testing and validation, using many data sets, are presented in section 4. In section 5 we make a comparison of different particular methods and present conclusions and further directions of study.

2   The Frame of the Hybrid Method

2.1   General Presentation

We started from the idea presented in [7]. We construct the multiple kernels using a hybrid method structured on two levels. The macro level is represented by a genetic algorithm which builds the multiple kernel. Multiple kernels are coded into chromosomes. In the micro level the quality of chromosomes is computed using a SVM algorithm. The fitness function is represented by the classification accuracy on a validation set of data. The optimal multiple kernel is computed in the genetic algorithm. The chromosome containing this kernel is decoded and then the cross-validation techniques is applied for obtaining the "Cross Validation Accuracy". A multiple kernel has, in this frame, a formal representation and a structural representation.

2.2   Formal Representation of Multiple Kernels

The multiple kernel can be formally represented using a tree, which terminal nodes contain a single kernel and the intermediate nodes contains operations.

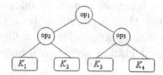

Fig. 1. Formal representation of multiple kernel

It is proved in the kernel theory that multiple kernels can be obtained using the set of operations $(+, *, exp)$, which preserve Mercer's conditions (1)-(3). If a node contains the operation exp only the left of its descendants is considered. The number of the terminal nodes is an input data of formal representation. The linear kernels are obtained when only the additive operation is used in the formal representation of multiple kernels. A formal representation of the multiple kernel $K = (K_1 \ op_2 \ K_2)op_1(K_3 \ op_3 \ K_4)$ is given in figure 1.

2.3   Structural Representation of Multiple Kernels

The structural representation of multiple kernels represents the structure of chromosome in the genetic algorithm from the macro level of the frame method. We propose two types of structural representation:

1. Tree representation
2. Linear representation

The details of structural representation differentiate the methods derived from the general frame. Other kind of representations could be added in order to enlarge our frame.

In [7,10] the chromosome is coded using a tree structure identical with the formal representation. In [11] we used maximum 4 polynomials kernels $(K_i^{d_i, r}; \ i = 1, \ldots, 4)$ and 3 operations. The chromosome has a linear structure composed by 34 genes. Each operation, op_j, $j = 1, 2, 3$ is represented using two genes, for a degree d_i are allocated 4 genes, and the variable r, which is the same for all the simple kernels, is represented using 12 genes. If one of the operations is exp, only the first 12 from the 16 genes allocated for the degrees d_j are representative. In [13,14] we used maximum 4 single kernels, which can be polynomial, RBF, or sygmoidal, and 3 operations. The chromosome has a linear structure composed by 78 genes: 2 genes for each operation (op_i, $i = 1, 2, 3$), 2 genes for the kernel's type, t_i, $i = 1, 2, 3$. If the single kernel K_i is polynomial, we use 4 genes for the degree parameter d_i and 12 genes for r_i. If the associated kernel is not polynomial, the last 16 genes are used to represent the real value of parameter γ_i. The linear structures of the chromosome, presented above, are represented in fig. 2.

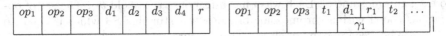

op_1	op_2	op_3	d_1	d_2	d_3	d_4	r

op_1	op_2	op_3	t_1	d_1	r_1	t_2	$\dots$
				γ_1			

Fig. 2. Linear representations of multiple kernel-model

2.4 Genetic Algorithm

One of the elements that influence the behavior of the hybrid algorithm which built the SVM kernel is the genetic algorithm used in the macro level. We implemented and tested many kind of genetic algorithms. The difference between them is made by the mutation operators. Three type of mutation operators were used:

1. Classical mutation operators (like the algorithms implemented in [5]).
2. Co-mutation operators.
3. Improved co-mutation operators using a wasp-based computational scheme.

We used the co-mutation operator M_{ijn} defined in [9], instead of classical mutation and cross-over operator. The M_{ijn} operator realizes a mutation of a number of adjacent bits at the same time. It does not perform only a flip-mutation, it mutates substrings in order to save the implicit information contained in the adjacency property of these bits.

We also used the co-mutation operator $LR - M_{ijn}$ that we defined in [12]. This co-mutation operator finds the longest sequence of bits, situated in the left or in the right of a position p, randomly chosen. If the longest sequence is in the left of p, the $LR - M_{ijn}$ operates identical with M_{ijn}, otherwise it operates on the set of bits starting from p and going to the right. Both co-mutation operators we considerate have about the same capabilities of local search as ordinary mutation, but allows long jumps, to reach far regions in the search space which cannot be reached by the classical bit- flip mutation.

The single kernels chosen and the operations between them are coded inside a chromosome. In order to realize an equilibrium between the changing of the operations and the changing of single kernels' parameters we improved the mutation and co-mutation operators using a wasp-based computational scheme.

Wasp based computational models are used for solving dynamic repartition of tasks. The wasp natural behavior is governed by a stimulus - response mechanism. The response threshold of an individual wasp for each zone of the nest together with the stimulus from brood located in this zone determine the engagement of the wasp in the task of foraging for this zone. A wasp based computational model uses a pair of stimulus - response for computing the probability of some actions. One or many rules for updating the response threshold may be established in order to adapt the model to particular requirements ([2,4]). Let consider in our case a mutation or a co-mutation operator, denoted by M. The number of genes for operations' representation is much smaller than the number of genes for kernel's parameters. For increasing the probability of changing of the operations we use a wasp based computational model. We associate a wasp to each chromosome C. Each wasp has a response threshold θ_C. The set of operations coded within chromosome broadcasts a stimulus S_C which is equal to difference between maximum classification accuracy (100) and the actual classification accuracy obtained using the multiple kernel coded in the chromosome, $S_c = 100 - CA_C$. The probability that the operator M perform a mutation that will change the operations coded within chromosome is $P(\theta_C, S_C) = \dfrac{S_C^\gamma}{S_C^\gamma + \theta_C^\gamma}$, where γ is a system parameter. Good results were obtained for $\gamma = 2$.

The threshold update rule is defined as follows: $\theta_C = \theta_C - \delta$, $\delta > 0$, if the classification accuracy of the new chromosome C is lower than in the previous step, and $\theta_C = \theta_C + \delta$, $\delta > 0$, if the classification accuracy of the new chromosome C is greater than in the previous step.

A similar model can be used for increasing the probability for changing the type of single kernels which composed the multiple kernels. New methods for building multiple kernels can be obtained using other modifications to classical genetic algorithms.

2.5   SVM Algorithm

The fitness function for evaluation of chromosomes is the classification accuracy given by an SVM algorithm acting on a particular set of data. The data are divided into two subsets: the training subset, used for problem modeling and the test subset used for evaluation. The training subset is also randomly divided into a subset for learning and a subset for validation. The data from the learning subset are used for training and the data from the validation subset for computing the classification accuracy.

2.6   Model Evaluation

After the multiple kernel is found by the genetic algorithm we apply cross validation method for estimating the performance of our predictive model. In our frame we use K-fold cross validation.

One idea could be to use cross validation as fitness function in the genetic algorithm instead of classification accuracy. Our implementations and practical results proved that this is not a valid solution due to the huge time required.

3   Implementation Details

In order to implement one particular method derived from the general frame presented in section 2, we start from the classes implemented in *libsvm* [5] and modify them according to the chosen structural representation of multiple kernel. The classes *svm_parameter*, *svm_predict* and *Kernel* must be adapted to our particular model. The class *svm_predict* was extended with the *predict* method. The *Kernel* class is modified to accomplish the kernel substitution. A method for computing the hybrid multiple kernels is necessary. We construct a new method, that we named $k\_function$, for the computation of our simple kernel. Then, the simple kernels are combined using operation given in the linear structural model of the chromosome. In the genetic algorithm, the operations and all parameters assigned to a simple kernel (type of the simple kernels and all other parameters) are obtained from a chromosome, which is then evaluated using the result of the modified *predict* method. After the end of the genetic algorithm, the best chromosome gives the multiple kernel which can be evaluated on the test subset of data.

The cross-validation method is applied for the "optimal" multiple kernel obtained in the genetic algorithm. In order to obtain the cross validation accuracy, which characterize our predictive model given by the multiple kernel, we modify the class svm_train, introducing the method $do\_cross\_validation$ which takes into account the structural model of the multiple kernel.

4   Experimental Results

We used the "Leukemia" and "Vowel" data sets from the standard libsvm package [5] and different kinds of genetic algorithms: classical genetic approach (from [5]), approach using M_{ijn}co-mutation operator, approach using $LR - M_{ijn}$ co-mutation operator, approaches derived from the previous ones combined with a wasp based computational model. For each execution, dimension of population was 40 and the number of generations was 30.

The experimental results for the Leukemia data set are presented in Table 1. We denoted the approaches as follows, taking into account the genetic algorithm

Table 1. Table 1. Leukemia data set

Approach	Type of single kernels	Operations	Parameters of optimal kernel	Classif. accuracy	Cross Validation accuracy
1	POL, RBF, SYG	+,exp, +	$d_1 = 1$, $r_1 = 1887$, $\gamma_{1,1} = 1.399$,$\gamma_{1,2} = 1.890$	88.23%	86.73 %
2	POL, SYG, SYG	+,exp +	$d_2 = 3$, $r_2 = 2383$, $\gamma_{2,1} = 0.622$,$\gamma_{2,2} = 0.256$	90.03%	81.57 %
3	RBF, SYG, POL	exp,* +	$d_3 = 2$, $r_3 = 1503$, $\gamma_{3,1} = 0.294$,$\gamma_{3,2} = 0.065$	91.17%	92.10 %
1*	SYG, SYG, POL, POL	*,+ +	$\gamma_{4,1} = 1.456$,$\gamma_{4,2} = 1.245$, $d_{4,1} = 1$, $r_{4,1} = 1373$, $d_{4,2} = 3$, $r_{4,1} = 2165$	91.17%	81.57%
2*	RBF, POL RBF	+,exp +	$\gamma_{5,1} = 1.596$, $d_5 = 3$, $r_5 = 2198$, $\gamma_{5,2} = 1.309$	91.17%	81.57 %
3*	SYG, RBF, POL, SYG	+,* *	$\gamma_{6,1} = 0.016$, $\gamma_{6,2} = 1.997$, $d_6 = 1$, $r_6 = 1543$, $\gamma_{6,3} = 0.068$	94.11%	89.47 %

Table 2. Table 2- Vovel data set

Approach	Type of single kernels	Ope- rations	Parameters of optimal kernel	Classif. accuracy	Cross Validation accuracy
1	RBF, RBF, RBF	*,exp +	$\gamma_{1,1} = 0.824$, $\gamma_{1,2} = 0.943$, $\gamma_{1,3} = 0.048$	61.47%	97.34 %
2	RBF, SYG, RBF, SYG	*,* +	$\gamma_{2,1} = 0.654$, $\gamma_{2,2} = 0.445$, $\gamma_{2,3} = 0.298$,$\gamma_{2,4} = 0.017$	61.68%	98.86 %
3	SYG, SYG, RBF, RBF	*,+ *	$\gamma_{3,1} = 0.190$, $\gamma_{3,2} = 0.014$ $\gamma_{3,3} = 0.372$, $\gamma_{3,4} = 0.760$	62.33%	99.62 %
1*	RBF, SYG, RBF	*,exp *	$\gamma_{4,1} = 1.064$, $\gamma_{4,2} = 0.094$ $\gamma_{4,3} = 0.273$	61.53%	98.24%
2*	SYG, RBF, RBF,	*,exp *	$\gamma_{5,1} = 0.243$, $\gamma_{5,2} = 0.700$, $\gamma_{5,3} = 0.249$	61.73%	99.24 %
3*	RBF, SYG, RBF, SYG	*,+ +	$\gamma_{6,1} = 0.694$, $\gamma_{6,2} = 0.097$, $\gamma_{6,3} = 0.257$, $\gamma_{6,4} = 0.014$,	62.81%	99.62 %

used in the hybrid approach: 1 - classical genetic approach, 2 - M_{ijn} co-mutation genetic operator, 3 - $LR - M_{ijn}$ co-mutation genetic operator. The methods denoted with an additional * are improved methods using a wasp computational scheme. The terminal nodes of structural representation of multiple kernel (simple kernels) are given from left to right. The operations are given beginning with the last intermediate level. By example, the first multiple kernel in the table 1 is $(K_{POL} + K_{RBF}) + (exp(K_{SIG}))$.

The experimental results for the Vowel data set are presented in Table 2.

5   Conclusions and Further Direction of Study

In this article we introduce a general frame which allows us to obtain hybrid methods for building optimal multiple SVM kernels. Our scheme follows 5 steps: formal representation of the multiple kernels, structural representation, choice of genetic algorithm, SVM algorithm and model evaluation. Structural representation and genetic algorithm are most important in differentiate particular methods from this frame. We implemented many particular methods, for the data sets *Leukemia* and *Vowel*. Analyzing the results, presented in Table 1 and Table 2, we can conclude that utilization in the genetic algorithm of more performing co-mutation algorithms generally improves both the classification and cross validation accuracy. The results are strongly dependent on data sets. The existence of a frame, from which many methods for building optimal multiple kernels can be easily obtained, is very important. It makes possible a quick comparison of the performances of multiple kernels and allows the choice of the better method for a given data set. The performances of the particular methods implemented are promising. Our frame is an open one, it may be enlarged and offer possibility for further development.

References

1. Aizerman, M., Braverman, E., Rozonoer, L.: Theoretical foundations of the potential function method in pattern recognition learning. Automation and Remote Control 25, 821–837 (1964)
2. Bonabeau, E., Theraulaz, G., Demeubourg, J.I.: Fixed response thresholds and the regulation of division of labor in insect societies. Bull. Math. Biol. 60, 753–807 (1998)
3. Campbell, C.: An Introduction to Kernel Methods Radial Basis Function Network: Design and Applications, pp. 1–31. Springer, Berlin (2000)
4. Cicirelo, V.A., Smith, S.F.: Wasp-like Agents for Distributed Factory coordination. Autonomous Agents and Multi-Agent Systems 8(3), 237–267 (2004)
5. Chang, C.-C., Lin, C.-J.: LIBSVM: a library for support vector machines (2001), Software available at http://www.csie.ntu.edu.tw/cjlin/libsvm
6. Diosan, L., Oltean, M., Rogozan, A., Pecuchet, J.P.: Improving SVM Performance Using a Linear Combination of Kernels. In: Beliczynski, B., Dzielinski, A., Iwanowski, M., Ribeiro, B. (eds.) ICANNGA 2007. LNCS, vol. 4432, pp. 218–227. Springer, Heidelberg (2007)
7. Diosan, L., Oltean, M., Rogozan, A., Pecuchet, J.P.: Genetically Designed Multiple-Kernels for Improving the SVM Performance, portal VODEL (2008), http://vodel.insa-rouen.fr/publications/rfia
8. Minh, H.Q., Niyogi, P., Yao, Y.: Mercer's Theorem, Feature Maps, and Smoothing, http://people.cs.uchicago.edu/~niyogi/papersps/MinNiyYao06.pdf
9. De Falco, I., Iazzetta, A., Della Cioppa, A., Tarantino, E.: A new mutation operator for evolutionary airfoil design. In: Soft Computing - A Fusion of Foundations, Methodologies and Applications, vol. 3(1), pp. 44–51. Springer, Heidelberg (1999)
10. Simian, D.: A Model For a Complex Polynomial SVM Kernel. In: Proceedings of the 8th WSEAS Int. Conf. on Simulation, Modelling and Optimization, Santander Spain, within Mathematics and Computers in Science and Engineering, pp. 164–170 (2008)
11. Simian, D., Stoica, F.: An evolutionary method for constructing complex SVM kernels. In: Proceedings of the 10th International Conference on Mathematics and Computers in Biology and Chemistry, MCBC 2009, Recent Advances in Mathematics and Computers in Biology and Chemistry, Prague, Chech Republic, pp. 172–178. WSEAS Press (2009)
12. Stoica, F., Simian, D., Simian, C.: A new co-mutation genetic operator. In: Proceeding of the 9th Conference on Evolutionay Computing, Advanced Topics on Evolutionary Computing, Sofia, pp. 76–82 (2008)
13. Simian, D., Stoica, F.: Evaluation of a hybrid method for constructing multiple SVM kernels. In: Proceedings of the 13th WSEAS International Conference on Computers. Recent Advances in Computer Engineering Series, pp. 619–623. WSEAS Press (2009)
14. Simian, D., Stoica, F., Simian, C.: Optimization of Complex SVM Kernels Using a Hybrid Algorithm Based on Wasp Behaviour. In: Lirkov, I., Margenov, S., Waśniewski, J. (eds.) LSSC 2009. LNCS, vol. 5910, pp. 361–368. Springer, Heidelberg (2010)
15. Sonnenburg, S., Rätsch, G., Schäfer, C., Scholkopf, B.: Large scale multiple kernel learning. Journal of Machine Learning Research 7, 1531–1565 (2006)
16. Vapnik, V.: The Nature of Statistical Learning Theory. Springer, Heidelberg (1995), http://www.csie.ntu.edu.tw/~cjlin/libsvmtools/datasets

Part VI

Environmental Modeling

Modeling of Toxic Substances in the Atmosphere – Risk Analysis and Emergency Forecast

A. Brandiyska[1], K. Ganev[1], D. Syrakov[2], M. Prodanova[2], and N. Miloshev[1]

[1] National Institute of Geophysics, Geodesy and Geography,
Bulgarian Academy of Sciences, Sofia, Bulgaria
kganev@geophys.bas.bg
[2] National Institute of Meteorology and Hydrology,
Bulgarian Academy of Sciences, Sofia, Bulgaria

Abstract. The present paper describes the activities and results achieved in the development of a modelling system for operational response to accidental releases of harmful gases in the atmosphere. The main envisaged functions of the system are:

1. Perform highly accurate and reliable risk analysis and assessment for selected "hot spots";
2. Provide the national authorities and the international community with operational short-term local-to regional scale forecast of the propagation of harmful gases;
3. Perform, in an off–line mode, a more detailed and comprehensive analysis of the possible longer-term impacts on the environment and human health.

The system is based on the following models: *WRF*, used as meteorological pre-processor; *SMOKE* - the emission pre-processor; *CMAQ* - the Chemical Transport Model (CTM) of the system.

For the needs of the emergency response preparedness mode the risk is defined as probability the national regulatory threshold values for toxic gases to be exceeded. Maps of the risk around potential sources of emergency toxic gas releases are constructed and demonstrated in the current paper.

Some examples of the system "operational mode" results are demonstrated as well.

1 Introduction

For emergency management, the adage "It's better to do something than nothing" is not true since the wrong response can be very costly and potentially as dangerous as the threat itself. Scientific and technical information is critical for helping emergency managers to make sound decisions with regards to response to critical threats. The Balkan Peninsula is a region with complex topography, which causes significant disturbances of the air flows. These mesoscale disturbances may have a great influence not only on the local pollution transport and

I. Lirkov, S. Margenov, and J. Waśniewski (Eds.): LSSC 2011, LNCS 7116, pp. 267–274, 2012.

hence on the detailed pollutionpattern, but also on the trans-boundary transport of harmful substances (see for example [8,7,6]. Using simplified approaches and tools for the calculation of the atmospheric dispersion would inevitably lead to inappropriate response to the threat.

The present work demonstrates some preliminary results produced by the Bulgarian system for emergency response. The planned modelling system has the potential to assist emergency managers in three stages:

In preparedness mode, "risk analysis" is performed. It results in a set of risk assessments for different emergency scenarios for selected "hot spots". These assessments can be of a direct use for the relevant decision making bodies for developing strategies for immediate emergency response (for example evacuation of people from the pollution exposed regions, proper assignment of medical teams) in order to minimize the pollution impact on human health. They could give valuable information for optimisation of the air quality monitoring network.

In the operational mode the system produces fast short-term forecast of the pollutant propagation in local and regional scale. This information will help the authority decisions about the immediate measures and activities to be carried out. This information will also warn the international community of possible trans-boundary harmful pollutant transport.

In the off-line mode (after an accident happens) the modelling system will produce a more detailed and comprehensive analysis of the possible longer-term impact of the harmful releases on the environment and human health in local to regional scales. This information will help the formulation of long–term strategic measures and activities for abatement of the caused damages and gradual restoration of the environment.

The system is based on up-to date and complex meteorological and pollution transport models with proved high-quality simulation performance, high-spatial resolution and options for two way nesting: WRF ([10], used as meteorological pre-processor; SMOKE ([4]) – the emission pre-processor. SMOKE currently supports BEIS (Biogenic Emissions Inventory System) mechanism, versions 3.13 ([11]). An alternative method for calculating biogenic emissions is suggested, for example, in [9]; CMAQ ([2,3]) – the Chemical Transport Model of the system. A chlorine chemical mechanism has been added to CMAQ based on [12].

As the simulation, especially in the risk analysis and assessment mode of the system require huge amount of computer resources, the grid computing technology is applied. The Computational Grid, or shortly, the Grid, is a computing environment which enables the unification of widely geographically distributed computing resources into one big (super)-computer ([1,5,6]).

The work on the system is still going on, yet some results from the system preparedness and on-line mode will be demonstrated in this paper.

2   Some Examples of Risk Assesment Simulations

The risk may generally be defined as product of probability (recurrence) of happening of a given event and the impact of the same event (in this particular case

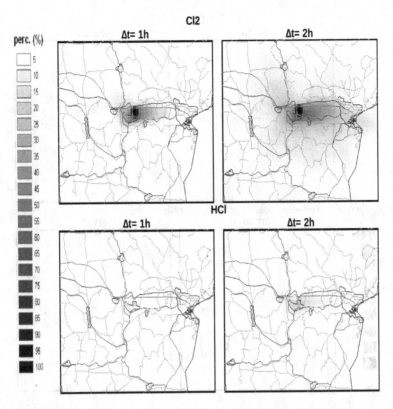

Fig. 1. Annual fields of probability surface Cl_2 and HCl concentrations to exceed the threshold value for times $\Delta t = 1h$ and $2h$ after the release (2 o'clock). The annual risk for HCl pollution for $\Delta t = 1h$ is zero.

on human health). The impact should be evaluated by some metrics directly giving the effect of toxic gas on human health – in this case the national regulatory threshold values (defined for Cl_2 and HCl), so the risk is simply the probability these thresholds to be exceeded.

Following the above definition of risk it is clear that a large number of simulations of the toxic gases (primary as well as secondary) dispersion around the potentially dangerous site should be made under comprehensive set of meteorological conditions and for different accidental release times. The averaged over this ensemble concentration fields should be treated as an assessment of the potential risk.

The NCEP Global Analysis Data with $1° \times 1°$ resolution was used as meteorological forcing and the WRF and CMAQ nesting capabilities were used (the simulations are carried out in three nested domains with resolutions 25, 5, 1km, covering respectively the regions of South-Eastern Europe, Bulgaria, and the area surrounding the particular site) for downscaling the problem to a horizontal resolution of 1 km around the site. It should be noted that the "regular"

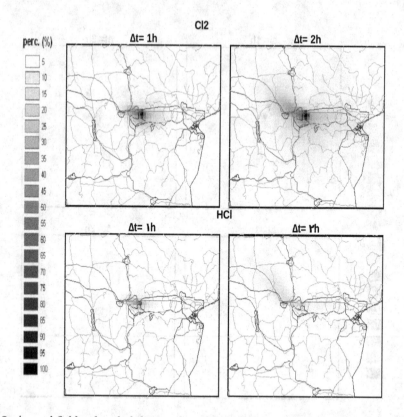

Fig. 2. Annual fields of probability surface Cl_2 and HCl concentrations to exceed the threshold value for times $\Delta t = 1h$ and $2h$ after the release (14 o'clock)

emissions are also taken into account not only in the innermost domain, but also in the domain with spatial resolution of 5 km (Bulgaria) thus providing appropriate boundary conditions for the innermost domain.

The simulations were carried out day by day for several hours after the release and thus a large ensemble of simulations was obtained. For each individual run it was checked up where the respective concentration exceeds the regulatory threshold value and so the impact is set equal to 1, and in the points where the concentrations are below the regulatory threshold, the impact value is set equal to zero. Then a simple statistics over this ensemble gives the probability the regulatory threshold to be exceeded – the risk.

Risk simulations for an instantaneous release of 25 t chlorine (Cl_2) released at the site of "VEREJA-HIM" Jambol in 3 o'clock in the morning were demonstrated in ([13]). Another industrial site, for which the risk analysis results will be shown here is "POLIMERI-DEVNYA" located close to the Black sea. The simulations were carried out for two release times — 2 o'clock and 14 o'clock local time.

The probabilities surface concentration fields of Cl2 (the primary pollutant) and HCl (one of the secondary pollutants) to exceed the respective threshold values are calculated for the four seasons and for the whole year (due to the limited volume of the paper only annual risk plots will be shown — Figs. 1,2). There are several things about the probability fields that should be mentioned:

1. The area with non-zero probabilities could cover a big domain around the site. This is another good demonstration why the risk assessment is necessary. It can be seen that significant risk could happen far from the site and with temporal delay.

2. The probability fields are quite different for the different seasons, which yet again demonstrates the importance of the meteorological conditions for the pollution field pattern. The probability field pattern is pretty complex for both Cl_2 and HCl, still a tendency for spreading the pollution in E-W direction could be seen, which reflects the local circulation patterns. In summer, due to the vicinity of the Black Sea, there is a breeze circulation, so during daytime the plume moves inland, while during night time the plume moves towards the sea.

3. The closest neighbouring cities are under threat – Devnya (large probability during warm daytime conditions) and Varna (during the night or in winter).

4. Some of the roads at certain time could be rather risky, so they should not be used for evacuation of the population. Such roads are both the highway and the T1 road from Devnya to Varna, if the accident happens at night-time or in winter, or the same roads in the opposite direction (Devnya–Shumen) if it happens during daytime in spring and summer.

5. The secondary pollutant concentrations may also exceed the regulatory threshold – for example HCl concentrations are high during spring and summer.

3   Some Examples of the System "Fast Decision" Mode

It is clear that the amount and nature of the accidental released gases will most probably not be known at the moment of the accident. That is why the concept of developing operational forecasting system, which follows the propagation of tracers (no photolisis, no chemistry, no aerosols, no cloud processes) from selected sites, was accepted.

A script, which starts the system, is run every 6 hours. First, the input data is downloaded from NCEP and the WRF input files are prepared. Then the meteorological model is run for 15 hours, with a nesting chain down to a 2 km resolution over the territory of Bulgaria. The first 6 hours of the simulation are a spin-up and the next 9 hours are processed to become meteorological input for the CMAQ chemical transport model. The corresponding emission files are generated. The CMAQ model is run for 9 hours and the output file with 15-min average surface concentrations is kept for archive in the OUTPUT folder. At the end images are produced, showing the 11 tracer concentrations. The units are provisional, since the amount and the coverage of the toxic spill, as well as the molar mass of the pollutant and the evaporation rate, are hypothetical.

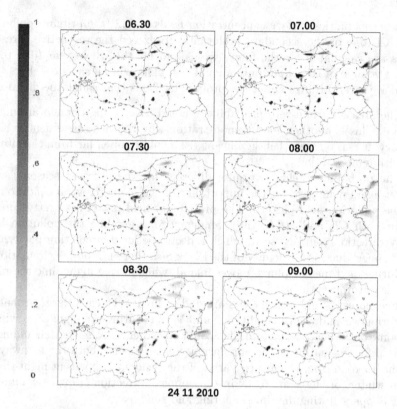

24 11 2010

Fig. 3. 3-hours tracer evolution for a release in 6 o'clock

Example of the simulated tracer evolution is shown in Fig. 3.

The main characteristics of the plume propagation, which were outlined by the numerical experiments, carried out so fat are the following:

1. The shape of the plumes and the direction of their movement are governed by the meteorological conditions. Since there are no chemical reactions, the changes in the pollutant concentrations are only due to dispersion and deposition.

2. The meteorological wind circulation can be very different on the local scale and on the mesoscale. This yet again confirms the need of high resolution. This is especially true for areas with complex terrain, like southern Bulgaria. The breeze circulation at the coast of the Black Sea during the warm months is also important – it can contribute to cleansing the air (when the wind is blowing from the west) or to high pollutant concentrations inland (when the wing is blowing from the east).

3. The simulations show, that if there are strong winds or the thermodynamic conditions are unstable, the surface concentrations drop 5 times in 2 hours, but this is a rare case. In most simulations the concentrations remain high at least 3 hours after the release. For some sites in some simulations the toxic

plumes remain almost stationary and the surface concentrations remain high for more than 2 hours.

Some of the sites are very close to big cities (Varna, Burgas, Rousse) and major roads or railway lines, so the emergency simulations could be really useful, showing if the plume approaches a region with dense population. Even cities, that are more than 20km away can be under threat: on 24.11.2010, 18h release — the city of Plovdiv is threatened by the plume originating from "Velichkovo" site.

4 Conclusions

The shown results are very preliminary. In order to achieve a real assessment of the risk simulations should be made for a much larger set of meteorological conditions (probably several years in order to account for the seasonal variability) and for much more release times in order to account for diurnal variability. The on-line emergency mode of the system is still not fully operational. At this stage of the system development, however, it can be concluded that:

1. The chosen modelling tools are probably suitable for this particular task and the obtained preliminary results are quite realistic and promising;
2. The Grid computing technologies are very appropriate for approaching tasks like risk assessment, which require performing of a very large number of simulations;
3. The shown numerical results very well demonstrate the practical value of both the preparedness and the operational mode of the system.

Acknowledgments. The present work is supported by the NATO SfP program (project ESP.EAP.SFPP 981393) and the Bulgarian National Science Fund (grants No. D002–161/16.12.2008 and DCVP–02/1/29.12.2009).

The work of the young scientist A. Brandiyska is supported by the ESF project No. BG51PO001-3.3.04-33/28.08.2009.

A. Brandiyska is a World Federation of Scientists grant holder.

Deep gratitude is due to US EPA and US NCEP for providing free-of-charge data and software. Special thanks to the Netherlands Organization for Applied Scientific research (TNO) for providing the study with the high-resolution European anthropogenic emission inventory.

References

1. Atanassov, E., Gurov, T., Karaivanova, A.: Computational Grid: structure and Applications. Journal Avtomatica i Informatica (in Bulgarian), 40–43 (September 2006) ISSN 0861-7562
2. Byun, D., Young, J., Gipson, G., Godowitch, J., Binkowski, F.S., Roselle, S., Benjey, B., Pleim, J., Ching, J., Novak, J., Coats, C., Odman, T., Hanna, A., Alapaty, K., Mathur, R., McHenry, J., Shankar, U., Fine, S., Xiu, A., Jang, C.: Description of the Models-3 Community Multiscale Air Quality (CMAQ) Modeling System. In: 10th Joint Conference on the Applications of Air Pollution Meteorology with the A&WMA, Phoenix, Arizona, January 11-16, pp. 264–268 (1998)

3. Byun, D., Ching, J.: Science Algorithms of the EPA Models-3 Community Multiscale Air Quality (CMAQ) Modeling System. EPA Report 600/R-99/030, Washington DC (1999)
4. Sparse Matrix Operator Kernel Emission (SMOKE) Modeling System, University of Carolina, Carolina Environmental Programs, Research Triangle Park, North Carolina
5. Foster, J., Kesselmann, C.: The Grid: Blueprint for a New Computing Infrastructure. Morgan Kaufmann (1998)
6. Ganev, K., Syrakov, D., Prodanova, M., Miloshev, N., Jordanov, G., Gadjev, G., Todorova, A.: Atmospheric composition modeling for the Balkan region. In: SEE-GRID-SCI USER FORUM 2009, Istanbul, 9-10, pp. 77–87 (December 2009) ISBN:978-975-403-510-0
7. Melas, D., Zerefos, C., Rapsomanikis, S., Tsangas, N., Alexandropoulou, A.: The war in Kosovo: Evidence of pollution transport in the Balkans during operation 'Allied Force'. Environmental Science and Pollution Research 7(2), 97–104 (2000)
8. Rappenglük, B., Melas, D., Fabian, P.: Evidence of the impact of urban plumes on remote sites in the Eastern Mediterranean. Atmospheric Environment 37(13), 1853–1864 (2003)
9. Symeonidis, P., Poupkou, A., Gkantou, A., Melas, D., Devrim Yay, O., Pouspourika, E., Balis, D.: Development of a computational system for estimating biogenic NMVOCs emissions based on GIS technology. Atmospheric Environment 42(8), 1777–1789 (2008)
10. Shamarock, et al.: A description of the Advanced Research WRF Version 2 (2007), http://www.mmm.ucar.edu/wrf/users/docs/arw_v2.pdf
11. Schwede, D., Pouliot, G., Pierce, T.: Changes to the Biogenic Emissions Invenory System Version 3 (BEIS3). In: Proc. of 4th Annual CMAS Models-3 Users's Conference, September 26-28, Chapel Hill, NC (2005)
12. Tanaka, P.L., Allen, D.T., McDonald-Buller, E.C., Chang, S., Kimura, Y., Yarwood, G., Neece, J.D.: Development of a chlorine mechanism for use in the carbon bond IV chemistry model. Journal of Geophysical Research 108, 41–45 (2003)
13. Todorova, A., Ganev, K., Syrakov, D., Prodanova, M., Georgiev, G., Miloshev, N., Gadjhev, G.: Bulgarian emergency response system for release of hazardous pollutants — design and first tests. In: Proceedings of the 13th International Conference on Harmonisation within Atmospheric Dispersion Modelling for Regulatory Purposes, Paris, France, June 1-4, pp. 495–499 (2010) ISBN: 2-8681-5062-4

Some Aspects of Impact in the Potential Climate Change on Ozone Pollution Levels over Bulgaria from High Resolution Simulations

Hristo Chervenkov

National Institute of Meteorology and Hydrology, branch Plovdiv - Bulgarian
Academy of Sciences, "Russki" 139, 4000 Plovdiv, Bulgaria
hristo.tchervenkov@meteo.bg

Abstract. According to the European Environment Agency (EEA), the
ground-level ozone is one of the most serious air pollutants in Europe
today. High levels of ozone can affect the respiratory system and increases morbidity and mortality, particularly in sensitive groups of the
population. Ozone also damages vegetation, reduces crop yields and
corrodes technological materials. Ozone pollution is pronounced in regions with strong photochemical activity, such as the Mediterranean
basin and Balkan Peninsula. Due to its central location in the second
region, Bulgaria may be considered as a hot-spot for ozone and representative of ozone effects on Balkan ecosystems. Ozone concentrations
are highly dependent on environmental conditions, including temperature. It is thought to be likely that long-term changes in climate will
affect levels of future ozone pollution. Based on the calculated and accumulated in NIMH - Bulgaria during the CECILIA WP7-program data,
scope of the presented work is to investigate the changes in ozone pollution levels, expressed with some exposure indices, due to climate change
in Southeastern Europe.

Keywords: Ozone pollution, exposure index, ecological impact, climate
change.

1 Introduction

Ozone is formed in the atmosphere through chemical reactions between nitrogen
oxides (NO_x) and volatile organic compounds (VOC) in the presence of sunlight
during a timescale from hours to days. Ozone is lost from the atmosphere through
dry deposition to surfaces (including the human respiratory tracts), uptaken by
vegetation and chemical reactions in the atmosphere with a timescale of hours to
weeks. Meteorological conditions have a decisive influence on the concentration
of ground-level ozone. Net formation of ozone requires solar ultra violet (UV)
radiation, high temperatures increase the efficiency of the photochemical formation and surface drought reduces deposition to the ground. Furthermore ozone
formation is related to precursor emissions in a non-linear fashion. Thus, the
influence on ozone by the prevailing meteorological conditions in Europe from

I. Lirkov, S. Margenov, and J. Waśniewski (Eds.): LSSC 2011, LNCS 7116, pp. 275–282, 2012.

one year to another can easily mask the changes in ozone caused by precursor emission changes. Although emissions from human activities thought to be a major cause of climate change, climate change is, in turn, affecting the concentration and spread of pollutant in the atmosphere resulting in a feedback loop between climate and chemistry in the atmosphere. The presented work is based on the primary results, obtained in modelling study performed in the Bulgarian NIMH in the frame of the EC FP6 Project "Central and Eastern Europe Climate Change Impact and Vulnerability Assessment" (CECILIA) program. Main aim is to reveal and assess the trend in the pollution levels of ozone expressed with the most used and special for this purpose calculated exposure indexes due only to the (possible) climate change in the future decays.

2   Concept and Methodology

The influence of meteorological variability on surface ozone levels is well known. Secondary pollutants, like ozone and particulate matter are controlled by the interplay of emissions and meteorology [7]. There are several processes responsible for this link. Increased temperatures speed up most of the chemical reactions in the atmosphere producing ozone; high-pressure situations may create soil drought and thereby reduce surface deposition of ozone; elevated temperatures may enhance biogenic emissions leading to higher ozone production; and subsidence may trap the photochemical pollution in the planetary boundary layer for an extended time [19]. The relationships between ozone and meteorological variability are complex and in reality require a chemical transport model (CTM) to provide meaningful relationships. Already, numerous studies dealt with future air quality projections on global scale using coupled climate-chemistry models and most of them were focusing on ozone on global scale ([11,5,19,17,18]). Fewer studies provide information on regional and local scale air quality changes ([21,14,15]). These studies focused on the average species concentration change, but only few or no results were given on the impact on the occurrence of extreme air pollution events. A recent report issued by the U.S. Environmental Protection Agency assessing the regional impact of climate change on U.S. air quality, indicated that the number of high ozone situations will increase in the future, potentially causing air quality alerts earlier in spring and later in the autumn [8]. The rise in the frequency of the occurrence of elevated ozone levels due to higher temperatures in the future climate is shown by many others as well ([3,13,23]).

Specific tasks of the activities in CECILIA's workpackage 7 (WP7) was to study (based on the high (10 km) resolution downscaling results) the impacts of climate change on health and air quality (photochemistry of air pollution, aerosols). For this purpose special modellig system was build in NIMH - BAS Sofia, described in details in [20]. A future climate has been imposed using the IPCC SRES A1B scenario. To isolate the effect of climate change on air quality, present day (year 2000) anthropogenic emissions were used in all simulations, while climate-sensitive biogenic emissions were allowed to vary with the simulated climate. To study anticipated climate impacts on air quality, simulations were performed for three decades: the present decade 1991-2000, the decade 2041-2050 and 2091-2100, which we will refer to as control run (CR), near (NF) and far (FF) future decades, respectively. The choice of decadal time slices was caused by the huge computer time and storage requirements of RCM-AQM model simulations. However, the comparison of driving climate parameters for the selected 10-year time slices (adopted in this study) with 30-year time slices usually adopted in climate studies, shows that the selected 10-year time slices, 2041-2050 and 2091-2100, are comparable with the 30-year time slices 2021-2050 and 2071-2100, respectively [12].

3 Ozone Metrics

The impact of surface ozone on ecosystem and human health can be assessed using suitable measures. Four among the most frequently used ([16,6]) for research and regulatory purposes to protect vegetation and human health, are chosen and summarized below.

AOT40 (expressed in $(\mu g/m^3)$.hours) means the sum of the difference between hourly concentrations greater than 80 $\mu g/m^3$ (= 40 parts per billion (ppb)) and 80 $\mu g/m^3$ over a given period using only the 1 hour values measured between 8:00 and 20:00 Central European Time (CET) each day ([10]). Two modifications of this exposure index are taken into consideration:

AOT40c for protection of agricultural crops. For this parameter the integration window is between May 1 and August 1.

AOT40f for protection of (deciduous) forests. The integration period is between the April 1 and September 1.

The target value, set by the EC for the year 2010 is 18000 $\mu g/m^3$.h (9000 ppb.h) and 20000 $\mu g/m^3$.h (10000 ppb.h) for AOTc and AOTf correspondingly. Additionally for the AOTc is set also a long-term objective of 6000 $\mu g/m^3$.h (3000 ppb.h) ([10]).

The impact on the human health is assessed with two parameters. Even though current evidence was insufficient to derive a level below which ozone has no effect on mortality, a cut-off at 35 ppb, considered as a daily maximum 8-hour mean ozone concentration, is used ([22]). This means that for days with ozone concentration above 35 ppb as maximum 8-hour mean, only the increment exceeding 35 ppb is used to calculate effects.

The exposure parameter called SOMO35 (sum of means over 35) and is the sum of excess of daily maximum 8-h means over the cut-off of 35 ppb calculated for all days in a year ([1]).

This parameter, the SOMO35, is a measure of accumulated high exposure. No limit values have been established for SOMO35, but a threshold of 3000 ppb days is consistent with air quality limits currently in use in North America and Europe ([5,8]).

The last metrics used in this study is the $N_{8hDM120}$ also noted as NOD60 ([20,23]):

The metric called $N_{8hDM120}$ is equal to the Number of days with a maximum eight hours running average ozone concentration exceeding 120 µg/m^3. This target value may be exceeded up to 25 days a year, as an average over the three preceding years ([10]). The directive [10] recommends to use 24 8-hour running intervals between 17:00 on the previous day to 24:00 on the current they in the obtaining of the SOMO35 and NOD60, but we have use only the intervals from the current day. Our opinion is that this difference in the methodology can not lead to significant underestimation of the SOMO35, because in the late afternoon and night hours (the omitted intervals) the ozone concentrations are small due to the weak photochemical activity. Due to this reason no difference in the values of the NOD60 obtained in the both ways is expected also.

4   Overview of the Results

The values of the four ozone metrics for each year in all three time slices 1991-2000, 1951-2060 and 2091-2100 are calculated. The averaged quantities are treated as representatives for the corresponding decade.

First we evaluated the AOT40 for crops plotted in Figures 1a-1c. The values range from 15000 to 30000 ppbv.h, being high over the southerwestern part of the country. AOT40 for forests, plotted in Figures 2a-2c, reaches higher values due to longer integration period and has maxima around 33000 ppbv.h, again in small area in southerwestern Bulgaria and over the Black sea. The spatial distribution of the parameter SOMO35 for the three decades, shown on Figures 3a-3c, is similar to these of the both AOT40: On the bigger part of the model domain it has value below 3300 ppb.h, with broad maxima over the above mentioned regions reaching in the last decade 5000 ppb.h. The last parameter, the NOD60, is relatively homogeneous distributed with values in nearly whole domain below 10 days in year.

The comparison between the figures for the three time slices show that the AOTs will mostly increase. The difference between the last and the first decade can be judged very roughly up to 2000 ppb.h or 20% for the AOTc and up to 3000 ppb.h or again around 20% for the AOTf. The trends by the SOMO35 shows clearly positive tendency, better expressed by the comparison between the far and near future. Quantitatively this increase, again after rough judgment, is around 300 ppb.h, or relatively 9%. The NOD60 in all decades remains over the bigger part of the model domain below 10 days and the tendency, although positive obtained by the comparison of the small areas around the maxima, is very weak.

5   Summary and Discussion

Most important results from the study is the significant overrun of the values of all exposure indexes (AOTc, AOTf and SOMO35) in comparison with the specified (or recommended) in the legislative directives objectives and its significant increase especially in the last decade, the far future. Obviously elevated

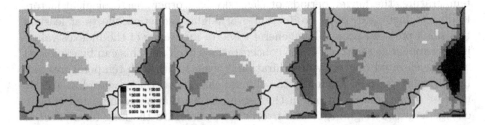

Fig. 1. Spatial distribution of the AOT for crops (AOTc) for the control run (left panel, fig. 1a), near future (centre panel, fig. 1b) and far future (right panel, fig. 1c). The units are ppb. hours and the common grey scale is shown in left panel.

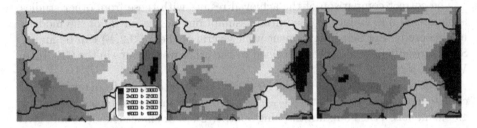

Fig. 2. Same as Fig.1, but for AOT for forests (AOTf)

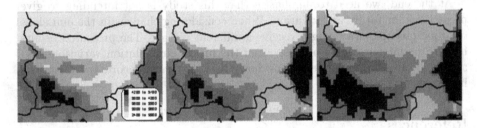

Fig. 3. Spatial distribution of the SOMO35 for the control run (left panel, fig. 3a), near future (centre panel, fig. 3b) and far future (right panel, fig. 3c). The units are ppb.days and the common grey scale is shown in left panel.

Fig. 4. Spatial distribution of the NOD60 for the control run (left panel, fig. 4a), near future (centre panel, fig. 4b) and far future (bottom panel, fig. 4c). The units are number of days and the common gray scale is shown in left panel.

temperature, the longer periods of clear sky and other meteorological factors leading to increased ozone production overweights reduction effects at least in this small part of Europe. The general trend of increasing of the indexes values is more robust as the trend of the concentrations itself and this can by explained with the elevated frequency and duration of events with high temperatures and bright sunny periods. Evidently the climate change has the potential to greatly increase ozone concentrations in the region. Given that the committed climate change is already significant and irrevocable, we regard the long-term effect of climate change a very serious impediment in reaching and retaining acceptable ozone levels in continental Europe during the 21st century. The climate change will to some extent counteract the gains in the indexes achieved through reducing ozone precursor emissions in Europe ([2]).

The correct comparison of the achieved results with these from other studies is hampered due mainly to the initial assumptions by the set-up of the models, but generally a principal agreement with several of them ([9,2,4]) can be found. Similar results have also been obtained by the other workgroups in the CECILIA WP7 project.

In our study we used one particular CTM operating on data from one regional climate model simulating one CO_2 emission scenario and forced by one particular global model on the boundaries. Using another CTM or climate model would give different details, but the overall effect of regional climate change on European ozone in this study, as already mentioned, is similar to other recent studies.

At the end, we need to emphasize that this study is not intending to give predictions for future air pollution. When considering changes in the emissions of ozone precursors, we might receive a different picture. The presented results have to be interpreted as one aspect of the future air pollution variations and a step towards better understating how the long-term meteorology variations influences air pollution.

References

1. Amann, M., Bertok, I., Cofala, J., Gyarfas, F., Heyes, C., Klimont, Z., et al.: Baseline scenarios for the clean air for Europe (CAFE) programme. Final Report, p. 79. IIASA, Laxenburg (2005)
2. Andersson, C., Engardt, M.: European ozone in a future climate: Importance of changes in dry deposition and isoprene emissions. J. Geophys. Res. 115, D02303 (2010), doi:10.1029/2008JD011690
3. Bell, M.L., Goldberg, R., Hogrefe, C., Kinney, P.L., Knowlton, K., Lynn, B.: Climate change, ambient ozone, and health in 50 US cities. Clim. Change 82, 61–76 (2007)
4. Carvalho, A., Monteiro, A., Solman, S., et al.: Climate-driven changes in air quality over Europe by the end of the 21st century, with special reference to Portugal. Environmental Science & Policy 13, 445–458 (2010)

5. Dentener, F., Stevenson, D., Ellingsen, K., van Noije, T., Schultz, M., Amann, M., Atherton, C., Bell, N., Bergmann, D., Bey, I., Bouwman, L., Butler, T., Cofala, J., Collins, B., Drevet, J., Doherty, R., Eickhout, B., Eskes, H., Fiore, A., Gauss, M., Hauglustaine, D., Horowitz, L., Isaksen, I.S.A., Josse, B., Lawrence, M., Krol, M., Lamarque, J.F., Montanaro, V., Muller, J.F., Peuch, V.H., Pitari, G., Pyle, J., Rast, S., Rodriguez, J., Sanderson, M., Savage, N.H., Shindell, D., Strahan, S., Szopa, S., Sudo, K., Van Dingenen, R., Wild, O., Zeng, G.: The Global Atmospheric Environment for the Next Generation. Environ. Sci. Technol. 40(11), 3586–3594 (2006), doi:10.1021/es0523845
6. EEA Report 2/2007. Air pollution in Europe 1990-2004, ISSN 1725-9177
7. EEA Technical report 7/2009, Assessment of ground-level ozone in EEA member countries, with a focus on long-term trends, ISSN 1725-2237
8. EPA: Climate Change Impacts on Regional Air Quality Report, U.S. Environmental Protection Agency, EPA (2009)
9. Engardt, M., Bergstrom, R., Andersson, C.: Climate and Emission Changes Contributing to Changes in Near-surface Ozone in Europe over the Coming Decades: Results from Model Studies Ambio 38(8), 452–458 (2009)
10. EC: Directive 2002/3/EC of the European Parliament and the Council relating to ozone in ambient air ('Third Daughter Directive'). OJ L 67, 14–30 (March 09, 2002)
11. Hauglustaine, D.A., Lathiere, J., Szopa, S., Folberth, G.A.: Future tropospheric ozone simulated with a climate-chemistry biosphere model. Geophys. Res. Lett. 32, L24807 (2005), doi:10.1029/2005GL024031
12. Juda-Rezler, K., Zanis, P., Syrakov, D., Reizer, M., Chervenkov, H., Huszar, P., Melas, D., Krueger, B., Trapp, W., Halenka, T.: On the effect of climate change on regional air quality over Europe: concept, evaluation and future projections. Clim. Res., CECILIA Special Issue (in press, 2012)
13. Mahmud, A., Tyree, M., Cayan, D., Motallebi, N., Kleeman, M.J.: Statistical downscaling of climate change impacts on ozone concentrations in California. J. Geophys. Res. 113, D21103 (2008), doi:10.1029/2007JD009534
14. Meleux, F., Solmon, F., Giorgi, F.: Increase in summer European ozone amounts due to climate change. Atmos. Environ. 41(35), 7577–7587 (2007)
15. Nolte, C.G., Gilliland, A.B., Hogrefe, C., Mickley, L.J.: Linking global to regional models to assess future climate impacts on surface ozone levels in the United States. J. Geophys. Res. 113, D14307 (2008), doi:10.1029/2007JD008497
16. Paoletti, E., De Marco, A., Racalbuto, S.: Why Should We Calculate Complex Indices of Ozone Exposure? Results from Mediterranean Background Sites Environ Monit Assess 128, 19–30 (2007), doi:10.1007/s10661-006-9412-5
17. Racherla, P.N., Adams, P.J.: Sensitivity of global tropospheric ozone and fine particulate matter concentrations to climate change. J. Geophys. Res. 111, D24103 (2006), doi:10.1029/2005JD006939
18. Racherla, P.N., Adams, P.J.: The response of surface ozone to climate change over the Eastern United States. Atmos. Chem. Phys. 8, 871–885 (2008)
19. Solberg, S., Hov, Ø., Søvde, A., Isaksen, I.S.A., Coddeville, P., De Backer, H., Forster, C., Orsolini, Y., Uhse, K.: European surface ozone in the extreme summer. J. Geophys. Res. 113, D07307 (2008)
20. Syrakov, D., Prodanova, M., Miloshev, N., Ganev, K., Jordanov, G., Spiridonov, V., Bogatchev, A., Katragkou, E., Melas, D., Poupkou, A., Markakis, K.: Climate Change Impact Assessment of Air Pollution Levels in Bulgaria. In: Lirkov, I., Margenov, S., Waśniewski, J. (eds.) LSSC 2009. LNCS, vol. 5910, pp. 538–545. Springer, Heidelberg (2010) ISSN: 0302-9743

21. Szopa, S., Hauglustaine, D.A., Vautard, R., Menut, L.: Future global tropospheric ozone changes and impact on European air quality. Geophys. Res. Lett. 33, L18805 (2006), doi:10.1029/2006GL25860
22. WHO: WHO air quality guidelines for particulate matter, ozone, nitrogen dioxide and sulfur dioxide. Global update, summary of risk assessment. WHO/SDE/PHE/OEH06.02 (2006)
23. Zlatev, Z.: Impact of future climatic changes on high ozone levels in European suburban areas. Clim. Change (2009), doi:10.1007/s10584-009-9699-7

New Parallel Implementation of an Air Pollution Computer Model – Performance Study on an IBM Blue Gene/P Computer

Krassimir Georgiev[1], Tzvetan Ostromsky[1], and Zahari Zlatev[2]

[1] Institute of Information and Communication Technologies,
Bulgarian Academy of Sciences, Sofia, Bulgaria
{georgiev,ceco}@parallel.bas.bg
[2] National Environmental Research Institute, Aarhus University
Frederiksborgvej 399, P.O. Box 358, DK-4000 Roskilde, Denmark
zz@dmu.dk

Abstract. A new parallel version of the Danish Eulerian model for long transport of air pollutants over the territory of Europe (UNI–DEM) is presented. It is based on the domain partitioning of the space domain both in the Ox and Oy directions. This new approach gives possibilities to use large number of processors (or cores) on the IBM BlueGene/P computer. The new version of the parallel code of the UNI–DEM is created by using MPI standard library and appears to be highly portable and shows good efficiency and scalability. Discussions according to the performance, speed-ups and efficiency achieved in the first testing runs of the new parallel code on an IBM Blue Gene/P computer are presented.

1 Introduction

The modern community is needed of sustainable and robust strategies control to keep the level of the concentrations of some dangerous air pollutants under some safety levels. The process of finding the optimal solution is in general a long process, and therefore the efficiency of the corresponding computer codes used is highly desirable. Computations in real time are needed in many typical situations, e.g. in connection with the EU ozone directives [4] and/or when the global climate changes are studied. High ozone concentrations occur in large parts in Europe during the summer season, and they can cause damages. Reliable mathematical models have to be coupled with weather forecasting models, and they have to be run operationally in order to predict the appearance of high concentrations and to warn the community in such dangerous situations.

The research on the process of the long-range transport of air pollutants requires using large mathematical models in which at least the most of the atmospheric physical and chemical processes are adequately described. The Danish Eulerian Model (DEM) is an appropriate model for investigating both the transport on long distances and the transformations in the atmosphere of pollutants, which are potentially harmful. This mathematical model is described by a system

I. Lirkov, S. Margenov, and J. Waśniewski (Eds.): LSSC 2011, LNCS 7116, pp. 283–290, 2012.
© Springer-Verlag Berlin Heidelberg 2012

of partial differential equations and was developed at the National Environmental Research Institute (NERI, Roskilde, Denmark, [17]). The spatial domain of the model covers Europe and some neighbouring parts of this continent belonging to the Atlantic Ocean, Asia and Africa. More details about DEM and some of the computer implementation can be found in [6,7,14,15,16,18,19,20]. Large–scale models, as DEM, can efficiently be run only on high–performance parallel computer architectures. The parallel code of DEM (here we refer to the last unified version popular as UNI–DEM) is created by using the Message Passing Interface (MPI) standard library [9].

The rest of the paper is organized as follows. Some very short mathematical background of DEM is discussed in Section 2. Section 3 is devoted to the parallelization strategies – the old (based on partitioning of the computational domain in the Oxy plane on strips in south – north direction) and the new one (based on partitioning of the computational domain in the Oxy plane on rectangles). In Section 4 some characteristics of the computer used, analysis of the performed runs of the UNI–DEM on an IBM Blue Gene/P supercomputer and some preliminary results on the performance, speed ups and efficiency achieved are presented. Finally, some concluding remarks can be found.

2 Short Mathematical Background of the Danish Eulerian Model

The Danish Eulerian model is a large air pollution model for studying different phenomena connected with transport of air pollutants in the atmosphere. The space domain of the model contains the whole of Europe[17]. All important physical processes (advection, diffusion, deposition, emissions and chemical reactions) are represented in the mathematical formulation of the model. All important chemical species (sulphur pollutants, nitrogen pollutants, ammonia-ammonium, ozone, as well as many radicals and hydrocarbons) can be studied by the model. The chemical reactions are described by using the well-known condensed CBM IV scheme. The space domain is discretized by using (32×32), (96×96), (288×288) or (480×480) grid in the two–dimensional version of the model and $(96 \times 96 \times 10)$, $(288 \times 288 \times 10)$ or $(480 \times 480 \times 10)$ grid in the three–dimensional version. The horizontal discretization is performed by using equidistant grid-squares, correspondingly $(150km \times 150km)$, $(50km \times 50km)$, $(16.7km \times 16.7km)$ or $(10km \times 10km)$, while non–equidistant grid is used in the vertical direction (finer resolution close to the surface).

The three dimensional version of DEM is represented mathematically by the following system of partial differential equations (PDE):

$$\frac{\partial c_s}{\partial t} = -\frac{\partial(uc_s)}{\partial x} - \frac{\partial(vc_s)}{\partial y} - \frac{\partial(wc_s)}{\partial z}$$

$$+\frac{\partial}{\partial x}\left(K_x\frac{\partial c_s}{\partial x}\right) + \frac{\partial}{\partial y}\left(K_y\frac{\partial c_s}{\partial y}\right) + \frac{\partial}{\partial z}\left(K_z\frac{\partial c_s}{\partial z}\right) \qquad (1)$$

$$+E_s - (\kappa_{1s} + \kappa_{2s})c_s + Q_s(c_1, c_2, \ldots, c_q) \; ; \quad s = 1, \ldots, q,$$

where the following notations are used:

- $c_s = c_s(t, x, y, z)$ is the concentration of the chemical species s at point (x, y, z) of the space domain and at time t of the time-interval;
- $u = u(t, x, y, z), v = v(t, x, y, z)$ and $w = w(t, x, y, z)$ are the wind velocities along Ox, Oy and Oz directions respectively at point (x, y, z) and time t;
- $K_x = K_x(t, x, y, z)$, $K_y = K_y(t, x, y, z)$ and $K_z = K_z(t, x, y, z)$ are diffusion coefficients;
- E_s are the emissions;
- $\kappa_{1s} = \kappa_{1s}(t, x, y, z)$ and $\kappa_{2s} = \kappa_{2s}(t, x, y, z)$ are the coefficients for dry and wet deposition of the chemical species s at point (x, y, z) and at time t of the time-interval;
- $Q_s(t, x, y, z, c_1, c_2, \ldots, c_q)$ are nonlinear expressions that describe the chemical reactions under consideration.

Numerical methods are used to find approximate values of the solution at the grid–points. Most often the classical variant of the Finite Element Method (linear or bilinear conforming triangles) is used in the numerical treatment of DEM. Some experiments were done with bilinear nonconforming finite elements (see [3]). It is also appropriate to split the model into several sub-models (sub-systems) that are in some sense simpler. There is another advantage when some splitting procedure is applied: the different sub–systems have different mathematical properties which can be used in the selection of the best numerical method for each of the sub-systems. The splitting procedure proposed in [11,12] is used in the UNI–DEM. It leads to five submodels representing the horizontal advection (2), the horizontal diffusion (3), the chemistry and the emissions (4), the depositions (dry and wet) (5) and the vertical exchange (6).

$$\frac{\partial c_s^{(1)}}{\partial t} = -\frac{\partial (u c_s^{(1)})}{\partial x} - \frac{\partial (v c_s^{(1)})}{\partial y} \tag{2}$$

$$\frac{\partial c_s^{(2)}}{\partial t} = \frac{\partial}{\partial x}\left(K_x \frac{\partial c_s^{(2)}}{\partial x}\right) + \frac{\partial}{\partial y}\left(K_y \frac{\partial c_s^{(2)}}{\partial y}\right) \tag{3}$$

$$\frac{dc_s^{(3)}}{dt} = E_s + Q_s(c_1^{(3)}, c_2^{(3)}, \ldots, c_q^{(3)}) \tag{4}$$

$$\frac{dc_s^{(4)}}{dt} = -(\kappa_{1s} + \kappa_{2s})c_s^{(4)} \tag{5}$$

$$\frac{dc_s^{(5)}}{dt} = -\frac{\partial (w c_s^{(5)})}{\partial z} + \frac{\partial}{\partial z}\left(K_z \frac{\partial c_s^{(5)}}{\partial z}\right). \tag{6}$$

These five systems of PDE's are solved successively at each time step. The concentrations obtained for each submodel at a given time step are used as initial conditions for the next submodel. Finally, solving the sub–system (6) at the time step $t + \Delta t$ with initial conditions the solution obtained solving the subsystem (5) the solution c_s at a time step $(t + \Delta t)$ is obtained.

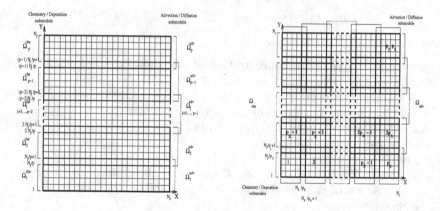

Fig. 1. Domain partitioning - **OLD** (left), **NEW** (right)

3   Briefly about the Parallelization Strategies

The parallelization strategy which is used in the computer treatment of DEM is based on the MPI standard library routines. The MPI (Message Passing Interface) (see e.g. [9]) was initially developed as a standard communication library for distributed memory computers. Later, having proved to be efficient, portable and easy to use, it became one of the most popular parallelization tools for application programming. Now it can be used on much wider classes of parallel systems, including shared-memory computers and clustered systems (each node of the cluster being a separate shared-memory machine). Thus it provides high level of portability of the code.

Our MPI parallelization is based on a *space domain partitioning*. The space domain is divided into several sub-domains (see Fig. 1). The number of the sub-domains is being equal to the number of MPI tasks. In the more often used now version of the UNI–DEM (this version has been implemented in almost all different kinds of parallel computers) the model parallelization is based on partitioning of the computational domain in the Oxy plane on strips in south–north direction (see Fig. 1, left part). Further on we will call this version as *OLD* or *Algorithm 1*. This version has proved to have very good parallel properties on a relatively small amount of processors, up to 120. The possibility to use the IBM Blue Gene/P computer and similar computers with a large amount of processors, $10^4 - 10^5$ and even more was the reason to develop a new parallel version of the UNI–DEM based on partitioning of the computational domain in the Oxy plane on rectangles (see Fig. 1, right part). Further on we will call this version as *NEW* or *Algorithm 2*.

Each MPI task works on its own sub–domain. At each time step there is no data dependency between the MPI tasks on both the chemistry and the vertical exchange stages. In the advection–diffusion part the spatial grid partitioning between the MPI tasks requires overlapping of the inner boundaries and exchange of certain boundary values on the neighboring subgrids for proper treatment of the boundary conditions. This leads to the following two main consequences: (*i*) certain computational overhead and load imbalance in the advection–diffusion stage in comparison with the chemistry and the vertical transport stages; (*ii*) communication necessity for exchanging boundary values on each time step (done in a separate communication step).

The main computational steps, performed on each time step, are:

- *Horizontal advection and diffusion.* The horizontal advection and diffusion submodels are treated together. In fact several independent advection–diffusion subproblems with Dirichlet boundary conditions arise on each time step after the splitting procedure, one for every chemical compound on every layer (in the 3-D version). These tasks are often too large to exploit efficiently the fast cache memory of the processors. This may have dramatic impact on the performance in some runs where the number of the processors are relatively small according to the grid size of the discrete problem.
- *Chemistry and deposition.* The calculations of these two processes can be carried out independently for each grid-point. At opposite to the advection stage, there are many small independent tasks (their number is equal to the number of grid-points in the spatial domain). In order to achieve good data locality, the smaller (low-level) tasks are grouped in chunks for more efficient cache utilization.
- *Vertical exchange.* The vertical exchange submodel splits into independent relatively simple advection–diffusion subproblems (along each vertical grid-line). The number of such tasks is relatively large and they are not very big. Like in the chemistry–deposition stage, the tasks can be grouped in chunks to improve the cache utilization.

In addition to the main stages, two extra procedures (a pre-processing and a post-processing) are used for scattering the input data and gathering the partial results at the beginning and at the end of each month during the run (see [5]).

4 Some Main Characteristics of an IBM Blue Gene/P Computer and Numerical Tests with the UNI–DEM Model

The first Bulgarian supercomputer IBM Blue Gene/P is used for numerical experiments with the new partitioning algorithm. As it is reported, this machine consists of two racks, 2048 Power PC 450 based compute nodes, 8192 processor cores and a total of 4TB random access memory (see http://www.bgsc.acad.bg/). Each processor core has a double–precision, dual pipe floating–point core accelerator. Sixteen I/O nodes are connected via fibre optics to a 10 Gb.s Ethernet switch. The smallest partition size, available currently, is 128 compute nodes (512 processor cores). The theoretical performance of the computer is 27.85 Tflops, while the maximum LINPACK performance achieved is 23.42 Tlops ($\approx 84\%$).

Some results of the test runs of the UNI–DEM for one–year period with the two parallel algorithms are presented in Table 1 for *Algorithm 1* and in Table

2 for *Algorithm 2*. In the space and time discretization the following values are used: (a) Advection step: 150 s.; (b) Chemical step: 150 s., 480×480 grid in the *Oxy* plane ($10km \times 10km$ cells).

The subdomains which appeared after the partitioning of the computational domain are usually too large to fit into the fast cache memory of the target processor, even for *Algorithm 2*. In order to achieve good data locality, the smaller low–level tasks are grouped in chunks where appropriate for more efficient cache utilization. A special parameter (called CHUNKSIZE) is provided in the chemical–emission part of the UNI–DEM code. This parameter should be tuned with respect to the cache size of the target machine. All runs reported here were performed by using chunks of length 48, which was obtained to be optimal for the cache size of the IBM Blue Gene/P computer.

Let us note that the speed–ups and the efficiency achieved are not classical one but according to the minimal number of processors used (15 for *Algorithm 1* and 4 for *Algorithm 2*).

Table 1. Results of some runs of the *Algorithm 1* on the IBM Blue Gene/P computer

No. of proc.	CPU time (in sec.)	Speedup	Efficiency (in %)	Comm. time in sec (%)
15	2803	-	-	316.8 (11.4)
30	1476	1.90	95	212.4 (15.3)
60	913	3.07	77	211.2 (24.8)
120	540	5.19	65	211.8 (39.2)

Table 2. Results of some runs of the *Algorithm 2* on the IBM Blue Gene/P computer

No. of proc.	CPU time (in sec.)	Speedup	Efficiency (in %)	Comm. time in sec (%)
4	8371	-	-	272.01 (3.2)
8	4303	7.97	99.6	138.00 (3.3)
16	1757	19.06	119.1	71.60 (4.1)
32	879	38.09	119.0	43.62 (5.0)
64	471	71.09	111.1	27.62 (5.9)
128	253	132.35	103.4	15.88 (6.3)
256	131	255.60	99.8	10.42 (8.0)
512	73	458.68	89.6	8.96 (12.3)
1024	52	643.92	62.9	10.78 (20.7)

One can see from the reported results that *Algorithm 2* shows better parallel properties. The communication time is more than two times less (compare e.g. the row four, the last column in Table 1 with the row six, the last column in Table 2). Nevertheless, it is to large in percent when the number of the processors grow (e.g. when the number of the processors used is 1024 the communication time is more than 20% of the total CPU time). It can be seen that the new algorithm shows to be very efficient up to 512 processors and its efficiency goes down dramaticly to 62% when 1024 processors are used. Another observation is the superlinear speed–up when the number of the processors is very small.

5 Conclusions

The unified parallel code of DEM, UNI–DEM, created by using MPI standard library, appears to be highly portable and shows good efficiency and scalability on the IBM Blue Gene/P machine for the new *Algorithm 2*. Some additional work can be done in order to avoid the superlinear speed–ups and for decreasing the communication time when the number of processors grows.

Acknowledgments. This research is supported in part by grants DCVP-02/1 and DO 02-161/08 from the Bulgarian NSF and the Bulgarian National Center for Supercomputing Applications (NCSA) giving access to the IBM Blue Gene/P computer.

References

1. Abdalmogith, S., Harrison, R.M., Zlatev, Z.: Intercomparison of inorganic aerosol concentrations in the UK with predictions of the Danish Eulerian Model. Journal of Atmospheric Chemistry 54, 43–66 (2006)
2. Alexandrov, V., Owczarz, W., Thomsen, P.G., Zlatev, Z.: Parallel runs of large air pollution models on a grid of SUN computers. Mathematics and Computers in Simulations 65, 557–577 (2004)
3. Antonov, A., Georgiev, K., Komsalova, E., Zlatev, Z.: Implementation of Bilinear Nonconforming Finite Elements in an Eulerian Air Pollution Model: Results Obtained by Using the Rotational Test. In: Dimov, I.T., Lirkov, I., Margenov, S., Zlatev, Z. (eds.) NMA 2002. LNCS, vol. 2542, pp. 379–386. Springer, Heidelberg (2003)
4. Directive 2002/3/EC of the European Parliament and the Council of 12 February 2002 relating to ozone in ambient air. Official Journal of the European Communities L67, 14–30 (2002)
5. Georgiev, K., Ostromsky, T.: Performance results of a large-scale air pollution model on two parallel computers. Int. J. of Environment and Pollution 22(1/2), 43–50 (2004)
6. Georgiev, K., Zlatev, Z.: Running an Advection-Chemistry Code on Message Passing Computers. In: Alexandrov, V., Dongarra, J. (eds.) PVM/MPI 1998. LNCS, vol. 1497, pp. 354–363. Springer, Heidelberg (1998)
7. Georgiev, K., Zlatev, Z.: Parallel sparse matrix algorithms for air pollution models. Parallel and Distributed Computing Practices 2, 429–442 (2000)
8. Georgiev, K., Zlatev, Z.: Runs of UNI–DEM Model on IBM Blue Gene/P Computer and Analysis of the Model Performance. In: Lirkov, I., Margenov, S., Waśniewski, J. (eds.) LSSC 2009. LNCS, vol. 5910, pp. 188–196. Springer, Heidelberg (2010)
9. Gropp, W., Lusk, E., Skjellum, A.: Using MPI: Portable programming with the message passing interface. MIT Press, Cambridge (1994)

10. Hov, Ø., Zlatev, Z., Berkowicz, R., Eliassen, A., Prahm, L.P.: Comparison of numerical techniques for use in air pollution models with non-linear chemical reactions. Atmospheric Environment 23, 967–983 (1988)
11. Marchuk, G.I.: Mathematical modeling for the problem of the environment. Studies in Mathematics and Applications, vol. 16. North-Holland, Amsterdam (1985)
12. McRae, G.J., Goodin, W.R., Seinfeld, J.H.: Numerical solution of the atmospheric diffusion equations for chemically reacting flows. J. Comp. Physics 45, 1–42 (1984)
13. Ostromsky, T., Owczarz, W., Zlatev, Z.: Computational Challenges in Large-scale Air Pollution Modelling. In: Proc. 2001 International Conference on Supercomputing in Sorrento, pp. 407–418. ACM Press (2001)
14. Ostromsky, T., Zlatev, Z.: Parallel Implementation of a Large-Scale 3-D Air Pollution Model. In: Margenov, S., Waśniewski, J., Yalamov, P. (eds.) LSSC 2001. LNCS, vol. 2179, pp. 309–316. Springer, Heidelberg (2001)
15. Ostromsky, T., Zlatev, Z.: Flexible Two-Level Parallel Implementations of a Large Air Pollution Model. In: Dimov, I.T., Lirkov, I., Margenov, S., Zlatev, Z. (eds.) NMA 2002. LNCS, vol. 2542, pp. 545–554. Springer, Heidelberg (2003)
16. Owczarz, W., Zlatev, Z.: Running a large-scale air pollution model on fast supercomputer. In: Chok, D.P., Carmichael, G.R. (eds.) Atmospheric Modelling, pp. 185–204. Springer, Heidelberg (2000)
17. WEB-site of the Danish Eulerian Model,
    http://www.dmu.dk/AtmosphericEnvironment/DEM
18. Zlatev, Z.: Computer treatment of large air pollution models. Kluwer (1995)
19. Zlatev, Z., Dimov, I.: Computational and Environmental Challenges in Environmental Modelling. Elsevier (2006)
20. Zlatev, Z., Dimov, I., Georgiev, K.: Three-dimensional version of the Danish Eulerian Model. Zeitschrift für Angewandte Mathematik und Mechanik 76, 473–476 (1996)

Simulation of the 2009 Harmanli Fire (Bulgaria)

Georgi Jordanov[1], Jonathan D. Beezley[2], Nina Dobrinkova[3],
Adam K. Kochanski[4], Jan Mandel[2], and Bedřich Sousedík[2]

[1] Institute of Geophysics, Bulgarian Academy of Sciences, Sofia
gjordanov@geophys.bas.bg
[2] Department of Mathematical and Statistical Sciences,
University of Colorado Denver, Denver, CO
{jon.beezley.math,jan.mandel,bedrich.sousedik}@gmail.com
[3] Institute of Information and Communication Technologies
Bulgarian Academy of Sciences, Sofia
ninabox2002@gmail.com
[4] Department of Meteorology, University of Utah, Salt Lake City, UT
adam.kochanski@utah.edu

Abstract. We use a coupled atmosphere-fire model to simulate a fire
that occurred on August 14–17, 2009, in the Harmanli region, Bulgaria.
Data was obtained from GIS and satellites imagery, and from standard
atmospheric data sources. Fuel data was classified in the 13 Anderson
categories. For correct fire behavior, the spatial resolution of the models
needed to be fine enough to resolve the essential micrometeorological
effects. The simulation results are compared to available incident data.
The code runs faster than real time on a cluster. The model is available
from **openwfm.org** and it extends WRF-Fire from WRF 3.3 release.

1 Introduction

Research in the southern member states of European Union (EU) in the last 30
years noted very high increase of the forest fires on their territories and Bulgarian
statistic shows a similar trend [4]. A team from Bulgarian Academy of Sciences
has started a literature-based analysis on the available models in 2007 and a
wildland-fire modeling initiative in 2008, and they have selected the Weather
Research and Forecasting model with fire behavior module, WRF-Fire [5, 6]. In
2009, it was decided to select a real fire from the national database maintained
by the Ministry of Agriculture, Forest and Food, administrative division Forests
and Forest Protection, for a demonstration of the model. The objective of the
present work is to demonstrate the simulation capabilities of the model with real
input data.

There were 108 forest fires in 2009, with total 18,105 hectares of forest burned.
In most of the cases, the fires have started from the surrounding area with tall
grasses and different types of bushes. 15,072 hectares of these non forest areas
burned, which showed to the authorities that without prevention by prescribed
burns, burning grass with two-three years old layers can easily turn into a forest

I. Lirkov, S. Margenov, and J. Waśniewski (Eds.): LSSC 2011, LNCS 7116, pp. 291–298, 2012.

fire. We have chosen as the case study a large forest fire close to town Harmanli, Bulgaria, on August 14–17, 2009. This fire is said to be caused by a barbecue on land with tall grasses and bushes near to the forest, which turned into a three-day forest fire. The area of the fire is on the south border of the protected zone "Ostar Kamak," a part of the network NATURA 2000, near Bulgaria-Greek-Turkey border.

The aim of this paper is to describe how input data, obtained in Bulgaria, may be used to simulate fire behavior, and to compare the results with the observed fire. WRF-Fire was used to simulate a wild fire in Bulgaria [4] before with partly ideal data. This is the first use with real data and an assessment its capability for prediction. If further validation is satisfactory, this model might be used in forecast mode in future.

2   Summary of the Model

The model couples the mesoscale atmospheric code WRF-ARW [9] with a fire spread module, based on the Rothermel model [8] and implemented by the level set method. It has grown out of CAWFE [2, 3], which couples the Clark-Hall atmospheric model with fire spread implemented by tracers. The atmospheric model supports refined meshes, called domains. Only the finest domain is coupled with the fire model. See [5, 6] for further details and references.

3   Data Sources

Collecting input data and making it usable for the model is a major component of the work necessary to simulate fire behavior. A WRF-Fire simulation requires input data from a variety of sources from meteorological initial and boundary conditions to static surface properties. Because WRF is a mesoscale meteorological model, typical data sources are only available at resolutions ranging from 10 to 100 km, while our simulations occur on meshes of resolution 1, 000 times finer. For this simulation of the Harmanli fire, we employ the highest resolution available to us. As more accurate and higher resolution data become available in the future, more detailed simulations will become possible.

For the meteorological inputs, we use a global reanalysis from the U.S. National Center for Environmental Protection (NCEP). This data is given on a 1 degree resolution grid covering the entire globe with 6 hour reanalysis cycle. The data is freely available and can be downloaded automatically over HTTP using a simple script. The data is downloaded as gridded binary (GRIB) files, which are extracted into an intermediate format using a utility called ungrib included in the WRF preprocessing system (WPS). Although the resolution of the NCEP global analysis is limited, it still may be useful as data source of data for model initialization in multi-domain setups. While there are many local sources of finer

Table 1. Fuel categories from satellite imagery and CORINE code (in parentheses)

Category	Description
1	Artificial, non-agricultural vegetated areas (141,142)
2	Sport Complex,Irrigated Cropland and Pasture,Bare Ground Tundra, Arable land (211,212,213), Open spaces with little or no vegetation (331,332,333,334,335)
3	Cemeteries, Dryland Cropland and Pasture, Grassland, Permanent crops(221,222,223), Pastures (231), Heterogeneous agricultural areas(241,242,243,244), Scrub and/or herbaceous vegetation associations (321,322,323,324)
4	Herbaceous Tundra, Parks
5	Wooded Wetland
6	Wooded Tundra, Orchard
7	Mixed Forest
8	Deciduous Needleleaf Forest, Forests (311,312,313)
9–13	N/A
14	Urban fabric (111,112), Industrial, commercial, and transport units (121,122,123,124), Mine, dump and construction sites (131,132,133), Wetlands(411,412,421,422,423), Water bodies (511,512,521,522,523)

resolution meteorological data available, they must be obtained individually for each region and combined into a single source. This process is labor intensive and cannot be automated on demand. Because our ultimate goal is to run the model in real time to forecast ongoing events, the use of such data is impractical.

Creating the simulation also requires a number of static data fields describing the surface properties of the domain. All such data is available as part of a standard global dataset for WRF. The fields in this dataset are available at various resolutions ranging from about 1 km to 10 km, which is sufficient for most mesoscale weather modeling purposes. Each field is stored in a unique format consisting of a series of simple binary files described by a text file. The geogrid utility in WPS interpolates the data in these files onto the model grid and produces an intermediate NetCDF file used in further preprocessing steps. While the standard geogrid dataset is sufficient for most weather forecasting applications, it lacks two high resolutions fields. These fields, surface topography and fuel information, are essential for accurately modeling fire behavior because they directly affect the rate of spread of the fire front inside the model.

While topography data is provided to WPS from the standard source, it comes from the USGS 30 arc second resolution global dataset (GTOPO30), which lacks enough detail to be used for our purposes. As a result, much more detailed source for the area of Harmanli from the Shuttle Radar Topography Mission (SRTM) at http://eros.usgs.gov is used, which provides topography at a resolution of about 90 m.

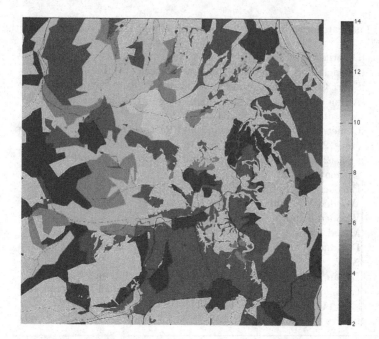

Fig. 1. Fuel map of the simulation area. The colors are the 13 Anderson fuel categories.

The data received from the server is a GIS raster format (DTED), which must first be processed and converted to `geogrid`'s binary data format. Using the powerful and open-source Quantum GIS (`www.qgis.org`), we open the downloaded raster and fill in the missing data with bilinear interpolation and project the data onto the Lambert Conformal Conic projection used in the model. Finally, we export the raster into a `GeoTIFF` file, which can be converted into the `geogrid` format using a utility included with the extended source distribution from `openwfm.org`, called `convert_geotiff`. This procedure is explained in detail at `http://www.openwfm.org/wiki/How_to_run_WRF-Fire_with_real_data` .

The final piece of surface data needed for input into `geogrid` is a categorical field describing the properties of the fuels. In the U.S., this data is readily available from the USGS; however, no such data exists for the Harmanli region. Instead, we create this field using data from the Corine Landcover Project (financed by the European Environment Agency and the member states). This project provides landcover data for Bulgaria with 100 m resolution with a 25 ha minimum mapping unit (`http://www.eea.europa.eu/data-and-maps/data/corine-land-cover -2006-raster`). We use downloaded data along with orthophoto data from the geoportal of the Ministry of Regional Development and Public Works (MRDPW) of Bulgaria to estimate the fuel behavior throughout the domain. All rivers, lakes, villages and forest areas have been vectorized using the orthophoto images combined with CORINE2006 into a GIS vector shape file. The vectorized file provides very high accuracy of representation for non burning areas like rivers and lakes as well as areas with high burning fuel level like woods. We assess the fuel

Fig. 2. Burned area after the first four hours of the fire

behavior of every land cover category using data from the MRDPW and assign each area a fuel category using the standard 13 Anderson fuel models [1]. Table 1 gives a description of the fuel categories used in the Harmanli simulation.

This fuel level data combined with the vectorized landcover areas gives us a final shape file with attributes for each polygon fuel level. We again use Quantum GIS to rasterize the shape file into a 10 m resolution `GeoTIFF` file, which we convert into the `geogrid` format as described above. Along with WPS's standard global datasets, we place the newly created files in the WPS working directory and run the `geogrid` binary. The resulting input files contain all the standard WRF fields along with several additional variables generated from the high resolution topography and fuel categories.

4   Simulation Results

The atmospheric model was run with two domains. The outer one with 250m resolution consists of $180 \times 180 \times 41$ grid points, while the inner one with 50m resolution consists of $221 \times 221 \times 41$ grid points. The fire model (coupled with the inner domain) runs with mesh step 5m. The time step for the inner domain and the fire model was 0.3s, while for the outer domain it was 1.5s.

The fuel data is in Fig. 1. The simulated fire spread is shown in Fig. 2. The actual final burn area is shown in Fig. 3 for comparison. Fig. 4 shows the atmospheric flow above the fire.

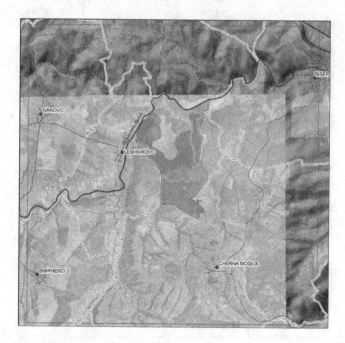

Fig. 3. Map of the burned area and the fire perimeter

Table 2. Execution times divided by simulation time for increasing number of processor cores. The fraction is given separately for the 3 components of the simulation: fire model and the two atmospheric model domains. The outer domain time includes communication between the atmospheric domains.

Cores	6	12	24	36	60	120	240	360	480	720	960	1200
Fire	1.91	1.08	0.50	0.34	0.22	0.13	0.08	0.06	0.06	0.04	0.10	0.04
Inner domain	6.76	7.05	2.90	2.06	1.20	0.73	0.45	0.32	0.26	0.23	0.24	0.17
Outer domain	0.00	0.00	0.00	0.02	0.02	0.04	0.04	0.06	0.06	0.08	0.07	0.15
Total	10.59	9.21	3.91	2.75	1.64	0.99	0.61	0.44	0.37	0.31	0.44	0.26

5   Parallel Performance

The speed of the simulation is essential, because only a model that is faster than real time can be used for forecasting. We have performed computations on the Janus cluster at the University of Colorado. The computer consists of nodes with dual Intel X5660 processors (total 12 cores per node), connected by QDR InfiniBand. Table 2 and Figure 5 show that the coupled model is capable of running faster than real time, and the performance scales well with the number of processors. The model runs slightly faster than real time (1s of simulation time takes 0.99s to compute) on 120 cores.

Fig. 4. The heat flux (red is high), the burned area (black), and the atmospheric flow (purple is over 10m/s). Note the updraft caused by the fire. Ground image from Google Earth.

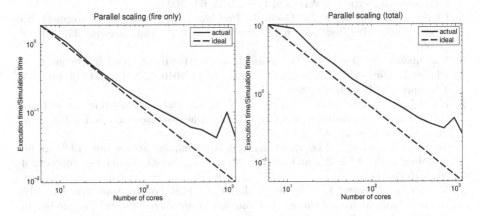

Fig. 5. Parallel scalability data plotted from Table 2

6   Conclusion

We have demonstrated wildfire simulation based on real data in Bulgaria from satellite measurement and existing GIS databases. While the simulation provides a reasonable reproduction of the fire spread, further refinement is needed. Data assimilation [5, 7] will also play an important role. As seen from comparison of Figures 2 and 3, there is a good, but not not complete, agreement of the simulated and the real burned area. Despite this, the simulation showed correct fire line propagation, and it can give forecast and valuable information for future firefighting actions in different areas with different meteorological conditions. The model can perform faster than real time at the required resolution, thus satisfying one basic requirement for a future use for prediction.

Acknowledgments. This work was partially supported by the National Science Fund of the Bulgarian Ministry of Education, Youth and Science under Ideas Concourse Grant DID-02-29 "Modelling Processes with Fixed Development Rules (ModProFix)," ESF project number BG51PO001-3.3.04-33/28.08.2009, U.S. National Science Foundation (NSF) under grant AGS-0835579, and by U.S. National Institute of Standards and Technology Fire Research Grants Program grant 60NANB7D6144. Computational resources were provided by NSF grant CNS-0821794, with additional support from UC Boulder, UC Denver, and NSF sponsorship of NCAR.

References

1. Anderson, H.E.: Aids to determining fuel models for estimating fire behavior. General Technical Report INT-122, US Department of Agriculture, Forest Service, Intermountain Forest and Range Experiment Station (1982),
   http://www.fs.fed.us/rm/pubs_int/int_gtr122.html
2. Clark, T.L., Coen, J., Latham, D.: Description of a coupled atmosphere-fire model. International Journal of Wildland Fire 13, 49–64 (2004)
3. Clark, T.L., Jenkins, M.A., Coen, J., Packham, D.: A coupled atmospheric-fire model: Convective feedback on fire line dynamics. J. Appl. Meteor. 35, 875–901 (1996)
4. Dobrinkova, N., Jordanov, G., Mandel, J.: WRF-Fire Applied in Bulgaria. In: Dimov, I., Dimova, S., Kolkovska, N. (eds.) NMA 2010. LNCS, vol. 6046, pp. 133–140. Springer, Heidelberg (2011)
5. Mandel, J., Beezley, J.D., Coen, J.L., Kim, M.: Data assimilation for wildland fires: Ensemble Kalman filters in coupled atmosphere-surface models. IEEE Control Systems Magazine 29, 47–65 (2009)
6. Mandel, J., Beezley, J.D., Kochanski, A.K.: Coupled atmosphere-wildland fire modeling with WRF 3.3 and SFIRE 2011. Geoscientific Model Development 4, 591–610 (2011)
7. Mandel, J., Beezley, J.D., Kondratenko, V.Y.: Fast Fourier transform ensemble Kalman filter with application to a coupled atmosphere-wildland fire model. In: Gil-Lafuente, A.M., Merigo, J.M. (eds.) Computational Intelligence in Business and Economics, Proceedings of MS 2010, pp. 777–784. World Scientific (2010), also available as arXiv:1001.1588
8. Rothermel, R.C.: A mathematical model for predicting fire spread in wildland fires. USDA Forest Service Research Paper INT-115 (1972),
   http://www.treesearch.fs.fed.us/pubs/32533
9. Skamarock, W.C., Klemp, J.B., Dudhia, J., Gill, D.O., Barker, D.M., Duda, M.G., Huang, X.Y., Wang, W., Powers, J.G.: A description of the Advanced Research WRF version 3. NCAR Technical Note 475 (2008),
   http://www.mmm.ucar.edu/wrf/users/docs/arw_v3.pdf

A Computational Approach for Remediation Procedures in Horizontal Subsurface Flow Constructed Wetlands

Konstantinos Liolios[1], Vassilios Tsihrintzis[1], Konstantinos Moutsopoulos[1], Ivan Georgiev[2], and Krassimir Georgiev[3]

[1] Democritus University of Thrace, Department of Environmental Engineering, Laboratory of Ecological Engineering and Technology, Xanthi, Greece
`kliolios@env.duth.gr`
[2] Bulgarian Academy of Sciences, Institute of Mathematics and Informatics, Acad. G. Bonchev Str., Bl. 8, 1113 Sofia, Bulgaria
`john@parallel.bas.bg`
[3] Bulgarian Academy of Sciences,
Institute of Information and Communication Technologies,
Acad. G. Bonchev Str., Bl. 25a, 1113 Sofia, Bulgaria
`georgiev@parallel.bas.bg`

Abstract. A large-scale computational approach for groundwater flow and contaminant transport and removal in porous media is presented. Emphasis is given to remediation procedures in horizontal subsurface flow constructed wetlands. For the numerical procedure, the MODFLOW computer code family is used. Application is made for the simulation of horizontal subsurface flow wetlands pilot-scale units, constructed and operated in Democritus University of Thrace, Xanthi, Greece. The effects of the inlet and outlet recharge positions to the optimum contaminant removal are also numerically investigated.

Keywords: Computational Groundwater Flow, Contaminant Transport and Removal, Constructed Wetlands, Remediation Procedures.

1 Introduction

As Environmental Engineering praxis has shown [1–3], numerical simulations of groundwater flow and contaminant transport and removal through porous media require usually large-scale computational procedures. This holds also for constructed wetlands (CW), which are recently used [4, 5] as a good alternative solution for small settlements in order to treat municipal wastewater and to remedy contaminated soils. So it seems necessary, in parallel to experimental research, to investigate computationally the optimal design characteristics of CW, in order to maximize the removal efficiency and keep their area and construction cost to a minimum. In a previous study [15], the optimal contaminant inlet positions in horizontal subsurface flow CW has been numerically investigated. In the present paper, a numerical simulation of underground flow and contaminant removal in porous media is also presented. Emphasis to horizontal subsurface flow

I. Lirkov, S. Margenov, and J. Waśniewski (Eds.): LSSC 2011, LNCS 7116, pp. 299–306, 2012.

in constructed wetlands under Mediterranean conditions is given. The objective here is to study numerically the removal of BOD (Biochemical Oxygen Demand) in an already contaminated unit, and so to obtain an estimate for the effectiveness of remediation procedures [1, 14]. First the mathematical modelling of the problem is formulated. Next, for the numerical simulation, the Visual MODFLOW computer family code [10] is used. Further, the numerical procedure is applied for the simulation of pilot-scale units of horizontal subsurface flow wetlands. These pilot-scale units were constructed and operated in the Laboratory of Ecological Engineering and Technology, Department of Environmental Engineering, Democritus University of Thrace (DUTh), in Xanthi, Greece, see [6, 7]. A control of the simulation in the wetlands is done by comparison to existing experimental results. Finally, the effects of the inlet and outlet recharge positions to the optimum remediation procedure have been numerically investigated.

2   Methods and Materials

2.1   The Mathematical Modeling

As well known - see e.g. [3, 8, 9, 15] - the partial differential equation describing the fate and transport of contaminants of species k in 3-D, transient groundwater flow systems can be written, using tensorial notation ($i, j = 1, 2, 3$) as follows:

$$\frac{\partial(\theta C^k)}{\partial t} = \frac{\partial}{\partial x_i} \left(\theta D_{ij} \frac{\partial C^k}{\partial x_j} \right) - \frac{\partial}{\partial x_i} \left(\theta v_i C^k \right) + q_v C_s^k + \sum R_n \qquad (1)$$

where: θ = porosity of the subsurface medium, dimensionless; C^k = dissolved concentration of species k, in $[ML^{-3}]$; D_{ij} = hydrodynamic dispersion coefficient tensor in $[L^2 T^{-1}]$; v_i = seepage or linear pore water velocity in $[LT^{-1}]$, it is related to the specific discharge or Darcy flux through the relationship, $v_i = q_i \theta$; q_v = volumetric flow rate per unit volume of aquifer representing fluid, sources (positive) and sinks (negative) in $[T^{-1}]$; C_s^k = concentration of the source or sink flux for species k in $[ML^{-3}]$; ΣR_n = chemical reaction term in $[ML^{-3}T^{-1}]$. This last term, in the simplest linear case, depends on the first-order removal coefficient λ and is equal to $(-\lambda \theta C)$.

The above Equation (1) is the governing equation underlying in the transport model and contaminant removal. This required velocity v_i is computed through the Darcy relationship:

$$v_i = -\frac{K_{ij}}{\theta} \frac{\partial h}{\partial x_j} \qquad (2)$$

where: K_{ij} = a component of the hydraulic conductivity tensor in $[LT^{-1}]$; h = hydraulic head in $[L]$.

The hydraulic head $h = h(x_i; t)$ is obtained from the solution of the three-dimensional groundwater flow equation:

$$\frac{\partial}{\partial x_i} \left(K_i \frac{\partial h}{\partial x_j} \right) + q = S_y \frac{\partial h}{\partial t} \qquad (3)$$

where: S_y = the specific yield of the porous materials; q = the volumetric flow rate per unit area of aquifer representing fluid sources (positive) and sinks (negative) in $[LT^{-1}]$, when precipitation and evapotranspiration effects are taken in to account, respectively.

So, the above equations (1)-(3), combined with appropriate initial and boundary conditions, describe the 3-dimensional flow of groundwater and the transport and removal of contaminants in a heterogeneous and anisotropic medium. Thus, for the case of one only ($k = 1$) pollutant species, the unknowns of the problem are the following five space-time functions: - the hydraulic head: $h = h(x_i; t)$, - the three velocity components: v_i and the concentration: $C = C(x_i; t)$.

2.2   The Numerical Simulations

The constructed wetlands usually have a rectangular scheme. So, for the numerical solution of the above problem, the Finite Difference Method is chosen among the other numerical methods. This method is the basis for the computer code family MODFLOW (Modular Finite-Difference Ground-Water Flow Model), [10]. This family code, accompanied by packages, is widely used for the simulation of groundwater flow and mass transport, see e.g. [1–3].

In the present study, use is made of the MODFLOW code accompanied by the effective computer package MT3DMS module.(Modular 3-D Multi-Species Transport Model for Simulation of Advection, Dispersion and Chemical Reactions of Contaminants in Griundwater Systems). This combination is used for the analysis of the pilot-scale units in DUTh. Double discretization, in space and in time, takes place. At each time-step, a large system of linear equations has to be solved. For this purpose, the Strongly Implicit Procedure (SIP) is used. For more details see e.g. [1–3].

3   The Investigated Case of Pilot-Scale Wetland Units in DUTh

3.1   DUTh Pilot-Scale Units Description

In the Laboratory of Ecological Engineering and Technology, Department of Environmental Engineering, Democritus University of Thrace (DUTh), Xanthi, Greece, five similar pilot-scale horizontal subsurface flow constructed wetlands (CW) have been constructed and are in operation [6, 7]. A schematic view of the experimental layout is shown in Figure 1 and Figure 2. They are rectangular tanks made of steel, with dimensions $L = 3m$ long, $0.75m$ wide and $1m$ deep. The wetland units are equipped with inlet and outlet hydraulic structures, similar to those used in real systems. These five pilot-scale units were operated continuously from January 2004 until January 2006 in parallel experiments, in order to investigate the effect of temperature, hydraulic residence time (HRT), vegetation type and porous media material and grain size on the performance of horizontal subsurface flow (HSF) constructed wetlands treating wastewater.

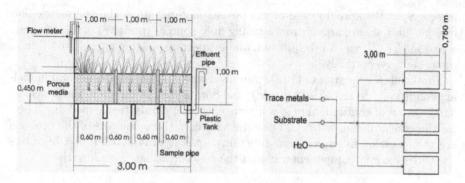

Fig. 1. Schematic section along one wetland **Fig. 2.** General layout of the facility
tank   (plan)

Three of the above five units contained medium gravel (MG) obtained from
a quarry. The other two units contained fine gravel (FG) the first and cobbles
(CO) the second, both obtained from a river bed. The two units (MG) were
planted: The first with common reeds (R, Phragmites australis), the second one
with cattails (C, Typha latifolia) and the third one was kept unplanted (Z). The
other two units were planted with common reeds. Planting and porous media
combinations were appropriate for comparison of the effect of vegetation and
media type on the function of the system. Synthetic wastewater was introduced
in the units. For more details concerning the experimental results see [6, 7].

3.2   Representative Results of the Numerical Simulation

The proposed numerical approach, based on MODFLOW family codes, has been
applied to the tank (MG-R). Available literature and laboratory estimates for
various hydraulic parameters and parameters of mass transfer, see e.g. [11], and
proper boundary and initial conditions, are used. The BOD concentration at
selected points in the tank MG-R are evaluated. So, for this tank, with length
L = 3 m, the following data are used: Confined aquifer storage: $S_s = 10^{-5} l/m$;
Unconfined aquifer storage: $S_y = 0.37$; Effective and total porosity: Eff.Por. =
Tot.Por. = 0.37; Hydraulic conductivity: $K = 0.345 m/s$; Diffusion coefficient:
Diff. Coef. = $0.0000036 m^2/hr$; Longitudinal dispersion: Long. Disp. = $0.027 m$;
Horizontal / Longitudinal dispersion = Vertical / Longitudinal dispersion =
0.01566; Initial hydraulic head: $0.45 m$; Finally, the first-order reaction rate for
BOD removal, depending on temperature, is $\lambda = RC1 = 0.125 (days)^{-1}$; as
estimated in Liolios [12].

3.3   Calibration Procedure

First a sensitivity analysis by applying and calibrating MODFLOW has been realized in [12]. The relative optimal space discretization has resulted in 8400 cells.

For each cell, values of the previously mentioned (in section 2.1) five quantities, i.e. h, C, v_i, (where $i = 1, 2, 3$) have to be computed. So, in each time step, a system of $5 \times 8400 = 42000$ unknowns has to be solved. This is obtained by using the iterative SIP procedure (Strong Implicit). The total time (hydraulic residence time, HRT) was one month, i.e. $30 \times 24 = 720$ hours, and the selected time step was $\Delta t = 15$ minutes.

Next, based on the above experimental results [6, 7] for both units, (MG-R) and (MG-Z), an estimation for the optimal range of the values for the first order reaction rate λ for BOD removal, depending on temperature and hydraulic retention time, has been obtained in Liolios [12] by using inverse problem methodologies [13].

Further, the calibration for MODFLOW code concerning the tank (MG-R) is realized by taking as boundary inlet condition for the BOD input concentration: $C_{in} = 360mg/L$ and as initial background concentration a zero one $(C = 0)$.

Representative numerical results are first shown in Figure 3 and Figure 4. These results concern inlet $(C_{in} = 360mg/L)$ and outlet concentrations (final $C_{out} = 59.84mg/L$) versus time (in hours, hr). Concentration values are computed at selected unit points, for which experimental results are available. These points were the places of observation wells at the tank entrance $x_0 = 0$, at distances $x_1 = L/3$, $x_2 = 2L/3$ along the tank length $L = 3m$ from the entrance, and at the tank outlet $x_L = 3m$. At the entrance, the monitoring positions A, B, C were in the corresponding levels: $z_A = 44cm$, $z_B = 22cm$, $z_C = 1cm$. At distances x_1 and x_2, the monitoring positions had levels of 22 cm. At the outlet, the monitoring position had a level of 1 cm.

The above computed results are in good agreement with the experimental measured ones, given in [1, 2] and shown in Figure 5. So, the effectiveness of the proposed numerical procedure is verified.

3.4   Remediation Procedure

Further, in order to investigate the remediation effectiveness, the tank (MG-R) is considered as already fully contaminated by initial $300mg/L$ BOD and subjected to a continuous inlet concentration $300mg/L$ BOD.

The effects of the inlet and outlet recharge positions to the optimum remediation effectiveness have been computed. For this purpose, the following scenarios concerning the inlet places of the pollutant have been treated:

Case 1: Inlet of the whole pollutant at the entrance of the tank: $x_{in} = 0.0m$

Case 2: Inlet of the 60% of pollutant at the entrance of the tank and 40% at the place: $x_1 = 1m(L/3)$.

Case 3: Inlet of the 60% of pollutant at the entrance of the tank, 25% at the place $x_1 = 1m(L/3)$ and 15% at the place $x_2 = 2m(2L/3)$.

Case 4: Inlet of the 80% of pollutant at the entrance of the tank and 20% at the place: $x_1 = 1m(L/3)$.

Case 5: Inlet of the 90% of pollutant at the entrance of the tank and 10% at the place: $x_1 = 1m(L/3)$.

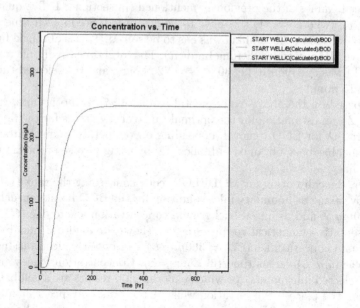

Fig. 3. Calibration procedure: Concentrations at the entrance (upper curve, monitoring position: $z_A = 44$ cm, middle curve: $z_B = 22$ cm, lower curve: $z_C = 1$ cm)

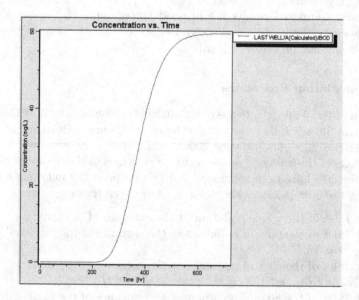

Fig. 4. Calibration procedure: Concentrations at the outlet (monitoring position: $z_C = 1$ cm)

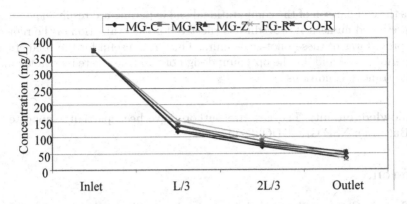

Fig. 5. Calibration procedure: Experimental average values BOD along the pilot units

Table 1. Remediation procedure: Concentrations C_{out} (in mg/L) at the tank outlet

λ	0.10	0.125	0.15	0.17	0.19	0.20
Case 1	163.12	140.341	120.811	104.044	89.723	77.442
Case 2	165.65	143.452	124.203	107.852	93.849	81.627
Case 3	167.64	145.833	127.312	111.382	97.747	85.945
Case 4	163.81	141.271	121.772	105.180	90.702	78.592
Case 5	163.68	140.671	121.241	104.471	90.202	77.815

The range of values $\lambda = 0.10 - 0.20$ (in days^{-1}) for the first order reaction rate for BOD removal, depending on temperature and hydraulic residence time (HRT), has been and estimated in Liolios [12].

Next Table 1 summarizes the results concerning the outlet concentration for the above Cases 1–5. As concluded by the Table values, the Case 1 is the optimal remediation procedure. The above results concern the case of a typical pilot-scale unit, which here had dimensions $3.00 \times 0.75 \times 0.45$, in meters. For this case, as mentioned, a system of 42000 unknowns it was to be solved in each time step. Obviously, for the case of a real horizontal subsurface flow CW, which has usually dimensions of several decades meters (e.g. $100 \times 50 \times 0.60$, in meters), the system to be solved in each time step will have a greater number of unknowns. Therefore, large-scale computation procedures are needed for such realistic Environmental Engineering applications .

4   Concluding Remarks

Large-scale computational procedures have been used for simulating remediation procedures in constructed wetlands of horizontal subsurface flow. After the mathematical modelling of the problem, the Visual MODFLOW computer code family, based on the finite difference method and combined with the MT3DMS module, has been used for the numerical simulation of rectangular pilot-scale units, constructed and operated in DUTh.

As the comparison of the computational and experimental results have shown, the presented numerical procedure is effective for the evaluation of the remediation procedures of these pilot-scale units. Thus, it is useful for the prediction of the performance and for the optimum design of real constructed wetlands under Mediterranean conditions.

Acknowledgement. The last two authors have been partially supported by the Bulgarian NSF Grant DO-02-147/2008.

References

1. Bear, J., Cheng, A.H.-D.: Modeling Groundwater Flow and Contaminant Transport. Springer, Berlin (2010)
2. Batu, V.: Applied Flow and Solute Transport Modelling in Aquifers. Taylor & Francis, CRC Press, New York (2006)
3. Zheng, C., Bennett, G.D.: Applied Contaminant Transport Modelling, 2nd edn. Wiley, New York (2002)
4. Kadlec, R., Wallace, S.: Treatment Wetlands, 2nd edn. CRC Press, New York (2009)
5. Angelakis, A.N., Tschobanoglous, G.: Wastewater: Treatment and Reuse. Physical Systems. Crete University Editions, Herakleion, Greece (1995) (in greek)
6. Akratos, C.: Optimization of Design Parameters for subsurface Flow Constructed Wetlands by using Pilot-scale Units. Doctoral Dissertation, Department of Environmental Engineering, Democritus University of Thrace, Xanthi, Greece (2006) (in greek)
7. Akratos, C.S., Tsihrintzis, V.A.: Effect of temperature, HRT, vegetation and porous media on removal efficiency of pilot - scale horizontal subsurface flow constructed wetlands. Ecological Engineering J. 29, 173–191 (2007)
8. Bear, J.: Hydraulics of Groundwater. McGraw-Hill, New York (1979)
9. de Marsily, G.: Quantitative Hydrogeology. Academic Press, New York (1986)
10. Visual MODFLOW v.4.2. User's Manual. Waterloo Hydrogeologi. In (2006)
11. Spitz, K., Moreno, J.: A Practical Guide to Groundwater and Solute Transport Modeling. Wiley, New York (1996)
12. Liolios, K.: Simulation of Flow and Performance Factors for Contaminant Removal in Horizontal Subsurface Flow Constructed Wetlands. Master Thesis, Departments of Civil and Environmental Engineering, Democritus University of Thrace, Xanthi, Greece (2008) (in greek)
13. Sun, N.-Z.: Inverse Problems in Groundwater Modeling. Springer, Berlin (1994)
14. Voudrias, E.A.: Pump-and-treat remediation of groundwater contaminated by hazardous waste: Can it really be achieved? Global Nest: the International J. 3(1), 1–10 (2001)
15. Liolios, K., Tsihrintzis, V., Radev, S.: A Numerical Investigation for the Optimal Contaminant Inlet Positions in Horizontal Subsurface Flow Wetlands. In: Dimov, I., Dimova, S., Kolkovska, N. (eds.) NMA 2010. LNCS, vol. 6046, pp. 167–173. Springer, Heidelberg (2011)

Parallel Computation of Sensitivity Analysis Data for the Danish Eulerian Model

Tzvetan Ostromsky[1], Ivan Dimov[1], Rayna Georgieva[1], and Zahari Zlatev[2]

[1] Institute of Information and Communication Technologies,
Bulgarian Academy of Sciences,
Acad. G. Bonchev str., bl. 25-A, 1113 Sofia, Bulgaria
{ceco,rayna}@parallel.bas.bg, ivdimov@bas.bg
http://www.bas.bg/iict/
[2] National Environmental Research Institute,
Department of Atmospheric Environment, Frederiksborgvej 399 P.O. Box 358,
DK-4000 Roskilde, Denmark
zz@dmu.dk
http://www.dmu.dk/AtmosphericEnvironment

Abstract. Sensitivity Analysis of the Danish Eulerian Model requires
an extensive amount of output data from computationally expensive numerical experiments with a specially adapted for the purpose version
of the model, called SA-DEM. It has been successfully implemented and
run on the most powerful parallel supercomputer in Bulgaria - IBM Blue-
Gene/P. A new enhanced version, capable of using efficiently the full capacity of the mashine, has recently been developed. It will be described
in this paper together with some performance analysis and numerical
results. The output results are used to construct some mesh-functions of
ozone concentrations ratios to be used further in sensitivity analysis of
the model by using Monte Carlo algorithms.

1 Introduction

The Unified Danish Eulerian Model (UNI-DEM) is a powerful air pollution
model, used to calculate the concentrations of various dangerous pollutants and
other species over a large geographical region (4800 × 4800 km), covering the
whole of Europe, the Mediterranean and some parts of Asia and Africa. It takes
into account the main physical, chemical and photochemical processes between
the studied species, the emissions, the quickly changing meteorological conditions. This large and complex task is not suitable for direct numerical treatment.
For the purpose of numerical solution it is split into submodels, which represent
the main physical and chemical processes. The sequential splitting [5] is used in
the production version of the model, although other splitting methods have also
been considered and implemented in some experimental versions [1,4]. Spatial
and time discretization makes each of the above submodels a huge computational
task, challenging for the most powerful supercomputers available nowadays. That
is why the parallelization has always been a key point in the computer implementation of DEM since its very early stages. A coarse-grain parallelization strategy

I. Lirkov, S. Margenov, and J. Waśniewski (Eds.): LSSC 2011, LNCS 7116, pp. 307–315, 2012.

based on partitioning of the spatial domain appears to be the most efficient and well-balanced way on widest class of nowadays parallel machines (with not too many processors), although some restrictions apply. Other parallelizations are also possible and suitable to certain classes of supercomputers [7,8].

In the chemical submodel there is a number of parameters for control on the speed of the corresponding chemical reactions. By introducing some regular perturbations in these parameters we produce the necessary data to be used later in a new adaptive Monte Carlo approach to variance-based sensitivity analysis. In general, sensitivity analysis can help us to find out which simplifications can be done without significant loss of accuracy. It is also important to analyze the influence of variations of the chemical rate coefficients (and other parameters on a later stage), as there is always a certain level of uncertainty for their values. This knowledge can show us which parameters are most critical for a certain set of output results.

A special parallel version (SA-DEM) of the UNI-DEM has been created for this purpose [2,6] and efficiently implemented on the IBM BlueGene/P, the most powerful parallel machine in Bulgaria. Nevertheless, the sensitivity analysis task remains a huge computational problem, which requires enormous resources of storage and CPU time. Essential improvements of this version are made by introducing two new levels of parallelism (top-level(MPI) and bottom-level(OpenMP) respectively) in SA-DEM. They allow us to shorten many times the necessary computing time for obtaining the sensitivity analysis results and to use efficiently the IBM BlueGene/P machine up to its full capacity.

The general concept of sensitivity analysis and our utilization for the above problem is described briefly in Section 2. The mathematical background of Danish Eulerian Model and the scheme of its numerical solution are described in Section 3. Some details on parallelization of the improved SA-DEM version, performance and scalability results obtained on the IBM BlueGene/P are presented in the rest of this paper.

2   Sensitivity Analysis Concept — Sobol's Approach

Sensitivity analysis (SA) is the study of how much the uncertainty in the input data of a model (due to any reason: inaccurate measurements or calculation, approximation, data compression, etc.) is reflected in the accuracy of the output results [9]. Two kinds of sensitivity analysis are present in the existing literature, local and global. Local SA studies how much some small variations of inputs around a given value can change the value of the output. Global SA takes into account all the variation range of the input parameters, and apportions the output uncertainty to the uncertainty in the input data. Subject to our study in this paper is the global sensitivity analysis.

Several sensitivity analysis techniques have been developed and used throughout the years [9]. In general, these methods rely heavily on special assumptions connected to the behaviour of the model (such as linearity, monotonicity and additivity of the relationship between input and output parameters of the model).

Among the quantitative methods, variance-based methods are most often used. The main idea of these methods is to evaluate how the variance of an input or a group of inputs contributes to the variance of model output.

Assume that a model is represented by the following model function: $u = f(\mathbf{x})$, where the input parameters $\mathbf{x} = (x_1, x_2, \ldots, x_d) \in U^d \equiv [0,1]^d$ are independent (non-correlated) random variables with a known *joint probability distribution function*. In this way the output u becomes also a random variable (as it is a function of the random vector $\mathbf{x}$) and let $\mathbf{E}$ be its mathematical expectation. Let $\mathbf{D}[\mathbf{E}(u|x_i)]$ be the variance of the conditional expectation of u with respect to x_i and $\mathbf{D}_u$ - the total variance according to u. This indicator is called *first-order sensitivity index* by Sobol [10] or sometimes *correlation ratio*.

Total Sensitivity Index (TSI) [10] of an input parameter x_i, $i \in \{1, \ldots, d\}$ is the sum of the complete set of mutual sensitivity indices of any order (main effect, two-way interactions (second order), three-way interactions (third order), etc.):

$$S_{x_i}^{tot} = S_i + \sum_{l_1 \neq i} S_{il_1} + \sum_{l_1, l_2 \neq i, l_1 < l_2} S_{il_1 l_2} + \ldots + S_{il_1 \ldots l_{d-1}}, \tag{1}$$

where $S_{il_1 \ldots l_{j-1}}$ – j^{th} order sensitivity index for the parameter x_i ($1 \leq j \leq d$), $j = 1$: S_i – the "main effect" of x_i. According to the values of their total sensitivity indices, the input parameters are classified in the following way: very important ($0.8 < S_{x_i}^{tot}$), important ($0.5 < S_{x_i}^{tot} < 0.8$), unimportant ($0.3 < S_{x_i}^{tot} < 0.5$), irrelevant ($S_{x_i}^{tot} < 0.3$). In most practical problems the high dimensional terms can be neglected, thus reducing significantly the number of summands in (1).

The Sobol's method is one of the most often used variance-based methods. It is based on a unique decomposition of the model function into orthogonal terms (summands) of increasing dimension and zero means. Its main advantage is computing in a uniform way not only the first order indices, but also the higher order indices (in quite a similar way as the computation of the main effects). The total sensitivity index can then be calculated with just one Monte Carlo integral per factor.

The Sobol's method for global SA, applied here, is based on the so-called $HDMR^1$ (2) of the model function f (integrable) in the d-dimensional factor space:

$$f(\mathbf{x}) = f_0 + \sum_{s=1}^{d} \sum_{l_1 < \ldots < l_s} f_{l_1 \ldots l_s}(x_{l_1}, x_{l_2}, \ldots, x_{l_s}), \tag{2}$$

where f_0 is a constant. The representation (2) is not unique. Sobol has proven that under the conditions (3) for the right-hand-side functions

$$\int_0^1 f_{l_1 \ldots l_s}(x_{l_1}, x_{l_2}, \ldots, x_{l_s}) \, dx_{l_k} = 0, \quad 1 \leq k \leq s, \quad s = 1, \ldots, d \tag{3}$$

---

[1] High Dimensional Model Representation.

the decomposition (2) is unique and is called $ANOVA^2$-HDMR of the model function $f(\mathbf{x})$. Moreover, the functions of the right-hand side can be defined in a unique way by multidimensional integrals [11].

3   The Danish Eulerian Model

In this section we describe shortly the Danish Eulerian Model (DEM) [13] and its current production version UNI-DEM [12]. It is mathematically represented by the following system of partial differential equations, in which the unknown concentrations of a large number of chemical species (pollutants and other chemically active components) take part. The main physical and chemical processes (advection, diffusion, chemical reactions, emissions and deposition) are represented in that system.

$$\frac{\partial c_s}{\partial t} = -\frac{\partial(uc_s)}{\partial x} - \frac{\partial(vc_s)}{\partial y} - \frac{\partial(wc_s)}{\partial z} +$$

$$+ \frac{\partial}{\partial x}\left(K_x\frac{\partial c_s}{\partial x}\right) + \frac{\partial}{\partial y}\left(K_y\frac{\partial c_s}{\partial y}\right) + \frac{\partial}{\partial z}\left(K_z\frac{\partial c_s}{\partial z}\right) + \qquad (4)$$

$$+ E_s + Q_s(c_1, c_2, \dots c_q) - (k_{1s} + k_{2s})c_s, \quad s = 1, 2, \dots q \,.$$

where

- c_s – the concentrations of the chemical species;
- u, v, w – the wind components along the coordinate axes;
- K_x, K_y, K_z – diffusion coefficients;
- E_s – the emissions;
- k_{1s}, k_{2s} – dry / wet deposition coefficients;
- $Q_s(c_1, c_2, \dots c_q)$ – non-linear functions describing the chemical reactions between species under consideration.

The above rather complex system (4) is split (by using the most straightforward sequential splitting scheme) according to the major physical and chemical processes. Finaly, the following 3 submodels are formed:

$$\frac{\partial c_s^{(1)}}{\partial t} = -\frac{\partial(uc_s^{(1)})}{\partial x} - \frac{\partial(vc_s^{(1)})}{\partial y} + \frac{\partial}{\partial x}\left(K_x\frac{\partial c_s^{(1)}}{\partial x}\right) + \frac{\partial}{\partial y}\left(K_y\frac{\partial c_s^{(1)}}{\partial y}\right) = A_1c_s^{(1)}(t)$$

horizontal advection & diffusion

$$\frac{\partial c_s^{(2)}}{\partial t} = E_s + Q_s(c_1^{(2)}, c_2^{(2)}, \dots c_q^{(2)}) - (k_{1s} + k_{2s})c_s^{(2)} = A_2c_s^{(2)}(t)$$

chemistry, emissions & deposition

$$\frac{\partial c_s^{(3)}}{\partial t} = -\frac{\partial(wc_s^{(3)})}{\partial z} + \frac{\partial}{\partial z}\left(K_z\frac{\partial c_s^{(3)}}{\partial z}\right) = A_3c_s^{(3)}(t)$$

vertical transport

---

[2] ANalysis Of VAriances.

Spatial and time discretization of the above submodels on the EMEP[3] grid or its refinements (see Table 1) makes each of them a huge computational task. Thus the high performance and parallel computing become vital for the real-time numerical solution of the model.

The following methods are used in the numerical solution of the submodels:

- **Advection-diffusion part:** Finite elements, followed by predictor-corrector schemes with several different correctors.
- **Chemistry-deposition part:** An improved version of the QSSA (Quazi Steady-State Approximation)
- **Vertical transport:** Finite elements, followed by theta-methods.

4 UNI-DEM, the Improved Sensitivity Analysis Version SA-DEM and Their Parallel Implementation Features

The development and improvements of DEM throughout the years has lead to a variety of different versions with respect to the grid-size/resolution, vertical layering (2D or 3D model respectively) and the number of species in the chemical scheme. The most prospective of them have been united in the packege UNI-DEM. The available up-to-date versions, the selecting parameters and their optional values are shown in Table 1.

A coarse-grain parallelization strategy based on partitioning of the spatial domain in strips or blocks is currently used in UNI-DEM. For the purpose of this study, the strip-based distributed memory parallelization of the model via MPI is used [3,7,14]. It is based on partitioning of the horizontal grid, which implies certain restrictions on the number of MPI tasks and requires communication on each time step. Improving the data locality for more efficient cache utilization is achieved by using *chunks* to group properly the small tasks in the chemistry-deposition and vertical exchange stages. Additional pre-processing and post-processing stages are needed for scattering the input data and gathering the results, causing some overhead.

SA-DEM is a modification of UNI-DEM, specially adjusted to be used in the first stage of our sensitivity analysis concept (see [2]). There are additional input

Table 1. User-determined parameters for selecting an appropriate UNI-DEM version

Parameter	Description	Optional values		
NX = NY	Grid size (Grid step)	96 × 96 (50 km)	288 × 288 (16.7 km)	480 × 480 (10 km)
NZ	# layers (2D/3D)	1 or 10		
NEQUAT	# chem. species	35, 56 or 168		

---

[3] European Monitoring and Evaluation Programme.

parameters in the main program, allowing the user to set some changes of the parameters subject to sensitivity analysis deeply in the code. These are constants in the original model and normaly there is no direct user access to their values. In our particular sensitivity analysis study regular perturbations have to be done on some chemical rate coefficients in the chemistry submodel. These coefficients must be modified on the course of the SA experiments, either separately or in groups in dependence with the dimension of the particular sensitivity analysis study. That is a typical SIMD[4] task, if considering the coarsest possible level of the strusture of our algorithm. By using it we introduce a new, higher level of parallelism in SA-DEM on the top of the grid-partitioning level, the basis for distributed-memory MPI parallelization in UNI-DEM.

Our target hardware can optionally offer a limited amount of shared memory parallelism. In order to exploit it efficiently, we introduced an additional (finer-grain) level of parallelism in our algorithm by using OpenMP standard directives.

All three levels of parallelism can be used efficiently in the calculations of the necessary data for sensitivity analysis on a powerful Blue Gene/P computing system. This is shown by experiments in the next section. Finally, for extracting the ozone mean monthly concentrations and computing the necessary mesh functions an additional program was developed. The last task is much simpler and not computationally intensive, so currently we left it beyond the scope of our highly parallel supercomputer implementation.

5 Numerical Experiments on the IBM Blue Gene/P

In this section we present some execution times and speed-ups in order to show the scalability of SA-DEM on the Bulgarian IBM Blue Gene/P , the main computing platform used in our sensitivity analysis study. The IBM Blue Gene/P is a state-of-the-art high-performance system with 8192 CPU in total and theoretical peak performance more than 23 TFLOPS. It consists of 2048 compute cards (nodes), each of them being a quard core PowerPC 450 (4 CPU, 850 MHz, 2 GB RAM). A single compute card is in fact a 4-CPU shared-memory computational unit with possible multithreading support via OpenMP. It can be used in 3 different modes: VN, DUAL and SMP. With respect to the MPI parallelism there are 4 MPI processes per node in VN mode, 2 - in DUAL mode, and one in VN mode. Thus, in the last two cases the machine offers limited, but natural from hardware viewpoint shared memory parallelism, exploited on the lowest (finer-grain) level in the new implementation of SA-DEM, as mentioned above. There is 8 MB L3 cache per node, 32 KB L1 cache per CPU (private).

The results of 20-sample one-year experiments with the SA-DEM (on the 2D medium resolution spatial grid (96 x 96 x 1)), executed on the Blue Gene/P are presented in Table 2 below.

The load managing policy of this huge parallel system is based on allocating whole number of planes per job (a multiple of 128 nodes). Therefore it does not encourage submission of long jobs that use considerably less nodes, as this

---

[4] Single Instruction Multiple Data, according to Flynn's taxonomy (1966).

Table 2. Time (T) in seconds and **speed-up (Sp)** of SA-DEM with MPI parallelism on the Bulgarian IBM Blue Gene/P (in VN mode)

# CPU	Time and **speed-up** of SA-DEM on the IBM Blue Gene/P (96 × 96 × 1) grid, 35 species, CHUNKSIZE=48							
	Advection		Chemistry		Comm.	I/O	TOTAL		
	time [s]	**(Sp)**	time [s]	**(Sp)**	time [s]	time [s]	time [s]	**(Sp)**	E [%]
40	3410	(**40**)	15925	(**40**)	94	1116	20733	(**40**)	100%
80	1715	(**79**)	7948	(**80**)	99	1151	11000	(**75**)	94%
120	1154	(**118**)	5291	(**120**)	138	1051	7664	(**108**)	90%
160	870	(**157**)	3983	(**160**)	137	1076	6204	(**134**)	84%
240	586	(**233**)	2643	(**241**)	140	1107	4562	(**182**)	76%
320	464	(**294**)	1974	(**323**)	153	1131	3810	(**218**)	68%
480	344	(**396**)	1321	(**482**)	221	1651	3659	(**227**)	47%
640	283	(**482**)	985	(**647**)	176	1973	3473	(**239**)	37%
960	206	(**662**)	656	(**971**)	172	1972	3114	(**266**)	28%

Table 3. Time (T) in seconds and **speed-up (Sp)** of SA-DEM with both MPI and OpenMP parallelism on the Bulgarian IBM Blue Gene/P

# CPU	MPI p-s × OMP thr.	MODE	Time and **speed-up** of SA-DEM (MPI+OpenMP) on the IBM Blue Gene/P (96 × 96 × 1) grid, 35 species, CHUNKSIZE=48						
			Advection		Chemistry		TOTAL		
			T [s]	**(Sp)**	T [s]	**(Sp)**	T [s]	**(Sp)**	E [%]
40	40 × 1	VN	3410	(**40**)	15925	(**40**)	20733	(**40**)	100%
80	40 × 2	DUAL	1778	(**77**)	7972	(**80**)	11295	(**73**)	92%
160	80 × 2	DUAL	889	(**153**)	3960	(**161**)	6153	(**135**)	84%
240	120 × 2	DUAL	647	(**211**)	2655	(**240**)	4712	(**176**)	73%
320	160 × 2	DUAL	502	(**271**)	1978	(**322**)	4006	(**207**)	65%
480	240 × 2	DUAL	358	(**381**)	1329	(**479**)	3418	(**243**)	51%
640	160 × 4	SMP	223	(**612**)	997	(**639**)	2768	(**300**)	47%
960	480 × 2	DUAL	218	(**626**)	659	(**967**)	2684	(**309**)	32%
960	240 × 4	SMP	158	(**863**)	667	(**955**)	2292	(**362**)	38%
1280	320 × 4	SMP	122	(**1118**)	499	(**1277**)	2109	(**393**)	31%
1920	480 × 4	SMP	99	(**1378**)	338	(**1885**)	2568	(**323**)	17%
2560	640 × 4	SMP	83	(**1643**)	332	(**1919**)	2182	(**380**)	15%
3840	960 × 4	SMP	58	(**2352**)	168	(**3792**)	1653	(**502**)	13%

would be a waste of resourse. In our experiments the runs start from 40 CPU. In order to obtain comparable figures and correct scalability results we calculate the speed-up in the next tables by using the following formula (assuming that the speed-up on 40 processors is 40):

$$Sp(n) = 40 \, \frac{T(n)}{T(40)} \qquad (5)$$

where n is the number of processors (given in the first column). The time and the **speed-up (Sp)** of the main computational stages and in total are given in separate columns. The last column contains also the total efficiency E (in percent), where $E = 100\, Sp(n)/n\,\%$.

The total time includes also the MPI communication time as well as the time for some I/O procedures, which are not parallelizable. Moreover, the larger the number of MPI tasks, the more I/O device conflicts arise, which results in a significant drop-down in the total efficiency. I/O device access appear to be the performance bottleneck in this case, partially avoided by using the lowest level OpenMP parallelisation (see Table 3). On the other hand, the computational stages scale pretty well, even the speed-up of the chemistry stage tends to be slightly superlinear (due to the cache memory effects).

6   Conclusions and Plans for Future Work

We consider a 3-stage variance-based sensitivity analysis method. For the purpose of sensitivity analysis of the Danish Eulerian Model with respect to variation of certain chemical rate coefficients, a special version of the model has been developed and implemented efficiently on the IBM Blue Gene/P (called SA-DEM). Experiments, showing its scalability and efficiensy on a huge parallel system (IBM Blue Gene/P) are presented in this paper.

The first stage of our 3-stage sensitivity analysis method is completed by extracting from the output results some mean monthly concentrations of the ozone and producing the necessary mesh functions. The second stage of this sensitivity analysis research includes approximation of the mesh functions by polynomials of 3-rd / 4-th degree or by cubic B-spline functions. A Monte Carlo integration method is furtherly applied to these functions on the third stage. The results of the last two stages are presented in another paper.

Our near future plans include:

- Optimization of the I/O operations in order to overcome the bottleneck, causing a significant efficiency dropdown;
- Extending the abilities of SA-DEM (including experiments with more chemical species and on finer resolution grids (storage-permitting);
- Extending the scope of the sensitivity analysis study with respect to the emission levels and the boundary conditions.

Acknowledgments. This research is partly supported by the Bulgarian NSF grants Bulgarian NSF Grants DCVP02/1/2010 (SuperCA++) and DTK 02/44/ 2009.

References

1. Dimov, I., Faragó, I., Havasi, Á., Zlatev, Z.: Operator splitting and commutativity analysis in the Danish Eulerian Model. Math. Comp. Sim. 67, 217–233 (2004)
2. Dimov, I., Georgieva, R., Ivanovska, S., Ostromsky, T., Zlatev, Z.: Studying the Sensitivity of the Pollutants Concentrations Caused by Variations of Chemical Rates. Journal of Computational and Applied Mathematics 235(2), 391–402 (2010)
3. Dimov, I., Georgiev, K., Ostromsky, T., Zlatev, Z.: Computational challenges in the numerical treatment of large air pollution models. Ecological Modelling 179, 187–203 (2004)
4. Dimov, I., Ostromsky, T., Zlatev, Z.: Challenges in using splitting techniques for large-scale environmental modeling. In: Faragó, I., Georgiev, K., Havasi, Á. (eds.) Advances in Air Pollution Modeling for Environmental Security. NATO Science Series, vol. 54, pp. 115–132. Springer, Heidelberg (2005)
5. Marchuk, G.I.: Mathematical modeling for the problem of the environment. Studies in Mathematics and Applications, vol. 16. North-Holland, Amsterdam (1985)
6. Ostromsky, T., Dimov, I., Georgieva, R., Zlatev, Z.: Sensitivity Analysis of a Large-Scale Air Pollution Model: Numerical Aspects and a Highly Parallel Implementation. In: Lirkov, I., Margenov, S., Waśniewski, J. (eds.) LSSC 2009. LNCS, vol. 5910, pp. 197–205. Springer, Heidelberg (2010)
7. Ostromsky, T., Zlatev, Z.: Parallel Implementation of a Large-Scale 3-D Air Pollution Model. In: Margenov, S., Waśniewski, J., Yalamov, P. (eds.) LSSC 2001. LNCS, vol. 2179, pp. 309–316. Springer, Heidelberg (2001)
8. Ostromsky, T., Zlatev, Z.: Flexible Two-Level Parallel Implementations of a Large Air Pollution Model. In: Dimov, I.T., Lirkov, I., Margenov, S., Zlatev, Z. (eds.) NMA 2002. LNCS, vol. 2542, pp. 545–554. Springer, Heidelberg (2003)
9. Saltelli, A., Tarantola, S., Campolongo, F., Ratto, M.: Sensitivity Analysis in Practice: A Guide to Assessing Scientific Models. Halsted Press, New York (2004)
10. Sobol, I.M.: Sensitivity estimates for nonlinear mathematical models. Mathematical Modeling and Computational Experiment 1, 407–414 (1993)
11. Sobol, I.M.: Global Sensitivity Indices for Nonlinear Mathematical Models and Their Monte Carlo Estimates. Mathematics and Computers in Simulation 55(1-3), 271–280 (2001)
12. WEB-site of the Danish Eulerian Model, http://www.dmu.dk/AtmosphericEnvironment/DEM
13. Zlatev, Z.: Computer treatment of large air pollution models. Kluwer (1995)
14. Zlatev, Z., Dimov, I., Georgiev, K.: Three-dimensional version of the Danish Eulerian Model. Zeitschrift für Angewandte Mathematik und Mechanik 76(S4), 473–476 (1996)

Implementation of Two Different Shadow Models into EULAG Model: Madrid Case Study

R. San Jose[1], J.L. Perez[1], and R.M. Gonzalez[2]

[1] Environmental Software and Modelling Group, Computer Science School,
Technical University of Madrid (UPM),
Campus de Montegancedo, Boadilla del Monte, 28660 Madrid, Spain
roberto@fi.upm.es
[2] Department of Meteorology and Geophysics, Faculty of Physics,
Complutense University of Madrid (UCM),
Ciudad Universitaria, 28040, Madrid, Spain

Abstract. Two shadow models are compared focusing on the energy fluxes. The first one uses a simplified geometry of the building, this approach is included in the Urban Canopy Model (UCM). The second one is a new three-dimensional urban solar radiation model (SHAMO) which has been developed by the authors of the present contribution. It has been developed to calculate short wave radiation over urban high resolution grids.

The data produced by the urban solar radiation model has been used in large scale numerical experiments to simulate turbulent fluxes for urban areas; in this contribution over Madrid (Spain) city. We have applied a modified version of the EULAG (UCAR) micro scale model (CFD) which includes an energy balance equation to obtain the urban atmosphere/biosphere energy exchange. Results of the micro scale simulations and sensitivity analysis related to the solar radiation approach will be presented in this paper.

1 Introduction

The short wave radiation is affected by the shadows from buildings and reflections. In some cases the net radiation between buildings could be substantially higher than that received in the roof surface, demonstrating that the radiation is trapped inside of the canyons. These effects can be simulated with a simplified geometry of the buildings, this approach is included in the Urban Canopy Model (UCM), National Centre for Atmospheric Research (NCAR, US) [1]. The simplified urban geometry consists of a two-dimensional approach with symmetrical street canyons with infinite length.

The UCM approach is a good initial and ideal approach to develop micro urban meteorological simulations over urban domains. However, the real urban structures are not simply geometric shaped and have no uniformity. Micro urban simulations require high detailed information about spatial and temporal distribution of short wave radiation taking into account the three-dimensional

I. Lirkov, S. Margenov, and J. Waśniewski (Eds.): LSSC 2011, LNCS 7116, pp. 316–323, 2012.

building morphology of the city and the change of solar positions, as well as the effects of multiple reflections and shading in an urban canopy.

In order to include the building morphology and all solar radiation reflections and interactions in the urban environment with a very high spatial resolution in the WRF-UCM structure, we have used a new model specifically devoted for Computational Fluid Dynamics (CFD) applications named EULAG. The computational effort needed to simulate 3D building morphology with a few meters spatial resolution by using WRF-UCM is prohibited. So, the boundary information is produced by WRF-UCM with high spatial resolution (200 m) and sent to EULAG simulation (after including our flux balance and SHAMO model) to produce detailed (4 m spatial resolution) surface flux maps.

The EULAG CFD model will be used in combination with a mesoescale meteorological model, Weather Research and Forecasting Model (WRF [10]), to provide boundary and initial conditions.

The urban land surface models UCM and NOAH have been coupled with the CFD model EULAG [2]. EULAG model does not include the flux calculation, so, it is necessary to include the energy fluxes into the system in order to have surface patterns of sensible, ground and latent heat fluxes with high spatial resolution in an urban environment, so we have developed this section. When the grid cell has a non-urban land-use type (natural), the NOAH land-surface model [11] approach is applied into EULAG and in case of a urban grid cell, the UCM scheme is used to calculate the fluxes.

2   Fluxes Calculations into EULAG

In this section, we explain how the fluxes are integrated into the EULAG CFD model. Meteorological, radiation information and surface exchange coefficient are provided by the mesoescale model WRF. The urban/natural surface scheme (NOAH/UCM) is applied over high resolution grid cells using detailed data of winds and temperature from EULAG.

New skin temperature is estimated applying the Similarity Theory and the Energy balance equation:

$$Rn - H - L - G = 0, \tag{1}$$

where, Rn is the net radiation which includes the short and long wave; H is the sensible heat flux; L is the latent heat flux and G is the ground heat flux.

$$H = \rho \; cp \; k \frac{u^*}{\psi h}(\Omega - \Omega_0) \tag{2}$$

where

Ω_0 =Air temperature at roughness length height;
ρ =air density;
Ω =Air temprature (K);
cp = Heat capacity of dry air;
$k = 0.4$ (von Karman constant);

ψh =Heat integrated universal function;
u^* =Friction velocity.

From Similarity Theory [4,6,7,9], we obtain $u*$ and ψh using two levels, $Zwrf - Zeulag$. The latent heat flux is calculated as:

$$L = \rho \; E1 \; k \frac{u^*}{\psi h}(q - q_0) \tag{3}$$

where
$\rho =$ air density;
$q0 =$ WRF surface specific humidity (at roughness length);
$El =$ heat capacity of vaporation;
$q = 0.622 * ES(PS(saturation pressure) - 0.378 * ES(saturation vapor pressure))$.

Ground heat flux:

$$G = Thermal \quad conductivity * \left(\frac{\Omega(surface) - \Omega(layer\,1)}{(depth/2)} \right) \tag{4}$$

Radiation budget:

$$Rn = (1 - \alpha)short\_wave\_radiation + \text{emissivity} * long\_wave\_radiation - \sigma\Omega^4 \tag{5}$$

Short wave and long wave radiation is affected by the shadows and reflections.

3   Shadow Models

To calculate the downward short wave flux, two shadow models are available. The first one, the simple shadow model (CANYONSHADOW) has been implemented by using an ideal canyon street. The short wave radiation has been calculated in each grid cell as:

$$SWR = SWR_i + SWR_r \tag{6}$$

SWR_i is the incident short wave radiation and SWR_r is the reflected short wave radiation. In this simple approach we have the h parameter: building height, the w parameter: street width and the ls parameter: shadow length. Each grid cell is associated to the closest building in order to obtain the h parameter. We have used the City 3D shape files with the buildings morphology in the GIS file. The shadow length is calculated as:

$$ls = h * \tan \sigma_z * \sin \sigma_n, \tag{7}$$

where σ_z is the solar zenith angle (as received from WRF-UCM) and σ_n is the street orientation (from GIS data set) in case ls $\geq$ w. If ls $\leq$ w we force ls = w since all street is under the shadow. The short wave radiation coming from WRF model is split into direct (d) and diffuse (q) solar radiation as:

$$SWR_d = SWR\_WRF * 0.8 \tag{8}$$

$$SWR_q = SWR\_WRF * 0.2,$$

where SWR_d is the direct solar radiation and SWR_q is the diffuse solar radiation. The incident solar radiation is calculated as:

$$SWR_i = SWR_d\frac{(w - ls)}{w}(1 - \alpha) + SWR_q F_{sky}(1 - \alpha) \tag{9}$$

The reflected solar radiation is calculated as:

$$SWR_r = SWR_d\frac{ls}{2h}\alpha_w F_{walltoground}(1 - \alpha) + SWR_q F_{sky}\alpha_w F_{walltoground}(1 - \alpha) \tag{10}$$

where, α is the ground albedo, α_w is the wall albedo and $F_{sky} = 1 - F_{walltoground}$.

$$F_{walltoground} = \int_{o}^{H\_building} 1 - \frac{h}{\sqrt{h^2 + w^2}} \tag{11}$$

In the case of a detailed shadow model, (MIROSYS SHADOW, SHAMO) has been developed. A new 3D urban solar radiation model has been developed by the authors, the final solar radiation has been calculated as.

$$FSR_{i,m} = \sum FSR_{k,m-1} * ALB * FREFLECTION \tag{12}$$

where, $FSR_{i,m}$ is the final solar radiation in the grid cell i, after m reflections; $FSR_{k,m-1}$ is the final solar radiation in the grid cell j after m-1 reflections; the ALB is the surface albedo and finally, FREFLECTION is the reflection factor. If m=0 the initial solar radiation is ISR (No reflections yet). The variables to be calculated to have a full description of the total short wave solar radiation in every grid cell are: sun location (x, y, z), shadows, sky view factor, direct & diffuse solar radiation, reflections and finally total solar radiation in every grid cell. The process continues by an iterative process until FSR $\leq$ 1% ISR. The ISR is calculated as:

$$ISR = FSHADOW * DIRSR + FSKY * DIFSR, \tag{13}$$

where, FSHADOW is the building shadow effect. The FSHADOW=1 in case the grid cell is receiving full solar radiation and the SHADOW=0 in case the grid cell is full in a shadow; DIRSR is the direct solar radiation (W/m2), DIFSR is the diffuse solar radiation (W/m2) and FSKY is the sky view factor (0-1). From every grid cell, a straight line between sun and the grid cell is traced. From the grid cell, we follow the straight line and if a building is found, the FSHADOW is set to 0, if not, FSHADOW is set to 1.

In Figure 1, we show an example of the Madrid shadows in 1 km x 1 km domain (our experiment) with 4 m spatial resolution on June, 28, 2008, 07:00 GMT (left) and 17:00 GMT (right) using our MICROSYS shadow module. White areas are buildings, gray areas are shadow and black are sun areas.

The FSKY & FREFLECTION variables are calculated using a 3D tracking method and casting rays from a hemisphere located on the center of each grid cell.

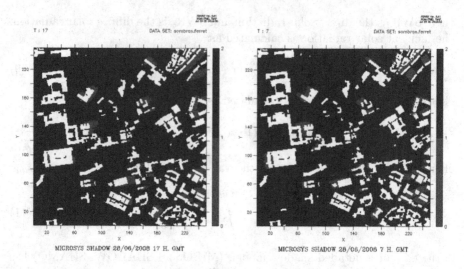

Fig. 1. Shadow module (SHAMO MODEL) results on June, 28, 2008 over a domain of 1 km x 1 km with 4 m spatial resolution. Left 7H and right 17H. The white areas are buildings.

The global total radiation is calculated as the direct plus the diffuse solar radiation. The diffuse solar radiation is calculated as the global radiation obtained from the WRF/UCM model multiplied by the turbidity factor. It is defined as the relation between extraterrestrial solar radiation and the incoming solar radiation over the horizontal plane. The turbidity factor is calculated as follows:

$$TF = MIN(1, \ 1/A) \tag{14}$$

$$A = MAX(0.1, \ B) \tag{15}$$

$$B = 2.1 - 2.8 * LOG(LOG(SOLTOP/GLOBAL)) \tag{16}$$

$$SOLTOP = 1370 * DAYFACTOR * \cos(zenith) \tag{17}$$

The ratio DIRECT/DIFFUSE solar radiation depends on the solar zenith angle and the sky conditions (cloudiness). Under high cloudiness conditions and small solar angles the ratio is close to 0.5 and with large solar angles, the ratio is close to 0.2.

4   Results and Discussion

The sensitivity study tries to assess the differences in model outputs about energy fluxes (surface results) by a simple shadow model (CANYONSHADOW) and the new 3d shadow model (SHAMO) implemented. The EULAG domain is limited to 1 km x 1 km with 100 m in height. Our objective is to have a 4 m surface resolution simulation [3,5,8]. The vertical resolution is also fixed to 4 m. The EULAG model estimates the wind and temperature at 2 m in height (4 m cubes).

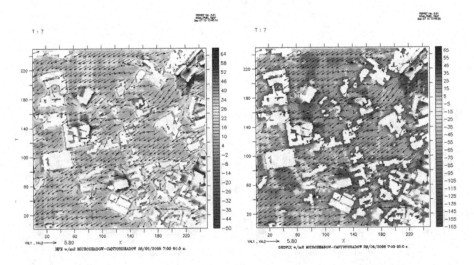

Fig. 2. Differences between the sensible heat fluxes(left) and groun heat flux (righ) obtained by the detailed shadow model (MICROSYS SHADOW) (SHAMO) and the simple shadow model (CANYONSHADOW) on June, 28, 2010 at 07h00 GMT after 60 s simulation time of EULAG model.

We have performed two simulations, one with the simple shadow model (SHADOW CANYON) and other with the new 3D shadow model (MICRO SHADOW SHAMO) to test the sensitivity of fluxes to the shadow model.

The results show important differences between different grid cells in the experiment model domain. The Northwest area of the domain has a building density lower than the Southeast area of the model domain, so that when running our shadow models the results show higher differences in the latter area.

We will show the results for two significant hours of June 28, 2008. On Fig.2 results for 7:00 AM (sunrise) are presented while on Fig.3 results for 6:00 PM (sunset) are shown.

After the results analysis, we can conclude that the positive differences are predominant for the sensible heat flux analysis. These are located around the buildings. At 7:00 AM the differences can be seen in the eastern part of the buildings and at 18:00 on the west part, which is consistent with the actual sun movement. The maximum differences reach values up to 60 W/m2, this is because some areas are being considered as shadow places by the simple shadow model, with values around -20 W/m^2. But the new 3d shadow model implemented considers that some rays of sun are coming at this area, with sensible heat flux values of about 40 W/m^2.

The ground heat flux differences follow the same procedure for the sensible heat flux, taking into account the sign convention changes. Positive ground heat flux means that the heat is moving from the ground into the subsoil. The differences are larger than in the case of the sensible heat flux differences because

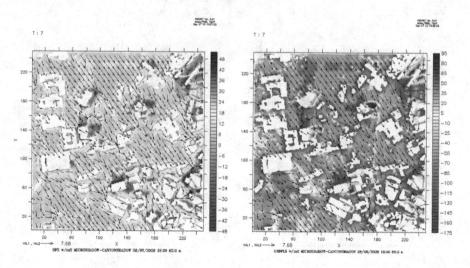

Fig. 3. Differences between the sensible heat fluxes(left) and groun heat flux (righ) obtained by the detailed shadow model (MICROSYS SHADOW) (SHAMO) and the simple shadow model (CANYONSHADOW) on June, 28, 2010 at 18h00 GMT after 60 s simulation time of EULAG model.

ground heat flux is more sensible to the solar radiation finally reaching each grid cell, which depends on the calculated shadow degree by the shadow model.

The simple shadow model uses street canyons to calculate shadows. If you apply this methodology over an urban area where there is not a clear canyon street, it introduces an uncertainty in the fluxes model, which can be solved by using the new 3D micro- shadow model implemented.

We have implemented the turbulent energy fluxes (sensible, latent and ground) into the EULAG code. We have developed a detailed shadow model to take into account the different reflections of the short wave solar radiation over the buildings in a typical urban environment. The approaches have been developed for a high resolution runs of EULAG in an urban environment (in our experiment with 4 m spatial resolution grid cells). The code has been developed according to the structure of EULAG and receiving the information (nesting approach) from the WRF/UCM model. Two different shadow models have been developed: a) a simple shadow model based on a so-called canyon street approach which calculates the short wave radiation in different locations based on height of the buildings and the width of the street; b) a second shadow model has been developed based on a much more detailed approach. This approach is based on N reflections of the solar rays over the different building faces until reaching the surface. Casting different rays from every grid cell (4 m x 4m) to sun produces information related to the sky view factor for every grid cell. The results show – as expected – a high sensitivity to the shadow model. The turbulent fluxes reflect a high sensitivity to the shadow model which has been used to determine the amount of short wave radiation reaching every grid cell in surface and building faces.

Acknowledgment. We would like to acknowledge INDRA ESPACIO for the remote data provided from different satellites and sources and for partially funding this research. We also would like to acknowledge the BRIDGE EU project FP7/2007-2013, under grant agreement 211345 which partially funded this research. Authors thankfully acknowledge the computer resources, technical expertise and assistance provided by the Centro de Supercomputación y Visualización de Madrid (CeSVIMa) and the Spanish Supercomputing Network (BSC).

References

1. Lorenc, A.C.: Analysis methods for numerical weather prediction. Q. J. R. Meteorol. Soc. 112, 1177–1194 (1986)
2. Todling, R., Cohn, S.E.: Suboptimal schemes for atmospheric data assimilation based on the Kalman filter. Mon. Wea. Rev. 122, 2530–2557 (1994)
3. Pielke, R.A.: Mesoscale Meteorological Modeling, 1st edn. Academic Press, New York (1984)
4. Grell, G.A., Emeis, S., Stockwell, W.R., Schoenemeyer, T., Forkel, R., Michalakes, J., Knoche, R., Seidl, W.: Application of the multiscale, coupled MM5/chemistry model to the complex terrain of the VOTALP valley campaign. Atmospheric Environment 34, 1435–1453 (2000)
5. Skamarock, W.C., Klemp, J.B., Dudhia, J., Gill, D., Barker, D., Wang, W., Powers, J.G.: A description of the advanced research WRF Version 2. NCAR Technical Note NCAR/TN-468+STR (2005)
6. Flanner, M.G.: Integrating anthropogenic heat flux with global climate models. Geophys. Res. Lett. 36, L02801 (2009)
7. Kusaka, H., Kondo, H., Kikegawa, Y., Kimura, F.: A simple-layer urban canopy model for atmospheric models: comparison with multi-layer and slab models. Boundary-Layer Meteorology 101, 329–358 (2001)
8. San José, R., Pérez, J.L., Morant, J.L., González, R.M.: Elevated PM10 and PM2.5 concentrations in Europe: a model experiment with MM5-CMAQ and WRF/CHEM. WI Transactions on Ecology and the Environment 116, 3–12 (2008) ISSN: 1743-3541
9. Chimklai, P., Hagishima, A., Tanimoto, J.: A computer system to support albedo calculation in urban areas. Building and Environment 39, 1213–1221 (2004)
10. Michalakes, J., Chen, S., Dudhia, J., Hart, L., Klemp, J., Middlecoff, J., Skamarock, W.: Development of a Next Generation Regional Weather Research and Forecast Model (2001)
11. Chen, F., Mitchell, K., Schaake, J., Xue, Y., Pan, H., Koren, V., Duan, Y., Ek, M., Betts, A.: Modeling of land-surface evaporation by four schemes andcomparison with FIFE observations. J. Geophys. Res., 7251–7268 (1996)

Model Simulation of Air Pollution
Due to April 2010 Iceland Volcano Eruption

D. Syrakov and M. Prodanova

National Institute of Meteorology and Hydrology,
Bulgarian Academy of Sciences, Sofia, Bulgaria
dimiter.syrakov@meteo.bg

Abstract. The eruption of Iceland volcano Eyjafjallajokull in April
2010 caused enormous disruption to air transport over Europe over a
long period of time. The losses and inconveniences for air companies,
common business and usual passengers are difficult to estimate but in
any case are rather considerable. There are few centres devoted to observation and forecast of such events. The ENSEMBLE consortium led
by European JRC in Ispra, Italy, which is aimed at elaborating ensemble forecast on the base of individual forecasts of almost all European
Emergency Response Systems (ERS) in case of nuclear accident, decided
to launch a series of exercises, devoted to simulation of the first week air
pollution dilution caused by Iceland volcano eruption. Bulgarian ERS
(BERS) was upgraded to be able to take part in these exercises. This
work presents the results of BERS participation in the exercises and
comparison with other model results.

1 Introduction

The severe atmospheric events like powerful volcano eruption do not cause serious problems for air quality at ground surface except for the neighboring territories. But sometimes they can cause serious economic problems. The disruption in
ground transport caused by snow, floods etc. paled into insignificance compared
with the havoc caused by the ash cloud spewed out by Eyjafjallajokull volcano
in Iceland. The event started on 14 April 2010. It lasted with different intensity
more than a month and ended on 23 May. The eruption has been characterized
by two main phases of intense ash emissions spanning April 14-21 and May 1-10.
    Volcanic eruptions are often referred to as violent explosions which spew huge
rocks and hot lava. Though these can be extremely damaging, a stealthier threat
lies in the air. This is because volcanic eruptions can create very large clouds of
ash that can be quite dangerous. These ash clouds contain microscopic particles
of rock and glass. When they pass through the extremely hot jet engines, they
melt into glass which coats any exposed areas. This has the potential to cause
the engines to malfunction, including total engine failure. A detail description
of these effects can be found in the report of the Institute for Risk and Disaster
Reduction at University College London [14]. Latest investigations on the matter
are presented in [9].

I. Lirkov, S. Margenov, and J. Waśniewski (Eds.): LSSC 2011, LNCS 7116, pp. 324–332, 2012.
© Springer-Verlag Berlin Heidelberg 2012

Due to the damaging effect of ash particles to aircraft engines, during the Eyjafjallajokull eruption a lot of airports all over Europe have been closed sporadically for different periods of time. Even at open airport conditions many flights have been cancelled or delayed according to the volcanic ash distribution. The cancelled flights are estimated to more than 100 000. Millions of passengers were stranded not only because of the dust clouds, but due to the ultra-strict safety rules which banned flying if there was even a slightest trace of volcanic ash. The European Union's executive body estimates the losses from the first two weeks of the disaster to 1.5–2.5 billion Euros.

2   Scientific Activities during and after the Event

All around the world, *nine* Volcanic Ash Advisory Centers (VAACs) are set, responsible for advising international aviation of the location and movement of clouds of volcanic ash. The Icelandic volcano was handled by the VAAC in London, England [15]. Advisory charts have been issued several times per day during the entire event. Quite good description of London VAAC activity connected with Eyjafjallajokull eruption is presented in[13].

Some met-services succeeded to set operational volcanic ash dispersion systems just in connection with south Iceland volcano eruption. Their forecasts with different time horizons used to be available for common use by Internet [16].

It must be emphasized that the modeling of volcanic ash distribution suffers from serious uncertainties, mainly due to the uncertainty of the source term. The main results of eruption monitoring are made by Icelandic Coast Guard and Iceland Met Office whish is responsible for monitoring of seismic and volcano activity on the island [17]. One of the most important parameters is the evolution of the height of ash release. Poor information is available about the concentration of particles and gases in the release column, i.e. the release rate and its vertical profile.

The London and some other VAACs use the dependence of emission rates on plume rise following the achievements in [6,7,8]. This dependence is worked out on the base of data from multiple volcano eruptions all over the world. This approach is used also for the simulations presented by USA at the ICAO meeting in September 2010 [4]. The SILAM group at the Finish Meteorological Institute (FMI) elaborated its own release profile on the base of multiple runs and comparing the modeling results with satellite measurements of the optical depth. This profile can be found in [18], justification given in [19].

In the same period, measurement activities took place by means of dedicated flights, remote sensing by satellites, lidars, ceilometers etc. Unfortunately, this information is quite insufficient and suffers from different shortcomings – it is quite rare in time and space, very often the data obtained by different approaches is inconsistent, it is often columnar.An example of this inconsistency is the estimates of total mass released during the event. According to [9], the volcano ejected about 40 Mt of ash mass and 10 Mt of SO_2 during the whole eruption period. Another estimate performed in the French HotVolc Observation System ([20]) by [5], amounts these masses to about 2.5 Mt and 0.35 Mt, respectively.

An intensive measuring activity took place in Germany. Several publications in discussion phase can be found in the ACP Special Issue [21]. Generally, a few scientific publications are released so far. Most of the works referred here are presentations at different forums, web archives of different organizations, or pre-publication sites of scientific journals.

The modeling activity performed in real or near-real time during the event suffers from different uncertainties. In [13], special attention is paid on it. The main feature of the modeling activities just described is that only individual forecasts are issued. That is why the ENSEMBLE consortium (described in the next paragraph) decided to launch series of modeling exercises in attempt to obtain more realistic simulations of this event. Ensemble-type modeling and modeling of other pollutants (mainly SO2) are in the planned works of the London VAAC ([13]).

3   ENSEMBLE Project

Ensemble modeling, originally developed for weather prediction, is lately being extended to atmospheric dispersion applications ([1,2]. Ensemble dispersion modeling can be defined as statistical analysis and treatment of several dispersion simulations of a common case study. Once an ensemble of dispersion simulations is obtained, a series of methods can be applied for analysis. The ensemble results contain more information than single-model results. This aspect becomes more evident when no monitoring data can be used for the evaluation and validation of the modeling results.

Following this ideology the EU 5FP project ENSEMBLE was launched as a natural continuation of ETEX project ([3]) and of the follow-up RTMOD project ([1]). After its completing, the European Commission, recognizing the importance of the project achievements, converted its consortium to a permanent structure lead by JRC in Ispra, Italy (see the ENCSEMBLE web-site [22]).

In the frame of ENSEMBLE a web-based decision support system has been created to collect atmospheric dispersion forecasts produced by various participants. Model results are uploaded in the specific format via Internet and can be displayed in real time as single model output or grouped (ensembles) for ad-hoc statistical treatments. ENSEMBLE system can be accessed by means of a simple web-browser, through which the analysis of uploaded information is calculated and displayed dynamically. The tool provides also a number of various plots and allows users to perform on-line ensemble analysis.

The ENSEMBLE Consortium decided to use all abilities of ENSEMBLE system for analyzing the dilution of volcano materials during the first week of the event (14-21 April 2010). For the exercise the vertical structure was changed from 5 to 10 levels. Four cases had to be modeled:

Case 1: Tracer (no removal); source term the FMI profile.
Case 2: Tracer; source term provided by Montreal VAAC (following [6,7]).
Case 3: Aerosol (137CS–like), source term by Montreal VAAC.
Case 4: SO_2, source from FMI.

Table 1. Models participating in the ENSEMBLE volcano exercise (Case 3)

ENS No.	Abbr.	Institution	Meteorological model	Disp. model
00100	UK1	UK MetOffice	UNIFIED (UM) MODEL (UM)	NAME III
00300	SE1	SMHI, Sweden	HIRLAM 5.0.0	MATCH
00400	DE1	DWD, Germany	GME	DWD–LPDM
00600	AT1	ZAMG, Austria	ESMWF T319L50	TAMOS
00700	GR1	Democritos, Greece	ESMWF	DIPCOT
01001	PL2	IAE, Poland	GFS/COSMO	HYSPLIT
01500	US1	Savanah River Site, USA	RAMS43	LPDM
02700	BG1	NIMH, Bulgaria	DWD-GME	EMAP
03000	CA2	Environment Canada	GEM Global	CANERM
04800	FR1-1	IRSN, France	ARPEGE	
02500	FR2	IPSl, France	ARPEGE	CHIMERE

The simulation starts at 06:00 UTC on 14 April 2010 (further denoted as T_0) and ends at 06:00 UTC on 21 April.

4   Participating Parties and Models

Eleven institutions, presented in Table 1, took part in the Case 3 of the ENSEMBLE volcano exercise. Many of the models used, especially the meteorological ones, are world wide known. It is important to stress on the fact that the used models and combination of models differ essentially, so the formed ensemble is of kind 5, according to the classification given in [2]. The use of dispersion fields produced by different models based on different meteorology is particularly suitable when no measurements are available for model validation and tuning.

Bulgarian Emergency Response System (BERS) in case of nuclear accident works operationally since 2001. Its creation and development was highly stimulated by the ETEX experiment. NIMH took part in all activities of ETEX with a puff-type dispersion model. Lately, the new 3D model EMAP was tested; performing better than the puff model and now it is the core of BERS. Description of the system and of EMAP is given in [11,12].

For ENSEMBLE applications EMAP is realized in a domain of 250×210 points with 20 km resolution covering the whole Europe on stereographic map projection. The participation of BERS in volcano exercise laid essential upgrades of EMAP and the whole system. The number of EMAP levels was increased and set as required by the exercise's technical documents. The operational met-data transferred from German Met Office (DWD) is quite limited: only wind and temperature at 850 hPa level, surface temperature and precipitation amounts every 12 hours. The necessary data was downloaded from the free NCEP data base [23]. This information is global with six–hour frequency. All this brought to a mixture of met-data but the final results occur quite promising.

5   BERS' Simulation Results

BERS took part in all ENSEMBLE volcano exercise cases; results encrypted and uploaded to the ENSEMBLE web server. From all statistical parameters supported by ENSEMBLE system the only "overlap in space" is used and respective images are presented here. It is done because of the quite illustrative character of FMS (Figure of Merit in Space) indicator. This is the percentage of overlapping area to the common area of both plumes. For demonstrating EMAP simulation ability, only part of BERS results versus ensemble ones will be presented. All images are produced by ENSEMBLE system.

Many simulations were made by BERS in attempt to tune the EMAP model. First was the decrease of the horizontal diffusion coefficient from 5×10^4 to $1 \times 10^4 m^2/s$. The different shapes of EMAP's and the ensemble plums were the next problem, best seen in the beginning of the release. The attentive analysis of these images leads to the conclusion that the low time resolution of the met-information is the main reason for the displacement. New six–hours met file was produced that increased the simulation considerably as seen in Fig.1.

The next two figures demonstrate the EMAP simulation abilities in comparison with an ensemble composed by the other 10 models from Table 1. In Fig. 2, the concentration spots of both simulations are shown for two particular periods. EMAP plume coincides with ensemble one in big extent.

In Fig. 3, the accumulated deposition fields (dry and wet) are presented for three days after T_o. The main difference between both deposition patterns is the most fragmental character of the wet deposition fields (following the character of precipitation field).

In order to investigate the simulation ability of EMAP model in comparison with the other models in the ensemble, Table 2 was created. It presents the percentage of overlapping of all 11 models taking part in ENSEMBLE volcano exercise versus the results of the ensemble composed by the other 10 models. The data for only 7 particular hours (the end of the respective days of simulation)

Fig. 1. Influence of met-data frequency on simulation results (first 24 hours integration): Case 3, concentration at 6000 m height, threshold $1\mu g/m^3$

Fig. 2. Forecasted time evolution of EMAP simulation versus ensemble one: Case 3, concentration at 6000 m height, threshold $1\mu g/m^3$

Fig. 3. Forecasted time evolution of EMAP simulation versus ensemble one: Case 3, accumulated dry (left) and wet (right) deposition, threshold $1000\mu g/m^3$

Table 2. FMS estimates (% of overlapping) of all models versus the ensemble composed by the remaining 10 models: Case 3, concentration at 6000 m height, threshold $1\mu g/m^3$

Model	\multicolumn{8}{c}{Simulation period}							
	24 h	48 h	72 h	96 h	120 h	144 h	168 h	mean	rank
UK1	61	62	59	51	43	38	32	49	4
SE1	69	81	69	74	75	74	79	74	1
DE1	8	5	8	7	10	9	6	8	11
AT1	20	23	27	22	15	22	27	22	8
GR1	4	15	24	28	29	34	41	25	7
PL2	10	9	12	15	14	13	11	12	10
US1	21	20	17	17	16	16	15	17	9
BG1	52	56	39	36	38	28	19	38	5
CA2	36	34	35	42	43	37	34	37	6
FR1-1	76	77	68	69	65	67	65	70	2
FR2	56	70	65	72	73	70	74	69	3

are shown in attempt to give the simulation ability of each model in evolution. The mean percentage of each model is displayed as well. In the last column, ranking of the mean percents is shown in attempt to evaluate the agreement of every model with the ensemble. EMAP model (BG1) presents quite good ranking regardless its simplicity and quite poor meteorological input.

6   Conclusions

In the paper a short review of the activities (mainly modeling ones) related to April 2010 Iceland volcano eruption is presented. The incompatibility of measurements during the events and the lack of enough volume of measured concentrations do not allow full verification of the modeling tools used on- and off-line. Until reliable data base is assembled, the only way to obtain reliable modeling results is the ensemble approach. The ENSEMBLE consortium decided to launch a volcano exercise. Eleven members of the consortium took part in case 3 of this exercise and uploaded its results to the ENSEMBLE web-based analyzing system.

The Bulgarian Emergency Response System was one of the members of this ensemble.

The main conclusion is that despite many unfavorable circumstances as different source of meteorological information for low and high layers, met-data set low time resolution, simplest numerical schemes etc., EMAP produces quite satisfactory results. All main features of volcanic ash distribution are caught by the model. The comparison of the levels of agreement of all models that took part in the exercise versus the ensemble of the remaining 10 models shows that EMAP results are in the middle of the ranking. This entire means that NIMH with its BERS is a useful member of ENSEMBLE consortium.

References

1. Galmarini, S., Bianconi, R., Bellasiob, R., Graziani, G.: Forecasting the consequences of accidental releases of radionuclides in the atmosphere from ensemble dispersion modelling. Journal of Enveronmental Radioactivity 57, 203–219 (2001)
2. Galmarini, S., Bianconi, R., Klug, W., Mikkelsen, T., Addis, R., Andronopoulos, S., Astrup, P., Baklanov, A., Bartniki, J., Bartzis, J.C., Bellasio, R., Bompay, F., Buckley, R., Bouzom, M., Champion, H., D'Amours, R., Davakis, E., Eleveld, H., Geertsema, G.T., Glaab, H., Kollax, M., Ilvonen, M., Manning, A., Pechinger, U., Persson, C., Polreich, E., Potemski, S., Prodanova, M., Saltbones, J., Slaper, H., Sofiev, M.A., Syrakov, D., Sorensen, J.H., Van der Auwera, L., Valkama, I., Zelazny, R.: Ensemble dispersion forecasting - Part I: concept, approach and indicators. Atmospheric Environment 38(28), 4607–4617 (2004)
3. Girardi, F., Graziani, G., van Velzen, D., Galmarini, S., Mosca, S., Bianconi, R., Bellasio, R., Klug, W., Fraser, G.: ETEX - The European tracer experiment. EUR 18143 EN, 1-108. Office for Official Publications of the European Communities, Luxembourg (1998)
4. IVATF: Effect of Volcanic Input on Uncertainties in Forecasting Volcanic Ash Concentrations in the Atmosphere, Report presented by USA, International Civil Aviation Organization (ICAO), International Volcanic Ash Task Force (IVATF) First Meeting, Montréal, July 27-30 (2010),
   http://www2.icao.int/en/anb/metaim/met/ivatf/Documents/WP.15.pdf

5. Labazuy, P., Gouhier, M., Harris, A., Guéhenneux, Y., Hervo, M., Bergès, J.C., Cacault, P., Rivet, S.: Near Real-Time Monitoring of the April-May 2010 Eyjafjöll's Ash Cloud. In: 13th International Conference on Harmonisation within Atmospheric Dispersion Modelling for Regulatory Purposes, June 1-4, Paris, France (2010), http://www.harmo.org/Conferences/Proceedings/_Paris/publishedSections/H13-E02-abst.pdf

6. Mastin, L., Guffanti, M., Servranckx, R., Webley, P., Barsotti, S., Dean, K., Durant, A., Ewert, J.W., Neri, A., Rose, W.I., Schneider, D., Siebert, L., Stunder, B., Swanson, G., Tupper, A., Volentik, A., Waythomas, C.F.: A multidisciplinary effort to assign realistic source parameters to models of volcanic ash-cloud transport and dispersion during eruptions. Journal of Volcanology and Geothermal Research (2009a), http://dx.doi.org/10.1016/j.jvolgeores.2009.01.008

7. Mastin, L., Guffanti, M., Ewert, J.W., Spiegel, J.: Spreadsheet of eruption source parameters for active volcanoes of the world, in U.S. Geological Survey open-file report 2009-1133 (2009b), http://pubs.usgs.gov/of/2009/1133/

8. Mastin, L., Schwaiger, H., Denlinger, R.: Why do models predict such large ash clouds? An investigation using data from the Eyjafjallajökull eruption, Iceland. Poster Presented at the Atlantic Conference on Eyjafjallajokull and Aviation, Keflavik Airport, Iceland, September 15-16 (2010), http://en.keilir.net/static/files/Aviation/PDF/Poster_Presentations _Abstracts.pdf

9. Schumann, U., Weinzierl, B., Reitebuch, O., Schlager, H., Minikin, A., Forster, C., Baumann, R., Sailer, T., Graf, K., Mannstein, H., Voigt, C., Rahm, S., Simmet, R., Scheibe, M., Lichtenstern, M., Stock, P., Rüba, H., Schäuble, D., Tafferner, A., Rautenhaus, M., Gerz, T., Ziereis, H., Krautstrunk, M., Mallaun, C., Gayet, J.-F., Lieke, K., Kandler, K., Ebert, M., Weinbruch, S., Stohl, A., Gasteiger, J., Olafsson, H., Sturm, K.: Airborne observations of the Eyjafjalla volcano ash cloud over Europe during air space closure in April and May 2010. Atmos. Chem. Phys. Discuss. 10, 22131–22218 (2010), http://www.atmos-chem-phys-discuss.net/10/22131/2010/acpd-10-22131-2010.pdf

10. Sigfusson, T., Thorbjörnsson, I., Gudmundsson, T., Sigurjónsson, B., Gunnarsson, G., Matthíasson, J., Jóhannesson, B., Hilmarsson, H., Arason, A., Gudmundsson, A., Gudlaugsson, G.: First Results of Experiments with the Effect of Volcanic Ash from Eyjafjallajokull Volcano on Aircraft Turbines, A Report to the Icelandic Civil Aviation Administration, Innovation Center Iceland (2010), http://www.nmi.is/files/First%20Results%20of%20Experiements%20with%20 the%20Effect%20of%20Volcanic%20Ash%20Oct25_563345673.pdf

11. Syrakov, D., Prodanova, M., Slavov, K.: Description and performance of Bulgarian Emergency Response System in case of nuclear accident (BERS). Int. J. Environment and Pollution 20, 1–6 (2003)

12. Syrakov, D., Veleva, B., Prodanova, M., Popova, T., Kolarova, M.: The Bulgarian Emergency Response System for dose assessment in the early stage of accidental releases to the atmosphere. Journal of Environmental Radioactivity 100, 151–156 (2009)

13. Thomson, D.: The Met Office Volcanic Ash Advisory Centre (VAAC) Response to the Eruption. In: Annual Science Conference National Centre for Earth Observation, Univ. of Leicester, UK (September 29, 2010), http://www.nceo.ac.uk/documents/18.%20Dave%20Thomson.pdf

14. UCL, Volcanic Hazard from Iceland: Analysis and Implications of the Eyjafjal-lajökull Eruption, Report of the Institute for Risk and Disaster Reduction, University College London (2010),
https://www.ucl.ac.uk/rdr/publications/icelandreport
15. http://www.metoffice.gov.uk/aviation/vaac/
16. http://silam.fmi.fi/AQ_forecasts/v4/forecasts_2010/index.html
17. http://en.vedur.is/earthquakes-and-volcanism/articles/nr/1884
18. http://silam.fmi.fi/AQ_forecasts/v4/forecasts_2010/source_rate_20100510_web.pdf
19. http://silam.fmi.fi/AQ_forecasts/v4/forecasts_2010/source_calibration_20100510_web.pdf
20. http://wwwobs.univ-bpclermont.fr/SO/televolc/hotvolc/
21. http://www.atmos-chem-phys-discuss.net/special_issue122.html
22. http://ensemble2.jrc.ec.europa.eu/
23. http://dss.ucar.edu/datasets/ds083.2/

Automatic Data Quality Control
of Environmental Data

A. Tchorbadjieff

Institute for Nuclear Research and Nuclear Energy–BAS,
72 Tzarigradsko chaussee, Blvd. 1784 Sofia, Bulgaria
assen@inrne.bas.bg

Abstract. The modern physics needs large uninterrupted data for advanced study. The research data must be reliable, precise, adequate and available on time. It can happen only with working advanced information systems. Those systems must implement the most advanced technologies, algorithms and knowledge of informatics, programming and mathematics. This article describes an automatic data quality control of large amount of real time uninterrupted data, implemented in the Institute for Nuclear Research and Nuclear Energy (INRNE) at the BEO Observatory at Moussala.

1 Introduction

The Basic Environmental Observatory (BEO) on Moussala has very turbulent history. It was established in early fifties in twenties century as cosmic rays observatory. But fire destroyed observatory in 1983 after almost thirty years very active and successful scientific work. Later, the Institute of Nuclear Research and Nuclear Energy (INRNE) managed to restore it after almost one decade. The new profile is extended to complex environmental observatory, with atmospheric research facility in addition to cosmic ray study [1];. The very important task in the process of re-establishment of observatory was information system development. It was designed for real time data acquisition and storage in INRNE data centre server. In the initial data system was stored more than 1.29GB scientific data until the end of 2010. The expansion of measured data volumes was very fast and missed to implement adequate data quality verification. Because of that, the new system implementation is in process in BEO observatory. It is part of support from UBA[1], Germany, during the process of World Meteorological organization (WMO) membership of BEO since 2009.

The new data acquisition system, called DAQAS, includes raw data acquisition, data transfer to the main database server in Sofia and storage of metadata (station log book) and processed intermediate data. The dedicated software client application with special data drivers enables easy, quick and on time access to all available raw data [2];.

---

[1] UBA - Federal Environmental Agency (Umwelt Bundes Amt).

I. Lirkov, S. Margenov, and J. Waśniewski (Eds.): LSSC 2011, LNCS 7116, pp. 333–340, 2012.
© Springer-Verlag Berlin Heidelberg 2012

The directly acquired data from scientific measurements may contain very different type of errors. The main role of data quality process is to filter erroneous data from whole dataset and to provide clean and directly usable data for scientific research. The number of observed parameters in BEO observatory expanded largely to 38 during past ten years. Many of them have different nature, pattern and scientific equipment characteristics. But as parameters number grows, the process of error finding consumes enormous work and is prone to miss any problems. Because of that implementation of automatic data quality system is very urgent and important.

2   Measurement Data

The observed parameters in BEO observatory have very different nature. The origin of observed data is from different sources - particle astrophysics, atmosphere particle physics, atmospheric chemistry, radiation and meteorology. Every parameter produces time-ordered univariate time series denoted by:

$$y_{t=1}, y_{t=2}, \cdots, y_{t=n}, \cdots. \tag{1}$$

Every measurement, assumed as a part of long infinite time period t, can be decomposed in three types of patterns: trends in time T (t); seasonal cycle S(t), and random and irregular pattern $\xi(t)$:

$$y_t = T_t + S_t + \xi_t. \tag{2}$$

The trend of random variable y(t) in time is regular change over time of y estimation:

$$T_t = Ey_t \quad t = 1, 2, \cdots, n. \tag{3}$$

The time period of natural observations trend vary from parameter to parameter and depends on physical properties of parameter. The trend could be local for very short time period t, as a result of a temporal event. The most extreme example for temporal local trend is detected in December 2006 Forbush effect, shown with Muons Telescope graphics (Fig.1). It is very well known phenomena when detected in the ground secondary cosmic ray particles decline quickly as result of Solar flares. Then it follows a period of consecutive trend of value increase to normal values levels. The most popular example for long period trend in natural science is CO_2 annual mean values grow at CO_2 data from Mauna Loa (Fig.1).

The seasonal cycle of data describes repeating data patterns over regular time intervals. The pattern can be approximated with a periodic function S(t) with time period d: $S(t) = S(t + d)$.The period d is different for every measured parameter. The most of observed parameters have composite seasonal circle - cycles involved in cycles. The secondary cosmic rays have observable seasonal pattern for daily, 27 days, annual and 11 years cycles. Compare to that CO_2 has "only" two reasonable patterns - diurnal and annual. The annual cycles of CO_2 can be noticed in Mauna Loa graphics (Fig1). The comparison of measured data

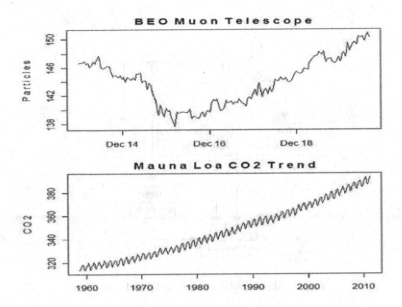

Fig. 1. The plot above shows raw data for Forbush effect detected with Muons Telescope in BEO in December 2006. The graphic below shows increasing long time trend from NOAA Mauna Loa CO_2 measurements. The data source is original NOAA data publication site: http://www.esrl.noaa.gov/gmd/ccgg/trends/.

to the theoretical seasonal templates produces important data quality procedure for long term data sets.

During the normal measurement process the random pattern ξ_t is independent and identically distributed (iid) noise with $WN(0, \sigma^2)$. But random pattern turns very easily to non-normal distributed variable when something gone wrong in measurement process. The result is biased error data within y_t value that mask seasonal and trend properties of data. The process of data quality is to detect and classify data records with biased random pattern and to separate them from the general data set. Because of large data volumes, the process of random pattern must be automated.

3   Initial Data Quality

The new Data Acquisition process in BEO observatory was set up in cooperation with Schneefernerhaus observatory in UBA Germany. The installed DAQAS system packet consists of SAP Database server, client and server applications. The server application manages measurement instruments, data acquisition process, data storage system and technical data classification in database. All procedures are classified as Layer 0. The stored data has two levels - raw and operational type. Raw data is direct data from instruments. The operational data allows predefined parameters as combination of related raw data records. The client

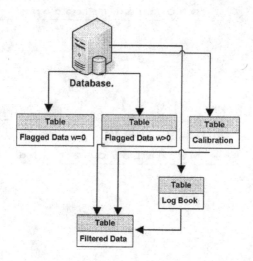

Fig. 2. Abstract data model in Data Layer 0

application provides user interface portal for data access, visual presentation, and system control as definition of calibration procedures, data flagging, and station book metadata. The first step for data quality estimation in database is flagging. Every record is classified by integer parameters '0', '1', '2' and '3' for operationally state following internal temperature, current, control values according technical specifications. This filters data in terms of operational classification. Flag '0' means that instrument does not work according the requirements. The rest flags provide classification of data, as weight parameter, classifying data as partially good. Flag '2' is equal to weight parameter $w = 1/2$, as well as '1' means 1, and '3' means 1/3. The weight functions provide stability flagging in averaging procedures:

$$Mean = \frac{1}{\sum w_i} \sum_{i=1}^{m} (w_i \cdot y_i). \tag{4}$$

In addition, it is provided additional flagging for every separate calibration procedures. This separates calibration data from general data set with predefined automatic or manual procedures for gas analyzers. The process of data filtering in Layer 0 is completed with adding metadata, as log book, which enables registering events related to measurements, as mountain flares in nearby area that concern gas analyzers measurements. The workflow of separation is described in Fig.2.

The successful data filtering in Data Layer 0 depends on adequate knowledge for technical and artificial events. But it leakages in cases when data in meta information is not correct, or has unknown nature. The noise in produced time series grows with time period of irregularities. In this case the process of data acquisition must include additional computational procedures for data filtering. This procedures must upgrade quality information about data inside the database.

The proposed solution is to implement automatic data quality procedure to observe real time data. The automatic filtering have to request Data Layer 0 data directly, to recognize nonnormal distributed ξ_t values, split them from general data set and store it to additional data Layer, numbered with 1.

4 Automatic Filtering Model

Let us assume very small time interval d' within a single seasonal period d with frequency $\omega = \dfrac{d'k\pi}{d}$ and seasonal function S(t) approximates to linear trend window $S'(t)$. The value of $S'(t)$ corresponds to the spectral density of frequency in time interval d' (see e.g. in [3]). For example, during diurnal solar activity, the spectral frequency for the most parameters results in higher $S'(t)$, conversely to night period. Unfortunately, it is impossible to compute a complete data spectral analysis in a real time filtering. This is the main reason to follow another approach to the problem.

The solution for seasonal parameter linearization is usage of local linear filter within interval $(\omega - d'/2, \omega + d'/2)$, where d' is small enough for linear approximation of $S'(t)$. In addition, to avoid sharp $S'(t)$ spectral differences, the interval D of filtering is very slowly moving time window. It heads on with only one measurement forward in every filtering procedure, removing the last value with the newer that has passed filtering in previous filtering step.

One of the simplest ways for implementation of linear filter is to construct a time series X_t, strictly stationary and $(Ex^2 < \infty)$, that smoothes an independent and identically distributed (iid) sequence of random variables. The series X_t is a moving-average process of order q if

$$X_t = Z_t + \theta_1 Z_{t-1} + \cdots \theta_q Z_{t-q}, \tag{5}$$

where $Z_t \sim N(0, \sigma^2)$ and $\theta_1 \cdots \theta_q$ are constants.

We say that the constructed stationary series X_t are q-dependent, i.e., that X_s and X_t are independent whenever $|t - s| > q$. In our case this dependence must be limited between local regions of data. Constructed by this model time series relates to short time characteristics of measured parameters and can be interpreted as a time of parameter perturbations [3].

The criterion of data filtering is the property of the $MA(q)$ as linear filter. The data is modelled by comparison of measured values in data block with size T with prediction estimation of next in time order values. This prediction in l-step for stationary time series X_t in time interval T is

$$y_{T+l} = \sum_{k=0}^{l+1} \theta_{Tk} Z_{T+l-k}. \tag{6}$$

The indexed parameters θ are generated recursively in straighforward recursive substitution in (5) following innovative algorithm in [3]. Then the prediction intervals assume normality of the errors with $100(1 - \alpha)\%$ confidence interval

$$y_{T+l} = \bar{y}_{T+l} \pm (t\text{-}value)s \sqrt{\sum_{k=0}^{l} \bar{\theta}_k^2}. \tag{7}$$

The t-$value$ is $(1-\alpha/2)$-th percentile from a t-distribution with degree of freedom $df = (T - q)$, where s is the standard deviation and $\bar{\theta}$ is the coefficient error estimation. The model separates the records with values outside of the interval of y_l in step l from the general time series.

The very serious concern of proposed model is possible data gaps as result of missing measurements during filtering in Layer 0 or linear filtering in (7). The model assumes missed values in data set as measurements with value 0. That values is excluded from $MA(q)$ and future prediction processes. The filtering is extended to next value, with step $l + 1$ and missing value changes the t-value degree of freedom to $T - q - r$, where r is the number of zeroed values. In case of long series of missing in row values, data threshold regions are possible because of time variability. Because of that it is assumed that computation must stop when measuring distance of data interruption (r) exceeds some level of acceptance. This limits the count of missing values to T and q values.

5   Practical Implementation

The proposed model was tested with several atmospheric and high energy physics parameters. In all tests the most challenging was CO_2 filtering. The tested measurements have some serious errors and defects. The very interesting time series is data from 13 to 16 of January 2011. The period is well observed, all interfering events are very well documented and period is exactly after gas analyzer calibration. During that period, all instruments on the roof were freezing, including gas analyzer inlet, very fast because cold and high humidity. This was a reason why regular work of cleaning nearby and on the gas analyser inlet was undergone. The consequence of those activities is sharp unexpected picks in measured data (CO_2 Raw data plot on Fig 3). The filtering program is Java application computing with R statistics engine. The data is acquired by SQL queries to filtered data from Layer 0 database. But, for demonstration and model testing the original raw data was selected to test extreme cases without any basic filtering. The tested model implements window size of 40 records for 1 minute data. It approximates $\pi/36$ part from daily seasonal pattern. The selected values for, q and d are $q = 10$ and $\max d = 10$. This corresponds to ten step backshift recursion with only allowed 10 minutes time interval for consecutive missing values and errors. The confidence interval was accepted to be 95%.

The proposed model was tested successfully. The smaller data peaks were separated from general dataset. When the first long row of wrong data (higher peak) was detected, the first 10 consecutive error records were separated and the program stopped computation with message showing error. The model was resumed after problem analyses and data trend returns to normal values. The filtered series are plotted in (CO_2 Raw data filtered plot on Fig 3). That plot

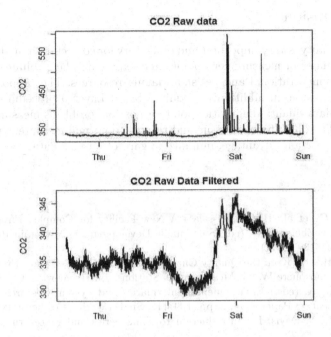

Fig. 3. The data plot of raw and filtered CO_2 data from 13 Jan 2011 12:00:00 to 16 Jan 2011 24:00:00 GMT. The highest value is 1.837987 times bigger than lower one. The filtered data consists of merged outputs after long error series caused interruption.

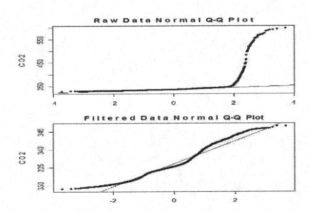

Fig. 4. The tested CO_2 series $Q-Q$ plot for raw and filtered (below) data. The straight line draws normal distribution. Large tail in raw data represents errors. The filtered data is fitted successfully.

shows that every serious error is detected. The simplest graphical test with $Q-Q$ plot confirms obvious data quality improvements (Fig. 4).

6   Conclusion

The data quality is very important and obligatory for correct data modeling. The growing number of measurement points and expanding data volumes, requires additional computational and decision making resources. The proposed model of data diagnostic, in addition to required based Layer 0 procedures, enables automatic data quality diagnostic, pointing possible problems measurements in real time. The quick fixing of issue and resume to normal measurement work is the most important advantage, minimizing gaps of missing data.

References

1. Angelov, C., et al.: BEO Moussala - A New Facility for Complex Environmental Studies. In: Zhelezov, G. (ed.) Sustainable Development in Mountain Regions, ch. 11, pp. 123–139 (2011)
2. Ries, L.: How to Keep the Quality Chain, Work Flow-Oriented Data Processing for Global Atmosphere Watch Measurement Stations. In: Hryniewicz, O., Studzinski, J., Szediw, A. (eds.) Environmental Informatics and System Research Workshop and Application Papers, vol. 2, pp. 109–112. Shaker Verlag, Aachen (2007)
3. Brockwell, P., Davis, R.: Introduction to Time Series and Forecasting. Springer, Heidelberg (2002)

Part VII

Large-Scale Computing on Many-Core Architectures

Computing Boundary Element Method's Matrices on GPU

Gundolf Haase[1], Martin Schanz[2], and Samar Vafai[1]

[1] Institute of Mathematics and Scientific Computing, University of Graz
[2] Institute of Applied Mechanics, Graz University of Technology

Abstract. Matrices resulting from standard boundary element methods are dense and computationally expensive. To speed up the computational time, the matrix computation is done on a GPU. The parallel processing capability of the Graphics Processing Unit (GPU) allows us to divide complex computing tasks into several thousands of smaller tasks that can be run concurrently. We achieved an acceleration of $31-36$ in comparison to a computation performed on the CPU, serially.

1 Introduction

Wave propagation is an important topic in engineering sciences, e.g., in the field of solid mechanics. Applications of wave phenomena can be found in nearly every field of engineering. In non-destructive testing, disturbances of travelling waves are measured to identify cracks or inclusions in the material. In the field of mining, blasting introduces intense stress waves to burst rocks or parts of it. Seismic waves are used to study the interior construction of the earth. Waves produced by earthquakes can cause tremendous destruction in buildings or other man made constructions. Therefor, knowledge is necessary how waves propagate in soil to prevent buildings or dams from destruction [1].

This short, certainly incomplete listing shows the importance of wave propagation problems in engineering mechanics. Tackling such problems correctly will lead to an improvement of constructions and higher quality of living by protecting houses from tremors [1].

In this study, the wave propagation in the elastic material (elastodynamics) is taken into account. This physical phenomena can be well described by the Lamé-Navier equation. To solve this equation numerically, the boundary element method is implemented. The resulting linear system of equations needs to be set up and solved several times, for different frequencies. The large, dense matrices appearing in this linear system of equations are computationally expensive.

To speed up the computational time, the matrix computation is done on GPU. The parallel processing capability of the Graphics Processing Unit (GPU) allows us to divide complex computing tasks into several thousands of smaller tasks that can be run concurrently. We achieved an acceleration of $31-36$ in comparison to a computation performed on the CPU, serially.

I. Lirkov, S. Margenov, and J. Waśniewski (Eds.): LSSC 2011, LNCS 7116, pp. 343–350, 2012.

2   Problem Setting

In an elastic body $\Omega \subset \mathbb{R}^3$ with a Lipschitz boundary $\Gamma = \overline{\Gamma}_D \cup \overline{\Gamma}_N$ and a fixed final time $T \in \mathbb{R}^+$ the following mixed initial boundary value problem has to be solved:

$$-(\mu + \lambda)\nabla\nabla.\mathbf{u}(\mathbf{x},t) - \mu\Delta\mathbf{u}(\mathbf{x},t) + \rho\frac{\partial^2\mathbf{u}}{\partial t^2}(\mathbf{x},t) = 0 \quad (\mathbf{x},t) \in \Omega \times (0,T) \quad (1)$$

$$\mathbf{u}(\mathbf{y},t) = \mathbf{g}_D(\mathbf{y},t) \quad (\mathbf{y},t) \in \Gamma_D \times (0,T)$$
$$\mathbf{t}(\mathbf{y},t) := \tau_y u(\mathbf{y},t) = g_N(\mathbf{y},t) \quad (\mathbf{y},t) \in \Gamma_N \times (0,T)$$
$$\mathbf{u}(\mathbf{x},0) = \tfrac{\partial\mathbf{u}}{\partial t}(\mathbf{x},0) = 0 \quad \mathbf{x} \in \Omega$$

The surface displacements $\mathbf{u}(\mathbf{x},t)$ and tractions $\mathbf{t}(\mathbf{x},t)$ are prescribed by some given data $\mathbf{g}_D(\mathbf{x},t)$ on Γ_D and $\mathbf{g}_N(\mathbf{x},t)$ on Γ_N, respectively. The traction operator τ_x reads as

$$(\tau_x\mathbf{u})(\mathbf{x},t) = (\sigma.\mathbf{n})(\mathbf{x},t) \quad (2)$$

with the stress tensor $\sigma(\mathbf{x},t)$ incorporating Hooke's law and the outward normal vector $\mathbf{n}(\mathbf{x})$ on the boundary Γ [3].

3   Numerical Schemes

The boundary element method is implemented on equation (1). To do the space discretization the collocation technique and for time discretization the Convolution Quadrature Method (CQM) are applied, and the following final discretized linear system of equations is achieved, [1,3]:

$$V(s_l)\mathbf{t}_l - K(s_l)\mathbf{u}_l = C\mathbf{u}_l . \quad (3)$$

The equation (3) consists of N elliptic problems for the complex 'frequency' s_l, $l = 0, 1, ..., N-1$, where N is the total number of time steps.

The entries of the matrices $V(s_l)$ and $K(s_l)$ come from the following formulas:

$$V(y, node_e^f, s_l) = \int_{\Gamma_e} \hat{U}_{ij}^*(x,y,s_l)N_e^f(x)\,d\Gamma_e \quad (4a)$$

$$K(y, node_e^f, s_l) = \int_{\Gamma_e} \hat{T}_{ij}^*(x,y,s_l)N_e^f(x)\,d\Gamma_e \quad (4b)$$

which are the discretized single and double layer potential at all elements.

According to the boundary element method, the boundary surface Γ is discretized by E elements, Γ_e, $e = 1, ..., E$, where polynomial spatial shape functions $N_e^f(x)$ with F nodes are defined, $f = 1, ..., F$. In this problem, the boundary surface elements are triangles. The $\hat{U}_{ij}^*(x,y,s_l)$ and $\hat{T}_{ij}^*(x,y,s_l)$ are the displacement and traction fundamental solutions, respectively. In $3D$, these fundamental solutions are 3×3 tensors. y is the collocation point. In order to obtain the linear system of equations, all nodes of the shape functions $N_e^f(x)$ are chosen as the collocation points. Each column of the given matrices are presented by $node_e^f$, the f_{th} node belonging to mesh element e. C is an integral free term. It is a diagonal matrix corresponding to the singularity appears when formulating the integral equations and all its entries are added to the diagonal elements of the matrix K.

The integrals which appear in equations $(4a, 4b)$ must be computed numerically. All regular integrals are performed with Gaussian quadrature formulas. The strong singular integrals are performed with the method from [2] and the weak singular ones with polar coordinate transformation.

For simplicity of presentation, we restrict ourselves in the numerical example to the case of regular integrations. The implementation of the singular integrals will be similar with respect to GPU acceleration. We also restrict ourselves to 79 Gauss points per element, the highest number of points needed for regular quadrature on triangles, when the highest order of complete polynomial 20 is desirable. These rules are taken from [4].

After implementing the Gauss quadrature method on equations $(4a, 4b)$, we will have the following formulas:

$$V(y, node_e^f, s_l) = \sum_{g=1}^{G} \hat{U}_{ij}^*(x_g, y, s_l) N_e^f(x_g) W_g |Gram| \qquad (5a)$$

$$K(y, node_e^f, s_l) = \sum_{g=1}^{G} \hat{T}_{ij}^*(x_g, y, s_l) N_e^f(x_g) W_g |Gram| \qquad (5b)$$

where G, x_g, W_g and $|Gram|$ are the total number of Gauss points in each element, the Gauss point, the quadrature weight, and the Gram determinant, respectively.

4   Parallel Computing Approach

In the large scale problems, where the same computations must be done several times and at the same time these computations are computationally expensive, the parallel computing plays a role to overcome these bottlenecks.

Parallel computing is a form of computation in which many calculations are carried out simultaneously, operating on the principle that large problems can often be divided into smaller ones, which are then solved concurrently ("in parallel"). Among different possible parallel computing approaches, here GPGPU programming is adopted.

For the purpose of GPU programming, the CUDA hardware and software architecture is implemented. The graphical card in this case is NVIDIA GeForce GTX 480 [5].

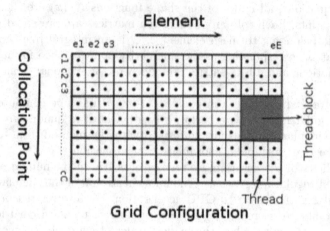

Fig. 1. The grid of thread blocks: it shows how the grid configuration looks like in the given parallel algorithm

4.1   Parallel Algorithm

To compute the matrices $V(s_l)$ and $K(s_l)$, appearing in the equations (5a, 5b), on GPU, the 2D grid is taken into account. In this case, the thread block is also defined in 2D. If, as an example, the number of threads in each thread block is chosen as 64, 8 in each direction x and y, the number of thread blocks in the x-direction will be equal to the number of elements (on the geometrical mesh) divided by 8. The same is true for the number of thread blocks in the y-direction which is derived by splitting the number of collocation points into 8 equal parts. The schematic grid configuration is depicted in figure 1.

All these computations $\sum_{g=1}^{G} \hat{U}_{ij}^*(x_g, y, s_l) N_e^f(x_g) W_g |Gram|$ and
$\sum_{g=1}^{G} \hat{T}_{ij}^*(x_g, y, s_l) N_e^f(x_g) W_g |Gram|$ appeared in equations (5a, 5b) are done on each thread. In other words, all the computations corresponding to the pair combination of one element and one collocation point are performed on one thread. In addition, the same collocation point is used for all the thread blocks on the same row. The blocks on the same columns deals with the computations of the same element but different collocation points.

At the end, 3 number of 3×3 matrices are computed on each thread corresponding to each vertex of the element (triangle in this case). Then, these results must be stored in the correct place in the given matrices $V(s_l)$ and $K(s_l)$. To do that, the values computed on each thread corresponding to the same node on the geometrical mesh are summed up and stored in the position corresponding to that node in the given matrix on the global memory. The global memory has enough storage space, so the $V(s_l)$ and $K(s_l)$ matrices are stored there.

Since all the threads in the same thread blocks need to read the local coordinates of the Gauss points and the quadrature weights repeatedly in a loop over the Gauss points and with respect to the fact that access to the shared memory is much more faster than to the global memory, to gain benefit out of it, all these values are stored in the shared memory.

To clarify what was explained earlier, part of the code is put here:

```
int main ()
{
    ...
    //2D block — 64 threads in each block
    dim3 dimBlock (8,8);
    //num_1: number of elements
    //number of blocks in the x−direction
    const int num_block_x = ceil(num_1/(float)8);
    //num: number of collocation points
    //number of blocks in the y−direction
    const int num_block_y = ceil(num/(float)8);
    dim3 dimGrid (num_block_x,num_block_y);  //2D Grid

    //on the device (GPU)
    matrix_element<<<dimGrid,dimBlock>>>(A_d,num,num_1,...,
            coords_d,connectivity_matrix_d,x_d,quad_weight_d);

    cudaThreadSynchronize ();
    ...
    return(0);
}
```

```
__global__ void matrix_element
(cuDoubleComplex A[],int N,int E,...,double coords_d[],
int connectivity_matrix_d[],double x_d[],
double quad_weight_d[])
{
    #define L 79

    //collocation point
    int ii = (blockIdx.y*blockDim.y)+threadIdx.y;
    //element
    int jj = (blockIdx.x*blockDim.x)+threadIdx.x;

    //storing the Gauss points on the shared memory
    __shared__ double x[L*2];
    for(int idx=((threadIdx.y*blockDim.x)+threadIdx.x);
            idx<(L*2); idx += (blockDim.x*blockDim.y))
        {x[idx] = x_d[idx]};

    //storing the quadrature weights on the shared memory
    __shared__ double quad_weight[L];
```

```
for(int idx=((threadIdx.y*blockDim.x)+threadIdx.x);
           idx<L; idx += (blockDim.x*blockDim.y))
  {quad_weight[idx] = quad_weight_d[idx]};

if ( !( (ii<N) && (jj<E) ) ) {return;}

cuDoubleComplex U[9];
cuDoubleComplex sum_1[9];
...
for(int i=0; i<3; i++)
   {
     for(int j=0; j<3; j++)
       {
         sum_1[i*3+j] =  make_cuDoubleComplex(0.0,0.0);
         ...
         U[i*3+j] =  make_cuDoubleComplex(0.0,0.0);
       }
   }

//Gram determinant —— is the mapping function
//from the reference element to the real element
double gram_y;
Gram_determinant(gram_y,coords_d,connectivity_matrix_d);
gram_y *= 0.5;

double shape_fun_0, shape_fun_1, shape_fun_2;
cuDoubleComplex result_complex;

__syncthreads();

//Loop over all Gauss points in each element
for(int l=0; l<L; l++)
  {
    //Single layer fundamental solution
    //Tensor with 9 elements in 3D—U
    //x: Gauss point —— y: Collocation point(coords_d)
    evaluateFundSol_SLP(U,coords_d,connectivity_matrix_d,
           x,l,shape_fun_0,shape_fun_1,shape_fun_2,...);

    //change from double to cuDoubleComplex
    result_complex = make_cuDoubleComplex
               (shape_fun_0*quad_weight[l]*gram_y,0.0);

    //Multiplication of all the elements taking part into
    //the computations:
```

```
for(int  i=0;  i<3;  i++)
  {
    for(int  j=i;  j<3;  j++)
      {
      sum_1[i*3+j]  =  cuCadd(cuCmul(U[i*3+j],
                        result_complex),sum_1[i*3+j]);
      }
  }
sum_1[3]  =  sum_1[1];
sum_1[6]  =  sum_1[2];
sum_1[7]  =  sum_1[5];
  ...
}

//A  is  the  final  square  matrix  on  global  memory.
//plug  the  value  of  sum  into  the  A  matrix
//The  value  of  the  same  node/vertex  on  different  threads
//summed  up  together  and  insert  into  the  corresponding
//matrix  element:
int  count  =  0;
int  node  =  connectivity_matrix_d[jj*3];
  for(int  J=((node-1)*3);  J<((node-1)*3+3);  J++)
    {
      for(int  I=((ii)*3);  I<((ii)*3+3);  I++)
        {
          A[(3*N)*I+J]=cuCadd(sum_1[count],A[(3*N)*I+J]);
          count  +=  1;
        }
    }
  __syncthreads();
    ...
};
```

5   Numerical Examples

Equation (1) is solved for the special case where a steel rod, which is fixed from one end and free from the other side, is excited by pressure P in its longitudinal direction. The code runs 10 times with a time step of $\Delta t = 0.01$ sec. on both CPU and GPU, separately and the elapsed time for each case corresponding to different number of mesh elements is measured, see in Table 1.

The elapsed time for both GPU and CPU is the time for doing matrix computations. The methods which are used on CPU and GPU to compute the matrix elements are not the same. On CPU, the matrix is divided into four parts which are computed sequentially, but on GPU, the whole matrix is computed in one step.

Table 1. Elapsed time (seconds)-Number of time steps= 10, $\Delta t = 0.01$

Number of elements	CPU	GPU
12	0	0.00231
112	0.16	0.00512
822	9.95	0.275237
3288	162.04	4.46914

6   Conclusion

By performing all the computations corresponding to one element and one collocation point on one thread and putting all those data used repeatedly by all threads of the same thread blocks in the shared memory, we tried to increase this ratio $\frac{number\ of\ arithmetic}{memory\ transfer}$. As it can be seen in Table 1, the computations done on GPU are $31 - 36$ times faster than on one CPU core.

At this step, our goal was to see whether it is paid off to compute the boundary element method's matrices on GPU, so in our computations just the regular integrals with 79 number of Gauss points on each element were considered. However, on the next step, the singular integrals will also be taken into account and the number of Gauss points on each element will be selected depending on the relative position of the element with respect to the collocation point.

In this case, the number of threads in each thread block is chosen to be 64, but it is going to be increased to 128 and 256, respectively. At the same time, the code will run for different arrangement of the threads in a thread block.

To save the memory space and reduce the memory transfer, the symmetric 3×3 sum matrices computed on each thread will be declared as matrices with 6 elements, instead.

References

1. Schanz, M.: Wave Propagation in Viscoelastic and Poroelastic Continua: A Boundary Element Approach. Lecture Notes in Applied Mechanics, vol. 2 (2001)
2. Guiggiani, M., Gigante, A.: A general algorithm for multidimensional Cauchy principal value integrals in the boundary element method. J. of Appl. Mech. 57, 906–915 (1990)
3. Schanz, M.: On a Reformulated Convolution Quadrature Based Boundary Element Method. Computer Modeling in Engineering and Sciences 58, 109–128 (2010)
4. Dunavant, D.A.: High degree efficient symmetrical Gaussian quadrature rules for the triangle. IJNME 21, 1129–1148 (1985)
5. NVIDIA Corporation, NVIDIA CUDA C Programming Guide Version 4.0 (June 2011), http://developer.download.nvidia.com/compute/DevZone/docs/html/C/doc/CUDA_C_Programming_Guide.pdf

A Parallel Algorithm with Improved Performance of Finite Volume Method (SIMPLE-TS)

Kiril S. Shterev[1], Stefan K. Stefanov[1], and Emanouil I. Atanassov[2]

[1] Institute of Mechanics, Bulgarian Academy of Sciences,
Acad. G. Bonchev Str., Block 4, Sofia 1113, Bulgaria
{kshterev,stefanov}@imbm.bas.bg
http://www.imbm.bas.bg
[2] Institute of Information and Communication Technologies,
Bulgarian Academy of Sciences,
Acad. G. Bonchev Str., Block 2, Sofia 1113, Bulgaria
emanouil@parallel.bas.bg
http://www.iict.bas.bg

Abstract. In this paper a parallel version of the finite volume method SIMPLE-TS for calculation of two-dimensional compressible, viscous gas flows with improved performance is presented. It is demonstrated on a problem regarded to micro gas flows, taking place in Micro-Electro-Mechanical Systems (MEMS). The reorganisation of the parallel algorithm improve the algorithm performance, when more cores are used for calculations on computational grids with relatively small number of nodes or cells. The reorganisation is two-fold: first to reduce the number of communications between the processes, and second to reorder the calculation of some variables in such a way that increases the number of calculations during the communications between the processes. The comparison of speed-up between previous and new parallel versions of SIMPLE-TS was performed on two types of clusters with regard to the communication hardware: the first uses specialised cards with low latency for the interconnections between the computers and the other uses conventional cards for the interconnections. The clusters are a part of the GRID-infrastructure of the European Research Area (ERA).

Keywords: finite volume method, SIMPLE-TS, gas microflows, parallel algorithms, GRID.

1 Introduction

The computational analysis of fluid dynamics problems depends strongly on the computational resources [8]. The computational demands are related mainly to: the CPU performance and the memory size. In this paper we consider the problem of calculation of a two-dimensional unsteady state gaseous flow past a particle moving with supersonic speed in a planar microchannel, which is a typical example of problem requiring very big amount of computational resources.

I. Lirkov, S. Margenov, and J. Waśniewski (Eds.): LSSC 2011, LNCS 7116, pp. 351–358, 2012.

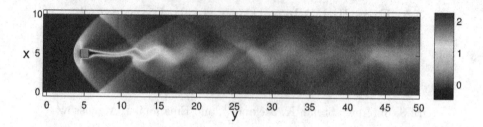

Fig. 1. Horizontal velocity field calculated by parallel version of SIMPLE-TS

We consider an unsteady supersonic flow with Mach number equal to 2.43. The shock wave formed in front of the particle reflects from the channel walls and interacts with the Karman vortex street (see Fig. 1) behind it. The shock wave have significant gradients of velocities, pressure and temperature. Thus, an accurate calculation of the flow requires the use of a very fine or adaptive grid. The steady state calculations have been carried out for a set of gradually refined meshes. Finally, a mesh with 8000x1600 cells was found to give stable and accurate enough results [6]. In this paper speedups of the parallel version of SIMPLE-TS published in [5] and proposed parallel version of SIMPLE-TS are compared. Hereafter the parallel version of SIMPLE-TS, published in paper [5], is referred as SIMPLE-TS [5], while the proposed in this paper is referred as SIMPLE-TS.

2 Continuum Model Equations

A two dimensional system of equations describing the unsteady flow of viscous, compressible, heat conductive fluid can be expressed in a general form as follows:

$$\frac{\partial \rho}{\partial t} + \frac{\partial (\rho u)}{\partial x} + \frac{\partial (\rho v)}{\partial y} = 0 \tag{1}$$

$$\frac{\partial (\rho u)}{\partial t} + \frac{\partial (\rho u u)}{\partial x} + \frac{\partial (\rho v u)}{\partial y} = \rho g_x - A\frac{\partial p}{\partial x} + B\left[\frac{\partial}{\partial x}\left(\Gamma\frac{\partial u}{\partial x}\right) + \frac{\partial}{\partial y}\left(\Gamma\frac{\partial u}{\partial y}\right)\right]$$
$$+ B\left\{\frac{\partial}{\partial x}\left(\Gamma\frac{\partial u}{\partial x}\right) + \frac{\partial}{\partial y}\left(\Gamma\frac{\partial v}{\partial x}\right) - \frac{2}{3}\frac{\partial}{\partial x}\left[\Gamma\left(\frac{\partial u}{\partial x} + \frac{\partial v}{\partial y}\right)\right]\right\} \tag{2}$$

$$\frac{\partial (\rho v)}{\partial t} + \frac{\partial (\rho u v)}{\partial x} + \frac{\partial (\rho v v)}{\partial y} = \rho g_y - A\frac{\partial p}{\partial y} + B\left[\frac{\partial}{\partial x}\left(\Gamma\frac{\partial v}{\partial x}\right) + \frac{\partial}{\partial y}\left(\Gamma\frac{\partial v}{\partial y}\right)\right]$$
$$+ B\left\{\frac{\partial}{\partial y}\left(\Gamma\frac{\partial v}{\partial y}\right) + \frac{\partial}{\partial x}\left(\Gamma\frac{\partial u}{\partial y}\right) - \frac{2}{3}\frac{\partial}{\partial y}\left[\Gamma\left(\frac{\partial u}{\partial x} + \frac{\partial v}{\partial y}\right)\right]\right\} \tag{3}$$

$$\frac{\partial (\rho T)}{\partial t} + \frac{\partial (\rho u T)}{\partial x} + \frac{\partial (\rho v T)}{\partial y}$$
$$= C^{T1}\left[\frac{\partial}{\partial x}\left(\Gamma^\lambda\frac{\partial T}{\partial x}\right) + \frac{\partial}{\partial y}\left(\Gamma^\lambda\frac{\partial T}{\partial y}\right)\right] + C^{T2}.\Gamma.\Phi + C^{T3}\frac{Dp}{Dt} \tag{4}$$

$$p = \rho T \tag{5}$$

where:

$$\Phi = 2\left[\left(\frac{\partial u}{\partial x}\right)^2 + \left(\frac{\partial v}{\partial y}\right)^2\right] + \left(\frac{\partial v}{\partial x} + \frac{\partial u}{\partial y}\right)^2 - \frac{2}{3}\left(\frac{\partial u}{\partial x} + \frac{\partial v}{\partial y}\right)^2 \tag{6}$$

u is the horizontal component of velocity, v is the vertical component of velocity, p is pressure, T is temperature, ρ is density, t is time, x and y are coordinates of a Cartesian coordinate system. Parameters A, B, g_x, g_y, C^{T1}, C^{T2}, C^{T3} and diffusion coefficients Γ and Γ^λ, given in Eqs. (1)-(5), depend on the gas model and the equation non-dimensional form. A first order upwind scheme is used for the approximation of the convective terms, and a second order central difference scheme is employed for the approximation of the diffusion terms.

The Navier-Stokes-Fourier equations (1) - (5) are given in general form. For gaseous microflow description we use the model of a compressible, viscous hard sphere gas with diffusion coefficients determined by the first approximation of the Chapman-Enskog theory for low Knudsen numbers [7]. The Knudsen number (Kn), a nondimensional parameter, determines the degree of appropriateness of the continuum model. It is defined as the ratio of mean free path ℓ_0 to macroscopic length scale of the physical system L $(Kn = \ell_0/L)$. For the calculated case the Knudsen number is equal to $Kn = 0.001$. For a hard-sphere gas, the viscosity coefficient μ and the heat conduction coefficient λ (first approximations are sufficient for our considerations) read as follows:

$$\mu = \mu_h\sqrt{T}, \ \mu_h = (5/16)\rho_0\ell_0 V_{th}\sqrt{\pi} \tag{7}$$

$$\lambda = \lambda_h\sqrt{T}, \ \lambda_h = (15/32)c_p\rho_0\ell_0 V_{th}\sqrt{\pi} \tag{8}$$

The Prandtl number is given by $Pr = 2/3$, $\gamma = c_p/c_v = 5/3$. The dimensionless system of equations (1) - (5) is scaled by the following reference quantities, as given in [7]: molecular thermal velocity $V_0 = V_{th} = \sqrt{2RT_0}$ for velocity, for length - square size a (Fig. 1), for time - $t_0 = a/V_0$, the reference pressure (p_0) is pressure at the inflow of the channel, the reference temperature (T_0) is equal to the channel walls, reference density (ρ_0) is calculated using equation of state (5). The corresponding non-dimensional parameters in the equation system (1) - (5) are computed by using the following formulas:

$$A = 0.5, \ B = \frac{5\sqrt{\pi}}{16}Kn, \ \Gamma = \Gamma^\lambda = \sqrt{T}$$

$$C^{T1} = Kn\sqrt{\pi\frac{225}{1024}}, \ C^{T2} = \frac{\sqrt{\pi}}{4}Kn, \ C^{T3} = \frac{2}{5} \tag{9}$$

3   Parallel Organization

Parallel organizations of FVM algorithms are discussed in many papers (see, e.g., [9,4]). In this paper we present a modification of the parallel organization of algorithm SIMPLE-TS. Details for a conventional domain decomposition (data partitioning) approach are given in [5]. The realization was accomplished by using standard MPI (Message Passing Interface) [3] instructions. There are a lot of communications in one iteration of SIMPLE-TS. To reach high parallel efficiency non-blocking communications are used and communications are overlapped with calculations.

The corresponding serial algorithm, coefficients of numerical equations and base ideas of the algorithm SIMPLE-TS are presented in details in [6]. Here is placed briefly the parallel algorithm SIMPLE-TS [5], which follows the steps of the serial algorithm SIMPLE-TS:

The SIMPLE-TS [5] algorithm
Initialize variables.
Start loop 1:

Set the initial condition for the calculated time step.
Start loop 2 (calculating a state for a new time step):
   Calculate convective and diffusion fluxes.
   Calculate pseudo velocities (velocities, without pressure term), coefficients for pressure equation and start send/receive messages between the processes.
   Wait to complete communications between the processes.
   Start loop 3:
      Solve the coupled equations for energy and pressure and start send/receive messages between processes (there is exchange between data of boundary conditions and *maximum residuals for pressure and temperature of sub-domains in order to determine maximum residuals in the computational domain*).
      Wait to complete communications between processes.
   Stop loop 3. In most cases two iterations are sufficient.
   Calculate velocities using pseudo velocities and pressure (calculated within loop 3) and start send/receive messages between processes (here are exchanged data of boundary conditions and maximum residuals for velocities of sub-domains to determine maximum residuals in computational domain).
   Compute density, using pressure and temperature calculated within loop 3.
   Wait to complete communications between processes.
   Convergence of loop 2: Check for convergence of the iteration process for the current time step.
  Convergence of loop 1: If the final time is not reached continue.

The idea of parallel algorithm SIMPLE-TS with improved performance, is to improve the parallel scalability, when hundreds of cores are used. The number of iterations of loop 3 is fixed. This reduces the number of communications needed to determine the maximum residual for pressure and energy equations within the loops and gives the possibility to make a lot of useful calculations during the communication process. The main bottleneck of the parallel algorithm SIMPLE-TS [5] is avoided with this approach. Another new part of the presented algorithm are the calculations of convective and diffusion fluxes at the end of the loop 2, increasing the number of calculations, which mask communications time between processes. The underlined, bold and italic text in the presented algorithms (SIMPLE-TS [5] and SIMPLE-TS) denotes the differences between them.

The SIMPLE-TS algorithm with improved performance
Initialize variables.
Start loop 1:

Set the initial condition for the calculated time step.
Start loop 2 (calculating a state for a new time step):
Calculate pseudo velocities (velocities, without pressure term), coefficients for pressure equation and start send/receive messages between processes.
Wait to complete communications between processes.
Start loop 3:
Solve the coupled equations for energy and pressure and start send/receive messages between processes (here are exchanged data of boundary conditions).
Wait to complete communications between processes.
Stop loop 3. Make fixed number of iterations.
Start send/receive maximum residuals for pressure and tempera-
ture of sub-domains in order to determine maximum residuals
in the computational domain.
Calculate velocities using pseudo velocities and pressure (calculated within loop 3) and start send/receive maximum residuals for velocities of sub-domains to determine maximum residuals in computational domain.
Compute density, using pressure and temperature calculated within loop 3.
Calculate convective and diffusion fluxes.
Wait to complete communications between processes.
Convergence of loop 2: Check for convergence of the iteration process for the current time step.
Convergence of loop 1: If the final time is not reached continue.

4   Speedup Analysis

The comparison of speedup of both parallel algorithms SIMPLE-TS and SIMPLE-TS [5] for calculations of gas microflows, was performed on two clusters. The first cluster uses DDR InfiniBand [1] interconnection (BG01-IPP). InfiniBand is a switched fabric communications link used in high-performance computing and enterprise data centres [2]. High throughput and low latency of InfiniBand are the most important features for achieving good speedup of the algorithm SIMPLE-TS. The second cluster, uses conventional Gigabit Ethernet cards for interconnection (BG03-NGCC). The worker nodes of both clusters contain two quad core processors. The characteristics of the clusters are shown in Table 1. The first cluster is part of the HP-SEE project infrastructure, while the latter is a GRID cluster, accessible via X509 certificate authentication.

Table 1. The characteristics of the clusters BG01-IPP and BG03-NGCC

Cluster:	BG01-IPP	BG03-NGCC
Numbers of cores:	300 (600 Hyper-Threading enabled)	200
Numbers of cores per worker node:	8 (16 Hyper-Threading enabled)	8
CPU model:	Intel(R) Xeon(R) CPU X5560	Intel(R) Xeon(R) CPU E5430
CPU GHz:	2.80 GHz	2.66 GHz
Cache size:	8 MB	6 MB
RAM per core:	6 GB (3 Hyper-Threading enabled)	2 GB
Interconnection:	InfiniBand	Conventional cards

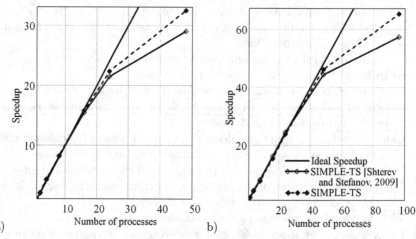

Fig. 2. The speedup of the parallel algorithm SIMPLE-TS [5] and parallel algorithm SIMPLE-TS presented in this paper on cluster BG01-IPP, for meshes: a) 500x100 cells and b) 1000x200 cells

The speedup is calculated as $S_n = T_s/T_{par}$, where n is the number of cores (CPUs), S_n is the speedup, when n-cores are used, T_s is the time of a run on a single core, T_{par} is the time of a run performed on n-cores.

The test calculations were carried out on two meshes: 500x100 and 1000x200 cells. The final state is a periodic unstedy state flow. Determine of unsteady state regime of gas microfow for the case 500x100 cells, takes an hour on 24 cores on BG01-IPP. The same calculation on one core would needs about 23 hours Fig. 2 a).

The results concerning the speedup of both parallel algorithms are shown in Fig. 2 and Fig. 3. In the proposed parallel version of SIMPLE-TS all processes are synchronised at the end of loop 2 or in other two places: at the end of loop 3 and loop 2. The parallel algorithm becomes unstable, when no synchronisation is used. The command MPI_Barrier() is used for synchronisation. SIMPLE-TS

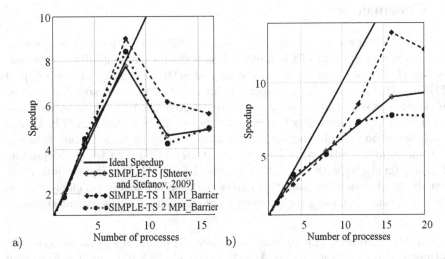

Fig. 3. The speedup of the parallel algorithm SIMPLE-TS [5] and parallel algorithm SIMPLE-TS presented in this paper with one synchronisation and two synchronisations of pal processes on cluster BG03-NGCC, for meshes: a) 500x100 cells and b) 1000x200 cells

[5] synchronises all processes in two places: at the end of loop 3 and loop 2. SIMPLE-TS synchronises all processes at the end of loop 3 and loop 2, when it was run on cluster BG01-IPP. A linear speedup was reached on the cluster BG01-IPP for the case with mesh 500x100 cells for less than 24 processes, Fig. 2. a). The speedup for 48 processes of SIMPLE-TS is higher than speedup of SIMPLE-TS [5]. For a finer mesh (1000x200 cells) the speedup, Fig. 2. b), is very close to ideal. Here the speedup of SIMPLE-TS is also higher than speedup of SIMPLE-TS [5] for more than 48 processes. The results show that scalability of the parallel algorithm SIMPLE-TS with improved performance is better than the algorithm SIMPLE-TS [5].

Three speedups are compared on cluster BG03-NGCC. First speedup is of SIMPLE-TS [5]. The second is of SIMPLE-TS, which make one synchronisation of all processes at the end of loop 2. The third is of SIMPLE-TS, which makes synchronisations of all processes at the end of loop 3 and loop 2. A linear speedup was reached on the cluster BG03-NGCC for the case with mesh 500x100 cells and for less than 8 processes (one worker node), Fig. 3. a), and decrease, when more than 8 processes (more than one worker node) are used. The reason for the speedup decreasing is regarded to the communications between two worker nodes. The speedup of SIMPLE-TS with one synchronisation point is better than the others. The speedups of SIMPLE-TS [5] and SIMPLE-TS with two synchronisation points are close. For a finer mesh (1000x200 cells) the speedup of SIMPLE-TS, which makes one synchronisation, Fig. 3. b), is very good. The speedup of SIMPLE-TS [5] is better than speedup of SIMPLE-TS, with two synchronisation points.

5   Conclusions

The non-blocking communications give the possibility to reach excellent speedup. The speedup of SIMPLE-TS is improved on a cluster with InfiniBand interconnection (BG01-IPP) compared to speedup of SIMPLE-TS [5]. The speedup of SIMPLE-TS with one synchronisation point is better than the speedup reached on the cluster with conventional cards (BG03-NGCC). In other case, the algorithm SIMPLE-TS [5] shows better results, when in algorithm SIMPLE-TS two synchronisations of all processes are made. The algorithm with one synchronisation point is stable, which is important for reaching an improved parallel efficiency on cluster with conventional cards (BG03-NGCC). These results show that the good parallel organization makes the proposed algorithm efficient even, when it is run on clusters with conventional cards, used for interconnections.

Acknowledgments. The authors appreciate the financial support by the NSF of Bulgaria under Grant (SuperCA++)- 2009 No DCVP 02/1 and the European Comission 7FP HP-SEE - 2010, grant No 261499.

References

1. InfiniBand trade association, http://www.infinibandta.org
2. Introduction and hardware charchteristics of InfiniBand, http://en.wikipedia.org/wiki/InfiniBand
3. The official MPI (Message Passing Interface) standards documents, errata, and archives of the mpi forum, http://www.mpi-forum.org
4. Bui, T.T.: A parallel, finite-volume algorithm for large-eddy simulation of turbulent flows. Computers & Fluids 29(8), 877–915 (2000), http://dx.doi.org/10.1016/S0045-7930(99)00040-7
5. Shterev, K.S., Stefanov, S.K.: A Parallelization of Finite Volume Method for Calculation of Gas Microflows by Domain Decomposition Methods. In: Lirkov, I., Margenov, S., Waśniewski, J. (eds.) LSSC 2009. LNCS, vol. 5910, pp. 523–530. Springer, Heidelberg (2010), http://www.springeronline.com/978-3-642-12534-8
6. Shterev, K.S., Stefanov, S.K.: Pressure based finite volume method for calculation of compressible viscous gas flows. Journal of Computational Physics 229(2), 461–480 (2010), http://dx.doi.org/10.1016/j.jcp.2009.09.042
7. Stefanov, S., Roussinov, V., Cercignani, C.: Rayleigh-bénard flow of a rarefied gas and its attractors. I. Convection regime. Physics of Fluids 14(7), 2255–2269 (2002)
8. Versteeg, H.K., Malalasekra, W.: An Introduction to Computational Fluid Dynamics: The Finite Volume Method, 2nd edn. Prentice Hall, Pearson (2007)
9. Wang, P., Ferraro, R.D.: Parallel multigrid finite volume computation of three-dimensional thermal convection. Computers & Mathematics with Applications 37(9), 49–60 (1999), http://dx.doi.org/10.1016/S0898-1221(99)00113-3

Towards Distributed Heterogenous High-Performance Computing with ViennaCL

Josef Weinbub[1], Karl Rupp[1,2], and Siegfried Selberherr[1]

[1] Institute for Microelectronics, TU Wien, Vienna, Austria
[2] Institute for Analysis and Scientific Computing, TU Wien, Vienna, Austria

Abstract. One of the major drawbacks of computing with graphics adapters is the limited available memory for relevant problem sizes. To overcome this limitation for the ViennaCL library, we investigate a partitioning approach for one of the standard benchmark problems in High-Performance Computing (HPC), namely the dense matrix-matrix product. We apply this partitioning approach to problems exceeding the available memory on graphics adapters. Moreover, we investigate the applicability on distributed memory systems by facilitating the Message Passing Interface (MPI). Our approach is presented in detail and benchmark results are given.

1 Introduction

In the last couple of years, graphics adapters have been increasingly used for computations in the field of scientific computing, especially for linear algebra problems [1,2,6], but also to distribute data on several computing nodes powered by a graphics adapter [3]. The first major push forward has been accomplished by NVIDIAs CUDA library [14]. However, the CUDA library solely relies on NVIDIA products and therefore excludes other computing resources, like, graphics adapters from ATI/AMD and CPUs in general. OpenCL, on the other hand, overcomes this drawback as it offers a unified parallel programming model for a multitude of targets. The general concept of utilizing graphics adapters as processing units is referred to as General-Purpose computation on Graphics Processing Units (GPGPU).

To utilize OpenCL for linear algebra, ViennaCL [16] has been developed. The aim of this library is to provide a convenient means to access the vast computing resources of GPUs and multi-core CPUs.

Aside from the high-performance capabilities of graphics adapters, the limited memory of such computing targets is a drawback which hinders problems of considerable size to be computed on the graphics adapters. To further increase the execution performance, a distribution approach is required which facilitates the computational capabilities of several computing nodes.

We base our investigations on the dense matrix-matrix product within the ViennaCL library. We introduce an approach to overcome the memory restrictions on graphics adapters. Moreover, we distribute the workload on several computing nodes by MPI.

I. Lirkov, S. Margenov, and J. Waśniewski (Eds.): LSSC 2011, LNCS 7116, pp. 359–367, 2012.

The structure of this paper is as follows: Section 2 provides a short overview of the ViennaCL library. Section 3 discusses the memory constraints on graphics adapters. Section 4 rigorously discusses our approach to overcome the memory restrictions on a single graphics adapter. Moreover, a distributed approach based on MPI is presented. In Section 5 results are presented and discussed in detail.

2   ViennaCL

The Vienna Computing Library (ViennaCL) provides standard data types for linear algebra operations on GPUs and multi-core CPUs [4,5]. The library is based on OpenCL [12], from which ViennaCL inherits the unified parallel programming approach for personal computers, servers, handhelds, and embedded devices. The ViennaCL API is compatible to the Boost uBlas [9] library, which is a generic template class library with BLAS level 1, 2, 3 support. Therefore, existing uBlas implementations can be conveniently adapted to utilize the vast computing resources of, for example, GPUs. By now, additional support is provided for Eigen [10] and the Matrix Template Library 4 [15]. ViennaCL aims to provide a convenience layer for developing GPU accelerated applications.

3   Challenges of GPGPU

Although OpenCL and thus ViennaCL provide access to various computing targets, we focus our investigations on graphics adapters for two reasons. First, graphics adapters provide a massively parallelized environment, which can be facilitated for tasks in the field of linear algebra. Second, the memory constraints of graphics adapters introduce the need for special treatment. Let us consider these restrictions based on the dense matrix-matrix product (Equation 1).

$$C = A \times B; \quad A \in \mathbb{R}^{n \times m}, B \in \mathbb{R}^{m \times p}, C \in \mathbb{R}^{n \times p} \tag{1}$$

For this product the memory requirements (Equation 2) and the number of operations are as follows (Equation 3).

$$\texttt{memory} : (n \cdot m + m \cdot p + n \cdot p) \cdot \texttt{sizeof}(double) \tag{2}$$

$$\texttt{operations} : (2 \cdot n \cdot m \cdot p) \tag{3}$$

If we consider $n = m = p =: N$, it can clearly be seen that the memory complexity is $\mathcal{O}(N^2)$ and the complexity in regard to the required operations is $\mathcal{O}(N^3)$. For example, $N = 10\,000$ would yield $2.4 \cdot 10^9$ Bytes (~ 2.3 GBytes) of memory and $2 \cdot 10^{12}$ floating point operations.

On the contrary, the matrix-vector product offers the same complexity in the computation as in the memory requirements, which would be $\mathcal{O}(N^2)$ for $n = m = p =: N$. Consequently, this makes the matrix-matrix product much more interesting for investigations in regard to computational performance.

The memory requirements hinder the immediate application of large scale computations on graphics adapters, as the available memory of mere consumer level graphics adapters is typically much smaller, for example, the NVIDIA Geforce GTX 570 (1.2 GB) or the AMD Radeon HD 6970 (2 GB).

To tackle this fact, high-end workstation solutions are available which offer a considerable larger memory, like the NVIDIA Tesla C2050 (3 GB) or the Tesla C2070 (6 GB). These products naturally help to ease the problem, but still, they cannot even remotely compete with the system memory of workstations, let alone computing clusters. These computing environments typically offer system memory in the range from 32 GB to 128 GB per node. Although the memory restrictions limit the immediate application of large scale problems, the computing power of graphics adapters is due to the massive parallelized architecture of interest to the scientific community. Graphics adapters offer a high-performance per value in contrast to expensive cluster systems. Consequently, this introduces the need to overcome the memory restrictions to access the vast computing resources of graphics adapters which, however, requires special treatment of algorithms, for example, a matrix decomposition approach for the dense matrix-matrix product.

4   Our Approach

This section presents our approach which is based on two steps. First, we introduce a decomposition for the dense matrix-matrix product to overcome the memory restrictions of graphics adapters for a single computing node. This approach enables computing problem sizes with ViennaCL, which would normally be impossible due to the memory limitations. Second, a distribution approach based on MPI is presented, which enables to facilitate heterogenous computing nodes.

4.1   Overcoming the Memory Restrictions

To be able to process matrix products of large matrix sizes on memory restricted graphics adapters, a partitioning approach is used. These partitioned parts are then sequentially fed to the ViennaCL matrix product algorithms. Figure 1 depicts this approach.

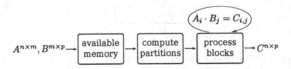

Fig. 1. Overview of the partitioning approach. The problem is partitioned into blocks based on the input matrix sizes and the available memory. Each block is processed by the ViennaCL matrix product algorithm and therefore contributes a subset to the overall solution matrix.

First, the available memory of the OpenCL device, in our case the graphics adapter, is computed. Then, the required partitions of the input matrices A and B are computed according to the available memory and the problem size. Blocks are extracted from the input matrices A_i, B_j and processed sequentially, based on the partitioning information. The computed partial results $C_{i,j}$ are collected and finally returned as overall result C. In the following the three different modules depicted in Figure 1 are discussed in some detail.

To be able to partition an input problem for a specific computing target, the available memory must be known. Unfortunately, OpenCL does not provide functionality which directly allows to determine the available memory of the device. OpenCL provides a means to extract two different memory values of the device by the `clGetDeviceInfo()` function, namely the global memory and the maximum allocable memory accessed by the `CL_DEVICE_GLOBAL_MEM_SIZE` and the `CL_DEVICE_MAX_MEM_ALLOC_SIZE` flag, respectively. However, both values cannot be interpreted as the allocable memory available on the device. The global memory refers to the total amount of memory physically available on the graphics adapter. This value cannot be allocated as the global memory exceeds the OpenCL internal limitations for maximum memory allocations and the host system allocates some of the graphics adapters memory too. On the contrary, the provided maximum allocable memory value yields an upper bound for a single allocation, while several allocations easily exceed this value. In other words, the OpenCL-provided maximum allocable memory value does not actually represent the maximum allocable memory.

Therefore, we use a simple algorithm which evaluates a reasonable allocable memory size specifically for the matrix-matrix product. A fraction of the maximum allocable memory is used as the initial value of this algorithm which is iteratively incremented until the memory can indeed not be allocated anymore. The last allocable memory value is than used as the available memory value of the OpenCL device. Figure 2 depicts the principle of our algorithm.

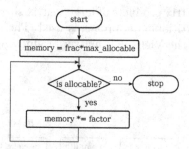

Fig. 2. Flowchart of our memory algorithm. The algorithm starts from a low value and iteratively increases the memory by a certain factor. The value which is last known to be allocable is returned.

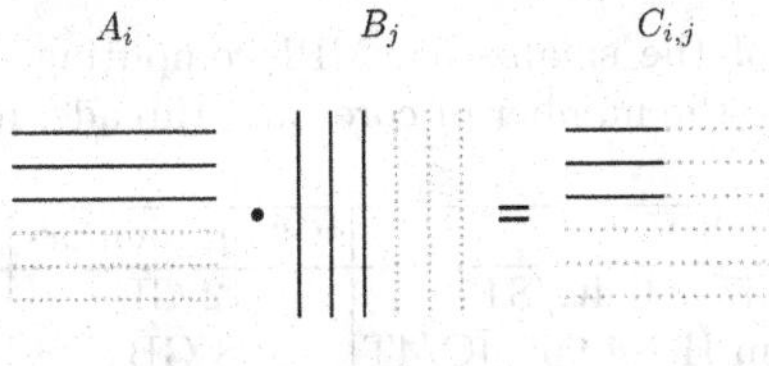

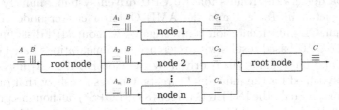

A_i B_j $C_{i,j}$

Fig. 3. The matrices A, B, C are partitioned to fit on the graphics adapters memory

The problem can be partitioned in such a way that it fits into the memory of the graphics adapter, based on the computed available memory estimate. To maximize the performance of our ViennaCL implementation the problem is partitioned in regard to maximum memory utilization, as the execution performance increases with the problem size.

Aside from the memory constraints, the partitioning algorithm has to obviously respect the mathematical rules of matrix multiplication. Therefore, the rows of the matrix A and the columns of the matrix B are partitioned as depicted in Figure 3.

4.2   Distributed Computing

When considering larger problems, the workload for the matrix products can be distributed on available nodes. Each node is governed by a ViennaCL implementation capable of dealing with matrix products of considerable size, as introduced in Section 4.1. The distribution of the matrices is realized using Boost MPI. The matrix A for both matrix products is distributed evenly between the nodes by partitioning the rows. For our numerical experiments, the matrix B is distributed as a whole to all nodes. Note that in the case that B does not fit on the node, additional partitioning has to be applied in accordance to the memory availability on the respective computing nodes. Figure 4 depicts the MPI distribution approach in which, without loss of generality, we assume that the total number of entries in A is larger or equal to that of B.

Fig. 4. Overview of the matrix distribution for different computation nodes. The matrix B is transferred as a whole to all nodes, whereas the matrix A is row-wise partitioned and evenly distributed.

Table 1. Specifications of the small-scale MPI computing environment. The CPU properties C and T denote the number of cores and threads, respectively.

MPI node	CPU	system memory	GPU
root	INTEL i7 960 4C/8T	12 GB	-
1	AMD Phenom II X4 965 4C/4T	8 GB	NVIDIA Tesla C2050
2	AMD Phenom II X4 955 4C/4T	8 GB	NVIDIA GTX 470
3	INTEL Core2 Q9550 4C/4T	4 GB	NVIDIA GTX 580

For the matrix storage we use a linear memory approach based on the standard C++ Standard Template Library (STL) vectortype, for which tuned Boost MPI transmission implementations are available.

The matrices A and B are stored in row-major and column-major order, respectively, since these memory layouts both favor the MPI based transmission efficiency and the memory layout of the OpenCL devices.

5   Results

This section discusses the gathered results based on our approach, for two different computing environments. The first environment is composed of consumer level computing targets as depicted in Table 1. All systems are driven by a Funtoo Linux 64-bit distribution and NVIDIA drivers of version 260.19.36.

The second environment consists of one node of our professional MPI environment. The node is powered by four AMD Opteron 8435 CPUs, each providing six cores yielding a total number of 24 cores, and 128 GB of system memory. Our benchmarks are based on the double precision floating point datatype.

It is important to note that due to the OpenCL back-end, the same implementation is used to run on the CPU and the GPU driven systems. To be able to use CPUs as OpenCL computing targets, the AMD Accelerated Parallel Processing Standard Development Kit (APP SDK) 2.3 is used [8].

Figure 5a depicts the results for two CPU driven systems, namely the root node as depicted in Table 1 and the AMD Opteron cluster node. The standalone ViennaCL implementation is investigated without MPI distribution. We compare our CPU based results with reference implementations based on Boost uBLAS [9] and on libflame [13,7] driven by the GotoBLAS2 [11] library which is specifically tuned for the individual targets. Our results show that our implementation executed on the INTEL i7 target (8 GFLOPS[1]), although significantly slower than the reference libflame/GotoBLAS2 implementation (25 GFLOPS), scales adequately in reference to the uBLAS implementation (1 GFLOPS). Note that although the GotoBLAS2 library has been compiled to support 8 threads, it only utilized 4 threads on the INTEL i7 CPU, omitting the 4 additional threads provided by the Hyperthreading technology which is due to the missing support

---

[1] Giga Floating Point Operations per Second.

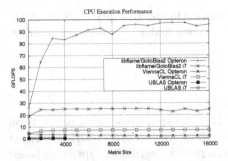
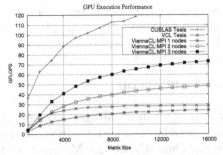

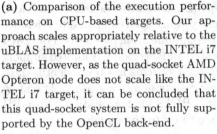

(a) Comparison of the execution performance on CPU-based targets. Our approach scales appropriately relative to the uBLAS implementation on the INTEL i7 target. However, as the quad-socket AMD Opteron node does not scale like the INTEL i7 target, it can be concluded that this quad-socket system is not fully supported by the OpenCL back-end.

(b) Comparison of the execution performance on GPU-based targets. The scaling of our MPI distribution approach as well as the MPI overhead can be identified. The reference CUBLAS implementation outperforms our standalone approach. However, it is due to the memory requirements restricted to matrix sizes of approximately 10000.

Fig. 5. Comparison of the execution performance on CPU- and GPU-based targets

for this specific technology by the GotoBLAS2 library. The quad-socket AMD Opteron cluster node is not fully utilized by the OpenCL back-end provided by the APP SDK. The uBLAS implementation yields 0.4 GFLOPS which is, compared to the INTEL i7 CPU, largely due to the significantly lower clock rate.

The ViennaCL implementation, however, reaches a mere 3 GFLOPS which, considering the higher performance achieved by the INTEL i7 target, indicates that the quad-socket AMD Opteron system is not efficiently utilized, because when restricted to six threads approximately the same performance can be achieved.

Apparently, such multi-processor computing targets are not fully supported by the APP SDK. Although our MPI distributed ViennaCL computation is capable of being executed on our cluster, it is skipped for benchmarking investigations due to the current state of support as depicted.

Figure 5b presents benchmark results based on GPU targets. We compare our approach on a single ViennaCL node with the MPI distribution approach as depicted in Table 1 and with a reference implementation based on CUBLAS [14]. The CUBLAS and our standalone ViennaCL implementations are executed on the root node (Table 1). The MPI distribution overhead can be identified when comparing MPI node 1 (crosses) with the standalone implementation (stars). The scaling behavior of our distribution approach can be recognized when analyzing the execution performance of one to up to three nodes. Although the reference CUBLAS implementation performs significantly faster than our standalone ViennaCL implementation, it is restricted to matrices of approximately the size of 10000 due to the required memory. On the contrary both our standalone variant as well as our MPI variants, are capable of processing matrices of significantly larger problem sizes.

6   Conclusion

We discussed the challenges of computing large dense matrix-matrix products on graphics adapters. We introduced an approach to overcome this restriction based on the ViennaCL library. Moreover, we presented a distribution approach based on MPI to facilitate the power of several OpenCL computing nodes. We provided benchmark results for CPU and GPU driven systems, each compared to reference implementations. The scaling as well as the communication overhead of our MPI based distribution approach has been presented and discussed. A limited support of the APP SDK for quad-socket AMD Opteron systems has been identified. In regard to the CPU based investigations our results depict a reasonable scaling behavior relative to the reference single-core uBLAS implementation. Our GPU powered MPI distribution scales well for one to up to three nodes.

Acknowledgments. This work has been supported by the European Research Council through the grant #247056 MOSILSPIN. Karl Rupp gratefully acknowledges support by the Graduate School PDETech at the TU Wien. The authors thank NVIDIA for providing a Tesla C2050.

References

1. Agullo, E., et al.: Numerical Linear Algebra on Emerging Architectures: The PLASMA and MAGMA Projects. Journal of Physics: Conference Series 180 (2009)
2. Bell, N., Garland, M.: Efficient Sparse Matrix-Vector Multiplication on CUDA. Tech. Rep. NVR-2008-004, NVIDIA (2008)
3. Lawlor, O.S.: Message Passing for GPGPU Clusters: cudaMPI. In: IEEE Cluster PPAC Workshop (2009)
4. Rupp, K., Rudolf, F., Weinbub, J.: ViennaCL - A High Level Linear Algebra Library for GPUs and Multi-Core CPUs. In: Proceedings International Workshop on GPUs and Scientific Applications (GPUScA), pp. 51–56 (2010)
5. Rupp, K., Weinbub, J., Rudolf, F.: Automatic Performance Optimization in ViennaCL for GPUs. In: Proceedings Parallel/High-Performance Object-Oriented Scientific Computing Workshop, POOSC (2011)
6. Tomov, S., Dongarra, J., Baboulin, M.: Towards Dense Linear Algebra for Hybrid GPU Accelerated Manycore Systems. Parallel Computing 36, 232–240 (2010)
7. Zee, F.G.V., et al.: The libflame Library for Dense Matrix Computations. Computing in Science and Engineering 11, 56–63 (2009)
8. AMD Accelerated Parallel Processing SDK,
   http://developer.amd.com/gpu/amdappsdk/
9. Boost uBLAS, http://www.boost.org/libs/numeric/ublas/
10. Eigen, http://eigen.tuxfamily.org
11. GotoBLAS2, http://www.tacc.utexas.edu/tacc-projects/gotoblas2/

12. Khronos OpenCL, http://www.khronos.org/opencl/
13. libflame, http://z.cs.utexas.edu/wiki/flame.wiki/libflame/
14. NVIDIA CUDA, http://www.nvidia.com/cuda/
15. SimuNova Matrix Template Library 4, http://www.simunova.com
16. ViennaCL, http://viennacl.sourceforge.net

High-Throughput-Screening of Medical Image Data on Heterogeneous Clusters

Peter Zinterhof

Salzburg University, Austria
peter.zinterhof3@sbg.ac.at

Abstract. Non-invasive medical imaging by means of computed tomography (CT) and fMRI helps clinicians to improve diagnostics and - hopefully - treatment of patients. Due to better image resolutions as well as ever increasing numbers of patients who undergo these procedures, the amount of data that have to be analyzed puts great strain on radiologists. In an ongoing development with SALK (Salzburger Landeskrankenhaus) we propose a system for automated screening of CT data for cysts in the patient's kidney area. The proper detection of kidneys is non-trivial, due the high variance of possible size, location, levels of contrast and possible pathological anomalies a human kidney can expose in a CT slice. We employ large-scale, semi-automatically generated dictionaries (based on 10^7 training images) to be used in injunction with principal component analysis (PCA). Heterogeneous clusters of CPU-, GPGPU-, and Cell BE-processors are used for high-throughput-screening of CT data. For data-parallel programming CUDA, OpenCL and the IBM Cell SDK have been used. Task parallelism is based on OpenMPI and a dynamic load-balancing scheme, which demonstrates very low latencies by means of double-buffered, multi-threaded queues.

1 Introduction

In this work we focus on the application of principal component analysis (PCA) for automated detection of kidneys within computed tomography (CT) data. Our main motivation for the development of such a system is given by the option of further automating medical diagnostic routines, such as detecting cysts in the kidneys of a patient. Another motivation is based on the fact, that radiology departments are legally constrained to store medical imaging data for many years after creation of the data, thus building up vast libraries of potentially valuable data. In general, these data cannot be reassessed easily on a large scale by clinicians due to time constraints. With respect to medical research, automated analytical systems seem to be a potential way to answer newly arising questions by screening large scale medical image databases. In machine learning and pattern recognition two problems regularly arise. First, definition and production of proper training samples can be a difficult task as badly chosen samples may compromise the quality of the final system. One answer to this problem can be to provide large numbers of samples, with the intention of statistically averaging out sub-optimal samples. Secondly, under the premise of a large training set, one

I. Lirkov, S. Margenov, and J. Waśniewski (Eds.): LSSC 2011, LNCS 7116, pp. 368–377, 2012.

has to question the stability of a classification algorithm and also its computational feasibility. In this work we show how to generate very large training sets semi-automatically and describe a way for screening large numbers of CT slices in parallel with the help of currently available Cell- and GPU-accelerators.

2   Related Work

Several approaches for the automated segmentation and registration of organ tissue have been proposed, not necessarily restricted to kidney tissue. These include contour-based [1], graph-based [3], texture-based, and statistical [7] methods. The latter seems especially promising, because of its ability to cope with cysts, that may distort the kidney and make the detection even more difficult. Due to a lack of appropriate data, we could not yet test our system with images of pathologically distorted kidneys. Petkov [4] describes a system of n parallel 1-dimensional Kohonen self-organizing maps for general image classification purposes. Kohonen self-organizing maps are also demonstrated in [10], who accomplish kidney detection on the base of features that are generated by means of an evolutionary strategy. The texton-based approach [6] also bears a rather close resemblance with our approach, in that features (textons) are directly derived from visual sample data. A more general approach is taken by Pinto et al. [5] who describe a very efficient method of high-throughput screening of good visual representations on Cell and GPU hardware. Currently, we are not aware of approaches for automated kidney detection on Cell or GPU hardware except for the work of [10].

3   Training

In short, training consists of building the covariance matrix M over all image samples S_i in the training data-set. Eigenvalues and corresponding eigenvectors of M are sorted by absolute value. As experiments show no degradation of classification quality, we do not subtract the mean vector from sample images, thus accelerating computation. In the last step we obtain a description vector of each sample image based on the linear combination of the top 32 eigenvectors. Finally, the resulting description vectors are annotated with the corresponding image classes for each vector and the result set is treated as a dictionary throughout the classification process. A good overview of the process can be found in [8]. The provision of appropriate numbers of samples often proves difficult and literature gives different solutions. For instance, [1] increase the limited set of 60.000 samples in the MNIST data-set by artificially distorting them geometrically. Our approach is based on a sliding-window[1] algorithm, that samples each CT slice in two dimensions. By varying the stepping of the window, we can control the resulting number of samples. Each sample is automatically labeled to either class 0 (non-kidney) or 1 (kidney). This automatic process makes use of a map, which has to be created manually for each CT slice of the training data set. The map describes kidney areas in the CT slice, thus enabling rapid checks for all pixel positions in question during the sliding-window algorithm. The resulting set of correctly labeled images (64x64 pixels) serves as training set for the subsequent PCA method. In order to enable computation of large training sets in a reasonable time, we rely on GPU-acceleration. The GPU is utilized for the computation of millions of inner products that constitute the elements of the covariance matrix M. Due to the usually large size of the training set, images cannot be stored in full on current GPUs. We therefore employ a blocked algorithm, which iteratively takes n image vectors at a time, computes their inner products and adds results to matrix M.

Consider the pseudo-code of the standard (non-blocked) algorithm for computing M:

```
#define SAMPLES  10000000

char images[SAMPLES][64*64];  // each row holds another image of 4096 pixels
double M[4096][4096];  // covariance matrix M

for (y=0; y < 4096; y++) {
  for (x=0; x <= y; x++) {
    temp = 0.0;
    for (nr = 0; nr < SAMPLES; nr++) {
      temp = temp + images[nr][y] * images[nr][x];
    }
    M[y][x] = temp;
    M[x][y] = temp;  // due to symmetry of M
  }
}
```

A blocked version of the code provides higher data locality by using only P images at each iteration, thus preventing high cache-miss rates on the CPU. On the GPU blocking is the only way to compute large image data sets, because only small portions of the data fit into device memory and constant re-transfer (e.g. by using page-locked or 'pinned' host memory) of memory chunks would severely degrade performance. So blocking proves to be a necessary means on both architectures.

---

[1] According to the convolution theorem it is possible to replace the sliding window filtering process by a single application of the filter in the frequency domain of both the filter mask and the input CT-image. FFT-based filtering would deliver the same results much more efficiently but we chose not to include it in this stage of our work. As the described load-balancing scheme effectively hides all communications overheads, as well as the execution of the sliding-window filtering process, we cannot expect any further speedup from using an FFT-based approach in our current setup. This assumption will not hold for substantially larger clusters of GPUs (which we currently do not have access to), so the introduction of FFT-based filtering will definitely make a viable upgrade path for our set of algorithms.

```
#define SAMPLES   10000000
#define BLOCKSIZE 10000
char images[SAMPLES][64*64];  // each row holds another image of 4096 pixels
double M[4096][4096];  // covariance matrix M
char block_of_images[BLOCKSIZE][64*64];  // working copy of images-array

void ComputeBlock () {
for (y=0; y < 4096; y++) {
  for (x=0; x <= y; x++) {      temp = 0.0;
    for (nr = 0; nr < BLOCKSIZE; nr++) {
      temp = temp + block_of_images[nr][y] * block_of_images[nr][x];
    }
    M[y][x] = M[y][x] + temp;  // add local sum to global array M
    M[x][y] = M[x][y] + temp;  // due to symmetry of M
  } } }  // end function ComputeBlock

for (nr = 0; nr < SAMPLES; nr++) {
  if ((nr>0)&&(nr MOD BLOCKSIZE == 0)) { ComputeBlock(); }
  for (t = 0; t < 64*64; t++) {
    block_of_images[nr MOD BLOCKSIZE][t] = images[nr][t];  // copy image nr into working memory
  } }
```

3.1 Kidney Detection

PCA-based[2] detection of kidneys boils down to a nearest-neighbor (Euclidean metric) search of target vectors T_{pos} within a dictionary D of n pre-computed and classified dictionary vectors. In order to detect kidneys in a given CT slice (512 x 512 pixels resolution), a window of 64 x 64 pixels is moved across each position $P_{x,y}, 0 < x, y < 512 - 64$ of the slice. The pixel data (4096 values) are then being projected onto the PCA-feature space, constituting a target vector $T_{x,y}$ of length 32. The use of the top 32 eigenvectors of M gives good results in the classification process and also has nice computational- and memory-layout-properties on GPU and Cell. In the last step, the minimum euclidean distance min_d of target T and all dictionary entries D_i with $d_i = \|(D_i - T_{x,y}\|, 0 < i < n$ is computed, along with the corresponding vector index V_x of that minimum. By looking up index V_x in the dictionary, the result of the classification (class 'kidney' or class 'non-kidney', respectively) is directly available. Due to the high dimensionality (32) of the dictionary vectors, there arise major problems ("curse of dimensionality") when applying standard search[9] and optimization techniques such as space partitioning and clustering. Despite any potential ordering of the vector space, every vector might be the nearest neighbor to a given vector. According to [9] this phenomenon usually arises at a dimensionality of 10 and worsens with higher dimensions. Thus, exact nearest-neighbor solutions, which form the base of our described approach, unfortunately are bound to linear search.

3.2 Data Parallelism

Finding the minimum Euclidean distance of some vector within a large matrix is very well suited to the data parallel programming paradigm and to GPU hardware. Rows of large (dictionary-) matrices can efficiently be scanned by thousands of threads in parallel. Due to data independence of matrix rows, explicit synchronization of threads is not necessary until the final step, in which the compute kernel has to merge the local minima of the participating work-groups and determine the global minimum, along with its corresponding (row-) position. For portability reasons we chose OpenCL 1.0, which, other than OpenCL 1.1, does not offer an atomic exchange operation. Hence we could not rely on mutexes on the GPU and therefore we moved this functionality back to the CPU portion of the code and let the CPU compute the final global minimum. This has no significant impact on performance[2]. Considering the problem dimension of some 40.000 pixels to be compared against 10^7 dictionary entries, we do not only face huge numbers of floating point operations, but also very high memory-bandwidth requirements. Despite the widely acclaimed memory bandwidth of many current GPU cards, we cannot let the GPU kernel compute one target-vector at a time as this would result in 40.000 repetitions of moving some 1.3 GB of device memory into the MPs. Instead, we utilize shared memory on the MP, in which 32 target-vectors can be stored and computed by separate threads in parallel. This reduces memory transfers by a factor of 32. The IBM OpenCL framework, which is intended for their Cell-based blade systems only, can also be installed on the Playstation3 (PS3) gaming console (YellowDog Linux 6.1, gcc 4.1.2). Our OpenCL-kernels can then be executed on the PS3 without modification, nevertheless the resulting performance seems to indicate an issue with the scheduling mechanism, which utilizes only one out of six accessible SPEs on the Cell processor. Several experiments with varying work-group allocations did not change this rather odd behavior of OpenCL on the Cell, which is why we switched back to the standard IBM Cell-SDK 1.2 (SPE Runtime Management Library) and created the necessary kernel, which also provides a scheme for double-buffered data transfer. This scheme (listing 3) works analogously to the queuing system for load-balancing (described in chapter 3.3), in that data transfers and computations are interleaved and - most importantly - accomplished by separate threads.

```
// listing 3:  SPU (pseudo-) code / double buffering scheme
char local_buffer[2][4096];
void fetch_data (int nr) {
        mfc_get(&local_buffer[nr % 2][0], base.ui[1], BLOCKSIZE*(32*4), buf_idx, 0, 0); }
void process_data (int nr)  {
        wait_for_completion(1<<(nr % 2));  // data should be in place already
        compute_minimum (); }
  fetchdata(0);      // start transfer of first data block
  for (n=0; n < samples-1; n++)
    {
    fetchdata(n + 1);  // asynchronously initiate transfer of next data block (n+1)
    process_data (n);  // process data block n
    }
  process_data (samples-1);
```

Hereby we gained total RAM transfer rates of 19.9 GB/s for six SPUs (DMA blocksize 16KB) and an overall floating point performance of 23 Gflop/s. This is well below the theoretical peak of 6 x 25.6 Gflop/s. Screening of a CT slice takes

---

[2] Experiments with OpenCL 1.1 and CUDA, which both offer atomic functions on GPU demonstrated levels of performance, that have been comparable to our mixed GPU/CPU (OpenCL 1.0) version, so we do not consider missing atomic functions in OpenCL 1.0 a drawback.

116 s (dictionary of 669.089 vectors) on the PS3. Unfortunately, computation of the minimum distance requires extensive use of conditional clauses, which are expensive on the Cell. Using manual branch prediction pragmas in order to assist the compiler and also employing special spu_cmpgt- and spu_sel-functions [10] did not significantly improve performance.

3.3 Task Parallelism

As there are no data-dependencies within CT-slices in the course the classification process, we implemented a standard master/worker parallelization scheme (OpenMPI), in which workers asynchronously request work-items and return results to the master. This simple hand-shake protocol works well on small to medium sized clusters in combination with coarsely grained work-loads, but it introduces some latency during which workers may be forced to idle-state because they do not receive new work items on time. This problem is likely to increase with growing cluster size as the probability of request collisions increases. As we aim for high performance on a variety of hardware, e.g. GPUs of different individual performance, there is a need for optimal allocation of resources and load-balancing. This can be achieved by choosing a fine granularity in the work load, which increases the number of requests and hence, latency. Our solution to the latency problem is to create copies of work items remotely. Whenever a worker is in need of a new work item (e.g. a block of 32 target vectors), instead of issuing a request at the network level it can directly fetch some work item from a locally managed queue. The same holds for result sets, which are being copied to a local out-queue. In this scheme all participating processes maintain two local queues, one for new work items and one for finished work items. A separate thread is assigned for managing both queues. This thread then acts as a proxy between its local queues and the two globally accessible queues maintained by the master. It handles incoming MPI messages and actively maintains a minimum level (> 1) of locally available work items at all times. This last point is crucial for proper operation since latency elimination is based on readily available items in the in-queue of each worker process. This also implies the capability of the master process to supply work items at a rate that matches or

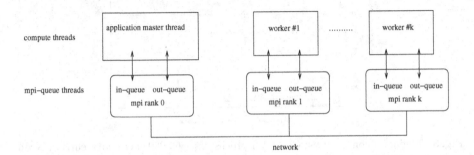

Fig. 1. Overview MPI-queuing system

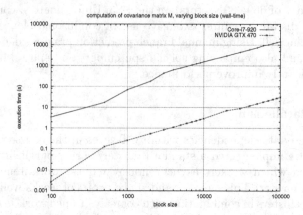

Fig. 2. Timing: computation of covariance matrix

exceeds the rate at which workers are capable of producing results. The master process is also greatly simplified. It can produce new work items asynchronously, putting them in the out-queue. In the next step it simply collects result after result from the in-queue. There is no need to keep track which worker requests are pending or how many tasks each worker should receive. The load is evenly balanced on a first-come/first-served base. Figure 1 shows the proposed decoupling of compute- and communications-threads. On a cluster of Intel-based systems (Q6600, i7-920) we encountered remote work item access latencies (average delay for fetching a work item) below of $2\mu s$ on standard GBIC interconnects. This compares very well to the intrinsically much faster Infiniband (DDR) interconnect, but we want to point out, that our system offers this kind of extra-low latency even for large numbers of compute nodes.

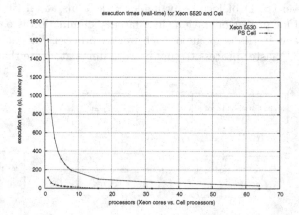

Fig. 3. Timing: Xeon 5520 vs. Cell (PS) processors, 669.089 dictionary entries, Xeon numbers denote single cores, Cell numbers denote Cell processors with six active SPUs per node

4   Results

Generating the covariance matrix M during training is a data-intensive (approximately 80 GB of data have to be processed in a typical run) and time consuming process, which can be sped up dramatically by using a GPU accelerator. Figure 2 gives an overview of execution times for different block-sizes. As can be seen, the NVIDIA GTX 470 provides a speedup in the range of 100 - 120 compared to the baseline Intel Core-i7-920 system. CPU performance can be leveraged by choosing small block-sizes, which enables effective use of L2 and L3 caches. The GPU code is based on the shared memory of each multiprocessor, which can be interpreted as a special form of blocking as well. Nevertheless, total run-time on the GPU does not correlate with blocksize and it is safe to choose the largest possible blocksize in order to reduce the number of kernel calls.

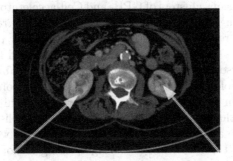

Fig. 4. Sample CT slice with marked kidney area

Table 1. Execution times (wall-time) for classification of one CT slice, dictionary size 6.000.000 entries, GTX295 numbers denote single GPUs (a GTX 295 card holds two GPUs). Systems given in each row of the table operate in parallel.

X5520	GTX 295	GTX 470	GTX 480	Tesla 2050	latency (ms)	run-time (s)
2					2.2e-03	7207.4 s
4					1.8e-03	3492.8 s
8					1.7e-03	1737.6 s
16					1.8e-03	876.04 s
32					1.7e-03	444.39 s
64					1.8e-03	237.14 s
		1			8.3e-04	146.7
		2			8.4e-04	73.6
			1		1.3e-03	122.1
				1	1.6e-03	171.8
				2	1.8e-03	85.0
	4	2	3	4	7.23e-01	17.7

Based on 10^7 samples, building the covariance matrix takes 50 minutes on the NVIDIA GTX470 GPU (448 CUDA cores) and 16 days on the Core-i7-920 CPU.

Run-times for classification of a single CT slice vary considerably on CPU, Cell and GPU architectures (table 1, figure 3). It seems noteworthy, that a cluster of 8 Cell processors at 3.2 GHz (48 SPUs total) is able to outperform a cluster of 64 Xeon cores at 2.3 GHz. As can be seen in table 1, latency for task fetches is very low throughout the experiments. Also, a GTX 480 GPU is able to classify a CT slice (based on 669k entries) in 13.5 s, while this takes a cluster of 8 PS3 consoles 15.6 s and a cluster of 64 Xeons (5520) 31.92 s.

Figure 4 displays the typical result of a classification run (with the cores of both kidneys being marked), based on a dictionary of 10^7 training samples.

5   Conclusions

The proposed system for medical image classification has been successfully implemented on heterogeneous clusters of GPU- and Cell-accelerators. Its performance allows for high-throughput screening of CT data, bringing real time assistance for clinicians within reach. The Cell processor shows decent performance but as its system memory is limited, it can only be recommended for relatively small feature matrices. Also, as the NVIDIA GTX480 GPU is able to outperform a cluster of 8 PS3 consoles, it cannot be recommended to rely on PS3 systems only. The proposed system for load-balancing reduces work fetching-induced latency to very low levels, thus enabling fine grained task parallelism while maximizing utilization of the accelerators. On the algorithmic side we have shown that the computation of the mean image during training is not strictly necessary regarding the application of principle component analysis for kidney detection. This both simplifies and accelerates training. The system offers very good performance on clusters of GPUs, along with good scalability. Future work will include introduction of our Cell code's double-buffered memory transfer mechanism into GPU kernels.

References

1. Ghosh, P., Mitchell, M.: Segmentation of medical images using a genetic algorithm. In: Proceedings of the 8th Annual Conference on Genetic and Evolutionary Computation, GECCO 2006, pp. 1171–1178. ACM, New York (2006)
2. Jolliffe, I.T.: Principal Component Analysis. Springer, New York (2002)
3. Lai, C.C., Chang, C.Y.: A hierarchical evolutionary algorithm for automatic medical image segmentation. Expert Syst. Appl. 36(1), 248–259 (2009)
4. Petkov, N.: Biologically motivated computationally intensive approaches to image pattern recognition. Future Generation Computer Systems 11(4-5), 451–465 (1995)
5. Pinto, N., Doukhan, D., DiCarlo, J.J., Cox, D.D.: A high-throughput screening approach to discovering good forms of biologically inspired visual representation. PLoS Comput. Biol. 5(11), e1000579 (2009), http://dx.doi.org/10.1371%2Fjournal.pcbi.1000579
6. Johnson, M., Shotton, J.: Semantic Texton Forests. In: Cipolla, R., Battiato, S., Farinella, G.M. (eds.) Computer Vision. SCI, vol. 285, pp. 173–203. Springer, Heidelberg (2010)

7. Touhami, W., Boukerroui, D., Cocquerez, J.P.: Fully Automatic Kidneys Detection in 2D CT Images: A Statistical Approach. In: Duncan, J., Gerig, G. (eds.) MICCAI 2005. LNCS, vol. 3749, pp. 262–269. Springer, Heidelberg (2005)
8. Turk, M., Pentland, A.: Eigenfaces for Recognition. Journal of Cognitive Neuroscience 3(1), 71–86 (1991), http://dx.doi.org/10.1162/jocn.1991.3.1.71
9. Weber, R.: Similarity Search in High-Dimensional Data Spaces. Ph.D. thesis, Grundlagen von Datenbanken (1998)
10. Zinterhof, P.: Distributed Computation of Feature-Detectors for Medical Image Processing on GPGPU and Cell Processors. In: Guarracino, M.R., Vivien, F., Träff, J.L., Cannatoro, M., Danelutto, M., Hast, A., Perla, F., Knüpfer, A., Di Martino, B., Alexander, M. (eds.) Euro-Par-Workshop 2010. LNCS, vol. 6586, pp. 339–347. Springer, Heidelberg (2011)

Part VIII

Multiscale Industrial, Enviromental and Biomedical Problems

Preconditioning of Linear Systems Arising in Finite Element Discretizations of the Brinkman Equation

P. Popov

Institute of Information and Communication Technologies,
Bulgarian Academy of Sciences, Sofia, Bulgaria
ppopov@parallel.bas.bg

Abstract. In this work we present a preconditioner for the pressure Schur complement of the linear system, resulting from finite element discretizations of the Stokes-Brinkman equation. The work is motivated by the need to solve numerically the Stokes-Brinkman system. The particular focus are two specific applications: industrial filtration problems and vuggy subsurface flows. The first problem features complex filtering media, coupled to a free flow (Stokes) domain. In vuggy subsurface flows one has free flow inclusions of various connectivity, embedded in highly heterogeneous diffusive media. The Birnkman equation provides a new modeling path to both problems, which warrants the search for efficient methods of solving the resulting linear systems. We consider a block-preconditioning approach for the pressure Schur complement. The starting point is an Incomplete Cholesky factorization of the velocity block. Based on it, an approximate pressure Schur complement is constructed and applied using Preconditioned Conjugate Gradient (PCG). The key in this scheme is an efficient preconditioning of this approximate Schur complement. This is achieved by introducing a second approximation of the pressure Schur complement based on an incomplete back-substitution scheme, followed by a second IC factorization. Numerical examples, illustrating the efficiency of this approach are also presented.

1 Introduction

Consider a domain Ω in $\mathbb{R}^n$, which, in general consists of two parts, a porous Ω_p and free flow one Ω_s, that is $\Omega = \Omega_p \cup \Omega_s$. In this work, we model the flow a slowly moving Newtonian fluid by the Stokes-Brinkman equation which is written as follows (c.f. eg. [2,7,10]):

$$\mu \mathbf{K}^{-1}\mathbf{v} + \nabla p - \mu \Delta \mathbf{v} = \mathbf{f} \qquad \text{in } \Omega, \qquad (1a)$$

$$\nabla \cdot \mathbf{v} = 0 \qquad \text{in } \Omega. \qquad (1b)$$

where $\mathbf{v}$ and p are the fluid velocity and pressure, respectively. In the above system of equations, $\mathbf{K}$ is a permeability tensor, which in Ω_p is equal to the Darcy permeability of the porous media. In either domain μ is the physical viscosity of the fluid.

I. Lirkov, S. Margenov, and J. Waśniewski (Eds.): LSSC 2011, LNCS 7116, pp. 381–389, 2012.

The Stokes-Brinkman equations (1) provide a unified approach in modeling porous media and Stokes flow. In free flow regions, e.g. Ω_s, one sets $\mathbf{K} = \infty$ and as a result the above system is reduced to the Stokes equation [6]. In porous regions, the permeability is finite. When the permeability is small it dominates the laplace term in (1a) and the above system can be thought of as a perturbed Darcy equation in Ω_p. However, for larger values of $\mathbf{K}$, which are generally associated with high porosity, the Brinkman model is a better approximation than Darcy [7,10]. It should be noted that, following [10], the viscosity in the porous part Ω_p is taken to be the fluid viscosity.

The coupling on the interface of Ω_s and Ω_p, implied in (1), is one of continuous velocity. When the permeability is small, this implies the development of a boundary layer, which depends on the flow regime near the interface, e.g. normal, tangential, etc [7]. Thus by resolving the boundary layer, e.g. by adaptive refinement, one can obtain a physically consistent interface model.

Numerical discretizations of the Brinkman system involve large jumps in the coefficients, as well as substantial local refinement near interfaces. Thus, preconditioning the linear systems, generated by various discretizations, is not simple. In this work we consider a mixed Finite Element discretization of (1) (Section 2) and we use a Preconditioned Conjugate Gradient (PCG) iteration on the pressure Schur complement as a linear solver (Section 3). The preconditioner described in Section 4 is suitable for moderately sized problems of several hundred thousand degrees of freedom, featuring large coefficient jumps in complex geometries with small features and extensive use of local refinement.

2   Finite Element Discretization

Consider a mixed finite element discretization in the primal variables $\mathbf{v}$ and p over the finite element spaces V_h and W_h (which we assume satisfy the discrete inf-sup condition, c.f. e.g [4]) and the following bilinear forms:

$$\mathcal{A}(\mathbf{v}, \mathbf{w}) = \int_\Omega \mu \nabla \mathbf{v} \cdot \nabla \mathbf{w} dx \qquad\qquad \mathbf{v}, \mathbf{w} \in V_h, \qquad (2)$$

$$\mathcal{M}(\mathbf{v}, \mathbf{w}) = \int_\Omega \mu \left(\mathbf{K}^{-1}\mathbf{v}\right) \cdot \mathbf{w} dx \qquad\qquad \mathbf{v}, \mathbf{w} \in V_h, \qquad (3)$$

$$\mathcal{C}(p, \mathbf{w}) = \int_\Omega -p \nabla \cdot \mathbf{w} dx \qquad\qquad p \in W_h \text{ and } \mathbf{w} \in V_h. \qquad (4)$$

Also, let us denote the standard L^2 linear form on V_h by

$$(\mathbf{f}, \mathbf{w}) = \int_\Omega \mathbf{f} \cdot \mathbf{w} dx, \quad \mathbf{w} \in V_h.$$

With this notation, the Stokes-Brinkman system (1) can be rewritten in weak form to obtain the following discrete system of equations: Find $\mathbf{v} \in V_h$ and $p \in W_h$, such that:

$$\mathcal{A}(\mathbf{v}, \mathbf{w}) + \mathcal{M}(\mathbf{v}, \mathbf{w}) + \mathcal{C}(p, \mathbf{w}) = (\mathbf{f}, \mathbf{w}) \qquad\qquad \text{for } \forall \mathbf{w} \in V_h, \qquad (5)$$

$$\mathcal{C}(q, \mathbf{v}) = 0 \qquad\qquad \text{for } \forall q \in W_h. \qquad (6)$$

In matrix form, after factoring in essential boundary conditions, the last system is written as

$$\begin{pmatrix} \mathbf{B} & \mathbf{C} \\ \mathbf{C}^T & \mathbf{0} \end{pmatrix} \begin{pmatrix} \mathbf{u} \\ \mathbf{p} \end{pmatrix} = \begin{pmatrix} \mathbf{F} \\ \mathbf{G} \end{pmatrix} \tag{7}$$

where $\mathbf{B} = \mathbf{A} + \mathbf{M}$. Observe, that both $\mathbf{A}$ and $\mathbf{M}$ are symmetric, positive definite matrices. However, the entire system (7), while still symmetric, is indefinite.

3   The Schur Complement Approach

In this work we solve the system (7) by using the pressure Schur complement formulation, a popular approach in the numerical Stokes community [12]. The system (7) is transformed to a symmetric, positive definite system by block-elimination of the velocity block $\mathbf{B}$:

$$-\mathbf{C}^T \mathbf{B}^{-1} \mathbf{C} \mathbf{p} = \mathbf{G} - \mathbf{C}^T \mathbf{B}^{-1} \mathbf{F} \tag{8}$$

The pressure Schur complement $\mathbf{S} = \mathbf{C}^T \mathbf{B}^{-1} \mathbf{C}$ is clearly a symmetric matrix. It can be shown (c.f. e.g. [1]) that $\mathbf{S}$ is also a positive definite matrix. Therefore one can apply the Conjugate Gradient (CG) method [11] to solve (8).

The CG method is applied to (8) by implementing the action of $\mathbf{B}^{-1}$ implicitly, via a second (nested) CG iteration. The reason for this is that computing explicitly $\mathbf{S}$ is both expensive and not necessary. Indeed, denote by N_v the dimension of V_h and by N_p the dimension of W_h. First, computing $\mathbf{B}^{-1}$ involves $\mathcal{O}(N_v^3)$ operations. Secondly, $\mathbf{B}^{-1}$ is generally a dense matrix and, as a consequence, so is $\mathbf{S}$. Finally, the CG iteration requires only the action of $\mathbf{S}$, once per iteration. Therefore, the matrix-vector multiplication with $\mathbf{S}$ can be implemented without the need for explicit construction of $\mathbf{B}^{-1}$. Every time the action of $\mathbf{B}^{-1}$ on a vector is needed one solves a linear system with $\mathbf{B}$ instead.

The most costly operation in a CG iteration with $\mathbf{S}$ is the linear solve with $\mathbf{B}$, required at each iteration. This implies that an efficient implementation needs good preconditioners for both $\mathbf{B}$ and $\mathbf{S}$. Preconditioning $\mathbf{B}$ for problems of moderate dimension (up to a few hundred thousand degrees of freedom) can be achieved easily by a number of standard methods, such as incomplete factorizations. Preconditioning $\mathbf{S}$, however is a different matter. For the pure Stokes equation, the condition number of $\mathbf{S}$ is independent of the mesh parameter h and can be substantially reduced by simple preconditioners, such as a pressure mass matrix, c.f. e.g. [12]. This, however, is not the situation with the Brinkman system. Not only does the condition number of $\mathbf{S}$ depend on h, but it is adversely affected by jumps in $\mathbf{K}$. Since a typical simulation with free flow and porous part entails large jumps in the permeability, preconditioning $\mathbf{S}$ by standard methods that work for the Stokes equation is not possible. In particular, one cannot adapt preconditioners for the time-dependent Stokes system (e.g. [1]).

4   Preconditioning

The starting point in the construction is an Incomplete Cholesky (IC) factorization [11] of $\mathbf{B}$. Let $\mathbf{L} = IC_{m_L}(\mathbf{B})$ be the lower factor of an IC factorization of $\mathbf{B}$ with no more than m_L entries are allowed in each row of $\mathbf{L}$ (i.e. limited fill-in):

$$\mathbf{B} \approx \mathbf{LL}^T. \tag{9}$$

This factorization follows the standard row-wise construction of a Cholesky factorization. Once a row is computed, the largest m_L elements by absolute value are selected and retained and the rest are dropped. Note that the dropping criteria can be efficiently implemented using a modified quick-sort algorithm [11].

This preconditioner was found to be quite efficient in solving linear systems with $\mathbf{B}$ using PCG. It also defines an approximation for the Schur complement. Indeed, let $\mathbf{Q} = \mathbf{L}^{-1}\mathbf{C}$ and consider

$$\tilde{\mathbf{S}} := \mathbf{C}^T \left(\mathbf{LL}^T\right)^{-1} \mathbf{C} = \left(\mathbf{L}^{-1}\mathbf{C}\right)^T \left(\mathbf{L}^{-1}\mathbf{C}\right) = \mathbf{Q}^T\mathbf{Q} \approx \mathbf{S} \tag{10}$$

Assuming that $\mathbf{LL}^T$ approximates $\mathbf{B}$ well, then the above should be a good approximation to $\mathbf{S}$. To be an effective preconditioner however, the action of $\tilde{\mathbf{S}}^{-1}$ should be easily computable. Now $\tilde{\mathbf{S}}$ is in general dense and it cannot be constructed explicitly in linear time, because of the required back-substitution (constructing $\mathbf{Q}$ explicitly). However, the action of $\tilde{\mathbf{S}}$ is easily computable in linear time. Therefor a PCG iteration with $\tilde{\mathbf{S}}$ is taken as preconditioner to $\mathbf{S}$.

Now, this shifts the difficulty in preconditioning $\tilde{\mathbf{S}}$. This is done as follows. Consider the factor $\mathbf{Q} = \mathbf{L}^{-1}\mathbf{C}$. $\mathbf{L}$ is a lower triangular matrix with limited fill-in. The explicit computation of $\mathbf{Q}$ by back-substitution with $\mathbf{L}$ will lead to a dense matrix. Traditionally, back-substitution will result in the *columns* of $\mathbf{Q}$ being sequentially computed. However, the process can be reorganized so that the *rows* of the product matrix $\mathbf{Q}$ are sequentially computed. This allows to limit the fill-in as each row is computed, resulting in a sparse matrix $\hat{\mathbf{Q}}$, which approximates $\mathbf{Q}$. In summary, we have:

Algorithm 1. *Incomplete back-substitution with limited fill-in* $IBS_{m_Q}(\mathbf{L}^{-1}\mathbf{C})$

1. *For $i = 1...N_\mathbf{v}$ do*
2. *Compute the row $\hat{\mathbf{Q}}(i,:)$ by back substitution.*
3. *Keep only the largest (by absolute value) m_Q elements of $\hat{\mathbf{Q}}(i,:)$*
4. *End For*

Note that step 2 of the above algorithm uses only rows 1 through $i-1$ of $\hat{\mathbf{Q}}$ and since $\mathbf{L}$ is a sparse matrix the operation involves at most $m_L m_Q$ flops. Therefore, the cost of the algorithm is at most $N_\mathbf{v} m_L m_Q$ flops. Moreover, it can be implemented efficiently using standard compressed row storage for the matrices $\mathbf{L}$ and $\mathbf{C}$. Thus we obtain $\hat{\mathbf{S}}$ as an approximation to $\tilde{\mathbf{S}}$:

$$\hat{\mathbf{S}} = \hat{\mathbf{Q}}^T\hat{\mathbf{Q}}, \qquad \text{where } \hat{\mathbf{Q}} = IBS_{m_Q}(\mathbf{L}^{-1}\mathbf{C}) \tag{11}$$

where $\hat{\mathbf{Q}}$ is a sparse, $N_v \times N_p$ matrix with maximum m_Q elements per row. The product $\hat{\mathbf{S}} = \hat{\mathbf{Q}}^T\hat{\mathbf{Q}}$ is now generally a sparse matrix, although the sparsity pattern is denser than that of its two factors. Since $\hat{\mathbf{S}}$ can be computed explicitly in linear time, we then construct its Incomplete Cholesky factorization, $IC_{m_S}\left(\hat{\mathbf{S}}\right)$. Thus the final preconditioner for $\tilde{\mathbf{S}}$ is $IC_{m_S}\left(\hat{\mathbf{S}}\right)$. In summary, each PCG iteration for the pressure Schur Complement $\mathbf{S}$ involves

1. Applying the action of $\mathbf{B} = (\mathbf{A} + \mathbf{M})$ on the residual via
   - CG preconditioned with $\mathbf{LL}^T$.
2. Preconditioning the residual by applying $\tilde{\mathbf{S}}^{-1}$ via
   - CG preconditioned with $IC\left(\hat{\mathbf{Q}}^T\hat{\mathbf{Q}}\right)$.

It should be noted that this scheme has three different parameters. The first one, m_L controls the quality of preconditioning of $\mathbf{B}$ and consequently $\mathbf{S}$. The two parameters m_Q and m_S control the preconditioning of $\tilde{\mathbf{S}}$. As such they do not have an effect on the global number of iterations but on the internal PCG iterations that are used to compute the action of $\tilde{\mathbf{S}}^{-1}$.

5   Numerical Examples

In this section we present two numerical examples which illustrate the behavior of the preconditioner. Both examples come from upscaling certain subsurface flows in karst reservoirs with fractures, vugs and caves [9]. The first example (Section 5.1) involves isolated elliptical inclusions (vugs) in a porous matrix. This problem tests the effect of coefficient jumps in the permeability. The second problem (Section 5.2) also features fractures, connecting the vugs. In that setting local refinement near the fractures is necessary. These two examples are illustrative of many situations that arise in modeling subsurface flows. Both were discretized with Taylor-Hood finite elements on triangular meshes [12]. This choice of elements was done for simplicity. However, the framework outlined can be applied to other choices, e.g. [8,3,5].

5.1   Isolated Vugs Embedded in Porous Matrix

The simulation is set-up as follows: the domain containing the matrix and the vugs has dimensions $20 \times 20m$ (Figure 1). The Stokes-Brinkman model (1) is discretized by Taylor-Hood elements on a mesh with 6458 triangles. The total number of degrees freedom is 29579. The porous matrix has a uniform permeability of $\mathbf{K} = k_r\mathbf{I}$ and the vugs are free flow zones ($\mathbf{K} = \infty$). The flow is driven by distributed body force acting in the x direction and periodic boundary conditions are used.

The problem was run for several different values of k_r in the range $[10^{-12}, 10^4]$. At the lower end of permeabilities ($k_r = 10^{-12}$) one has perturbed Darcy flow in the porous matrix. As the flow is mostly normal to the interface with the

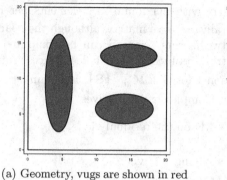
(a) Geometry, vugs are shown in red

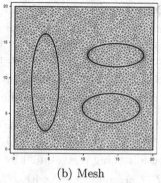
(b) Mesh

Fig. 1. Three elliptical, free flow vugs (a), are embedded in a porous matrix of permeability k_r. The mesh used in the Taylor-Hood discretization is shown in (b).

Table 1. Example 1, number of iterations required to reach relative precision is 10^{-13}

k_r	$m_L = 16$		$m_Q = m_S$		$m_L = 32$		$m_Q = m_S$		$CG(\mathbf{S})$
	PCG	PCG	$PCG\left(\tilde{\mathbf{S}}\right)$, avrg.		PCG	PCG	$PCG\left(\tilde{\mathbf{S}}\right)$, avrg.		
	$(\mathbf{S})$	$(\mathbf{B})$	16	32	$(\mathbf{S})$	$(\mathbf{B})$	32	64	
10^{-12}	13	16.1	108.3	76.8	8	9.6	86.7	61.2	6976
10^{-8}	13	15.6	86.3	61.0	8	9.1	68.0	48.0	8216
10^{-4}	13	15.0	62.8	44.1	8	9.0	48.3	34.3	2107
10^{0}	16	24.0	36.0	36.3	10	13.0	31.7	26.7	61
10^{4}	25	49.9	32.8	26.5	15	27.0	32.6	27.2	60

vugs, no significant boundary layers are observed. At intermediate k_r, the flow is essentially in the Brinkman regime. For $k_r > 1$ one has essentially Stokes flow.

All simulations were repeated for several different values of m_L, m_Q and m_S and the number of iterations are reported in Table 1. As noted in Section 4, m_L affects the global number of PCG iterations and the number of internal iterations needed to apply the action of $\mathbf{B}^{-1}$. These numbers are reported for two different values of m_L: 16 and 32 in the columns labeled $PCG(\mathbf{S})$ and $PCG(\mathbf{B})$, respectively. The number of internal PCG iterations executed when computing the action of $\tilde{\mathbf{S}}^{-1}$ is controlled by m_Q and m_S. So, for each m_L we used two different sets for m_Q and m_S ($m_Q = m_S = 16, 32$ for $m_L = 16$ and $m_Q = m_S = 32, 64$ for $m_L = 32$). The number of internal PCG iterations needed to compute the action of $\tilde{\mathbf{S}}^{-1}$, averaged over the global PCG iteration is reported in the columns labeled $PCG(\tilde{\mathbf{S}})$. For reference, the unpreconditioned global pressure Schur CG iteration is reported in the last column of the table.

The numerical examples demonstrate that the global preconditioned CG iteration for the pressure Schur complement is very insensitive to k_r in essentially its entire spectrum: from the Darcy regime, through intermediate Brinkman regimes to essentially Stokes flow. What is surprising is that even with very little fill-in

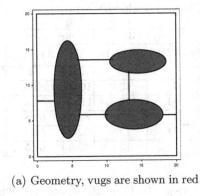
(a) Geometry, vugs are shown in red

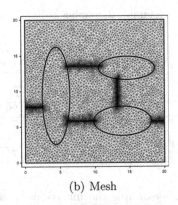
(b) Mesh

Fig. 2. Three elliptical, free flow vugs are connected by fractures of permeability k_f and embedded in a porous matrix of permeability $k_r = 10^{-12}$ (shown in a). The mesh (b) used in the Taylor-Hood discretization is locally refined near the fractures.

($m_L = 16$), smaller than the typical Taylor-Hood sparsity pattern, it took just 13-25 $PCG(\mathbf{S})$ iterations for the global pressure Schur PCG iteration to converge to a precision of 10^{-13}. The average number of PCG iterations for the action of $\mathbf{B}^{-1}$ was in the range 15-50 ($m_L = 16$), the higher value being the Stokes regime. These numbers compare very favorably with a global unpreconditioned CG iteration, reported in the last column of Table 1.

The cost of applying $\tilde{\mathbf{S}}^{-1}$ is also competitive. The action of $\tilde{\mathbf{S}}^{-1}$ was achieved by a reasonable number of iterations in all settings and was very mildly dependent on k_r. It should also be noted that the smaller number of internal iterations achieved by allowing higher fill-in ($m_Q = m_S = 32$) was offset by the longer time it took to construct explicitly $\hat{\mathbf{Q}}$, $\hat{\mathbf{Q}}^T\hat{\mathbf{Q}}$ and $IC\left(\hat{\mathbf{Q}}^T\hat{\mathbf{Q}}\right)$. In all cases the total computational time was less than a minute on a $1.3GHz$ Intel $SU4100$ dual-core processor. For example, the case $k_r = 10^{-12}$ with $m_L = m_Q = m_S = 16$ took $6s$ to set-up all preconditioners and $32s$ to solve the problem. With $m_L = m_Q = m_S = 32$ the times were $21s$ for the set-up and $24s$ for the solve.

In terms of the mesh parameter h, the preconditioner is not uniform with the number of iterations for $PCG(\mathbf{S})$ roughly depending as $h^{0.8}$. Nevertheless, for moderate sized problems the numbers are small and the set-up times reasonable.

5.2   Vugs Connected by Fractures

This example is set-up in the same way as the previous one with the difference that now the vugs are connected by a fracture network. The fractures (the bold lines connecting the vugs in Figure 2) are straight and have aperture in the range $1.1 - 1.7cm$. Due to the aperture size, the mesh (Figure 2(b)) was accordingly refined near the fractures. It contained 27493 triangles with a total of 124314 degrees of freedom for the global Taylor-Hood space.

Table 2. Example 2, number of iterations required to reach relative precision is 10^{-13}

k_f	$m_L = 16$		$m_Q = m_S$		$m_L = 32$		$m_Q = m_S$	
	$PCG\,(\mathbf{S})$	$PCG\,(\mathbf{B})$	$PCG\left(\tilde{\mathbf{S}}\right)$, avrg.		$PCG\,(\mathbf{S})$	$PCG\,(\mathbf{B})$	$PCG\left(\tilde{\mathbf{S}}\right)$, avrg.	
			16	32			32	64
10^{-12}	17	23.1	242.8	141.6	11	13.8	166.3	100.1
10^{-8}	17	23.0	143.1	99.5	11	13.8	112.8	84.8
10^{-4}	17	22.9	99.0	71.8	11	13.8	82.5	61.8
10^{0}	17	22.9	98.4	71.6	11	13.8	82.1	61.5
10^{4}	17	22.9	98.5	71.6	11	13.8	82.0	61.5

In this example the matrix permeability is set to $k_r = 10^{-12}$ and the vugs are again free flow zones. The fracture permeability $\mathbf{K} = k_f \mathbf{I}$ is varied in the range $10^{-12} - 10^4$. The main goal of this example was to test the behavior of the method for domains with small features. Also, the case of nearly free flow fractures ($k_f > 1$) introduced boundary layers in the solution.

We used the same set-up with respect to the fill-in parameters m_L, m_Q and m_S as in the previous example. The results are summarized in Table 2. For the case $m_L = 16$ the global pressure Schur PCG took 17 to converge, independent of the fracture permeability. For $m_L = 32$ the number was 11. To illustrate the effects of mesh refinement, on can compare the case $k_f = 10^{-12}$ with the case $k_r = 10^{-12}$ from the previous example. It is seen that refining heavily the mesh in certain locations leads to a very moderate increase in global pressure Schur PCG iterations (17 instead of 13 for $m_L = 16$ and 11 instead of 8 for $m_L = 32$).

Similarly to the previous example, the averaged number of internal iterations needed to apply $\tilde{\mathbf{S}}^{-1}$ was also mildly dependent on k_f. Again, computations took a small number of minutes. The case $k_f = 10^{-18}$ with $m_L = m_Q = m_S = 16$ took $30s$ to set-up and $250s$ to solve the problem. With $m_L = m_Q = m_S = 32$ the times were $93s$ to set-up and $186s$ to solve the problem. Iteration count is not reported for unpreconditioned pressure Schur CG iterations as these were much more than the previous example and the process would take days.

Acknowledgements. The author acknowledges the partial support provided by the China Petroleum & Chemical Corporation (SINOPEC), US National Science Foundation grant NSF-DMS-0811180 and European Commission grant FP7-PEOPLE-IRG-230919.

References

1. Bramble, J., Pasciak, J.: Iterative techniques for time dependent stokes problems. Computers and Mathematics with Applications 33(1/2), 13–30 (1997)
2. Brinkman, H.C.: A calculation of the viscous force exerted by a flowing fluid on a dense swarm of particles. Applied Scientific Research Section A-Mechanics Heat Chemical Engineering Mathematical Methods 1, 27–34 (1947)

3. Burman, E., Hansbo, P.: A unified stabilized method for stokes' and darcy's equations. Journal of Computational and Applied Mathematics 198, 35–51 (2007)
4. Girault, V., Raviart, P.: Finite Element Methods for Navier-Stokes Equations. Springer, Berlin (1986)
5. Gulbransen, A., Hauge, V., Lie, K.: A multiscale mixed finite element method for vuggy and naturally fractured reservoirs. SPE Journal 15, 395–403 (2010)
6. Gurtin, M.E.: An Introduction to Continuum Mechanics. Academics Press, San Diego (1981)
7. Laptev, V.: Numerical solution of coupled flow in plain and porous media. Ph.D. thesis, Technical University of Kaiserslautern, Germany (2003)
8. Mardal, K., Tai, X., Winther, R.: A robust finite element method for darcy–stokes flow. SIAM J. Numer. Anal. 40, 1605–1631 (2005)
9. Popov, P., Efendiev, Y., Qin, G.: Multiscale modeling and simulations of flows in naturally fractured karst reservoirs. Communications in Computational Physics 6, 162–184 (2009)
10. Rajagopal, K.: On a hierarchy of approximate models for flows of incompressible fluids through porous solids. Mathematical Models and Methods in Applied Sciences 17, 215–252 (2007)
11. Saad, Y.: Iterative Methods for Sparse Linear Systems. PWS, New York (1996)
12. Turek, S.: Efficient Solvers for Incompressible Flow Problems: An Algorithmic and Computational Approach. Springer, New York (1999)

Part IX

Efficient Algorithms of Computational Geometry

Blending Functions for Hermite Interpolation by Beta-Function B-Splines on Triangulations*

Børre Bang, Lubomir T. Dechevsky, Arne Lakså, and Peter Zanaty

Priority R&D Group for Mathematical Modelling,
Numerical Simulation & Computer Visualization,
Faculty of Technology, Narvik University College
2 Lodve Lange's Str., P.O.B. 385, N-8505 Narvik, Norway

Abstract. In the present paper we compute for the first time Beta-function B-splines (BFBS) achieving Hermite interpolation up to third partial derivatives at the vertices of the triangulation. We consider examples of BFBS with uniform and variable order of the Hermite interpolation at the vertices of the triangulation, for possibly non-convex star-1 neighbourhoods of these vertices. We also discuss the conversion of the local functions from Taylor monomial bases to appropriately shifted and scaled Bernstein bases, thereby converting the Hermite interpolatory form of the linear combination of BFBS to a new, Bezier-type, form. This conversion is fully parallelized with respect to the vertices of the triangulation and, for Hermite interpolation of uniform order, the load of the computations for each vertex of the computation is readily balanced.

1 Introduction

Generalized expo-rational B-splines (GERBS) [1,2] are a generalization of expo-rational B-splines (ERBS) [3,4] which includes the polynomial simplified modifications of ERBS, termed Euler Beta-function B-splines (BFBS) in [1,2], and which also includes other basis/blending functions with the relevant properties such as minimal support (as a 1st-degree piecewise affine B-spline), value 1 at the center knot, value zero at all other knots, and (at least one) derivative equal zero at all knots. Examples are provided, e.g., by a trigonometric basis function $\sin^2$ introduced in [5] and a rational basis function proposed in [6]. Compared to these other 'bell-shaped' functions, however, smooth GERBS (and, in particular, BFBS and ERBS) have the property to provide minimally supported convex partitions of unity over triangulated polygonal domains in $\mathbb{R}^2$. The minimality of the supports of BFBS is in the sense that these supports coincide with the supports of the respective piecewise-affine B-splines. At the same time, unlike piecewise-affine B-splines, BFBS are smooth in the vertices of the triangulation and can be used to provide not only Lagrange interpolation but also Hermite

* Research partially supported by the 2010 and 2011 Annual Research Grants of the Priority R&D Group for Mathematical Modelling, Numerical Simulation & Computer Visualization at Narvik University College.

I. Lirkov, S. Margenov, and J. Waśniewski (Eds.): LSSC 2011, LNCS 7116, pp. 393–401, 2012.

Fig. 1. A GERBS triangular patch is a blending of three local triangular patches where the blending functions are the GERBS basis functions. The GERBS triangular patch interpolates (position and derivatives of) each local triangular patch at its respective corner.

interpolation there. This is achieved by upgrading the constant coefficients in a linear combination of piecewise-affine B-splines to functional coefficients (local functions) in a linear combination of BFBS, where the local functions are Taylor polynomials. The degree of the Taylor polynomial corresponds to the order of Hermite interpolation in the given vertex of the triangulation.

ERBS were introduced in 2003, and a first comprehensive study of their properties was conducted in [3,4]; their applications to geometric modelling of curves and surfaces were discussed in [7,4,8,9], while [10] was dedicated to a first discussion of the potential applications of multilevel ERBS to the theory and design of spline multiwavelets.

BFBS were introduced in 2007 as an important simple instance of GERBS which is essentially complementary to ERBS [1,2].

2 GERBS on Triangulations

The general definition of a simplicial GERBS (and, in particular, a simplicial ERBS) for every dimension $n = 1, 2, 3, 4, \ldots$ of the simplex was proposed in [1] as part of one of the four new constructions of multivariate GERBS providing smooth convex resolution of unity on general, possibly highly irregular, domain partitions in $\mathbb{R}^n$ and/or Hermite interpolation scattered-point sets in $\mathbb{R}^n$.

Triangular GERBS patches are in general of the form

$$C(u, v, w) = \sum_{i=1}^{3} c_i(u, v, w) B_i(u, v, w), \quad u + v + w = 1, \tag{1}$$

where $c_i(u, v, w)$, $i = 1, 2, 3$, are local triangular patches, and where $B_i(u, v, w)$, $i = 1, 2, 3$, are GERBS blending functions.

In Figure 1 a GERBS triangular patch and its local triangular patches (here all three chosen to be affine) are shown.

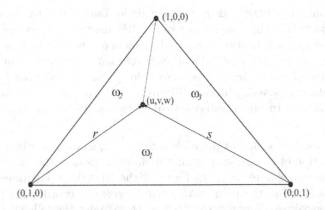

Fig. 2. A triangle defined by the tree point (in barycentric coordinates): $(1,0,0)$, $(0,1,0)$, and $(0,0,1)$. Inside the triangle there is a point (u,v,w) and two vectors $r = (u, v-1, w)$ and $s = (u, v, w-1)$.

3   Construction of GERBS Basis Functions on Triangulations

The exposition in this section follows closely [1] where this new construction was proposed for the first time, for every dimension $n = 1, 2, 3, 4, \ldots$ of the simplectified domain. Here we provide the details for the case $n = 2$, for triangulated domains in $\mathbb{R}^2$, which is the case relevant to geometric modelling of parametric curves and surfaces in 3D. The details and proofs in the general case of the construction for arbitrary $n = 1, 2, 3, 4, \ldots$ proposed in [1] are considerably more technically involved and spacious, and will not be discussed here.

The general construction of the GERBS basis functions in [1] is based on two key ideas which are essentially new for the theory of spline (and B-spline) functions.

- The first one of these ideas was initially proposed in [3] for the new at that time ERBS and discussed in the main theoretical paper [4] on ERBS, and later upgraded in [1] and the main theoretical paper [2] for GERBS: *Upgrade the classical concept of* **Lagrange interpolation by a linear combination of piecewise affine B-splines with constant coefficients** *to* **Hermite interpolation by a linear combination of upgraded smooth B-splines with the same supports as piecewise B-splines and with functional coefficients**.

  This idea was 'the red thread' going through all constructions based on ERBS and GERBS, and its impact drastically increases with increasing the dimension of the problem and the irregularity of the data sets to be processed. In particular, it was also key common component for all four new constructions in [1] of smooth resolution of unity for possibly irregular domain partitions and/or Hermite interpolation at possibly irregularly scattered points sets.

- The second one of these ideas was specific for only one of the four new constructions in [1] – the construction of GERBS upgrades of piecewise affine B-splines on simplectifications (triangulations in our present context of $n = 2$): *Upgrade the characteristics of a simplex relevant to the definition of a piecewise affine B-spline from 'geometrical' to 'physically-weighted geometrical'* by introducing **possibly non-uniform density of mass distribution on each triangle in the triangulation** of the parametric domain.

Here we discuss the second of these ideas for $n = 2$, as follows. Fix a triangle in the triangulation of the global parametric domain. Denote the local parameter domain of the triangle to be Ω. In Figure 2 the triangle is partitioned in three sub-triangles (sharing the point with local barycentric coordinates (u, v, w) as their common vertex) denoted Ω_1, Ω_2 and Ω_3. We get the following definition for the values of the three GERBS, each of which is 'centered' at 'its' respective vertex of the triangle [1]:

$$B_i(u, v, w) = \frac{mass(\Omega_i)}{mass(\Omega)}, \quad i = 1, 2, 3, \tag{2}$$

where B_i is the GERBS centered at the i-th vertex of the triangle, $i = 1, 2, 3$. The density of mass distribution function $\Psi(u, v, w)$ proposed in [1] is:

- for ERBS:

$$\Psi(u, v, w) = \exp\left(\frac{-1}{uvw}\right); \tag{3}$$

- for BFBS:

$$\Psi(u, v, w) = u^{k_1} v^{k_2} w^{k_3}, \ k_i \in \mathbb{N}, \ i = 1, 2, 3. \tag{4}$$

In view of the fact that the GERBS values in (2) are defined via a rational form, in each of the above-said two cases the density is defined modulo a constant multiplicative factor.

Let us note that in the indexation in (2) here we deviated a little from the definition of GERBS on a triangulation, as given in [1]. Here we gave a triangle-by-triangle definition, while the formulation of the definition in [1] was in the classical way, vertex by vertex, GERBS by GERBS. In the latter case, formula (2) was given for every triangle in the star-1 neighbourhood of the j-th vertex in the triangulation, where the local index i was taking only the value corresponding to the j-th vertex in the triangulation as a vertex in the triangle considered. Clearly, the two formulations of the definition are equivalent.

4   The BFBS Blending Functions for Triangles

According to (2), to compute the blending functions B_i, $i = 1, 2, 3$, we have to calculate

$$B_1(u, v, w) = \frac{\int_{\Omega_1} \Psi(u,v,w) d\Omega_1}{\int_{\Omega} \Psi(u,v,w) d\Omega}$$

$$B_2(u, v, w) = \frac{\int_{\Omega_2} \Psi(u,v,w) d\Omega_2}{\int_{\Omega} \Psi(u,v,w) d\Omega} \quad (5)$$

$$B_3(u, v, w) = \frac{\int_{\Omega_3} \Psi(u,v,w) d\Omega_3}{\int_{\Omega} \Psi(u,v,w) d\Omega}$$

where the domains Ω_i, $i = 1, 2, 3$, are shown in Figure 2 and Ω is the parameter domain of the whole triangle.

The computation proceeds by reducing the area (i.e., double) integrals in (5) to 2-fold ones, as follows.

$$B_1(u, v, w) = \frac{\int_0^u \int_{\frac{v}{u}\mu}^{1 - \frac{1-v}{u}\mu} \Psi(u, v, w) \, dv d\mu}{\int_0^1 \int_0^{1-\mu} \Psi(u, v, w) \, dv d\mu}. \quad (6)$$

The limits in the integrals in the denominator and the first integral in the numerator are self explanatory. The more complex limits to explain is the lower and the upper limit in the second integral in the numerator. They follow by using the respective vertex point and the vectors r and s in figure 2 scaled by $\frac{\mu}{u}$. We get

$$(0, 0, 1) + \frac{\mu}{u} ((u, v, w) - (0, 0, 1)) = \left(\mu, \frac{v}{u}\mu, 1 - \frac{1-w}{u}\mu \right), \quad (7)$$

where we have used the second component as the lower limit, and

$$(0, 1, 0) + \frac{\mu}{u} ((u, v, w) - (0, 1, 0)) = \left(\mu, 1 - \frac{1-v}{u}\mu, \frac{w}{u}\mu, \right) \quad (8)$$

where we have used the second component as the upper limit. This completes the computation of B_1, as soon as the respective single integrals of polynomial integrands are calculated; B_2 and B_3 can then be computed, based on symmetry argument, by cyclic shifting of the barycentric variables u, v, w.

The respective results are:

- For $k_0 = k_1 = k_2 = 0$:

$$B_1(u, v, w) = u$$
$$B_2(u, v, w) = v \quad (9)$$
$$B_3(u, v, w) = w$$

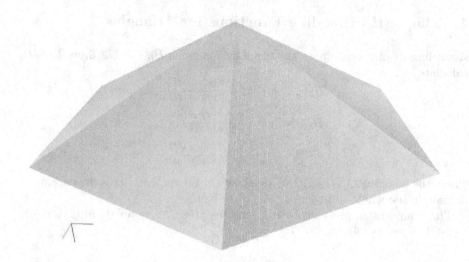

Fig. 3. The entire graph of an order-0 BFBS blending functions over the star-1 neighbourhood of the vertex at which the blending function is 'centered'. Outside this star-1 neighbourhood the blending function is identically zero. Overall, the function is C^0-regular (i.e., continuous) including the vertices and edges.

– For $k_0 = k_1 = k_2 = 1$:

$$
\begin{aligned}
B_1(u, v, w) &= (6vw - 2u + 3)u^2 \\
B_2(u, v, w) &= (6uw - 2v + 3)v^2 \\
B_3(u, v, w) &= (6uv - 2w + 3)w^2
\end{aligned}
\tag{10}
$$

– For $k_0 = k_1 = k_2 = 2$:

$$
\begin{aligned}
B_1(u, v, w) &= (30vw(3vw - u + 1) + 6u^2 - 15u + 10)u^3 \\
B_2(u, v, w) &= (30uw(3uw - v + 1) + 6v^2 - 15v + 10)v^3 \\
B_3(u, v, w) &= (30uv(3uv - w + 1) + 6w^2 - 15w + 10)w^3
\end{aligned}
\tag{11}
$$

The entire graphs of the order-0, order-1 and order-2 BFBS basis functions for the same given vertex are plotted in Figures 3, 4 and 5, respectively, for a domain containing in its interior the entire star-1 neighbourhood of the vertex at which all the BFBS are 'centered'.

The computation of the derivatives is straightforward, and will not be given explicitly here.

5   Some Properties and Fields of Applications

1. The Hermite interpolation problem using the so-defined GERBS, and BFBS, in particular, has generalized Vandermonde determinant which is always readily in Jordan normal block-diagonal form; moreover, if the order of Hermite multiplicity is uniform, i.e., the multiplicities are the same for each of

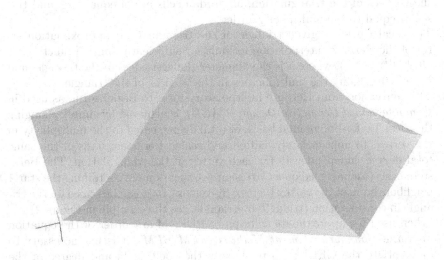

Fig. 4. The entire graph of an order-1 BFBS blending functions over the star-1 neighbourhood of the vertex at which the blending function is 'centered'. Outside this star-1 neighbourhood the blending function is identically zero. The function is C^1-smooth at the vertices.

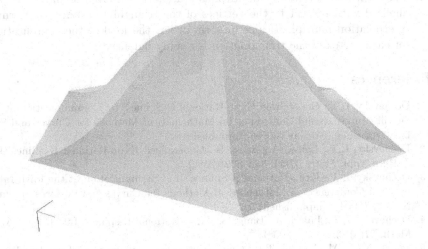

Fig. 5. The entire graph of an order-2 BFBS blending functions over the star-1 neighbourhood of the vertex at which the blending function is 'centered'. Outside this star-1 neighbourhood the blending function is identically zero. Overall, the function is C^2-smooth at the vertices.

the vertices of the triangulation, all Jordan cells are of equal size, and this size is equal to the multiplicity/order.

2. The local (inside a given triangle of the triangulation) approximation error of the Hermite interpolation formula for sufficiently smooth functions is $O(h^{\min\{k_0,k_1,k_2\}})$, where h is the diameter (longest edge) of the triangle, and k_j, $j = 0, 1, 2$, are the multiplicities in the vertices of the triangle.

3. The conversion from Hermite interpolatory form to Bezier form, as used in *Computer-aided Geometric Design (CAGD)*, is achieved by simply changing the local Taylor polynomial basis, of total degree equal to the multiplicity in the vertex, to appropriately shifted and scaled Bernstein basis of the same total degree, independently for each vertex of the triangulation. The Bernstein basis can be scaled on a Cartesian-product domain containing the star-1 neighbourhood of the vertex (vertex-by-vertex), or it can be scaled on the triangle in consideration (triangle-by-triangle, see the example in Figure 2).

4. When used for approximate solution of PDEs of any order of the equation in *Finite/Boundary Element Methods (FEM/BEM)*, it is not necessary to differentiate the GERBS; instead, only the coefficients and degree of the Taylor polynomial coefficients are modified to correspond to the respective (partial) derivative. The high approximation error in item 2 ensures that this is good approximation to the respective derivative. The respective stiffness matrices of the FEM/BEM which need inversion are in block-band-limited form, where the bandwidth is narrow and depends only on the triangulation and not on the order of the PDE; the order of the PDE influences only the size of the blocks in the block-band-limited form.

5. The respective relevant computations in Items 1–4 above are all readily parallelized with respect to the vertices of the triangulation and, for Hermite interpolation multiplicities of uniform order, the load of the computations for each vertex of the triangulation is readily balanced.

References

1. Dechevsky, L.T.: Generalized Expo-Rational B-Splines. In: Communication at the Seventh International Conference on Mathematical Methods for Curves and Surfaces, Tønsbeg, Norway (2008) (unpublished)
2. Dechevsky, L.T., Lakså, A., Bang, B.: Generalized Expo-Rational B-splines. Int. J. Pure Appl. Math. 57(1), 833–872 (2009)
3. Dechevsky, L.T.: Expo-Rational B-Splines. In: Communication at the Fifth International Conference on Mathematical Methods for Curves and Surfaces, Tromsø, Norway (2004) (unpublished)
4. Dechevsky, L.T., Lakså, A., Bang, B.: Expo-Rational B-splines. Int. J. Pure Appl. Math. 27(3), 319–369 (2006)
5. Szivasi-Nagy, M., Vendel, T.P.: Generating curves and swept surfaces by blended circles. Comput. Aided Geom. Design 17(2), 197–206 (2000)
6. Hartmann, E.: Parametric G^n blending of curves and surfaces. The Visual Computer 17, 1–13 (2001)
7. Lakså, A., Bang, B., Dechevsky, L.T.: Exploring Expo-Rational B-splines for Curves and Surfces. In: Dæhlen, M., Mørken, K., Schumaker, L. (eds.) Mathematical Methods for Curves and Surfaces, pp. 253–262. Nashboro Press (2005)

8. Dechevsky, L.T., Lakså, A., Bang, B.: NUERBS form of Expo-Rational B-splines. Int. J. Pure Appl. Math. 32(1), 11–332 (2006)
9. Lakså, A.: Basic properties of Expo-Rational B-splines and practical use in Computer Aided Geometric Design, Doctor Philos. Dissertation at Oslo University, Unipub, Oslo, 606 (2007)
10. Dechevsky, L.T., Quak, E., Lakså, A., Kristoffersen, A.R.: Expo-rational spline multiwavelets: a first overview of definitions, properties, generalizations and applications. In: Truchetet, F., Laligant, O. (eds.) Wavelet Applications in Industrial Processing V, SPIE Conference Proceedings, vol. 6763, article 676308 (2007)

Index Mapping between Tensor-Product Wavelet Bases of Different Number of Variables, and Computing Multivariate Orthogonal Discrete Wavelet Transforms on Graphics Processing Units*

Lubomir T. Dechevsky[1], Jostein Bratlie[1], and Joakim Gundersen[1,2]

[1] R&D Group for Mathematical Modelling,
Numerical Simulation & Computer Visualization,
Faculty of Technology, Narvik University College
2 Lodve Lange's Str. , P.O.B. 385, N-8505 Narvik, Norway
[2] Nordnorsk Havkraft AS,
P.O.B. 55, N-8501 Narvik, Norway

Abstract. An algorithm for computation of multivariate wavelet transforms on graphics processing units (GPUs) was proposed in [1]. This algorithm was based on the so-called *isometric conversion between dimension and resolution* (see [2] and the references therein) achieved by mapping the indices of orthonormal tensor-product wavelet bases of different number of variables and a tradeoff between the number of variables versus the resolution level, so that the resulting wavelet bases of different number of variables are with different resolution, but the overall dimension of the bases is the same.

In [1] we developed the algorithm only up to mapping of the indices of *blocks* of wavelet basis functions. This was sufficient to prove the consistency of the algorithm, but not enough for the *mapping of the individual basis functions* in the bases needed for a programming implementation of the algorithm. In the present paper we elaborate the full details of this 'book-keeping' construction by passing from block-matrix index mapping on to the detailed index mapping of the individual basis functions. We also consider some examples computed using the new detailed index mapping.

1 Introduction

Denote by $\mathbf{V}_{j-1} = \mathbf{V}_j \oplus \mathbf{W}_j$, $j \in \mathbb{Z}$, the nested $\mathbf{V}$-subspaces and $\mathbf{W}$-spaces of an orthonormal multiresolution analysis in $L_2(\mathbb{R}^n, \mathbb{R}^m)$ (see, e.g., [3,7]), where "$\oplus$" is the direct sum of orthogonal subspaces.

* Research partially supported by the 2009, 2010 and 2011 Annual Research Grants of the Priority R&D Group for Mathematical Modeling, Numerical Simulation and Computer Visualization at Narvik University College, Norway.

I. Lirkov, S. Margenov, and J. Waśniewski (Eds.): LSSC 2011, LNCS 7116, pp. 402–410, 2012.

In [2] a new method was introduced for matching the bases of scaling functions in the V-spaces of orthonormal multiresolution analyses in L_2-spaces of n-variate m-dimensional vector-fields. In comparison with the Cantor diagonal type method proposed earlier in [5,6], the method proposed in [2] was less general, but algorithmically simpler. In [2] this new simpler method was used for solving systems of nonlinear equations, with application to the solution of computational-geometric intersection problems in n-dimensions, $n = 2, 3, 4, \ldots$.
In [1] the approach proposed in [5,6], endowed with the new method introduced in [2], was applied to solve another problem within the rapidly expanding field of computational imaging: the parallel computation of n-variate orthogonal discrete wavelet transforms (DWTs) on graphics processing units (GPUs).

In [1] the new algorithm was developed only up to mapping of the indices of *blocks* of wavelet basis functions. This was sufficient to prove the consistency of the algorithm, but not enough for the *mapping of the individual basis functions* in the bases needed for a GPU-programming implementation of the algorithm. In the present paper we provide explicit formulas for the direct and inverse basis matching in the 'book-keeping' construction in [1], by passing from block-matrix index mapping on to the detailed, explicit, index mapping of the individual basis functions.

The exposition of the paper assumes knowledge of the preliminary notation and results about isometric Hilbert spaces, orthogonal operators and their adjoints, orthogonal wavelet expansions in one and several variables and sequence norms in terms of wavelet coefficients, as introduced in [2, section 2]. The notation used is fully consistent with [1].

Part of the present work, together with [5,6,2,1], was part of the research of the R&D Group for Mathematical Modelling, Numerical Simulation and Computer Visualization at Narvik University College within two consecutive Strategic Projects of the Norwegian Research Council –'GPGPU[1] – Graphics Hardware as a High-end Computational Resource' (2004-2007) and 'Heterogeneous Computing' (2008-2010).

The organization of the paper is, as follows. In section 2 we provide a concise summary of the main results of [1] which serve as problem setup and provide the reader with the necessary background needed for the exposition in the next section. Section 3 contains the main new results; these results are a development and upgrade of the main results of [1] (see section 2). In section 4 we provide a first graphical visualization of the results of the new explicit basis-matching algorithm on a benchmark image. Section 5 is comprised of some concluding remarks.

2   The Algorithm Developed in [1]

In [1] we considered in detail the case $n = d$, $m = 1$ (the d-dimensional DWT, i.e., the DWT for processing d-variate scalar-valued fields). The case of m-dimensional

---

[1] *General Purpose* computation on *Graphics Processing Units* (GPGPU): http://gpgpu.org/

vector-fields, $m = 2, 3, \ldots$, follows easily by applying the results of [1, section 2] to each coordinate of the vector-field. We shall continue making use of this reduction also in the present paper.

With the latter reduction in mind, consider the Hilbert spaces with respect to the standard inner products in $L_2(\mathbb{R}^d, \mathbb{R}^1)([0,1]^d)$ and $L_2(\mathbb{R}^1, \mathbb{R}^1)([0,1]^1)$, respectively, $\mathbf{V}_{d,J} = \underbrace{\mathbf{V}_J \otimes \mathbf{V}_J \otimes \ldots \otimes \mathbf{V}_J}_{d-times}$, with dimension (number of variables) d of the integration variable, resolution level J and, $\mathbf{V}_{1,dJ}$, and with dimension 1 (univariate) integration variable, resolution level dJ, where the operator $\otimes$ corresponds to tensor-product of spaces spanned over scaling function bases, as in [3,4,1].

Following [1], the bases in $\mathbf{V}_{d,J}$ and $\mathbf{V}_{1,dJ}$ are denoted by $\boldsymbol{\Phi}_{d,J}$ and $\boldsymbol{\Phi}_{1,Jd}$, respectively, with $\dim \boldsymbol{\Phi}_{d,J} = 2^{Jd} = \dim \boldsymbol{\Phi}_{1,Jd}$.

Since the spaces $\mathbf{V}_{d,J}$ and $\mathbf{V}_{1,dJ}$ have the same dimension, and each one of them has a selected fixed basis in it, there is a "canonical" isometry $\mathbf{A}_{d,J}$ between them which maps $1-1$ the fixed basis of $\mathbf{V}_{d,J}$ onto the fixed basis of $\mathbf{V}_{1,dJ}$, see [1].

This choice of the canonical isometry is unique modulo right composition with an orthogonal operator $\sigma_{d,J}$ and modulo left composition with an orthogonal operator $\sigma_{1,dJ}$ corresponding to a permutation of the basis in $\mathbf{V}_{d,J}$ and $\mathbf{V}_{1,dJ}$, respectively. Any such permutation operator would be represented by a permutation matrix (orthogonal matrix with only one non-zero element in every row and every column, the non-zero element being equal to 1). See the mutually inverse commutative diagrams in [1, Fig.1].

Let $d \in \mathbb{N}$. Without loss of generality, in [1] it was assumed that $J \geq 1$ and $0 \leq j = j_d \leq J - 1$.

Choosing and fixing j_0: $0 \leq j_0 \leq J - 1$, and setting $j = j_d = j_0$, $j_1 = j_0 \cdot d$, in [1] $\mathbf{V}_{d,J}$ and $\mathbf{V}_{1,Jd}$ were decomposed into direct sums of $\mathbf{V}$- and $\mathbf{W}$-spaces, as follows:

$$\mathbf{V}_{d,J} \text{ isometric to } \bar{\mathbf{V}}_{d,J,j_0} = \mathbf{V}_{d,j_0} \oplus \bigoplus_{\mu=j_0}^{J-1} \mathbf{W}_{d,\mu},$$

$$\mathbf{V}_{1,Jd} \text{ isometric to } \bar{\mathbf{V}}_{1,Jd,j_1} = \mathbf{V}_{1,j_0 d} \oplus \bigoplus_{\nu=j_0 d}^{Jd-1} \mathbf{W}_{1,\nu}.$$

Next, in [1] an isometry was constructed between $\bar{\mathbf{V}}_{d,J,j_0}$ and $\bar{\mathbf{V}}_{1,Jd,j_1}$ by proposing an appropriate $1-1$ bijection between their bases. Namely, rewriting the $\oplus$-term in the RHS of $\bar{\mathbf{V}}_{1,Jd,j_1}$ to match RHS in $\bar{\mathbf{V}}_{d,J,j_0}$ as

$$\bigoplus_{\nu=j_0 d}^{Jd-1} \mathbf{W}_{1,\nu} = \bigoplus_{\mu=j_0}^{J-1} \underbrace{\bigoplus_{\nu=\mu d}^{(\mu+1)d-1} \mathbf{W}_{1,\nu}}_{=\mathbf{W}_{1,\mu}^d} = \bigoplus_{\mu=j_0}^{J-1} \mathbf{W}_{1,\mu}^d,$$

in [1] it was obtained that

$\mathbf{V}_{d,J}$ isometric to $\bar{\mathbf{V}}_{d,J,j_0} = \mathbf{V}_{d,j_0} \oplus \bigoplus_{\mu=j_0}^{J-1} \mathbf{W}_{d,\mu},$

$\mathbf{V}_{1,Jd}$ isometric to $\bar{\mathbf{V}}_{1,Jd,j_1} = \mathbf{V}_{1,j_0d} \oplus \bigoplus_{\mu=j_0}^{J-1} \mathbf{W}_{1,\mu}^d.$

The next step in [1] was, therefore, to design appropriate bijections between the orthonormal bases of $\mathbf{W}_{d,\mu}$ and $\mathbf{W}_{1,\nu}^d$, $\mu = j_0, \dots, J-1$, given by

$$\mathbf{W}_{d,\mu} = \bigoplus_{l=1}^{2^d-1} (\mathbf{Z}_{1,\mu}^{\sigma_{1,l}} \otimes \mathbf{Z}_{1,\mu}^{\sigma_{2,l}} \otimes \cdots \otimes \mathbf{Z}_{1,\mu}^{\sigma_{d,l}})$$

and

$$\mathbf{W}_{1,\nu}^{d,\mu} = \bigoplus_{n=0}^{d-1} \mathbf{W}_{1,n+\mu d}.$$

Here $\sigma_{k,l} \in \{0,1\}$, $k = 1, \dots, d$, $l = 1, \dots, 2^d - 1$; $\sigma_{k,l}$, $k = 1, \dots, d$, are uniquely determined by the binary representation of l; $l = \sum_{k=1}^{d} \sigma_{k,l} 2^{d-k}$ and

$$\mathbf{Z}_{1,\mu}^{\sigma_{k,l}} = \begin{cases} \mathbf{V}_{1,\mu}, & \sigma_{k,l} = 0 \\ \mathbf{W}_{1,\mu}, & \sigma_{k,l} = 1 \end{cases}$$

$k = 1, \dots, d$, $l = 1, \dots, 2^d - 1$, $\mu = j_0, j_0 + 1, \dots, J - 1$.

Next, in [1] $W_{d,\mu}$ was mapped 1-1 onto $W_{1,\mu}^d$ for every $\mu = j_0, \dots, J-1$, in terms of blocks of basis functions of dimension $2^{\mu d}$ each, as follows, see [1, Fig. 2]:

1. $\mathbf{V}_{d,j_0}$ maps 1-1 onto $\mathbf{V}_{1,j_0d}$ by the isometric operator $\mathbf{A}_{d,j_0}$.
2. For every $\mu = j_0, \dots, J-1$, $\mathbf{W}_{d,\mu}$ will be mapped to the $2^d - 1$ corresponding blocks of basis functions $\mathbf{W}_{1,n+\mu d}$ for $n = 0, \dots, d-1$.
3. Every corresponding block of basis functions in $\mathbf{W}_{d,\mu}$ (with dimension d and 2^μ) is then being mapped 1-1 onto a respective block of basis functions $\mathbf{W}_{1,\mu}$ (with dimension 1 and size $2^{\mu d}$) by the isometric operator $\mathbf{A}_{d,\mu}$. The selection of the operator $\mathbf{A}_{d,\mu}$ can be made in many different ways, for example, by the original Cantor diagonal method developed in [5,6], or by the method proposed in [2].

In the present paper we opt for the latter method which is simpler to implement, thanks to which it will lead us to the derivation of explicit formulae in the next section 3. Namely (see [2, subsection 3.1]), $\mathbf{A}_{d,\mu} \colon (k_1, \dots, k_d) \in \{0, 1, \dots, 2^\mu - 1\}^d \mapsto n \in \{0, 1, \dots, 2^{\mu d} - 1\}$ is defined by

$$n = \sum_{i=1}^{d} k_i \cdot 2^{\mu(d-i)}.$$

Its inverse (and, in view of its orthogonality, also its adjoint)

$\mathbf{A}_{d,J}^T$: $n \in \{0, 1, \ldots, 2^{\mu d} - 1\} \mapsto (k_1, \ldots, k_d) \in \{0, 1, \ldots, 2^\mu - 1\}^d$ is given by

$$
\begin{aligned}
\left[\frac{n}{2^{\mu(d-1)}}\right] &= k_1 \ , \ n - k_1 \cdot 2^{\mu(d-1)} = n_1, \\
\left[\frac{n_1}{2^{\mu(d-2)}}\right] &= k_2 \ , \ n_1 - k_2 \cdot 2^{\mu(d-2)} = n_2, \\
&\vdots \qquad\qquad\qquad \vdots \\
\left[\frac{n_d}{2^{\mu(d-d)}}\right] &= k_d \ , \ n_{d-1} - k_d \cdot 2^0 = 0.
\end{aligned}
$$

where $[a]$ is the customary Gaussian bracket (integer part) of a: $[a] = n \in \mathbb{Z}$, where n is the only integer solution of the system of inequalities $n : n \leq a < n+1$.

The explicitness of the definition of the so-defined operator $\mathbf{A}_{d,\mu}$ and its inverse $\mathbf{A}_{d,J}^T$ will allow us to obtain explicit formulae for the direct and inverse matching of the indices of the bases in the respective d-variate and univariate tensor-product wavelet bases, as follows in the next section.

3 Main Results

This section contains the main new results in the present paper, where we go one crucial step further in the development of the basis-matching algorithm proposed in [1]: we pass from block-matrix index mapping in [1, Fig. 2] on to the detailed index mapping of the individual basis functions, by obtaining explicit formulae for the direct and inverse matching of the elements of the index sets for the d-variate and univariate case. Our new results are formulated in the following theorem.

Theorem 1. *Consider the index sets by means of which a d-variate tensor-product orthonormal wavelet basis and a corresponding univariate orthonormal wavelet basis are being enumerated:*

$$(d; j; i; k_1, \ldots, k_d) \leftrightarrow (\gamma, \varkappa),$$

respectively.

Assume that the ranges of the parameters in the index sets are consistently matched, as follows:

$$j = j_0, \ldots, J - 1; \quad i = 1, \ldots, 2^{d-1} - 1; \quad k_l = 0, \ldots, 2^j - 1, \quad l = 1, \ldots, d, \quad (1)$$

for the d-variate case, and

$$\gamma = j_0 d, j_0 d + 1, \ldots, J d - 1; \quad \varkappa = 0, 1, \ldots, 2^\gamma - 1, \quad (2)$$

for the univariate case, respectively.

Then,

1. *The direct (d-variate to 1-variate) index mapping*

$$\mathcal{I}_{d \to 1} : (j, i, k_1, \ldots, k_d) \mapsto (\gamma, \varkappa)$$

is computed by

$$\gamma = jd + [\log_2 i] ; \tag{3}$$

$$\varkappa = 2^{jd} \left(i - 2^{[\log_2 i]} \right) + \sum_{l=1}^{d} k_l \, 2^{j(d-l)}. \tag{4}$$

2. *The inverse (1-variate to d-variate) index mapping*

$$\mathcal{I}_{1 \to d} : (\gamma, \varkappa) \mapsto (j, i, k_1, \dots, k_d)$$

is computed by

$$j = \left[\frac{\gamma}{d} \right] ; \tag{5}$$

$$i = 2^{\gamma - [\frac{\gamma}{d}]d} + \left[\frac{\varkappa}{2^{[\frac{\gamma}{d}]d}} \right] ; \tag{6}$$

$k_l, \, l = 1, \dots, d \; :$

$$\left[\frac{n}{2^{\mu(d-1)}} \right] = k_1, \; n - k_1 \times 2^{\mu(d-1)} = n_1 \tag{7}$$

$$\left[\frac{n_1}{2^{\mu(d-2)}} \right] = k_2, \; n_1 - k_2 \times 2^{\mu(d-2)} = n_2 \tag{8}$$

$$\vdots \tag{9}$$

$$\left[\frac{n_d}{2^{\mu(d-d)}} \right] = k_d, \; n_{d-1} - k_d \times 2^0 = 0, \tag{10}$$

where

$$n = \varkappa_0 = \varkappa - \left[\frac{\varkappa}{2^{[\frac{\gamma}{d}]d}} \right] \times 2^{[\frac{\gamma}{d}]d}, \tag{11}$$

$$\mu = \left[\frac{\gamma}{d} \right] . \tag{12}$$

3.

$$\mathcal{I}_{1 \to d} \mathcal{I}_{d \to 1} = \mathcal{I}_d, \quad \mathcal{I}_{d \to 1} \mathcal{I}_{1 \to d} = \mathcal{I}_1, \tag{13}$$

where $\mathcal{I}_d$ and $\mathcal{I}_1$ are the identities on the ranges of the index sets for the d-variate and univariate case, respectively, and, in this sense,

$$\mathcal{I}_{1 \to d} = \mathcal{I}_{d \to 1}^{-1}, \quad \mathcal{I}_{d \to 1} = \mathcal{I}_{1 \to d}^{-1} \tag{14}$$

hold true.

Theorem 1 can be seen as a new detailed closed-form formulation, by explicit formulae for the indices, of the graphical block-by-block outline of the algorithm in [1, Figure 2].

Due to space constraints, the proof of the theorem cannot be provided here, but in section 5 we include a brief discussion how the general case of $1 - 1$ mapping between d_1-variate and d_2-variate wavelet bases, $d_1, d_2 \in \mathbb{N}$, can be reduced to Theorem 1.

Fig. 1. The index matching used on the benchmark "Lena" image (section 4)

4   Graphical Visualization

In this section we provide graphical visualization of the results of the new index-matching algorithm. In Figure 1 we analyze the original benchmark "Lena" image, by converting, as customary, the greyscale values in the pixels to values of scaling-function coefficients for resolution level J chosen so, as to fit the support of the tensor-product scaling function to the size of the pixel.

The following description is consistent with the commutative diagram in [1, Figure 1]:

- Image representation:
    - *Top left:* Representation in $V_{2,J} = V_{1,J} \otimes V_{1,J}$.
    - *Bottom left:* Representation in $V_{1,2J}$.
    - *Top right:* Representation in $V_{2,J-1} \oplus W_{2,J-1}^{[1]} \oplus W_{2,J-1}^{[2]} \oplus W_{2,J-1}^{[3]}$.
    - *Bottom right:* Representation in $V_{1,2J-2} \oplus W_{1,2J-2} \oplus W_{1,2J-1}$.
- Transformation between image representations:
    - *Top left to top right:* 2-dimensional DWT.
    - *Top left to bottom left:* cartesian algorithm from [2] for matching scaling-function bases.
    - *Bottom left to bottom right:* 1-dimensional DWT.
    - *Bottom right to top right:* The new (inverse) algorithm for matching scaling-function and wavelet bases, as developed in [1] and the present paper.

The consecutive composition of the transformations in the latter 3 items (in this order) is equivalent to the transformation in the first item (cf. [1, Figure 1]).

5 Concluding Remarks

Remark 1. Currently, we are working on extending Theorem 1 to the more general case of matching the tensor-product wavelet bases of d_1 and d_2 variables, for appropriate selection of the respective resolutions. In this more general context, our present result should coincide with the partial case $d_1 = d$, $d_2 = 1$.

Remark 2. Prospective applications of the results in Theorem 1 include, but are not limited to, the following.

1. In *computational imaging*: for computing multivariate DWTs via GPGPU-programming.
2. In *telecommunications*: for transferring large volumes of multidimensional geometric data using any of the standard technologies for transferring information in the form of univariate signals.
3. In *data encryption*: if a more general multi-wavelet construction is used, possible encryption keys may include the number of scaling functions and the filter coefficients generating them.
4. In *functional analysis*: for exploring new trace and retract theorems for function spaces of different numbers of variables, such as the Besov and Triebel-Lizorkin space scales.

References

1. Dechevsky, L.T., Gundersen, J., Bang, B.: Computing n-Variate Orthogonal Discrete Wavelet Transforms on Graphics Processing Units. In: Lirkov, I., Margenov, S., Wasniewski, J. (eds.) LSSC 2009. LNCS, vol. 5910, pp. 730–737. Springer, Heidelberg (2010)
2. Dechevsky, L.T., Bang, B., Gundersen, J., Lakså, A., Kristoffersen, A.R.: Solving Non-linear Systems of Equations on Graphics Processing Units. In: Lirkov, I., Margenov, S., Waśniewski, J. (eds.) LSSC 2009. LNCS, vol. 5910, pp. 719–729. Springer, Heidelberg (2010)
3. Cohen, A.: Wavelet methods in numerical analysis. In: Ciarlet, P.G., Lions, J.L. (eds.) Handbook of Numerical Analysis, vol. VII, pp. 417–712. Elsevier (2000)
4. Dechevsky, L.T.: Atomic decomposition of function spaces and fractional integral and differential operators. In: Rusev, P., Dimovski, I., Kiryakova, V. (eds.) Transform Methods and Special Functions, Part A. Fractional Calculus & Applied Analysis, vol. 2(4), pp. 367–381 (1999)
5. Dechevsky, L.T., Gundersen, J.: Isometric conversion between dimension and resolution. In: Dæhlen, M., Mørken, K., Schumaker, L.L. (eds.) Mathematical Methods for Curves and Surfaces: Tromsø 2004, pp. 103–114. Nashboro Press, Brentwood TN (2005)

6. Dechevsky, L.T., Gundersen, J., Kristoffersen, A.R.: Wavelet-based isometic conversion between dimension and resolution and some of its applications. In: Proceedings of SPIE: Wavelet Applications in Industrial Processing V, Boston, Massachusetts, USA, vol. 6763 (2007)
7. Mallat, S.: A Wavelet Tour of Signal Processing, 2nd edn. Acad. Press (1999)

Interpolation of Curvature and Torsion Using Expo-Rational B-Splines*

Lubomir T. Dechevsky[1] and Georgi H. Georgiev[2]

[1] R&D Group for Mathematical Modelling,
Numerical Simulation & Computer Visualization,
Faculty of Technology, Narvik University College
2 Lodve Lange's Str. , P.O.B. 385, N-8505 Narvik, Norway
LubomirT.Dechevsky@hin.no
[2] Faculty of Mathematics and Informatics, Konstantin Preslavsky University
115 Universitetska Str., BG-9712 Shumen, Bulgaria
g.georgiev@fmi.shu-bg.net

Abstract. Expo-rational B-splines (ERBS) are well adapted for Hermite interpolation of any prescribed order [1]. This property of ERBS can be used to interpolate and approximate a variety of differential geometric structures of smooth manifolds, such as length, curvature, torsion, area, volume, etc. In this paper we solve the simplest of these problems: finding the explicit formulae for ERBS-based interpolation of curvature and torsion of unit-speed 3D-space curves and the order of the rate of approximation it provides.

1 Introduction

Hermite interpolation up to certain order of the derivatives of univariate and multivariate smooth functions can be used for interpolation of differential geometric structures of smooth manifolds, where the order of the derivatives involved in the structure does not exceed the order of Hermite interpolation in any of the interpolation knots. When Hermite-interpolating with the interpolation and approximation apparatus of *expo-rational B-splines* (ERBS) [1], it is possible to achieve Hermite interpolation of any prescribed order without having to change the ERBS basis providing it. Moreover, the fact that ERBS simultaneously provide respective smooth convex partitions of unity guarantees that the Hermite interpolation property of ERBS generates good approximation of the interpolated function. Together, these two properties of ERBS make them a good tool for solving computational problems in differential geometry of curves, surfaces, volume deformations and higher dimensional manifolds, with potential applications

* The research of Dechevsky and Georgiev was partially supported, respectively, by the 2010 and 2011 Annual Research Grant of the Priority R&D Group for Mathematical Modeling, Numerical Simulation and Computer Visualization at Narvik University College, and by the Bulgarian Ministry of Education, Youth and Science under EEA Grant No. D02-780/20.10.2010.

I. Lirkov, S. Margenov, and J. Waśniewski (Eds.): LSSC 2011, LNCS 7116, pp. 411–419, 2012.

in computer-aided geometric design, computational physics and fluid dynamics, as well as for solving approximation problems involving vector calculus and tensor analysis. The above-said properties of ERBS will be used to interpolate and approximate length, curvature, torsion, area, volume, and other intrinsic invariants in differential geometry. In the present introductory paper we shall address the simplest one of these problems: finding formulae for ERBS-based interpolation of curvature and torsion of unit-speed 3D-space curves, together with the rate of approximation provided by this interpolation.

The organization of the paper is, as follows. In section 2 we provide the necessary minimum of information concerning the differential geometry of smooth space curves (in subsection 2.1) and the concept of ERBS (in subsection 2.2). Section 3 contains the main results. In particular, we study ERBS-based interpolation of arc-length, curvature, and torsion of normally parametrized (unit-speed) regular curves in subsections 3.1, 3.2, and 3.3, respectively. In section 4 we identify the class of regular smooth space curves which can be exactly reproduced (modulo congruence) via the new interpolation formulae in section 3; we also provide some model instances of unit-speed curves which can be used as examples for numerical testing of the results of the paper. Section 5 contains some concluding remarks concerning the continuation of this research.

2   Preliminaries

2.1   Smooth Regular Curves in the Euclidean 3-Space

Norm of the tangent vector, natural parametrization, curvature, and torsion are among the main concepts in differential geometry of curves. Their numerical approximations provide important tools in computational geometry of parametric space curves. Consider a curve with parametrization

$$\overrightarrow{p} : [a, b] \to \mathbb{E}^3$$

given by

$$\overrightarrow{p}(t) = \big(x(t), y(t), z(t)\big), \quad t \in [a, b].$$

We assume that $\overrightarrow{p} \in C^m$, $m \geq 1$, as well as that the curve is regular, i.e., $\overrightarrow{p}(t) = \frac{d}{dt}\overrightarrow{p}(t)$ is nonzero everywhere on $[a, b]$.

For all relevant information used in the paper on differential geometry of curves, we refer to [3] or [2].

2.2   Hermite Interpolation with ERBS

Consider knots

$$a = t_1 < t_2 < \ldots < t_N = b,$$

and suppose that

$$h = \max(t_k - t_{k-1}).$$

Then, the Hermitian interpolation formula for smooth function is determined by

$$H(f;t) = \sum_{k=1}^{N} \left[\sum_{j=0}^{r_k-1} \frac{f^{(j)}(t_k)}{j!}(t-t_k)^j \right] B_k(t), \quad t \in [a,b],$$

where $f^{(j)}(t) = \dfrac{d^j}{dt^j}f(t)$, r_k is the multiplicity of the knot t_k, $B_k(t)$ is the ERBS associated with the strictly increasing adjacent knots t_{k-1}, t_k and t_{k+1} (see [1] and [4] for all relevant information on the definition and properties of ERBS). Note that, when using ERBS, the Hermite interpolation can even be transfinite: some or all of the multiplicities r_k can be equal to $+\infty$, in which case the interpolated function is assumed analytic on (t_{k-1}, t_{k+1}) and the functional coefficient multiplying the respective B_k is a Taylor series.

We may consider approximation by the Hermitian interpolant for the curve $\vec{p}$ of class C^m, $m \in \mathbb{N}$, as follows:

$$\vec{p}(t) = \sum_{k=1}^{N} \left[\sum_{j=0}^{m-1} \frac{\vec{p}^{(j)}(t_k)}{j!}(t-t_k)^j \right] B_k(t) + O_h(h^m), \tag{1}$$

where $\vec{p}^{(j)}(t) = \dfrac{d^j}{dt^j}\vec{p}(t)$. Here and in the remaining part of the paper the error remainders are in Peano form.

3 Main Results

3.1 The Square of the Norm of the Tangent Vector

The tangent vector $\dot{\vec{p}}$ can be approximated by the same type of Hermite-interpolating formula as $\vec{p}$ in (1):

$$\dot{\vec{p}}(t) = \sum_{k=1}^{N} \left[\sum_{j=1}^{m-1} \frac{\vec{p}^{(j)}(t_k)}{(j-1)!}(t-t_k)^{j-1} \right] B_k(t) + O_h(h^{m-1}). \tag{2}$$

We may now approximate the square of the norm of the tangent vector $\|\dot{\vec{p}}\|^2 = \langle \dot{\vec{p}}, \dot{\vec{p}} \rangle$, as follows.

Theorem 1. *For the curve $\vec{p}(t)$ of class C^m it is fulfilled*

$$\|\dot{\vec{p}}(t)\|^2 = \|B_{k-1}(t)\dot{\vec{p}}(t_{k-1}) + B_k(t)\dot{\vec{p}}(t_k)\|^2 + \mathcal{R}_h^{(1)}, \tag{3}$$

where $t \in (t_{k-1}, t_k)$,

$$\mathcal{R}_h^{(1)} = \begin{cases} o_h(1), & h \to 0^+, \text{ if } \vec{p} \in C^1; \\ O(h), & h \to 0^+, \text{ if } \vec{p} \in C^m, m \geq 2. \end{cases} \tag{4}$$

Proof. First, we calculate

$$
\langle \vec{\rho}, \vec{\rho} \rangle
$$

$$
= \Big\langle \sum_{k_1=1}^{N} \Big[\sum_{j_1=1}^{m-1} \frac{\vec{\rho}^{(j_1)}(t_{k_1})}{(j_1-1)!} (t-t_{k_1})^{j_1-1} \Big] B_{k_1}(t) + O_h(h^{m-1}),
$$

$$
\sum_{k_2=1}^{N} \Big[\sum_{j_2=1}^{m-1} \frac{\vec{\rho}^{(j_2)}(t_{k_2})}{(j_2-1)!} (t-t_{k_2})^{j_2-1} \Big] B_{k_2}(t) + O_h(h^{m-1}) \Big\rangle
$$

$$
= \sum_{k_1=1}^{N} \sum_{k_2=1}^{N} A_{k_1,k_2}(t) B_{k_1}(t) B_{k_2}(t) + O_h(h^{m-1}),
$$

where

$$
A_{k_1,k_2}(t) = \sum_{j_1=1}^{m-1} \sum_{j_2=1}^{m-1} \frac{(t-t_{k_1})^{j_1-1}(t-t_{k_2})^{j_2-1}}{(j_1-1)!(j_2-1)!} \Big\langle \vec{\rho}^{(j_1)}(t_{k_1}), \vec{\rho}^{(j_2)}(t_{k_2}) \Big\rangle.
$$

From the minimal support (over $[t_{k-1}, t_k]$) of $B_k(t)$ and the Taylor interpolation form of its local function it follows that

$$
\langle \vec{\rho}, \vec{\rho} \rangle = \sum_{k_1=1}^{N} \sum_{k_2=\max\{0,k_1-1\}}^{\min\{N,k_1+1\}} A_{k_1,k_2}(t) B_{k_1}(t) B_{k_2}(t).
$$

Replacing k_1 with k and restricting the range of t onto (t_{k-1}, t_k), we obtain

$$
\langle \vec{\rho}, \vec{\rho} \rangle = A_{k-1,k-1}(t) B_{k-1}(t) B_{k-1}(t) + A_{k-1,k}(t) B_{k-1}(t) B_k(t)
$$
$$
+ A_{k,k-1}(t) B_k(t) B_{k-1}(t) + A_{k,k}(t) B_k(t) B_k(t) + O_h(h^{m-1}).
$$

Finally, the square of the norm of the tangent vector can be written in the form

$$
\langle \vec{\rho}, \vec{\rho} \rangle = [B_{k-1}(t)]^2 \|\vec{\rho}(t_{k-1})\|^2 + 2B_{k-1}(t) B_k(t) \Big\langle \vec{\rho}(t_{k-1}), \vec{\rho}(t_k) \Big\rangle
$$
$$
+ [B_k(t)]^2 \|\vec{\rho}(t_k)\|^2 + \mathcal{R}_h^{(1)} + O_h(h^{m-1})
$$
$$
= \Big\| B_{k-1}(t) \vec{\rho}(t_{k-1}) + B_k(t) \vec{\rho}(t_k) \Big\|^2 + \mathcal{R}_h^{(1)}.
$$

This completes the proof of (3), since the approximation rate $O_h(h^{m-1})$ is higher than or equal to $\mathcal{R}_h^{(1)}$ (see also Remark 4).

The error-rate estimate of $\mathcal{R}_h^{(1)}$ in (4) follows from the properties of a Taylor expansion with its error remainder taken in the o_h- or O_h-form of Peano, depending on the type of regularity of the highest order derivative of ρ.

$\square$

Remark 1. As it is well known, the arc-length function $s(t)$ and the norm of the tangent vector are related by

$$
s(t) = \int_a^t \sqrt{\|\vec{\rho}(\theta)\|^2} \, d\theta. \tag{5}
$$

The parametrization with respect to s is customarily called *a natural parametrization* and the derivatives with respect to s are denoted by $'$, $''$ etc. The regular curves endowed with natural parametrization are called *unit-speed* curves, and for them the formula for computing their curvature is particularly simple. We shall use this in subsection 3.2.

Remark 2. Note the important fact that in (3) $B_{k-1}(t)$ and $B_k(t)$ form a *smooth convex partition of unity* for every $t \in (t_{k-1}, t_k)$.

3.2   The Curvature of Unit-Speed Space Curves

In this and the next subsection we assume that the parameter t is natural (with some minor abuse of notation with respect to Remark 1). Clearly, any unit-speed curve $\vec{p}$ is regular, so if $\vec{p} \in C^m$ with $m \geq 2, \ldots$, then, together with (1) and (2), we may also express the Hermite interpolant for ρ'':

$$\vec{p}''(t) = \sum_{k=1}^{N} \left[\sum_{j=2}^{m-1} \frac{\vec{p}^{(j)}(t_k)}{(j-2)!} (t - t_k)^{j-2} \right] B_k(t) + O_h(h^{m-2}). \tag{6}$$

Now we can approximate the square of the curvature $\kappa^2 = \langle \vec{p}'', \vec{p}'' \rangle$, as follows.

Theorem 2. *For the regular space curve $\vec{p}(t)$ of class C^m, $m \geq 2, \ldots$, with natural parametrization, it holds that*

$$\kappa^2 = \| B_{k-1}(t) \vec{p}''(t_{k-1}) + B_k(t) \vec{p}''(t_k) \|^2 + \mathcal{R}_h^{(2)}, \tag{7}$$

where $t \in (t_{k-1}, t_k)$ and

$$\mathcal{R}_h^{(2)} = \begin{cases} o_h(1), & h \to 0^+, \text{ if } \vec{p} \in C^2; \\ O(h), & h \to 0^+, \text{ if } \vec{p} \in C^m, m \geq 3. \end{cases} \tag{8}$$

Proof. (Outline.) The proof of Theorem 2 is analogous to the proof of Theorem 1: here one only needs to replace the first derivative with respect to an arbitrary parameter with the second derivative with respect to a natural (an arc-length) parameter (and, thus, formula (6) is used in the proof instead of formula (2)). See also Remark 2.   □

3.3   The Torsion of Unit-Speed Curves

Consider a regular curve of class C^m, $m \geq 3, \ldots$,

$$\rho : [a, b] \to \mathbb{E}^3$$

with natural parametrization. Then, together with (1), (2) and (6), the Hermite interpolant for ρ''' can be written as

$$\vec{p}'''(t) = \sum_{k=1}^{N} \left[\sum_{j=3}^{m-1} \frac{\vec{p}^{(j)}(t_k)}{(j-3)!} (t - t_k)^{j-3} \right] B_k(t) + O_h(h^{m-3}). \tag{9}$$

Now we can approximate the torsion

$$\tau = \frac{\vec{p}' \vec{p}'' \vec{p}'''}{\langle \vec{p}'', \vec{p}'' \rangle} \tag{10}$$

of the curve, as follows.

Theorem 3. *For the regular space curve $\vec{\rho}(t)$ of class C^m, $m \geq 3$, with natural parametrization it holds that*

$$
\begin{aligned}
\tau = & \frac{[B_{k-1}(t)]^2 \langle \vec{\rho}''(t_{k-1}), \vec{\rho}'''(t_{k-1}) \rangle (t - t_{k-1})}{\|B_{k-1}(t)\vec{\rho}''(t_{k-1}) + B_k(t)\vec{\rho}''(t_k)\|^2 + \mathcal{R}_h^{(2)}} \\
& + \frac{2B_{k-1}(t)B_k(t) \langle \vec{\rho}''(t_{k-1}), \vec{\rho}'''(t_k) \rangle (t - t_k)}{\|B_{k-1}(t)\vec{\rho}''(t_{k-1}) + B_k(t)\vec{\rho}''(t_k)\|^2 + \mathcal{R}_h^{(2)}} \\
& + \frac{[B_k(t)]^2 \langle \vec{\rho}''(t_k), \vec{\rho}'''(t_k) \rangle (t - t_k) + \mathcal{R}_h^{(3)}}{\|B_{k-1}(t)\vec{\rho}''(t_{k-1}) + B_k(t)\vec{\rho}''(t_k)\|^2 + \mathcal{R}_h^{(2)}}
\end{aligned}
\tag{11}
$$

where $t \in (t_{k-1}, t_k)$,

$$
\mathcal{R}_h^{(2)} = O(h), \quad h \to 0^+
\tag{12}
$$

and

$$
\mathcal{R}_h^{(3)} = \begin{cases} o_h(h), & h \to 0^+, \ \text{if } \vec{\rho} \in C^3; \\ O(h^2), & h \to 0^+, \ \text{if } \vec{\rho} \in C^m, m \geq 4. \end{cases}
\tag{13}
$$

Proof. Let us consider the cross product

$$
\begin{aligned}
\sigma(t) &= \vec{\rho}'(t) \vec{\rho}''(t) \vec{\rho}'''(t) \\
&= \sum_{k_1=1}^{N} \sum_{k_2=1}^{N} \sum_{k_3=1}^{N} A_{k_1,k_2,k_3}(t) B_{k_1}(t) B_{k_2}(t) B_{k_3}(t) + r_h ,
\end{aligned}
$$

where the function $A_{k_1,k_2,k_3}(t)$ can be expressed as

$$
\sum_{j_1=1}^{m-1} \sum_{j_2=2}^{m-1} \sum_{j_3=3}^{m-1} \frac{(t-t_{k_1})^{j_1-1}(t-t_{k_2})^{j_2-2}(t-t_{k_3})^{j_3-3}}{(j_1-1)!(j_2-2)!(j_3-3)!} \vec{\rho}^{(j_1)}(t_{k_1}) \, \vec{\rho}^{(j_2)}(t_{k_2}) \, \vec{\rho}^{(j_3)}(t_{k_3})
$$

and the error remainder r_h depends on the error remainders in (2), (6), and (9). (The rate of r_h will be computed below.) The range of supports of the involved ERBS implies

$$
\sigma(t) = \sum_{k_1=1}^{N} \sum_{k_2=\max\{0,k_1-1\}}^{\min\{N,k_1+1\}} \sum_{k_3=\max\{0,k_1-1\}}^{\min\{N,k_1+1\}} A_{k_1,k_2,k_3}(t) B_{k_1}(t) B_{k_2}(t) B_{k_3}(t) .
$$

Hence, reducing the range of the parameter t onto the interval (t_{k-1}, t_k) yields

$$
\begin{aligned}
\sigma(t) = & A_{k-1,k-1,k-1}(t) B_{k-1}(t) B_{k-1}(t) B_{k-1}(t) \\
& + A_{k-1,k-1,k}(t) B_{k-1}(t) B_{k-1}(t) B_k(t) \\
& + A_{k-1,k,k-1}(t) B_{k-1}(t) B_k(t) B_{k-1}(t) \\
& + A_{k,k-1,k-1}(t) B_k(t) B_{k-1}(t) B_{k-1}(t) \\
& + A_{k-1,k,k}(t) B_{k-1}(t) B_k(t) B_k(t) \\
& + A_{k,k-1,k}(t) B_k(t) B_{k-1}(t) B_k(t) \\
& + A_{k,k,k-1}(t) B_k(t) B_k(t) B_{k-1}(t) \\
& + A_{k,k,k}(t) B_k(t) B_k(t) B_k(t) + r_h ,
\end{aligned}
$$

and $r_h = O_h(h^{m-2})$. From here, it follows that

$$\sigma(t) = [B_{k-1}(t)]^2 \langle \overrightarrow{\rho}''(t_{k-1}), \overrightarrow{\rho}'''(t_{k-1}) \rangle (t - t_{k-1})$$
$$+ 2B_{k-1}(t)B_k(t) \langle \overrightarrow{\rho}''(t_{k-1}), \overrightarrow{\rho}'''(t_k) \rangle (t - t_k)$$
$$+ [B_k(t)]^2 \langle \overrightarrow{\rho}''(t_k), \overrightarrow{\rho}'''(t_k) \rangle (t - t_k) + \mathcal{R}_h^{(3)} + r_h \,,$$

which, together with Theorem 2, implies (11), since the rate r_h is higher or equal to the rate $\mathcal{R}_h^{(3)}$ (see also Remark 4).

The estimates in (12) and (13) follow from the properties of the respective Taylor expansion with error remainder in Peano form, under the corresponding assumptions of smoothness of $\overrightarrow{\rho}$ and its derivatives.

$\square$

Remark 3. Note that the observation made in Remark 2 concerning (3) implies an analogous important observation in (11): $[B_{k-1}(t)]^2$, $2B_{k-1}(t)B_k(t)$ and $[B_k(t)]^2$ also form a *smooth convex partition of unity* for every $t \in (t_{k-1}, t_k)$.

Remark 4. For the smallest m considered in Theorems 1–3 the O_h-rate in (2), (6), (9) can be replaced by o_h-rate.

4 Inverse Results and Some Examples

Let us now recall *the fundamental theorem of curves in differential geometry*, which, concisely formulated, states that *any regular curve with non-zero curvature has its shape (and size) completely determined by its intrinsic invariants: curvature and torsion.* Considering the main-term ansatz in (11) together with the main-term one in the collective numerator of (11), and using the aforementioned fundamental theorem, it is possible to obtain an inverse result, in the sense of identifying (modulo congruence) the exact subclass of unit-speed curves on which the interpolation formulae in Theorems 2 and 3 are *exact*. We shall not be discussing this in detail here, since it seems more appropriate to study this topic when investigating curves with general, not necessarily natural, parametrization (see also section 5), but we note here that the subclass of such curves with natural parametrization is non-void, since it contains, e.g., the following two elements:

- *Example 1.* The *planar circle*, with natural parametrization, constant non-zero curvature and zero torsion.
- *Example 2.* A non-planar generalization of Example 1: the *helix*, with natural parametrization, constant non-zero curvature and constant non-zero torsion.

In principle, the present Theorems 2 and 3 are applicable to arbitrary regular curves with sufficiently smooth, possibly not unit-speed, general parametrization, since any such curve has a natural parametrization which can be computed, at least numerically, with any prescribed precision. As we shall remark in the next section, however, in subsequent research we intend to generalize Theorems 2 and 3 for arbitrary, not necessarily unit-speed, parametrizations. Nevertheless, there is a relatively narrow class of generally parametrized regular curves for which the following two statements hold simultaneously.

1. The natural arc-length parameter $s = s(t)$ can be computed in terms of elementary (or specific classes of special) functions, when the integral in (5) can be solved in closed-form quadratures (e.g., by symbolic integration in *Maple* or *Mathematica*).
2. The equation $s = s(t)$ obtained in the previous item can be solved with respect to t: $t = t(s)$ in terms of elementary (or specific classes of special) functions (e.g., by symbolic computation in *Maple* or *Mathematica*).

Here we have provided one model example of a curve (non-planar and not a helix) which satisfies both of the above requirements.

Example 3.

$$\vec{p}(t) = \left(e^t, \frac{2\sqrt{2}}{3}e^{3t/2}, \frac{1}{2}e^{2t}\right), \quad t \in [t_0, T], \quad -\infty \le t_0 < T \le +\infty. \tag{14}$$

For this example, (5) becomes

$$s = s(t) = \frac{1}{2}e^{2t} + e^t - \frac{1}{2}e^{2t_0} - e^{t_0} \tag{15}$$

which, after inversion, becomes

$$t = t(s) = \ln\left(\sqrt{2s + e^{2t_0} + 2e^{t_0} + 1} - 1\right). \tag{16}$$

Now (16) can be plugged back into (14) to derive the needed natural parametrization.

Example 3 can be used as a reference case for geometrical investigation of the approximation properties of the interpolants of the curvature and torsion designed in Theorems 2 and 3. We shall not conduct this study here, since we intend to include it as part of a similar study for the generalizations of Theorems 2 and 3 for general parametrization (see the next section).

5   Concluding Remarks

Forthcoming research involving ERBS on this topic includes:

1. Generalizing Theorems 2 and 3 for the case of arbitrary, not necessarily unit-speed parametrizations.
2. Conducting an analogous study for the area element, the normal vector and the curvature of surfaces.
3. Study of the interpolation of higher-order intrinsic geometric invariants and higher-dimensional smooth manifolds.

References

1. Dechevsky, L.T., Lakså, A., Bang, B.: Expo-Rational B-splines. International Journal of Pure and Applied Mathematics 27(3), 319–369 (2006)
2. do Carmo, M.P.: Differential Geometry of Curves and Surfaces. Prentice-Hall (1976)
3. Gray, A.: Modern Differential Geometry of Curves and Surfaces with Mathematica, 2nd edn. CRC Press (1998)
4. Lakså, A., Bang, B., Dechevsky, L.T.: Exploring expo-rational B-splines for curves and surfaces. In: Dæhlen, M., Mørken, K., Schumaker, L.L. (eds.) Mathematical Methods for Curves and Surfaces: Tromsø 2004, pp. 253–262. Nashboro Press, Brentwood TN (2005)

Hermite Interpolation Using ERBS with Trigonometric Polynomial Local Functions[*]

Lubomir T. Dechevsky[1] and Rumen Uluchev[2]

[1] Narvik University College, P.O.B. 385, N-8505 Narvik, Norway
ltd@hin.no
[2] University of Transport (VTU), Sofia 1574, Bulgaria
rumenu@vtu.bg

Abstract. Given a sequence of knots $t_0 < t_1 < \cdots < t_{n+1}$, an expo-rational B-spline (ERBS) function $f(t)$ is defined by

$$f(t) = \sum_{k=1}^{n} \ell_k(t) B_k(t), \quad t \in [t_1, t_n],$$

where $B_k(t)$ are the ERBS and $\ell_k(t)$ are local functions defined on (t_{k-1}, t_{k+1}). Consider the Hermite interpolation problem at the knots $0 \leq t_1 < t_2 < \cdots < t_n < 2\pi$ of arbitrary multiplicities. In [3] a formula was suggested for Hermite-interpolating ERBS function with $\ell_k(t)$ being algebraic polynomials. Here we construct Hermite interpolation by an ERBS function with trigonometric polynomial local functions. We provide also numerical results for the performance of the new trigonometric ERBS (TERBS) interpolant in graphical comparison with the interpolant from [3]. Potential applications and some topics for further research on TERBS are briefly outlined.

1 Introduction and Preliminaries

A new type of B-splines – the Expo-Rational B-splines (ERBS) – were introduced in [3]. ERBS are the asymptotic limits of polynomial B-splines letting the degree of the polynomial B-splines to tend to infinity. The new B-splines hold some important properties where they outperform usual B-splines. An important example is constructing a minimally supported C^∞-partition of unity in one and several dimensions, with potential applications, among others, in 2D/3D geometric modeling.

We proceed to give the definition and a brief description of basic properties of ERBS.

[*] The research of Dechevsky and Uluchev was partially supported, respectively, by the 2010 and 2011 Annual Research Grant of the Priority R&D Group for Mathematical Modeling, Numerical Simulation and Computer Visualization at Narvik University College, and by the Bulgarian Ministry of Education, Youth and Science under Grant No. DDVU 02/30, 2010.

I. Lirkov, S. Margenov, and J. Waśniewski (Eds.): LSSC 2011, LNCS 7116, pp. 420–428, 2012.

Definition. Given a strictly increasing sequence of knots $t_0 < t_1 < \cdots < t_{n+1}$, an *expo-rational B-spline* associated with the knots t_{k-1}, t_k, and t_{k+1} is defined, as follows:

$$B_k(t) = \begin{cases} \int_{t_{k-1}}^{t} \psi_{k-1}(s)\,ds / \int_{t_{k-1}}^{t_k} \psi_{k-1}(s)\,ds, & \text{if } t \in (t_{k-1}, t_k], \\ \int_{t}^{t_{k+1}} \psi_k(s)\,ds / \int_{t_k}^{t_{k+1}} \psi_k(s)\,ds, & \text{if } t \in (t_k, t_{k+1}), \\ 0, & \text{if } t \notin (t_{k-1}, t_{k+1}), \end{cases}$$

where

$$\psi_k(t) = e^{-\frac{\left(t - \frac{t_k + t_{k+1}}{2}\right)^2}{(t - t_k)(t_{k+1} - t)}}.$$

Basic properties of ERBS

P1. $B_k(t) > 0$, $t \in (t_{k-1}, t_{k+1})$, and $B_k(t) = 0$, $t \notin (t_{k-1}, t_{k+1})$;

P2. $\sum_{k=1}^{n} B_k(t) = 1$, i.e., $B_k(t) + B_{k+1}(t) = 1$, $t \in (t_k, t_{k+1}]$, $k = 1, \ldots, n-1$;

P3. $B_k(t_k) = 1$, $k = 1, \ldots, n$;

P4. $\dfrac{d^j}{dt^j} B_k(t_i) = 0$, $j = 1, 2, \ldots$, $k = 1, \ldots, n$;

P5. $B_k \in C^\infty(\mathbb{R})$ and B_k is analytic on $\mathbb{R} \setminus \{t_{k-1}, t_k, t_{k+1}\}$, $k = 1, \ldots, n$.

Given a sequence of knots $t_0 < t_1 < \cdots < t_{n+1}$, an *ERBS function* $f(t)$ is defined by

$$f(t) = \sum_{k=1}^{n} \ell_k(t) B_k(t), \quad t \in [t_1, t_n],$$

where $B_k(t)$ are expo-rational B-splines and $\ell_k(t)$ are local functions defined on (t_{k-1}, t_{k+1}). A formula for Hermite interpolation by ERBS function with $\ell_k(t)$ being algebraic polynomials was suggested in [3].

Here, based on a result of Du, Han, and Jin [5], we construct a new Hermite-interpolating ERBS function using trigonometric polynomial local functions. In the sequel, the resulting new trigonometric ERBS interpolant will be referred to as TERBS, for short.

After formulating and proving our new result, we briefly outline some potential applications and relevant topics for further research.

2   Hermite Interpolation by Trigonometric Polynomials and TERBS

The problem of interpolation at nodes with multiplicities by trigonometric polynomials was studied, e.g., by Delvos [1], Dryanov [4], Kress [6], Salzer [7], etc. However, a complete constructive solution of the problem for arbitrary nodes and multiplicities was given only recently by Du, Han, and Jin [5]. We sketch their approach.

Denote by $\mathcal{T}_N$ the linear space of trigonometric polynomials with real coefficients of up to degree N of the form

$$\tau(x) = a_0 + \sum_{k=1}^{N} (a_k \cos kx + b_k \sin kx), \tag{1}$$

and by $\mathcal{T}_{N+1/2}$ the linear space of up to half-integer degree $N+1/2$ trigonometric polynomials with real coefficients of the form

$$\tau(x) = \sum_{k=0}^{N} \left[a_k \cos\left(k + \frac{1}{2}\right)x + b_k \sin\left(k + \frac{1}{2}\right)x \right], \tag{2}$$

$\mathcal{T}_N$ and $\mathcal{T}_{N+1/2}$ containing all respective cases with $a_N^2 + b_N^2 \neq 0$ in (1) and (2), as well as all respective trigonometric polynomials of lower degree, where $a_N^2 + b_N^2 = 0$.

Consider the following

Hermite interpolation problem. Let $0 \leq t_1 < t_2 \cdots < t_n < 2\pi$ be given points, $\lambda_1, \lambda_2 \ldots, \lambda_n$ be positive integers, $\{d_{ij}\}_{i=1,\, j=0}^{n,\ \lambda_i-1}$ be arbitrary real numbers and $\Lambda = \lambda_1 + \cdots + \lambda_n$.

Find a trigonometric or half-degree trigonometric polynomial of minimal degree, which satisfies the interpolation conditions

$$\tau^{(j)}(t_i) = d_{ij}, \qquad i = 1, \ldots, n, \quad j = 0, \ldots, \lambda_i - 1. \tag{3}$$

Note the two extreme special cases:

1. *Taylor-type interpolation:*
$$n = 1, \quad \lambda_1 \in \mathbb{N}; \tag{4}$$

2. *Lagrange-type interpolation:*
$$n \in \mathbb{N}, \quad \lambda_i = 1, \ i = 1 \ldots, n. \tag{5}$$

Following the idea and notations in [5], we consider the analytic function

$$\left(\frac{x}{2\sin\frac{x}{2}}\right)^{m+1} = \sum_{j=0}^{+\infty} a_{2j}(m) x^{2j}, \qquad |x| < 2\pi, \tag{6}$$

where m is a non-negative integer. For integer m and s, $0 \leq s \leq m$, the functions

$$h_{s,m}(t) = \sin^{m+1}\frac{t}{2} \cdot \frac{(-1)^{m-s}2^m}{s!} \sum_{\ell=0}^{\left[\frac{m-s}{2}\right]} \frac{a_{2\ell}(m)}{(m-s-2\ell)!} \cdot D^{m-s-2\ell} \cot\frac{t}{2}$$

satisfy

$$D^\ell h_{s,m}(0) = \delta_{\ell,s}, \qquad 0 \le \ell, s \le m,$$

$\delta_{\ell,s}$ being the Kronecker's symbol and $a_{2\ell}(m)$ is the coefficient in the Taylor series in (6). Finally, we denote

$$\Delta_n(t) = \prod_{j=1}^n \sin^{\lambda_j}\left(\frac{t - t_j}{2}\right), \qquad \Delta_{n,r}(t) = \prod_{j=1,\, j\ne r}^n \sin^{\lambda_j}\left(\frac{t - t_j}{2}\right),$$

$$\phi_{r,k}(t) = \frac{(t - t_r)^k}{k!\,\Delta_{n,r}(t)}, \qquad \phi_{r,k}^s = D^s \phi_{r,k}(t_r),$$

and

$$T_{r,k}(t) = \frac{2^{\lambda_r - 1}\Delta_n(t)}{\sin\frac{t - t_r}{2}} \left\{ \frac{1}{(\lambda_r - 1)!} \left(\phi_{r,k}^{\lambda_r - 1} - \sum_{\ell=k}^{\lambda_r - 2} \phi_{r,k}^\ell\, D^{\lambda_r - 1} h_{\ell, \lambda_r - 2}(0) \right) \right.$$
$$\left. + \sum_{s=k}^{\lambda_r - 2} \frac{(-1)^{\lambda_r - s}\phi_{r,k}^s}{2s!} \sum_{\ell=0}^{[(\lambda_r - s)/2] - 1} \frac{a_{2\ell}(\lambda_r - 2)}{(\lambda_r - 2 - s - 2\ell)!}\, D^{\lambda_r - 2 - s - 2\ell} \cot\frac{t - t_r}{2} \right\}. \tag{7}$$

In fact, $T_{r,k}(t)$ defined in (7) are the fundamental polynomials for interpolation problem (3) satisfying

$$T_{r,k}^{(j)}(t_i) = \delta_{i,r}\delta_{j,k}, \qquad i, r = 1, \ldots, n, \quad j, k = 0, \ldots, \lambda_i - 1. \tag{8}$$

In [5] it was shown that the solution to the Hermite interpolation problem (3) can be obtained in parallel for the mutually complementary and alternating cases of trigonometric and half-integer degree trigonometric polynomials, in the sense that the following theorem holds true.

Theorem 1 ([5]). *Let*

$$\tau(t) = \sum_{i=1}^n \sum_{j=0}^{\lambda_i - 1} d_{ij}\, T_{i,j}(t),$$

where $T_{i,j}(t)$ is given in (7). Then, $\tau(t) \in \mathcal{T}_{\Lambda/2}$ is the unique solution of the problem (3). Moreover, $\tau(t)$ is a trigonometric polynomial if Λ is odd and $\tau(t)$ is a half-integer degree trigonometric polynomial if Λ is even.

Our main result follows: we construct Hermite interpolation ERBS function

$$f(t) = \sum_{i=1}^n \ell_i(t) B_i(t), \qquad t \in [t_1, t_n],$$

with trigonometric polynomial local functions $\ell_i(t)$. In analogy with the case of algebraic polynomial local curves in [3], here $\ell_i(t)$ will be the solutions to a Taylor-type interpolation problem (4).

Theorem 2. *Let* $0 \le t_0 < t_1 < \cdots < t_n < t_{n+1} \le 2\pi$ *be given points,* $\lambda_1, \lambda_2 \ldots, \lambda_n$ *be positive integers,* $\{d_{ij}\}_{i=1, j=0}^{n, \; \lambda_i-1}$ *be arbitrary real numbers and*

$$f(t) = \sum_{i=1}^{n} \left[\sum_{j=0}^{\lambda_i-1} d_{ij} T_{i,j}(t) \right] B_i(t), \quad t \in [t_1, t_n], \tag{9}$$

where $T_{i,j}(t)$ *are the fundamental polynomials for case (4) of problem (3), as defined in (7). Then, the ERBS function* $f(t)$ *satisfies the Hermite interpolation conditions*

$$f^{(k)}(t_r) = d_{rk}, \qquad r = 1, \ldots, n, \quad k = 0, \ldots, \lambda_r - 1. \tag{10}$$

Proof. Let us set

$$\ell_i(t) = \sum_{j=0}^{\lambda_i-1} d_{ij} T_{i,j}(t), \qquad i = 1, \ldots, n,$$

for the local functions in representation (9) of the ERBS function $f(t)$. By the construction of the fundamental polynomials $T_{i,j}(t)$, (8) and (10) it follows that

$$\ell_i^{(k)}(t_r) = \sum_{j=0}^{\lambda_i-1} d_{ij} T_{i,j}^{(k)}(t_r) = d_{ik} T_{i,k}^{(k)}(t_r) = d_{ik}\delta_{i,r} = \begin{cases} d_{rk}, & \text{if } i = r, \\ 0, & \text{if } i \ne r. \end{cases}$$

Now using properties P3 and P4 of the ERBS we have

$$f^{(k)}(t_r) = \sum_{i=1}^{n}\sum_{s=0}^{k} \binom{k}{s} \ell_i^{(s)}(t_r) B_i^{(k-s)}(t_r) = \ell_r^{(k)}(t_r) B_r(t_r) = d_{kr},$$

for all $r = 1, \ldots, n$ and $k = 0, \ldots, \lambda_r - 1$. ∎

3   Numerical Experiments

In this section we present numerical results from the study of two model examples. We interpolate a function $g(t)$ at knots with multiplicities by ERBS functions with local trigonometric (TERBS) and local algebraic polynomial (AERBS) functions. For each one of the two examples, plots of $g(t)$ and the two interpolants are shown, together with plots of the local functions $\ell_i(t)$ for each one of the two interpolants.

Based on the respective formulae in Section 2, the main steps of the algorithm for computing the ERBS Hermite interpolant using trigonometric polynomial local functions can be outlined, as follows.

Algorithm:

1. Compute the coefficients $a_{2\ell}(m)$ and store in a matrix. They do not depend on the data (knots, multiplicities, derivative values of $g(t)$).

2. Compute the derivative values $\phi_{r,k}^s = D^s\phi_{r,k}(t_r)$.
3. Compute the derivative values $D^{\lambda_r-1}h_{\ell,\lambda_r-2}(0)$.
4. Compute the fundamental trigonometric polynomials $T_{r,k}(t)$.
5. Compute the trigonometric polynomial local functions

$$\ell_i(t) = \sum_{j=0}^{\lambda_i-1} g^{(j)}(t_i)T_{i,j}(t), \qquad i = 1,\ldots,n.$$

6. Compute the ERBS function using trigonometric polynomial local functions

$$f(t) = \sum_{i=1}^{n} \ell_i(t)B_i(t).$$

The functions to be interpolated and the knot-vectors with respective multiplicities used in the numerical experiments are given in Examples 1 and 2 below. The programming implementation of the algorithm is in *Mathematica*.

Example 1. We interpolate the function $g(t) = e^{t/2}\sin 5t$ with the following data:

ERBS knots	0	1	2	3	4
Interpolation knots t_i		1	2	3	
Multiplicities λ_i		3	3	3	

The resulting graphs are shown in Figures 1, 2 and 3.

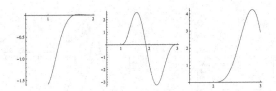

Fig. 1. The local functions of TERBS interpolant

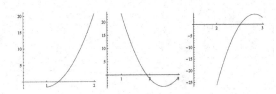

Fig. 2. The local functions of AERBS interpolant

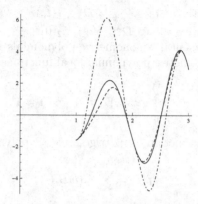

Fig. 3. Function $g(t)$ (solid), TERBS interpolant (dashed) and AERBS interpolant (dotdashed)

Example 2. We interpolate the function $g(t) = \dfrac{\cos(t^2/5)}{(t - 0.7)(t - 6)}$ with the data:

ERBS knots	0	1	2.236	3.234	3.972	4.591	5	6
Interpolation knots t_i		1	2.236	3.234	3.972	4.591	5	
Multiplicities λ_i		3	3	3	2	3	1	

In this case, the knots $t_2 = 2.236$, $t_3 = 3.234$, $t_4 = 3.972$ and $t_5 = 4.591$ are chosen to be the local extrema of $g(x)$. The plots are shown in Figures 4, 5 and 6.

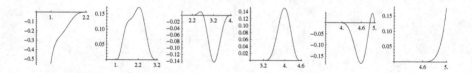

Fig. 4. The local functions of TERBS interpolant

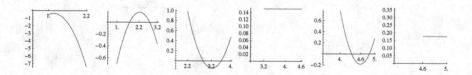

Fig. 5. The local functions of AERBS interpolant

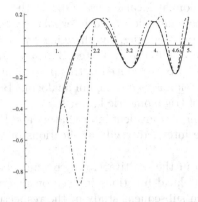

Fig. 6. Function $g(t)$ (solid), TERBS interpolant (dashed) and AERBS interpolant (dotdashed)

One of the observations that can be made from the graphs in this section is that, compared to the ERBS-based Hermite interpolant with algebraic polynomial local functions, the ERBS function with trigonometric polynomial local functions is closer to the graph of the original function most of the time. In our opinion, this is mainly due to the choice of trigonometric functions as model functions to be interpolated in Examples 1 and 2, which choice is in line with the prospective applications discussed in the next section.

4   Concluding Remarks

Remark 1. (Prospective applications.) The importance of using local functions belonging to classes other than algebraic polynomials was already discussed in [3] and [2]. Here we briefly continue this discussion.

Consider, for example, the families of eigenfunctions of Sturm-Liouville boundary-value problems. One general observation is that important properties of these eigenfunctions, such as, e.g., positivity, monotonicity, convexity, pattern of oscillation, and so on, depend on the type of the underlying differential equation. As a rule, the elliptic and parabolic case vary considerably from the hyperbolic case. We see the prospective applications of the new TERBS construction mainly in relevance to hyperbolic problems. In the first place, TERBS are expected to perform well in processing sound signals and, in particular, music. The multivariate extensions of TERBS can be expected to perform well in collocation, Galerkin and other projection-based numerical methods for solving partial differential equations describing wave processes and other processes with finite speed of propagation.

Remark 2. (Ongoing and forthcoming research.)

1. The currently ongoing research of the authors in this direction is mainly based on studying TERBS by exploring the analogy with the case of ERBS

with algebraic polynomial local curves. One of the important features in the latter case is the conversion from interpolatory to Bezier form for the purposes of geometric modeling, by changing the local Taylor monomial bases to appropriately shifted and dilated Bernstein polynomial bases. The representation of TERBS introduced in this paper is interpolatory; our first new task in this direction is to develop an analogous Bezier-type representation in the context of trigonometric bases.

2. After the development of equivalent interpolatory and Bezier-type representations of TERBS, we intend to study applications of the new apparatus to sound processing.

3. A natural next step in the multivariate case would be to introduce multivariate trigonometric local functions in the construction of a multivariate ERBS function, with subsequent study of the respective interpolatory and Bezier-type representations.

References

1. Delvos, F.: Hermite interpolation with trigonometric polynomials. BIT 33, 113–123 (1993)
2. Dechevski, L., Bang, B., Lakså, A.: Generalized Expo-rational B-splines. Int. J. Pure Appl. Math. 57(6), 833–872 (2009)
3. Dechevski, L., Lakså, A., Bang, B.: Expo-rational B-splines. Int. J. Pure Appl. Math. 27(3), 319–369 (2006)
4. Dryanov, D.: Quadrature formulae with free nodes for periodic functions. Numer. Math. 47, 441–464 (1994)
5. Du, J., Han, H., Jin, G.: On trigonometric and paratrigonometric Hermite interpolation. J. Approx. Theory 131, 74–99 (2004)
6. Kress, R.: On general Hermite trigonometric interpolation. Numer. Math. 20, 125–138 (1972)
7. Salzer, H.: New formulas for trigonometric interpolation. J. Math. Phys. 39, 85–96 (1960)

Triangular Beta-Function B-Spline Finite Elements: Evaluation and Graphical Comparisons

Lubomir T. Dechevsky and Peter Zanaty

R&D Group for Mathematical Modelling,
Numerical Simulation & Computer Visualization,
Faculty of Technology, Narvik University College
2 Lodve Lange's Str. , P.O.B. 385, N-8505 Narvik, Norway

Abstract. This work is dedicated to the computation of Euler Beta-function B-spline (BFBS) finite elements (FE) on triangulations, and to comparative visualization of their graphs. BFBS are a particular type of generalized expo-rational B-splines (GERBS) [2] and provide a piece-wise polynomial modification of the true expo-rational B-splines (ERBS) [3]. The organization of the exposition is, as follows. First, we derive new formulae for triangular BFBS FE having C^r smoothness at the vertices $r \in \mathbb{N}$. Second, we provide visualization of their graphs. Third, we compare the interpolatory and fitting properties of the new triangular BFBS FE of different polynomial degrees on two model surfaces used as a benchmark manifold.

Keywords: finite elements, bivariate interpolant, scattered data, incomplete Euler Beta function, expo-rational, Hermite interpolation, Bernstein-Beziér form.

1 Introduction

The present paper is a continuation of [4], in which a construction of simplicial finite elements on triangulated two-dimensional polygonal domains were discussed, having $C^i, i = 0, 1, 2$, smoothness at the vertices of the triangles. In this paper, we extend the model to be C^r-smooth at the vertices, for any non-negative integer r.

The use of bivariate interpolation over triangles is an important tool for many applications related to Computer Aided Geometric Design (CAGD) and FE. In this paper we focus on bivariate interpolations applying position and derivative data only on (arbitrarily scattered) given vertices, which have been triangulated (typically via Delaunay triangulation, but our results here are valid for any non-degenerate triangulation).

Before discussing the results, let us introduce a few terms, which will be used throughout this paper. For convenience, we will use a notation similar to the one in [6, Chapter 2.].

I. Lirkov, S. Margenov, and J. Waśniewski (Eds.): LSSC 2011, LNCS 7116, pp. 429–436, 2012.

Definition 1 (B-form). *Given a triangle Ω with its vertices V_1, V_2, V_3, every bivariate polynomial p of degree d can be uniquely written as*

$$p(v) = \sum_{i+j+k=d} c_{ijk} B_{ijk}^{\Omega}(v), \tag{1}$$

where $B_{ijk}(v)$ is the so-called Bernstein basis polynomial, which is defined as

$$B_{ijk}^{\Omega}(v) = \frac{(i+j+k)!}{i!j!k!} b_1^i b_2^j b_3^k, \tag{2}$$

where $b_i, i = 1, 2, 3$-s are the barycentric coordinates of the point v, such as $v = \sum_{i=1}^{3} b_i V_i$ and $\sum_{i=1}^{3} b_i = 1$. Let us denote this projection from the Euclidean space to the barycentric affine space with ξ^{Ω}, such as $b_i = \xi_i^{\Omega}(v)$, $i = 1, 2, 3$.

A common technique to achieve Hermite interpolation over triangles relies on the Bernstein-Beziér-form of polynomials. The degree of the interpolant is chosen to match the number of known derivatives (i.e. if derivatives up to, including k, are all known at the vertices, the degree is $d = 2k+1$), then the B-form representation (see Definition 1) allows certain coefficients to be computed exactly, resulting in a polynomial, which approximates the given data with various degrees of freedom. Dealing with the degrees of freedom is referred to as condensation of parameters. In this paper we will use a condensation scheme which produces interpolants which are accurate on polynomials of maximal possible degree. From now on, we will refer to these types of bivariate polynomial interpolations as BB interpolants.

2   Hermite Interpolating BFBS and ERBS on Triangles

The classes of univariate GERBS are discussed in [2]; the multivariate construction was introduced in [1]. Our construction of simplicial elements is the following.

Definition 2. *Given a triangle Ω with its vertices V_1, V_2, V_3, we construct a finite element on this triangle as*

$$H(\Omega, T, v) = \sum_{k=1}^{3} \Gamma_k(v) T_{V_k}(v), \tag{3}$$

where $T_{V_k}(v) = T_{\Omega, V_k}(v)$ is the functional coefficient at the vertex V_k and $\Gamma_k(\vartheta)$ is the respective base. In this paper, we are discussing Hermite interpolation, hence, the functional coefficients are Taylor expanding polynomials around the respective vertex up to a given order

$$T_{V_k}(x, y) = \sum_{i+j \le \phi_k} \frac{\partial^{i+j}}{\partial x^i \partial y^j} f(V_k) \frac{(x - V_{k,x})^i (y - V_{k,y})^j}{i!j!}. \tag{4}$$

The bases are constructed as an integral of a density function over a partition of the triangular domain as introduced in [1].

Definition 3. *Given a triangle Ω, with its vertices V_1, V_2, V_3, a density function $\rho(v)$, and a point $v \in \Omega$. The kth base with multiplicity φ is defined as:*

$$\Gamma_k^\varphi(v) = \frac{\int_{\Omega_k} \rho(w)dw}{\int_\Omega \rho(w)dw}, \tag{5}$$

where $\Omega_1 = \sum_2(v, V_2, V_3), \Omega_2 = \sum_2(V_1, v, V_3), \Omega_3 = \sum_2(V_1, V_2, v)$ are the three sub-triangles generated by v.

2.1 BFBS

As in [4], the construction of Hermite interpolatory multivariate FE uses the Bernstein polynomial density which is corresponding to the degrees of the interpolation $\varphi_1, \varphi_2, \varphi_3$ at the vertices of the triangle Ω

$$\rho(v) = B_{\varphi_1 \varphi_2 \varphi_3}^\Omega(v). \tag{6}$$

Now, let us state the explicit form of the base Hermite interpolatory BFBS FE base functions.

Theorem 1. *Given a triangle Ω, with its vertices V_1, V_2, V_3 and the corresponding Hermite multiplciity $\varphi = (\varphi_1, \varphi_2, \varphi_3)$, the bases can be formulated as*

$$\Gamma_1^\varphi(v) = \xi_1^\Omega(v) \sum_{i=\varphi_1}^{\varphi_1} \sum_{j=0}^{\varphi_2} \sum_{k=0}^{\varphi_3} B_{ijk}^\Omega(v) \tag{7a}$$

$$\Gamma_2^\varphi(v) = \xi_2^\Omega(v) \sum_{i=0}^{\varphi_1} \sum_{j=\varphi_2}^{\varphi_2} \sum_{k=0}^{\varphi_3} B_{ijk}^\Omega(v) \tag{7b}$$

$$\Gamma_3^\varphi(v) = \xi_3^\Omega(v) \sum_{i=0}^{\varphi_1} \sum_{j=0}^{\varphi_2} \sum_{k=\varphi_3}^{\varphi_3} B_{ijk}^\Omega(v). \tag{7c}$$

In order to prove formulas (7a-7c), we will recite three theorems relevant to the B-form of bivariate polynomials from [6, Chapter 2.].

Lemma 1 (Theorem 2.33. of [6]). *Given a triangle Ω, let p be a polynomial of total degree d written in B-form. Then*

$$\int_T p(x,y)dxdy = \frac{A_\Omega}{\binom{d+2}{d}} \sum_{i+j+k=d} c_{ijk}, \tag{8}$$

where A_T is the area of T.

Lemma 2 (Theorem 2.10. of [6]). *The coefficients in the de Casteljau algorithm for point v, which has barycentric coordinates $b = \xi^\Omega(v)$, are given by*

$$c_{ijk}^{(\ell)}(b) = \sum_{\mu+\nu+\kappa=\ell} c_{i+\mu,j+\nu,k+\kappa} B_{\mu\nu\kappa}(v), \quad i+j+k = d-\ell. \tag{9}$$

Lemma 3 (Theorem 2.38. of [6]). *Given a triangle Ω, let v be a point in the interior of Ω with its barycentric coordinates $a = \xi^\Omega(v)$ and $\Omega_i, i = 1, 2, 3,$ being the sub-triangles in Definition 3. Then, for any polynomial p of degree d, we have*

$$p(w) = \begin{cases} \sum\limits_{i+j+k=d} c_{0jk}^{(i)}(a) B_{ijk}^{\Omega_1}(w), & w \in \Omega_1 \\ \sum\limits_{i+j+k=d} c_{i0k}^{(j)}(a) B_{ijk}^{\Omega_2}(w), & w \in \Omega_2 \\ \sum\limits_{i+j+k=d} c_{ij0}^{(k)}(a) B_{ijk}^{\Omega_3}(w), & w \in \Omega_3. \end{cases} \qquad (10)$$

Proof (of Theorem 1). We have the inventory ready to prove Theorem 1. From Definition 3 we have

$$\Gamma_1^\varphi(v) = \frac{\int_{\Omega_1} \rho(w)dw}{\int_\Omega \rho(w)dw} \qquad (11)$$

$$= \frac{\int_{\Omega_1} B_{\varphi_1\varphi_2\varphi_3}^\Omega(w)dw}{\int_\Omega B_{\varphi_1\varphi_2\varphi_3}^\Omega(w)dw} \qquad (12)$$

using Lemma 3, this can be written as

$$= \sum_{i+j+k=d} \frac{\int_{\Omega_1} c_{0jk}^{(i)}(a) B_{ijk}^{\Omega_1}(w)dw}{\int_\Omega B_{\varphi_1\varphi_2\varphi_3}^\Omega(w)dw}. \qquad (13)$$

Now, by applying Lemma 1, we get

$$= \frac{A_{\Omega_1}}{A_\Omega} \sum_{i+j+k=d} c_{0jk}^{(i)}(a) \qquad (14)$$

$$= \xi_1^\Omega(v) \sum_{i+j+k=d} c_{0jk}^{(i)}(a). \qquad (15)$$

From Lemma 2, this is equal to

$$= \xi_1^\Omega(v) \sum_{i+j+k=d} \sum_{\mu+\nu+\kappa=i} c_{\mu,j+\nu,k+\kappa} B_{\mu\nu\kappa}^\Omega(v) \qquad (16)$$

$$= \xi_1^\Omega(v) \sum_{\mu+j+\nu+k+\kappa=d} c_{\mu,j+\nu,k+\kappa} B_{\mu\nu\kappa}^\Omega(v). \qquad (17)$$

Noting, that the only nonzero B-form coefficient in (17) is $c_{\varphi_1\varphi_2\varphi_3} = 1$, we finally arrive to

$$\Gamma_1^\varphi(v) = \xi_1^\Omega(v) \sum_{\mu=\varphi_1}^{\varphi_1} \sum_{\nu=0}^{\varphi_2} \sum_{\kappa=0}^{\varphi_3} B_{\mu\nu\kappa}^\Omega(v). \qquad (18)$$

Remark 1. It can be seen, that all bases are obtained from (18) via respective cyclic transpositions of the parameters involved.

Two examples of the resulting bases are plotted in Figure 1.

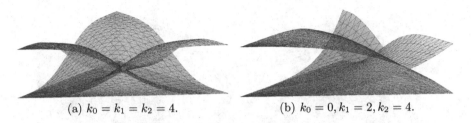

(a) $k_0 = k_1 = k_2 = 4$.   (b) $k_0 = 0, k_1 = 2, k_2 = 4$.

Fig. 1. Graphs of the two BFBS bases for different multiplicities. Notice how the bases blend the functional coefficients if the scheme is not symmetrical.

3   Numerical Results

Here we will present some numerical and graphical results comparing the BFBS, ERBS, and the classical BB interpolants. Two exemplary manifolds are used, the first one being a torus, described by

$$x(u,v) = \left(\frac{1}{3} + \frac{1}{5} \cos v \right) \cos u \tag{19a}$$

$$y(u,v) = \left(\frac{1}{3} + \frac{1}{5} \cos v \right) \cos v \tag{19b}$$

$$x(u,v) = \frac{1}{5} \sin v, \tag{19c}$$

the second example is Franke's functional surface [5], given on the unit square $[0,1] \times [0,1]$ by

$$f(x,y) = \frac{3}{4} e^{-((9x-2)^2+(9y-2)^2)/4} + \frac{3}{4} e^{-(9x+1)^2/49-(9y+1)/10)/4}$$
$$+ \frac{1}{2} e^{-((9x-7)^2+(9y-2)^3)/4} - \frac{1}{5} e^{-((9x-4)^2+(9y-7)^3)/4}. \tag{20}$$

Figure 2 provides a graphical visualization of the result of BB and BFBS interpolation for the torus. As the approximations become visually indistinguishable at this resolution even at low degrees, only the first few interpolants (i.e. which still have visible deviations from the torus) were plotted.

Tables 1 and 2 present the numerical data of average errors obtained by the approximation of classical BB and BFBS interpolants for the first and second benchmark model examples, respectively. The error is measured as an average of the L_2-distance between the interpolant and the interpolated manifold.

Figures 3 and 4 present the L_2-error of the BB and BFBS interpolations for the first and the second model example, respectively. The errors are plotted in a truncated semi-logarithmic manner. Based on the truncation threshold Ψ, the logarithm of the error of the interpolation is plotted as functional surface, defined by

$$\Delta(u,v) = \begin{cases} 1 + \log_{\frac{1}{\Psi}} (|f(v) - H_f(v)|), & |f(v) - H_f(v)| > \Psi \\ 0, & \text{otherwise.} \end{cases} \tag{21}$$

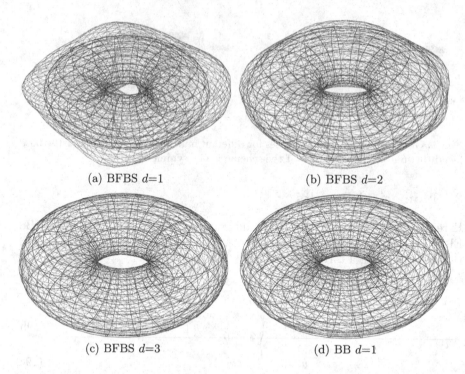

(a) BFBS $d=1$ (b) BFBS $d=2$

(c) BFBS $d=3$ (d) BB $d=1$

Fig. 2. BFBS and BB interpolation of the torus on a triangulation of the uniform 4×4 grid of the unit square is shown for the first few degrees

Table 1. Numerical results: average L_2-error on the torus

Method	d	4×4	5×5	6×6	7×7	8×8	9×9
BB	1	7.820658e-1	6.456645e-1	6.154056e-1	5.986260e-1	5.950372e-1	5.913668e-1
BB	2	2.320363e-2	6.458763e-3	2.302709e-3	9.511990e-4	4.437263e-4	2.284408e-4
BB	3	2.018414e-3	6.094913e-4	2.212516e-4	9.229894e-5	4.284313e-5	2.163905e-5
BB	4	6.585538e-5	9.896403e-6	2.137236e-6	5.906688e-7	1.958328e-7	7.548639e-8
BFBS	1	2.415451	1.665822	1.226894	9.235570e-1	7.262199e-1	5.802837e-1
BFBS	2	8.445627e-1	3.891506e-1	2.040861e-1	1.170928e-1	7.238259e-2	4.760752e-2
BFBS	3	1.644066e-1	8.633400e-2	4.753908e-2	2.786822e-2	1.720148e-2	1.112569e-2
BFBS	4	8.321772e-2	2.522363e-2	9.302814e-3	3.970121e-3	1.896154e-3	9.896265e-4
BFBS	5	1.064981e-2	3.591457e-3	1.391344e-3	6.021768e-4	2.858082e-4	1.464316e-4
BFBS	6	4.604780e-3	9.118412e-4	2.375760e-4	7.572531e-5	2.810590e-5	1.173973e-5
BFBS	7	5.296442e-4	1.061089e-4	2.767954e-5	8.664038e-6	3.120174e-6	1.255970e-6
BFBS	8	1.619771e-4	2.084811e-5	3.834115e-6	9.118729e-7	2.627810e-7	8.782046e-8
BFBS	9	1.763331e-5	2.128035e-6	3.735963e-7	8.444205e-8	2.303980e-8	7.277805e-9

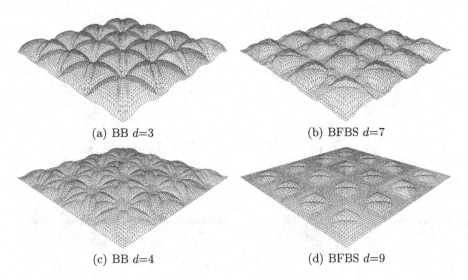

(a) BB $d=3$ (b) BFBS $d=7$

(c) BB $d=4$ (d) BFBS $d=9$

Fig. 3. Logarithmic plots of errors of BB and BFBS interpolants on the torus on a triangulation of the uniform 4×4 grid of the unit square. Errors below 10^{-15} have been truncated.

Table 2. Numerical results: Average L_2-error on Franke's function

Method				Grid Size		
	d	4×4	5×5	6×6	7×7	8×8	9×9
BB	1	1.674185e-2	1.477192e-2	1.415778e-2	1.379736e-2	1.375453e-2	1.362808e-2
BB	2	1.311981e-3	5.441518e-4	2.208328e-4	7.573881e-5	5.534924e-5	2.633061e-5
BB	3	4.526312e-4	1.293705e-4	3.822375e-5	1.356023e-5	7.783664e-6	3.629180e-6
BB	4	2.132594e-4	4.470613e-5	1.022222e-5	3.050981e-6	1.157867e-6	4.534583e-7
BFBS	1	1.260340e-2	1.040526e-2	7.690138e-3	6.139326e-3	5.019522e-3	4.073451e-3
BFBS	2	1.131285e-2	6.270927e-3	3.694548e-3	2.158332e-3	1.405364e-3	8.814113e-4
BFBS	3	5.665282e-3	2.284858e-3	1.026595e-3	5.419032e-4	3.616880e-4	2.381736e-4
BFBS	4	4.434235e-3	1.525956e-3	5.942543e-4	2.864580e-4	1.541490e-4	7.806419e-5
BFBS	5	3.037121e-3	7.950973e-4	2.521999e-4	7.652658e-5	4.834835e-5	2.412451e-5
BFBS	6	2.164649e-3	4.690393e-4	1.088383e-4	4.240184e-5	1.679686e-5	7.489437e-6
BFBS	7	1.702230e-3	2.747375e-4	6.653708e-5	1.519939e-5	7.421848e-6	3.300495e-6
BFBS	8	1.196474e-3	1.571107e-4	2.529921e-5	7.206004e-6	2.129230e-6	6.918364e-7
BFBS	9	8.649081e-4	9.450772e-5	1.658151e-5	3.073790e-6	9.365283e-7	3.759385e-7

4   Conclusion

The graphical and numerical results in Section 3 suggest that the classical polynomial BB-interpolants often outperform the comparable BFBS interpolants on the same interpolation data; while, by increasing the degree of the functional coefficients over each vertex (which requires the use of BFBS of higher degree), it is possible outperform any *a priori* given classical BB-interpolant on the class of all sufficiently smooth functions.

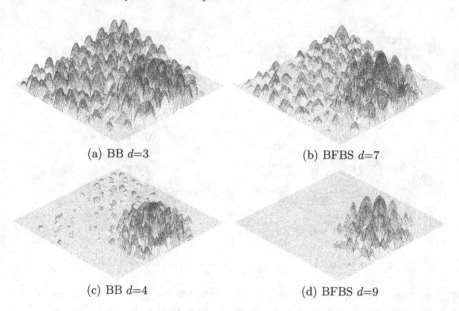

    (a) BB d=3   (b) BFBS d=7

    (c) BB d=4   (d) BFBS d=9

Fig. 4. Logarithmic plots of errors of BB and BFBS interpolants on Franke's function on a triangulation of the uniform 8×8 grid of the unit square. Errors below 10^{-15} have been truncated.

The advantage of using BFBS interpolants becomes clear when the BFBS simplicial elements are grouped into star-1 neighborhoods of the scattered interpolation vertices. Then, the B-splines generated via this construction will be entirely supported by this star-1 neighborhood alone, freeing us from dealing with additional topology (i.e. star-2, star-3, etc., neighborhoods). Furthermore, the functional coefficients are only depending on the data at the corresponding vertex, whereas in the classical polynomial case they are influenced by the triangle as well.

References

1. Dechevsky, L.T.: Generalized Expo-Rational B-Splines. In: Communication at the 7th International Conference on Mathematical Methods for Curves and Surfaces (2008)
2. Dechevsky, L.T., Bang, B., Lakså, A.: Generalized expo-rational B-splines. Int. J. Pure Appl. Math. 57(6), 833–872 (2009)
3. Dechevsky, L.T., Lakså, A., Bang, B.: Expo-rational B-splines. Int. J. Pure Appl. Math. 27(3), 319–369 (2006)
4. Dechevsky, L.T., Zanaty, P., Lakså, A., Bang, B.: First Instances of Generalized Expo-Rational Finite Elements on Triangulations. To appear in Proceedings of the American Institute of Physics (2011)
5. Franke, R.: Scattered data interpolation: Tests of some method. Mathematics of Computation 38(157), 181–200 (1982), http://www.jstor.org/stable/2007474
6. Lai, M.J., Schumaker, L.L.: Spline functions on Triangulations. Encyclopedia of Mathematics and its Applications. Cambridge University Press, New York (2007)

Part X

High-Performance Monte Carlo Simulations

Sensitivity Study of Heston Stochastic Volatility Model Using GPGPU

Emanouil I. Atanassov and Sofiya Ivanovska

Institute of Information and Communication Technlogies,
Bulgarian Academy of Sciences
Acad. G. Bonchev St., Bl.25A, 1113 Sofia, Bulgaria
{emanouil,sofia}@parallel.bas.bg

Abstract. The focus of this paper is on effective parallel implementation of Heston Stochastic Volatility Model using GPGPU. This model is one of the most widely used stochastic volatility (SV) models. The method of Andersen provides efficient simulation of the stock price and variance under the Heston model. In our implementation of this method we tested the usage of both pseudo-random and quasi-random sequences in order to evaluate the performance and accuracy of the method.

We used it for computing Sobol' sensitivity indices of the model with respect to input parameters. Since this method is computationally intensive, we implemented a parallel GPGPU-based version of the algorithm, which decreases substantially the computational time. In this paper we describe in detail our implementation and discuss numerical and timing results.

1 Introduction

One of the most popular models for option pricing is the Heston stochastic volatility model. The typical workflow when trying to price a more complex (exotic) option involves calibration of the model parameters using market prices of European options of the same underlying asset and then simulating the evolution of the underlying using some Monte Carlo or quasi-Monte Carlo scheme. The calibration process has some inherent uncertainty since some of the prices that serve as input are known only approximately and as a result there is certain level of freedom in how to achieve the "best" fit. In our work we study the sensitivity of the price of the complex (exotic) option with respect to the model parameters, where we consider as an example a particular knock-out barrier option. We use the generic scheme of Andersen for efficient simulation of Heston-type stochastic volatility models, with the martingale correction enabled. This scheme is numerically efficient and can be applied for pricing other path-dependent options. Our aim is to demonstrate how one can use scrambled low-discrepancy sequences on GPU hardware in order to compute such quantities in an efficient manner. In the next section we describe the problem under consideration, while in the following sections we describe our algorithm and the GPU implementation. Then we proceed to apply the algorithm for computing the Sobol' sensitivity coefficients for

I. Lirkov, S. Margenov, and J. Waśniewski (Eds.): LSSC 2011, LNCS 7116, pp. 439–446, 2012.
© Springer-Verlag Berlin Heidelberg 2012

the Heston model, presenting the numerical results that were obtained. In the last section we present some conclusions and directions for future work.

2   The Problem of Option Pricing

One of the most important problems of Financial Mathematics is that of pricing of financial assets. The so-called derivatives are financial instruments which are derived from other forms of assets. The financial options are a complex type of contract, but in our work we concentrate on options with a single underlying asset S, where at a specified time in the future (option expiration time) T, the owner of the option is entitled to receive payout, which is a function F of the evolution of the price of the asset S during the time interval $[0, T]$. More extensive information about the various aspects of option pricing can be found in [6] and [7]. Here we outline only the formulas and notations concerning our research. The payout $max(0, S(T) - K)$ corresponds to European call options, where the owner has the right to buy the underlying asset at a fixed price K at expiration. The European options are extensively studied and widely traded type of options. The American options offer the possibility to buy or sell the underlying asset at any time before expiration, which is an important difference if the underlying assets offers dividend payments within the considered time interval. We consider the price of the underlying as a stochastic process and we denote by Q the unique risk-neutral measure for the model, under which the discounted prices are martingales:

$$\exp^{-rt} S_t = E_Q[\exp^{-rT} S_T | \mathcal{F}_t].$$

Under these assumptions today's fair value of the option is $\exp(-rT)E_Q F$.

In order to obtain numerically the price of an option under these assumptions, we must choose a model for the process S_t and then estimate the payout via Monte Carlo or other techniques [5]. For many underlying assets the market of European or American call options is liquid and thus can provide a basis for evaluating the validity of our model and numerical schemes. For most models the computation of European or American call options is numerically efficient and accurate and thus serves as a basis to calibrate the various parameters of the model before undertaking the evaluation of more complex, also called "exotic", options. The Black-Scholes formula for the prices of European options introduced in [4] is still used in theory and practice. Under the Black-Scholes model the price of the asset is lognormally distributed:

$$S_t = s_0 \exp((r - \frac{1}{2}\sigma^2)t + \sigma W_t),$$

where r is the risk-free interest rate and σ^2 is the "volatility". For each maturity and strike price one can obtain the value of volatility for which the Black-Scholes formula gives the observed price. For each time to expiration and strike price, we obtain data points for constructing the "implied volatility surface" (see Fig. 1).

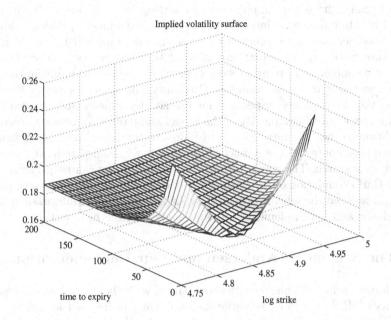

Fig. 1. Implied volatility surface

The Black-Scholes model has been extended in many directions. One can look at [9] for comparison of several popular models when evaluating various types of options. It can be seen that when the underlying assets and the types of options vary, the "best" model also varies. Models, which consider the volatility of the underlying asset as a stochastic process, are called stochastic volatility models. One of the advantages of such models is their ability to fit the observed implied volatility surface without introduction of too many parameters (see, e.g., [6]). The implied volatility is obtained by finding the value of volatility which fits the Black-Scholes formula to the observed price of an European option and the volatility surface is the result of computing the implied volatilities for all certain number of option strike values and expiration times. One of the most important stochastic volatility models is the Heston model [8], where the price of the underlying $X(t)$ and the instantaneous volatility $V(t)$ satisfy the following:

$$dX(t)/X(t) = \sqrt{V(t)}dW_X(t)$$

$$dV(t) = \kappa(\theta - V(t))dt + \varepsilon\sqrt{V(t)}dW_V(t)$$

where $\kappa, \theta, \varepsilon$ are constants, W_X and W_V are Brownian motions with correlation ρ: $< dW_X(t), dW_V(t) >= \rho dt$. The initial conditions are $X(0) = X_0$ and $V(0) = V_0$.

The prices of European call options under Heston model can be obtained efficiently, e.g. using Fourier transform, and thus model calibration using optimisation procedures like Adaptive Simulated Annealing (ASA), is fast. We note that we make the usual simplification of setting interest rate to 0, since the case of constant non-zero interest rate poses only technical problems. Models which consider stochastic processes for the interest rate will be considered in our future work. An important extension of the Heston model is obtained by introducing one or more jump processes, which correspond to jumps in the price of the underlying or in the volatility. Obviously this allows for better fit of the implied volatility surface, especially for the shorter times before expiration at the expense of added complexity of the model and more parameters to consider. In addition to that the introduction of jumps to the model presents difficulties for the quasi-Monte Carlo approach. That is why we decided to leave jumps for our future research. The computation of an exotic option can be performed via Monte Carlo simulation of the path of the underlying or some other schemes. In our work we used the scheme proposed by Andersen [1]. More information about this scheme and our implementation will be given in the next section.

3   The Scheme of Andersen and Our Implementation

The Monte Carlo methods, sometimes extended with the use of low-discrepancy sequences [11], is the most versatile method that can be used for option pricing. There are several schemes for simulating the Heston model, of which the Andersen scheme is considered one of the fastest while maintaining small bias. The Andersen scheme models the evolution of the price of the underlying $\hat{X}(t)$ and the instantaneous variance $\hat{V}(t)$ along time step Δ by computing certain treshold $\psi = \frac{s^2}{m^2}$ and then:

- if $\psi \leq 1.5$ set $\hat{V}(t+\Delta) = a(b+Z_V)^2$ where Z_V is sampled from $N(0,1)$, for example by $Z_V = \Phi^{-1}(U_V)$, where U_V is uniformly distributed in $[0,1]$.
- otherwise $\psi > 1.5$ set $\hat{V}(t+\Delta) = \Psi^{-1}(U_V; p, \beta)$, where U_V is uniformly distributed in $[0,1]$.

$$\ln \hat{X}(t+\Delta) = \ln \hat{X}(t) + K_0^* + K_1 \hat{V}(t) + K_2 \hat{V}(t+\Delta) + \sqrt{(K_3 \hat{V}(t) + K_4 \hat{V}(t+\Delta))} Z,$$

where Z is a standard Gaussian random variable, independent of $\hat{V}$.

We refer the reader to the original work of Andersen for the exact definitions of the various quantities in these formulae. For each time-step we need to sample two independent random numbers with appropriate distribution. Without loss of generality we assume that we sample uniformly distributed random number, using the inverse function method for obtaining normally distributed numbers where necessary. Thus our problem of pricing a financial option with this scheme becomes an integration problem over the unit cube of dimension $2s$, where s is the number of time steps. When using quasi-random sequences [10], we use one point of the sequence for sampling one trajectory and we use the consecutive coordinates of the point for sampling U_V and Z_V, using the inverse function method for obtaining variables with normal distribution.

Our implementation of the Andersen scheme uses GPU computing. The General Purpose GPU computing utilizes powerful graphics cards for power- and cost-efficient computations. State of the art graphics cards have large number (hundreds, even thousands) of cores. For NVIDIA cards one can use CUDA for parallel computations. Parallel processing on such cards is based upon splitting the computations between grids of threads. For Monte Carlo or quasi-Monte Carlo computations it is natural to split the computations among trajectories. We use threadsize of 256, which is optimal taking into account the relatively large number of registers that we need. By varying the gridsize we determined that it should not be overextended, because it is related to the overall usage of graphics memory. That is why we fixed the gridsize also at 256, so that one invocation of the graphics "kernel" function yields 65536 results. We also found that we have to add the option "–maxrregcount 64" in order to reduce the usage of registers by the compiler and prevent abnormal behaviour in some cases. The Sobol' sequence with Owen scrambling is generated by scrambling consequtive digits, using previous digits to generate "random trees" that serve to permute the next digits (see our previous works [2,3]). For the Monte Carlo variant of the program we used the random number generator CURAND provided by the CUDA SDK.

As an example of an exotic option that is path-dependent we select a knockout barrier call option. This option entitles the buyer of an asset with current price 1 to receive payout, which is $min(max(0, S(T) - 1), 0.2)$ on the condition that the asset never exceeds 1.5. Otherwise the payout is 0. The parameters of the Heston model are calibrated using current option prices of a particular asset - share prices of SPDR Gold Trust ETF, which can be considered as proxy for the price of gold. We found that the knock-out condition is an important factor for the performance of the quasi-Monte Carlo method because of the discontinuity. Table 1 shows comparison between mean-square errors with CURAND [13] versus scrambled Sobol' sequence. This comparison would be even more favorable if the knock-out condition is removed.

When using smaller number of time steps, e.g. 12, we again observe improvements from using the scrambled Sobol' sequence. Results from such computations are shown in Table 2.

The generation of the scrambled Sobol' sequence is computationally expensive, especially on a CPU. Most of the time is spent in the scrambling process. On an Intel Xeon X5560 (2.8 GHz) CPU the evaluation of the above option with 1000 timesteps and 2097152 samples takes more than 1620 seconds. We performed the same computation on two dual GPU cards, i.e. cards that combine two graphics devices in one envelope. In both cases we only used one of the available CUDA devices. On an NVIDIA GTX 295, using 240 cores, the same test completed in 155 seconds, while on a Tesla M2050 card from AMAZON EC2, using 120 cores, the test completed in 66 seconds. The advantage of the Tesla card is due

Table 1. Errors when using Scrambled Sobol'

Number of trajectories	2^{13}	2^{14}	2^{15}	2^{16}	2^{17}
Error	7.9978e-04	5.2838e-04	3.3107e-04	2.6362e-04	1.4171e-04

Table 2. Errors for sampling 12 timesteps (24 dimensions)

Number of trajectories	Scrambled Sobol'	CURAND
2^{14}	2.1582e-04	7.5218e-04
2^{15}	1.2762e-04	5.0740e-04
2^{16}	7.9609e-05	3.2131e-04
2^{19}	5.9588e-05	2.5437e-04
2^{20}	3.9779e-05	7.1946e-05

to its superiour performance in double precision computations. Computations using the CURAND function are substantially faster (23 seconds for the same computation on NVIDIA GTX 295), but one should take into account that to achieve twice more accuracy one must perform 4 times more computations. The logic of the option under consideration does not produce substantial amount of additional operations, otherwise we can expect that the difference in time will diminish further.

4   Sobol' Sensitivity Indices

It is obvious that once the model is fixed, there is some degree of uncertainty about what the "right" parameters are, not only due to numerical errors, but also due to missing data or difficulty in interpretation of some input data. For example, for each option at any given moment in time we have 3 prices - the last price, the bid and the offer price. The last price can be outdated, especially for options with low open interest. Thus we consider the middle between bid and offer as a more reasonable value for the price of the option. However, these prices are offered usually by market-makers and have a tendency to be rounded, while the price of the underlying is known in any moment in time with much greater degree of accuracy due to higher liquidity. A valid approach would be to take closing prices in all cases, although some aberrations can still occur. Now let us consider the problem of a large institution that desires to offer to their clients a type of complex option contract instead of simple deposit with fixed rate. Since there will be some period of validity of such offer, the bank would have to estimate not only the price of the option that it is effectively selling, based on the current data, but also the sensitivity of the price with respect to changes in the market conditions. Thus we can consider the model parameters as random variables and look at the sensitivity of the resulting price of the exotic option. The Sobol' global sensitivity indices provide quantitative estimate of model sensitivity in such a setting. Let us consider a function F with parameters $x_1, \ldots, x_s$. The Sobol' global sensitivity indices estimate the variability of the model result that is dependent on particular variables. We use the following definition [12]:

Definition 1. *Total Sensitivity Index of input parameter* x_i, $i \in \{1, \ldots, d\}$:

$$S_{x_i}^{tot} = S_i + \sum_{l_1 \neq i} S_{il_1} + \sum_{l_1, l_2 \neq i, l_1 < l_2} S_{il_1 l_2} + \ldots + S_{il_1 \ldots l_{d-1}},$$

Table 3. First order(S_i) and total(S_{Ti}) sensitivity indices for the first scenario

	ε	κ	ρ	θ_0	θ
μ	0.6224	5.1566	0.3464	0.0196	0.0564
σ	0.1134	2.0442	0.0004	0.0013	0.0084
S_i	0.162578	0.808468	5.62E-05	3.91E-05	0.013896
S_{Ti}	0.178835	0.8276	0.000572	0.000357	0.019321

Table 4. First order(S_i) and total(S_{Ti}) sensitivity indices for the second scenario

	ε	κ	ρ	θ_0	θ
μ	1.5963	16.5136	0.2205	0.0332	0.0551
σ	0.9733	46.8082	0.1189	0.0109	0.0180
S_i	0.019638	0.000964	0.79929	0.020213	0.006157
S_{Ti}	0.059721	0.090854	0.941485	0.037	0.020939

where S_i - is the main effect (first-order sensitivity index) of x_i and
$S_{il_1...l_{j-1}}$ – j^{th} order sensitivity index for parameter x_i $(2 \leq j \leq d)$.

Let $\mathbf{x} = (\mathbf{y}, \mathbf{z}) \in \mathbb{R}^d, \mathbf{y} = (x_{k_1}, \ldots, x_{k_m}) \in \mathbb{R}^m, K = (k_1, \ldots, k_m)$.
Variance of the subset $\mathbf{y}$: $\mathbf{D_y} = \sum_{n=1}^{m} \sum_{(i_1 < ... < i_n) \in K} \mathbf{D}_{i_1,...,i_n}$.

We fix all other conditions and consider the contribution of 5 coefficients to the option price: $\varepsilon, \kappa, \rho, \theta_0, \theta$. By repeatedly performing estimation of model coefficients based on current European options prices, we found two realistic scenarious for the distribution of the coefficients of the Heston model. The first one roughly corresponds to variation within one day of the SPRD GLD shares, while the second one correspond to variation in a larger time frame of 3 weeks. By μ and σ we denote the mean and standard deviation for each parameter. The computations of the Sobol' sensitivity indices were carried out using procedure from Simlab [14] and the results are shown in the next two tables. Each of the tables shows the results after performing 8000 computations for combinations of those parameters. We can see that the most important contribution to the mean comes from different parameters in the two cases. Indeed during the smaller timeframe there has been little change in the parameter ρ and this justifies its small sensitivity indices. However, the situation in the larger timeframe has been different. We believe that such study of Sobol' sensitivity indices should be standard procedure when it is desirable for the option's price to be stable for a certain period of time.

5   Conclusions and Future Work

An efficient implementation of the Andersen scheme on GPU has been proposed, using pseudorandom numbers or scrambled Sobol' sequences. The scheme has good efficiency when compared with CPU implementation and the overhead of scrambling is acceptable. The algorithm is applicable for approximating the value of any type of path-dependent options like barrier options, Bermuda options, Asian options, etc. We used the algorithm to compute efficiently the Sobol' sensitivity indices in the setting of the Heston model. Our implementation can also be adapted to model jumps in the stock price or volatility. Tests with other low-discrepancy sequences are under way with initial positive results.

Acknowledgments. This work is supported by the National Science Fund of Bulgaria under Grants DCVP 02/1 and DO 02-215/2008.

References

1. Andersen, L.B.G.: Efficient Simulation of the Heston Stochastic Volatility Model. Banc of America Securities, http://ssrn.com/abstract=946405
2. Atanassov, E.: A New Efficient Algorithm for Generating the Scrambled Sobol' Sequence. In: Dimov, I.T., Lirkov, I., Margenov, S., Zlatev, Z. (eds.) NMA 2002. LNCS, vol. 2542, pp. 83–90. Springer, Heidelberg (2003)
3. Atanassov, E., Karaivanova, A., Ivanovska, S.: Tuning the Generation of Sobol Sequence with Owen Scrambling. In: Lirkov, I., Margenov, S., Waśniewski, J. (eds.) LSSC 2009. LNCS, vol. 5910, pp. 459–466. Springer, Heidelberg (2010)
4. Black, F., Scholes, M.S.: The pricing of options and corporate liabilities. Journal of Political Economy 81(3), 637–654 (1973)
5. Caflisch, R.: Monte Carlo and quasi-Monte Carlo methods. Acta Numerica 7, 1–49 (1998)
6. Gatheral, J.: The Volatility Surface: A Practitioner's Guide. Wiley Finance (2006)
7. Glasserman, P.: Monte Carlo Methods in Financial Engineering. Springer, New York (2003)
8. Heston, S.: A closed-form solution for options with stochastic volatility. Review of Financial Studies 6, 327–343 (1993)
9. Christoffersen, P., Heston, S.L., Jacobs, K.: The Shape and Term Structure of the Index Option Smirk: Why Multifactor Stochastic Volatility Models Work so Well. Management Science - Management 55(12), 1914–1932 (2009)
10. Niederreiter, H.: Random Number Generations and Quasi-Monte Carlo Methods. SIAM, Philadelphia (1992)
11. Owen, A.B.: Scrambling Sobo'l and Niederreiter-Xing points. Journal of Complexity 14, 466–489 (1998)
12. Sobol, I.M.: Global Sensitivity Indices for Nonlinear Mathematical Models and Their Monte Carlo Estimates. Mathematics and Computers in Simulation 55(1-3), 271–280 (2001)
13. CUDA, http://developer.nvidia.com/category/zone/cuda-zone
14. SIMLAB, http://simlab.jrc.ec.europa.eu/

A Monte Carlo Simulator for Non-contact Mode Atomic Force Microscopy

Lado Filipovic[1,2] and Siegfried Selberherr[1]

[1] Institute for Microelectronics, Technische Universität Wien,
Gußhausstraße 27–29/E360, A-1040 Wien, Austria
{filipovic,selberherr}@iue.tuwien.ac.at
[2] Christian Doppler Laboratory for Reliability Issues in Microelectronics at the
Institute for Microelectronics, Wien, Austria

Abstract. Nanolithography using Non-Contact Mode Atomic Force Microscopy (NCM-AFM) is a promising method for the manufacture of nanometer sized devices. Compact models which suggest nanopatterned oxide dots with Gaussian or Lorentzian profiles are implemented in a Monte Carlo simulator in a level set environment. An alternative to compact models is explored with a physics based Monte Carlo model, where the AFM tip is treated as a point charge and the silicon wafer as an infinite conducting plane. The strength of the generated electric field creates oxyions which accelerate towards the silicon surface and cause oxide growth and surface deformations. A physics based model is presented, generating an oxide dot based on the induced surface charge density. Comparisons to empirical models suggest that a Lorentzian profile is better suited to describe surface deformations when compared to the Gaussian profile.

1 Introduction

Atomic Force Microscopy (AFM) [1] was developed in 1986 to measure protuberances and depressions on a nanometer sized section of a desired surface. For this purpose AFM has been a useful tool in physics, chemistry, biology, biochemistry, as well as in the semiconductor industry. This method utilizes the Van der Waals interaction between the tip and a sample surface to determine the surface properties [11]. Several years after the development of AFM, it was noted that this method can also be implemented to pattern semiconductor or metal surfaces with nanometer scale precision for processes such as local anodic oxidation of silicon [4]. With the ability to pattern nanoscale devices, the importance of AFM in the semiconductor industry increased significantly. There are various modes in which AFM can operate. The most commonly used for local anodic oxidation of semiconductor surfaces are contact mode (CM-AFM) and non-contact mode (NCM-AFM) [2–4]. NCM-AFM allows narrow patterns to be generated at relatively high speed, making it the preferred method for local anodic nanooxidation of silicon surfaces [7].

I. Lirkov, S. Margenov, and J. Waśniewski (Eds.): LSSC 2011, LNCS 7116, pp. 447–454, 2012.

Over the past decades the use of NCM-AFM as a tool for the local anodic oxidation of semiconductor surfaces has significantly increased, generating a demand for a simulation tool which predicts semiconductor surface deformation due to AFM processing. Conventional photolithographic methods are unable to accurately describe the processing steps necessary to generate a nanoscale device using AFM. Having the ability to simulate the AFM process is essential in order to understand the extent of devices which can be manufactured using this method. In this work, we present an analytical model implemented using a Monte Carlo simulator in a level set environment [6]. The level set simulator is an existing silicon process simulator which describes surface deformations after conventional processing steps are performed on a silicon surface [5]. The final deformation of a silicon surface after local anodic nanooxidation using NCM-AFM can either have a Gaussian [10] or a Lorentzian [8] profile.

We further explore an alternative to this compact model by developing a physics based Monte Carlo model, where the AFM needle tip is treated as a point charge which generates an electric field towards the silicon surface. The silicon surface is treated as an infinite conductive plane, since the silicon wafer extends to lengths much longer than the nanoscale AFM tip distance. With this physics based Monte Carlo model, the validity of the available analytical model is explored and the final shape of a silicon wafer dot after applying AFM nanolithographic processing steps is generated. Using the results, a suggestion will be made whether to use a Gaussian or Lorentzian profile for the analytical model. The processing steps for a physics based Monte Carlo approach are introduced.

2   Non-Contact Mode Atomic Force Microscopy

AFM is a relatively modern lithographic technique, capable of generating patterns on semiconductor surfaces through the application of an electric field. It is a very useful tool for local anodic nanooxidation of silicon surfaces. Fig. 1 shows the schematic of NCM-AFM, where the tip is placed near the silicon surface and a negative bias voltage, relative to the silicon potential, is applied. An electric field is generated between the AFM tip and the silicon wafer. The high electric field near the AFM tip causes the creation of oxyions (O^- and OH^-) from the air ambient, which are then accelerated towards the silicon surface, away from the negatively charged AFM tip. The oxyions react with the silicon substrate, resulting in the generation of silicon dioxide which is simultaneously expanded into the silicon substrate and the ambient. A detailed explanation of the AFM method and its use to physically or chemically modify surfaces at nanometer scales can be found in [11].

3   Empirical Atomic Force Microscopy Model

An empirical model for the generation of an oxide dot using local anodic nanooxidation of silicon surfaces with the NCM-AFM method was presented in [2].

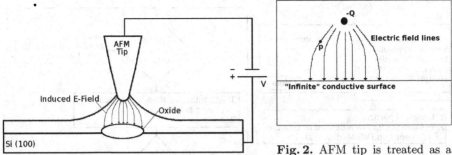

Fig. 1. Basic schematic for the local anodic nano-oxidation of silicon surfaces using NCM-AFM

Fig. 2. AFM tip is treated as a point charge in order to create a physics based Monte Carlo simulation model

It is suggested that the width and height of an oxide dot have a linear dependence on the applied bias voltage, while a logarithmic dependence exists for the pulse duration. An empirical equation which governs the height of the oxide dot produced with NCM-AFM mode has been presented in [2]

$$h(t, V) = h_0(V) + h_1(V) \ln(t), \tag{1}$$

where $h_0(V) = -2.1 + 0.5V - 0.006V^2$ and $h_1(V) = 0.1 + 0.03V - 0.0005V^2$, and the size, voltage, and time are expressed in nanometers, volts, and seconds, respectively. Similarly, the equation which governs the width of the oxide dot, represented as the full width at half maximum (FWHM) is

$$w(t, V) = w_0(V) + w_1(V) \ln(t), \tag{2}$$

where $w_0(V) = 11.6 + 9V$ and $w_1(V) = 2.7 + 0.9V$, and the size, voltage, and time are expressed in nanometers, volts, and seconds, respectively.

3.1 Monte Carlo Simulation Approach

An analytical simulator for NCM-AFM was created in the level set environment described in [6] with a Monte Carlo method as shown in Fig. 3. The Monte Carlo technique is performed by generating a desired number of particles, distributed using a selected probability distribution along a plane parallel to the silicon surface, ensuring that the FWHM fits with (2). Each particle is accelerated towards the silicon surface. After a collision with the surface, the silicon dioxide is expanded by a height determined by (1) and the total number of particles. After all the particles have been simulated, the result is a surface topography of a nanodot with a height and width dependent on (1) and (2), respectively. An example from [2] is simulated at a bias voltage of 20V and a pulse time of 0.125ms, resulting in a height and width of 1nm and 5.6nm, respectively, shown in Fig. 4.

The results of the experiments from [2] and the analytical model were used in order to construct a physics based model for NCM-AFM in Section 4.

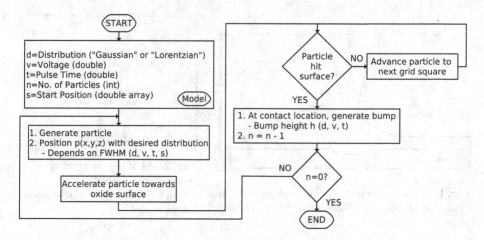

Fig. 3. Flow chart for the analytical Monte Carlo NCM-AFM simulator

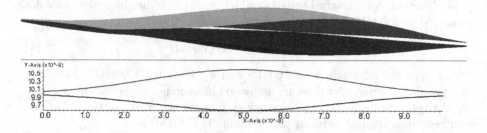

Fig. 4. AFM simulation: bias voltage is 20V and pulse width is 0.125ms. Top surface is the oxide-ambient interface, while the bottom surface is the oxide-silicon interface.

3.2 Lorentzian versus Gaussian Profiles

There are a very limited number of simulators which can deal with NCM-AFM nanooxidation, therefore there is not a clear consensus on the exact structure of the final oxide dot. The important feature of the nanodot is its height and FWHM. However, in order to create a simulator for NCM-AFM, more information is required. Some literature states that the final oxide dot approximately follows a Gaussian curvature [10], while some suggest a Lorentzian profile [8]. The differences between the two options are shown in Fig. 5 where the cross-section of a NCM-AFM simulation is depicted. The simulation is performed with the bias voltage set to 20V and a pulse time of 0.125ms. According to the model described in Section 3.1 a height of 1nm and a width of 5.6nm are achieved.

From Fig. 5 it is evident that both curves have the same height and width, since the points of intersect for the two curves are at $\pm\frac{1}{2}FWHM$. The difference is that the Lorentzian profile suggests that the oxide expands laterally near the silicon interface, while the Gaussian profile has a steeper curve, resulting in slightly more oxide in the top half of the nanodot.

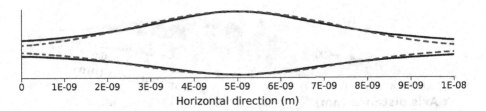

0   1E-09   2E-09   3E-09   4E-09   5E-09   6E-09   7E-09   8E-09   9E-09   1E-08
Horizontal direction (m)

Fig. 5. Cross-section of the oxide pattern when using Gaussian (red-dashed line) distribution and Lorentzian (blue-solid line) distribution

4 Electric Field and Ambient Profiles

In order to simulate the AFM oxidation process with a physical model, the tip of the cantilever from Fig. 1 is treated as a negatively charged stationary dot, while the silicon wafer is treated as an infinitely long conducting plane, shown in Fig. 2. The generated oxyions (particle p from Fig. 2) will move through the ambient towards the silicon surface aided by the electric field. It is assumed that the silicon surface is on the x-y plane, at z=0, the origin p(0,0,0) is located on the silicon surface directly below the charged dot, and the charged dot is at a distance d away from the silicon surface. The equation for the strength of the electric field at a point p(x,y,z) is well known. It is calculated by assuming the presence of an oppositely charged stationary dot located at (0,0,-d) and implementing the image charge method [9]. This method is used in order to calculate the effective charge of the AFM needle tip, the electric field between the AFM tip and the silicon surface, as well as the surface charge density on the silicon surface. The equation which governs the electric potential at a point p(x,y,z) is

$$V\left(\overrightarrow{p}\right) = k\left[\frac{Q}{\left(x^2+y^2+(z-d)^2\right)^{1/2}} - \frac{Q}{\left(x^2+y^2+(z+d)^2\right)^{1/2}}\right], \tag{3}$$

where $k = 1/(4\pi\epsilon_0)$, Q is the charge at a distance d from the surface. The voltage at p(0,0,d) is known, since this is the applied bias voltage. Using (3), we can find the effective dot charge Q to simulate the AFM tip.

$$E_x = kQ\left[\frac{x}{\left(x^2+y^2+(z-d)^2\right)^{3/2}} - \frac{x}{\left(x^2+y^2+(z+d)^2\right)^{3/2}}\right]$$

$$E_y = kQ\left[\frac{y}{\left(x^2+y^2+(z-d)^2\right)^{3/2}} - \frac{y}{\left(x^2+y^2+(z+d)^2\right)^{3/2}}\right] \tag{4}$$

$$E_z = kQ\left[\frac{z-d}{\left(x^2+y^2+(z-d)^2\right)^{3/2}} - \frac{z+d}{\left(x^2+y^2+(z+d)^2\right)^{3/2}}\right]$$

The induced surface charge density is represented as $\sigma\left(x,y,0\right) = \epsilon_0 E_z\left(x,y,0\right)$, leading to the expression

$$\sigma\left(x,y,0\right) = -\frac{dQ}{2\pi\left(x^2+y^2+d^2\right)^{3/2}}. \tag{5}$$

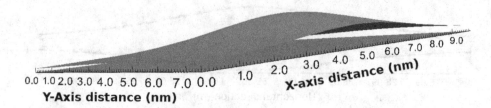

Fig. 6. Shape of the induced surface charge density above the silicon surface

Using the example from [2], modeled with the empirical simulator from Section 3, (3), and (5) we developed a physics based Monte Carlo model for the final pattern of the oxide nanodot.

4.1   Surface Charge Density Modeling

In order to obtain the final pattern for an oxide nanodot generated using NCM-AFM, the physical equations of Section 4 will be implemented for the example from Section 3.1. The initial steps used to generate an oxide nanodot were:

1. Calculate Q based on the applied voltage (20V) and (3) in order to obtain a dot charge of -2.2×10^{-19}C.
2. Using (5) and the known value of the FWHM (5.6nm), calculate the effective location of the dot charge above the silicon surface, d=3.7nm.
3. Using the effective location of the dot charge above the silicon surface, generate an induced surface charge density curve, shown in Fig. 8 and Fig. 9.
4. Use the Monte Carlo rejection technique in order to distribute particles around the AFM tip according to the induced surface charge density, as shown in Fig. 6 and further detailed in Section 5.

5   Monte Carlo Model Implementation

The implementation of the Monte Carlo rejection technique in order to distribute a desired number of particles according to the electric field strength and induced surface charge density is as follows:

1. Generate an evenly distributed particle at position $p_0(x_0, y_0, z_0)$, located on a plane parallel to the silicon surface. x_0 and y_0 are evenly distributed random variables, while $z_0 = d$ is the effective vertical position of the static dot charge.
2. Calculate the induced surface charge density $\sigma(x, y, 0)$ at $p_0(x_0, y_0, 0)$ using (5).
3. Calculate the maximum induced surface charge density $\sigma_{max}(0, 0, 0)$, at position $p_{max}(0,0,0)$.
4. Generate an evenly distributed random number ρ between 0 and σ_{max}.
5. If $\rho > \sigma(x, y, 0)$ remove the particle and repeat the procedure from step 1. If $\rho \leq \sigma(x, y, 0)$ continue to the next step.

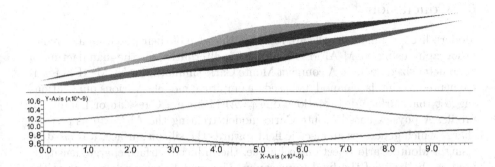

Fig. 7. Surface topography obtained with the presented model

6. Accelerate the particle towards the silicon surface along the vertical direction, until it collides with the surface.
7. At the impact location, advance the ambient-oxide interface towards the ambient while advancing the oxide-silicon interface towards the silicon.
8. If the number of particles is 0 the simulation is complete. Otherwise reduce the number of particles by 1 and repeat the procedure from step 1.

The surface topography, following the above simulation steps, is shown in Fig. 7, where the top and bottom surfaces are the final oxide-ambient and oxide-silicon interfaces, respectively. The induced surface charge density curve is compared to empirical simulations using Gaussian and Lorentzian distributions in Fig. 8 and Fig. 9, respectively. It is immediately evident that the Lorentzian distribution is a better fit to the physics based model. The Gaussian distribution has a sharp slope, which fails to properly show the lateral extension of the oxide nanodot.

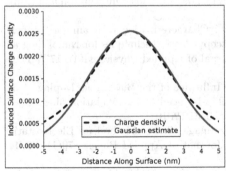

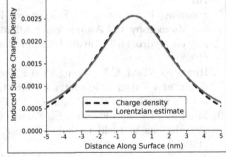

Fig. 8. Comparison to simulation with a Gaussian distribution

Fig. 9. Comparison to simulation with a Lorentzian distribution

6   Conclusion

Modern lithographic methods are reaching the edge of their potential and nano-lithography using NCM-AFM is a promising alternative for the manufacture of nanometer sized devices. A compact Monte Carlo simulator for NCM-AFM is implemented to simulate nanodots based on empirical models. Various publications suggest that surface deformations due to AFM have a Gaussian or Lorentzian profile. A physics based Monte Carlo model, treating the AFM tip as a static charge which generates an electric field towards the silicon surface, was created. Using a Monte Carlo rejection technique, the induced surface charge density is used as the basis of the final topography. The model shows a better fit to the Lorentzian distributed empirical model when compared to the Gaussian model.

References

1. Binning, G., Quate, C.F., Gerber, C.: Atomic Force Microscopy. Physical Review Letters, 930 (1986)
2. Calleja, M., García, R.: Nano-Oxidation of Silicon Surfaces by Noncontact Atomic-Force Microscopy: Size Dependence on Voltage and Pulse Duration. Applied Physics Letters 76(23), 3427–3429 (2000)
3. Dagata, J.A., Perez-Murano, F., Abadal, G., Morimoto, K., Inoue, T., Itoh, J., Yokoyama, H.: Predictive Model for Scanned Probe Oxidation Kinetics. Applied Physics Letters 76(19), 2710–2712 (2000)
4. Dagata, J., Schneir, J., Harary, H., Evans, C., Postek, M., Bennett, J.: Modification of Hydrogen-Passivated Silicon by a Scanning Tunneling Microscope Operating in Air. Applied Physics Letters 56, 2001–2003 (1990)
5. Ertl, O., Selberherr, S.: A Fast Level Set Framework for Large Three-Dimensional Topography Simulations. Computer Physics Communications 180(8), 1242–1250 (2009)
6. Filipovic, L., Ceric, H., Cervenka, J., Selberherr, S.: A Simulator for Local Anodic Oxidation of Silicon Surfaces. In: IEEE Canadian Conference on Electrical and Computer Engineering 2011 (CCECE 2011), Niagara Falls, Ontario, Canada (May 2011)
7. Fontaine, P., Dubois, E., Stiévenard, D.: Characterization of Scanning Tunneling Microscopy and Atomic Force Microscopy-Based Techniques for Nanolithography on Hydrogen-Passivated Silicon. Journal of Applied Physics 84(4), 1776–1781 (1998)
8. Huang, J., Tsai, C.L., Tseng, A.A.: The Influence of the Bias Type, Doping Condition and Pattern Geometry on AFM Tip-Induced Local Oxidation. Journal of the Chinese Institute of Engineers 33(1), 55–61 (2010)
9. Mesa, G., Dobado-Fuentes, E., Saenz, J.: Image Charge Method for Electrostatic Calculations in Field Emission Diodes. Journal of Applied Physics 79(1), 39–44 (1996)
10. Notargiacomo, A., Tseng, A.: Assembling Uniform Oxide Lines and Layers by Overlapping Dots and Lines Using AFM Local Oxidation. In: 9th IEEE Conference on Nanotechnology, pp. 907–910 (July 2009)
11. Tang, Q., Shi, S.Q., Zhou, L.: Nanofabrication with Atomic Force Microscopy. Journal of Nanoscience and Nanotechnology 4(8), 948–963 (2004)

Numerical Integration Using Sequences Generating Permutations

Sofiya Ivanovska[1], A. Karaivanova[1], and N. Manev[2]

[1] Institute of Information and Communication Technologies, BAS
Acad. G. Bonchev St., Bl.25A, 1113 Sofia, Bulgaria
{sofia,anet}@parallel.bas.bg
[2] Institute of Mathematics and Informatics, BAS
Acad. G. Bonchev St., Bl.8, 1113 Sofia, Bulgaria
nlmanev@math.bas.bg

Abstract. In this paper we propose a new class of pseudo random number generators based on a special linear recursions modulo m. These generators produce sequences which are permutations of the elements of a $\mathbb{Z}_m$. These sequences have been developed for other applications but the analysis of their statistical properties and the experiments described in this paper show that they are appropriate for multiple integration. Here we present some results from numerical tests comparing the performance of the two proposed generators with Mersenne Twister.

Keywords: Monte Carlo methods, Pseudo Random Number Generators, multidimensional integration.

1   Introduction

Stochastic simulations use sequences of numbers with certain statistical properties. In particular Monte Carlo methods need *independent and identically distributed* random variables having uniform distribution over the interval $(0, 1)$ (*i.i.d. $U(0, 1)$ variables*). If it is necessary, these variables are later transformed to random variables with the desired distribution.

Real-valued random variables $\{u_j\}_{j=0}^{\infty}$ are i.i.d. $U(0, 1)$, if for all integers $i \geq 0$ and $t > 0$, the vectors $(u_i, u_{i+1}, \ldots, u_{i+t-1})$ are uniformly distributed over the t-dimensional hypercube $(0, 1)^t$.

The sequences used in computer simulations are usually generated by so called Pseudo Random Number Generators (PRNG). A short introduction to them is given in Section 2.

In section 3 we describe in brief a new class of PRNGs based on a special linear recursions modulo prime power. These recursions have been originally constructed and studied in order to be applied to areas far from stochastic simulations, but their statistical properties have inspired us to test whether these recursions can be utilized by Monte Carlo methods.

The numerical experiments that we have carried out by using two generators from the proposed new class in computing multidimensional integrals are

I. Lirkov, S. Margenov, and J. Waśniewski (Eds.): LSSC 2011, LNCS 7116, pp. 455–463, 2012.
© Springer-Verlag Berlin Heidelberg 2012

described in Section 4. The obtained results are compared with the results of integrations based on using Mersenne Twister random number generator.

2   Preliminaries

Definition 1 (L'Ecuyer [5]). *A PRNG is a structure* (S, μ, f, U, g)*, where*

- *S is a finite set called **the state space;***
- *μ is the probability distribution on S;*
- *$f : S \to S$ is a function called **transition function;** starting with a given initial state s_0, which is selected according to the distribution μ, all elements of S are generated according to $s_i = f(s_{i-1})$.*
- *U is the **output space;***
- *$g : S \to U$ is the **output function:** $u_i = g(s_i)$.*

For the specific PRNGs used in most Monte Carlo computations we have

- $S = \mathbb{Z}_m^k$, where $\mathbb{Z}_m$ is the ring of integers modulo m,
- the output set U is i.i.d. $U(0, 1)$ (or $U(0, 1)^t$),
- $u_i = s_i/m$.

To some extend, the quality requirements for PRNGs depend on applications to which they are applied. But there are some properties that are relevant to and required by any application. The most important such properties are listed below.

Efficiency. The generator has to be implementable by a deterministic polynomial-time algorithm, i.e., to run in time bounded by a polynomial of the length of the initial state. The implementation has to be realized by as few as possible arithmetical operations and use little memory.

Long period. The period T of the generator has to be a square, or sometimes a cube of the required number of points, that is, even for modest applications we need $T > 10^{18}$.

Repeatability and Portability. These properties guarantee the ability to exactly generate the same sequence of random numbers on different machines and at different times. For the purposes of testing and development these properties are very important.

Uniformity and Independence. They are relevant to the ability to generate i.i.d. $U(0, 1)$ sequences.

Ability "to skip ahead". This property characterizes the ability of calculating u_k for large k without generating all values $u_0, u_1, \ldots, u_{k-1}$. It is a property important for parallel realizations.

Pseudorandomness. This is very important characteristic for random number generators. Informally pseudorandomness means "The generators output has to look random". This is quite vague statement and different trends in its understanding can be observed. Indeed there are three main approaches to formalization of pseudorandomness:

- **Probabilistic** (Shannon): Shannon's information theory considers perfect randomness as the extreme case and it is associated with a unique distribution the uniform one.

- **Computational Complexity** (Kolmogorov, Chaitin, Solomonov): This approach is based on the Kolmogorov's computational complexity [4,8].

- **Computational Indistinguishability** (Blum, Goldwasser, Micali,Yao, Goldreich): A distribution is pseudorandom if no efficient procedure can distinguish it from the uniform distribution [3,2].

We will not enter into details since the discussion on such topics is far from the goals of this paper. Such considerations are of greater concern in cryptography. For Monte Carlo integration, for example, it is not so important if the next generated value is unpredictable. We refer the interested reader to the cited literature.

As examples of generator often used in Monte Carlo applications we give the next PRNGs which are included in the GNU Library:

1. Fifth-order multiple recursive with period $\approx 10^{46}$:
   $x_n = 107374182 x_{n-1} + 104480 x_{n-5} \,. (\mathrm{mod}\ 2^{31} - 1)$.

2. Combined multiple recursive $z_n = x_n - y_n \ (\mathrm{mod}\ 2^{31} - 1)$, where $\{x_n\}$ and $\{y_n\}$ are 3rd order linear recurrent sequences modulo $2^{31} - 1$ and 2145483479, respectively. Its period is $\approx 2^{185}$.

3. Generalized (lagged) Fibonacci: $y_n = y_{n-s} \theta y_{n-r}$, $r > s$, where θ is $+, -, *$ modulo m, or xor. If θ is the addition modulo 2^k, then the period is $(2^r - 1) 2^{k-1}$.

4. **Mersenne Twister generator**. It is equi-distributed in 623 dimensions and has period $2^{19937} - 1 \approx 10^{6000}$ [10].

The last generator has recently become popular for simulation and it has been installed as the default PRNG for the most used mathematical packages. That is why we have chosen to compare our random number generator with the Mersenne Twister generator.

3 A New Class of Congruential Generators

Definition 2 ([9]). *Let* $\mathbf{S} = \{s_n\}_{n \geq 0}$ *be a sequence with terms in a finite ring* R. *The sequence* $\mathbf{S}$ *is called* ***strictly balanced (in short SB) sequence***, *if it is periodic and each element of* R *occurs equal number of times in one period of the sequence. If each element of* R *appears exactly once in a period, the sequence* $\mathbf{S}$ *is called* ***sequence generating permutation (in short SGP)***.

The period of an SGP sequence is equal to the cardinality $|R|$ of the ring.

Both from algebraic and practical points of view, the most important case is $R = \mathbb{Z}_{p^m}$.

Recall that a kth-*order homogeneous linear recurrence sequence,* $\mathbf{S} = \{s_n\}$, *with constant terms in* R is defined by the recursion

$$s_{n+k} = a_{k-1}s_{n+k-1} + a_{k-2}s_{n+k-2} + \cdots + a_1 s_1 + a_0 s_n \qquad (1)$$

and *initial terms* $s_0, s_1, \ldots, s_{k-1}$. It is obvious that any such sequence over a finite ring is periodic.

Theorem 1 ([9]). *Let* $\{s_n\}$ *be a second-order sequence defined by*

$$s_{n+2} = as_{n+1} + bs_n, \quad (\bmod \ p^m). \qquad (2)$$

It is an SGP sequence if and only if

$$\mu(x^2 - ax - b) = (x - 1)^2, \qquad (3)$$

where $\mu : \mathbb{Z}_{p^m}[x] \to \mathbb{Z}_p[x]$ *for* $p > 2$ *and* $\mu : \mathbb{Z}_{2^m}[x] \to \mathbb{Z}_4[x]$ *for* $p = 2$.
In the case $p = 3$ *the condition*

$$u + v \not\equiv 2 \pmod 3, \qquad \text{where } a = 2 + 3u, \quad b = -1 + 3v,$$

has to be added to (3) in order the theorem to be true.

The higher order case is more complicated and a necessary and sufficient condition cannot be formulated in a simple form. But if $f(x) = x^k - a_{k-1}x^{k-1} - a_{k-2}x^{k-2} - \cdots - a_1 x - a_0$ is the minimal polynomial of the recursion of order $k \geq 3$ with $M = p^m$, the following conditions are sufficient for generating an SGP sequence:

- $f(x) \equiv (x - 1)^k \pmod p$

- $s_0, s_1, \ldots, s_{k-1}$ are different modulo p, e.g., $0, 1, 2, \ldots, k - 1$.

Using the Chinese Reminder Theorem we can construct an SGP sequence for any $M = p_1^{e_1} p_2^{e_2} \ldots p_k^{e_k}$.

Given an SGP sequence $\{s_n\}$ over $\mathbb{Z}_M$ we can transform it into a sequence of numbers in $[0, 1)$ dividing each element by M. Hence the resulting sequence of M elements is a permutation of the numbers

$$\left\{ \frac{i}{M} \ \middle| \ i = 0, 1, 2, \ldots, M - 1 \right\}. \qquad (4)$$

We consider the following two generators of the proposed class:

Generator M1
Consider the second-order sequence with terms in $\mathbb{Z}_M$, where $M = 5^{19}$ defined by

$$s_{n+2} = (5^{10} + 2)s_{n+1} - 5556s_n$$

We start with $s_0 = 0$ and $s_1 = 1$ (or any $s_1 \equiv 1 \pmod 5$).

Similarly, the coefficients $a_1 = 5^{10} + 2$ and $a_0 = -5556$ can be replaced with any $a_1 \equiv 2$, $a_0 \equiv -1 \pmod 5$.

Generator M2

Consider the third-order sequence with terms in $\mathbb{Z}_M$ ($M = 2^{61}$) defined by

$$s_{n+3} = (2^{10} - 1)s_{n+2} + (2^8 + 2^6 + 1)s_{n+1} + s_n \pmod{2^{61}}.$$

Any elements s_0, s_1, s_2 of $\mathbb{Z}_M$, such that $s_0 \equiv 0$, $s_1 \equiv 1$, $s_2 \equiv 2 \pmod 4$ are suitable for initial parameters. The coefficients a_2, a_1, a_0 can be replaced with any $a_2 \equiv -1$, $a_1 \equiv 1$, and $a_0 \equiv 1 \pmod 4$, but they should be taken with large absolute values in order to improve uniformity of the distribution of segments $(u_{n+1}, u_{n+2} \ldots, u_{n+t})$.

Bellow some features which characterize the proposed class of PRNGs are listed:

– *Good lattice structure can be arranged.*
  We can make the lower bound $\left(\sum_{i=0}^{k-1} a_i^2\right)^{-1/2}$ for the minimal distance between hyperplanes, d_t, sufficiently small by choosing large coefficients.

– *The generators modulo 2^m are very efficient.*
  For instance, the implementation of Example 2 gives a two times faster generator than Mersenne Twister. The generation is realized by shifting and addition and only in the last step of the algorithm a multiplication by $1/2^m$ is used.

– *There is no theoretical limit for the period*
  But in order to keep the advantage in speed and low complexity of the considered class of PRNGs the period should be less than 2^{128} in practical implementation. This is relatively short period in comparison to one of Mersenne Twister but enough long for many applications. Indeed we can lengthen the period without enlarging M by permitting the repetition of the elements of (4).

– *The proposed class of PRNGs is significantly different* from algebraic point of view. In contrast to other congruential generators the minimal polynomial of each generator is purely inseparable polynomial (this corresponds to purely inseparable extensions of the basic field).

4   Computational Experiments

4.1   Description of the Experiments

We have carried out our computational experiments with the following d-dimensional test integrals:

$$I_1 = \int_{(0,1)^d} F_1(\mathbf{x})d\mathbf{x} \quad \text{and} \quad I_2 = \int_{(0,1)^d} F_2(\mathbf{x})d\mathbf{x},$$

where $\mathbf{x} = (x_1, x_2, \dots, x_d)$ and

$$F_1(\mathbf{x}) = \prod_{i=1}^{d} (x_i^3 + 0.75) \quad \text{and} \quad F_2(\mathbf{x}) = \prod_{i=1}^{d} |4x_i - 2|.$$

The values of I_1 and I_2 are both equal to 1.

The first of these test functions is taken from Schmid and Uhl (2001). I_2 is known as Roos and Arnold's example and it is suggested as a test function by Owen (2003).

It is straightforward to calculate the variances σ_1^2 and σ_2^2 of F_1 and F_2:

$$\sigma_1 = \sigma[F_1] = \sqrt{\left(\frac{121}{112}\right)^d - 1} \quad \text{and} \quad \sigma_2 = \sigma[F_2] = \sqrt{\left(\frac{4}{3}\right)^d - 1} \quad (5)$$

In particular we have

$$d = 10: \quad \sigma_1 = 1.07984949546134, \quad \sigma_2 = 4.09362023566092$$
$$d = 20: \quad \sigma_1 = 1.92142671333385; \quad \sigma_2 = 17.72954751823117;$$
$$d = 30: \quad \sigma_1 = 3.02703898016620; \quad \sigma_2 = 74.82423185191648;$$

As it is well known the error of integration tends asymptotically to

$$e_N[F_i] \approx \frac{\sigma_i}{\sqrt{N}} \nu,$$

where ν is a standard normal ($N(0,1)$) random variable and σ_i^2 is the variance of F_i, $i = 1, 2$.

Parameters of our tests are:

- Tested generators: *Mersenne Twister, M1, and M2.*

- Dimensions: $d = 10, 20, 30$.

- Number of points: $N = 2^m$, where $m = 10, 11, \dots, 20$.

- 200 calculations have been done for each generator and for each pair (d, m). The presented value of the error is the average over these 200 calculations of the absolute values of the error for each pair (d, m).

The described experiments can be considered as a continuation of the research given in [1].

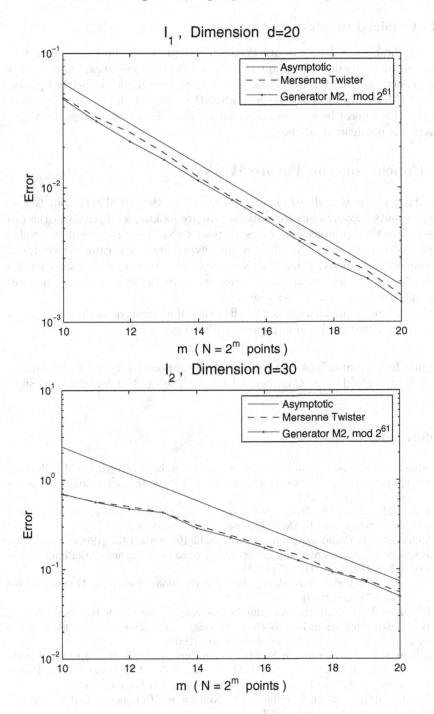

Fig. 1. Graphical presentations of the experimental results

4.2   Obtained Results

Two of the obtained results are graphically represented in Fig. 1. The term "asymptotic" is used for the graph of $\sigma_i/\sqrt{N}$ ($N = 2^m$). The experiments show that generators from the new class demonstrate even modestly better behavior than the Mersenne Twister. The generators (Examples 1 and 2 and many others not described here) have been chosen at random. Hence we believe that their behavior is intrinsic to all class.

5   Conclusion and Future Work

The study of the considered class of PRNGs is at the initial stage but the obtained results encourage us to continue. We are looking for PRNGs in the proposed class that produce sequences of points that demonstrate high level of uniform distribution in high dimensional hypercubes. We cannot recommend yet concrete parameters (i.e., concrete generators) since we do not have enough knowledge about discrepancy and lattice structure of the generated sequences. This is the subject of a further research.

Also, we are going to analyze the efficiency of generators modulo p^m, $p > 2$, that are implemented by Montgomery arithmetic.

Acknowledgments. This work was partially supported by the Bulgarian National Science Fund under Contracts DO02-146/2008 and DCVP02/1 CoE Super CA++.

References

1. Atanassov, E., Karaivanova, A., Gurov, T., Ivanovska, S., Durchova, M., Dimitrov, D.: Quasi-Monte Carlo integration on the grid for sensitivity studies. Earth Sci. Inform. 3, 289–296 (2010)
2. Blum, M., Micali, S.: How to Generate Cryptographically Strong Sequences of Pseudo-Random Bits. SIAM J. Comput. 13(4), 850–864 (1984)
3. Goldreich, O.: Pseudorandomness. Notices 46(10), 1209–1216 (1999)
4. Kolmogorov, A.N.: Three approaches to the concept of the amount of information. Probl. Inf. Transm. 1(1), 1–7 (1965)
5. L'Ecuyer, P.: Uniform random number generation. Annals of Operations Research 53, 77–120 (1994)
6. L'Ecuyer, P.: Uniform random number generation. In: Henderson, S.G., Nelson, B.L. (eds.) Simulation, Handbooks in Operations Research and Management Science, ch. 3, pp. 55–81. Elsevier, Amsterdam (2006)
7. L'Ecuyer, P.: Pseudorandom Number Generators. In: Cont, R. (ed.) Encyclopedia of Quantitative Finance, pp. 1431–1437. John Wiley, Chichester (2010); Platen, E., Jaeckel, P. (eds.) Simulation Methods in Financial Engineering
8. Li, M., Vitanyi, P.: An Introduction to Kolmogorov Complexity and Its Applications. Springer, New York (1993)
9. Manev, N.L.: Sequences generating permutations. Applied Mathematics and Computation 216(3), 708–718 (2010)

10. Matsumoto, M., Nishimura, T.: Mersenne twister: a 623-dimensionally equidistributed uniform pseudo-random number generator. ACM Transactions on Modeling and Computer Simulation (TOMACS) 8(1), 3–30 (1998)
11. Owen, A.: The dimension distribution and quadrature test functions. Stat. Sin. 13, 1–17 (2003)
12. Schmid, W., Uhl, A.: Techniques for parallel quasi-Monte Carlo integration with digital sequences and associated problems. Math. Comp. Sim. 55, 249–257 (2001)

Optimization of Intermolecular Interaction Potential Energy Parameters for Monte-Carlo and Molecular Dynamics Simulations*

Dragan Sahpaski[1], Ljupčo Pejov[2], and Anastas Misev[1]

[1] Institute of Informatics, Faculty of Natural Sciences and Mathematics,
University "Ss. Cyril and Methodius", Skopje, Macedonia
[2] Institute of Chemistry, Faculty of Natural Sciences and Mathematics,
University "Ss. Cyril and Methodius", Skopje, Macedonia

Abstract. Derivation of high-quality intermolecular potentials for molecular dynamics (MD) and Monte Carlo (MC) simulations is crucial for efficient modeling of molecular systems. Despite their overall complexity, the interactions potentials have often been derived in a semiempirical manner, though in certain cases, also *ab initio* techniques have been involved in their construction. In the present study, we aim to construct optimized intermolecular interaction potentials to be used for MD and MC simulations of pure molecular liquids and their mixtures. We have focused on one of the simplest forms of the potentials, namely the Lennard-Jones (LJ) + Coulomb electrostatic terms. Interaction between each pair of atoms in the molecular liquids has thus been characterized by the LJ parameters + atomic charges. The optimization has been performed by genetic algorithms. An in-depth analysis of the performances of both the standard, widely used (*i.e.* non-optimized), and the optimized interaction potentials was carried out. This analysis was carried out from various aspects related to their performances.

Keywords: optimization, genetic algorithms, Monte Carlo, molecular dynamics, intermolecular interaction potentials.

1 Introduction

Condensed-phase systems are of substantial importance in both fundamental natural sciences and technology. Theoretical modeling of such systems has been shown to be of crucial importance for a thorough understanding of their properties. It has been often demonstrated that the theoretical model can be complementary to the experimental studies of condensed phases. On one hand, theory can certainly enable a better understanding of the experimental observations related to a variety of properties of the mentioned systems. On the other hand, theory can sometimes predict certain properties or systems' behavior which hasn't

---

* This paper is based on the work done in the framework of the SEE-GRID-SCI FP7 EC funded project, with partial support from NSFB grant DO02 - 146/2008.

I. Lirkov, S. Margenov, and J. Waśniewski (Eds.): LSSC 2011, LNCS 7116, pp. 464–471, 2012.

been observed yet (or, in certain cases, is not even observable with the current experimental techniques). The most widely used theoretical methods for modeling of condensed phases are Monte Carlo and molecular dynamics techniques [1]. Though the *ab initio* variants of both of these techniques are exact and extremely useful, they are computationally highly demanding as well. The usefulness of the classical counterparts thereof, on the other hand, while being much less demanding from computational aspect, depends highly on the chosen interaction potential. Though interaction potentials can be constructed from the first principles, the construction procedures often contain certain semiempirical elements. The final forms of the widely used interaction potentials are therefore, more or less semiempirical. Semiempirically constructed potentials can be quite useful with respect to prediction of certain condensed phase properties, but, at the same time, may fail completely for others. In the present paper, we aim to propose a general methodology (approach) for optimization of the interaction potentials, using genetic algorithms. We analyze in details the performance and drawbacks of non-optimized potentials and emphasize the need for a very careful construction of general-purpose potentials. As a particular example, we focus our attention on liquid carbon tetrachloride (CCl4).

2   Computational Details and Algorithms

2.1   Monte Carlo Simulations

To generate the structure of liquid, first a series of Monte-Carlo (MC) simulations were performed, using the statistical mechanics code DICE [2]. All MC simulations were performed in the isothermal-isobaric (NPT) ensemble, implementing the Metropolis sampling algorithm, at T = 300 K, P = 1 atm, using the experimental density of liquid carbon tetrachloride (CCl_4) of 1.5867 gcm^{-3} at these conditions as a starting point. We carried out series of MC simulations of 500 carbon tetrachloride molecules placed in a cubic box with side length of 43.36 Å, imposing periodic boundary conditions. These simulation conditions correspond to the experimental data for the basic experimental parameters for this liquid system. Long-range corrections (LRC) to the interaction energy were calculated for interacting atomic pairs between which the distance is larger than the cutoff radius defined as half of the unit cell length. The Lennard-Jones (LJ) contribution to the interaction energy beyond this distance was estimated assuming uniform density distribution in the liquid (i.e. $g(r) \approx 1$), while the electrostatic contribution was estimated by the reaction field method involving the dipolar interactions. In all MC simulations carried out in the present study, intermolecular interactions were described by a sum of Lennard-Jones 12-6 site-site interaction energies plus Coulomb terms:

$$U_{ab} = \sum_i^a \sum_j^b 4\varepsilon_{ij} \left(\left(\frac{\sigma_{ij}}{r_{ij}} \right)^{12} - \left(\frac{\sigma_{ij}}{r_{ij}} \right)^6 \right) + \frac{q_i q_j e^2}{4\pi\varepsilon_0 r_{ij}} \tag{1}$$

where i and j are sites in interacting molecular systems a and b, r_{ij} is the interatomic distance between sites i and j, while e is the elementary charge.

The following combination rules were used to generate two-site Lennard-Jones parameters ε_{ij} and σ_{ij} from the single-site ones:

$$\varepsilon_{ij} = \sqrt{\varepsilon_i \varepsilon_j} \tag{2}$$

$$\sigma_{ij} = \sqrt{\sigma_i \sigma_j} \tag{3}$$

The main aim of the present study was to optimize the intermolecular interaction energy potentials to be used in sequential statistical physics — quantum mechanical methods for studying molecular solvation processes. For that purpose, we want to point out at certain subtle inadequacies of the presently implemented and widely used "standard" interaction potentials for liquids. Therefore, we have also carried out a Monte Carlo simulation of pyrrole solution in CCl_4. This system is important for modeling the solvent effects on the vibrational N-H stretching band of pyrrole in liquid carbon tetrachloride, for which relatively accurate experimental data are available. The Monte Carlo simulations were this time carried out for 1 pyrrole molecule solvated by 412 carbon tetrachloride molecules placed in a cubic box with side length of 40.67 Å, imposing periodic boundary conditions. The LJ intermolecular interaction energy parameters are tabulated in Table 1. Atomic charges of CCl_4 and pyrrole were computed by fitting the point charges to the molecular electrostatic potential calculated from the $MP2/6 - 311 + +G(d, p)$ wavefunction (for the optimized geometry on $MP2/6 - 311 + +G(d, p)$ potential energy surface) using the CHELPG point selection algorithm [3]. MC simulations were performed with the DICE statistical mechanics Monte Carlo suite of codes [2], while all quantum-chemical calculations were performed with the Gaussian03 series of codes [4].

2.2   The Optimization Problem

Let q_{Cl}, ε_{Cl}, σ_{Cl}, q_C, ε_{Cl}, and σ_C be charge, epsilon and sigma values of the Cl and C atoms respectively and let ρ, α_P, β_T, and $C_{P,m}$ be the density, thermal expansion coefficient, isothermal compressibility and molar specific heat at constant pressure of the system as calculated by the simulation procedure run with the DICE software. The optimization problem is defined as follows. Find a set of values for $S = \{q_{Cl}, \varepsilon_{Cl}, \sigma_{Cl}, q_C, \varepsilon_{Cl}, \sigma_C\}$, such that the cost function

$$Cost(S) = c_1 \cdot relerr(D) + c_2 \cdot relerr(D) + c_3 \cdot relerr(D) + c_4 \cdot relerr(D) \tag{4}$$

is minimal. The function $relerr$ gives the relative error of the parameter computed by the simulation procedure and the experimental value for the parameters ρ, α_P, β_T and $C_{P,m}$ (vide supra). c_1, c_2, c_3 and c_4 are integer constants defining the weights in which each relative error affects the cost function.

2.3   The Optimization Procedure

We now describe the optimization procedure for obtaining the minimum Cost function using genetic algorithms as defined in the previous section.

Table 1. Lennard-Jones interaction potential parameters used initially in Monte-Carlo simulations

Atom	q/e	$\varepsilon/kcal\,mol^{-1}$	$\sigma/\text{Å}$
N	-0.208	0.170	3.250
H	0.307	0.000	0.000
C	-0.181	0.070	3.550
C	-0.181	0.070	3.550
C	-0.137	0.070	3.550
C	-0.137	0.070	3.550
H	0.413	0.030	2.420
H	0.413	0.030	2.420
H	0.413	0.030	2.420
H	0.413	0.030	2.420
C	0.248	0.050	3.800
Cl	-0.062	0.266	3.470
Cl	-0.062	0.266	3.470
Cl	-0.062	0.266	3.470
Cl	-0.062	0.266	3.470

Algorithm 1. The optimization procedure using a Genetic Algorithm

Generate a random starting population
i=0
repeat
    Calculate the cost function for each chromosome in the population using DICE
    Apply genetic operators and obtain the new population
    i=i+1
until repeat until i = maximum number of generations

A genetic algorithm (GA) [5] is a randomized search method for finding approximate solutions to optimization problems. The design of the genetic algorithm is inspired by concepts in evolutionary biology such as mutation, selection, crossover, and survival of the fittest. Candidate solutions to a given problem are also called chromosomes and are represented most commonly as bit strings, arrays of integers, floating point precision numbers, but also other encodings are possible. The individual bits, numbers etc. are called genes where each gene is an abstraction of some atomic entity in the problem domain. The algorithm starts from a population of randomly generated solutions (chromosomes) and continues to evolve the population in different iterations (*i.e.* generations). In each generation, the cost of every solution in the population is evaluated, multiple solutions are selected from the current population based on their cost, and they are modified (recombined and possibly randomly mutated) to form a new population. The new population is then used in the next iteration. The algorithm terminates when either a maximum number of generations has been produced, or a solution with satisfactory cost has been found.

2.4   Representation of the Solution

The solution is represented as an array of six genes represented as double precision (64 bit) numbers corresponding to: q_{Cl}, ε_{Cl}, σ_{Cl}, q_C, ε_{Cl}, and σ_C. Each gene is bounded by an interval of values determined by the physically acceptable values of the corresponding parameter. One chromosome represents one possible solution to the problem. A single point crossover operator is used, which chooses a random position from two parent chromosomes, *i.e.* solutions, and then performs a swap of that number and all subsequent numbers between the two parent chromosomes, in order to obtain two new offspring chromosomes. The mutation operation is performed over each gene of a chromosome and mutates them with a given probability. Because the genes are represented as double precision floating point numbers, a mutation of a gene means flipping the value of a random bit in the binary representation of the number with a given probability. We use a natural selection operator where a chromosome is selected for survival in the next generation with a probability proportional to the cost of the solution represented by the chromosome. The termination of the GA is done after a fixed number of iterations. The experiments were conducted using the Java Genetic Algorithms and Genetic Programming Package — JGAP [6].

3   Results and Discussion

3.1   Results with the Standard LJ Parameters

First of all, we compared the results obtained from the "classical" MC simulations, using the standard (*i.e.* non-optimized) LJ parameters for the *pure liquid carbon tetrachloride*. We have chosen the following physical quantities as representative to test the quality of the used LJ potential energy parameters: the average density of the liquid (ρ), the thermal expansion coefficient (α_P), isothermal compressibility (β_T) as well as the molar heat capacity at constant pressure ($C_{P,m}$) of the liquid. We have calculated these data from the MC simulations of the pure liquid, using the following basic relation for the average enthalpy H:

$$\langle H \rangle = \frac{3+\nu}{2} NkT + \langle U \rangle + P\langle V \rangle \tag{5}$$

where ν is the number of molecular degrees of freedom other than translations, N is the number of molecules in the box, T is thermodynamic temperature, k is the Boltzman constant, U is the internal energy of the system, while P and V are pressure and volume, correspondingly. In 5, as well as in other following relations, $\langle \rangle$ denotes ensemble averaging. Molar heat capacity was calculated by:

$$\langle C_P \rangle = \left(\frac{\partial H}{\partial T} \right)_P = \frac{3+\nu}{2} Nk + \frac{\partial H_c^2}{kT^2} \tag{6}$$

where

$$\langle \partial H_c^2 \rangle = \langle H_c^2 \rangle - \langle H_c \rangle^2 \tag{7}$$

The thermal expansion coefficient was, on the other hand, computed by the expression:

$$\alpha_P = \frac{1}{\langle V \rangle} \left(\frac{\partial \langle V \rangle}{\partial T} \right)_P = \frac{1}{T} + \frac{\langle \partial V \partial H_c \rangle}{\langle V \rangle k T^2} \tag{8}$$

In the last equation,

$$\langle \partial V \partial H_c \rangle = \langle V H_c \rangle - \langle V \rangle \langle H_c \rangle \tag{9}$$

Finally, the isothermal compressibility was calculated by:

$$\beta_T = \frac{1}{\langle V \rangle} \left(\frac{\partial \langle V \rangle}{\partial P} \right)_T = \frac{\langle \partial V^2 \rangle}{\langle V \rangle k T} \tag{10}$$

In Table 2, the calculated values of these four parameters are compared to the available experimental data, along with the corresponding relative errors. As it can be seen, though the experimental density and isothermal compressibility are well-reproduced with the current set of LJ potential parameters, the relative errors in molar heat capacity at constant pressure and thermal expansion coefficient are very high. This means that though the widely used LJ parameters show satisfactory overall performance, they are not properly chosen to enable modeling of a wider variety of physical quantities characterizing the liquid.

Table 2. Comparison of the density, thermal expansion coefficient, molar heat capacity at constant pressure and isothermal compressibility of liquid carbon tetrachloride computed from the MC simulation with the standard (non-optimized) LJ potential parameters with the available experimental data

Parameter	MC-opt	Experimental	Relative error %
$\rho/(g cm^{-3})$	1.5697	1.5867	1.07
α_P/K^{-1}	80.65	129.35	37.6
β_T/Pa^{-1}	$1.126 \cdot 10^{-9}$	$1.034 \cdot 10^{-9}$	8.9
$C_{P,m}/J K^{-1} mol^{-1}$	$4.6199 \cdot 10^{-3}$	$1.236 \cdot 10^{-3}$	273.8

3.2 The Optimized LJ Parameters

The optimized parameters for the CCl_4 solvent by the genetic algorithm are shown in Table 3. As it can be seen, the charge separation between C and Cl atoms within the CCl_4 molecule is larger as predicted by this algorithm in comparison to the corresponding value obtained by fitting to the molecular electrostatic potential. Also, the σ values for the Cl atoms are notably smaller than those initially used.

Table 4 summarizes the relative errors in the density, thermal expansion coefficient, molar heat capacity at constant pressure and isothermal compressibility of liquid carbon tetrachloride computed from the MC simulation with the optimized LJ potential parameters. As it can be seen, while the improvement in

Table 3. The optimized Lennard-Jones for CCl4 parameters by the genetic algorithm

Atom	q/e	$\varepsilon/kcalmol^{-1}$	$\sigma/\text{Å}$
C	0.413	0.025	3.328
Cl	-0.103	0.374	0.149
Cl	-0.103	0.374	0.149
Cl	-0.103	0.374	0.149
Cl	-0.103	0.374	0.149

Table 4. Comparison of the density, thermal expansion coefficient, molar heat capacity at constant pressure, and isothermal compressibility of liquid carbon tetrachloride computed from the MC simulation with the optimized LJ potential parameters with the available experimental data

Parameter	MC-opt	Experimental	Relative error %
$\rho/(gcm^{-3})$	1.5884	1.5867	0.1
α_P/K^{-1}	122.13	129.35	5.6
β_T/Pa^{-1}	$3.459 \cdot 10^{-12}$	$1.034 \cdot 10^{-9}$	99.6
$C_{P,m}/JK^{-1}mol^{-1}$	$3.352 \cdot 10^{-3}$	$1.236 \cdot 10^{-3}$	171.2

the liquid's density, molar heat capacity and the thermal expansion coefficient is significant, the isothermal compressibility value obtained by this simulation is actually worse than the initial one. However, having in mind the extremely small values of the last parameters, the present error, though seemingly large, is physically acceptable.

4   Conclusions and Directions for Future Work

In the present study, we have efficiently implemented a genetic algorithm to optimize the interaction potential energy parameters of liquid CCl_4 to be used in statistical physics simulations of the pure liquid, as well as of various solutions thereof. We have demonstrated that it is possible to improve the values of certain parameters characterizing the static and dynamical properties of the liquid by the approach that we have adopted. We will further apply the optimized LJ parameters to study salvation of pyrrole as well as other molecular systems in this widely used organic solvent, with an emphasis on solute's vibrational spectroscopic properties. It is also tempting to apply such novel approach to the problem of construction and optimization of intermolecular interaction energy parameters for various types of simulations of a number of molecular liquid systems.

References

1. Allen, M.P., Tildesley, D.J.: Computer Simulation of Liquids. Oxford University Press, New York (1997)
2. Coutinho, K., Canuto, S.: DICE: a Monte Carlo Program for Molecular Liquid Simulation. University of São Paulo, São Paulo (2003)
3. Breneman, C.M., Wiberg, K.B.: J. Comp. Chem. 11, 361 (1990)
4. Frisch, M.J., Trucks, G.W., Schlegel, H.B., Scuseria, G.E., Robb, M.A., Cheeseman, J.R., Zakrzewski, V.G., Montgomery, J.A., Stratmann, R.E., Burant, J.C., Dapprich, S., Millam, J.M., Daniels, A.D., Kudin, K.N., Strain, M.C., Farkas, O., Tomasi, J., Barone, V., Cossi, M., Cammi, R., Mennucci, B., Pomelli, C., Adamo, C., Clifford, S., Ochterski, J., Petersson, G.A., Ayala, P.Y., Cui, Q., Morokuma, K., Malick, D.K., Rabuck, A.D., Raghavachari, K., Foresman, J.B., Cioslowski, J., Ortiz, J.V., Stefanov, B.B., Liu, G., Liashenko, A., Piskorz, P., Komaromi, I., Gomperts, R., Martin, R.L., Fox, D.J., Keith, T., Al-Laham, M.A., Peng, C.Y., Nanayakkara, A., Gonzalez, C., Challacombe, M., Gill, P.M.W., Johnson, B.G., Chen, W., Wong, M.W., Andres, J.L., Head-Gordon, M., Replogle, E.S., Pople, J.A.: Gaussian 1998 (revision A.11). Gaussian, Inc., Pittsburgh PA (1998)
5. Holland, J.H.: Adaptation in Natural and Artificial Systems. University of Michigan Press (1995)
6. Meffert, K.: JGAP - Java Genetic Algorithms and Genetic Programming Package, http://jgap.sf.net
7. Coutinho, K., Canuto, S.: Adv. Quantum Chem. 28, 89 (1997)
8. Becke, A.D.: Phys. Rev. A 38, 3098 (1988)
9. Lee, C., Yang, W., Parr, R.G.: Phys. Rev. B 37, 785 (1988)

Phonon-Induced Decoherence
in Electron Evolution

Philipp Schwaha[2,1], Mihail Nedjalkov[1,3],
Siegfried Selberherr[1], and Ivan Dimov[3]

[1] Institute for Microelectronics, TU Wien
Gußhausstraße 27-29/E360, A-1040 Vienna, Austria
[2] Shenteq s.r.o., Záhradnícka 7, 811 07 Bratislava, Slovak Republic
[3] Institute of Information and Communication Technlogies
Bulgarian Academy of Sciences
Acad. G. Bonchev, Bl25A, 1113 Sofia, Bulgaria

Abstract. A Monte Carlo analysis of the evolution of an electron interacting with phonons is presented in terms of a Wigner function. The initial electron state is constructed by a superposition of two wave packets and a pronounced interference term. The results show that phonons effectively destroy the interference term. The initial coherence in wave vector distribution is pushed towards the equilibrium distribution. Phonons hinder the natural spread of the density with time and advance it towards a classical localization. The decoherence effect due to phonons, which brings about the transition from a quantum to a classical state, is demonstrated by the purity of the state, which decreases from its initial value of 1, with a rate depending on the lattice temperature.

1   Introduction

Quantum computational and communication processes rely on the fundamental physical notions of superposition, entanglement, uncertainty, and interference. The idea for such processes is related to the fundamental physical limits of computation, which are foreseeable due to the saturation in down-scaling the feature sizes of transistors, the basic elements of today's computing engines. Today, features are already characterized by the nanometre scale, where few tens of atom layers represent the active region of devices. As the physical laws at such scale are inherently quantum mechanical in nature, the idea for quantum computations arises in a natural way. The research on quantum computing is mainly concerned with the possible speed-up, quantum complexity bounds, and construction of optimal quantum algorithms [1].

The foundations for quantum algorithms rest on the basic quantum units of information (qubits) and the basic logical manipulations provided by quantum gates. The qubit is a quantum state which may be conveniently presented by the states 0 or 1 of the classical bit forming the basis $|0\rangle$ $|1\rangle$. Any normalized superposition

$$|\psi\rangle = \alpha|0\rangle + \beta|1\rangle \tag{1}$$

I. Lirkov, S. Margenov, and J. Waśniewski (Eds.): LSSC 2011, LNCS 7116, pp. 472–479, 2012.

of such states is a legitimate qubit. A single classical register can store the states only one at a time, while the quantum counterpart stores superpositions of them. In general, a quantum computer with n qubits can be in an arbitrary superposition of up to 2^n different states simultaneously, whereas a normal computer can only be in one of these 2^n states at any given time. The difficult task, however, is to retrieve this information efficiently. The complex numbers α and β can only be measured statistically, which is related to the very nature of quantum mechanics as will be seen below. The property (1) also gives rise to quantum teleportation, since the states can be highly non-local, and to quantum cryptography. In quantum communication it is easy to detect, if the state has been subject to undesired observation, since measurements disturb quantum states, due to the entanglement of $|\psi\rangle$ with the states of the detector. The realizations of all these novel and fascinating scientific ideas rely on the condition that (1) remains coherent, i.e. is a subject of unitary evolution, which is equivalent to say 'remains quantum', since measurements and processes of interaction with the environment try to turn the quantum system into a classical one, a process known as decoherence. To clarify the difference we recall that expectation values $\langle A \rangle$ of physical quantities A, presented by a Hermitian operator $\hat{A}$ are obtained by the trace operation:

$$\langle A \rangle = Tr(\hat{A}\hat{\rho}) = \sum_{i=0,1} \langle i|\hat{A}\hat{\rho}|i\rangle; \qquad Tr(\hat{\rho}) = 1. \tag{2}$$

The density operator $\hat{\rho}$ is defined with the help of (1):

$$\hat{\rho} = |\psi\rangle\langle\psi| = |\alpha|^2|0\rangle\langle 0| + |\beta|^2|1\rangle\langle 1| + \alpha\beta^*|0\rangle\langle 1| + \alpha^*\beta|1\rangle\langle 0| \tag{3}$$

The physical quantity 'expectation value for the first basis state' in (1), for example, expressed by the operator $\hat{\rho}_0 = |0\rangle\langle 0|$, evaluates to $|\alpha|^2$ using the trace operation, while for the second state it yields $|\beta|^2$. The relation $|\alpha|^2 + |\beta|^2 = 1$ allows to interpret these values as probabilities, which implies the normalization of (1) and ensures the last equality in (2).

In order to measure quantities we need to prepare a detector which discriminates states $|0\rangle$ from $|1\rangle$ by virtue of their orthogonality. There are two peculiarities of this process. The measurement disturbs the superposition state (1) by leaving it in the measured state. The probability of finding the system in the alternative state after the measurement is zero. Moreover, the last two terms in (3) remain unobservable for such a detector. However, they can be observed by other kind of detectors and actually reveal the quantum character (1): they account for the superposition of the amplitudes which lead to interference effects.

If these interference effects are neglected, the density operator $\hat{\rho}$ reduces to $\hat{\rho}_{cl} = |\alpha|^2\hat{\rho}_0 + |\beta|^2\hat{\rho}_1$. This density operator again provides the values $|\alpha|^2$ and $|\beta|^2$ for the basis states, however, it corresponds to an entirely different set-up of the system. The latter measures the register by generating *either* the state 0 with a probability $|\alpha|^2$, *or* the complementary state with the complementary

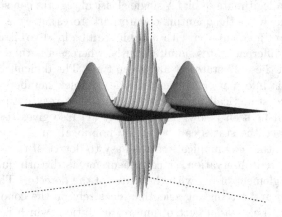

Fig. 1. Entangled wave packets used as initial condition

probability. The collapse of (1) into one of the two possible basis states as a result of a measurement gives rise to the same $\hat{\rho}_{cl}$, which is a return to the principles of the classical computer. Accordingly, $\hat{\rho}_{cl}$ can be called classical density operator.

Decoherence destroys the unitary evolution of the coherent state and is thus the biggest enemy of an effective practical realization of the aforementioned ideas. The system interacts with the environment so that system and environment states entangle into a common, usually macroscopic state. The system state now is obtained by applying a trace on the additional variables, which precludes certain correlations. The theory of decoherence addresses the manner in which some quantum systems become classical due to such entanglement with the environment. The latter monitors certain observables in the system, destroying coherence between the states corresponding to their eigenvalues. Only preferred states survive consecutive 'measurements' by the environment as in the above example. The remainder of states which actually comprises a major part of the Hilbert space is eliminated. Many of the features of classicality are actually induced in quantum systems by their environment [2]. The role of scattering has been intensively studied by different models describing quantum Brownian motion. Peculiar for the equation governing the evolution of the density matrix in the spatial coordinate representation $\langle x|\hat{\rho}|x'\rangle$ is a term giving rise to an exponential damping in time with a rate Λ of the off-diagonal elements ($x \neq x'$). Thus the initial wave packet of an electron does not follow the natural process of spreading due to the coherent evolution, but shrinks around the line $x = x'$ revealing a classical localization [3].

Recently the problem has been reformulated in phase space giving rise to a Wigner equation with a Fokker-Planck term describing the diffusion in the phase space [4]. The analysis of the equation provides an alternative interpretation of the process of decoherence in phase space. Quantum coherence effects as a rule give rise to rapid oscillations of the Wigner function. The diffusion

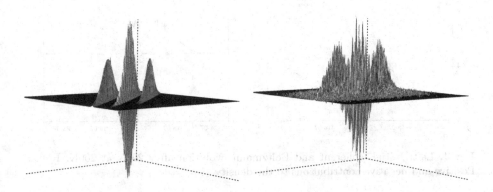

Fig. 2. Left: The coherent evolution leaves the basic structure of the entangled wave packets intact even after 900fs. Right: Scattering mechanisms destroy the initial structure of the Wigner-function as shown here after 300fs.

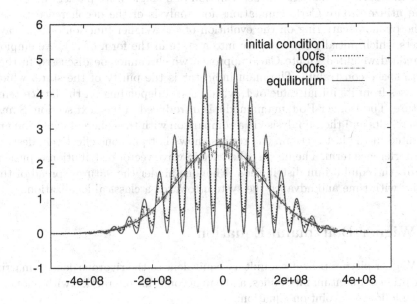

Fig. 3. The momentum distribution of the initial condition decays to at a temperature $T = 200K$ the thermal equilibrium in approximately 1ps. Without scattering the initial distribution remains frozen in place.

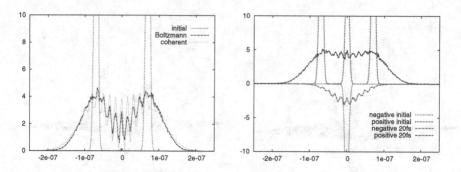

Fig. 4. Left: Initial, coherent and Boltzmann evolution after 200fs at 200K. Right: Positive and negative contributions to the density.

term destroys these oscillations thus effectively suppressing quantum coherence effects. Furthermore, this model has been compared with the Wigner-Boltzmann equation showing that the latter reduces to the former provided that the wave vector of the lattice vibrations becomes much smaller than the electron counterpart [4]. It follows that decoherence effects can definitely be expected as a result of the scattering by phonons. These effects have been well demonstrated by Monte Carlo simulations of the evolution of a single wave packet [4].

We utilize Monte Carlo simulations for analysis of the decoherence caused by the phonon scattering on the evolution of the Wigner function of two wave packets which initially superimpose into a state in the form of (1). We employ a standard weighted Monte Carlo approach which cannot be discussed further due to space constraints. The main indicator is the purity of the state, which decreases from its initial value of 1, with a speed depending on the lattice temperature. The Wigner-Boltzmann model is introduced in the next section. Simulation results and their analysis via a comparison with the coherent evolution are presented in the last section. The results show that phonons effectively destroy the interference term. The initial coherence in wave vector distribution is pushed towards the equilibrium distribution. Phonons hinder the natural spread of the density with time and advance the system towards a classical localization.

2   Wigner-Boltzmann Equation

The Wigner picture provides a unitary equivalent to the rigorous density matrix description of quantum mechanics, and can account for interaction with phonons via the following evolution equation:

$$\left(\frac{\partial}{\partial t} + \frac{\hbar k_x}{m}\frac{\partial}{\partial x}\right) f_w(x, \mathbf{k}, t) = \int dk_x' V_w(x, k_x' - k_x) f_w(x, k_x', \mathbf{k}_{yz}, t) \tag{4}$$

$$+ \int d\mathbf{k}' f_w(x, \mathbf{k}', t) S(\mathbf{k}', \mathbf{k}) - f_w(x, \mathbf{k}, t)\lambda(\mathbf{k})$$

Here, the phase space is formed by a single position and three wave vector coordinates. Quantum correlations are described by the arguments x and k_x of the Wigner potential V_w. Phase-breaking processes are accounted for by the Boltzmann scattering operator with $S(\mathbf{k}, \mathbf{k}')$, the scattering rate for a transition from $\mathbf{k}$ to $\mathbf{k}'$. $\lambda(\mathbf{k}) = \int d\mathbf{k}' S(\mathbf{k}.\mathbf{k}')$ is the total out-scattering rate. The Wigner function f_w is a real quantity. Physical averages are obtained according to $\langle A \rangle = \int d\mathbf{k} dx A(x, \mathbf{k}) f_w$, where A is a generic dynamical function in phase space. Thus, f_w resembles the classical distribution function. However, in contrast to the latter, it allows negative values. Actually, the only positive Wigner function is the equilibrium Maxwell-Boltzmann distribution f_{MB}. Moreover, this is the only function which equates the two terms in the scattering operator and thus remains unchanged by scattering.

In the following, we consider the case without electric potential, so that the V_w term disappears. At first glance (4) reduces to the classical field less Boltzmann equation in this particular case. This, however, is not true, since the classification of the equation depends on the initial condition. If it is non-negative and normalized to unity, this is indeed a legitimate classical distribution function. Alternatively, a phase space function f_w^0 may be chosen, which corresponds to a fully coherent initial system. In this case the uncertainty relation is manifested by the shape of the function: the shape is such that the density matrix $\langle x|\hat{\rho}|x' \rangle$ obtained from the function must be a product of the type $\psi(x)\psi(x')$ where ψ is a quantum state function [5]. Then equation (4) describes the evolution of f_w^0 due to processes of scattering. In the next section we demonstrate that this evolution transforms f_w^0 towards a classical distribution by destroying all incorporated coherence effects.

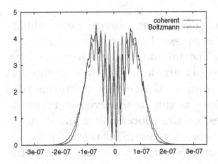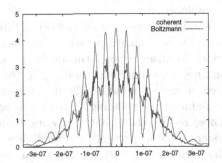

Fig. 5. The spatial broadening of the wave packet is hindered by scattering processes. The coherent wave packet is slightly broader after 200fs (left hand side); with this trend continuing so that the coherent wave packet begins to reach beyond the simulation domain after 500fs, while the wave packet experiencing scattering still exists completely within the confines of the simulation.

3   Simulations and Analysis

The chosen initial condition is the superposition (1) of two Gaussian wave packages: $e^{\frac{-(x\pm a)^2}{2\sigma^2}}e^{\imath bx}$. The corresponding initial Wigner function

$$f_w^0(x, k_x) = Ne^{-(k_x-b)^2\sigma^2}\left[e^{-\frac{(x-a)^2}{\sigma^2}} + e^{-\frac{(x+a)^2}{\sigma^2}} + e^{-\frac{x^2}{\sigma^2}}\cos\left((k_x-b)2a\right)\right] \quad (5)$$

comprised by two Wigner wave packets and oscillatory term is shown in Figure 1. Equilibrium is assumed in the other two directions of the wave space, so that $\frac{\hbar^2}{2\pi mkT}e^{-\frac{\hbar^2(k_y'^2+k_z'^2)}{2mkT}}$ multiplies (5) to give $f_w^0(x, \mathbf{k})$. A GaAs semiconductor with a single Γ valley and scattering mechanisms given by elastic acoustic phonons and inelastic polar optical phonons is considered, while setting the parameter $a = 70$nm. The choice of $2\sigma^2 = \frac{\hbar^2}{2mkT}$ along with $b = 0$ gives rise to $f_{MB}(\mathbf{k})$, which minimizes the effect of the phonons on the change in the shape of the wave vector distribution. The SI units [m], $[m^{-1}]$ and [s] are used.

During a coherent evolution the initial structure of the Wigner function remains intact, as shown in the left hand side of Figure 2. The oscillatory term corresponding to the off-diagonal elements of the density matrix is responsible for the coherence of the state, since the $\mathbf{k}$ distributions of the other two components (in k_y and k_z) remain unchanged. Thus the oscillatory term is most affected by scattering as seen on the right hand side of Figure 2. Indeed, the shape of the initial momentum distribution, $f(k_x) = \int dx dk_y dk_z f_w$, in Figure 3 is due entirely to the oscillatory term, as the other two components of $\mathbf{k}$ are distributed according to thermal equilibrium, which is indicated by the thin line. The initial shape remains frozen during coherent evolution, while as it is deducible from the figure, scattering destroys the coherence in about $1ps$ and forces the distribution to equilibrium.

Figure 4 shows the density $n(x) = \int dk_x dk_y dk_z f_w$. The coherent curve exhibits pronounced oscillations, which are being suppressed in the Boltzmann curve which localizes around the initial peaks. The initially well balanced positive and negative contributions of f_w to the density are destroyed by scattering, as seen on the right hand side of Figure 4. Another effect is that scattering tries to reduce the spreading of the wave packets as can be observed in Figure 5. These results show that scattering induces a spatial localization and destroys coherence, thus preventing reversibility in time. A measure for this behaviour is the purity $p = \int dx dk_x dk_y dk_z f_w^2$. For coherent evolution it remains 1 while the loss of information in the initial state is given by its decrease. An increase of the temperature leads to increase of the electron-phonon coupling and thus an accelerated drop of purity, as depicted in Figure 6.

It is concluded that phonons are an important cause of decoherence. The transition from a quantum to a classical electron state occurs at a picosecond time scale, which acts as a limit for the speed of operation of future semiconductor quantum computers.

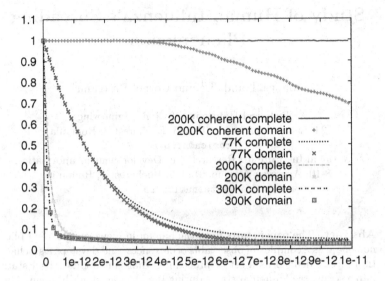

Fig. 6. Evolution of purity at different temperatures. Pairs of lines are obtained by neglecting particles which leave the simulation domain (domain) and by a complete record of all particles (complete). The former case may lead to an artificial indication for loss of coherence.

Acknowledgement. This work has been supported by the Austrian Science Fund Project FWF-P21685-N22, the Österreichische Forschungsgemeinschaft (ÖFG), Project MOEL453, and by the Bulgarian Science Fund Project DTK 02/44.

References

1. Heinrich, S.: Quantum Complexity of Numerical Problems. In: Cucker, F., et al. (eds.) London Mathematical Society Lecture Note Series, vol. 312, pp. 76–95. Cambridge University Press (2004)
2. Zurek, W.H.: Decoherence, Einselection, and the Quantum Origins of the Classical. Review of Modern Physics 75, 715–775 (2003)
3. Joos, E., et al.: Decoherence and Appearance of the Classical World in the Quantum Theory. Springer, Heidelberg (2003)
4. Querlioz, D., Dollfus, P.: The Wigner Monte Carlo Method for Nanoelectronic Devices. Wiley (2010)
5. Tatarskii, V.I.: The Wigner Representation of Quantum Mechanics. Sov. Phys. Usp. 26, 311–327 (1983)

Study of Human Influenza's Spreading Phenomenon

Romică Trandafir[1] and Cornel Resteanu[2]

[1] Technical University of Civil Engineering,
24, Lacul Tei Avenue, 020396, Bucharest 2, Romania
romica@utcb.ro
[2] National Institute for Research and Development in Informatics,
8-10, Averescu Avenue, 011455, Bucharest 1, Romania
resteanu@ici.ro

Abstract. The paper starts from the human influenza's spreading phenomenon, as a complex of observable occurrences, and develops a stochastic process, defined as a set of procedures that convert its initial state into a sequence of different states during the phenomenon's lifespan. The Monte Carlo simulation method for a stochastic discrete event system is used. This system is completely described in terms of: entities, sequential states, transition tables of states, sets of input/output events, internal/external transition function, events/time advance function, input/output parameters. The simulation can encompass several contagion schema and health policy responses. Finally, some information about the software in the field is given.

Keywords: Spreading Diseases, RNA Viruses, Stochastic Discrete Event Systems, Monte Carlo Simulation, Analysis and Prognosis.

1 Introduction

Human influenza's spreading phenomenon concerns, on the one hand, the symptoms of a specific group of individuals caused by a set of viruses (RNA viruses of the Orthomyxoviridae family: types A, B C, Isa, Thogoto, etc.) in precise climate conditions, in a given time period, and on the other hand, the response of medical organizations [5,6]. The symptoms include chills, fever, nasal secretions, watering eyes, sore throat, muscle pain, severe headache, coughing, nausea, vomiting, diarrhea, abdominal pain, weakness/fatigue and general discomfort, sometimes leading to death. The group of individuals is defined by number, age, sex, general state of health, vaccinations, topological placement in subgroups, distance between the subgroups, interacting coefficient, and shedding. The climate conditions combine sunlight, air-temperature, humidity, wind, and aerosols. This paper attempts to provide information that would aid with the second aspect of human influenza's spreading phenomenon, the response of the medical organizations.

There are several contagion schema, such as direct (from person to person) or indirect (airborne or through contact with a contaminated surface), leading from

I. Lirkov, S. Margenov, and J. Waśniewski (Eds.): LSSC 2011, LNCS 7116, pp. 480–486, 2012.

limited spreading to epidemics or even to pandemics. Obviously, the health policy response implies several direct and indirect costs. A study of the phenomenon and its consequences using mathematics and information technology techniques has merit and is the purpose of this paper. Modeling the human influenza's spreading phenomenon can be a valuable tool in analysis and prognosis.

The paper is organized as follows: the next section proposes a contagion scheme for human influenza's spreading phenomenon that may be utilized in the numerical simulation. One presents afterwards a general methodology for describing the stochastic discrete event system and its application for the proposed contagion scheme. The corresponding algorithm to solve this problem given in the next section. A general presentation of the software in the field is made in the conclusions' section. Some suitable references are also present.

2 Numerical Simulation of Contagion Schema

Contagion schemes can be modeled using the framework of chains with complete connections [10] or stochastic discrete-event systems [4,9,1]. In this paper we use the last approach.

First, a simple human influenza's spreading problem is stated, then an algorithm for numerical simulation of the contagion scheme is developed, and then some improvements of this algorithm are presented.

The analysis and prognosis of the spreading phenomenon of the virus is performed using computer simulation and based on a mathematical and logical model describing the behavior of real systems.

Let $\mathcal{O}$ be a human population of cardinality No distributed on a region of K communities and let $\mathcal{V}$ be a set of virus types of cardinality Nv. Suppose that for every virus type $v \in \mathcal{V}$ there is an incubation period ($v = 0$ means no virus), represented by an uniform random variable on $[l_v, u_v]$. For the sake of simplicity, suppose that an individual contaminated with a virus type cannot be contaminated with another virus type. Every community is characterized by following attributes: the number of healthy individuals (Nh_i), the number of contaminated individuals for every virus type ($N_{v,i}$), its Cartesian coordinates (x_i, y_i), the proportion of commuters ($F_i = (pc_i)$, $i = \overline{1, K}$) and the radius of its area (r_i). Denote with $\mathcal{S}$ the set of states characterizing an individual $o \in \mathcal{O}$, i.e. $\mathcal{S} = \{H$ – healthy; Hi – healthy and immune; Ck – contaminated, aware of it; Ct – contaminated, under treatment; $Cknt$ – contaminated, aware but not under treatment; D – dead$\}$, and with $\mathcal{O}c$ the set of contagious individuals ($\mathcal{O}c = \{Ck, Ct, Cknt\}$).

Suppose that:

– all previous input data are known at time t_0;
– at time t_1, an commuter from L_i can meet η_M individuals from L_i. If the commuter is infected with virus type v, ξ_c among the individuals met will be contaminated. Here η_M is a realization of a Poisson random variable with parameter α_i ($Po(\alpha_i)$), and ξ_c is a realization of a binomial random variable with parameters (η_M, q_v) ($Bi(\eta_M, q_v)$);

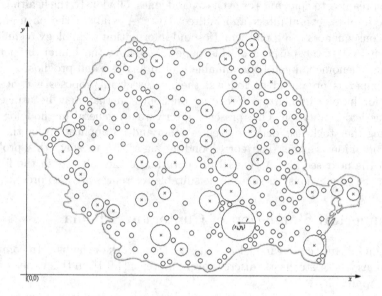

Fig. 1. Communities in the system built for the study on human influenza spreading

- the ξ_c contaminated individuals are distributed as follows: $\xi_k \in Ck$, where ξ_k is a realization of a binomial random variable $Bi(\xi_c, q_k)$; $\xi_t \in Ct$, where ξ_t is a realization of a binomial random variable $Bi(\xi_c, q_t)$; $\xi_h \in Hi$, where ξ_h is a realization of a binomial random variable $Bi(\xi_c, q_h)$; $\xi_{knt} \in Cknt$, where ξ_{knt} is a realization of a binomial random variable $Bi(\xi_c, q_{knt})$; and $\xi_D = \xi_c - \xi_t - \xi_h - \xi_{knt}$.

The purpose is to determine the average cardinality of the subsets of S and the average treatment costs for each virus type v and for every community. The corresponding algorithm needs to be executed multiple times in order to obtain these output data. The parameters of the random variable distributions involved are provided by a specialized company.

3   Stochastic Discrete-Event System

The formal framework for modeling and simulating discrete-event systems, DEVS, and stochastic discrete-event system specifications, STDEVS, is presented in [2,7,11,12,8].

A STDEVS model $\mathcal{M}$ representing the proposed contagion scheme has the following components:

- X: set of events of type "an individual infected with a certain virus type has contact with non-contaminated individuals";
- Y: the set of averages of the random variables giving the cardinalities of the sets of individual states and of the average costs at the end of the simulation;

- ta: a random variable indicating the contamination time for a virus type;
- $\delta_{ext} : Q \times X \to S$: the function giving the number of individuals that switch into state $s \in S$ (note that δ_{ext} is a random variable);
- $\delta_{int} : S \to S$: the function giving the number of individuals that switch from one state to another (e.g., $H \to Ck$, $Ckt \to Hi$, $Cknt \to D$, etc.), characterized by the distribution table $\mathcal{H}_{v,i}$;
- $Q = \{(s, ta)\}$;
- $\lambda : S \to Y$: the exit function, which calculates the averages of the cardinalities of the sets of states and the average costs.

4   Simulation Algorithm

In order to analyze this contagion scheme with some type of distribution for the involved input random variables, a simulation model needs to be constructed. An algorithm produces artificial experiments via computer runs. These experiments (in fact realizations of output variables) are processed and analyzed, yielding solutions for the management of actual contagion scheme described by the simulation model. The instruments used to construct the simulation model (i.e. algorithm) are the clock time and the agenda. The clock time has a dual purpose: to record the time elapsed in the system and to track the correct order of the events produced by the simulation model. The agenda is a concept related to recording the events produced by the simulation. The clock time increases by a finite number. After each increment of the clock, the simulation algorithm processes the events occurred at that moment in time (these events define the agenda of current events, ACE). When the ACE is empty, the clock is incremented with some value and the process is repeated for the new events from ACE. When an event is processed, it can produce another event (at a future moment, in which case the set of these events is the agenda of future events, AFE) or can cancel some events from ACE (this is the set of canceled events, CE). Therefore the agenda A has a dynamic evolution in the form

$$A = ACE + AFE - CE$$

The clock time can be incremented in two ways: with variable increments (called variable increment clock time) or with constant increments (called constant increment clock time). Initially the clock is zero. In the case of variable increments, the clock time is increased up to the time of occurrence of the first event in AFE. In the case of constant increments, the clock is increased by a constant c. After incrementing the clock, the main cycle of the simulation algorithm selects the events from AFE with time of occurrence equal to the current clock time and places them in ACE. Then, the events from ACE are processed; when ACE is empty, the clock time is advanced again and the process is repeated until the clock time takes the value TS, where TS is the input value for the end time of the simulation. As an alternative, the simulation ends when the number of simulated events of a specific type takes a given value $Nmax$. Sometimes, this equivalent rule (the next-event rule) is used instead of the variable clock time; in this case, the end of the simulation is determined by $Nmax$.

Algorithm 1. Main algorithm

Input: K, Nv, $N_{v,i}$, l_v, u_v, $pc_{v,i}$, x_i, y_i, d_i, α_i, $q_{v,i}$, qk_i, qt_i, qh_i, $qknt_i$, $i = \overline{1,k}$, $v = \overline{1, Nv}$, $Dist \in \mathcal{M}_{K,K}(R)$, $Nmax$, Ts

Initialize: $N = 0$; {count of current iteration} $n = 1$; {index in agenda} $Clock = 0$;

repeat

    Generate $i \sim \mathcal{U}(1, K)$;   {a community}

    Step 1: Generate $u \sim \mathcal{U}(0, 1)$; {determine if the current individual is commuter}

    if $u < pc_i$ **then**

        begin

        Generate $d \sim \mathcal{U}(0, d_{max})$ {$d_{max} = \max_{1 \leq i \leq K} d_i$}

        if $d < r_i$ **then**

            $I = i$ {the individual remains in his community}

        else

            begin

            Determine $\mathcal{L} = \{L_k | Dist_{i,k} \leq d_i, k = \overline{1,K}\}$, $|\mathcal{L}| = L$

            Generate $I \sim \mathcal{U}(1, L)$ {determine the destination community}

            Generate $v \sim F_i$ {a virus type}

            if $v = 0$ **then**

                go to Step 1   {$v = 0$ means no virus}

            Generate $\eta_M \sim Po(\alpha_I)$ {a value of the random variable representing the number of individuals met by an individual contaminated with virus type v}

            Generate $\xi_c \sim Bi(\eta_M, q_{v,I})$ {a value of the random variable representing the number of individuals from η_M contaminated after the contact with the individual contaminated with virus type v}

            for $i = 1$ **to** ξ_c **do**

                begin

                Generate $T_{v,i} \sim \mathcal{U}([l_v, u_v])$ {a value of the random variable representing the contamination period of individual o, contaminated with virus type v}

                $n = n + 1$

                $T_n = Clock + T_{v,i}$

                end

            $N_{v,I} = N_{v,I} + \xi_c$

            $|C_{v,I}| = |C_{v,I}| + \xi_c$

            $Clock = T_o$ {time of the last processed event/individual}

            {determine the state of the current individual at time T according to the probability distribution $\mathcal{H}$}

            if $o_n \in Ck$ **then**

                Compute $|Ck_{v,I}| = |Ck_{v,I}| + 1$

            if $o_n \in Ct$ **then**

                Compute $|Ct_{v,I}| = |Ct_{v,I}| + 1$ and the Cost;

            if $o_n \in Cknt$ **then**

                Compute $|Ckknt_{v,I}| = |Cknt_{v,I}| + 1$;

Algorithm 1. *(Continued)*

>   if $o_n \in Hi$ then
>   Compute $|Hi_{v,I}| = |Hi_{v,I}| + 1$;
>   if $o_n \in D$ then
>   Compute $|D_{v,I}| = |D_{v,I}| + 1$;
> $N = N + 1$ {next event; the first event/individual having $T_n > Clock$}
>   end
> end
> **until** $N = Nmax$ or $Clock > Ts$
>
> {Outputs of simulation}
>
> **for** $i = 1$ **to** K **do**
>   **for** $v = 1$ **to** Nv **do**
>   **begin**
>   Calculate $E[|Hi_{v,i}|] = \frac{|Hi_{v,i}|}{N}$, $E[|Ci_{v,i}|] = \frac{|Ci_{v,i}|}{N}$, $E[|Ck_{v,i}|] = \frac{|Ck_{v,i}|}{N}$,
> $E[|Ct_{v,i}|] = \frac{|Ct_{v,i}|}{N}$, $E[|Cknt_{v,i}|] = \frac{|Cknt_{v,i}|}{N}$, $E[|D_{v,i}|] = \frac{|D_{v,i}|}{N}$, $Cost_{v,i}$
>   **end**

Classical algorithms can be used to generate samples of the different types of random variables involved in this algorithm, such as the Poisson, Binomial and discrete distributions [3].

It is assumed that the transition of a contaminated individual from one state to another, for each community, is according to the distribution table $\mathcal{H}$:

$$\tau_i : \begin{pmatrix} Hi & Ckt & Cknt & D \\ p_{1,i} & p_{2,i} & p_{3,i} & p_{D,i} \end{pmatrix}$$

and that the virus types have the distribution table

$$\zeta_i : \begin{pmatrix} 1 & 2 & \cdots & Nv \\ p'_{1,i} & p'_{2,i} & \cdots & p'_{Nv,i} \end{pmatrix}$$

The proposed contagion scheme would be closer to reality if contaminated individuals with virus type v do not become immune to that virus type after healing, but rather can be contaminated with any virus type. Also, individuals contaminated with one virus type could be contaminated with another virus type during their contamination period. In this case, the exact cause of death is not known if the individual dies. In the situation when there are sufficient data to determine suitable distribution functions for the random variables involved in this model, the corresponding algorithms can be used to generate samples for these variables.

5   Conclusions

The human influenza phenomenon is still studied by scientists in various disciplines. The medical field regularly faces new variants of the virus, such as bird influenza or H4N1, and has to find new treatments. The mathematics or computer science fields try to improve the systems that mimic the behavior of viruses in its spreading phenomenon. This paper is an example of an attempt to simulate the spreading of a virus as close to reality as possible. The chosen framework-model is STDEVS, which is very complex and can represent the entire phenomenon. The technique chosen for simulations is Monte Carlo, which produces good results for a sufficiently large number of iterations.

The experience of the authors in this area shows that the carefully made simulations with the HIS software (Human Influenza Spreading software) yield useful results for analysis and forecast. There are several programs used to model and solve stochastic discrete event systems, such as SimEvents, DevSim++, adevs, PowerDES etc, but it seems more appropriate to consider software especially built for this particular problem. The proposed algorithm would provide such a specialized software. The software designers could allow the users to change certain parameters, such as the target region or the distribution function of commuters and of infected individuals for every community in the system.

References

1. Banks, J., Carson, J.S., Nelson, B.L., Nicol, D.M.: Discrete-Event System Simulation, 5th edn. Prentice Hall (July 2009)
2. Castro, R., Kofman, E., Wainer, G.: A formal framework for stochastic discrete event system specification modeling and simulation. Simulation 86(10), 587–611 (2010), http://sim.sagepub.com/content/86/10/587.abstract
3. Devroye, L.: Non-uniform random variate generation. Springer, Heidelberg (1986)
4. Fishman, G.S.: Monte Carlo: concepts, algorithms, and applications. Springer, Heidelberg (1996)
5. Goldsmith, C.: Influenza. Twenty-First Century Books (August 2010)
6. Kamps, B.S., Hoffmann, C., Preiser, W. (eds.): Influenza Report. Flying Publisher (2006)
7. Kofman, E., Lapadula, M., Pagliero, E.: PowerDEVS: a DEVS-Based environment for hybrid system modeling and simulation (2003), http://citeseerx.ist.psu.edu/viewdoc/download?doi=10.1.1.73.4537&rep=rep1&type=pdf
8. Sobh, M., Owen, J.C., Valavanis, K.P., Gracanin, D.: A subject-indexed bibliography of discrete event dynamic systems. IEEE Robotics and Automation Magazine 1, 14–20 (1994)
9. Söderström, T.: Discrete-time stochastic systems: estimation and control. Springer, Heidelberg (2002)
10. Trandafir, R.: On some contagious schema. In: Despotis, D.K., Zapounidis, K.C. (eds.) Proceedings of the 5th International Conference of the Decision Sciences Institute, Athens, Greece, vol. 2, pp. 1910–1912 (1999)
11. Zeigler, B.P., Praehofer, H., Kim, T.G.: Theory of modeling and simulation: integrating discrete event and continuous complex dynamic systems. Academic Press (2000)
12. Zimmermann, A.: Stochastic discrete event systems: modeling, evaluation, applications. Springer, Heidelberg (2008)

Part XI

Voxel-Based Computations

Multilevel Solvers with Aggregations for Voxel Based Analysis of Geomaterials

Radim Blaheta[1] and Vojtěch Sokol[2]

[1] Institute of Geonics of the Academy of Sciences of the Czech Republic,
Ostrava, Czech Republic
blaheta@ugn.cas.cz
[2] Department of Applied Mathematics, FEECS,
VŠB - Technical University of Ostrava, Czech Republic
vojtech.sokol@vsb.cz

Abstract. Our motivation for voxel based analysis comes from the investigation of geomaterials (geocomposites) arising from rock grouting or sealing. We use finite element analysis based on voxel data from tomography. The arising finite element systems are large scale, which motivates the use of multilevel iterative solvers or preconditioners. Among others we concentrate on multilevel Schwarz preconditioners with aggregations. The aggregations are efficient even in the case of problems with heterogeneity, coefficient oscillations and large coefficient jumps if the aggregations include a proper handling of the strong couplings.

1 Introduction

Our work is motivated by the investigation of processes in heterogeneous materials, mainly in the rocks and geocomposites arising from rock or soil grouting. The considered materials allow to distinguish at least two scales, which will be called macroscale and microscale. The characteristic length of the macroscale l_M corresponds to engineering objects, samples or representative volumes, the characteristic length of the microscale l_m corresponds to grains, microfractures and other structural details. The processes to be analysed are mechanical behaviour (elastic, inelastic) and flow in porous media.

The final goal is to be able to analyse the processes in the macroscale thereby taking into account influences of the microscale. Such analysis can be done physically in laboratory (or even in-situ) by loading the material samples and observing the response. Such testing can be also done numerically by simulation of these tests. It provides well known homogenization or upscaling techniques, which are especially useful if the scales are well separated ($l_m \ll l_M$). In these numerical tests Ω will denote the body of a sample or a representative volume.

In the case of *linear material behaviour*, we are interested in obtaining macroscale (apparent) constitutive relation operators, i.e. elasticity tensor $\mathcal{C}_M$ or permeability tensor $\mathcal{K}_M$, These tensors can be calculated from the expressions

$$\langle \sigma \rangle = \mathcal{C}_M \langle \varepsilon \rangle, \quad \langle \phi \rangle = \mathcal{K}_M \langle \psi \rangle, \tag{1}$$

I. Lirkov, S. Margenov, and J. Waśniewski (Eds.): LSSC 2011, LNCS 7116, pp. 489–497, 2012.

where ε, σ and ϕ, ψ are strains, stresses and pressure gradients, fluxes in the microstructure, respectively, and $\langle \cdot \rangle$ denotes the componentwise averaging operator.

The strains, stresses, gradients, fluxes in the microstructure are computed by solving the following boundary value problems

$$\left.\begin{array}{c} -\mathrm{div}\,\sigma = f \\ \sigma = \mathcal{C}_m \varepsilon \\ \varepsilon = \frac{1}{2}\left(\nabla u + \nabla^T u\right) \end{array}\right\} \text{ in } \Omega \quad (2) \qquad \left.\begin{array}{c} -\mathrm{div}\,\phi = q \\ \phi = \mathcal{K}_m \psi \\ \psi = \nabla p \end{array}\right\} \text{ in } \Omega \quad (3)$$

Above $\mathcal{C}_m = \mathcal{C}_m(x)$ and $\mathcal{K}_m = \mathcal{K}_m(x)$ denote local constitutive tensors in microstructure, which can change rapidly (oscillatorily) with $x \in \Omega$ and, moreover, their components can be jumping by many orders of magnitude. Normally, the inner forces or sources are taken as zero ($f \equiv 0$, $q \equiv 0$) and the loading is given by boundary conditions. Standard choices are the pure Dirichlet or pure Neumann boundary conditions, [7].

For inelastic material behaviour, the investigation of the influence of microstructure to macrostructure behaviour can be done in a similar way - via numerical simulation of laboratory tests and evaluation of macro stresses/strains by averaging computed micro stresses/strains, but the exploited material models can be much more difficult. As an example, we can mention the investigation involving damage behaviour in the microstructure, see e.g. [7] and [4] describing applications of continuum damage mechanics in structural mechanics and biomechanics. Note that local nonlinear or damage behaviour parameters can be difficult to determine a priori but can be obtained by a proper calibration using a physical test data.

2   Voxel Analysis

To perform the analysis described in the previous section, we use finite element discretization of the domain Ω representing the material sample. The input information about the microstructure is frequently based on tomography and converting the tomography outputs (CT data) to information about material heterogeneity in the microstructure. As the tomography information is organized in voxels, it is natural to use regular voxel based grids for the finite element analysis.

In both cases of linear and nonlinear material behaviour, we exploit the finite element discretization, which finally leads to the solution of linear systems. These systems can be large if we want to keep a high ratio of the sample size l_M to the microstructure details corresponding to the voxel size. The dimensions of the systems can be millions or billions of equations/unknowns, see [1]. The voxel elastic analysis of geocomposites is described in [3]. The arising systems can also be ill conditioned due to strong heterogeneity and high coefficient jumps. This aspect will be discussed in the following sections.

3   Multilevel Iterative Methods

Multilevel iterative methods (geometric and algebraic multigrid, AMLI, multilevel Schwarz type preconditioners, etc.) are know for their efficiency (and even optimality) when solving regular boundary value problems, typically with constant or piecewise constant non-oscillating coefficients. Their application to problems with material heterogeneity addresses the same questions as multiscale analysis of material, i.e. how to define the coarse (macro) level problem and how to transfer information between the micro and macro scales to obtain an efficient multilevel method.

To be specific, let us consider a two-level algorithm for solving the fine level problem $A_f u = b_f$, where the k-th iteration $u^k \rightarrow u^{k+1}$ involves the steps:

$(S0)$ $u \leftarrow u^k$

$(S1)$ perform ν smoothing steps $u \leftarrow S_f(u, A_f, b_f)$

$(S2)$ compute coarse level correction $w \leftarrow A_c^{-1} R_c r_f(u)$

$(S3)$ $u \leftarrow u + R_c^T w$

$(S4)$ perform ν smoothing steps $u \leftarrow S_f(u, A_f, b_f)$

$(S5)$ $u^{k+1} \leftarrow u$

Above, S_f can be a Jacobi-like smoothing method $S_f(u, A_f, b_f) = u + \omega_f r_f(u)$, $r_f(u) = (b_f - A_f u)$, R_c is a restriction operator (matrix) from the fine (f) to the coarse (c) level, R_c^T is the prolongation, ω_f is a parameter. Note that the smoothing step (S4) is for symmetrization of the iteration operator.

We shall illustrate the effect of different choice of transfer and coarse level operators on a simple 1D problem with periodic microstructure. Let us consider the boundary value problem

$$- (ku')' = f \text{ in } \Omega = [0, 1] , \ u(0) = u(1) = 0, \tag{4}$$

with an uniform coarse grid discretization $\mathcal{T}_H$ and fine grid discretization $\mathcal{T}_h$, $h = H/2$ and assume that the coefficient k is discontinuous on each coarse grid element; if $E = \langle x_i, x_{i+1} \rangle \in \mathcal{T}_H$, $x_i = iH$, then

$$k(x) = \begin{cases} k_1 = 10^0 \text{ for } x \in \langle x_i, x_{i+1/2} \rangle \\ k_2 = 10^n \text{ for } x \in \langle x_{i+1/2}, x_{i+1} \rangle \end{cases}.$$

Further, we introduce the operators

$$R_1 = \begin{bmatrix} z & & & \\ 1/2 & 1 & 1/2 & \\ & 1/2 & 1 & 1/2 \\ & & & \ddots \end{bmatrix}, \quad R_2 = \begin{bmatrix} 0 & & & \\ c & 0 & -c & \\ & c & 0 & -c \\ & & & \ddots \end{bmatrix}$$

$$R_3 = \begin{bmatrix} z & & & \\ 1 & 1 & & \\ & 1 & 1 & \\ & & & \ddots \end{bmatrix}.$$

Table 1. Numbers of iterations for solving the problem (4) by two-level methods, $\nu = 1$, relative accuracy $\epsilon = 10^{-6}$

		$k_2/k_1 = 10^n$, n=	0	1	2	3	4
(1)	$R_c = R_1$	$A_c = A_H$(Voigt)	7	46	451	-	-
(2)	$R_c = R_1$	$A_c = A_H$(homog.)	7	27	-	-	-
(3)	$R_c = R_1 + R_2$	$A_c = A_H$(homog.)	7	7	7	7	7
(4)	$R_c = R_3$	A_c aggregation	23	10	8	8	8

Above, z stands for the position of boundary condition and can be taken as zero, the value of $c = \frac{k_2 - k_1}{2(k_1 + k_2)}$ is the corrector.

Note that $\phi_i^{(1)} = R_1^T e_i$ represents a standard nodal basis function, $\phi_i^{(2)} = (R_1 + R_2)^T e_i$ is a multiscale basis function with optimal energy, $\phi_i^{(3)} = R_3^T e_i$ is aggregation basis function which is a limit case of $\phi_i^{(2)}$ for $k_2/k_1 \to \infty$. In Table 1, we report the efficiency of different two-grid methods.

Note that $A_H = R_1 A_f R_1^T$ is the standard coarse matrix. It is also obtained by using the standard FE basis and the averaged coefficient (in theory of homogenization representing the Voigt bound, see [13]). $A_H = (R_1 + R_2)A_f(R_1 + R_2)^T$ corresponds to the coarse matrix obtained by using the standard basis and the homogenized coefficient (here equal to the Reuss bound [13]). In Table 1, the method (2) is the only one not using the Galerkin relation $A_c = A_c^G = R_c A_f R_c^T$.

From Table 1, we can see that for problems with high jumps in coefficients it is possible to preserve the efficiency of two-level method by proper choice of transfer operators (which may be more difficult for 2D and 3D) or by use of aggregation that puts together strongly connected degrees of freedom.

4   Aggregation Techniques

Motivated by the results of the previous section, our aim is to construct the coarse space by adaptive aggregation, which respects strong connections between degrees of freedom. To this end, we shall describe and test several techniques described in the literature, namely [8], [12], [10]. Note that all of the mentioned papers use a node-wise aggregation technique. On the contrary [6] used element-wise aggregation, which seems to be very natural for two-level methods, but the extension to multi-level case seems to be more problematic.

Here, we consider the aggregation technique for solving the system

$$Au = b, \quad u, b \in R^n \tag{5}$$

arising from discretization of the scalar problem (3). The case of elasticity (2) will be discussed in the conclusions. The aggregation then divides the set of indices (nodes, degrees of freedom) $N = \{1, \ldots, n\}$ into disjoint subsets G_i of aggregates of unknowns, so that $N = \bigcup_{i=1}^{k} G_i$, $G_i \bigcap_{i \neq j} G_j = \emptyset$. Then the prolongation and restriction operators are defined by the boolean matrix R_c:

$$(R_c)_{i,j} = \begin{cases} 1 & \text{if } j \in G_i, \\ 0 & \text{otherwise.} \end{cases} \tag{6}$$

4.1 Aggregation Algorithms

In this subsection we focus on aggregation techniques that exploit the information directly stored in the matrix A_h, these include algorithms by Vaněk et al. [12], by Scheichl and Vainikko [10] and by Notay [8].

The algorithm by Vaněk et al. starts with defining strongly-connected neighbourhood $S_i = S_i(\varepsilon)$ of a node i with threshold parameter ε:

$$S_i(\varepsilon) = \left\{ j \in N : |a_{ij}| \geq \varepsilon \sqrt{a_{ii} a_{jj}} \right\}, \tag{7}$$

and then separates nodes that are not strongly connected to any other nodes. Then main part of the algorithm can be briefly described as follows (details can be found in [12]):

step 1: form set of unaggregated nodes U, initially $U = N \setminus \{\text{isolated DOF's}\}$

step 2: for $j = 1, \ldots$ form initial aggregates G_j^0 from strong neighbourhoods, i.e. if $S_i \subset U$ then ($G_j^0 = S_i$ & $U = U \setminus G_j^0$ & $j = j + 1$)

step 3: enlarge aggregates G_j^0 to G_i with respect to strong connection

step 4: process unaggregated nodes.

Given this aggregation, [12] creates tentative prolongation (6), which is further smoothed to get the final prolongation and restriction operators. We omit the smoothing here, as in case of high contrast in coefficients and aggregation in the stiff parts of the domain the situation is similar to the model problem in Section 3 – the energy of the basis functions created by plain aggregation is close to minimal and smoothing has not much to improve.

The next aggregation technique is that of Scheichl and Vainikko [10]. The algorithm uses a strongly-connected graph r-neighborhood $S_i(r, \varepsilon)$ defined as follows:

$$S_i(1, \varepsilon) = S_i(\varepsilon) = \left\{ \{i\} \cup \left\{ j \in N : \left| \hat{A}_{ij} \right| \geq \varepsilon \max_{k \neq i} \left| \hat{A}_{ik} \right| \right\} \right\}, \tag{8}$$

$$S_i(r, \varepsilon) = S_i(r - 1, \varepsilon) \cup S_i^+, \ S_i^+ = \bigcup_{j \in S_i(r-1, \varepsilon)} S_j(\varepsilon) \ \text{for } r = 2, 3 \ldots. \tag{9}$$

Above $\hat{A} = (\text{diag} A_h)^{-\frac{1}{2}} A_h (\text{diag} A_h)^{-\frac{1}{2}}$ is a scaled matrix and ε is a thresholding parameter for strong connections. The algorithm creates aggregates by finding a strongly-connected graph r-neighbourhood of a chosen seed node. To choose a good seed nodes an advancing front in the graph induced by nodes and edges of triangulation $\mathcal{T}_h$ is used.

The last aggregation algorithm by Notay [8] uses strong negative connectivity of nodes. For a node i, we define S_i as the set of all strongly negative connected nodes:

$$S_i(\varepsilon) = \left\{ j \in N : j \neq i, a_{ij} < -\varepsilon \max_{a_{ik}<0} |a_{ik}| \right\}, \tag{10}$$

where parameter ε is used as threshold for strong coupling. The sets $S_i(\varepsilon)$ are used to construct pairs of nodes that are most strongly negative connected. Recursively, the pairs can be aggregated into pairs of pairs (quadruples) etc.

5   Model Problem and Multilevel Schwarz Preconditioner

The aggregation techniques will be tested on a model problem, which is the Darcy flow problem described as follows:

$$\nabla \cdot v = f, \, v = -k\nabla u \quad \text{in} \quad \Omega = [0,2] \times [0,1] \tag{11}$$

with mixed boundary conditions: zero normal flux on the bottom and top sides and $u = 0, 1$ on the left and right vertical sides, respectively. The problem is discretized by linear triangular finite elements on a uniform grid.

The heterogeneity is represented by the permeability coefficient k. In our model problem, the coefficient is stochastically generated in such a way that

$$\log(k) \in N(0, \sigma)$$

which means that $\log(k)$ has a normal distribution with the mean $\mu = 0$ and the variance σ^2.

The linear system arising from the finite element discretization is solved by the conjugate gradient method with multilevel additive Schwarz preconditioner. It uses decomposition of the computational domain Ω into overlapping subdomains Ω_i^δ and a coarse space problem created by aggregations. This provides the following decomposition of the finite element space V_h:

$$V_h = V_0 + V_1 + \ldots + V_k$$
$$V_i = \left\{ v \in V_h, v \equiv 0 \text{ in } \Omega \setminus \Omega_i^\delta \right\}, \forall i \in \{1 \ldots k\} \,, \, V_0 = \text{Range}(R_c^T)$$

where $R_c = R_0$ is constructed by aggregation as described in the previous section. If R_i are restrictions to spaces V_i then the two-level additive Schwarz preconditioner (B_{AS}) can be written as

$$B_{AS} = \sum_{i=0}^{k} R_i^T A_i^{-1} R_i,$$

where $A_i = R_i A R_i^T$ are FE matrices corresponding to problems on subdomains and on the coarse space. More details about Schwarz type preconditioners can be found e.g. in [11].

Table 2. PCG iterations and sizes of coarse spaces, 32385 DOFs, $H_s = 16h$

(a) Notay, $\varepsilon = 0.1$, averages

jump	PCG it. 4lvl	3lvl	2lvl	coarse space size c_1	c_2	c_3
10^0	55	39	23	8097	2025	510
10^2	61	48	29	8103	2030	510
10^4	64	48	29	8235	2077	532
10^6	65	46	26	8648	2257	609
10^8	65	44	23	9140	2532	720
10^{10}	61	38	21	9535	2775	822

(b) Scheichl, $\varepsilon = 0.5, r = 2$, averages

jump	PCG it. 4lvl	3lvl	2lvl	coarse space size c_1	c_2	c_3
10^0	64	53	29	3655	242	18
10^2	69	58	30	5505	495	57
10^4	68	52	26	8939	1626	280
10^6	68	45	23	10751	2650	487
10^8	65	39	20	11921	3513	831
10^{10}	62	35	19	12833	4324	1115

(c) Vaněk, $\varepsilon = 0.16$, averages

jump	PCG it. 4lvl	3lvl	2lvl	coarse space size c_1	c_2	c_3	fill-in c_1	c_2	c_3
10^0	63	54	31	5471	638	85	6.9	6.7	6.2
10^2	68	55	31	6169	952	129	6.3	6.6	6.2
10^4	68	51	27	8135	1469	209	6.3	6.7	6.3
10^6	67	45	24	9683	2132	326	6.3	6.7	6.4
10^8	65	39	21	10867	2809	500	6.2	6.7	6.4
10^{10}	60	34	18	11670	3345	672	6.2	6.7	6.4

(d) Standard deviations

PCG it. 4lvl	3lvl	2lvl	c. space sz. c_1	c_2	c_3
0	0	0	0	0	0
1	1	1	90	19	6
2	1	1	123	46	8
2	1	1	38	23	9
3	2	1	46	18	8
3	2	1	30	24	12

In the case of two-level Schwarz preconditioners, the system with matrix A_0 is solved directly, in the multilevel case the FE space V_0 is decomposed again and the operation $g = A_0^{-1} r$ is replaced by $g = B_{AS}^{(1)} r$, where $B_{AS}^{(l)}, l = 1$, corresponds to the Schwarz preconditioner on a coarser level. This procedure is repeated with $l = 1, 2, \ldots$ until a desired size of the coarsest FE matrix is attained, which can be efficiently solved directly. The multilevel preconditioner on the finest level $B_{AS}^{(0)} = B_{AS}$ then forms a V-cycle multilevel method.

6   Numerical Experiments

For the numerical simulations the domain $\Omega = [0,2] \times [0,1]$ was decomposed into stripes of width H_s and minimal overlap $\delta = h$. The permeability coefficient k was stochastically generated with log-normal distribution, see the previous section, and the variance σ^2 was set such that the coefficient jump $\frac{k_{max}}{k_{min}}$ hits the predefined value. The preconditioned CG are tested with the use of relative accuracy $\epsilon_{rel} = 10^{-6}$.

The performance of various aggregation techniques can be seen from Table 2. Tables 2(a) and 2(b) show averaged numbers of PCG iterations and mean sizes of coarse spaces over 8 runs of the algorithm with different stochastically generated permeability field k. In Table 2(c) we also added mean numbers of nonzero elements per row (column fill-in) for the coarse space matrices. The last Table 2(d) shows standard deviations for numbers of iterations and coarse space dimensions in Table 2(c). The reported deviations show that the used aggregation technique is sufficiently robust with respect to the generated heterogeneity. The same can be said on the other two aggregation techniques.

Note that all aggregation techniques were tested with different parameters. The reported results use parameters which produce good performance results and comparable numbers of iterations for all considered techniques.

7 Conclusions

The paper is concerned with tomography based voxel analysis of materials possessing microstructure, which influence the macroscale behaviour. We outline the approach of numerical simulation of the processes in microscale and show that such analysis requires also investigation of suitable iterative solvers. In investigation of flow problems, we meet very high heterogeneity and jumps in coefficients. For these problems we suggest to use multilevel (Schwarz) methods with adaptive aggregations. The numerical experiments show that such an approach is able to work well. Still, there are several possibilities for improvement. For example the V-cycle in the multilevel Schwarz can be replaced by a W-cycle or even better by K-cycle based on inner Krylov solver, cf. [9]. It can be seen that all tested aggregation techniques can perform similarly (using proper aggregation parameters). The investigation of mechanical properties is characterized by heterogeneity with not so high contrast in coefficients describing the local material behaviour. A reasonable efficiency has been obtained using additive Schwarz preconditioner with coarse level created by regular aggregation [3]. Nevertheless, some adaptive aggregation is also possible, see e.g. [12] for a discussion of one possible strength connection measure.

Acknowledgement. This work is supported by the grants GD103/09/H078 and GACR105/09/1830 of the Grant Agency of the Czech Republic.

References

1. Arbenz, P., van Lenthe, G.H., Mennel, U., Mller, R., Sala, M.: A scalable multi-level preconditioner for matrix-free μ-finite element analysis of human bone structures. Int. J. for Numerical Methods in Engineering 73(7), 927–947 (2008)
2. Blaheta, R.: Algebraic Multilevel Methods with Aggregations: An Overview. In: Lirkov, I., Margenov, S., Waśniewski, J. (eds.) LSSC 2005. LNCS, vol. 3743, pp. 3–14. Springer, Heidelberg (2006)
3. Blaheta, R., Jakl, O., Starý, J., Krečmer, K.: Schwarz DD method for analysis of geocomposites. In: Proceedings of the Twelfth Int. Conference on Civil, Structural and Environmental Engineering Computing, Paper 111. Civil-Comp Press (2009)
4. Charlebois, M., Jirásek, M., Zysset, P.K.: A Nonlocal Constitutive Model for Trabecular Bone Softening in Compression. Biomechanics and Modeling in Mechanobiology 9(5), 597–611 (2010)
5. Fish, J., Belsky, V.: Multigrid method for periodic heterogeneous media Part 1: Convergence studies for one-dimensional case. Computer Methods in Applied Mechanics and Engineering 126(1-2), 1–16 (1995)

6. Fish, J., Belsky, V.: Generalized Aggregation Multilevel solver. Int. J. for Numerical Methods in Engineering 40(23), 4341–4361 (1997)
7. Hain, M., Wriggers, P.: Numerical homogenization of hardened cement paste. Comput. Mech. 42(2), 197–212 (2008)
8. Notay, Y.: Aggregation-Based Algebraic Multilevel Preconditioning. SIAM J. Matrix Anal. Appl. 27(4), 998–1018 (2006)
9. Notay, Y., Vassilevski, P.S.: Recursive Krylov-based multigrid cycles. Numer. Liner. Alg. Appl. 15(5), 473–487 (2008)
10. Scheichl, R., Vainikko, E.: Additive Schwarz with Aggregation-Based Coarsening for Elliptic Problems with Highly Variable Coefficients. Computing 80(4), 319–343 (2007)
11. Toselli, A., Widlund, O.: Domain Decomposition Methods - Algorithms and Theory. Springer, Berlin (2005)
12. Vaněk, P., Mandel, J., Brezina, M.: Algebraic Multigrid by Smoothed Aggregation for Second and Fourth Order Elliptic Problems. Computing 56(3), 179–196 (1996)
13. Zohdi, T.I., Wriggers, P.: An Introduction to Computational Micromechanics. Springer, Berlin (2005)

A Highly Scalable Matrix-Free Multigrid Solver for μFE Analysis Based on a Pointer-Less Octree

Cyril Flaig* and Peter Arbenz

ETH Zürich, Chair of Computational Science, 8092 Zürich, Switzerland
cflaig@inf.ethz.ch

Abstract. The state of the art method to predict bone stiffness is micro finite element (μFE) analysis based on high-resolution computed tomography (CT). Modern parallel solvers enable simulations with billions of degrees of freedom. In this paper we present a conjugate gradient solver that works directly on the CT image and exploits the geometric properties of the regular grid and the basic element shapes given by the 3D pixel. The data is stored in a pointer-less octree. The tree data structure provides different resolutions of the image that are used to construct a geometric multigrid preconditioner. It enables the use of matrix-free representation of all matrices on all levels. The new solver reduces the memory footprint by more than a factor of 10 compared to our previous solver ParFE. It allows to solve much bigger problems than before and scales excellently on a Cray XT-5 supercomputer.

Keywords: micro-finite element analysis, voxel based computing, matrix-free, geometric multigrid preconditioning, pointer-less octree.

1 Introduction

Osteoporosis is a bone disease affecting millions of people around the world. The disease entails low bone quality and increases the risk of bone fracture. For a better understanding of bone structures and to improve the prediction of bone fractures, a precise estimation of its stiffness and strength is required. Micro finite element analysis (μFE) is a tool to this end [12,17]. It is based on high-resolution 3D images that are obtained by computed tomography (CT).

The high resolution scans produce computation domains of complicated shape composed of a huge number of voxels (3D pixels), cf. Fig 1. Since voxels directly translate into finite elements the resulting linear systems can have enormous numbers of degrees of freedom (dofs). A few years ago, we have developed a fully parallel state-of-the-art solver called ParFE [2,11] based on the conjugate gradient algorithm preconditioned by smoothed aggregation-based algebraic multigrid. This code exploits the geometric properties of the underlaying rectangular grid by avoiding the assembly of the system matrix. The largest realistic bone model solved with ParFE so far had a size of about 1.5 billion dofs [3].

---

* Corresponding author.

I. Lirkov, S. Margenov, and J. Waśniewski (Eds.): LSSC 2011, LNCS 7116, pp. 498–506, 2012.
© Springer-Verlag Berlin Heidelberg 2012

It is natural to represent the voxel-based domains by octrees [4, 6, 15]. Sampeth et al. [15] used the different tree levels to construct a geometric multigrid preconditioner.

In this paper, we present a solver based on a pointer-less octree-like data structure. Both finite elements and nodes are identified by a key corresponding to a space filling curve. This curve is equivalent to an octree. In contrast to [4, 6, 15] we deal with incomplete octrees due to the bone free space. In full space approaches [9, 10] the bone free space is modeled by very soft material and its unknowns are included in the computations. With the help of the new data structure the algorithm can exploit the sparse structure of the bone. This enables us to run the simulation with up to 6 times smaller memory footprint compared to the geometric multigrid that also stores the empty bone region [9]. Compared to matrix-free ParFE, the memory savings is more than a factor of 10.

2   The Mathematical Modeling of the Problem

The linear elasticity theory is used to analyse the bone strength. The weak formulation in 3D reads as follows [5]: Find the displacement field $\mathbf{u} \in [H^1_E(\Omega)]^3 = \{v \in [H^1(\Omega)]^3 : \mathbf{v}_{|\Gamma_D} = \mathbf{u}_S\}$ such that

$$\int_\Omega [2\mu\varepsilon(\mathbf{u}) : \varepsilon(\mathbf{v}) + \lambda \operatorname{div} \mathbf{u} \operatorname{div} \mathbf{v}]\, d\Omega = \int_\Omega \mathbf{f}^T v d\Omega + \int_{\Gamma_N} \mathbf{g}_S^T v d\Gamma \qquad (1)$$

for all $\mathbf{v} \in [H^1_0(\Omega)]^3$ with the volume forces $\mathbf{f}$, the boundaries traction $\mathbf{g}$ on the Neuman boundary, the linearized symmetric strain tensor

$$\varepsilon(\mathbf{u}) := \frac{1}{2}(\nabla\mathbf{u} + (\nabla\mathbf{u})^T),$$

and the Lamé constants

$$\lambda = \frac{E\nu}{(1+\nu)(1-2\nu)}, \qquad \mu = \frac{E}{2(1+\nu)}.$$

Here, E is the Young's modulus and ν the Poisson's ratio.

We use two different boundary conditions. The Neuman boundaries are traction free, $\mathbf{g}_S = 0$. On the top and bottom of the domain we have Dirichlet boundary condition with a fixed displacement. The engineers look for regions with high stresses and strains to determine the quality of the bone [17].

The displacements are discretized by trilinear hexahedral elements. These are converted one-to-one from the voxels of the CT image. Thus, all elements are cubes of the same size. In contrast to ParFE only the Young's modulus can vary in the domain. The Poisson's ratio ν must be constant. Bone mass has a typical Poisson's ratio $\nu = 0.3$. Applying this finite element discretization to (1) results in a symmetric positive definite linear system

$$A\mathbf{u} = \mathbf{f}.$$

The number of degrees of freedom can exceed 10^9. For symmetric positive definite linear systems of this size the preconditioned conjugate gradient algorithm is the solver of choice [13]. We use a geometric multigrid preconditioner.

We coarsen by aggregating $2 \times 2 \times 2$ voxels. A voxel of the coarser level $\ell+1$ gets its Young's modulus by averaging the Young's moduli of the eight aggregated smaller voxels of level ℓ,

$$E_{x,y,z}^{\ell+1} = \frac{1}{8} \sum_{i,j,k=0}^{1} E_{2x+i,2y+j,2z+k}^{\ell}, \tag{2}$$

where the Young's modulus of a non-existing child element is zero. If this procedure is applied to a homogeneous grid with the standard prolongation (interpolation) and restriction it corresponds to the Galerkin product [16]. For smoothing we use a Chebyshev polynomial [1]. This type of smoother was successfully used in ParFE [2] in the context of a smoothed aggregation-based algebraic multigrid preconditioner.

3   Implementation Details

The mesh, which is constructed from a 3D image, is stored in an octree. An octree divides each spatial dimension in two parts. This means that each tree node has eight children. Finite elements and nodes of the grid that lie in bone free space are not stored. In our application we iterate over all elements of a multigrid (or octree) level. These elements have the same size. Both, the nodes and the elements of each level are stored in one array. Each element is identified by the coordinate of its node with local number 0. If the data item has a weight $w \geq 0$ then it represents both an element with a Young's modulus of $E_{elem} = w \cdot 1\text{GPa}$ *and* the node of the element with local number 0. Plain nodes are characterized by a negative weight. The nodes and elements are sorted against their position in the depth-first traversal of the tree. This so-called Morton ordering corresponds to a space filling curve called Z-curve [14]. The Morton key can be computed easily from the three coordinates (**short int**) by interleaving their bits $key = z_{15}y_{15}x_{15} \cdots z_1 y_1 x_1 z_0 y_0 x_0$. This pointer-less storing scheme reduces the needed memory to hold the octree by 24 Byte (on 64-bit by 56 Byte) per node. The whole application needs only about 100 Bytes per degree of freedom. That is about 16 times less compared to the matrix-free ParFE code.

3.1   Accessing Nodes of an Element

In matrix-free finite element applications the nodes of the corresponding element must be accessed. Usually an element-to-node table is queried to get the indices of the corresponding nodes. With the octree data structure this corresponds to the search of the eight neighbours in positive x, y, z direction. The binary search corresponds to the travel of the root down to the leaves. Nodes with bigger

Algorithm 1. Optimized Search

```
function int SearchIndex(int start, t_octree_key key, t_tree tree)
  int count = 1;
  while key > tree[start + count].key do
    count = count · 8;
  end while
  return binarySearch(start + count/8, start + count, key, tree);
```

coordinates have always a bigger Morton key. The search has to be done from the index of the actual element to the end of the array.

A faster way to access the neighbouring nodes is to ascend in the tree and descend to the wanted node [7]. Ascending in the full octree is an exponential interval search by a factor of eight (see Algorithm 1). The binary search combined with an exponential interval search speeds up the application.

3.2 Matrix-Vector Multiplication

The first step is to store the prescribed values at the Dirichlet boundary points and zero the corresponding components of the source vector. This is done because the boundary conditions are not taken into account in the matrix. Afterwards we import the ghost nodes. Then all elements must be traversed in order to compute the matrix-free matrix-vector product. All corresponding displacements of the nodes are loaded. This involves the neighbour search described in Section 3.1. Then the local stiffness matrix is applied with a scaling parameter that corresponds to the Young's modulus of the element. The results of the local element are added into the appropriate places in the destination vector and the ghost nodes are exported. Finally the displacements at the Dirichlet boundary points are restored.

3.3 Prolongation and Restriction

Compared to the matrix-vector multiplication prolongation and restriction are procedures that involve two tree levels. Instead of traveling between the levels, the two different resolutions are traversed concurrently. The keys on the coarser level are computed from those of the finer level by a division by eight. Because we traverse the mesh in one direction, we can use the fast search described in Section 3.1. Algorithm 2 describes the prolongation. The restriction is implemented in a similar way.

3.4 Load Balancing

The domain partitioning is obtained by splitting the space filling curve in equal sized sets of contiguous elements. This avoids the use of a data structure to store the mapping from the nodes to the processes.

After reading the image data each process sorts its nodes and elements according the space filling curve. Afterwards, the key space is subdivided binary

---

Algorithm 2. Prolongation

function void Prolongate(Vector c, Vector f)
  ImportGhostNodes(c);
  $cindex\_tmp = 0$;
  for each i in $TreeFineLevel$ **do**
    $coarsekey = i.key/8$; $bits = i.key$ mod 8; $factor = $ FactorOfElem($bits$);
    $cindex_tmp = $ SearchIndex($cindex\_tmp, coarsekey, coarsetree$);
    $f[$IndexOf(i)$] \mathrel{+}= factor \cdot c[cindex\_tmp]$;
    $coarsekeylist = $ AddCoarseKeysIfBitInDimensionIsSet($coarsekey, bits$);
    for each $cnode$ in $coarsekeylist$ **do**
      $cindex = $ SearchIndex($cindex\_tmp, cnode, coarsetree$);
      $f[$IndexOf(i)$] \mathrel{+}= factor \cdot c[cindex]$;
    end for
  end for
  ZeroBoundaryNodes(f);

---

into buckets until each holds less data items than a defined upper limit. Each process gets a number of consecutive buckets until the average size of elements is reached. This results in a nearly balanced distribution, cf. Fig. 1.

4 Numerical Results

We performed a strong and a weak scalability test. We used the boundary conditions described in Section 2. In each test we used the following stopping criterion: $\|\mathbf{r}_k\|_{M^{-1}} \leq 10^{-6}\|\mathbf{r}_0\|_{M^{-1}}$. We used a W-cycle in the multigrid preconditioner M. On the finest level we used a Chebyshev smoother of degree 6. On each coarser level the degree was increased by one. On the coarsest level we solved the problem by a Jacobi preconditioned CG algorithm. We stopped CG after 20 iterations or if the residual norm was decreased by a factor of 10^7. Usually the first criterion was met. The timings were made on the Cray XT5 of the Swiss National Supercomputing Center [8]. The Cray XT5 is based on Opteron processors with six cores running at 2.4 GHz. Each core has 1.33 GiB main memory.

4.1 Weak Scalability

The solver for the bone analysis is designed such that it scales well on MPI-based supercomputers with big-sized meshes. We have tested the weak scalability with up to 8000 cores with two different meshes, cf. Table 1. The larger grids are generated by 3D mirroring [2] from a bone sample encased in a cube, cf. Fig. 1. We have used two base meshes:

- c240 is encased in a 240^3 cube with $6.9 \cdot 10^6$ degrees of freedom and $1.46 \cdot 10^6$ elements (porosity 10.6%).
- c320 is encased in a 320^3 cube with $11.8 \cdot 10^6$ degrees of freedom and $2.23 \cdot 10^6$ elements (porosity 6.83%).

Fig. 1. Load balancing with a space filling curve in a cubical bone sample. 16 partitions are used. On the left side all partitions are shown. On the right side the partitions numbered three to nine are displayed. Note that partitions need not be connected.

Table 1. Weak scalability timings. The meshing time includes also the time to read the image data.

		64	512	1728	5832	8000
c240	dofs	$445 \cdot 10^6$	$3.6 \cdot 10^9$	$12.0 \cdot 10^9$	$40.5 \cdot 10^9$	$55.5 \cdot 10^9$
	meshing time [s]	8.5	20.9	52.7	154	204
	setup time [s]	20.4	21.4	23.1	28.9	33.6
	GFlops	32.3	253	854	2888	3947
c320	dofs	$758 \cdot 10^6$	$6.1 \cdot 10^9$	$20.4 \cdot 10^9$	$69.1 \cdot 10^9$	$94.7 \cdot 10^9$
	meshing time [s]	16.6	37.1	92.3	273	804
	setup time [s]	34.6	36.0	37.7	44.8	51.0
	GFlops	31.8	252	856	2865	3921

The biggest mesh on 8000 cores has $94.7 \cdot 10^9$ dofs and is 62 times bigger than the largest problem solved with ParFE [3]. In these tests we always used 7 levels in the multigrid preconditioner.

In Fig. 2 we see that the solver scales nearly perfectly up to 8000 cores. With both meshes, above 125 cores the solving time increases only little. Also the setup time and the flop rate of the matrix vector product scale very well, cf. Table 1. However, the meshing time doesn't scale. This time includes the construction of the octree (meshing) and, most of all, the time to distribute the voxel data among the cores. The latter means the broadcast of about 250 MiB = $320^3 \cdot 8$ B of image data from the root core to all others cores, which is a costly procedure.

4.2 Strong Scalability

For the strong scalability test a mesh based on c320 was used with $320 \cdot 10^6$ dofs. This moderately sized problem could be solved on a machine that is affordable

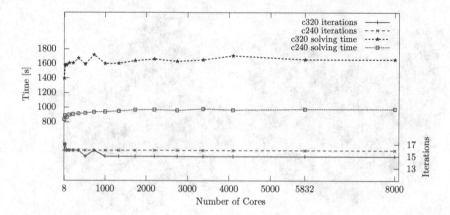

Fig. 2. Weak scaling with two different trabecular bone samples embedded in a 320^3 and a 240^3 regular grid. 3D mirroring is applied to generate the bigger meshes.

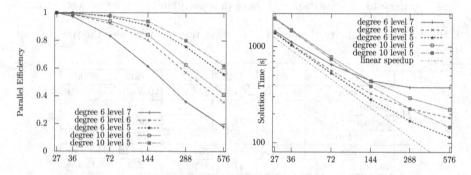

Fig. 3. Strong scaling with different smoother degrees and number of levels in the multigrid algorithm. c320 mesh three timed 3D mirrored used on initial 27 cores. On the left site the parallel efficiency. On the right side the solution time. The (yellow) dashed line in the bottom denotes linear speed up.

for a clinical institute. We have tested the scalability with different parameters to identify the limiting factors. The memory that is needed for solving this mesh forced us to use at least 27 cores.

Figure 3 shows that the application scales very well up to 576 cores. If the number of levels is chosen too big (red solid line with + sign) the parallel efficiency decreases and a configuration with a smoother of higher degree needs less time to solve with 144 cores. The reason is that the problem size on the coarser mesh gets very small and the communication dominates. With redistribution and using a smaller set of cores on coarser meshes the efficiency would be higher especially for large numbers of levels.

The higher smoother degree results in higher efficiency because on the fine meshes the matrix-vector product scales perfectly with the number of processors. However, on this mesh the smoother of degree ten needs more time to solve the problem than the smoother of degree six if the same number of levels is used.

5   Conclusions and Future Work

We have presented a highly parallel solver for voxel-based μFE bone analysis. The solver is based on the PCG method and uses a geometric multigrid preconditioner. Because the mesh is stored in a octree-like data structure all levels are implemented with matrix-free techniques. The minimal memory footprint enabled us to solve huge problems with more than $94 \cdot 10^9$ degrees of freedom. Solving these problems with the old solver ParFE would require 16 times as many processors! The solver also shows nearly perfect weak scalability up to 8000 of processors.

We plan to further improve the accessing of the element nodes by a low collision rate hashing. Further enhancements could be done with enabling repartitioning of the coarser level using a subset of processors. This would lower communication complexity and increase further the parallel efficiency.

Acknowledgments. The work of the first author has been funded in parts by the Swiss National Science Foundation project 205320_125114. The computations on the Cray XT5 have been performed in the framework of a Large User Project grant of the Swiss National Supercomputing Centre (CSCS).

References

1. Adams, M., Brezina, M., Hu, J., Tuminaro, R.: Parallel multigrid smoothing: polynomial versus Gauss–Seidel. J. Comput. Phys. 188(2), 593–610 (2003)
2. Arbenz, P., van Lenthe, G.H., Mennel, U., Müller, R., Sala, M.: A scalable multilevel preconditioner for matrix-free μ-finite element analysis of human bone structures. Internat. J. Numer. Methods Engrg. 73(7), 927–947 (2008)
3. Bekas, C., Curioni, A., Arbenz, P., Flaig, C., van Lenthe, G., Müller, R., Wirth, A.: Extreme scalability challenges in micro-finite element simulations of human bone. Concurrency Computat.: Pract. Exper. 22(16), 2282–2296 (2010)
4. Bielak, J., Ghattas, O., Kim, E.J.: Parallel octree-based finite element method for large-scale earthquake ground simulation. Comp. Model. in Eng. & Sci. 10(2), 99–112 (2005)
5. Braess, D.: Finite Elements: Theory, fast solvers and applications in solid mechanics, 2nd edn. Cambridge University Press, Cambridge (2001)
6. Burstedde, C., Wilcox, L.C., Ghattas, O.: p4est: Scalable algorithms for parallel adaptive mesh refinement on forests of octrees. accepted for publication in SIAM J. Sci. Comput.
7. Castro, R., Lewiner, T., Lopes, H., Tavares, G., Bordignon, A.: Statistical optimization of octree searches. Computer Graphics Forum 27(6), 1557–1566 (2008)
8. Swiss National Supercomputing Centre (CSCS), http://www.cscs.ch/

9. Flaig, C., Arbenz, P.: A Scalable Memory Efficient Multigrid Solver for Micro-Finite Element Analyses Based on CT Images. Parallel Computing 37(12), 846–854 (2011)
10. Margenov, S., Vutov, Y.: Comparative analysis of PCG solvers for voxel FEM systems. In: Proceedings of the International Multiconference on Computer Science and Information Technology, pp. 591–598 (2006)
11. The ParFE Project Home Page (2010), http://parfe.sourceforge.net/
12. van Rietbergen, B., Weinans, H., Huiskes, R., Polman, B.J.W.: Computational strategies for iterative solutions of large FEM applications employing voxel data. Internat. J. Numer. Methods Engrg. 39(16), 2743–2767 (1996)
13. Saad, Y.: Iterative Methods for Sparse Linear Systems, 2nd edn. SIAM, Philadelphia (2003)
14. Samet, H.: The quadtree and related hierarchical data structures. ACM Comput. Surv. 16, 187–260 (1984)
15. Sampath, R.S., Biros, G.: A parallel geometric multigrid method for finite elements on octree meshes. SIAM J. Sci. Comput. 32(3), 1361–1392 (2010)
16. Trottenberg, U., Oosterlee, C.W., Schüller, A.: Multigrid. Academic Press, London (2000)
17. Wirth, A., Mueller, T., Vereecken, W., Flaig, C., Arbenz, P., Müller, R., van Lenthe, G.H.: Mechanical competence of bone-implant systems can accurately be determined by image-based micro-finite element analyses. Arch. Appl. Mech. 80(5), 513–525 (2010)

Aliasing Properties of Voxels in Three-Dimensional Sampling Lattices

E. Linnér and R. Strand

Centre for Image Analysis, Uppsala University,
Box 337, 751 05 Uppsala, Sweden
{elisabeth,robin}@cb.uu.se

Abstract. It is well-known that band limited functions can be represented more efficiently on the body centered cubic (bcc) and face centered
cubic (fcc) lattices than on the standard cartesian (cubic) lattice. We
examine sampling properties of these lattices for non band-limited functions and see that, with respect to preserving the energy of the sampled
function, the bcc and fcc lattices perform better than the cubic lattice for
a given sampling density, indicating that sampling on these lattices will
produce less aliasing errors. We study the aliasing errors within a cross
section of a three dimensional, rotational invariant pattern through the
origin and how these errors depend on the normal of the cross section.
The results are in line with the theory, showing that the bcc and fcc
lattices have superior sampling qualities with respect to aliasing errors.

1 Introduction

Volume images can be classified based on the lattice on which they are sampled.
The de facto standard lattice for volume images is the cubic lattice. It has, however, been proven that the optimum, with respect to sampling properties, is to
use the face-centered cubic (fcc) lattice in the frequency domain, corresponding
to the body-centered cubic (bcc) lattice in the spacial domain [4, 10, 12]. The
Voronoi regions (voxels) of the cubic, bcc, and fcc lattices are cubes, truncated
octahedra and rhombic dodecahedra, respectively [4], and are shown in Fig. 1.

An n-dimensional lattice can be defined by an $n \times n$ matrix $\mathbf{V}$ such that each
lattice point can be expressed as a linear integer combination of the columns
of $\mathbf{V}$. The inverse transpose $\mathbf{V}^{-T} = \left(\mathbf{V}^{-1}\right)^{T}$ defines how the spectrum of a
function f sampled on $\mathbf{V}$ is replicated in the frequency domain [11]. Bcc and

Fig. 1. Voronoi regions in, from left to right, the cubic, bcc, and fcc lattices

I. Lirkov, S. Margenov, and J. Waśniewski (Eds.): LSSC 2011, LNCS 7116, pp. 507–514, 2012.

fcc are reciprocal lattices, meaning that if $\mathbf{V}$ defines a bcc lattice, $\mathbf{V}^{-T}$ defines an fcc lattice. The cubic lattice is defined by the identity matrix and is its own reciprocal lattice.

Three-dimensional volume images are used in a growing range of applications within research and medicine. Especially in the latter case, it is easy to relate to the importance of efficient sampling so as to capture fine details that may be crucial for posing a correct diagnosis. In [13], it is indicated that small tumors are more easily detectable in images sampled on a bcc lattice than on a cubic lattice of comparable resolution.

As the basis vectors of bcc and fcc lattices are not ortogonal, it is not straight forward to construct volume images on these lattices from a series of parallel 2D images. Bcc and fcc are thus most suitable for intrinsic three-dimensional image acquisition techniques, such as magnetic resonance imaging. Before a volume can be viewed on a screen, some volume rendering method needs to be applied. Many methods have been defined for the bcc and fcc lattices, such as splatting [12], Fourier domain rendering [2] and raycasting[6]. Raycasting, which we use for our experiments, is appreciated for its ability to render realistic images. Volume rendering is in general computationally intensive, but several methods, such as raycasting, are parallelizable and at least partly implementable on a GPU [6]. The computational costs of all methods depend on the sampling density of the volume.

Both the sampling of a volume on a lattice and the process of volume rendering produce aliasing errors for non band-limited signals. In applications and algorithm development it is important to be familiar with the nature of such errors so as to minimize their impact. It is well known that the bcc and fcc lattices have better sampling properties for band-limited signals than the cubic lattice[4, 11, 12]. In [11], sampling properties for 1D subsets of non band-limited signals are considered. In this paper, we study the directional dependency of such aliasing errors in 2D subsets of a non band-limited signal sampled on cubic, bcc and fcc lattices.

2   (Discrete Space) Fourier Transform

We use the following definitions of the Fourier transform $\mathcal{F}\left(f(\mathbf{x})\right)$ and inverse Fourier transform $\mathcal{F}^{-1}\left(F(\mathbf{u})\right)$ of a function $f(\mathbf{x})$

$$F(\mathbf{u}) = \int_{\mathbb{R}^n} f(\mathbf{x}) e^{-2\pi i \mathbf{u}^T \mathbf{x}} d\mathbf{x} \leftrightarrow f(\mathbf{x}) = \int_{\mathbb{R}^n} F(\mathbf{u}) e^{2\pi i \mathbf{x}^T \mathbf{u}} d\mathbf{u} \qquad \mathbf{x}, \mathbf{u} \in \mathbb{R}^n$$

A change of variables in spatial domain, see [11], gives the following Fourier pair:

$$f(\mathbf{V}\mathbf{x}) \longleftrightarrow \frac{1}{|\det \mathbf{V}|} F\left(\mathbf{V}^{-T}\mathbf{u}\right).$$

We define the Dirac comb, or Shah function, as

$$\text{III}_{\mathbf{V}}(\mathbf{x}) = \sum_{\mathbf{n} \in \mathbb{Z}^n} \delta\left(\mathbf{x} - \mathbf{V}\mathbf{n}\right),$$

where $\delta(\mathbf{x})$ is the n-dimensional Dirac delta function. The convolution theorem yields

$$f(\mathbf{x}) \mathrm{III}_{\mathbf{V}}(\mathbf{x}) \longleftrightarrow \frac{1}{|\det(\mathbf{V})|} (F(\mathbf{u}) * \mathrm{III}_{\mathbf{V}^{-T}}(\mathbf{u})),$$

where $*$ denotes convolution.

Let $\prod_{\mathbf{V}}$ be the Voronoi region of the lattice point $\mathbf{0}$ in the lattice defined by $\mathbf{V}$. Also, let

$$\sqcap_{\mathbf{V}}(\mathbf{x}) = \begin{cases} 1 \text{ if } \mathbf{x} \in \prod_{\mathbf{V}} \\ 0 \text{ otherwise.} \end{cases}$$

Define the ideal lowpass function $\phi_{\mathbf{V}}$ as:

$$\phi_{\mathbf{V}}(\mathbf{x}) = \mathcal{F}^{-1}(|\det(\mathbf{V})| \sqcap_{\mathbf{V}^{-T}}(\mathbf{u}).)$$

If f is bandlimited (i.e., F is zero outside $\prod_{\mathbf{V}^{-T}}$), then f is completely represented by its samples:

$$(f(\mathbf{x}) \cdot \mathrm{III}_{\mathbf{V}}(\mathbf{x})) * \phi_{\mathbf{V}}(\mathbf{x}) = \mathcal{F}^{-1}((F(\mathbf{u}) * \mathrm{III}_{\mathbf{V}^{-T}}(\mathbf{u})) \cdot \sqcap_{\mathbf{V}^{-T}}(\mathbf{u}))$$
$$= \mathcal{F}^{-1}(F(\mathbf{u})) = f(\mathbf{x}). \tag{1}$$

3   Sampling and Aliasing

Let $g(\mathbf{x})$ be the reconstructed function from the samples of f using (1):

$$g(\mathbf{x}) = f(\mathbf{x}) * \phi_{\mathbf{V}}(\mathbf{x})$$

As performance measure, the *aliasing error*, $\mathcal{E}$, (the relative error of the energy that is represented by the function reconstructed from the samples, see [9]), is used. The Plancherel theorem [5], saying that the L^2 norm is preserved by the Fourier transform, tells us that

$$\mathcal{E} = 1 - \frac{\int_{\mathbb{R}^n} |g(\mathbf{x})|^2 dx}{\int_{\mathbb{R}^n} |f(\mathbf{x})|^2 dx} = 1 - \frac{\int_{\mathbb{R}^n} |\mathcal{F}(g(\mathbf{x}))|^2 du}{\int_{\mathbb{R}^n} |\mathcal{F}(f(\mathbf{x}))|^2 du} = 1 - \frac{\int_{\prod_{\mathbf{V}^{-T}}} |F(\mathbf{u})|^2 du}{\int_{\mathbb{R}^n} |F(\mathbf{u})|^2 du}. \tag{2}$$

The Gauss function

$$f(x, y, z) = e^{-(x^2 + y^2 + z^2)}$$

is used as a test function. The total energy of f is

$$\int_{\mathbb{R}^n} |f(x, y, z)|^2 \, dx \, dy \, dz = \int_{\mathbb{R}^n} |F(u, v, w)|^2 \, du \, dv \, dw = \frac{\sqrt{2}}{4} \pi^{(3/2)}$$

The effect of sampling is simulated using (2). Cubic, fcc, and bcc sampling lattices with equal sampling density are compared by considering reciprocal lattices with equal volume of the Voronoi regions. The plot in Fig. 2 shows the aliasing error (y-axis), computed using Monte Carlo simulation, for sampling lattices of increasing density (x-axis). The volume of $\prod_{\mathbf{V}^{-T}}$, i.e., the period in frequency domain, needed to cover 98% of the relative energy for the cubic, fcc, and bcc lattices are 0.64, 0.56, and 0.55, respectively. This corresponds to spatial domain sampling densities 1.55, 1.80, and 1.81. For a sampling density of 1.55 in spatial domain, the fcc and bcc lattices cover 98.7% and 98.8% of the relative energy, respectively.

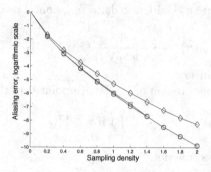

Fig. 2. The aliasing error as a function of sampling density when sampling $f(x, y, z) = e^{-(x^2+y^2+z^2)}$ on a cubic ($\diamond$), fcc ($\square$), and bcc ($\circ$) lattice

3.1   Rotational Dependency

In this section, aliasing errors of 2D subsets in spatial domain are analyzed for a number of rotations. We describe the formula when the 2D subset is the principal xy-plane. The rotation is obtained by rotating the domain of integration.

Analogous to the Fourier slice theorem [5], a slice in the spatial domain corresponds to a projection in frequency domain:

$$f(x, y, 0) \longleftrightarrow \int_{\mathbb{R}} F(u, v, w)dw. \tag{3}$$

The total energy in the 2D subset of a separable function $f(x, y, z) = f(x)f(y)f(z)$ with real-valued Fourier transform F is

$$\int_{\mathbb{R}^2} (f(x)f(y)f(0))^2 \, dx \, dy = \int_{\mathbb{R}^2} (F(u)F(v))^2 \left(\int_{\mathbb{R}} F(w)dw \right)^2 du \, dv.$$

The reconstructed function, i.e., the formula in (1), is now analyzed. We use (3) for the analysis in frequency domain.

The effect of sampling a two-dimensional subset on the three lattices is now considered. We will use the test function in (3) with $z = 0$, namely the pair

$$f(x, y) = e^{-(x^2+y^2)} \text{ and } F(u, v) = \pi e^{-\pi^2(u^2+v^2)}$$

with the total energy $\pi/2$. By the Fourier slice theorem, the energy in the 2D profile is

$$\int_{U_1}^{U_2} \int_{V_1(u)}^{V_2(u)} (F(u)F(v))^2 \left(\int_{W_1(u,v)}^{W_2(u,v)} F(w)dw \right)^2 dv \, du, \tag{4}$$

where $W_1, W_2, V_1, V_2, U_1, U_2$ bound (the rotated versions of) $\prod_{\mathbf{V}^{-T}}$.

By rewriting this integral, it can be computed numerically. We refer to the similar calculations in [11] for details.

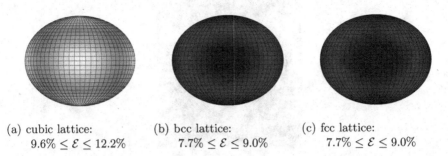

(a) cubic lattice:
    $9.6\% \leq \mathcal{E} \leq 12.2\%$

(b) bcc lattice:
    $7.7\% \leq \mathcal{E} \leq 9.0\%$

(c) fcc lattice:
    $7.7\% \leq \mathcal{E} \leq 9.0\%$

Fig. 3. Error distributions as function of direction. The same sampling density, 1.55, in 3D is used here. The spheres are viewed from the y-direction and the z-direction is up. A linear scale between 7.7% (black) and 12.2% (white) is used.

In Fig. 3, the aliasing error $\mathcal{E} = 1 - I/(\pi/2)$, where $\pi/2$ is the total energy and I is the integral (4), is shown for some different rotations. We see that the aliasing errors are most prominent along the x, y and z-axis for the cubic lattice and along the lines through the octant centers for the bcc and fcc lattices.

4   Raycasting

A raycaster imitates the functionality of a camera, tracing rays from a focus point through an image plane and into the volume that is to be depicted. Along each ray, we solve the volume-rendering integral to compute the radiance I, which is combined with a transfer function to decide what color to assign to the pixel of the image plane through which the ray is directed. In (5), we see the full version of the volume rendering integral, as it is stated in [3], where s_0 and D are the points where the ray enters and exits the volume, $T(s_1, s_2)$ is the transparency between points s_1 and s_2, and q is the emission coefficient.

$$I(D) = I_0 T(s_0, D) + \int_{s_0}^{D} q(s) T(s, D)\, ds \qquad (5)$$

The integral is evaluated at discrete intervals along the ray. Aliasing errors may arise during the raycasting process if the sampling density along the ray is too low, or if rays hit details in the volume that do not represent their neighborhood very well, providing a bad foundation for computing a suitable pixel value. The latter case can be avoided by casting several rays through each pixel of the image plane (known as supersampling, see [8]), or by using a more suitable interpolation method between the lattice points of the volume [6]. For the experiments described in this paper, we use the raycaster vuVolume [1], which is also used in [6].

Fig. 4. Cross section through the origin of the three-dimensional pattern P in (6). Aliasing errors in the form of circular artifacts may appear in printed versions of this paper due to insufficient resolution.

5   Experiments

To confirm the theory presented in section 3.1, we study the aliasing errors in the rotation invariant pattern

$$P = \cos\left(c^2\left(x^2 + y^2 + z^2\right)\right) + 1, \qquad c \in \mathbb{R},\ x, y, z \in [-1, 1]^3. \tag{6}$$

A cross section of (6) through the origin shows a pattern of concentric circles of increasing frequency, displayed in Fig. 4. .

The pattern in (6) is sampled on cubic, bcc anc fcc lattices of equal density, consciously chosen to produce aliasing errors. We extract part of the volume, corresponding to a plane through the origin convolved with a gaussian, which is a good approximation of the point spread function for some imaging methods [7], and visualize the result using the raycaster from [1]. The continuous volume is reconstructed by interpolation using quintic box splines, cubic box splines and cubic B-splines in the bcc, fcc and cubic lattices, respectively. The volume rendering integral is computed along each ray by numerical integration. The high order interpolation and a high sampling rate along the rays minimize the aliasing caused by the rendering process, so that any aliasing errors in the resulting images are likely to have arisen from the sampling of the volume. Using the GUI of [1], we apply a scalar transfer function to render voxels with high intensity to high opacity within the range $[0, 1]$. We study the pattern around cross sections of P with the normals $(1, 0, 0)$ and $(1, 1, 1)$, chosen from Fig. 3 as to represent the worst and best directions for the cubic lattice, and inversely the best and worst directions for the bcc and fcc lattices.

6   Results and Conclusions

Fig. 5 and 6 show the results of the experiment described in the previous section. The dark areas, ripples and circular artifacts are caused by aliasing. As a measure

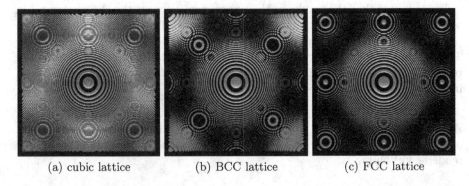

(a) cubic lattice   (b) BCC lattice   (c) FCC lattice

Fig. 5. Normal of intersecting plane: $(1,0,0)$. The horizontal and vertical axis in the image corresponds to the y and z axis, respectively.

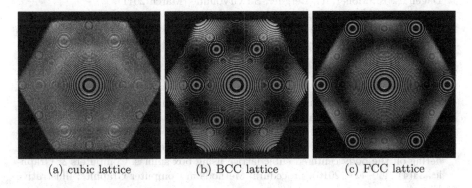

(a) cubic lattice   (b) BCC lattice   (c) FCC lattice

Fig. 6. Normal of intersecting plane: $(1,1,1)$. The horizontal and vertical axis in the image corresponds to $(1,-1,0)$ and $(-1,-1,2)$, respectively.

of the severity of the aliasing errors, we use the distance between the origin and the center of the first artifact, as well as the distance from the origin within which the rings of the pattern are undisturbed by aliasing.

As expected, we see that the bcc and fcc lattices outperform the cubic lattice in Fig. 5, while being comparable to each other. In Fig. 6, the performances of all three lattices are comparable.

For the cubic lattice, we notice that performance is better in Fig. 6, where we also see a higher degree of symmetry with respect to the distance between the origin and the center of the first artifact in different directions.

The performance of the fcc and bcc lattices is better in Fig. 5 than in Fig. 6, although the difference between the two cases is not as large as for the cubic lattice. In Fig. 5, we observe that the more severe aliasing errors appear in different directions for bcc and fcc. In Fig. 6, we observe a higher degree of symmetry in the appearance of the errors, and the orientation of the severe aliasing errors is the same for both lattices.

We conclude that the results of this experiment support the theory derived in section 3.1 about the directional dependency of the aliasing errors, and that the performances of the bcc and fcc lattices are superior to that of the cubic lattice. The directional dependency of the aliasing errors within the cross section of the pattern does not affect the energy of the reconstructed function, but it is still interesting and will be subject to further studies.

Acknowledgements. We would like to thank Dr Alireza Entezari, University of Florida, for providing us with valuable advice and support for the volume renderer vuVolume.

References

1. Dagenais, E., Möller, T., Bergner, S.: vuVolume (March 2011),
   http://sourceforge.net/projects/vuvolume/
2. Dornhofer, A.: A discrete fourier transform pair for arbitrary sampling geometries with applications to frequency domain volume rendering on the body-centered cubic lattice, diploma thesis (2003)
3. Engel, K., Hadwiger, M., Kniss, J., Rezk-Salama, C., Weiskopf, D.: Real-Time Volume Graphics. A K Peters, Ltd. (2006)
4. Entezari, A.: Towards computing on non-cartesian lattices, report (2006)
5. Feeman, T.G.: The Mathematics of Medical Imaging: A Beginner's Guide. Springer, Heidelberg (2010)
6. Finkbeiner, B., Entezari, A., Van De Ville, D., Möller, T.: Efficient volume rendering on the body centered cubic lattice using box splines. Computers & Graphics 34(4), 409–423 (2010), procedural Methods in Computer Graphics; Illustrative Visualization
7. Geets, X., Lee, J., Bol, A., Lonneux, M., Grégoire, V.: A gradient-based method for segmenting FDG-PET images: methodology and validation. European Journal of Nuclear Medicine and Molecular Imaging 34(9), 1427–1438 (2007)
8. Glassner, A. (ed.): An Introduction to Ray Tracing. Academic Press Inc. (1989)
9. Jackson, J.I., Meyer, C.H., Nishimura, D.G., Macovski, A.: Selection of a convolution function for fourier inversion using gridding. IEEE Transactions on Medical Imaging 10(3), 473–478 (1991)
10. Meng, T., Smith, B., Entezari, A., Kirkpatrick, A.E., Weiskopf, D., Kalantari, L., Möller, T.: On visual quality of optimal 3D sampling and reconstruction. In: Proceedings of Graphics Interface 2007 (GI 2007), pp. 265–272. ACM, New York (2007)
11. Strand, R.: Sampling and aliasing properties of three-dimensional point-lattices. In: Swedish Symposium on Image Analysis, Uppsala, Sweden, pp. 23–26 (2010), http://urn.kb.se/resolve?urn=urn:nbn:se:uu:diva-144097
12. Theussl, T., Möller, T., Gröller, E.: Optimal regular volume sampling. In: Proceedings of the IEEE Conference on Visualization 2001, San Diego, California. Washington, DC, USA, pp. 91–98 (2001)
13. Xu, F., Mueller, K.: Applications of optimal sampling lattices for volume acquisition via 3D computed tomography. In: Volume Graphics Symposium, Prague, Czech Republic, pp. 57–63 (2007)

Analysis of a Fast Fourier Transform Based Method for Modeling of Heterogeneous Materials

Jaroslav Vondřejc[1], Jan Zeman[1], and Ivo Marek[2]

[1] Czech Technical University in Prague,
Faculty of Civil Engineering, Department of Mechanics
vondrejc@gmail.com
[2] Czech Technical University in Prague,
Faculty of Civil Engineering, Department of Mathematics

Abstract. The focus of this paper is on the analysis of the Conjugate Gradient method applied to a non-symmetric system of linear equations, arising from a Fast Fourier Transform-based homogenization method due to Moulinec and Suquet [1]. Convergence of the method is proven by exploiting a certain projection operator reflecting physics of the underlying problem. These results are supported by a numerical example, demonstrating significant improvement of the Conjugate Gradient-based scheme over the original Moulinec-Suquet algorithm.

Keywords: Homogenization, Fast Fourier Transform, Conjugate Gradients.

1 Introduction

The last decade has witnessed a rapid development in advanced experimental techniques and modeling tools for microstructural characterization, typically provided in the form of pixel- or voxel-based geometry. Such data now allow for the design of bottom-up predictive models of the overall behavior for a wide range of engineering materials. Of course, such step necessitates the development of specialized algorithms, capable of handling large-scale voxel-based data in an efficient manner. In the engineering community, perhaps the most successful solver meeting these criteria was proposed by Moulinec and Suquet in [1]. The algorithm is based on the Neumann series expansion of the inverse of an operator arising in the associated Lippmann-Schwinger equation and exploits the Fast Fourier Transform to evaluate the action of the operator efficiently for voxel-based data. In our recent work [2], we have offered a new approach to the Moulinec-Suquet scheme, by exploiting the trigonometric collocation method due to Saranen and Vainikko [3]. Here, the Lippman-Schwinger equation is projected to a space of trigonometric polynomials to yield a non-symmetric system of linear equations, see Section 2 below. Quite surprisingly, numerical experiments revealed that the system can be efficiently solved using the standard Conjugate

I. Lirkov, S. Margenov, and J. Waśniewski (Eds.): LSSC 2011, LNCS 7116, pp. 515–522, 2012.

Gradient algorithm. The analysis of this phenomenon, as presented in Section 3, is at the heart of this contribution. The obtained results are further supported by a numerical example in Section 4 and summarized in Section 5.

The following notation is used throughout the paper. Symbols a, $\boldsymbol{a}$ and $\boldsymbol{A}$ denote scalar, vector and second-order tensor quantities, respectively, with Greek subscripts used when referring to the corresponding components, e.g. $A_{\alpha\beta}$. The outer product of two vectors is denoted as $\boldsymbol{a} \otimes \boldsymbol{a}$, whereas $\boldsymbol{a} \cdot \boldsymbol{b}$ or $\boldsymbol{A} \cdot \boldsymbol{b}$ represents the single contraction between vectors (or tensors). A multi-index notation is employed, in which $\mathbb{R}^N$ with $\boldsymbol{N} = (N_1, \ldots, N_d)$ represents $\mathbb{R}^{N_1 \times \cdots \times N_d}$ and $|\boldsymbol{N}|$ abbreviates $\prod_{\alpha=1}^{d} N_\alpha$. Block matrices are denoted by capital letters typeset in a bold serif font, e.g. $\mathbf{A} \in \mathbb{R}^{d \times d \times N \times N}$, and the superscript and subscript indexes are used to refer to the components, such that $\mathbf{A} = [A_{\alpha\beta}^{\boldsymbol{km}}]_{\alpha,\beta=1,\ldots,d}^{\boldsymbol{k},\boldsymbol{m}\in\bar{\mathbb{Z}}^N}$ with

$$\bar{\mathbb{Z}}^N = \left\{ \boldsymbol{k} \in \mathbb{Z}^d : -\frac{N_\alpha}{2} < k_\alpha \leqslant \frac{N_\alpha}{2}, \alpha = 1, \ldots, d \right\}.$$

Sub-matrices of $\mathbf{A}$ are denoted as

$$\mathbf{A}_{\alpha\beta} = \left[A_{\alpha\beta}^{\boldsymbol{km}}\right]^{\boldsymbol{k},\boldsymbol{m}\in\bar{\mathbb{Z}}^N} \in \mathbb{R}^{N \times N}, \quad \mathbf{A}^{\boldsymbol{km}} = \left[A_{\alpha\beta}^{\boldsymbol{km}}\right]_{\alpha,\beta=1,\ldots,d} \in \mathbb{R}^{d \times d}$$

for $\alpha, \beta = 1, \ldots, d$ and $\boldsymbol{k}, \boldsymbol{m} \in \bar{\mathbb{Z}}^N$. Analogously, the block vectors are denoted by lower case letters, e.g. $\mathbf{e} \in \mathbb{R}^{d \times N}$ and the matrix-by-vector multiplication is defined as

$$[\mathbf{Ae}]_\alpha^{\boldsymbol{k}} = \sum_{\beta=1}^{d} \sum_{\boldsymbol{m}\in\bar{\mathbb{Z}}^N} A_{\alpha\beta}^{\boldsymbol{km}} e_\beta^{\boldsymbol{m}} \in \mathbb{R}^{d \times N}, \tag{1}$$

with $\alpha = 1, \ldots, d$ and $\boldsymbol{k} \in \bar{\mathbb{Z}}^N$.

2   Problem Setting

Consider a composite material represented by a periodic unit cell

$$\mathcal{Y} = \prod_{\alpha=1}^{d} (-Y_\alpha, Y_\alpha) \subset \mathbb{R}^d.$$

In the context of linear electrostatics, the associated unit cell problem reads as

$$\boldsymbol{\nabla} \times e(\boldsymbol{x}) = 0, \qquad \boldsymbol{\nabla} \cdot \boldsymbol{j}(\boldsymbol{x}) = 0, \qquad \boldsymbol{j}(\boldsymbol{x}) = \boldsymbol{L}(\boldsymbol{x}) \cdot e(\boldsymbol{x}), \qquad \boldsymbol{x} \in \mathcal{Y} \tag{2}$$

where e is a $\mathcal{Y}$-periodic vectorial electric field, $\boldsymbol{j}$ denotes the corresponding vector of electric current and $\boldsymbol{L}$ is a second-order positive-definite tensor of electric conductivity. In addition, the field e is subject to a constraint

$$\langle e(\boldsymbol{x}) \rangle := \frac{1}{|\mathcal{Y}|} \int_{\mathcal{Y}} e(\boldsymbol{x}) \, \mathrm{d}\boldsymbol{x} = e^0, \tag{3}$$

where $|\mathcal{Y}|$ denotes the d-dimensional measure of $\mathcal{Y}$ and $e^0 \neq 0$ a prescribed macroscopic electric field.

The original problem (2)–(3) is then equivalent to the periodic Lippmann-Schwinger integral equation, formally written as

$$e(x) + \int_{\mathcal{Y}} \Gamma(x - y; L^0) \cdot \left(L(y) - L^0 \right) \cdot e(y) \, dy = e^0, \quad x \in \mathcal{Y}, \qquad (4)$$

where $L^0 \in \mathbb{R}^{d \times d}$ denotes a homogeneous reference medium. The operator $\Gamma(x, L^0)$ is derived from the Green's function of the problem (2)–(3) with $L(x) = L^0$ and can be simply expressed in the Fourier space

$$\hat{\Gamma}(k; L^0) = \begin{cases} 0 & k = 0 \\ \dfrac{\xi \otimes \xi}{\xi \cdot L^0 \cdot \xi} & \xi(k) = \left(\dfrac{k_\alpha}{Y_\alpha} \right)^d_{\alpha=1} \end{cases} ; k \in \mathbb{Z}^d \setminus 0. \qquad (5)$$

Operator $\hat{f} = \hat{f}(k)$ stands for the Fourier coefficient of $f(x)$ for the k-th frequency given by

$$\hat{f}(k) = \int_{\mathcal{Y}} f(x) \varphi_{-k}(x) \, dx, \quad \varphi_k(x) = |\mathcal{Y}|^{-\frac{1}{2}} \exp\left(i\pi \sum_{\alpha=1}^{d} \frac{x_\alpha k_\alpha}{Y_\alpha} \right), \qquad (6)$$

"i" is the imaginary unit ($i^2 = -1$). We refer to [2,4] for additional details. Note that the linear electrostatics serves here as a model problem; the framework can be directly extended to e.g. elasticity [5], (visco-)plasticity [6] or to multiferroics [7].

2.1 Discretization via Trigonometric Collocation

The numerical solution of the Lippmann-Schwinger equation is based on a discretization of a unit cell $\mathcal{Y}$ into a regular periodic grid with $N_1 \times \cdots \times N_d$ nodal points and grid spacings $h = (2Y_1/N_1, \ldots, 2Y_d/N_d)$. The searched field e in (4) is approximated by a trigonometric polynomial e^N in the form (cf. [3, Chapter 10])

$$e(x) \approx e^N(x) = \sum_{k \in \bar{\mathbb{Z}}^N} \hat{e}^k \varphi_k(x), \quad x \in \mathcal{Y}, \qquad (7)$$

where $\hat{e}^k = (\hat{e}^k_\alpha)_{\alpha=1,\ldots,d}$ designates the Fourier coefficients defined in (6). Notice that the trigonometrical polynomials are uniquely determined by a regular grid data, which makes them well-suited to problems with pixel- or voxel-based computations.

The trigonometric collocation method is based on the projection of the Lippmann-Schwinger equation (4) onto the space of the trigonometric polynomials

$$\mathcal{T}^N = \left\{ \sum_{k \in \bar{\mathbb{Z}}^N} c_k \varphi_k, c_k \in \mathbb{C} \right\}, \qquad (8)$$

leading to a linear system in the form, cf. [2]

$$(\mathbf{I} + \mathbf{B})\mathbf{e} = \mathbf{e}^0, \quad \mathbf{B} = \mathbf{F}^{-1}\hat{\mathbf{\Gamma}}\mathbf{F}(\mathbf{L} - \mathbf{L}^0), \tag{9}$$

where $\mathbf{e} = \left(e_\alpha^k\right)_{\alpha=1,\ldots,d}^{k\in\bar{\mathbb{Z}}^N} \in \mathbb{R}^{d\times N}$ is the unknown vector, $\mathbf{I} = \left[\delta_{\alpha\beta}\delta_{km}\right]_{\alpha,\beta=1,\ldots,d}^{km\in\bar{\mathbb{Z}}^N} \in \mathbb{R}^{d\times d\times N\times N}$ is the identity matrix, expressed as the product of the Kronecker delta functions $\delta_{\alpha\beta}$ and δ_{km}, and $\mathbf{e}^0 = \left(e_\alpha^0\right)_{\alpha=1,\ldots,d}^{k\in\bar{\mathbb{Z}}^N} \in \mathbb{R}^{d\times N}$.

All the matrices in (9) exhibit a block-diagonal structure. In particular,

$$\hat{\mathbf{\Gamma}} = \left[\delta_{km}\hat{\Gamma}_{\alpha\beta}^{km}\right]_{\alpha,\beta=1,\ldots,d}^{k,m\in\bar{\mathbb{Z}}^N}, \quad \mathbf{L} = \left[\delta_{km}\mathsf{L}_{\alpha\beta}^{km}\right]_{\alpha,\beta=1,\ldots,d}^{k,m\in\bar{\mathbb{Z}}^N}, \quad \mathbf{L}^0 = \left[\delta_{km}\mathsf{L}^0{}_{\alpha\beta}\right]_{\alpha,\beta=1,\ldots,d}^{k,m\in\bar{\mathbb{Z}}^N},$$

with $\hat{\Gamma}_{\alpha\beta}^{kk} = \hat{\Gamma}_{\alpha\beta}(k; L^0)$, $\mathsf{L}_{\alpha\beta}^{kk} = L_{\alpha\beta}(k)$ and $(\mathsf{L}^0)_{\alpha\beta} = L_{\alpha\beta}^0$. The matrix $\mathbf{F}$ implements the Discrete Fourier Transform and is defined as

$$\mathbf{F} = \left[\delta_{\alpha\beta}F^{km}\right]_{\alpha,\beta=1,\ldots,d}^{k,m\in\bar{\mathbb{Z}}^N}, \quad F^{km} = \frac{|\mathcal{Y}|^{\frac{1}{2}}}{\prod_{\alpha=1}^d N_\alpha} \exp\left(-\sum_{\alpha=1}^d 2\pi\mathrm{i}\frac{k_\alpha m_\alpha}{N_\alpha}\right), \tag{10}$$

with $\mathbf{F}^{-1}$ representing the inverse transform.

It follows from Eq. (1) that the cost of multiplication by $\mathbf{B}$ is dominated by the action of $\mathbf{F}$ and $\mathbf{F}^{-1}$, which can be performed in $O(|\mathbf{N}| \log |\mathbf{N}|)$ operations by the Fast Fourier Transform techniques. This makes the system (9) well-suited for applying some iterative solution technique. In particular, the original Fast Fourier Transform-based Homogenization scheme formulated by Moulinec and Suquet in [1] is based on the Neumann expansion of the matrix inverse $(\mathbf{I}+\mathbf{B})^{-1}$, so as to yield the m-th iterate in the form

$$\mathbf{e}^{(m)} = \sum_{j=0}^m (-\mathbf{B})^j \mathbf{e}^0. \tag{11}$$

As indicated earlier, our numerical experiments [2] suggest that the system can be efficiently solved using the Conjugate Gradient method, despite the non-symmetry of $\mathbf{B}$ evident from (9). This observation is studied in more detail in the next Section.

3 Solution by the Conjugate Gradient Method

We start our analysis with recasting the system (9) into a more convenient form, by employing a certain operator and the associated sub-space introduced later. Note that for simplicity, the reference conductivity is taken as $\mathbf{L}^0 = \lambda\mathbf{I}$.

Definition 1. *Given $\lambda > 0$, we define operator $\mathbf{P}_\mathcal{E} = \lambda\mathbf{F}^{-1}\hat{\mathbf{\Gamma}}\mathbf{F}$ and associated sub-space as*

$$\mathcal{E} = \left\{\mathbf{P}_\mathcal{E}\mathbf{x} \text{ for } \mathbf{x} \in \mathbb{R}^{d\times N}\right\} \subset \mathbb{R}^{d\times N}.$$

Lemma 1. *The operator $\mathbf{P}_\mathcal{E}$ is an orthogonal projection.*

Proof. First, we will prove that $\mathbf{P}_{\mathcal{E}}$ is projection, i.e. $\mathbf{P}_{\mathcal{E}}^2 = \mathbf{P}_{\mathcal{E}}$. Since $\mathbf{F}$ is a unitary matrix, it is easy to see that

$$\mathbf{P}_{\mathcal{E}}^2 = (\lambda \mathbf{F}^{-1}\hat{\mathbf{\Gamma}}\mathbf{F})(\lambda \mathbf{F}^{-1}\hat{\mathbf{\Gamma}}\mathbf{F}) = \mathbf{F}^{-1}(\lambda\hat{\mathbf{\Gamma}})^2\mathbf{F}. \tag{12}$$

Hence, in view of the block-diagonal character of $\hat{\mathbf{\Gamma}}$, it it sufficient to prove the projection property of sub-matrices $(\lambda\hat{\mathbf{\Gamma}})^{kk}$ only. This follows using a simple algebra, recall Eq. (5):

$$(\lambda\hat{\mathbf{\Gamma}})^{kk}(\lambda\hat{\mathbf{\Gamma}})^{kk} = \frac{\boldsymbol{\xi}(k) \otimes \boldsymbol{\xi}(k)}{\boldsymbol{\xi}(k) \cdot \boldsymbol{\xi}(k)} \cdot \frac{\boldsymbol{\xi}(k) \otimes \boldsymbol{\xi}(k)}{\boldsymbol{\xi}(k) \cdot \boldsymbol{\xi}(k)} = \frac{\boldsymbol{\xi}(k) \otimes \boldsymbol{\xi}(k)}{\boldsymbol{\xi}(k) \cdot \boldsymbol{\xi}(k)} = (\lambda\hat{\mathbf{\Gamma}})^{kk}.$$

The orthogonality of $\mathbf{P}_{\mathcal{E}}$ now follows from

$$\mathbf{P}_{\mathcal{E}}^* = \left(\lambda \mathbf{F}^{-1}\hat{\mathbf{\Gamma}}\mathbf{F}\right)^* = \lambda \mathbf{F}^*\hat{\mathbf{\Gamma}}^*\left(\mathbf{F}^{-1}\right)^* = \lambda \mathbf{F}^{-1}\hat{\mathbf{\Gamma}}\mathbf{F} = \mathbf{P}_{\mathcal{E}},$$

according to a well-known result of linear algebra, e.g. Proposition 1.8 in [8].   □

Remark 1. It follows from the previous results that the subspace $\mathcal{E}$ collects the non-zero coefficients of trigonometric polynomials $\mathcal{T}^N$ with zero rotation, which represent admissible solutions to the unit cell problem defined by (2). Note that the orthogonal space $\mathcal{E}^{\perp}$ contains the trigonometric representation of constant fields, cf. [4, Section 12.7].

Lemma 2. *The solution* $\mathbf{e}$ *to the linear system* (9) *admits the decomposition* $\mathbf{e} = \mathbf{e}^0 + \mathbf{e}_{\mathcal{E}}$, *with* $\mathbf{e}_{\mathcal{E}} \in \mathcal{E}$ *satisfying*

$$\mathbf{P}_{\mathcal{E}}\mathbf{L}\mathbf{e}_{\mathcal{E}} + \mathbf{P}_{\mathcal{E}}\mathbf{L}\mathbf{e}^0 = 0. \tag{13}$$

Proof. As $\mathbf{e} \in \mathbb{R}^{d \times N}$, Lemma 1 ensures that it can be decomposed into two orthogonal parts $\mathbf{e}_{\mathcal{E}} = \mathbf{P}_{\mathcal{E}}\mathbf{e}$ and $\mathbf{e}_{\mathcal{E}^{\perp}} = (\mathbf{I} - \mathbf{P}_{\mathcal{E}})\mathbf{e}$. Substituting this expression into (9), and using the identity $\mathbf{B} = \lambda \mathbf{F}^{-1}\hat{\mathbf{\Gamma}}\mathbf{F}\left(\frac{\mathbf{L}}{\lambda} - \mathbf{I}\right)$, we arrive at

$$\frac{1}{\lambda}\mathbf{P}_{\mathcal{E}}\mathbf{L}\mathbf{e}_{\mathcal{E}} + \mathbf{e}_{\mathcal{E}^{\perp}} + \frac{1}{\lambda}\mathbf{P}_{\mathcal{E}}\mathbf{L}\mathbf{e}_{\mathcal{E}^{\perp}} = \mathbf{e}^0. \tag{14}$$

Since $\mathbf{e}^0 \in \mathcal{E}^{\perp}$, we have $\mathbf{e}_{\mathcal{E}^{\perp}} = \mathbf{e}^0$ and the proof is complete.   □

With these auxiliary results in hand, we are in the position to present our main result.

Proposition 1. *The non-symmetric system of linear equations* (9) *is solvable by the Conjugate Gradient method for an initial vector* $\mathbf{e}_{(0)} = \mathbf{e}^0 + \tilde{\mathbf{e}}$ *with* $\tilde{\mathbf{e}} \in \mathcal{E}$. *Moreover, the sequence of iterates is independent of the parameter* λ.

Proof (outline). It follows from Lemma 2 that the solution to (9) admits yet another, optimization-based, characterization in the form

$$\mathbf{e} = \mathbf{e}^0 + \arg\min_{\tilde{\mathbf{e}} \in \mathcal{E}} \left[\frac{1}{2}\left(\mathbf{L}\tilde{\mathbf{e}}, \tilde{\mathbf{e}}\right)_{\mathbb{R}^{d \times N}} + \left(\mathbf{L}\mathbf{e}^0, \tilde{\mathbf{e}}\right)_{\mathbb{R}^{d \times N}}\right]. \tag{15}$$

The residual corresponding to the initial vector $\mathbf{e}_{(0)}$ equals to

$$\mathbf{r}_{(0)} = \mathbf{e}^0 - (\mathbf{I} + \mathbf{B})\left(\mathbf{e}^0 + \tilde{\mathbf{e}}\right) = -\frac{1}{\lambda}\mathbf{P}_{\mathcal{E}}\mathbf{L}\mathbf{e}^0 - \frac{1}{\lambda}\mathbf{P}_{\mathcal{E}}\mathbf{L}\tilde{\mathbf{e}} \in \mathcal{E}.$$

It can be verified that the subspace $\mathcal{E}$ is $\mathbf{B}$-invariant, thus $(\mathbf{I}+\mathbf{B})\mathcal{E} \subset \mathcal{E}$. Therefore, the Krylov subspace

$$\mathscr{K}_m(\mathbf{I} + \mathbf{B}, \mathbf{r}_{(0)}) = \text{span}\left\{\mathbf{r}_{(0)}, (\mathbf{I} + \mathbf{B})\mathbf{r}_{(0)}, \ldots, (\mathbf{I} + \mathbf{B})^m\mathbf{r}_{(0)}\right\} \subset \mathcal{E}$$

for arbitrary $m \in \mathbb{N}$. This implies that the residual $\mathbf{r}_{(m)}$ and the Conjugate Gradient search direction $\mathbf{p}_{(m)}$ at the m-th iteration satisfy $\mathbf{r}_{(m)} \in \mathcal{E}$ and $\mathbf{p}_{(m)} \in \mathcal{E}$. Since $\mathbf{B}$ is symmetric and positive-definite on $\mathcal{E}$, the convergence of the CG algorithm now follows from standard arguments, e.g. Theorem 6.6 in [8]. Observe that different choices of λ generate identical Krylov subspaces, thus the sequence of iterates is independent of λ.   □

Remark 2. Note that it is possible to show, using direct calculations based on the projection properties of $\mathbf{P}_{\mathcal{E}}$, that the Biconjugate Gradient algorithm produces exactly the same sequence of vectors as the Conjugate Gradient method, see [9].

4   Numerical Example

To support our theoretical results, we consider a three-dimensional model problem of electric conduction in a cubic periodic unit cell $\mathcal{Y} = \prod_{\alpha=1}^3(-\frac{1}{2}, \frac{1}{2})$, representing a two-phase medium with spherical inclusions of 25% volume fraction. The conductivity parameters are defined as

$$L(x) = \begin{cases} \rho I, & \|x\|_2 < \left(\frac{3}{16\pi}\right)^{\frac{1}{3}} \\ \begin{pmatrix} 1 & 0.2 & 0.2 \\ 0.2 & 1 & 0.2 \\ 0.2 & 0.2 & 1 \end{pmatrix}, & \text{otherwise} \end{cases}$$

where $\rho > 0$ denotes the contrast of phase conductivities. We consider the macroscopic field $e^0 = [1, 0, 0]$ and discretize the unit cell with $N = [n, n, n]$ nodes[1]. The conductivity of the homogeneous reference medium $L^0 \in \mathbb{R}^{d \times d}$ is parametrized as

$$L^0 = \lambda I, \qquad\qquad \lambda = 1 - \omega + \rho\omega, \qquad\qquad (16)$$

where $\omega \approx 0.5$ delivers the optimal convergence of the original Moulinec-Suquet Fast-Fourier Transform-based Homogenization (FFTH) algorithm [1].

We first investigate the sensitivity of Conjugate Gradient (CG) algorithm to the choice of reference medium. The results appear in Fig. 1(a), plotting the

---

[1] In particular, n was taken consequently as $16, 32, 64, 128$ and 160 leading up to $3 \cdot 160^3 \doteq 12.2 \times 10^6$ unknowns.

relative number of iterations for CG against the conductivity of the reference medium parametrized by ω, recall Eq. (16). As expected, CG solver achieves a significant improvement over FFTH method as it requires about 40% iterations of FFTH for a mildly-contrasted composite down to 4% for $\varrho = 10^3$. The minor differences visible especially for $\rho = 10^3$ can be therefore attributed to accumulation of round-off errors. These observations fully confirm our theoretical results presented earlier in Section 3.

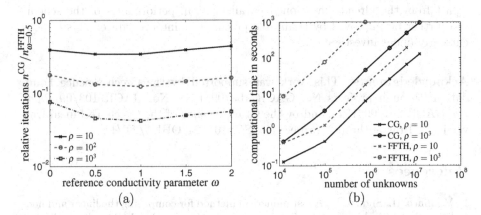

Fig. 1. (a) Relative number of iterations as a function of the reference medium parameter ω and (b) computational time as a function of the number of unknowns

In Fig. 1(b), we present the total computational time[2] as a function of the number of degrees of freedom and the phase ratio ρ. The results confirm that the computational times scales linearly with the increasing number of degrees of freedom for both schemes for a fixed ρ [2]. The ratio of the computational time for CG and FFTH algorithms remains almost constant, which indicates that the cost of a single iteration of CG and FFTH method is comparable.

In addition, the memory requirements of both schemes are also comparable. This aspect represents the major advantage of the short-recurrence CG-based scheme over alternative schemes for non-symmetric systems, such as GMRES. Finally note that finer discretizations can be treated by a straightforward parallel implementation.

5   Conclusions

In this work, we have proven the convergence of Conjugate Gradient method for a non-symmetric system of linear equations arising from periodic unit cell homogenization problem and confirmed it by numerical experiment. The important conclusions to be pointed out are as follows:

---

[2] The problem was solved with a MATLAB® in-house code on a machine Intel® Core™2 Duo 3 GHz CPU, 3.28 GB computing memory with Debian linux 5.0 operating system.

1. The success of the Conjugate Gradient method follows from the projection properties of operator $\mathbf{P}_\varepsilon$ introduced in Definition 1, which reflect the structure of the underlying physical problem.

2. Contrary to all available extensions of the FFTH scheme, the performance of the Conjugate Gradient-based method is independent of the choice of reference medium. This offers an important starting point for further improvements of the method.

Apart from the already mentioned parallelization, performance of the scheme can further be improved by a suitable preconditioning procedure. This topic is currently under investigation.

Acknowledgments. This work was supported by the Czech Science Foundation, through projects No. GAČR 103/09/1748, No. GAČR 103/09/P490, No. GAČR 201/09/1544, and by the Grant Agency of the Czech Technical University in Prague through project No. SGS10/124/OHK1/2T/11.

References

1. Moulinec, H., Suquet, P.: A fast numerical method for computing the linear and nonlinear mechanical properties of composites, Comptes rendus de l'Académie des sciences. Série II, Mécanique, physique, chimie, astronomie 318(11), 1417–1423 (1994)
2. Zeman, J., Vondřejc, J., Novák, J., Marek, I.: Accelerating a FFT-based solver for numerical homogenization of periodic media by conjugate gradients. Journal of Computational Physics 229(21), 8065–8071 (2010) arXiv:1004.1122
3. Saranen, J., Vainikko, G.: Periodic Integral and Pseudodifferential Equations with Numerical Approximation. Springer Monographs in Mathematics. Springer, Heidelberg (2002)
4. Milton, G.W.: The Theory of Composites. Cambridge Monographs on Applied and Computational Mathematics, vol. 6. Cambridge University Press, Cambridge (2002)
5. Šmilauer, V., Bittnar, Z.: Microstructure-based micromechanical prediction of elastic properties in hydrating cement paste. Cement and Concrete Research 36(9), 1708–1718 (2006)
6. Prakash, A., Lebensohn, R.A.: Simulation of micromechanical behavior of polycrystals: finite elements versus fast Fourier transforms. Modelling and Simulation in Materials Science and Engineering 17(6), 064010+ (2009)
7. Brenner, R., Bravo-Castillero, J.: Response of multiferroic composites inferred from a fast-Fourier-transform-based numerical scheme. Smart Materials and Structures 19(11), 115004+ (2010)
8. Saad, Y.: Iterative Methods for Sparse Linear Systems, 2nd edn. with corrections Edition. Society for Industrial and Applied Mathematics, Philadelphia (2003)
9. Vondřejc, J.: Analysis of heterogeneous materials using efficient meshless algorithms: One-dimensional study, Diploma thesis, Czech Technical University in Prague (2009), http://mech.fsv.cvut.cz/~vondrejc/download/ING.pdf

Part XII

Contributed Papers

Properties and Estimates of an Integral Type Nonconforming Finite Element

A.B. Andreev[1] and M.R. Racheva[2]

[1] Department of Informatics
Technical University of Gabrovo
5300 Gabrovo, Bulgaria
[2] Department of Mathematics
Technical University of Gabrovo
5300 Gabrovo, Bulgaria

Abstract. This paper is intended to provide an investigation to the application of an extended Crouzeix-Raviart (EC-R) nonconforming finite element. Integral degrees of freedom are used, which yields some superclose properties. The considered finite element basis contains an integral type bubble function. The approximate eigenvalues obtained by means of this nonconforming method give asymptotically lower bounds of the exact eigenvalues. It is considerable easier to obtain upper bounds for eigenvalues using variational numerical methods. That is why approximations from below are very valued and useful. Finally, computational aspects are discussed and numerical examples are presented.

Keywords: nonconforming element, postprocessing, supercloseness, lower bounds, eigenvalues, bubble function.

2010 MSC: 65N30, 65N25.

1 Introduction

Understandably, the theory of nonconforming finite element methods (FEM) has become an important area of research and has found application in an increasing number of scientific and commercial software products. These methods play a significant role in the numerical approximation of elliptic partial differential equations when conforming methods and others are too costly or unstable. Integral type nonconforming elements use integral values on their edges or/and on the element itself as degrees of freedom. Thus, the obtained finite element space is integrally continuous at the common edges between any two neighboring elements. This feature enables us to construct smoothing and convergence accelerating a posteriori procedures (see [1,2]). On the other hand, a key role of some postprocessing methods play by certain locally supported, nonnegative functions that are commonly referred to as a bubble functions [3]. We use an interior bubble function which is presented by integral value of this function on a single element. So, we consider triangular C-R elements extended by the interior

I. Lirkov, S. Margenov, and J. Waśniewski (Eds.): LSSC 2011, LNCS 7116, pp. 525–532, 2012.

bubble function (EC-R) which give slightly worse approximation compared to C-R elements, but they afford an advantages outlined into the last section.

Let Ω be a bounded polygonal domain in $\mathbf{R}^2$ with boundary $\partial\Omega$. Let also $H^m(\Omega)$ be the usual m-th order Sobolev space on Ω with a norm $\|\cdot\|_{m,\Omega}$ and seminorm $|\cdot|_{m,\Omega}$ and $(\cdot,\cdot)$ denotes the $L_2(\Omega)$-inner product.

Consider the following second-order model problem for $f \in L_2(\Omega)$:

$$-\Delta u + a_0 u = f \quad \text{in } \Omega,$$
$$u = 0 \text{ on } \partial\Omega, \tag{1}$$

where $a_0(x)$ is a nonnegative bounded function in Ω.

The weak formulation of the problem (1) is: find $u \in H_0^1(\Omega)$ such that:

$$a(u,v) = (f,v), \quad \forall v \in V \equiv H_0^1(\Omega), \tag{2}$$

where

$$a(u,v) = \int_\Omega (\nabla u \cdot \nabla v + a_0 uv) \, dx \, dy \quad \forall u,v \in H_0^1(\Omega).$$

By analogy of (2), we consider the variational elliptic eigenvalue problem (EVP): find $(\lambda, u) \in \mathbf{R} \times H_0^1(\Omega)$ such that:

$$a(u,v) = \lambda(u,v), \quad \forall v \in V. \tag{3}$$

Since the bilinear form $a(\cdot,\cdot)$ is symmetric and coercive on $H_0^1(\Omega)$, it is well known that the solution of (3) is given by a sequence of pairs (λ_j, u_j), with positive eigenvalues λ_j diverging to $+\infty$. Then we can assume that $0 < \lambda_1 \le \lambda_2 \le \dots$. The associated eigenfunctions u_j can be chosen to be orthonormal in $L_2(\Omega)$ and they constitute a Hilbert basis for V.

2   Nonconforming Finite Element Method

Let τ_h be family of regular finite element partitions of Ω which fulfill the usual shape regularity conditions [4]. We assume that any two triangles in τ_h share at most a vertex or an edge. Denote by h_K the diameter of the triangle K and

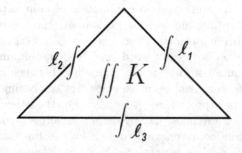

Fig. 1. The considered incomplete quadratic nonconforming FE

$h = \max_{K \in \tau_h} h_K$ is mesh parameter. The edges of any $K \in \tau_h$ are denoted by l_j, $j = 1, 2, 3$. We define nonconforming piecewise incomplete quadratic finite element space with integral type degrees of freedom (Fig. 1): $V_h = \{v \in V_{|K} \in \mathcal{P}_K$ is integrally continuous on the edges of K, for all $K \in \tau_h$ and $\int_l v\, dl = 0$ for every l being a side of the FE triangle that lies on the boundary $\partial\Omega\}$. Here, $\mathcal{P}_1 \subseteq \mathcal{P}_K \subseteq \mathcal{P}_2$, where $\mathcal{P}_1$ and $\mathcal{P}_2$ are the set of polynomials of degree less than or equal to 1 and 2, respectively. Obviously, any two adjacent triangles of τ_h have the same integral value $\int_l v_h\, dl$, $v_h \in V_h$ on their common edge l.

Let $T = \{(t_1, t_2) : t_1, t_2 \geq 0,\ t_1 + t_2 \leq 1\}$ be the reference element. Then, the shape functions of introduced EC-R element on T are:

$$\begin{aligned}
\varphi_1(t_1, t_2) &= -1 + 3(t_1^2 + t_2^2); \\
\varphi_2(t_1, t_2) &= 1 - 4t_1 - 2t_2 + 3(t_1^2 + t_2^2); \\
\varphi_3(t_1, t_2) &= 1 - 2t_1 - 4t_2 + 3(t_1^2 + t_2^2); \\
\varphi_4(t_1, t_2) &= 12t_1 + 12t_2 - 18(t_1^2 + t_2^2),
\end{aligned} \tag{4}$$

where the last function is the interior bubble function.

The functions φ_j, $j = 1, 2, 3, 4$ are obtained under conditions

$$\int_{l_i} \varphi_j\, dl = \delta_{ij},\ i = 1, 2, 3;\ j = 1, 2, 3, 4,$$

$$\int\int_T \varphi_j\, dt_1\, dt_2 = \delta_{4j},\ j = 1, 2, 3, 4.$$

We define the following bilinear form on $V_h + H_0^1(\Omega)$:

$$a_h(u, v) = \sum_{K \in \tau_h} \int_K (\nabla u \cdot \nabla v + a_0 uv)\, dx.$$

The nonconforming approximation problem of (2) is: find $u_h \in V_h$ such that:

$$a_h(u_h, v_h) = (f, v_h),\quad \forall v_h \in V_h. \tag{5}$$

Thus, it is easy to formulate and prove the following proposition:

Proposition 1. *(see [4]) The set of degrees of freedom is* $\mathcal{P}_K$*−unisolvent.*

We also consider the nonconforming FE approximation to the associated elliptic variational EVP: find $(\lambda_h, u_h) \in \mathbf{R} \times V_h$ with $\|u_h\|_{0,\Omega} = 1$, such that

$$a_h(u_h, v_h) = \lambda_h(u_h, v_h),\quad \forall v_h \in V_h. \tag{6}$$

This problem reduces to a generalized EVP involving positive definite symmetric matrices. It attains a finite number of eigenpairs $(\lambda_{h,j}, u_{h,j})$, $j = 1, \ldots, N_h$; $N_h = \dim V_h$ with positive eigenvalues which we assume increasingly ordered: $\lambda_{h,1} \leq \ldots \leq \lambda_{h,N_h}$.

Let us introduce the mesh-dependent norm and seminorm [5]. For any $v \in L_2(\Omega)$ with $v_{|K} \in H^m(K)$, for all $K \in \tau_h$ we define:

$$\|v\|_{m,h} = \left\{ \sum_{K \in \tau_h} \|v\|_{m,K}^2 \right\}^{1/2}, \quad |v|_{m,h} = \left\{ \sum_{K \in \tau_h} |v|_{m,K}^2 \right\}^{1/2}, \quad m = 0, 1.$$

Observe that for any $v_h \in V_h$, $\|v_h\|_{0,h} = \|v_h\|_{0,\Omega}$.

3   Some Approximation Properties

Let us introduce the interpolation operator i_h associated with the considered nonconforming element. The operator $i_h : L_2(\Omega) \to V_h$ can be defined using integral type conditions as degrees of freedom: for any $v \in L_2(\Omega)$ and for all $K \in \tau_h$,

$$\int_{l_j} i_h v \, dl = \int_{l_j} v \, dl \quad \text{and} \quad \int_K i_h v \, dx = \int_K v \, dx,$$

where l_j, $j = 1, 2, 3$ are the edges of K (Fig. 1).

Obviously, $i_h v \in V_h$, for all $v \in L_2(\Omega)$ and $i_h v \equiv v$, for all $v \in V_h$.

Our point of departure will be an essential result following by some interpolation properties (henceforth, C represents generic and positive constant):

Theorem 1. *Let u and u_h be the solution of (2) and (5), respectively. If the solution u belongs to $H^2(\Omega) \cap V$, then*

$$\|u - u_h\|_{s,h} \le Ch^{2-s}\|u\|_{2,\Omega}, \ s = 0, 1. \tag{7}$$

Proof. Besides the shape functions (4) we use the Lagrange interpolation operator π_h related to the linear nonconforming C-R element of integral type [6]. Its shape functions defined on the reference element T are:

$$\psi_1(t_1, t_2) = -1 + 2t_1 + 2t_2; \ \psi_2(t_1, t_2) = 1 - 2t_1; \ \psi_3(t_1, t_2) = 1 - 2t_2.$$

Using the shape functions φ_j, $j = 1, 2, 3, 4$ and ψ_j, $j = 1, 2, 3$, for any $v \in L_2(\Omega)$ we easily calculate:

$$(i_h v - \pi_h v)_{|T} = 3\left(3(t_1^2 + t_2^2) - 2t_1 - 2t_2\right) \left\{ \frac{1}{3} \sum_{j=1}^{3} \int_{l_j} v \, dl - \frac{1}{\text{meas } T} \int_T v \, dt \right\}.$$

This equality could be written in the form:

$$(i_h v - \pi_h v)_{|T} = 3A(t_1, t_2)L_T(v), \tag{8}$$

where $A(t_1, t_2)$ is a bounded function, namely $|A(t_1, t_2)| \le 1$ and $L_T(v)$ is a bounded linear functional in $H^2(T)$.

It is easy to verify that $L_T(v) = 0$ for any $v \in \mathcal{P}_1(T)$.

From the Bramble-Hilbert lemma [4] there exists constant $C > 0$ such that

$$|L_T(v)| \le C|v|_{2,T}.$$

Now, each triangular element may be regarded as the image of the reference triangle T under a continuous, affine, invertible transformation F_K. The transformation may be written in the form

$$F_K(t) = A_K t + b_K,$$

where A_K is a matrix and b_K is a vector. The regularity of the element K means that the A_K is nonsingular, and there exists a constant C such that:

$$
\begin{aligned}
\|A_K\| &\le Ch_K, \\
\|A_K^{-1}\| &\le C\rho_K^{-1}, \\
C\rho_K^2 &\le |\det(A_K)| \le Ch_K^2,
\end{aligned}
\tag{9}
$$

where $\|\cdot\|$ is a matrix norm and ρ_K is the diameter of the inscribed circle in K.

So, it follows that

$$|L_T(v)| \le Ch^2 |\det(A_K)|^{-1/2} |v|_{2,K}. \tag{10}$$

Thus, using that $\|i_h v - \pi_h v\|_{0,K} = |\det(A_K)|^{1/2} \|i_h v - \pi_h v\|_{0,T}$, from (8), (9) and (10) we obtain:

$$\|i_h v - \pi_h v\|_{0,h} = \left(\sum_{K \in \tau_h} \|i_h v - \pi_h v\|_{0,K} \right)^{1/2} \le Ch^2 |v|_{2,\Omega}. \tag{11}$$

Since $i_h v - \pi_h v$ is an incomplete quadratic polynomial on K, the explicit calculations in conjecture with the inverse inequality (see [4], pp. 133-137) we get:

$$\|\partial_i (i_h v - \pi_h v)\|_{|_T} \le Ch|v|_{2,K}, \quad i = 1, 2.$$

These inequalities and (11) give

$$\|i_h v - \pi_h v\|_{s,h} \le Ch^{2-s}|v|_{2,\Omega}, \quad s = 0, 1.$$

Using the estimate concerning integral type C-R linear element (see [5]), for any $v \in H^2(\Omega) \cap V$ it follows that:

$$\|v - \pi_h v\|_{s,h} \le Ch^{2-s}\|v\|_{2,\Omega}.$$

Finally, we obtain

$$\|v - i_h v\|_{s,h} \le \|v - \pi_h v\|_{s,h} + \|i_h v - \pi_h v\|_{s,h} \le Ch^{2-s}\|v\|_{2,\Omega}.$$

Therefore $\lim_{h \to 0} \|u - u_h\|_{1,h} = 0$ and from the FE interpolation theory (see Strang lemma [4] and (11)) we prove the estimate (7).

We can derive a superclose property of the interpolation operator i_h with respect to the elliptic a_h−form.

Lemma 1. *Let $v \in H^2(\Omega) \cap V$ and $i_h v$ be its interpolant constructed by means of extended C-R integral type nonconforming finite element. Then*

$$|a_h(v - i_h v, v_h)| \leq Ch^2 |v|_{2,\Omega} \quad \forall v_h \in V_h.$$

If in addition $a_0(x) \equiv 0$, then

$$a_h(v - i_h v, v_h) = 0 \quad \forall v_h \in V_h.$$

4   Lower Bounds for Eigenvalues

We analyze the approximation for the eigenvalues of the Laplace operator by the nonconforming EC-R integral type finite element. So, the bilinear form is

$$a_h(u, v) = \sum_{K \in \tau_h} \int_K \nabla u \cdot \nabla v \, dx, \quad \forall u, v \in V. \tag{12}$$

In the next theorem we shall prove asymptotically lower bounds of the exact second order eigenvalues by using defined above nonconforming finite element.

Theorem 2. *Let (λ_k, u_k) and $(\lambda_{h,k}, u_{h,k})$ be the solutions of (3) and (6), respectively, for any positive integer k and let also a_h be determined by (12). Assume that the eigenfunctions are normalized $\|u_k\|_{0,\Omega} = \|u_{h,k}\|_{0,\Omega} = 1$. Then, for small enough parameter h:*

$$\lambda_{h,k} \leq \lambda_k. \tag{13}$$

If in addition $u_k \in H^2(\Omega)$, then

$$\lambda_k - \lambda_{h,k} \leq Ch^2 \|u_k\|_{2,\Omega}^2. \tag{14}$$

Proof. It is a well-known fact that if λ_k is an exact eigenvalue of (3), the following inf-sup characterization is valid (see Courant and Hilbert [7]):

$$\lambda_k = \inf_{E^k \subset V} \sup_{v \in E^k} \frac{a_h(v, v)}{(v, v)},$$

where E^k denotes the $k-$dimensional subspace of $V \equiv H_0^1(\Omega)$ ($k \leq \dim V_h$).

Asymptotically lower bounds have been proved for eigenvalues when C-R integral type linear finite element is used. The considered cases are when Ω is nonconvex [8] and convex [6], respectively.

Let $\widetilde{\lambda}_{h,k}$ be approximate eigenvalue of λ_k using linear nonconforming C-R element [6,8]. Then

$$\widetilde{\lambda}_{h,k} = \inf_{E^k \subset \widetilde{V}_h} \sup_{v_h \in E^k} \frac{a_h(v_h, v_h)}{(v_h, v_h)},$$

where $\widetilde{V}_h$ is FE space obtained by integral type linear C-R nonconforming finite element.

In case of our considered element,

$$\lambda_{h,k} = \inf_{E^k \subset V_h} \sup_{v_h \in E^k} \frac{a_h(v_h, v_h)}{(v_h, v_h)}.$$

Having in mind that $\widetilde{V}_h \subset V_h$, the inf-sup conditions gives

$$\lambda_{h,k} \leq \widetilde{\lambda}_{h,k}. \tag{15}$$

The estimate (13) follows from the fact that the eigenvalues obtained by the linear C-R element approximate asymptotically the exact ones from below ([6,8]), i.e. $\widetilde{\lambda}_{h,k} \leq \lambda_k$.

It should be noted here that the considered nonconforming method produces a paradox of the EVP. Namely, in spite of the enhanced FE space V_h in comparison with $\widetilde{V}_h$, from (15) and (13) it follows that the more precision method is the nonconforming linear C-R FEM.

The estimate (14) could be obtained by using standard approaches of the elliptic solution operator corresponding to the Laplace operator. Then, assuming that the exact eigenfunction u_k belongs to $H^2(\Omega) \cap V$, from

$$\lambda_k - \lambda_{k,h} \leq C\|u_k - u_{k,h}\|_{1,h}^2$$

and the result of Theorem 1 we arrive to the estimate (14).

5   Numerical Example

The example is devoted to the EVP:

$$-\Delta u = \lambda u \quad \text{in} \ \ \Omega,$$

$$u = 0 \quad \text{on} \ \ \partial\Omega,$$

where Ω is the square $(0, \pi) \times (0, \pi)$. For this problem the exact eigenvalues are $\lambda_j = s_1^2 + s_2^2$, $s_{1/2} = 1, 2, 3, \ldots$, i.e. $\lambda_1 = 2$; $\lambda_2 = 5$; $\lambda_3 = 5$; $\lambda_4 = 8$; $\ldots$.

Table 1. Eigenvalues computed by means of C-R integral type nonconforming FEs (C-R) and using the proposed finite element (E C-R)

h		λ_1	λ_2	λ_3	λ_4
1/4	C-R	1.965475477	4.546038911	4.546162634	7.435539735
1/4	E C-R	1.903764937	4.236717491	4.259003007	6.732464172
1/8	C-R	1.991417651	4.888134342	4.888194004	7.910092786
1/8	E C-R	1.974736912	4.789539279	4.791231822	7.783327846
1/16	C-R	1.997857237	4.972126840	4.972177890	7.971004421
1/16	E C-R	1.994800036	4.931389746	4.933618833	7.939362469

Using uniform mesh, the domain is divided into $2n^2$ isosceles triangles and thus the mesh parameter is $1/n$, $n = 4, 8, 16$. In Table 1 the results from numerical experiments for the first four eigenvalues are given.

Finally, let us emphasize that in this case (Ω is a convex domain) both nonconforming FEMs approximate the exact eigenvalues from below.

6   Concluding Remarks

The considered integral type EC-R element has two main advantages by comparison with the usual linear (conforming and nonconforming) triangular finite elements:

- This integral type EC-R element is very appropriate to construct patch-recovery schemes [1]. The resulting macroelement (patch) contains 13 degrees of freedom. So, the postprocessing technique gives ultraconvergence results and it is relatively easy to implement.
- The availability of integral type bubble function makes the use of this finite element very suitable for elliptic problems defined on overlapping domains [9]. The same advantage could be emphasized concerning multicomponent elliptic problems with transition conditions.

Acknowledgement. This work is supported by the Bulgarian Ministry of Science under grants C 1001/2011 and DO02-147/2008.

References

1. Lin, Q., Yan, N., Zhou, A.: A rectangle test for interpolated finite elements. In: Proceedings of Systems Science & Systems Engineering, pp. 217–229. Culture Publish Co. (1991)
2. Andreev, A.B., Racheva, M.R.: Superconvergence of the interpolated quadratic finite elements on triangular meshes. Math. Balkanica, New Series, vol. 19, Fasc. 3–4, 385–404 (2005)
3. Ainsworth, M., Oden, J.T.: A Posteriori Error Estimation in Finite Element Analysis. Wiley Interscience, New York (2000)
4. Ciarlet, P.: Basic Error Estimates for the FEM, vol. 2, pp. 17–351. Elsevier, Amsterdam (1991)
5. Brenner, S., Scott, L.R.: The Mathematical Theory of Finite Element Methods. Springer, New York (1992)
6. Andreev, A.B., Racheva, M.R.: Lower Bounds for Eigenvalues by Nonconforming FEM on Convex Domain. In: AIP Conf. Proc., November 25, vol. 1301, pp. 361–369 (2010)
7. Courant, R., Hilbert, D.: Methods of Mathematical Physics, vol. II. Interscience, New York (1962)
8. Armentano, M.G., Duran, R.G.: Asymptotic lower bounds for eigenvalues by nonconforming finite element methods. Electron. Trans. Numer. Anal. 17, 92–101 (2004)
9. De Shepper, H.: Finite element analysis of a coupling eigenvalue problem on overlapping domains. J. Comput. Appl. Math. 132, 141–153 (2001)

Quadratic Finite Element Approximation of a Contact Eigenvalue Problem

A.B. Andreev[1] and M.R. Racheva[2]

[1] Department of Informatics
Technical University of Gabrovo
5300 Gabrovo, Bulgaria
[2] Department of Mathematics
Technical University of Gabrovo
5300 Gabrovo, Bulgaria

Abstract. In this paper we present a finite element method (FEM) to a nonstandard second-order elliptic eigenvalue problem defined on a two-component domain consisting of two intervals with a contact point. This vector problem involves a nonlocal (integral type) coupling condition between the solution components. By introducing suitable degrees of freedom for the quadratic finite element and a corresponding vector Lagrange interpolant we derive optimal order finite element approximation.

Some numerical aspects concerning the method implementation are considered. Illustrative example is given which shows the efficiency of the proposed method.

Keywords: contact problem, quadratic finite elements, integral type coupling condition, order of convergence.

2010 MSC: 65N30, 65N15, 65N25.

1 Introduction

In engineering society, engineers are interested in the approximation methods for some contact problems. The solution of these problems is significantly simplified if the function between contacting bodies is neglected [2]. When setting up expansion method for dynamic models, the different types eigenvalue problems (EVPs) arise here.

In this paper we will study an one-dimensional contact problem under the assumption that the contact region includes only a point. Here we could mention the boundary value problems modelling the heat exchange between two contacting rods [7].

Consider the following intervals: $\Omega_1 =] -1, 1[$ and $\Omega_2 =] 0, 1[$. Let the parameter $\varepsilon \in]0, 1[$ be fixed. Then the interval $\Omega_\varepsilon =] - \varepsilon, \varepsilon[$ is a subinterval of Ω_1 (fig. 1).

We define the vector space

$$L_2(\Omega) = L_2(\Omega_1) \times L_2(\Omega_2),$$

I. Lirkov, S. Margenov, and J. Waśniewski (Eds.): LSSC 2011, LNCS 7116, pp. 533–540, 2012.
© Springer-Verlag Berlin Heidelberg 2012

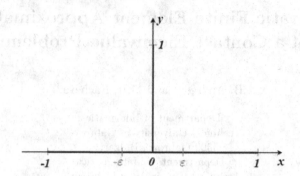

Fig. 1. The considered intervals $\Omega_1 =]-1,1[$, $\Omega_2 =]0,1[$ and $\Omega_\varepsilon =]-\varepsilon,\varepsilon[$

with its innerproduct $(u, v) = (u_1, v_1)_{L_2(\Omega_1)} + (u_2, v_2)_{L_2(\Omega_2)}$, for vector-valued functions $u = [u_1, u_2]$ and $v = [v_1, v_2]$.

We will use also the Sobolev spaces [3] $H^m(\Omega) = H^m(\Omega_1) \times H^m(\Omega_2)$, for nonnegative integer m, with their product norm $\|\cdot\|_{m,\Omega}$ and seminorm $|\cdot|_{m,\Omega}$.

Let us consider the model second order EVP consisting of the differential system: Find $[\lambda, u_1, u_2] \in \mathbf{R} \times H^3(\Omega)$ such that ($\chi_{(-\varepsilon,\varepsilon)}$ is the characteristic function of $]-\varepsilon,\varepsilon[$)

$$
\left.
\begin{aligned}
-\frac{d}{dx}\left(p_1 \frac{du_1}{dx}\right) + q_1 u_1 &= \lambda u_1 + \frac{1}{2\varepsilon}p_2(0)u_2'(0)\chi_{(-\varepsilon,\varepsilon)}, && \text{in } \Omega_1, \\
-\frac{d}{dy}\left(p_2 \frac{du_2}{dy}\right) + q_2 u_2 &= \lambda u_2, && \text{in } \Omega_2,
\end{aligned}
\right\}
\tag{1}
$$

with the classical Dirichlet conditions

$$
u_1(-1) = u_1(1) = 0, \ u_2(1) = 0,
\tag{2}
$$

as well as the following nonlocal coupling condition

$$
u_2(0) = \frac{1}{2\varepsilon}\int_{-\varepsilon}^{\varepsilon} u_1(x)\,dx.
\tag{3}
$$

The functions p_i, q_i are bounded and there exist constants $\bar{p}_i > 0 : p_i \geq \bar{p}_i$ and $q_i \geq 0$ in Ω_i, $i = 1, 2$.

Now we turn our attention to the corresponding variational problem. For any fixed $\varepsilon \in]0,1[$, the space of trial and test functions is

$$
V = \left\{ v = [v_1, v_2] \in V_1 \times V_2 : \ v_2(0) = \frac{1}{2\varepsilon}\int_{-\varepsilon}^{\varepsilon} v_1(x)\,dx \right\},
$$

where $V_1 = H_0^1(\Omega_1)$, $V_2 = \{v_2 \in H^1(\Omega_2) : v_2(1) = 0\}$.

Assuming that $p_i \in H^1(\Omega_i)$, $i = 1, 2$, we define the following EVP: Find $[\lambda, u_1, u_2] \in \mathbf{R} \times V$, $u \neq 0$ such that

$$
a(u, v) = \lambda(u, v), \ \forall v \in V,
\tag{4}
$$

where

$$a(u,v) = \int_{-1}^{1} (p_1 u_1' v_1' + q_1 u_1 v_1)\, dx + \int_{0}^{1} (p_2 u_2' v_2' + q_2 u_2 v_2)\, dy.$$

The problem (1)-(3) is considered and studied by De Schepper and Van Keer [6]. They recast this problem into the framework of abstract variational eigenvalue problems in Hilbert spaces. They also have proved that (4) and (1)-(3) are formally equivalent.

Clearly, $a(\cdot,\cdot)$ is bounded, symmetric and strongly coercive on $V \times V$, while V is dense in $L_2(\Omega)$ [6]. The next theorem proves the existence of the exact eigenpairs of (4).

Theorem 1. *([6], Theorem 1.1) The problem (4) has an infinite number of eigenvalues λ_l, all being strictly positive, having finite multiplicity. We arrange them as $0 < \lambda_1 \leq \lambda_2 \leq \ldots \to +\infty$. The corresponding eigenfunctions u_l can be chosen to be orthonormal in $L_2(\Omega)$, the sequence $\{u_l/\sqrt{\lambda_l}\}$ then being orthonormal with respect to $a(\cdot,\cdot)$. The eigenfunctions constitute a Hilbert basis for V.*

2 Finite Element Approximation

Consider families of regular finite element partitions of Ω_1 and Ω_2. Let $\{x_i\}_{i=0}^{2n_1+2n_2}$ be the set of points in $\overline{\Omega}_1$, where $-1 = x_0 < x_1 < \ldots < x_{2n_1+2n_2} = 1$, $x_{n_1} = -\varepsilon$ and $x_{n_1+2n_2} = \varepsilon$. We use one-dimensional quadratic finite elements. For that reason, let us introduce the spaces:

$$X_{1,h_1} = \left\{ v_1 \in C(\overline{\Omega}_1): \ v_1|_{[x_{i-1},x_i]} \in \mathcal{P}_2([x_{i-1},x_i]), \ i=1,2,\ldots,2n_1+2n_2 \right\},$$

$$V_{1,h_1} = \left\{ v_1 \in X_{1,h_1}: \ v_1(-1) = v_1(1) = 0 \right\},$$

where h_1 is the mesh parameter of the partition.

By analogy, let $\{y_i\}_{i=0}^{m}$ be the set of points in $\overline{\Omega}_2$, where $0 = y_0 < y_1 < \ldots < y_m = 1$. Consider

$$X_{2,h_2} = \left\{ v_2 \in C(\overline{\Omega}_2): \ v_2|_{[y_{i-1},y_i]} \in \mathcal{P}_2([y_{i-1},y_i]), \ i=1,2,\ldots,m \right\},$$

$$V_{2,h_2} = \left\{ v_2 \in X_{2,h_2}: \ v_2(1) = 0 \right\},$$

with a mesh parameter h_2. We assume that there exists a constant $\alpha > 0$ independent of the meshes such that $h_1/h_2 \leq \alpha$. Finally, let $h = \max\{h_1,h_2\}$. Then, the finite element space is defined by

$$V_h = \left\{ v_h = [v_{1h}, v_{2h}] \in X_{1,h_1} \times X_{2,h_2}: \ v_{2h}(0) = \frac{1}{2\varepsilon} \int_{-\varepsilon}^{\varepsilon} v_{1h}(x)\, dx \right\}.$$

The finite element approximation of the EVP (4) is: Find $[\lambda, u_{1h}, u_{2h}] \in \mathbf{R} \times V_h$, $u_h \neq 0$ such that

$$a(u_h, v_h) = \lambda_h(u_h, v_h), \quad \forall v_h \in V_h. \tag{5}$$

If the dimension of the finite element space V_h is $N = N(h)$, then the EVP (5) has N positive eigenvalues (see [6,5])

$$0 < \lambda_{1,h} \le \lambda_{2,h} \le \ldots \le \lambda_{N,h}.$$

The error analysis in [6] is based on a modification of the piecewise linear Lagrange interpolant (the so-called *imperfect* interpolant). Due to the nonlocal coupling condition (3) on the interface between Ω_1 and Ω_2, the use of standard finite elements leads to involved considerations in theoretical aspects. Also, the numerical implementation is complicated.

Herein, we propose a method which presents in a more natural way the characteristics of the problem. Namely, we use an integral type quadratic element in order to avoid construction of the imperfect interpolant.

Remark 1. We restrict ourselves to the case of second degree polynomials, but the proposed approach could be generalized to higher degree. $2D$ elements, including the linear nonconforming case, should be discussed separately.

Remark 2. The finite elements with integral degrees of freedom which are used herein could be applied to construct superconvergent patch-recovery technique (see [4,1]). This procedure can be obtained simultaneously with the finite element approximation solution.

Let us denote by π_h the quadratic interpolant corresponding to the one-dimensional finite element with an integral degree of freedom. Then, for any $v \in V$: $\pi_h \equiv [\pi_{1h_1} v_1, \pi_{2h_2} v_2] \in V_{1h_1} \times V_{2h_2}$ with

$$\pi_{1h_1} v_1(x_i) = v_1(x_i), \ i = 0, 1, \ldots, 2n_1 + 2n_2;$$

$$\int_{x_{i-1}}^{x_i} \pi_{1h_1} v_1(x) \, dx = \int_{x_{i-1}}^{x_i} v_1(x) \, dx, \ i = 1, \ldots, 2n_1 + 2n_2;$$

$$\pi_{2h_2} v_2(y_i) = v_2(y_i), \ i = 0, 1, \ldots, m;$$

$$\int_{y_{i-1}}^{y_i} \pi_{2h_2} v_2(y) \, dy = \int_{y_{i-1}}^{y_i} v_2(y) \, dy, \ i = 1, \ldots, m.$$

Obviously, this interpolant fulfils the condition $\pi_h v \in V_h$.

On the other hand, the classical three-point vector piecewise Lagrange interpolant on the mesh is defined by: $\Pi_h \equiv [\Pi_{1h_1} v_1, \Pi_{2h_2} v_2] \in V_{1h_1} \times V_{2h_2}, \ v \in V$ with

$$\Pi_{1h_1} v_1(x_i) = v_1(x_i), \ i = 0, 1, \ldots, 2n_1 + 2n_2;$$

$$\Pi_{1h_1} v_1 \left(\frac{x_{i-1} + x_i}{2} \right) = v_1 \left(\frac{x_{i-1} + x_i}{2} \right), \ i = 1, \ldots, 2n_1 + 2n_2;$$

$$\Pi_{2h_2} v_2(y_i) = v_2(y_i), \ i = 0, 1, \ldots, m;$$

$$\Pi_{2h_2} v_2 \left(\frac{y_{i-1} + y_i}{2} \right) = v_2 \left(\frac{y_{i-1} + y_i}{2} \right), \ i = 1, \ldots, m.$$

By the nonlocal coupling condition incorporated in V_h one has that $\Pi_h v \notin V_h$.

Now, we shall prove the main result:

Theorem 2. *Let (λ, u) be an exact eigensolution of (4) and $u = [u_1, u_2] \in V \cap H^3(\Omega)$. If (λ_h, u_h) is the corresponding approximate solution of (5), then there exist a constant $C = C(\Omega) > 0$, independent of h, such that*

$$\|u - u_h\|_{1,\Omega} \le Ch^2\|u\|_{3,\Omega},$$

$$|\lambda - \lambda_h| \le Ch^4\|u\|_{3,\Omega}^2. \tag{6}$$

Proof. First, we shall estimate the difference between both interpolants Π_h and π_h. Let $v = [v_1, v_2] \in V \cap H^3(\Omega)$. We introduce reference element T which coincides with the interval $[0,1]$. The basis functions of the Lagrange interpolant Π_h related to the argument $t \in T$ are

$$\psi_1(t) = 2t^2 - 3t + 1; \quad \psi_2(t) = -4t^2 + 4t; \quad \psi_3(t) = 2t^2 - t.$$

Accordingly, the basis functions corresponding to the integral type interpolant π_h are

$$\varphi_1(t) = 3t^2 - 4t + 1; \quad \varphi_2(t) = -6t^2 + 6t; \quad \varphi_3(t) = 3t^2 - 2t.$$

Any interval $K_i^{(1)} = [x_{i-1}, x_i]$ is transformed to T by means of linear function $L_i^{(1)}(x)$. Obviously, $|L_i^{(1)}(x)| = \mathcal{O}(\frac{1}{h})$, $i = 1, 2, \ldots, 2n_1 + 2n_2$. The considerations for $K_i^{(2)} = [y_{i-1}, y_i]$ are in the same manner.

We easily calculate

$$(\Pi_{1h_1} v_1 - \pi_{1h_1} v_1)_{|_T} = v_1(0)\,(\psi_1(t) - \varphi_1(t))$$

$$+ \left(v_1(\tfrac{1}{2})\psi_2(t) - \varphi_2(t) \int_0^1 v_1(t)\,dt\right) + v_1(1)\,(\psi_3(t) - \varphi_3(t))$$

$$= \varphi_2(t)\left(\frac{1}{6}v_1(0) + \frac{4}{6}v_1(\tfrac{1}{2}) + \frac{1}{6}v_1(1) - \int_0^1 v_1(t)\,dt\right).$$

The expression in the bracket could be presented by using error functional of quadrature formula [3]. Namely,

$$(\Pi_{1h_1} v_1 - \pi_{1h_1} v_1)_{|_T} = \varphi_2(t)E_T(v_1), \tag{7}$$

where error functional E_T represents the error of the Simpson quadrature formula on T. It is easy to verify that $E_T(v_1) = 0$ for all $v_1 \in \mathcal{P}_2(t)$. On the other hand $|\varphi_2(t)| \le 3/2$, $t \in T$.

From (7), using standard argument of Bramble-Hilbert lemma [3,5], for any element $K_i^{(1)}$, $i = 1, 2, \ldots, 2n_1 + 2n_2$ we obtain

$$|\Pi_{1h_1} v_1 - \pi_{1h_1} v_1\|_{K_i^{(1)}} \le Ch^3|v_1|_{3,K_i^{(1)}}.$$

From this inequality it follows the L_2-norm error estimate:

$$\|\Pi_{1h_1}v_1 - \pi_{1h_1}v_1\|_{0,\Omega_1} = \left(\sum_{i=1}^{2n_1+2n_2} \int_{K_i^{(1)}} |\Pi_{1h_1}v_1 - \pi_{1h_1}v_1|^2 \, dx \right)^{1/2} \tag{8}$$

$$\leq Ch^3\|v_1\|_{3,\Omega_1}.$$

Applying explicit calculations and the inverse inequality [3], we get

$$\left| \frac{d}{dx} \left(\Pi_{1h_1}v_1 - \pi_{1h_1}v_1 \right) \right|_{K_i^{(1)}} \leq Ch^2 |v_1|_{3,K_i^{(1)}}, \quad i = 1, 2, \ldots, 2n_1 + 2n_2.$$

Consequently, the H^1-norm error estimate is

$$\|\Pi_{1h_1}v_1 - \pi_{1h_1}v_1\|_{1,\Omega_1} \leq Ch^2\|v_1\|_{3,\Omega_1}.$$

This inequality and (8) give

$$\|\Pi_{1h_1}v_1 - \pi_{1h_1}v_1\|_{s,\Omega_1} \leq Ch^{3-s}\|v_1\|_{3,\Omega_1}, \quad s = 0, 1. \tag{9}$$

By analogy, if $v_2 \in V_2 \cap H^3(\Omega_2)$,

$$\|\Pi_{2h_2}v_2 - \pi_{2h_2}v_2\|_{s,\Omega_2} \leq Ch^{3-s}\|v_2\|_{3,\Omega_2}, \quad s = 0, 1. \tag{10}$$

Finally, from (9) and (10) we obtain

$$\|\Pi_h v - \pi_h v\|_{s,\Omega} \leq Ch^{3-s}\|v\|_{3,\Omega}, \quad s = 0, 1. \tag{11}$$

The Lagrange interpolant has an optimal order of convergence [3]. Taking into account that

$$\|v - \pi_h v\|_{s,\Omega} \leq \|v - \Pi_h v\|_{s,\Omega} + \|\Pi_h v - \pi_h v\|_{s,\Omega}, \quad s = 0, 1,$$

from (11) we have the estimate

$$\|v - \pi_h v\|_{s,\Omega} \leq Ch^{3-s}\|v\|_{3,\Omega}, \quad s = 0, 1. \tag{12}$$

Let us define the elliptic projector $\mathcal{R}_h : V \to V_h$ such that

$$a(v - \mathcal{R}_h v, v_h) = 0, \quad \forall v \in V, \ v_h \in V_h.$$

From the estimate (12) it follows that the finite element space $V_h \subset V$ satisfies the approximation property:

$$\inf_{v_h \in V_h} \{\|v - v_h\|_{0,\Omega} + h|v - v_h|_{1,\Omega}\} \leq Ch^3\|v\|_{3,\Omega}, \tag{13}$$

$$\|v - \mathcal{R}_h v\|_{1,\Omega} \leq Ch^2\|v\|_{3,\Omega}, \quad \forall v \in V \cap H^3(\Omega).$$

The theory of the finite element analysis could be adapted to the EVP (4) and its approximation (5), using the estimate (13) (see [5]). Namely, the order of convergence in (6) is determined by (13) which estimates are based on (12).

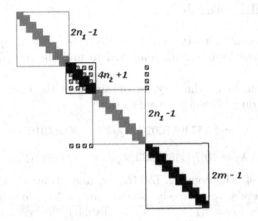

Fig. 2. Mass and stiffness matrix

3   Some Computational Aspects

It is evident, that for the approximate eigenfunction $u_h = [u_{1h}, u_{2h}]$ the following equality holds:

$$u_{2h}(0) = \frac{1}{2\varepsilon} \sum_{i=n_1}^{n_1+2n_2} \int_{x_{i-1}}^{x_i} u_{1h}(x)\, dx.$$

Due to this equality it is easy to determine a basis for V_h and to construct the mass and stiffness matrix.

Let $\{\widetilde{\varphi}_j\}_{j=0}^{4n_1+4n_2+1}$ be the canonical basis of X_{1h_1}, where $\widetilde{\varphi}_{2i}(x_j) = \delta_{ij}$ and $\int_{x_{j-1}}^{x_j} \widetilde{\varphi}_{2i-1}(x)\, dx = \delta_{ij}$, δ_{ij} is the Kronecker delta.

Similarly, let $\{\overline{\varphi}_j\}_{j=0}^{2m}$ be the canonical basis of X_{2h_2}, where $\overline{\varphi}_{2i}(x_j) = \delta_{ij}$ and $\int_{y_{j-1}}^{y_j} \overline{\varphi}_{2i-1}(y)\, dy = \delta_{ij}$.

Then, the basis of the space V_h consists of $4n_1 + 4n_2 + 2m - 2$ functions consecutively given below:

$$\Phi_j(x,y) = [\widetilde{\varphi}_j(x), 0], \qquad j = 1, \ldots, 2n_1;$$

$$\Phi_j(x,y) = [\widetilde{\varphi}_j(x), \frac{1}{2\varepsilon}\overline{\varphi}_0(y)], \quad j = 2k-1, k = n_1 + 1, \ldots, n_1 + 2n_2;$$

$$\Phi_j(x,y) = [\widetilde{\varphi}_j(x), 0], \qquad j = 2k, k = n_1 + 1, \ldots, n_1 + 2n_2 - 1;$$

$$\Phi_j(x,y) = [\widetilde{\varphi}_j(x), 0], \qquad j = 2n_1 + 4n_2, \ldots, 4n_1 + 4n_2 - 1;$$

$$\Phi_j(x,y) = [0, \overline{\varphi}_j(y)], \qquad j = 4n_1 + 4n_2, \ldots, 4n_1 + 4n_2 + 2m - 2.$$

In fig. 2 the structure of mass and stiffness matrix is shown.

4   Numerical Example

The example is devoted to the EVP (1)-(3) with $p_i = 1$, $q_i = 0$, $i = 1, 2$ and $\varepsilon = 0.2$. Calculations are implemented to find the first four approximate eigensolutions.

For this problem in [6] the way for obtaining of the exact eigenvalues and their dependence on ε is clearly given, so that:

$$\lambda_1 = 2.46740110027;\ \lambda_2 = 9.869604401089;$$

$$\lambda_3 = 10.7634821083;\ \lambda_4 = 22.2066099025.$$

Using uniform mesh, the domains $\Omega_1 \backslash \Omega_\varepsilon$, Ω_ε and Ω_2 are divided into $2n_1, 2n_2$ and m equal intervals, respectively. So that, $n_1 = 4n_2$, $m = n_1 + n_2$ and the mesh parameter is $h_1 = h_2 = 0.2/n_2$. In Table 1 the results from numerical experiments for the first four eigenvalues when for $n_2 = 1, 2, 4, 8$ are given.

n_2	$\lambda_{1,h}$	$\lambda_{2,h}$	$\lambda_{3,h}$	$\lambda_{4,h}$
1	2.46743431052	9.87169789914	10.7661731788	22.2299433588
2	2.46740318393	9.86973727943	10.7636545285	22.2082951278
4	2.46740123063	9.86961280498	10.7634938313	22.2070103253
8	2.46740110842	9.86960493939	10.7634830108	22.2067105097

Acknowledgement. This work is supported by the Bulgarian Ministry of Science under grants C 1001/2011 and DO02-147/2008.

References

1. Andreev, A.B., Racheva, M.R.: Superconvergence of the interpolated quadratic finite elements on triangular meshes. Math. Balkanica, New Series, vol.19, Fasc. 3–4, 385–404 (2005)
2. Galin, G.A.: Contact Problems: The Legacy of L.A. Galin. In: Gladwell, G.M.L. (ed.) Springer, Dordrecht (2008)
3. Ciarlet, P.: Basic Error Estimates for the FEM, vol. 2, pp. 17–35. Elsevier, Amsterdam (1991)
4. Lin, Q., Yan, N., Zhou, A.: A rectangle test for interpolated finite elements. In: Proceedings of Systems Science & Systems Engineering, pp. 217–229. Culture Publish Co. (1991)
5. Raviart, P.A., Thomas, J.M.: Introduction a l'Analyse Numerique des Equations aux Derivees Partielles, Masson Paris (1983)
6. De Shepper, H., Van Keer, R.: Finite Element Approximation of a Contact Vector Eigenvalue Problem. Applications of Mathematics 48(6), 559–571 (2003)
7. Shin, T.M.: Numerical Heat Transfer. Hemisphere Publ. Corp., Washington (1984)

Optimization Methods for Calibration
of Heat Conduction Models

Radim Blaheta, Rostislav Hrtus, Roman Kohut, and Ondřej Jakl

Institute of Geonics of the Academy of Sciences of the Czech Republic,
Ostrava, Czech Republic
{blaheta,hrtus,kohut,jakl}@ugn.cas.cz

Abstract. The paper provides a summary of techniques, which are
suitable for calibration of models like both stationary and nonstationary heat conduction. We assume that the PDE based models are discretized by finite elements and PDE coefficients are piecewise constant
on apriori given macroelements (subdomains). A special attention is
given to Gauss-Newton methods, evaluation of the derivatives and application of these methods to a heat evolution problem, which arose in
geoengineering.

1 Introduction

Problems of calibration of models by identification of parameters (mostly material parameters involved in constitutive relations) appear in many fields of science
and engineering. The main purpose of this paper is formulation of such calibra-
tion/identification problems and description of numerical optimization methods
for their solution. The identification problems are described as PDE–constrained
or unconstrained optimization. The gradient-based optimization methods were
implemented and tested on calibration of a nonstationary heat conduction model
arising in geomechanics.

2 Identification Problems — Stationary Case

Let us consider the elliptic heat flow problem in the form

$$- \operatorname{div} (k \operatorname{grad} u) = f \quad \text{in } \Omega, \tag{1}$$

$$u = \hat{u} \text{ on } \Gamma_D \quad \text{and} \quad k \operatorname{grad} u \cdot \nu = \hat{q} \quad \text{on } \Gamma_N, \tag{2}$$

where Ω is a domain with Lipschitz boundary, ν is the unit outer normal, $\partial\Omega = \Gamma_D \cup \Gamma_N$, $\Gamma_D \cap \Gamma_N = \emptyset$. We also assume discretization by the finite element
method given by division $\mathcal{T}_h$ of Ω into triangular/tetrahedral elements and
considering the space V_h of continuous piecewise linear functions. V_h can be
equipped with the nodal basis

$$\{\phi_1^h, \ldots, \phi_n^h\}, \quad \phi_i^h(x_j) = \delta_{ij} \text{ for nodes } x_j \in N_h.$$

I. Lirkov, S. Margenov, and J. Waśniewski (Eds.): LSSC 2011, LNCS 7116, pp. 541–548, 2012.

Further, the coefficient k is assumed to be constant on elements $E_i \in \mathcal{T}_h$ or on larger macroelements (subdomains) $\tilde{E}_j \in \mathcal{T}_H$. Here, each $\tilde{E}_j$ $(j = 1, \ldots, s)$ is a union of some elements from $\mathcal{T}_h$. Thus $s \leq nel = \text{card}(\mathcal{T}_h)$, but frequently $s \ll nel$.

This assumptions lead to introduction of the following admissible set of parameters (PDE coefficients),

$$K_H = \left\{ k = k(x) : \ \exists p \in R^s, \ k \mid_{\tilde{E}_j} = \exp(p_j) \quad \forall j = 1, \ldots, s \right\}.$$

For given $p \in R^s$, the *forward problem* is to find the solution of the discrete (FEM) approximation to the boundary value problem (1), (2). It means to compute the state variable (temperature)

$$u_h = u_h(p) = \sum \bar{u}_i \phi_i^h \ \equiv \ \bar{u} = (\bar{u}_1, \ldots, \bar{u}_n) \in \ R^n,$$

$$e = e(p, \bar{u}) \equiv A_h(p)\bar{u} - b_h = 0. \tag{3}$$

The *parameter identification problem* considers the coefficient k given by the vector $p \in R^s$ as unknown and attempts to determine $p \in R^s$ by exploiting given observation data

$$d_j = u_{obs}(x_j), \ x_j \in N_{obs} \subset N_h, \ \text{card}(N_{obs}) = m$$

and the objective function

$$F_{obs} = \frac{1}{2} \sum_{x_j \in N_{obs}} \mid d_j - \bar{u}_j(p) \mid^2. \tag{4}$$

The *identification problem* can be now formulated as follows

find $(p, \bar{u})$:

$$F(p, \bar{u}) = \min F(q, \bar{v}) \ \text{s.t.} \ e(q, \bar{v}) = 0, \ q \in R^s, \ \bar{v} \in R^n, \tag{5}$$

Above s.t. is an abbreviation for "subject to", $e(q, \bar{v}) = 0$ is the constraint by discretized PDE from (3) and

$$F(q, \bar{v}) = F_{obs}(q, \bar{v}) + \beta F_{reg}(q),$$

where F_{obs} is the observation error part (4) and F_{reg} is a possible regularization term with $\beta \geq 0$ being a regularization weight. The choice of F_{reg} is a specific important task [5,4].

The minimization problem (5) can be solved either as constrained minimization formulated as e.g. (augmented) Lagrangian problem (6) or the constraint can be involved in an unconstrained minimization (7), (8) approach.

The former approach can be based on the augmented Lagrangian functional

$$L(q, \bar{v}, \lambda) = F(q, \bar{v}) + \lambda e(q, \bar{v}) + \frac{c}{2} e^2(q, \bar{v}) \tag{6}$$

and the solution can use e.g. a modified Uzawa algorithm, see [4].

The unconstrained approach considers minimization of

$$\widetilde{F}(p) = F(p, \bar{u}(p)) \tag{7}$$

where evaluation of $\widetilde{F}$ involves the solution of the forward problem. The minimization problem then looks as

$$\text{find } p \in R^s \ : \ \widetilde{F}(p) = \min \widetilde{F}(q) \ : \ q \in R^s \ . \tag{8}$$

3 Identification Problems — Nonstationary Case

Now, we consider the parabolic heat flow problem

$$c\frac{\partial u}{\partial t} - \text{div} \ (k \ \text{grad} \ u) = f \quad \text{in } \Omega \times (0, T).$$

with boundary and initial conditions.

We shall consider the finite element discretization as in the previous section and a time discretization scheme, e.g. the implicit Euler method using uniform time discretization given by $t_i = i\Delta t$, $i = 1, \ldots, T/n_t$.

The problem coefficients now include not only the heat conductivity k but also the (volumetric) heat capacity $c = c_{spec}\rho$, where ρ denotes the specific weight. Similarly as in the previous section, we can construct the set of parameter vectors K_H,

$$K_H = \{k = k(x), \ c = c(x) : \ \exists p \in R^{2s} : \ k \,|_{\bar{E}_j} = \exp(p_j),$$
$$c \,|_{\bar{E}_j} = \exp \ (p_{j+s}) \quad \forall j = 1, \ldots, s\}$$

For given $p \in R^{2s}$, the *forward problem* is to find the sequence of vectors

$$\bar{u} = \{\bar{u}^k\} \in (R^n)^{n_t}, \quad \bar{e}(p, \bar{u}) = (\bar{e}_1(p, \bar{u}), \ldots, \bar{e}_{n_t}(p, \bar{u})) = 0, \tag{9}$$
$$\bar{e}_k(p, \bar{u}) \equiv [M_h(p) + \Delta t A_h(p)] \,\bar{u}^k - [\Delta t b_h^k + M_h(p)\bar{u}^{k-1}] \quad \forall k = 1, \ldots, n_t,$$

$\bar{u}^0$ is given by interpolation of the initial condition, A_h and M_h denote the FE and mass matrices, respectively, b_h^k is the source vector given by f and boundary values in the time level t_k.

The *parameter identification problem* considers the coefficients k, c represented by the vector $p \in R^{2s}$ as unknown and seek for optimal $p \in R^{2s}$ to fit the observation data

$$d_j^k = u_{obs}(x_j, t_k), \ x_j \in N_{obs} \subset N_h, \ \text{card}(N_{obs}) = m,$$
$$t_k \in T_{obs} \subset \{t_i : i = 1, \ldots, n_t\}, \ \text{card}(T_{obs}) = m_t$$

using the corresponding objective function

$$F_{obs} = \frac{1}{2} \sum_{x_j \in N_{obs}, t_k \in T_{obs}} | \ d_j^k - \bar{u}_j^k(p) \ |^2 \tag{10}$$

The identification problem can be then formulated as

$$\text{find } (p, \bar{u}) : \ p \in R^{2s}, \ \bar{u} \in (R^n)^{n_t} \tag{11}$$
$$F(p, \bar{u}) = \min F(q, \bar{v}) \quad s.t. \quad \bar{e}(q, \bar{v}) = 0, \ q \in R^s, \ \bar{v} \in (R^n)^{n_t}.$$

Above, $\bar{e}$ is the discretized PDE constraint from (9) and

$$F(q, \bar{v}) = F_{obs}(q, \bar{v}) + \beta F_{reg}(q),$$

where the construction of the regularization part is discussed e.g. in [5,7,4].

The minimization problem (11) can be again solved as an augmented Lagrangian problem or as an unconstrained minimization problem (with elimination of constrains). In the former approach, we define the Lagrangian functional

$$L(q, \bar{v}, \lambda) = F(q, \bar{v}) + \langle \lambda, \bar{e} \rangle + \frac{c}{2} \parallel e(q, \bar{v}) \parallel^2,$$

where $q \in R^{2s}, \bar{v} \in (R^n)^{n_t}$, $\lambda \in R^{n_t}$, $\langle , \rangle$ and $\parallel . \parallel$ denote the Eucledian inner product and the corresponding norm. Using this formulation, a modified Uzawa or another iterative methods can be used for the solution [4].

The direct minimization considers again the objective function

$$\widetilde{F}(p) = F(p, \bar{u}(p)), \tag{12}$$

which can be evaluated in the cost of the solution of the forward problem. The minimization can be done by different optimization problems, see the next section.

4   Unconstrained Minimization Techniques

The objective function $\widetilde{F}(p)$ from (7) and (12) has the nonlinear least squares structure given by (4) and (10), respectively. Generally, without regularization, we have

$$\widetilde{F}(p) = \frac{1}{2}(R(p))^T R(p)$$

where

$$p \in R^{m_p}, \ m_p = s \text{ or } m_p = 2s,$$
$$d \in R^{m_o}, \ m_o = m \text{ or } m_o = m \cdot m_t$$
$$\bar{u} \in R^{n_s}, \ n_s = n \text{ or } n_s = n \cdot n_t,$$

where m_p and m_o are different in stationary and non–stationary case. The residual $R(p)$ is defined as

$$R : R^{n_s} \to R^{m_o}, \ R(p) = S\bar{u}(p) - d$$

where $S : R^{n_s} \to R^{m_o}$ is the observation operator.

The optimization problem can be solved by direct methods like Nelder–Mead simplex method or genetic algorithms [2]. But here, we consider methods exploiting smoothness of the mapping $p \to R(p)$, i.e. the existence of Jacobian

$$J(p) = D_p R(p) = (J_{ij}(p)), \quad J_{ij}(p) = \frac{\partial R_i(p)}{\partial p_j} \, .$$

Having the Jacobian $J(p)$, the expressions

$$\text{Grad } \widetilde{F}(p) = J(p)^T R(p), \tag{13}$$

$$\text{Hess } \widetilde{F}(p) = J(p)^T J(p) + (D_p^2 R(p)) R(p) \sim J(p)^T J(p) = H_{GN}(p) \tag{14}$$

gives possibility of using iterative optimization methods as the steepest descent and conjugate gradient methods or Gauss–Newton and Levenberg-Marquardt type iterations. The neglecting of the second order term in (14) is mainly due to difficulty of its computation, but also this term is small if the residual goes to zero. A general solution algorithm can be written in the form

$$p^0 \text{ given }; \; p^{i+1} = p^i - \alpha_i z^i, \; i \geq 0, \tag{15}$$

where

$$z^i = g^i \text{ or } z^i = H_i^{-1} g^i, \quad g^i = J(p^i)^T R(p^i), \tag{16}$$

$$H_i = \nu_i I + J(p^i)^T J(p^i). \tag{17}$$

Note that $\alpha_i = 1$ or α_i are given by a line search/backtracking. Positive ν_i guarantee positive definiteness of H_i and regularize the problem but can destroy the local quadratic convergence. Therefore a strategy for choice of ν_i is needed, as e.g. $\nu_i = \min\{1; \; c \parallel J(p^i)^T R(p^i) \parallel\}$, where c is a positive constant, see e.g. [3].

The Jacobian $J(p)$ can be computed by differences

$$J_{ij}(p) = \frac{1}{\delta} [R_i(p + \delta e_j) - R_i(p)] \quad i = 1, \ldots, m_o, \; j = 1, \ldots m_p \tag{18}$$

at the cost of $m_p + 1$ evaluations of the residual $R(p)$, which involves $m_p + 1$ solutions of the forward problem. This evaluation can be sensitive to choice of $\delta > 0$ and costly, except having enough parallel computing power (the evaluation can be done fully in parallel). This approach is used especially for smaller m_p when $J(p)$ is computed from differences and systems with $J(p)^T J(p)$ are solved by a direct solution method.

Another way of computation of gradient (13) and implementation of matrix by vector computation with H_i (17) is as follows. In the stationary case, we have

$$R(p) = S A_h(p)^{-1} b_h - d \, ,$$

$$J(p) = -S A_h(p)^{-1} A'(p) A_h(p)^{-1} b_h \, ,$$

where $A'(p) = [D_{p_1} A(p), \ldots, D_{p_{m_p}} A(p)]$ contains the derivatives according to parameter components. These derivatives can be easily obtained as the elements of $A(p)$ are linearly depending on the coefficients (see local FE matrices). Now,

$$J(p)^T R(p)_i = \langle -SA_h(p)^{-1}(A'(p)e_i)A_h(p)^{-1}b_h, \, R(p)\rangle$$
$$= -\langle (A'(p)e_i)A_h(p)^{-1}b_h, \, A_h^T(p)^{-1}S^T(SA_h(p)^{-1}b_h - d)\rangle$$

and therefore $J(p)R(p)$ and similarly also the action of H_i (see [6]) can be computed at the cost of solution just one forward problem to get $A_h(p)^{-1}b_h$ and one forward adjoint problem to get $A^*(p)^{-1}w$. Moreover in our case $A^* = A$. This can be useful for implementing Newton–like method without change of H_i in each iteration or for the case, when the systems with H_i are solved iteratively, see [7]. The presented expressions can be also used for analysis of the iterative optimization methods.

This technique can be extended to non-stationary case by rewriting the timestepping process as the solution of the linear system $\mathcal{A}\bar{u} = \bar{b}$ of the following structure

$$\mathcal{A}(p) = \begin{bmatrix} I & & & \\ -M_h(p) & A_{ht}(p) & & \\ & & \ddots & \\ & & -M_h(p) & A_{ht}(p) \end{bmatrix}, \quad \bar{u} = \begin{bmatrix} \bar{u}^0 \\ \bar{u}^1 \\ \vdots \\ \bar{u}^{n_t} \end{bmatrix}, \quad \bar{b} = \begin{bmatrix} \bar{u}^0 \\ b_h^1 \\ \vdots \\ b_h^{m_t} \end{bmatrix} \tag{19}$$

where $A_{ht}(p) = M_h(p) + \Delta t A_h(p)$ and $\bar{u}^0$ given by the initial condition.

The expressions (19) allows to express the residual

$$R(p) = S\mathcal{A}(p)^{-1}b_h(p) - d$$

and derive its Jacobian in a similar way as above.

5 Model Problem and Numerical Results

The Äspö Pillar Stability Experiment (APSE) has been performed in natural geologic conditions with the aid of investigation of granite mass damage due to mechanical and thermal loading. The measured data are now used for validation of mathematical models for rock mass prediction within the DECOVALEX 2011 international project. APSE used electrical heaters to increase temperatures and induce stresses in a rock pillar between holes (Fig. 1) until its partial failure. To determine accurately the temperature changes, a heat flow model is formulated and monitored temperatures are used for model calibration, which should provide parameters taking into account water content and water flow. More details and another approach to the model calibration can be found in [2].

A FEM model of APSE, realized by GEM software [1], considers domain of $105 \times 125 \times 118$ m and tetrahedral 3D mesh with $99 \times 105 \times 59$ nodes. The grid is refined around the pillar, see Fig. 1. The heaters are producing heat which

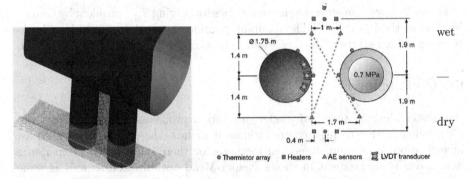

Fig. 1. The APSE model - detail of the FE grid around the pillar (GEM software [1]) and ground view on the pillar, holes, location of heaters and points of temperature measurement

varies in time. The model assumes original temperature $14.5°C$ on the outer boundaries, zero flux onto the tunnel and nonzero flux given the convection onto the holes. The initial condition is given again by the original temperature $14.5°C$.

Monitoring of the temperatures during two month heating phase of APSE is essential for calibration of the thermal model. There are 14 temperature monitoring positions and temperatures are measured in 12 time moments. Altogether 168 values of temperature measurement (vector d) are used for calibration, $m_o = 168$.

The material parameters are different conductivity k and heat capacity c. The simplest decomposition $\mathcal{T}_H^{(2)}$ decomposes the model domain just into two parts (dry and wet according to Fig.1 right). A further decomposition $\mathcal{T}_H^{(3)}$, used in [2], introduces also a damage material zone in the pillar. The choice of $\mathcal{T}_H$ can be refined with localization of calibration parameters just around the deposition holes. Also some other parameters like heat output due to water flow and convection constant in boundary values can be included in the model.

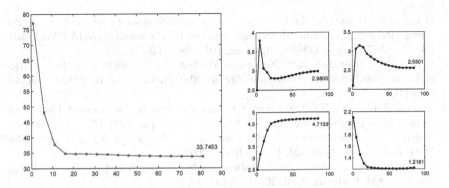

Fig. 2. Convergence of Gauss-Newton method with backtracking on the left and convergence of four parameters on the right (left top - k_{dry}, right top c_{dry}, left bottom k_{wet}, right bottom c_{wet})

Figure 2 shows convergence characteristics for solving $\mathcal{T}_H^{(2)}$ problem by Gauss–Newton method (15)–(18). The damping parameters are determined by a backtracking, which starts with $\alpha_i = 1$ and halves α_i if $F(p^{i+1}) < F(p^i)$ is violated.

6   Conclusions

Parameter identification and model calibration problems appear in many applications and due to high computational demands, there is a strong need for investigation of efficient numerical methods for their solution. Our experience now covers some gradient methods, Nelder-Mead optimization as well as some genetic algorithms [2], all applied only to problems with a small number of identification/calibration parameters.

Our future aims are in solving larger problems involving more parameters, to apply different techniques of parallel computing, using of semianalytic computation of gradient, testing of PDE constrained optimization approaches etc.

Acknowledgements. The work was conducted within the international DECOVALEX-2011 Project (DEvelopment of COdes and their VALidation against EXperiments). The authors are grateful to the Funding Organisations: CAS, JAEA, KAERI, POSIVA, SKB, TUL and WHU. The views expressed in the paper are however, those of the authors and are not necessarily those of the Funding Organisations. This work is also supported by the research plan AV0Z30860518 of the Academy of Sciences of the Czech Republic.

References

1. Blaheta, R., Jakl, O., Kohut, R., Starý, J.: GEM – A Platform for Advanced Mathematical Geosimulations. In: Wyrzykowski, R., Dongarra, J., Karczewski, K., Wasniewski, J. (eds.) PPAM 2009. LNCS, vol. 6067, pp. 266–275. Springer, Heidelberg (2010)
2. Blaheta, R., Kohut, R., Jakl, O.: Solution of Identification Problems in Computational Mechanics – Parallel Processing Aspects. In: Jónasson, K. (ed.) PARA 2010. LNCS, vol. 7134, pp. 399–409. Springer, Heidelberg (2012)
3. Dennis, J.E., Schnabel, R.B.: Numerical Methods for Unconstrained Optimization and Nonlinear Equations. Classics in Applied Mathematics, vol. 16. SIAM, Philadelphia (1996)
4. Nilssen, T.K., Tai, X.C.: Parameter estimation with the augmented Lagrangian method for a parabolic equation. J. Optim. Theory Appl. 124, 435–455 (2005)
5. Vogel, C.R.: Computational Methods for Inverse Problems. Frontiers in Applied Mathematics, vol. 23. SIAM, Philadelphia (2002)
6. Vogel, C.R.: Sparse matrix computations arising in distributed parameter identification. SIAM J. Matrix Anal. 20, 1027–1037 (1999)
7. Vogel, C.R., Wade, J.K.: Analysis of costate discretizations in parameter estimation for linear evolution equations. SIAM J. Control Optim. 33, 227–254 (1995)

Block-Preconditioners for Conforming and Non-conforming FEM Discretizations of the Cahn-Hilliard Equation[*]

P. Boyanova[1,2], M. Do-Quang[3], and M. Neytcheva[1]

[1] Uppsala University, Box 337, 751 05 Uppsala, Sweden
[2] Institute for Information and Communication Technology - BAS,
Acad. G. Bonchev Str., bl. 25A, 1113 Sofia, Bulgaria
[3] Department of Mechanics, Linné Flow Centre,
Royal Institute of Technology, SE-100 44 Stockholm, Sweden

Abstract. We consider preconditioned iterative solution methods to solve the algebraic systems of equations arising from finite element discretizations of multiphase flow problems, based on the phase-field model.

The aim is to solve coupled physics problems, where both diffusive and convective processes take place simultaneously in time and space. To model the above, a coupled system of partial differential equations has to be solved, consisting of the Cahn-Hilliard equation to describe the diffusive interface and the time-dependent Navier-Stokes equation, to follow the evolution of the convection field in time.

We focus on the construction and efficiency of preconditioned iterative solution methods for the linear systems, arising after conforming and non-conforming finite element discretizations of the Cahn-Hilliard equation in space and implicit discretization schemes in time. The nonlinearity of the phase-separation process is treated by Newton's method. The resulting matrices admit a two-by-two block structure, utilized by the preconditioning techniques, proposed in the current work. We discuss approximation estimates of the preconditioners and include numerical experiments to illustrate their behaviour.

1 Introduction

Multiphase flow problems are often modelled by the diffuse-interface phase-field model. Some phenomena, treated using this model, are coarsening kinetics of two-phase microstructures such as binary alloys, polymer systems, crystal growth, and spinodal decomposition; capillary phenomena, wetting (cf. [6]); modelling the dynamics of the biomass and the solvent components of a bacterial or other thin films (cf. [10]), see also [5] and the references therein.

[*] The work is supported by VR via the grant *Finite element preconditioners for algebraic problems as arising in modelling of multiphase microstructures*, 2009-2011. We also thank SNIC for resources provided through UPPMAX under Project p2009040, and the developers of the HSL MI-20 AMG package for kindly providing us with the AMG-`Matlab` interface.

I. Lirkov, S. Margenov, and J. Waśniewski (Eds.): LSSC 2011, LNCS 7116, pp. 549–557, 2012.

The major mathematical tool in the phase-field model is the Cahn-Hilliard equation, which in its original form is a fourth order nonlinear parabolic equation. In this paper we use an equivalent formulation, in terms of a coupled system of two second order partial differential equations (see, e.g. [7]), one of which is time-dependent, and the other one is nonlinear:

$$\eta - \Psi'(c) + \varepsilon^2 \Delta c = 0, \quad (\mathbf{x}, t) \in \Omega_T \equiv \Omega \times (0, T), \, \Omega \subset \mathbb{R}^2$$
$$\omega \Delta \eta - \frac{\partial c}{\partial t} - (\mathbf{u} \cdot \nabla)c = 0, \quad (\mathbf{x}, t) \in \Omega_T \tag{1}$$
$$\mathbf{n} \cdot \nabla c = -(\cos \alpha / \varepsilon^2) g'(c), \quad \mathbf{n} \cdot \nabla \eta = 0, \quad \mathbf{x} \in \partial\Omega; \quad c(x, 0) = c_0(x).$$

The unknown function $c(\mathbf{x}, t)$, also referred to as *phase-field*, represents the concentration of the fluids and serves to model the interface between them. It attains a distinct constant value in each bulk phase and rapidly, but smoothly, changes in the interface region. For a binary fluid, the usual assumption is that $c(\mathbf{x}, t)$ takes values between -1 and 1. The function $\Psi(c)$ is a double well potential with two stable minima at ± 1. Here we consider $\Psi(c) = \frac{1}{4}(c^2 - 1)^2$. The unknown function η is the so-called *chemical potential* and plays an auxiliary role, which we do not discuss further here. Within the phase-field theory the boundary condition that accounts for the free energy distribution between the different phases, sets the wetting boundary condition (α) for the interface. Here, $g(c)$ is a normalized function varying smoothly from 0 to 1. It is used to localize the surface energy of each phase on the energy system. In our simulation $g(c) = 0.5 - 0.75c + 0.25c^3$. For more details, see e.g. [12]. The problem parameter ε determines the thickness of the interface between the two phases. In order to resolve the interface in numerical simulations, the characteristic mesh size has to be chosen as $h = \varepsilon / r$, for a proper $r > 1$. The velocity field $\mathbf{u}$ is non-zero whenever the interface develops due to both diffusion and convection. To model such processes C-H is coupled to the Navier-Stokes equation. Usually, operator splitting is used for the coupled system, where $\mathbf{u}$ is assumed to be already computed when (1) needs to be solved. In non-convective models, $\omega = 1$. In the nondimensionalized formulation of the convective C-H equation, $\omega = 1/Pe$, where Pe is the Peclet number (e.g. [12]).

Section 2 outlines the considered discretizations and the treatment of the non-linearity. We propose some block-based preconditioning techniques for the resulting linear systems in Section 3. Section 4 includes numerical illustrations and convergence results. Concluding remarks are found in Section 5.

2 Discretization in Space and Time

Consider the weak formulation of (1) – find $c, \eta \in H^1(\Omega)$ such that for all $\nu \in H^1(\Omega)$,

$$(\eta, \nu) - (\Psi'(c), \nu) - \varepsilon^2 (\nabla c, \nabla \nu) = 0$$
$$\omega(\nabla \eta, \nabla \nu) + \left(\frac{\partial c}{\partial t}, \nu \right) + ((\mathbf{u} \cdot \nabla)c, \nu) = 0 \tag{2}$$

We use the finite element method (FEM) to discretize (2) in space, and assume that the same finite element space V_h is used for both variables. In this work we consider two choices of V_h, namely, Courant linear conforming FEs and Crouzeix-Raviart (CR) linear non-conforming FEs, defined as follows. For a triangulation $\mathcal{T}$ of the domain Ω, the nodes of the conforming FEM are the vertices of the triangles (elements) $e \in \mathcal{T}$ and $V_h = V_h^C$,

$$V_h^C = \{v_h(x) \in C(\Omega) : v_h(x)|_e \in \mathcal{P}_1(e), \forall e \in \mathcal{T}\}.$$

In the case of Crouzeix-Raviart elements the nodes are the midpoints of the edges of the triangles. The approximate solution is only guaranteed to be continuous in those points. We denote the finite element space $V_h = V_h^{CR}$,

$$V_h^{CR} = \{v_h(x) \in L^2(\Omega) : v_h(x)|_e \in \mathcal{P}_1(e), \forall e \in \mathcal{T}; v_h(x) \text{ - continuous at } \forall m_{i,e}\},$$

where $m_{i,e}$ $(i = 1, 2, 3)$ is the midpoint of the i-th edge of element e.

Note that V_h^C is a standard choice for the discretization of various PDE problems. The non-conforming V_h^{CR} space is of interest in computational fluid dynamics because of its local mass-conservation properties (see, e.g. [4]). On one hand, the number of degrees of freedom for the CR FE is larger than that for the Courant FE – about three times in 2D and about ten times in 3D. On the other hand, the sparsity structures of the matrices for non-conforming FE can be beneficial, including in parallel implementations. For example, one important property, used here, is that the CR mass matrix is diagonal.

To discretize (2) in time, we use the θ-method. Consider a sequence of time steps, $\{t_k\}, k = 0, 1, \ldots$, where $t_0 = 0, t_k = t_{k-1} + \Delta t_k$, and denote by $\mathbf{c}^{(k)} = \{c(\mathbf{x}_i, t_k)\}_{i=1}^N, \boldsymbol{\eta}^{(k)} = \{\eta(\mathbf{x}_i, t_k)\}_{i=1}^N$, where $\{\mathbf{x}_j\}_{j=1}^N$ are the finite element nodes. Then the fully discretized C-H system to be solved at each time step $k = 1, 2, \ldots$ has the following form:

$$\begin{aligned} M\boldsymbol{\eta}^{(k)} - f(\mathbf{c}^{(k)}) - \varepsilon^2 K \mathbf{c}^{(k)} = 0 \\ \theta\omega\Delta t_k K \boldsymbol{\eta}^{(k)} + M\mathbf{c}^{(k)} + \theta\Delta t_k W\mathbf{c}^{(k)} + \\ (1-\theta)(\omega\Delta t_k K\boldsymbol{\eta}^{(k-1)} + \Delta t_k W\mathbf{c}^{(k-1)}) - M\mathbf{c}^{(k-1)} = 0 \end{aligned} \quad (3)$$

where M, K, W are correspondingly the mass, stiffness matrices, and the matrix resulting from discretizing the convective term. The elements of the vector $f(\mathbf{c}^{(k)}) = \{f_i(\mathbf{c}^{(k)}\}_{i=1}^N$ are defined as $f_i(\mathbf{c}^{(k)}) = (\Psi'(\sum_{j=1}^N c(\mathbf{x}_j, t_k)\chi_j), \chi_i)$, for $\{\chi_j\}_{j=1}^N$ – the finite element nodal basis functions of V_h. The method parameter θ is in $(0, 1]$; for $\theta = 1$ we obtain the fully implicit backwards Euler scheme.

Due to the term $f(\mathbf{c}^{(k)})$, the systems to be solved at each time step are nonlinear. Here, we consider the classical Newton's method to obtain approximate solution of (3). The procedure is as follows. Using the solution at the previous time step as an initial approximation $\mathbf{y}^{k,0} = ((\mathbf{c}^{(k-1)})^T(\boldsymbol{\eta}^{(k-1)})^T)^T$, the consecutive Newton iterates are found as $\mathbf{y}^{k,s+1} = \mathbf{y}^{k,s} + \Delta\mathbf{y}^{k,s}$, where the updates $\Delta\mathbf{y}^{k,s}$ are computed by solving linear systems of the form

$$\begin{bmatrix} \theta M & -\theta J^{k,s} - \theta\varepsilon^2 K \\ \theta\omega\Delta t_k K & M + \theta\Delta t_k W \end{bmatrix} \Delta\mathbf{y}^{k,s} = A^{k,s}\mathbf{y}^{k,s} = \mathbf{b}(\mathbf{y}^{k,s}). \quad (4)$$

Here $J^{k,s}$ is the Jacobian of the nonlinear term $f(\mathbf{c})$, computed for the concentration at the previous Newton iterate $\mathbf{c}^{k,s}$, and the right-hand side $\mathbf{b}(\mathbf{y}^{k,s})$ can be straight-forwardly derived from (3). The block $J^{k,s}$ is assembled in the usual FEM manner from element matrices $J_e, e \in \mathcal{T}$ that have the form $J_e = j_e M_e$, where M_e are the corresponding element mass matrices and j_e depend on $\mathbf{c}^{k,s}$. Since $\mathbf{c}^{k,s} \in [-1,1]$, it can be shown that $-1 \le j_e \le 2$.

3   Preconditioning of the Linear Systems

The main computational kernel in the above numerical procedure is the solution of the linear systems (4) at each Newton step. To simplify the notations, we omit the sub- and superscripts and use the notations $\Delta t, J, A$ instead of $\Delta t_k, J^{k,s}, A^{k,s}$. Here we choose to utilize the natural two-by-two block form of the system matrix in (4), that results from the discretization of the system of PDEs (1). We consider preconditioners of the following block-factorized form

$$P = \begin{bmatrix} \theta\widehat{M} & 0 \\ \theta\omega\Delta t K & \widehat{S} \end{bmatrix} \begin{bmatrix} I & \widehat{M}^{-1}(-J - \varepsilon^2 K) \\ 0 & I \end{bmatrix}, \tag{5}$$

where $\widehat{M}, \widehat{S}$ are some approximations of M and the exact Schur complement $S = M + \theta\Delta t W + \theta\omega\Delta t K M^{-1}(J + \varepsilon^2 K)$, respectively.

It is important, that the pivot block of A is a mass matrix. In the case of CR FEM M is diagonal, and a solution of the original system can be done via exact block elimination of η and solution of the resulting Schur system for the concentration. Moreover, since M^{-1} is diagonal, S can be formed explicitly on low cost. When conforming FEM are used, the mass matrix M is no longer diagonal and its inverse is, in general, dense. Therefore, for approximate block-factorization preconditioners for A, both M and S have to be approximated.

3.1   SPD Schur Approximation

Clearly, A is non-symmetric, and so is S. However, we show here that for small enough time step Δt, the symmetric positive definite (SPD) matrix

$$\widetilde{S} = M + \theta\varepsilon^2\omega\Delta t K M^{-1}K \tag{6}$$

is a good approximation of S. We present a derivation of this result, using techniques as in [3] and [5]. Consider the error matrix

$$\widetilde{S}^{-1}(S - \widetilde{S}) = \theta\Delta t\widetilde{S}^{-1}W + \theta\omega\Delta t\widetilde{S}^{-1}KM^{-1}J. \tag{7}$$

Note first that by the Sherman-Morrison-Woodbury formula $\widetilde{S}^{-1} = M^{-1} - \theta\varepsilon^2\omega\Delta t M^{-1}K\widetilde{S}^{-1}KM^{-1}$, and since $M, K, \widetilde{S}$ are all positive (semi)definite, there holds that $\|\widetilde{S}^{-1}\| \le \|M^{-1}\|$. Then,

$$\|\theta\Delta t\widetilde{S}^{-1}W\| \le \theta\Delta t\|M^{-1}W\| = \theta O(\Delta t h^{-1}),$$

and, since $\theta \in (0,1]$, this term decays for Δt small enough with respect to h. For the other term in (7), we observe that due to the properties of J,

$\|M^{-1}J\| = O(1)$, thus, we only need to analyze further $\widetilde{S}^{-1}K$. Using the equivalent transformation $\widetilde{S}^{-1}KM^{-\frac{1}{2}}M^{\frac{1}{2}}$, we see, that the latter is similar to $M^{\frac{1}{2}}\widetilde{S}^{-1}KM^{-\frac{1}{2}} = (M^{-\frac{1}{2}}\widetilde{S}M^{-\frac{1}{2}})^{-1}M^{-\frac{1}{2}}KM^{-\frac{1}{2}} = (I + \theta\varepsilon^2\omega\Delta t\widetilde{K}^2)^{-1}\widetilde{K}$, where $\widetilde{K} = M^{-\frac{1}{2}}KM^{-\frac{1}{2}}$. We use the fact that $|a|/(1 + \beta^2a^2) \le 1/(2\beta)$ to obtain the estimate

$$\|\theta\omega\Delta t\widetilde{S}^{-1}KM^{-1}J\| \le \frac{\theta\omega\Delta t}{2\sqrt{\theta\varepsilon^2\omega\Delta t}}\|M^{-1}J\|.$$

We utilize the fact that $\varepsilon = rh$, $\theta \in (0, 1]$, and for $\omega = 1$ we see that the norm of the second term in (7) is of the order $O(\sqrt{\Delta t}/h)$. For problems with $\omega = 1/Pe < 1$, however, in order to get good resolution, the mesh size should be chosen as $h = \xi/Pe$, $\xi-$ small. Then we obtain that the norm of both terms in (7) is of order $O(\Delta t h^{-1})$, they decay for time steps Δt small compared to h and the error matrix goes to zero, which is confirmed by the numerical experiments.

The preconditioner $\widetilde{S}$ is an SPD matrix that, for a fixed mesh, does not change from one time step to the other. In the case of non-conforming FEM it can be computed explicitly. It can be further factorized via some exact or approximate method and used during whole time period. We note that, in general, $\widetilde{S}$ is not an M-matrix and methods that rely on such a property, such as, for example, incomplete Cholesky factorization, might fail in some (or all) occasions. However, preconditioners for general SPD matrices are applicable (e.g. [9]).

3.2 Diagonal Approximation of the Inverse Mass Matrix in $\widetilde{S}$

As already mentioned, in the case of conforming FEM the exact inverse of the mass matrix M is dense and is not cheap to form explicitly. Here, when constructing $\widetilde{S}$, we propose to approximate M^{-1} by D^{-1}, where $D = diag(M)$. It is known that the eigenvalues λ_D of $D^{-1}M$ are bounded independently of the mesh step h, see e.g. [13], and in two dimensions $1/2 \le \lambda_D \le 2$. We show that this estimate translates directly to the corresponding approximation $\widetilde{S}_D$ of $\widetilde{S}$.

To show the latter, consider an approximation of $\widetilde{S}$ in the form

$$\widetilde{S}_D = M + \theta\varepsilon^2\omega\Delta t K D^{-1}K. \tag{8}$$

The generalized eigenvalue problem $\lambda\widetilde{S}_D\mathbf{x} = \widetilde{S}\mathbf{x}$ can be transformed as

$$\lambda M^{\frac{1}{2}}(I + \theta\varepsilon^2\omega\Delta t\widetilde{K}\widetilde{D}\widetilde{K})M^{\frac{1}{2}}\mathbf{x} = M^{\frac{1}{2}}(I + \theta\varepsilon^2\omega\Delta t\widetilde{K}\widetilde{K})M^{\frac{1}{2}}\mathbf{x}, \tag{9}$$

where $\widetilde{D} = M^{\frac{1}{2}}D^{-1}M^{\frac{1}{2}}$. Denote by $\mathbf{y} = M^{\frac{1}{2}}\mathbf{x}/\|M^{\frac{1}{2}}\mathbf{x}\|$ and then from (9) it follows, $\lambda(1 + \theta\varepsilon^2\omega\Delta t k^2 d) = 1 + \theta\varepsilon^2\omega\Delta t k^2$, where $k = \mathbf{y}^T\widetilde{K}\mathbf{y}/(\mathbf{y}^T\mathbf{y})$, $d = \mathbf{y}^T\widetilde{D}\mathbf{y}/(\mathbf{y}^T\mathbf{y})$. Provided that $1 + \theta\varepsilon^2\omega\Delta t k^2 d \ne 0$, it holds that $\lambda = (1 + \theta\varepsilon^2\omega\Delta t k^2)/(1 + \theta\varepsilon^2\omega\Delta t k^2 d)$, and since the generalized eigenvalue problem $\mu M^{-1}\mathbf{v} = D^{-1}\mathbf{v}$ is equivalent to $\mu D\mathbf{v} = M\mathbf{v}$, in the 2D case $1/2 \le d \le 2$, and thus, $1/2 \le \lambda \le 2$.

3.3 Factorized Approximation of $\widetilde{S}$

The matrix $\widetilde{S}$ can be presented in the form $\widetilde{S} = \widetilde{S}_F - 2\varepsilon\sqrt{\theta\omega\Delta t}K$, where

$$\widetilde{S}_F = (M + \varepsilon\sqrt{\theta\omega\Delta t}K)M^{-1}(M + \varepsilon\sqrt{\theta\omega\Delta t}K). \tag{10}$$

It is shown in [11] that the eigenvalues of the generalized eigenvalue problem $\lambda\widetilde{S}_F\mathbf{x} = \widetilde{S}\mathbf{x}$ belong to the interval $1/2 \leq \lambda \leq 1$, thus (10) is a high-quality approximation of (6). The latter can be derived as well using results from [2], see also [5]. An advantage of the approximation $\widetilde{S}_F$ is that optimal order inner solvers can be straight-forwardly applied for systems with the factors $M + \varepsilon\sqrt{\theta\omega\Delta t}K$, such as multilevel or multigrid methods.

4 Numerical Experiments

We consider problem (1) with $\omega = 1$, $\epsilon = 0.0625$, $\alpha = \pi/2$, $\mathbf{u} = (0,0)$ in $\Omega = [-1/2, 1/2] \times [0,1]$, and random initial condition. There is no convection due to fluid flow, and the process of phase separation and coarsening takes place only due to diffusion. Note, that for this problem the estimate for the error matrix to approximate the exact Schur complement by $\widetilde{S}$ is more pessimistic for larger time steps. The numerical results below show that the preconditioner behaves well also for $\Delta t = O(h)$.

For the case of conforming FEM we apply Generalized Conjugate Gradient - Minimal Residual (GCG-MR) method, see e.g. [1], with preconditioner (5) to the non-symmetric systems with A. For CR FEM, GCG-MR is applied only when solving $\widetilde{S}$, preconditioned by $\widehat{S}$. We use a regular triangular mesh with a characteristic mesh size h. We apply Euler scheme in time. The stopping criterion for the Newton method at each time step is $||\Delta\mathbf{y}^{k,s}|| < 10^{-6}$. As an initial approximation in GCG-MR we use the solution at the previous time step, and terminate the iterative process when the norm of the residual is reduced by a factor 10^{-6} or the norm itself is smaller than 10^{-12}. Each table cell contains data in the format average number of Newton iterations per time step / average number of GCG-MR iterations per Newton iteration.

In Table 1 we present results from approximating the Schur complement by $\widetilde{S}_D$ as in (8). For CR FEs, $\widetilde{S}_D = \widetilde{S}$. For conforming FEM we use the diagonal of the mass matrix also for the pivot block in (5). We do the same for the experiments in Table 2 as well, where we also substitute M by D in the Schur approximation $\widetilde{S}_F$.

Table 1. $\widehat{S} = M + \varepsilon^2\omega\Delta t K D^{-1} K, \widehat{M} = D$, direct solves with $\widehat{S}$

	conforming FEM				non-conforming FEM		
Size	$\Delta t = h/4$	$\Delta t = h/10$	$\Delta t = h^2$	Size	$\Delta t = h/4$	$\Delta t = h/10$	$\Delta t = h^2$
8450	3 / 13	3 / 10	3 / 9	24832	3 / 6	3 / 5	3 / 4
33282	3 / 10	3 / 9	3 / 8	98816	3 / 5	3 / 4	3 / 3
132098	3 / 8	3 / 8	3 / 8	394240	3 / 4	3 / 4	3 / 3

Table 2. $\widehat{S} = (D + \varepsilon\sqrt{\omega\Delta t}K)D^{-1}(D + \varepsilon\sqrt{\omega\Delta t}K), \widehat{M} = D$, direct solves with $\widehat{S}$

conforming FEM				non-conforming FEM			
Size	$\Delta t = h/4$	$\Delta t = h/10$	$\Delta t = h^2$	Size	$\Delta t = h/4$	$\Delta t = h/10$	$\Delta t = h^2$
8450	3 / 17	3 / 13	3 / 12	24832	3 / 9	3 / 7	3 / 7
33282	3 / 13	3 / 11	3 / 11	98816	3 / 7	3 / 7	3 / 7
132098	3 / 11	3 / 10	3 / 12	394240	3 / 6	3 / 6	3 / 7

Table 3. $\widehat{S} = (D + \varepsilon\sqrt{\omega\Delta t}K)D^{-1}(D + \varepsilon\sqrt{\omega\Delta t}K), \widehat{M} = D$, inner solver using AMG

conforming FEM				non-conforming FEM			
Size	$\Delta t = h/4$	$\Delta t = h/10$	$\Delta t = h^2$	Size	$\Delta t = h/4$	$\Delta t = h/10$	$\Delta t = h^2$
8450	3 / 17 / 3	3 / 13 / 3	3 / 12 / 3	24832	3 / 10 / 3	3 / 8 / 3	3 / 7 / 3
33282	3 / 13 / 3	3 / 11 / 3	3 / 11 / 3	98816	3 / 8 / 3	3 / 7 / 3	3 / 7 / 3
132098	3 / 11 / 3	3 / 11 / 3	3 / 12 / 3	394240	3 / 7 / 3	3 / 7 / 3	3 / 7 / 3

In Tables 1 and 2, systems with $\widehat{S}$ are solved directly. In Table 3 we apply inner PCG iterations with Algebraic MultiGrid (AMG) preconditioner to the factors $M + \varepsilon\sqrt{\omega\Delta t}K$, with stopping tolerance 10^{-3}. The average number of the performed inner PCG iterations is indicated as a third term in the table cells.

We observe that for small enough time steps $\Delta t = O(h)$, the performance of the proposed preconditioners is stable with respect to both discretization parameters. Due to the additional approximation of M in the case of conforming FEM the GCG-MR iterations are almost twice more than in the non-conforming case. In both cases the convergence is (almost) not affected by inexact solutions with the Schur preconditioner $\widetilde{S}_F$.

Experiments, not included here, show that generally for smaller values of the problem parameter ε the linear solver needs more iterations to converge. Still, the number of iterations is always stable with respect to h and does not grow.

We also include some results for the case of a problem with nonzero convection. In Table 4 we present the iteration counts when solving a problem with velocity field $\mathbf{u} = (1, 1)$ and all other parameters the same as above. A comparison between the results in Tables 4 and 3 shows that the presence of the convective term does not have a significant effect on the convergence of the linear solver. We note that, in order for the nonlinear method to converge in the presence

Table 4. Problem with nonzero velocity field, $\mathbf{u} = (1, 1)$; $\widehat{S} = (D + \varepsilon\sqrt{\omega\Delta t}K)D^{-1}(D + \varepsilon\sqrt{\omega\Delta t}K), \widehat{M} = D$, inner solver using AMG

conforming FEM			
Size	$\Delta t = h/4$	$\Delta t = h/10$	$\Delta t = h^2$
8450	3 / 17 / 3	3 / 14 / 3	3 / 12 / 3
33282	3 / 13 / 3	3 / 12 / 3	3 / 12 / 3
132098	3 / 11 / 3	3 / 12 / 3	3 / 12 / 3

of stronger convection, the space and time discretization parameters h and Δt have to be small enough. Then, as expected, the proposed preconditioner also performs well.

5 Concluding Remarks

We propose block-factorized preconditioners for the C-H equation, where the Schur complement is approximated by a matrix $\widetilde{S}$ that does not change with time for fixed meshes. In the case of non-conforming FEM $\widetilde{S}$ is sparse and can be formed explicitly. For conforming FEM this can be achieved by additionally substituting the mass matrix with its diagonal in $\widetilde{S}$. Furthermore, an optimal preconditioner for $\widetilde{S}$ is available, that allows for multilevel or multigrid solvers to be applied. The presented experimental results confirm the numerical efficiency and the robustness of the proposed preconditioning techniques. The computational efficiency of the method in a slightly different setting of the preconditioner but using the same matrix ingredients is tested on large scale problems in [5]. There, some comparisons with other methods (MUMPS direct solver and an ILU preconditioner used in [14]) are also included. As expected, for large enough problems, over one million degrees of freedom in 2D, the preconditioned iterative solver is superior to a highly optimized direct sparse solver, both in terms of overall time and memory usage. The ILU-type preconditioners show lack of robustness for the important range of problem parameter values.

References

1. Axelsson, O.: Iterative Solution Methods. Cambridge University Press (1994)
2. Axelsson, O., Kucherov, A.: Real valued iterative methods for solving complex symmetric linear systems. Numer. Lin. Alg. Appl. 7, 197–218 (2000)
3. Axelsson, O., Neytcheva, M.: Operator splittings for solving nonlinear, coupled multiphysics problems with an application to the numerical solution of an interface problem, TR 2011-009, Institute for Information Technology, Uppsala Univ. (2011)
4. Bejanov, B., Guermond, J.-L., Minev, P.D.: A locally div-free projection scheme for incompressible flows based on non-conforming elements. Int. J. Numer. Meth. Fluids 49, 549–568 (2005)
5. Boyanova, P., Do-Quang, M., Neytcheva, M.: Solution methods for the Cahn-Hilliard equation discretized by conforming and non-conforming finite elements, TR 2011-004, Institute for Information Technology, Uppsala University (2011)
6. Do-Quang, M., Amberg, G.: The splash of a ball hitting a liquid surface: Numerical simulation of the influence of wetting. Journal Physic of Fluid (2008)
7. Elliott, C.M., French, D.A., Milner, F.A.: A second order splitting method for the Cahn-Hilliard Equation. Numer. Math. 54, 575–590 (1989)
8. Garcke, H.: On Cahn-Hilliard systems with elasticity. Proc. Roy. Soc. Edinburgh 133A, 307–331 (2003)
9. Kaporin, I.E.: High Quality Preconditioning of a General Symmetric Positive Definite Matrix Based on its $U^T U + U^T R + R^T U$-decomposition. Numer. Linear Algebra Appl. 5, 483–509 (1998)

10. Novick-Cohen, A.: The Cahn-Hilliard Equation: From Backwards Diffusion to Surface Diffusion, Draft. in preparation under contract with Cambridge University Press, http://www.math.technion.ac.il/~amync/
11. Pearson, J., Wathen, A.J.: A New Approximation of the Schur Complement in Preconditioners for PDE Constrained Optimization. The Mathematical institute, University of Oxford. Technical report, November 24 (2010), http://eprints.maths.ox.ac.uk/1021/
12. Villanueva, W., Amberg, G.: Some generic capillary-driven flows. International Journal of Multiphase Flow 32, 1072–1086 (2006)
13. Wathen, A.J.: Realistic eigenvalue bounds for the Galerkin mass matrix. IMA Journal of Numerical Analysis 7, 449–457 (1987)
14. Yue, P., Zhou, C., Feng, J.J., Ollivier-Gooch, C.F., Hu, H.H.: Phase-field simulations of interfacial dynamics in viscoelastic fluids using finite elements with adaptive meshing. Journal of Computational Physics 219, 47–67 (2006)

Comparison of Two Numerical Methods for Computation of American Type of the Floating Strike Asian Option

J.D. Kandilarov[1] and D. Ševčovič[2]

[1] Department of Mathematics, University of Rousse,
Studentska Str. 8, Rousse 7017, Bulgaria
ukandilarov@uni-ruse.bg
[2] Department of Applied Mathematics and Statistics, Comenius University
842 48 Bratislava, Slovak Republic
sevcovic@fmph.uniba.sk

Abstract. We present a numerical approach for solving the free boundary problem for the Black-Scholes equation for pricing American style of floating strike Asian options. A fixed domain transformation of the free boundary problem into a parabolic equation defined on a fixed spatial domain is performed. As a result a nonlinear time-dependent term is involved in the resulting equation. Two new numerical algorithms are proposed. In the first algorithm a predictor-corrector scheme is used. The second one is based on the Newton method. Computational experiments, confirming the accuracy of the algorithms, are presented and discussed.

1 Introduction

In this paper we consider the problem of pricing American style Asian options, analyzed in [1] (see also [11]). Asian options belong to the group of the so-called path-dependent options. Their pay-off diagrams depend on the spot value of the underlying asset during the whole or some part(s) of the life span of the option. Among path-dependent options, Asian option depend on the arithmetic or geometric average of spot prices of the underlying asset. During the last decade, the problem of solving the American option problem numerically has been subject for intensive research [1,6,9,10,13] (see also [11] for an overview). A comprehensive introduction to this topic can be found in [6]. Comparison of various analytical and numerical approximation methods of calculation of the early exercise boundary a position of the American put option paying zero dividends is given in [7]. An improvement of Han and Wu's algorithm [4] is described in [14]. Our goal is to propose and investigate two front-fixing numerical algorithms for solving free boundary value problems. The front-fixing method has been successfully applied to a wide range of applied problems arising from physics and engineering, cf. [3,8] and references therein. The basic idea is to remove the moving boundary by a transformation of the involved variables. Transformation techniques were used in the analysis and numerical computation of the early exercise boundary

I. Lirkov, S. Margenov, and J. Waśniewski (Eds.): LSSC 2011, LNCS 7116, pp. 558–565, 2012.

in the context of American style of vanilla options [10] as well as Asian floating strike options [1,11,12]. In comparison to the existing computational method [1] we do not replace the algebraic constraint by its equivalent integral form (see [1,12] for details) which is computationally more involved. In this paper we solve the corresponding parabolic equation with an algebraic constraint directly as it was proposed in [11]. The approach presented in [11] however suffered from the necessity of taking very small time discretization steps. Here we overcome this difficulty by proposing two new numerical approximation algorithms (see Section 4). They are based on the novel technique proposed in [5]. We extend this approach for American style of Asian options. In Section 5, a numerical example illustrating the capability of our algorithms are discussed.

2   The Free Boundary Problem

Following the classical Black-Scholes theory, the problem of pricing Asian options with arithmetically averaged strike price by means of a solution to a parabolic PDE with a free boundary $S_f(t, A)$ is analyzed in [1]:

$$\frac{\partial V}{\partial t} + \frac{\sigma^2}{2} S^2 \frac{\partial^2 V}{\partial S^2} + (r - q)S \frac{\partial V}{\partial S} + \frac{S - A}{t} \frac{\partial V}{\partial A} - rV = 0, \tag{1}$$

$0 < t < T, \ \ 0 < S < S_f(t, A)$, satisfying the boundary conditions

$$V(t, 0, A) = 0, \quad \text{for any } A > 0 \text{ and } 0 < t < T, \tag{2}$$

$$\frac{\partial V}{\partial S}(t, S_f(t, A), A) = 1, \ \ V(t, S_f(t, A), A) = S_f(t, A) - A, \tag{3}$$

and the terminal condition (terminal pay-off condition) at the maturity time T:

$$V(T, S, A) = \max(S - A, 0), \quad S, A > 0. \tag{4}$$

Here $S > 0$ is the stock price, $A > 0$ is the averaged strike price, $r > 0$ is the risk-free interest rate, $q > 0$ is a continuous dividend rate and $\sigma > 0$ is the volatility of the underlying asset returns. The arithmetically averaged price $A = A_t$ calculated from the price path $\{S_u, u \in [0, T]\}$ at the time t is defined as $A_t = \frac{1}{t} \int_0^t S_u \, du$. For floating strike Asian options, it is well known (see e.g. [6,2,1]) that one can perform a dimension reduction by introducing a new time variable $\tau = T - t$ and a similarity variable x defined as: $x = A/S$, $W(x, \tau) = V(t, S, A)/A$. The spatial domain for the reduced equation is given by $1/\rho(\tau) < x < \infty$, $\tau \in (0, T)$, $\rho(\tau) = S_f(T - \tau, A)/A$. Following [10,13,1], we can apply the Landau fixed domain transformation for the free boundary problem by introducing a new state variable ξ and an auxiliary function $\Pi(\xi, \tau) = W(x, \tau) + x \frac{\partial W}{\partial x}(x, \tau)$, representing a synthetic portfolio. Here $\xi = \ln(\rho(\tau)x)$. In [1,10,13] it is shown that under suitable regularity assumptions on the input data the free boundary problem (1)–(4) can be transformed into the initial boundary value problem for parabolic PDE:

$$\frac{\partial \Pi}{\partial \tau} + \alpha(\xi,\tau)\frac{\partial \Pi}{\partial \xi} - \frac{\sigma^2}{2}\frac{\partial^2 \Pi}{\partial \xi^2} + \beta(\xi,\tau)\Pi = 0, \quad \xi > 0, \quad \tau \in (0,T), \quad (5)$$

$$\Pi(0,\tau) = -1, \Pi(\infty,\tau) = 0, \quad \Pi(\xi,0) = \begin{cases} -1, & \text{for } \xi < \ln\rho(0), \\ 0, & \text{otherwise.} \end{cases} \quad (6)$$

The coefficients α and β are defined as follows:

$$\alpha(\xi,\tau) = \frac{\dot{\rho}(\tau)}{\rho(\tau)} + r - q - \frac{\sigma^2}{2} - \frac{\rho(\tau)e^{-\xi} - 1}{T - \tau}, \quad \beta(\xi,\tau) = r + \frac{1}{T-\tau}. \quad (7)$$

According to [1] the free boundary function $\rho(\tau)$ and the solution Π should fulfill the constraint:

$$\rho(\tau) = \frac{1 + r(T-\tau) + \frac{\sigma^2}{2}(T-\tau)\frac{\partial \Pi}{\partial \xi}(0,\tau)}{1 + q(T-\tau)}, \quad \rho(0) = \max\left(\frac{1+rT}{1+qT}, 1\right). \quad (8)$$

As for derivation of the initial free boundary position $\rho(0)$ in (8) we refer to [1] or [6,2]. A solution Π to the problem (5)-(8) is continuous for $t > 0$. The discontinuity appears only at the point $P^\star = (\ln(\rho(0)), 0)$. The derivatives of the solution exist and are sufficiently smooth in $[0,L] \times [0,T)$, outside of a small neighbourhood of $P^\star$. Another important fact to emphasize is that for times $t \to 0^+$ (i.e. when $\tau \to T$) the coefficients α, β become unbounded.

3   Finite Difference Schemes

In order to solve the problem (5)-(8) numerically, we introduce L which is sufficiently large upper limit of values of the ξ variable (a safe choice is to take L is equal to five times $\ln(\rho(0))$), where we prescribe $\Pi(L,\tau) = 0$. Next, for given positive integers N and M we define the uniform meshes: $\overline{\omega}_h = \{0\} \cup \{L\} \cup \omega_h$, $\omega_h = \{\xi_i = ih, \ i = 1, \ldots, (N-1), \ h = L/N\}$ and $\overline{\omega}_k = \{0\} \cup \{T\} \cup \omega_k$, $\omega_k = \{\tau_j = jk, \ j = 1, \ldots, (M-1), \ k = T/M\}$. Our goal is to define a finite difference method which is suitable for computing $y_i^j \approx \Pi(\xi_i, \tau_j)$ for $(\xi_i, \tau_j) \in \omega_h \times \omega_k$ and associated front position $z^j \approx \rho(\tau_j)$ for $\tau_j \in \omega_k$. The implicit difference scheme has the following form:

$$\frac{y_i^{j+1} - y_i^j}{k} + \alpha_i^{j+1}\frac{y_{i+1}^{j+1} - y_{i-1}^{j+1}}{2h} - \frac{\sigma^2}{2}\frac{y_{i+1}^{j+1} - 2y_i^{j+1} + y_{i-1}^{j+1}}{h^2} + \beta^{j+1}y_i^{j+1} = 0, (9)$$

$$y_0^{j+1} = -1, \quad y_N^{j+1} = 0; \quad y_i^0 = \begin{cases} -1, & \text{for } \xi_i \le \ln(\rho(0)), \\ 0, & \text{otherwise;} \end{cases} \quad (10)$$

$$\alpha_i^{j+1} = \frac{z^{j+1} - z^j}{kz^{j+1}} + r - q - \frac{\sigma^2}{2} - \frac{z^{j+1}\exp(-\xi_i) - 1}{T - \tau_{j+1}}, \quad \beta^{j+1} = r + \frac{1}{T - \tau_{j+1}}, (11)$$

$$z^{j+1} - \frac{1 + r(T - \tau_{j+1})}{1 + q(T - \tau_{j+1})} - \frac{\sigma^2}{2}\frac{T - \tau_{j+1}}{1 + q(T - \tau_{j+1})}\frac{-3y_0^{j+1} + 4y_1^{j+1} - y_2^{j+1}}{2h} = 0. (12)$$

For the initial condition for the free boundary we have $z^0 = \rho(0)$. An algebraic nonlinear system of equations can be derived from (9) for $i = 1, \ldots, N - 1$, (10) and (12). In [9] the authors apply implicit finite difference scheme, semi-implicit scheme and upwind explicit scheme for the American put option, combining with the penalty method. The time step parameter for the explicit case is very small, $k = 5.0 \cdot 10^{-6}$. Therefore in this work we consider the case of a fully implicit scheme. One can also apply a scheme of the Crank-Nicolson type.

4   Numerical Algorithms

In order to solve the nonlinear system of algebraic equations we developed the following two algorithms.

Algorithm 1. This algorithm is based on the *predictor-corrector* scheme and consists of the following steps, (see also [15,16] for the case of pricing American put options).

Step 1. Predictor. Let the solution and the free boundary position on the time level τ_j be known. Instead of (12) we use another approximation of (8) by introducing an artificial spatial node ξ_{-1}:

$$(1 + q(T - \tau_{j+1})) z^{j+1} = 1 + r(T - \tau_{j+1}) + \frac{\sigma^2}{2}(T - \tau_{j+1}) \frac{y_1^{j+1} - y_{-1}^{j+1}}{2h}. \quad (13)$$

An additional equation can be obtained from (5) by taking the limit $\xi \to 0$ and using the fact that $\partial_\tau \Pi(0, \tau) = 0$:

$$\alpha_0^{j+1} \frac{y_1^{j+1} - y_{-1}^{j+1}}{2h} - \frac{\sigma^2}{2} \frac{y_1^{j+1} - 2y_0^{j+1} + y_{-1}^{j+1}}{h^2} + \beta^{j+1} y_0^{j+1} = 0. \quad (14)$$

Using (13) we can express y_{-1}^{j+1} as:

$$y_{-1}^{j+1} = y_1^{j+1} - \left(q z^{j+1} - r + \frac{z^{j+1} - 1}{T - \tau_{j+1}}\right) \frac{4h}{\sigma^2}. \quad (15)$$

Inserting it into (14) we conclude the following equation for the value y_1^{j+1}:

$$y_1^{j+1} = \left(\frac{2\alpha_0^{j+1} h^2}{\sigma^4} + \frac{2h}{\sigma^2}\right) \left(q z^{j+1} - r + \frac{z^{j+1} - 1}{T - \tau_{j+1}}\right) - \frac{\beta^{j+1} h^2}{\sigma^2} - 1. \quad (16)$$

Instead of the implicit scheme (9) we make use of its explicit variant for $i = 1$ in order to derive

$$\frac{y_1^{j+1} - y_1^j}{k} + \alpha_1^{j+1} \frac{y_2^j - y_0^j}{2h} - \frac{\sigma^2}{2} \frac{y_2^j - 2y_1^j + y_0^j}{h^2} + \beta^{j+1} y_1^j = 0. \quad (17)$$

This way we obtain a nonlinear system (16), (17) for unknowns y_1^{j+1} and z^{j+1}. The system is indeed nonlinear as α_i^{j+1} depend on z^{j+1}. Now, by replacing $y_1^{j+1} \leftrightarrow \tilde{y}_1^{j+1}$ and $z^{j+1} \leftrightarrow \tilde{z}^{j+1}$ we construct the predictor value of $\tilde{z}^{j+1}$.

Step 2. Corrector. We again use Equation (9) in a slightly different form:

$$\frac{y_i^{j+1} - y_i^j}{k} + \widehat{\alpha}_i^{j+1} \frac{y_{i+1}^{j+1} - y_{i-1}^{j+1}}{2h} - \frac{\sigma^2}{2} \frac{y_{i+1}^{j+1} - 2y_i^{j+1} + y_{i-1}^{j+1}}{h^2} + \beta^{j+1} y_i^{j+1} = 0, \quad (18)$$

where approximation $\widehat{\alpha}_i^{j+1}$ takes into account the already constructed predictor value $\widetilde{z}^{j+1}$, i.e.

$$\widehat{\alpha}_i^{j+1} = \frac{\widetilde{z}^{j+1} - z^j}{k\widetilde{z}^{j+1}} + r - q - \frac{\sigma^2}{2} - \frac{\widetilde{z}^{j+1} \exp(-\xi_i) - 1}{T - \tau_{j+1}}. \quad (19)$$

Next we use the corrected solution y_i^{j+1} and Equation (12) in order to obtain the corrected value for the free boundary position z^{j+1} on the next time layer.

Algorithm 2. We now describe an algorithm based on the *Newton method*. A variant of this method was applied for an American Call option problem in [5].

Step 1. We eliminate the known boundary values $y_0^{j+1} = -1$ and $y_N^{j+1} = 0$ from (9). Taking into account (12) we obtain a nonlinear system for N unknowns: y_i^{j+1}, $i = 1, 2, ..., N-1$ and z^{j+1}. We denote by $\overset{l}{\mathbf{Y}}$ the vector of these N unknowns at the l-th iteration.

Step 2. We have to solve the equation $\overset{l}{\mathbf{F}} = 0$ with $\overset{l}{\mathbf{F}} = \left(\overset{l}{\mathbf{F}_1} \overset{l}{\mathbf{F}_2} \right)^T$ where $\overset{l}{\mathbf{F}_i}$, $i = 1, 2$, correspond to Equations (9) and (12), respectively. To this end, we apply the Newton method in the following form: $\overset{l}{\mathbf{J}} (\overset{l+1}{\mathbf{Y}} - \overset{l}{\mathbf{Y}}) = - \overset{l}{\mathbf{F}}$, with the Jacobi matrix defined by: $\overset{l}{\mathbf{J}} = (\overset{l}{\mathbf{J}}_{ij})_{i,j=1,2}$ where

$$\overset{l}{\mathbf{J}}_{11} = \begin{pmatrix} c_1^{j+1} & b_1^{j+1} & & & \\ a_2^{j+1} & c_2^{j+1} & b_2^{j+1} & & \\ & \ddots & \ddots & \ddots & \\ & & a_{N-2}^{j+1} & c_{N-2}^{j+1} & b_{N-2}^{j+1} \\ & & & a_{N-1}^{j+1} & c_{N-1}^{j+1} \end{pmatrix}, \quad \overset{l}{\mathbf{J}}_{12} = \begin{pmatrix} \frac{\partial a_1^{j+1}}{\partial z^{j+1}}(-1) + \frac{\partial b_1^{j+1}}{\partial z^{j+1}} y_2^{j+1} \\ \frac{\partial a_2^{j+1}}{\partial z^{j+1}} y_1^{j+1} + \frac{\partial b_2^{j+1}}{\partial z^{j+1}} y_3^{j+1} \\ \vdots \\ \frac{\partial a_{N-2}^{j+1}}{\partial z^{j+1}} y_{N-3}^{j+1} + \frac{\partial b_{N-2}^{j+1}}{\partial z^{j+1}} y_{N-1}^{j+1} \\ \frac{\partial a_{N-1}^{j+1}}{\partial z^{j+1}} y_{N-2}^{j+1} \end{pmatrix}$$

$\overset{l}{\mathbf{J}}_{21} = \left(\frac{-\sigma^2}{Dh}, \frac{\sigma^2}{4Dh}, 0, ..., 0 \right)$ where $D = q + 1/(T - \tau^{j+1})$ and $\overset{l}{\mathbf{J}}_{22} = 1$. Similarly $\overset{l}{\mathbf{Y}} = \left(\overset{l}{\mathbf{Y}_1} \overset{l}{\mathbf{Y}_2} \right)^T$, $\overset{l}{\mathbf{Y}_1} = \left(y_1^{j+1}, ..., y_{N-1}^{j+1} \right)$, $\overset{l}{\mathbf{Y}_2} = z^{j+1}$. As for the elements of the matrix $\overset{l}{\mathbf{J}}_{11}$ we have:

$$a_i^{j+1} = -\frac{1}{2h} \left(\frac{z^{j+1} - z^j}{kz^{j+1}} + r - q - \frac{\sigma^2}{2} \right) - \frac{\sigma^2}{2h^2} + d_i^{j+1},$$

$$c_i^{j+1} = \frac{1}{k} + \frac{\sigma^2}{h^2} + r + \frac{1}{T - \tau_{j+1}},$$

$$b_i^{j+1} = \frac{1}{2h} \left(\frac{z^{j+1} - z^j}{kz^{j+1}} + r - q - \frac{\sigma^2}{2} \right) - \frac{\sigma^2}{2h^2} - d_i^{j+1},$$

and $d_i^{j+1} = 1/(2h)(z^{j+1}\exp(-\xi_i) - 1)/(T - \tau_{j+1})$. The iteration process is repeated until the condition $\| \overset{l+1}{\mathbf{Y}} - \overset{l}{\mathbf{Y}} \| < tol$ is fulfilled.

Step 3. The solution on the $(j + 1)$-th time layer is considered as an initial iteration for the next time layer. For solving $\overset{l}{\mathbf{J}} (\overset{l+1}{\mathbf{Y}} - \overset{l}{\mathbf{Y}}) = - \overset{l}{\mathbf{F}}$ we perform the following stages. First, we solve the linear system of equations $\overset{l}{\mathbf{J}}_{11} \overset{l+1}{\mathbf{Y}}_1 = - \overset{l}{\mathbf{F}}_1$ $+ \overset{l}{\mathbf{J}}_{11}\overset{l}{\mathbf{Y}}_1 - \overset{l}{\mathbf{J}}_{12} \overset{l+1}{\mathbf{Y}}_2 + \overset{l}{\mathbf{J}}_{12}\overset{l}{\mathbf{Y}}_2$. Since the matrix $\overset{l}{\mathbf{J}}_{11}$ is tridiagonal we can apply the Thomas algorithm to find $\overset{l+1}{\mathbf{Y}}_1$. Next, we solve $\overset{l}{\mathbf{J}}_{12} \overset{l+1}{\mathbf{Y}}_1 + \overset{l}{\mathbf{J}}_{22} \overset{l+1}{\mathbf{Y}}_2 = - \overset{l}{\mathbf{F}}_2$.

Remark 1. In both algorithms we choose the last time step $k - \varepsilon$ with $\varepsilon = 10^{-7}$, i.e. $\tau_M = T - \varepsilon$. To overcome possible numerical instabilities of these methods for $\tau \to T$ (i.e. $t \to 0$) we use the so called upwind and downwind approximations of the term $\frac{z^{j+1}\exp(-\xi_i)-1}{T-\tau_{j+1}}\frac{\partial \Pi}{\partial \xi}$ depending of the sign of the term $z^{j+1}\exp(-\xi_i) - 1$.

5   Numerical Experiments

In this section we consider problem (1) with parameter values $r = 0.06$, $q = 0.04$, $\sigma = 0.2$, and $T = 50$, taken from examples presented in [1]. Since there exists no analytical solution to the proposed free boundary problem, we use the mesh refinement analysis with doubling the mesh size h. In Table 1 we present the position of the free boundary position $\rho(\tau)$ at different times τ constructed by the Newton method. We also present the difference between two consecutive values and the convergence ratio are presented. The results show nearly first order of accuracy for the free boundary and the CR increases with increasing τ (see Table 1). In Fig. 1a) a 3D plot of the portfolio function Π for $T = 50$, $N = 200$, $M = 500$ is presented. In Fig. 1b) the profiles of the function $\Pi(\xi, \tau)$ for $\tau = 0, 0.1, 10, 25, 50$ obtained by the Newton method are depicted.

Table 1. Mesh-refinement analysis and the convergence ratio (CR) of the Newton method

N	$\rho(\tau = 10)$	difference	CR	$\rho(\tau = 20)$	difference	CR	$\rho(\tau = 40)$	difference	CR
50	1.949988	-	-	1.991675	-	-	1.796663	-	-
100	1.955552	5.5640e-3	-	1.995525	3.8502e-3	-	1.803276	6.6133e-3	-
200	1.958037	2.4850e-3	1.16	1.996945	1.4194e-3	1.44	1.805149	1.8729e-3	1.82
400	1.959199	1.1617e-3	1.10	1.997515	5.7099e-4	1.31	1.805667	5.1799e-4	1.85
800	1.959758	5.5965e-4	1.05	1.997765	2.4919e-4	1.20	1.805813	1.4621e-4	1.82

In Fig. 2a) we show a comparison of the free boundary position $\rho(\tau)$ computed by our two algorithms (Predictor-corrector and Newton's based method) and by numerical methods from [1] (Bokes) and [2] (Kwok). It turns out that the Newton's based method gives nearly the same results as those of [1,2]. On the other hand, predictor-corrector methods slightly underestimates the free boundary position $\rho(\tau)$. In Fig. 2b) we show the free boundary position $x_f(t) = 1/\rho(T - t)$ for the original model variables $x = A/S$ and t. The continuation region and exercise region are also indicated.

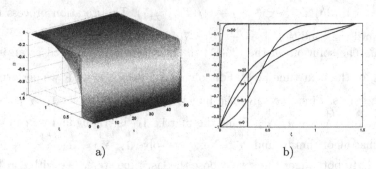

a)   b)

Fig. 1. (a) A 3D plot of the portfolio function Π for $T = 50$, $N = 200$, $M = 500$; (b) Profiles of the function $\Pi(\xi, \tau)$ for $\tau = 0$, $\tau = 0.1$, $\tau = 10$, $\tau = 25$, $\tau = 50$

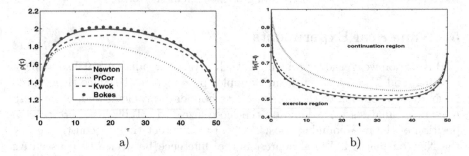

a)   b)

Fig. 2. a) Comparison of the free boundary $\rho(\tau)$ for various methods; b) the free boundary position $x_f(t) = 1/\rho(T - t)$ splitting the continuation and exercise region of American style of Asian call option

6   Conclusions

In this paper we have analyzed numerical algorithms for solving the free boundary value problem for American style of floating strike Asian options. To solve corresponding degenerate parabolic problem we have applied Landau's front fixing transformation method. We proposed two numerical algorithms based on the predictor-corrector scheme and the Newton's method. The predictor-corrector scheme is computationally faster when compared to the Newton method. It yields a good approximation close to expiry. However, its accuracy is decreased for times close to the initial time. The second algorithm based on Newton's method yields better approximation results over the whole time interval. Although all finite difference approximations are of second order, due to discontinuity of the initial datum and nonlinear behavior of the coefficients in all discrete equations, the results show nearly the first order rate of convergence.

Acknowledgments. The first author was supported by projects Bg-Sk-203/2008 and DID 02/37-2009. The second author was supported by the project APVV SK-BG-0034-08.

References

1. Bokes, T., Ševčovič, D.: Early exercise boundary for American type of floating strike Asian option and its numerical approximation. To appear in Applied Mathematical Finance (2011)
2. Dai, M., Kwok, Y.K.: Characterization of optimal stopping regions of American Asian and lookback options. Math. Finance 16(1), 63–82 (2006)
3. Gupta, S.C.: The Classical Stefan Problem: Basic Concepts, Modelling and Analysis. North-Holland Series in Applied Mathematics and Mechanics. Elsevier, Amsterdam (2003)
4. Han, H., Wu, X.: A fast numerical method for the Black-Scholes equation of American options. SIAM J. Numer. Anal. 41(6), 2081–2095 (2003)
5. Kandilarov, J., Valkov, R.: A Numerical Approach for the American Call Option Pricing Model. In: Dimov, I., Dimova, S., Kolkovska, N. (eds.) NMA 2010. LNCS, vol. 6046, pp. 453–460. Springer, Heidelberg (2011)
6. Kwok, J.K.: Mathematical Models of Financial Derivatives. Springer, Heidelberg (1998)
7. Lauko, M., Ševčovič, D.: Comparison of numerical and analytical approximations of the early exercise boundary of the American put option. ANZIAM Journal 51, 430–448 (2010)
8. Moyano, E., Scarpenttini, A.: Numerical stability study and error estimation for two implicit schemes in a moving boundary problem. Num. Meth. Partial Differential Equations 16(1), 42–61 (2000)
9. Nielsen, B., Skavhaug, O., Tveito, A.: Penalty and front-fixing methods for the numerical solution of American option problems. Journal of Comput. Finance 5(4), 69–97 (2002)
10. Ševčovič, D.: Analysis of the free boundary for the pricing of an American call option. Eur. J. Appl. Math. 12, 25–37 (2001)
11. Ševčovič, D.: Transformation methods for evaluating approximations to the optimal exercise boundary for linear and nonlinear Black-Sholes equations. In: Ehrhard, M. (ed.) Nonlinear Models in Mathematical Finance: New Research Trends in Optimal Pricing, pp. 153–198. Nova Sci. Publ., New York (2008)
12. Ševčovič, D., Takáč, M.: Sensitivity analysis of the early exercise boundary for American style of Asian options. To appear in International Journal of Numerical Analysis and Modeling, Ser. B (2011)
13. Stamicar, R., Ševčovič, D., Chadam, J.: The early exercise boundary for the American put near expiry: Numerical approximation. Canadian Applied Mathematics Quarterly 7(4), 427–444 (1999)
14. Tangman, D.Y., Gopaul, A., Bhuruth, M.: A fast high-order finite difference algorithms for pricing American options. J. Comp. Appl. Math. 222, 17–29 (2008)
15. Zhu, S.-P., Zang, J.: A new predictor-corrector scheme for valuing American puts. Applied Mathematics and Computation 217, 4439–4452 (2011)
16. Zhu, S.-P., Chen, W.-T.: A predictor-corrector scheme based on the ADI method for pricing American puts with stochastic volatility. To appear in Computers & Mathematics (2011)

A Kernel-Based Algorithm for Numerical Solution of Nonlinear PDEs in Finance

Miglena N. Koleva and Lubin G. Vulkov

University of Rousse,
8 Studentska Str., Rousse 7017, Bulgaria
{mkoleva,lvalkov}@uni-ruse.bg

Abstract. We present an algorithm for approximate solutions to certain nonlinear model equations from financial mathematics, using kernels techniques (fundamental solution, Green's function) for the linear Black-Scholes operator as a basis of the computation. Numerical experiments for comparison the accuracy of the algorithms with other known numerical schemes are discussed. Finally, observations are given.

1 Introduction

The solution of the famous (linear) Black-Scholes equation (Black and Scholes, 1973) provides both an option pricing formula for an European option and a hedging portfolio that replaces the contingent claim under restrictive assumptions, which are never fulfilled in the reality. To reflect the market properties in more details, several modifications of the Black-Scholes model have been derived. They take into account more realistic assumptions, such as transaction cost, risk from an unprotected portfolio, large investor's preference or illiquid markets, which may have an impact on the stock price, the volatility, the drift and the option price itself. This models defer from the classical Black-Scholes equation by a non-constant volatility term σ, which depends on time t, spot price S of the underlying and the second derivative (Greek Γ) of the option price $V(S,t)$. Hence, the model equation is the following nonlinear partial differential equation

$$V_t + \frac{1}{2}\sigma^2(t, S, V_{SS})S^2 V_{SS} + (r-q)SV_{SS} - rV = 0, \quad 0 \le S < \infty, \quad 0 \le t \le T, \quad (1)$$

with constant short rate r, dividend yield q, maturity T and $\sigma^2(t, S, V_{SS})$ depending on the particular model. For example, if the *transaction cost is taken into account*, $\sigma^2 = \sigma_L^2$ (Leland model, [15]) or $\sigma^2 = \sigma_{BS}^2$ (Barles-Soner model, [5]) or $\sigma^2 = \sigma_{JS}^2$ (Jandačka-Ševčovič, [14]) and

$$\sigma_L^2 = \sigma_0^2(1 + Le \cdot \text{sign}(V_{SS})), \quad Le = \sqrt{\frac{2}{\pi}} \frac{\kappa}{\sigma_0\sqrt{\delta t}},$$

$$\sigma_{BS}^2 = \sigma_0^2(1 + \Psi(e^{r(T-t)}a^2 S^2 V_{SS})), \quad \sigma_{JS}^2 = \sigma_0^2(1 + \mu(SV_{SS})^{\frac{1}{3}}),$$

where σ_0 is the volatility of the underlying asset, $Le \le 1$ is called *Leland number*, κ is the round trip transaction cost, δt denotes transaction frequency, a is a

I. Lirkov, S. Margenov, and J. Waśniewski (Eds.): LSSC 2011, LNCS 7116, pp. 566–573, 2012.

parameter to measure transaction cost and risk aversion, $\Psi(x)$ solves the ordinary differential equation

$$\Psi'(x) = (\Psi(x) + 1)/(2\sqrt{x\Psi(x)} - x), \quad x \neq 0, \quad \Psi(0) = 0 \tag{2}$$

and $\mu = 3(C^2 M/(2\pi))^{1/3}$, where $M \geq 0$ is the transaction cost measure, $C \geq 0$ is the risk premium measure. Here v^p with $v = SV_{SS}$ and $p = 1/3$ in σ_{JS}^2 stands for the signed power function, i.e. $v^p = |v|^{p-1}v$.

A second group of nonlinear Black-Scholes equations arises from the assumption that the *trading action of a large investor* will affect the price of the underlying asset, i.e. $\sigma^2 = \sigma_{AP}^2$ (Avellaneda-Parás, [4]) or $\sigma^2 = \sigma_{FP}^2$ (Frey-Patie, [11]) among others,

$$\sigma_{AP}^2 = \begin{cases} \sigma_{max}^2, & V_{SS} \leq 0, \\ \sigma_{min}^2, & V_{SS} > 0, \end{cases} \quad \sigma_{FP}^2 = \frac{\sigma_0^2}{1 - \rho\lambda(S)SV_{SS}}.$$

Here ρ is a parameter measuring the market liquidity and $\lambda(S)$ describes the liquidity profile in the dependence of the asset price. In Avellaneda-Parás model the volatility is not known exactly, but is assumed that lie between extreme values σ_{max} and σ_{min}.

We do not go further in describing other nonlinear model equations of type (1) for which our numerical method can be implemented.

We will study (1) for European Call option, i.e. the value $V(S,t)$ is the solution to (1), $q = 0$ on $0 \leq S < \infty, 0 \leq t \leq T$ with the following terminal and boundary conditions ($E > 0$ is the exercise price):

$$\begin{aligned} V(S,T) &= \max\{0, S - E\}, \quad 0 \leq S < \infty, \\ V(0,t) &= 0, \quad 0 \leq t \leq T \\ V(S,t) &= S - Ee^{-r(T-t)}, \quad S \to \infty, \end{aligned} \tag{3}$$

The models discussed above (and many other nonlinear modifications of the Black-Scholes equation) can be written as the backward parabolic fully nonlinear PDE

$$V_t + S^2 F(S, V_S, V_{SS}) = 0.$$

At some conditions on F, the most used of which are

$$F(S, p, r) \in C^2, \quad F_r(S, V_T'(S), V_T''(S)) > 0, \quad V(S,T) = V_T(S),$$

in [1,2,5,14,16] was obtained results for existence and uniqueness of solutions (classical or viscosity). It was checked that the model described above satisfies these conditions.

Since in general, a closed form solution to the nonlinear Black-Scholes equation does not exist (even in the linear case), these problems have to be solved numerically. Let us note that special exact solutions for some models of kind (1) are provided in [6].

There exists many possible discretizations, algorithms and some numerical methods for different versions of the non-linear Black-Scholes equation

[3,8,9,12,13,17]. Our goal is to present a *fast*, second order (in time and space) algorithm for solving a large class of nonlinear models in mathematical finance. We develop an implicit method, implementing the idea to use a kernel or Green's function to decouple the linear and nonlinear terms to allow for a two-stage solution process. A similar approach was applied to decouple the convection term from the diffusion operator in [10].

The remaining part of this paper is organized into three sections as follows. In Section 2, we describe our numerical method. In Section 3 some numerical experiments, concerning the Barles-Soner and Leland models are discussed. Finally, in Section 4, we make some concise conclusions.

2   Numerical Method

An often used approach to overcome a possible *degeneration* at $S = 0$ and to obtain a *forward* parabolic problem, is the following variable transformation [1,9,12]

$$x(S) = \log\left(\frac{S}{E}\right), \quad \tau(t) = \frac{1}{2}\sigma_0^2(T - t), \quad u(x, \tau) = e^{-x}\frac{V}{E}.$$

Now, denoting $K = 2r/\sigma_0^2$, the equation (1) transforms into

$$u_\tau - \tilde{\sigma}^2(\tau, x, u_x, u_{xx})(u_x + u_{xx}) - Ku_x = 0, \quad x \in \mathbb{R}, \quad 0 \le \tau \le \frac{\sigma_0^2 T}{2}, \quad (4)$$

where

$$\begin{aligned}
\tilde{\sigma}_L^2 &= 1 + f(u_x, u_{xx}), & f(u_x, u_{xx}) &= Le \cdot \text{sign}(u_x + u_{xx}), \\
\tilde{\sigma}_{BS}^2 &= 1 + f(x, \tau, u_x, u_{xx}), & f(x, \tau, u_x, u_{xx}) &= \Psi[e^{K\tau}a^2 Ee^x(u_x + u_{xx})], \\
\tilde{\sigma}_{JS}^2 &= 1 + f(x, u_x, u_{xx}), & f(x, u_x, u_{xx}) &= \mu[Ee^x(u_x + u_{xx})]^{1/3}, \\
\tilde{\sigma}_{AP}^2 &= f(u_x, u_{xx}), & f(u_x, u_{xx}) &= \begin{cases} \sigma_{max}^2, & u_x + u_{xx} \le 0, \\ \sigma_{min}^2, & u_x + u_{xx} > 0, \end{cases} \\
\tilde{\sigma}_{FP}^2 &= f(x, u_x, u_{xx}), & f(x, u_x, u_{xx}) &= [1 - \rho \cdot \lambda(Ee^x)(u_x + u_{xx})]^{-2}.
\end{aligned}$$

For computations we replace $\mathbb{R}$ by $\overline{\Omega} = [-L, L]$, $L > 0$. The problem (4) is computed by the following initial and boundary conditions, corresponding to (3)

$$u(x, 0) = u_0(x) = \max(0, 1 - e^{-x}), \quad u(-L, \tau) = 0, \quad u(L, \tau) = 1 - e^{-L - K\tau}. \quad (5)$$

Following the idea in [10] we first consider the case of zero Dirichlet boundary conditions. Let G be the Green's function for the linear part of (4), i.e. $\varphi(x, \tau) = \int_{-L}^{L} G(x, \tau; \xi)\varphi_0(\alpha)d\xi$ if and only if

$$\varphi_\tau = \nu\varphi_{xx} + (\nu + K)\varphi_x, \quad \varphi(-L, \tau) = \varphi(L, \tau) = 0, \quad \varphi(x, 0) = \varphi_0(x), \quad (6)$$

where $\nu = 1$ for $\tilde{\sigma}^2 = \{\tilde{\sigma}_L^2, \tilde{\sigma}_{BS}^2, \tilde{\sigma}_{JS}^2\}$ and $\nu = 0$ for $\tilde{\sigma}^2 = \{\tilde{\sigma}_{AP}^2, \tilde{\sigma}_{FP}^2\}$. Define the kernel integration for φ as $\mathcal{E}(\tilde{\tau})\varphi_0 = \varphi$:

$$(\mathcal{E}(\tilde{\tau})\varphi_0)(x, \tilde{\tau}) = \int_{-L}^{L} G(x, \tilde{\tau}; \xi)\varphi_0(\xi)d\xi.$$

Then we solve (4) by the variation of parameters formula

$$
u(x,\tau) = (\mathcal{E}(\tau)u_0)(x,\tau) + \left(\int_0^\tau \mathcal{E}(\tau - s) f(\cdot)[u_x(\cdot,s) + u_{xx}(\cdot,s)]ds \right)(x,\tau).
$$

Now, we replace the time interval $0 \le \tau \le \sigma_0^2 T/2$ by discrete times $\tau_n = n\Delta\tau$, $\Delta\tau = \sigma_0^2 T/(2N)$, $n = 0, \ldots, N$, denote by $u^n(\cdot) = u(\cdot, \tau_n)$ and write the solution at time τ by mean of the solution at time $\tau - \Delta\tau$

$$
u^{n+1}(x) = (\mathcal{E}(\tau)u^n(\cdot))(x,\tau) + \left(\int_{\tau-\Delta\tau}^\tau \mathcal{E}(\tau - s) f(\cdot)[u_x(\cdot,s) + u_{xx}(\cdot,s)]ds \right)(x,\tau).
$$

Taking into account the δ-function behavior of G, as the second arguments goes to 0 and using trapezoidal rule on the time integral we get

$$
u^{n+1}(x) - \frac{1}{2}\Delta\tau f^{n+1}(\cdot)[u_x^{n+1}(\cdot,s) + u_{xx}^{n+1}(\cdot,s)]
$$
$$
= (\mathcal{E}(\Delta\tau)(u^n(\cdot) + \frac{1}{2}\Delta\tau f^n(\cdot)[u_x^n(\cdot,s) + u_{xx}^n(\cdot,s)])) + \mathcal{O}(\Delta\tau^3). \tag{7}
$$

We need to approximate the operator $\mathcal{E}$. The spatial interval $[-L, L]$ is divided into M subintervals of length $h = 2L/M$, $x_i = -L + ih$, $i = 0, \ldots, M$. The discrete operator $\mathcal{E}_{h,\tau}$ is the solution operator for the linear system, generated from the approximation of the linear problem (6) and we can solve the problem, using (7). We will describe this algorithm in details for our model (4)-(5).

Algorithm for problem $(4) - (5)$

1. Given the solution u^n at previous time level, $n = 0, 1, \ldots, N$; $u^0 = u_0(x)$.

2. By Crank-Nicolson based scheme, solve the linear problem

$$
\tilde{u}_s = \nu\tilde{u}_{xx} + (\nu + K)\tilde{u}_x, \quad -L < x < L, \quad \tau - \Delta\tau < s < \tau,
$$
$$
\tilde{u}(-L, s) = 0, \quad \tilde{u}(L, \tau) = 1 - e^{-L-Ks}, \tag{8}
$$
$$
\tilde{u}(x, \tau - \Delta\tau) = u^n(\cdot) + \frac{1}{2}\Delta\tau f^n(\cdot)[u_x^n + u_{xx}^n].
$$

3. Solve the nonlinear problem, using a second order difference approximation

$$
u^{n+1}(x) - \frac{1}{2}\Delta\tau f^{n+1}(\cdot)[u_x^{n+1} + u_{xx}^{n+1}] = \tilde{u}^{n+1}, \quad -L < x < L, \quad \tau_n < s < \tau_{n+1}, \tag{9}
$$
$$
u^{n+1}(-L, s) = 0, \quad u^{n+1}(L, s) = 1 - e^{-L-Ks},
$$

where $\tilde{u}^{n+1}$ is computed at step 1.

4. Return to step 1.

Note, that the solution inverse matrix of the linear problem (8) is one and the same for all time levels. Moreover, the right hand side $\widetilde{u}^{n+1}$ of the nonlinear elliptic problem (9) is independent of u^{n+1} for all grid nodes x_i, thus simplifying the iteration of the nonlinear solver substantially.

The numerical stability condition for problem (8) is $\triangle \tau \leq 2h/(\nu+K)$ and for problem (9), solved by Newton's iterations: $\triangle \tau \leq 2h/|\alpha(f', f, u_x, u_{xx})|$, where the function α is computed at old iteration and depends on the model. Thus the restriction for $\triangle \tau$ is

$$\triangle \tau \leq \min \left\{ 2h/(\nu+K), 2h/|\alpha(f, f', u_x, u_{xx})| \right\},$$

while for the standard for example, Crank-Nicolson scheme combined with Newton's iteration we have

$$\triangle \tau \leq 2h/|\nu + K + \alpha(f, f', u_x, u_{xx})|.$$

3   Numerical Experiments

We will discuss two typical examples: Barles-Soner model and Leland model. The approach for the other problems is the same. Two groups of experiments are presented: test with exact solution and original solution computation. With appropriate initial and boundary conditions $u_{ex}(x, t) = e^{-Kt-x}$ is exact solution of (4). Thus the error $E_i = u_{ex}(x_i, T) - u_{num}(x_i, T)$, $i = 1, \ldots, M$ in maximal and L_2 discrete norms are given by

$$\|E^M\|_\infty = \max_{1 \leq i \leq M-1} |E_i|, \quad \|E^M\|_2 = \left[h \sum_{i=1}^{M-1} E_i^2 \right]^{1/2}$$

and the convergence rate is calculated using double mesh principle

$$CR_\infty = \log 2(\|E^M\|_\infty/\|E^{2M}\|_\infty), \quad CR_2 = \log 2(\|E^M\|_2/\|E^{2M}\|_2).$$

The average number of iterations (*iter*) at each time level and CPU time are also given. Iterations of the nonlinear solver continue until the maximal difference between two subsequent iterations is less than $tol = 10^{-12}$. The option parameters are: $r = 0.1$, $\sigma_0 = 0.2$, $E = 100$. The advantages of the Algorithm are illustrated for Barles-Soner model.

Example 1 (Barles-Soner model). The analysis of the ODE (2) by Barles and Soner [5] implies that $\lim_{x\to\infty} \Psi(x)/x = 1$ and $\lim_{x\to-\infty} \Psi(x) = -1$. This property allows the treatment of the function $\Psi(\cdot)$ as the identity for large arguments. Another approach is to solve (2) and use some interpolation to obtain the values of Ψ for arbitrary arguments [9,8]. In this case we obtain linear problem and the algorithm described in [10] can be applied. We take $\Psi(\rho) = \rho$. To solve the nonlinear system of equations, generated from (9), the classical Newton's method is used. Exact solution computations are performed with $a = 0.01$, $\triangle \tau = h/2$,

Table 1. Error, convergence rate, CPU time, and number of iterations

M	h	Algorithm				Algorithm		Standard meth.	
		$\|E^N\|_\infty$	CR_∞	$\|E^N\|_2$	CR_2	CPU	*iter*	CPU	*iter*
10	0.2	5.2873e-4		1.9012e-4		0.039	4.5	0.038	5.6
20	0.1	1.3215e-4	*2.0003*	3.3214e-5	*2.5171*	0.078	4.1	0.089	4.9
40	0.05	3.2837e-5	*2.0088*	5.8555e-6	*2.5039*	0.212	3.5	0.337	4.3
80	0.025	8.2161e-6	*1.9988*	1.0344e-6	*2.5010*	0.657	3.0	0.875	3.6
160	0.0125	2.0517e-6	*2.0016*	1.8283e-7	*2.5002*	2.265	2.6	2.852	3.3
320	0.00625	5.1293e-7	*2.0000*	3.2319e-8	*2.5001*	7.300	2.1	9.544	3.0
640	0.003125	1.2823e-7	*2.0000*	5.7132e-9	*2.5000*	26.804	2.0	33.454	2.9
1280	0.0015625	3.1933e-8	*2.0056*	1.0060e-9	*2.5056*	103.389	2.0	129.798	2.6

$L = 1$, $T = 2/\sigma_0^2$. For comparison we also solve the problem (4)-(5) using ordinary Crank-Nicolson scheme, also with classical Newton's iterations. Errors, convergence results (in maximal and L_2 discrete norms), number of iterations (*iter*) and CPU time are summarized in Table 1. The result shows that the presented Algorithm is much faster than the standard Crank-Nicolson method, especially for fine meshes. In the same manner the case $\sigma^2 = \{\sigma_{JS}^2, \sigma_{FP}^2\}$ can be treated.

We compute the Barles-Soner model with different transaction cost $a = \{0.02, 0.05\}$ and the corresponding model without transaction cost ($a = 0$). The payoff and boundary data are (5). The mesh parameters are: $L = 3$, $M = 160$, $N = 22$, $T = 1$. On Figure 1 (*left*) we plot $V(S,T)$ for different values of a (CPU=1.109 for $a = 0.05$). As the transaction cost increases we need a finer mesh in order to avoid oscillations. With the above mesh parameters and $a = 0.05$ the standard Crank-Nicolson scheme fails. For successful computations we need a little bit smaller time step, at least $N = 24$ (CPU=1.849 for $a = 0.05$).

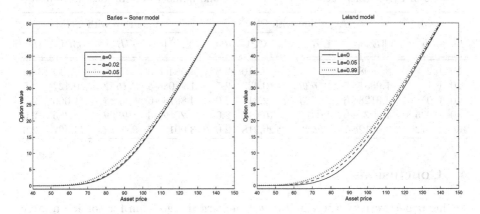

Fig. 1. Option value for Barles-Soner model (*left*) and Leland model (*right*)

Table 2. Error, convergence rate, CPU time and number of iterations

M	h	Algorithm with PI and QNM				Alg. with PI		Alg. with QNM	
		$\|E^N\|_\infty$	CR_∞	$\|E^N\|_2$	CR_2	CPU	iter	CPU	iter
10	0.2	5.2866e-4		1.8983e-4		0.036	3.1	0.014	6.9
20	0.1	1.3158e-4	2.0064	3.3098e-5	2.5199	0.069	3.0	0.058	6.9
40	0.05	3.2633e-5	2.0116	5.8203e-6	2.5076	0.178	3.2	0.112	7.4
80	0.025	8.0363e-6	2.0217	1.0088e-6	2.5285	0.620	3.3	0.267	7.2
160	0.0125	1.8747e-6	2.0999	1.6304e-7	2.5293	2.819	4.0	0.790	7.9

Example 2 (Leland model). The nonlinear problem (9) is solved by Picard-like (the term $f(\cdot)$ in the nonlinearity is computed at old iteration) iteration process (PI) and quasi-Newton method for nondifferentiable functions (QNM) [7]. For exact solution test we take $\delta t = 1$ (transaction frequency is one a week), $\kappa = 0.002$. The results (both with PI and QNM) with mesh parameters $L = 1$, $\triangle \tau = h/2$ at time $T = 2/\sigma_0^2$ are given in Table 2.

For great Le number the time step must be decreased, for example if $Le = 0.5$, then $\triangle \tau = h^2$ and in the case of critical Le number ($Le = 0.99$), we choose much smaller time step $\triangle \tau = h^2/2$. Results are summarized in Table 3 with $tol = 10^{-8}$. The iteration process QNM is more faster, because the solution at new iteration computes with explicit formula. On Figure 1 (*right*) we plot option value for different Le numbers and mesh parameters $L = 3$, $M = 22$, $N = 160$, $T = 1$.

In the case $\sigma^2 = \widetilde{\sigma}_{AP}^2$, there is a small difference between the parameters σ_{max}^2, σ_{min}^2 ($\sigma_{max}^2 = 0.25$, $\sigma_{min}^2 = 0.15$ [12]) and the same approach as for Leland model (with QNM) can be applied. We observe the same behavior of the solution as for Leland model.

Table 3. Error, convergence rate, CPU time and number of iterations, $Le = 0.99$

M	h	Algorithm with PI				Algorithm with QNM			
		$\|E^N\|_\infty$	CR_∞	CPU	iter	$\|E^N\|_\infty$	CR_∞	CPU	iter
10	0.2	8.8808e-5		0.110	2.8	8.8846e-5		0.314	35.9
20	0.1	1.9831e-5	2.1629	0.433	2.5	1.9808e-5	2.1652	0.427	9.5
40	0.05	4.8128e-6	2.0428	2.317	2.0	4.8073e-6	2.0428	1.066	2.4
80	0.025	1.2022e-6	2.0012	16.611	2.0	1.2019e-6	1.9999	6.385	2.0
160	0.0125	3.0012e-7	2.0020	129.218	2.0	3.0012e-7	2.0017	44.367	2.0

4 Conclusions

In this paper we have presented a *fast* numerical algorithm for solving a large class of nonlinear problems in mathematical finance. For this models, the real time period of computations is much more greater than $T = 1$. Thus the property of our algorithm to save computational time is very important.

The experiments show that presented method is asymptotically stable and second order accurate, with the stability being independent of the mesh parameters. It appears possible to modify the present numerical method to produce stable, second order accurate algorithms for nonlinear two-dimensional problems.

Acknowledgements. This work is supported by the Bulgarian National Fund of Science under Projects Sk-Bg-203 and DID 02/37-2009.

References

1. Abe, R., Ishimura, N.: Existence of solutions for the nonlinear partial differential equation arising in the optimal investment problem. Proc. Japan Acad., Ser. A 84, 11–14 (2008)
2. Agliardi, R., Popivanov, P., Slavova, A.: Nonhypoellipticity and comparison principle for partial differential equations of Black-Scholes type. Nonl. Anal.: Real Word Appl. 12(3), 1429–1436 (2011)
3. Ankudinova, J., Ehrhardt, M.: On tne numerical solution of nonlinear Black-Scholes equations. Int. J. of Comp. and Math. with Appl. 56, 779–812 (2008)
4. Avellaneda, M., Parás, A.: Managing the volatility risk of portfolios of derivative securities: the Lagrangian uncertain volatility model. Appl. Math. Fin. 3(1), 21–52 (1996)
5. Barles, G., Soner, H.M.: Option pricing with transaction costs and a nonlinear Black-Scholes equation. Finance and Stochastics 2(4), 369–397 (1998)
6. Bordag, L., Chmakova, A.: Explicit solutions for a nonlinear model of financial derivatives. Int. J. Theor. Appl. Finance 10, 1–21 (2007)
7. Chen, X., Yamamoto, T.: On the convergence of some quasi-Newton methods for nonlinear equations with nondifferentiable operators. Computing 49, 87–94 (1992)
8. Company, R., Navarro, E., Pintos, J., Ponsoda, E.: Numerical solution of linear and nonlinear Black-Scholes option pricing equations. Int. J. of Comp. and Math. with Appl. 56, 813–821 (2008)
9. Düring, B., Fournié, M., Jüngel, A.: Convergence of a high-order compact finite difference scheme for a nonlinear Black-Scholes equation. ESAIM: M2AN 38(2), 359–369 (2004)
10. Epperson, J.: A kernel-based method for parabolic equations with nonlinear convection terms. J. Comp. and Appl. Math. 36, 275–288 (1991)
11. Frey, R., Patie, P.: Risk management for derivatives in illiquid markets: a simulation-study. In: Sandmann, K., Schönbucher, P. (eds.) Adv. in Fin. and Stoch., pp. 137–159. Springer, Berlin (2002)
12. Heider, P.: Numerical Methods for Non-Linear Black-Scholes Equations. Appl. Math. Fin. 17(1), 59–81 (2010)
13. Ishimura, N., Koleva, M.N., Vulkov, L.G.: Numerical Solution of a Nonlinear Evolution Equation for the Risk Preference. In: Dimov, I., Dimova, S., Kolkovska, N. (eds.) NMA 2010. LNCS, vol. 6046, pp. 445–452. Springer, Heidelberg (2011)
14. Jandačka, M., Ševčovič, D.: On the risk-adjusted pricing-methodology-based valuation of vanilla options and explanation of the volatility smile. J. of Appl. Math. (3), 235–258 (2005)
15. Leland, H.E.: Option pricing and replication with transactions costs. J. of Finance 40(5), 1283–1301 (1985)
16. Ševčovič, D., Stehlíková, B., Mikula, K.: Analytical and Numerical Methods for Pricing Financial Derivatives. Nova Sci. Publ., NY (2010)
17. Seydel, R.: Tools for Computational Finance. Springer, Berlin (2003)

Improving the Efficiency of Parallel FEM Simulations on Voxel Domains

N. Kosturski, S. Margenov, and Y. Vutov

Institute of Information and Communication Technologies,
Bulgarian Academy of Sciences, Sofia, Bulgaria

Abstract. In this work, we consider large-scale finite element modeling on voxel grids. We are targeting the IBM Blue Gene/P computer, which features a 3D torus interconnect. Our previous parallelization approach was to divide the domain in one spatial direction only, which lead to limited parallelism. Here, we extend it to all three spatial directions in order to match the interconnect topology.

As a sample problem, we consider the simulation of the thermal and electrical processes, involved in the radio-frequency (RF) ablation procedure. RF ablation is a low invasive technique for the treatment of hepatic tumors, utilizing AC current to destroy the tumor cells by heating. A 3D voxel approach is used for finite element method (FEM) approximation of the involved partial differential equations. After the space discretization, the backward Euler scheme is used for the time stepping.

We study the impact of the domain partitioning on the performance of a parallel preconditioned conjugate gradient (PCG) solver for the arising large linear systems. As a preconditioner, we use BoomerAMG – a parallel algebraic multigrid implementation from the package Hypre, developed in LLNL, Livermore. The implementation is tested on the IBM Blue Gene/P massively parallel computer.

1 Introduction

This work is motivated by the need to improve the parallel efficiency of our supercomputer simulation of RF hepatic tumor ablation on the IBM Blue Gene/P massively parallel computer [5]. This simulation is based on a cubical computational domain, represented by a structured voxel grid. Here, different parallel partitioning strategies (illustrated on Figure 1) are given special attention.

Our previous implementation allowed only 1D partitioning of the computational domain among processors. The biggest limitation of this approach is that the maximum number of processors that can be utilized can never exceed the voxel image resolution in one direction. For higher voxel resolutions, the subdomains assigned to each processor may easily require more than the available amount of memory.

There is another disadvantage of the 1D partitioning strategy that is specific to the IBM Blue Gene/P computer. The communication patterns associated with the 1D partitioning cannot fully utilize the available hardware interconnect, which in this case is either a 3D mesh or a 3D torus, depending on the

I. Lirkov, S. Margenov, and J. Waśniewski (Eds.): LSSC 2011, LNCS 7116, pp. 574–581, 2012.

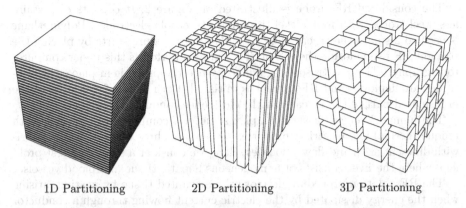

1D Partitioning   2D Partitioning   3D Partitioning

Fig. 1. Examples of Domain Partitioning Strategies

number of processors used. With a 3D partitioning, on the other hand, the communication patterns can be naturally mapped to the hardware interconnect topology. Therefore, this strategy is expected to provide the best scalability. The presented parallel tests fully confirm this.

2   Radio-Frequency Tumor Ablation

Let us turn our attention to the considered numerical simulation. RF ablation is an alternative, low invasive technique for the treatment of hepatic tumors, utilizing AC current to destroy the tumor cells by heating [6,7]. The destruction of the cells occurs at temperatures of 45°C–50°C. The procedure is relatively safe, as it does not require open surgery.

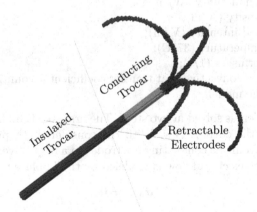

Fig. 2. The Structure of a Fully Deployed RF Probe

The considered RF probe is illustrated on Figure 2. It consists of a stainless steel trocar with four nickel-titanium retractable electrodes. Polyurethane is used to insulate the trocar. The RF ablation procedure starts by placing the straight RF probe inside the tumor. The surgeon performs this under computer tomography (CT) or ultrasound guidance. Once the probe is in place, the electrodes are deployed and RF current is initiated. Both the surfaces areas of the uninsulated part of the trocar and the electrodes conduct RF current.

The human liver has a very complex structure, composed of materials with unique thermal and electrical properties. There are three types of blood vessels with different sizes and flow velocities. Here, we consider a simplified test problem, where the liver consists of homogeneous hepatic tissue and blood vessels.

The RF ablation procedure destroys the unwanted tissue by heating, arising when the energy dissipated by the electric current flowing through a conductor is converted to heat. The bio-heat time-dependent partial differential equation [6,7]

$$\rho c \frac{\partial T}{\partial t} = \nabla \cdot k \nabla T + J \cdot E - h_{bl}(T - T_{bl}) \qquad (1)$$

is used to model the heating process during the RF ablation. The term $J \cdot E$ in (1) represents the thermal energy arising from the current flow and the term $h_{bl}(T - T_{bl})$ accounts for the heat loss due to blood perfusion.

The following initial and boundary conditions are applied

$$T = 37°C \text{ when } t = 0 \text{ at } \Omega,$$
$$T = 37°C \text{ when } t \geq 0 \text{ at } \partial\Omega. \qquad (2)$$

The following notations are used in (1) and (2):

- Ω – the entire domain of the model;
- $\partial\Omega$ – the boundary of the domain;
- ρ – density (kg/m^3);
- c – specific heat (J/kg K);
- k – thermal conductivity (W/m K);
- J – current density (A/m);
- E – electric field intensity (V/m);
- T_{bl} – blood temperature (37°C);
- w_{bl} – blood perfusion (1/s);
- $h_{bl} = \rho_{bl} c_{bl} w_{bl}$ – convective heat transfer coefficient accounting for the blood perfusion in the model.

The bio-heat problem is solved in two steps. The first step is finding the potential distribution V of the current flow. With the considered RF probe design, the current is flowing from the conducting electrodes to a dispersive electrode on the patient's body. The electrical flow is modeled by the Laplace equation

$$\nabla \cdot \sigma \nabla V = 0, \qquad (3)$$

with boundary conditions

$$V = 0 \text{ at } \partial\Omega,$$
$$V = V_0 \text{ at } \partial\Omega_{el}.$$

The following notations are used in the above equations:

- V – potential distribution in Ω;
- σ – electric conductivity (S/m);
- V_0 – applied RF voltage;
- $\partial\Omega_{el}$ – surface of the conducting part of the RF probe.

After determining the potential distribution, the electric field intensity can be computed from

$$E = -\nabla V,$$

and the current density from

$$J = \sigma E.$$

The second step is to solve the heat transfer equation (1) using the heat source $J \cdot E$ obtained in the first step.

For the numerical solution of both of the above discussed steps of the simulation the Finite Element Method (FEM) in space is used [2]. Linear conforming elements are chosen in this study. To apply the linear FEM discretization to the voxel domain, each voxel is split into six tetrahedra. To solve the bio-heat equation, after the space discretization, the time derivative is discretized via finite differences and the backward Euler scheme is used [3].

Let us denote with K^* the stiffness matrix coming from the FEM discretization of the Laplace equation (3). It can be written in the form

$$K^* = \left[\int_\Omega \sigma \nabla\Phi_i \cdot \nabla\Phi_j dx\right]_{i,j=1}^N,$$

where $\{\Phi_i\}_{i=1}^N$ are the FEM basis functions.

The system of linear algebraic equations

$$K^* X = 0 \tag{4}$$

is to be solved to find the nodal values X of the potential distribution.

The electric field intensity and the current density are then expressed by the partial derivatives of the potential distribution in each finite element. This way, the nodal values F for the thermal energy $E \cdot J$ arising from the current flow are obtained.

Let us now turn our attention to the discrete formulation of the bio-heat equation. Let us denote with K and M the stiffness and mass matrices from the finite element discretization of (1). They can be written as

$$K = \left[\int_\Omega k\nabla\Phi_i \cdot \nabla\Phi_j dx\right]_{i,j=1}^N,$$

$$M = \left[\int_\Omega \rho c\Phi_i\Phi_j dx\right]_{i,j=1}^N.$$

Let us also denote with Ω_{bl} the subdomain of Ω occupied by blood vessels and with M_{bl} the matrix

$$M_{bl} = \left[\int_\Omega \delta_{bl} h_{bl} \Phi_i \Phi_j d\mathbf{x} \right]^N_{i,j=1},$$

where

$$\delta_{bl}(x) = \begin{cases} 1 \text{ for } x \in \Omega_{bl}, \\ 0 \text{ for } x \in \Omega \setminus \Omega_{bl}. \end{cases}$$

Then, the parabolic equation (1) can be written in matrix form as:

$$M \frac{\partial T}{\partial t} + (K + M_{bl})T = F + M_{bl}T_{bl}. \tag{5}$$

If we denote with τ the time-step, with T^{n+1} the solution at the current time level, and with T^n the solution at the previous time level and approximate the time derivative in (5) we obtain the following system of linear algebraic equations for the nodal values of T^{n+1}

$$(M + \tau(K + M_{bl}))T^{n+1} = MT^n + \tau(F + M_{bl}T_{bl}). \tag{6}$$

The matrices of the linear systems (4) and (6) are ill-conditioned and very large, having around 10^8 rows. Since they are symmetric and positive definite, we use the PCG [1] method, which is the most efficient solution method in this case.

A parallel algebraic multigrid implementation is used as a preconditioner. The matrix $A = M + \tau(K + M_{bl})$ from (6) is assembled only once on the first time step and not varied after that. The corresponding AMG preconditioner is also constructed only on the first time step.

3   Parallel Tests

The results presented in this section are based on a high-resolution voxel-based representation of the computational domain. The domain consists of liver and tumor tissues, a large bifurcated blood vessel and the RF ablation probe (see Figure 3). We use three different domains (with sizes $127 \times 127 \times 127$, $255 \times 255 \times 255$, and $511 \times 511 \times 511$) to compare the performance and the weak scalability of the simulation using the three considered partitioning strategies.

Table 1 lists the thermal and electrical properties of the materials, which are taken from [6] as well as the blood perfusion coefficient $w_{bl} = 6.4 \times 10^{-3}$ 1/s. For the test simulations, an RF voltage of 10 V is applied for a duration of 8 minutes. A time step of $\tau = 5$ s is used.

Large-scale systems of linear algebraic equations arise from the FEM discretization of the considered problem, requiring an efficient parallel implementation. A parallel PCG solver is used here. The selected preconditioner is BoomerAMG [4,8] – a state of the art parallel preconditioner of optimal complexity. A relative PCG stopping criterion in the form

$$\mathbf{r}_k^T C^{-1} \mathbf{r}_k \leq \varepsilon^2 \mathbf{r}_0^T C^{-1} \mathbf{r}_0, \qquad \varepsilon = 10^{-6},$$

where $\mathbf{r}_k$ stands for the residual at the k-th step of the PCG method, is used.

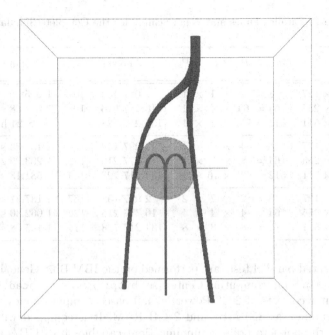

Fig. 3. High-Resolution 3D Voxel Representation of the Computational Domain

The settings for the BoomerAMG preconditioner were carefully tuned for maximum scalability in time. The selected coarsening algorithm is *Falgout-CLJP*. *Modified classical interpolation* is applied. The selected relaxation method is *hybrid symmetric Gauss-Seidel or SSOR*. To decrease the operator and grid complexities two levels of aggressive coarsening are used and the maximum number of elements per row for the interpolation is restricted to six. Smaller operator and grid complexities lead to faster iterations and reduced memory requirements, but can also affect the convergence rate of the solver. Thus, the values of the last two parameters must be carefully chosen to provide the best balance. With the above described setup, the solutions of the linear systems on each time step required 1–3 PCG iterations.

Table 1. Thermal and Electrical Properties of the Materials

Material	ρ (kg/m^3)	c (J/kg K)	k (W/m K)	σ (S/m)
Ni-Ti	6 450	840	18	1×10^8
Stainless steel	21 500	132	71	4×10^8
Liver	1 060	3 600	0.512	0.333
Blood	1 000	4 180	0.543	0.667
Polyurethane	70	1 045	0.026	10^{-5}

Table 2. Parallel Times and Weak Scaling for the Complete Simulation

Domain size	$N_p = P_x \times P_y \times P_z$	Unknowns	N_{it}	Time	Weak scaling
$127 \times 127 \times 127$	$8 = 8 \times 1 \times 1$	2 097 152	161	1 225.00 s	
$255 \times 255 \times 255$	$64 = 64 \times 1 \times 1$	16 777 216	128	5 951.08 s	21 %
$511 \times 511 \times 511$	$512 = 512 \times 1 \times 1$	134 217 728	—	> 24 h	< 2 %
$127 \times 127 \times 127$	$8 = 4 \times 2 \times 1$	2 097 152	167	1 137.83 s	
$255 \times 255 \times 255$	$64 = 8 \times 8 \times 1$	16 777 216	129	1 203.29 s	95 %
$511 \times 511 \times 511$	$512 = 32 \times 16 \times 1$	134 217 728	114	1 581.13 s	72 %
$127 \times 127 \times 127$	$8 = 2 \times 2 \times 2$	2 097 152	167	1 137.91 s	
$255 \times 255 \times 255$	$64 = 4 \times 4 \times 4$	16 777 216	128	1 062.30 s	107 %
$511 \times 511 \times 511$	$512 = 8 \times 8 \times 8$	134 217 728	114	1 155.08 s	99 %

The presented parallel tests are performed on the IBM Blue Gene/P machine at the Bulgarian Supercomputing Center (see http://www.scc.acad.bg/). This supercomputer consists of 2048 PowerPC 450 based compute nodes, each with four cores running at 850 MHz and 2 GB RAM. It is equipped with a torus network for the point to point communications capable of 5.1 GB/s and a tree network for global communications with a bandwidth of 1.7 GB/s. Our software is implemented in C++, using MPI for the parallelization. It is compiled using the IBM XL C++ compiler with the following options: "-O5 -qstrict".

The parallel times for the whole numerical simulation are presented in Table 2. The three parts of the table correspond to the three considered partitioning strategies. Here, N_p is the number of processors, P_x, P_y, P_z are the number of partitions in direction x, y, z respectively, and N_{it} is the number of PCG iterations performed during the simulation. The simulation in the 1D partitioning case for the largest domain could not finish in 24 hours, which is the hard limit for problems of this size on the Blue Gene/P machine in the Bulgarian Supercomputing Center. The weak scaling with respect to the smallest simulation domain is provided for each partitioning strategy. A big advantage of the 3D partitioning strategy over the other two is observed. The 1D partitioning is the least scalable of the considered three.

Equivalent computations are performed with each partitioning strategy, as each processor holds the same number of unknowns and the number of PCG iterations is almost the same. The communications, however, are quite different. In the case of 3D partitioning, the partition allocated to each processor has the same size ($64 \times 64 \times 64$ unknowns) for all domains. Typical communications involve the transfer of values for all the unknowns on the interfaces between neighboring partitions. Therefore, in the 3D partitioning case, the communication time should be independent of the domain size. Moreover, by mapping the partitioning to the underlying interconnect topology, we ensure that the communications in each direction can be performed in parallel. With the 1D partitioning strategy the size of the interfaces increases four times with each next domain size, which is the reason for the much lower scalability. The scalability of above 100 % can be explained by the smaller number of PCG iterations in this case.

4   Concluding Remarks

Parallel tests of a large-scale, time-dependent, voxel-based simulation of the RF ablation procedure are presented. Such simulations require very efficient use of supercomputer resources. Three approaches to the partitioning of the computational domain are presented and compared. Using a 3D domain partitioning leads to a substantial scalability improvement over our previous work, as clearly demonstrated by the parallel test results.

Our future plans include enhancing the model in order to simulate a more complex type of RF ablation probe, involving fluid injection during the procedure. Overcoming the previous restriction on the number of processors in our implementation as well as improving its scalability were crucial for enabling us to take the next step.

Acknowledgments. This work is partly supported by the Bulgarian NSF Grants DCVP 02/1 and DPRP7RP02/13. We also kindly acknowledge the support of the Bulgarian Supercomputing Center for the access to the IBM Blue Gene/P supercomputer.

References

1. Axelsson, O.: Iterative Solution Methods. Cambridge University Press (1996)
2. Brenner, S., Scott, L.: The mathematical theory of finite element methods. Texts in applied mathematics, vol. 15. Springer, Heidelberg (1994)
3. Hairer, E., Norsett, S.P., Wanner, G.: Solving ordinary differential equations I, II. Springer Series in Comp. Math. 2000 (2002)
4. Henson, V.E., Yang, U.M.: BoomerAMG: A parallel algebraic multigrid solver and preconditioner. Applied Numerical Mathematics 41(1), 155–177 (2002)
5. Kosturski, N., Margenov, S.: Supercomputer Simulation of Radio-Frequency Hepatic Tumor Ablation. In: AMiTaNS 2010 Proceedings. AIP CP, vol. 1301, pp. 486–493 (2010)
6. Tungjitkusolmun, S., Staelin, S.T., Haemmerich, D., Tsai, J.Z., Cao, H., Webster, J.G., Lee, F.T., Mahvi, D.M., Vorperian, V.R.: Three-dimensional finite-element analyses for radio-frequency hepatic tumor ablation. IEEE Transactions on Biomedical Engineering 49(1), 3–9 (2002)
7. Tungjitkusolmun, S., Woo, E.J., Cao, H., Tsai, J.Z., Vorperian, V.R., Webster, J.G.: Thermal-electrical finite element modelling for radio frequency cardiac ablation: Effects of changes in myocardial properties. Medical and Biological Engineering and Computing 38(5), 562–568 (2000)
8. Lawrence Livermore National Laboratory, Scalable Linear Solvers Project, http://www.llnl.gov/CASC/linear_solvers/

On the Robustness of Two-Level Preconditioners for Quadratic FE Orthotropic Elliptic Problems

J. Kraus[1], M. Lymbery[2], and S. Margenov[2]

[1] Johann Radon Institute for Computational and Applied Mathematics,
Austrian Academy of Sciences, Altenbergerstraße 69, A-4040 Linz, Austria
johannes.kraus@oeaw.ac.at
[2] Institute of Information and Communication Technologies,
Bulgarian Academy of Sciences, Acad. G. Bonchev Str., Bl. 25A,
1113 Sofia, Bulgaria
{mariq,margenov}@parallel.bas.bg

Abstract. We study the construction of subspaces for quadratic FEM orthotropic elliptic problems with a focus on the robustness with respect to mesh and coefficient anisotropy. In the general setting of an arbitrary elliptic operator it is known that standard hierarchical basis (HB) techniques do not result in splittings in which the angle between the coarse space and its (hierarchical) complement is uniformly bounded with respect to the ratio of anisotropy. In this paper we present a robust splitting of the finite element space of continuous piecewise quadratic functions for the orthotropic problem. As a consequence of this result we obtain also a uniform condition number bound for a special sparse Schur complement approximation. Further we construct a uniform preconditioner for the pivot block with optimal order of computational complexity.

1 Introduction

Let us consider the linear system of equations

$$A_h \mathbf{u}_h = F_h, \tag{1}$$

obtained after finite element discretization of the elliptic boundary value problem

$$-\nabla \cdot (a(x)\nabla u(x)) = f(x) \quad in \ \ \Omega, \tag{2a}$$
$$u = \ 0 \quad on \ \Gamma_D, \tag{2b}$$
$$(a(x)\nabla u(x)) \cdot \mathbf{n} = \ 0 \quad on \ \Gamma_N. \tag{2c}$$

The used notations are as follows: Ω - a polygonal convex domain in R^2; $f(x)$ - a given function in $L_2(\Omega)$, $a(x) = (a_{ij}(x))_{i,j=1}^2$ - a symmetric positive definite coefficient matrix, uniformly bounded in Ω; $\mathbf{n}$ - the outward unit vector normal to the boundary $\Gamma = \partial\Omega$, $\Gamma = \Gamma_D \cup \Gamma_N$; A_h - the stiffness matrix; F_h - the right hand side; h - the mesh parameter of the underlying partition $\mathcal{T}_h$ of Ω.

We assume that conforming quadratic finite elements have been used in the process of discretization of (2) and also that the partition of the domain Ω has

I. Lirkov, S. Margenov, and J. Waśniewski (Eds.): LSSC 2011, LNCS 7116, pp. 582–589, 2012.

been performed in such a way that over each element $e \in \mathcal{T}_h$ the functions $a_{ij}(x)$ are constant. Due to the partitioning of the domain Ω into finite elements, the stiffness matrix A_h can be written in the form

$$A_h = \sum_{e \in \mathcal{T}_h} R_e^T A_e R_e, \tag{3}$$

where A_e is the element stiffness matrix and R_e is the mapping that restricts a global vector to a given element $e \in \mathcal{T}_h$.

It can be proved, see, e.g., [8], that without loss of generality for the local analysis of (3) it is sufficient to consider an arbitrary anisotropic symmetric positive definite coefficient matrix $a(e)$ on the reference triangle $\tilde{e}$ with coordinates $(0,0), (1,0), (0,1)$, or, alternatively, the isotropic Laplace operator, i.e., $a(e) = I$, on a triangle of an arbitrary shape. The element stiffness matrix then can be written as

$$A_e = \begin{bmatrix} \dfrac{b+c}{2} & -\dfrac{2c}{3} & \dfrac{c}{6} & 0 & \dfrac{b}{6} & -\dfrac{2b}{3} \\[2mm] -\dfrac{2c}{3} & \dfrac{4(a+b+c)}{3} & -\dfrac{2c}{3} & -\dfrac{4b}{3} & 0 & -\dfrac{4a}{3} \\[2mm] \dfrac{c}{6} & -\dfrac{2c}{3} & \dfrac{a+c}{2} & -\dfrac{2a}{3} & \dfrac{a}{6} & 0 \\[2mm] 0 & -\dfrac{4b}{3} & -\dfrac{2a}{3} & \dfrac{4(a+b+c)}{3} & -\dfrac{2a}{3} & -\dfrac{4c}{3} \\[2mm] \dfrac{b}{6} & 0 & \dfrac{a}{6} & -\dfrac{2a}{3} & \dfrac{a+b}{2} & -\dfrac{2b}{3} \\[2mm] -\dfrac{2b}{3} & -\dfrac{4a}{3} & 0 & -\dfrac{4c}{3} & -\dfrac{2b}{3} & \dfrac{4(a+b+c)}{3} \end{bmatrix},$$

where in the first case $a = -a_{12}(e)$, $b = a_{12}(e) + a_{22}(e)$, $c = a_{12}(e) + a_{11}(e)$, while in the second a, b, c equal the cotangents of the angles in $e \in \mathcal{T}_h$.

One of the most popular iterative methods for solving systems of linear algebraic equations with large sparse symmetric and positive definite matrices is the Preconditioned Conjugate Gradient (PCG) method whose convergence rate crucially depends on the quality of the used preconditioner. The general strategy for efficient preconditioning requires that the spectral condition number $\kappa(C^{-1}A_h)$ of the preconditioned matrix is much less than the condition number of the original matrix and that the computational complexity $\mathcal{N}(C^{-1}v)$ of applying the preconditioner is much smaller than the complexity of solving the original problem. The symmetric and positive definite matrix C is referred to as an optimal preconditioner for the system (1) if $\kappa(C^{-1}A_h) = O(1)$ and $\mathcal{N}(C^{-1}v) = O(N)$, where N is the number of unknowns.

2   Two-Level Preconditioners

Let us consider two consecutive meshes obtained from a uniform refinement procedure and let us split the unknowns of the corresponding system into two groups where the first one consists of the nodes which do not belong to the coarse mesh and the second one contains all nodes that belong to the coarse mesh. Due to this two-level splitting we present the matrix $A = A_h$, and for convenience we skip the subscript h with the matrix A from now on, in the factorized form

$$A = \begin{bmatrix} A_{11} & A_{21}^T \\ A_{21} & A_{22} \end{bmatrix} = \begin{bmatrix} I & \\ A_{21}A_{11}^{-1} & I \end{bmatrix} \begin{bmatrix} A_{11} & A_{21}^T \\ & S \end{bmatrix}, \tag{4}$$

where the Schur complement is given by $S = A_{22} - A_{21}A_{11}^{-1}A_{21}^T$. Then we consider the following multiplicative two-level preconditioner

$$B = \begin{bmatrix} I & \\ B_{21}B_{11}^{-1} & I \end{bmatrix} \begin{bmatrix} B_{11} & B_{21}^T \\ & Q \end{bmatrix} \tag{5}$$

where Q is an approximation to the exact Schur complement S and B_{11} is spectrally equivalent to A_{11}.

Two-level preconditioners like (5) are the basis for constructing multilevel methods, which can be obtained by applying recursively a two-level method to the coarse-level matrix (Schur complement approximation). Optimal multilevel methods, however, additionally exploit certain stabilization techniques. For example, the so-called Algebraic Multi-Level Iteration (AMLI) algorithm stabilizes the condition number via certain matrix polynomials, cf. [3,4]. The AMLI preconditioner can be proven to be optimal if the degree v of the stabilization polynomial satisfies the condition

$$\left(1 - \gamma^2\right)^{-1/2} < v < \rho, \tag{6}$$

where γ is the constant in the strengthened Cauchy-Bunyakowski-Schwarz (CBS) inequality and ρ is the reduction factor of the number of degrees of freedom. For 2D problems and regular mesh refinement typically $\rho = 4$.

2.1 Hierarchical Basis Approximation

A commonly used approach for constructing two-level preconditioners involves the hierarchical basis, see e.g. [8,3,4]. Let $\mathcal{T}_H$ and $\mathcal{T}_h$ be two successive mesh refinements of the domain Ω and let $\{\phi_H^{(k)}, k = 1, 2, \cdots, N_H\}$ and $\{\phi_h^{(k)}, k = 1, 2, \cdots, N_h\}$ denote the corresponding standard finite element nodal basis functions. We assume that the diffusion tensor $a(x)$ is piecewise constant on the coarse mesh. Then we split the mesh points of $\mathcal{T}_h$ into two groups - coarse grid nodes, i.e. points from $\mathcal{T}_H$, and the rest, i.e. points from $\mathcal{T}_h \setminus \mathcal{T}_H$. Now we define the hierarchical two-level basis as follows:

$$\{\tilde{\phi}_h^{(i)}\}_{i=1}^{N_h} = \{\phi_h^{(m)} \text{ on } \mathcal{T}_h \setminus \mathcal{T}_H\} \cup \{\phi_H^{(l)} \text{ on } \mathcal{T}_H\}.$$

Obviously, the nodal basis matrix A and the hierarchical basis matrix $\tilde{A}$ are related to each other via $\tilde{A} = JAJ^T$ where J is a sparse matrix.

The hierarchical basis (multiplicative) two-level preconditioner is defined by

$$C = \begin{bmatrix} I & \\ \tilde{A}_{21}\tilde{A}_{11}^{-1} & I \end{bmatrix} \begin{bmatrix} \tilde{A}_{11} & \tilde{A}_{12} \\ & \tilde{A}_{22} \end{bmatrix}. \tag{7}$$

The pivot block and the Schur complement are invariant with respect to the above basis transformation, i.e., $\tilde{A}_{11} = A_{11}$ and $\tilde{S} = S$. For the second diagonal block, related to the coarse mesh unknowns, we have $\tilde{A}_{22} = A_H$.

Then the relative condition number with respect to the multiplicative two-level preconditioner C can be estimated by

$$\kappa(C^{-1}\tilde{A}) \leq \kappa(A_H^{-1}S) = \kappa(A_H^{-1}\tilde{S}) \leq 1/(1-\gamma^2) \tag{8}$$

where γ is the CBS constant related to the considered hierarchical two-level splitting. The bound for γ that we are going to present in Section 3 (Lemma 1) is in accordance with the numerical results in [6]. However, the experiments that were presented in [6] also show that for certain highly anisotropic elliptic problems $\gamma \to 1$ and thus the preconditioner (7) is not robust with respect to general operator and/or coefficient anisotropy.

2.2 Schur Complement Approximation

An alternative approach to construct a two-level preconditioner is to approximate the exact Schur complement S in (4) by a proper sparse matrix Q without changing the basis. The technique for computing such an approximation Q we are studying here has been introduced in [7]. The idea is first to compute all local macroelement Schur complements

$$S_E = A_{E:22} - A_{E:21}A_{E:11}^{-1}A_{E:12}, \qquad \forall E \in \mathcal{T}_H,$$

associated with a partition of the mesh into macro elements E and then to assemble the small-sized local Schur complement matrices S_E in order to obtain the global Schur complement approximation Q. That is, we define

$$Q := \sum_E R_E^T S_E R_E \tag{9}$$

where R_E is the mapping that restricts a global vector to a macro element $E \in \mathcal{T}_H$. Note that when replacing A_H with Q in (8) the estimate still holds true, i.e.,

$$\kappa(B^{-1}A) \leq \kappa(Q^{-1}S) \leq \frac{1}{1-\gamma^2}, \tag{10}$$

see [2] for details.

3   Orthotropic Elliptic Problem

Without loss of generality we can write the orthotropic elliptic operator $a(e)$ as

$$a(e) = \begin{bmatrix} 1 & 0 \\ 0 & \varepsilon \end{bmatrix}, \quad \varepsilon \in (0,1]. \tag{11}$$

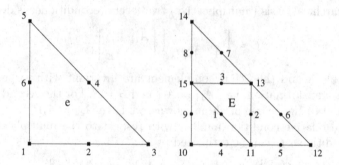

Fig. 1. Local numbering of the element and macroelement nodes

The element stiffness matrix for numbering of the nodes as in Fig. 1 looks like

$$
A_e = \begin{bmatrix}
(1+\varepsilon)/2 & -2/3 & 1/6 & 0 & \varepsilon/6 & -2\varepsilon/3 \\
-2/3 & 4(1+\varepsilon)/3 & -2/3 & -4\varepsilon/3 & 0 & 0 \\
1/6 & -2/3 & 1/2 & 0 & 0 & 0 \\
0 & -4\varepsilon/3 & 0 & 4(1+\varepsilon)/3 & 0 & -4/3 \\
\varepsilon/6 & 0 & 0 & 0 & \varepsilon/2 & -2\varepsilon/3 \\
-2\varepsilon/3 & 0 & 0 & -4/3 & -2\varepsilon/3 & 4(1+\varepsilon)/3
\end{bmatrix}. \quad (12)
$$

Then we can present the macroelement stiffness matrix assembled by four element stiffness matrices in the two-by-two block form

$$
A_E = \begin{bmatrix} A_{E:11} & A_{E:21}^T \\ A_{E:21} & A_{E:22} \end{bmatrix}, \quad (13)
$$

where all of its blocks depend only on one parameter - ε.

A well known result, see [3,8], is that the CBS constant γ can be estimated locally by the macroelement CBS constants γ_E

$$
\gamma \le \max_{E \in \mathcal{T}_h} \gamma_E, \qquad \gamma_E^2 = 1 - \mu_1 \quad (14)
$$

where μ_1 is the the minimal eigenvalue of the generalized eigenvalue problem

$$
S_E \mathbf{v}_{E:2} = \mu A_H \mathbf{v}_{E:2}, \qquad \mathbf{v}_{E:2} \ne const. \quad (15)
$$

If we denote

$$
B := \begin{bmatrix} A_{E:11}^{(k+1)} & A_{E:12}^{(k+1)} \\ A_{E:21}^{(k+1)} & A_{E:22}^{(k+1)} - \mu A_e^{(k)} \end{bmatrix} \quad (16)
$$

it can be proved, see [8], that it is enough to find a constant $\mu > 0$ for which B is symmetric positive semi-definite (SPSD) in order to provide a lower bound for μ in (15) and therefore an estimate for γ^2.

We can formulate now the following lemma:

Lemma 1. *The CBS constant γ corresponding to a hierarchical two-level splitting of the conforming quadratic finite element space is uniformly bounded with respect to the anisotropy ratio of the orthotropic operator, that is*

$$\gamma^2 \leq 3/4. \tag{17}$$

Proof. From (12), (13) one can see that the matrix B in (16) takes the form

$$B = (B_0 - \mu B_\mu) + \varepsilon(B_\varepsilon - \mu B_{\varepsilon\mu}) \tag{18}$$

where the matrices B_0, B_ε, B_μ and $B_{\varepsilon\mu}$ are SPSD and do not depend on any parameters. As we want B to remain SPSD for all $\varepsilon > 0$, we have to find the values of μ for which both $(B_0 - \mu B_\mu)$ and $(B_\varepsilon - \mu B_{\varepsilon\mu})$ remain SPSD.

First we solve the generalized eigenvalue problem

$$B_0 \mathbf{v} = \lambda B_\mu \mathbf{v}$$

and utilizing the Mathematica software tool we find its minimal eigenvalue $1/4$. This means that $(B_0 - \mu B_\mu)$ is SPSD for any $\mu \leq 1/4$.

In the same way we find that the necessary condition of $(B_\varepsilon - \mu B_{\varepsilon\mu})$ to be an SPSD matrix is $\mu \leq 1/4$, from where we conclude that $1/4$ is the sharp lower bound of the minimal eigenvalue μ_1 in (15). By gathering (14) with the last result we finally obtain

$$\gamma^2 \leq \max_{E \in \mathcal{T}_h} \gamma_E \leq 1 - \mu_1 \leq 1 - 1/4 = 3/4.$$

∎

Corollary 1. *The spectral condition number of the two-level preconditioner (9) is uniformly bounded with respect to the mesh size and the anisotropy ratio of the orthotropic elliptic operator, that is*

$$\kappa(Q^{-1}S) \leq 4. \tag{19}$$

Proof. Follows directly from (10) and (17). ∎

In Table 1 we list some values of the condition number of $Q^{-1}S$ for varying mesh size and anisotropy ratio.

3.1 Preconditioning of the Pivot Block

In this section we present the construction of a spectrally equivalent preconditioner to the pivot block A_{11} for the orthotropic elliptic operator that is based on the assembling of local preconditioners to the macroelement pivot blocks. If

$$B_{11} = \sum_E B_{E:11} \tag{20}$$

for the related spectral condition number we will have the estimate

$$\kappa(B_{11}^{-1} A_{11}) \leq \kappa(B_{E:11}^{-1} A_{E:11}). \tag{21}$$

Table 1. Relative condition number of sparse Schur complement approximation

$\kappa(Q^{-1}S)$	$\varepsilon = 10^0$	$\varepsilon = 10^{-1}$	$\varepsilon = 10^{-2}$	$\varepsilon = 10^{-3}$
$1/h = 8$	1.8755	2.2326	1.6447	1.1362
$1/h = 16$	1.8864	2.4259	2.2913	1.3415
$1/h = 32$	1.8911	2.4872	2.7174	1.9022
$1/h = 64$	1.8931	2.5061	2.8713	2.5305

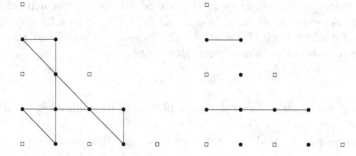

Fig. 2. Connectivity pattern of $A_{E:11}$ (left) and $B_{E:11}$ (right)

We construct the preconditioning matrix $B_{E:11}$ from $A_{E:11}$ by dropping all off-diagonal entries that correspond to vertical or diagonal edges in the graph representation of the macroelement pivot block, as shown in Fig. 2, and adding their value to the diagonal entry in the same row. With the right ordering the resulting global preconditioner B_{11} becomes a tridiagonal matrix and hence the number of arithmetic operations to solve a linear system with B_{11} is proportional to the number of unknowns in this system. Now we can formulate the following lemma:

Lemma 2. *The spectral condition number of the preconditioner (20) as defined above is uniformly bounded with respect to the mesh size and the anisotropy ratio of the orthotropic elliptic operator, that is*

$$\kappa(B_{11}^{-1}A_{11}) \leq 4. \tag{22}$$

Proof. In order to estimate $\kappa(B_{E:11}^{-1}A_{E:11})$, we consider the inequalities

$$\lambda_{min}v^T v \leq v^T B_{E:11}^{-1} A_{E:11} v \leq \lambda_{max} v^T v. \tag{23}$$

Our aim is to find a lower bound $\underline{\lambda}$ for λ_{min} and upper bound $\bar{\lambda}$ for λ_{max}. For that reason we rewrite the left side of (23) in the form

$$v^T(A_{E:11} - \lambda B_{E:11})v \geq 0. \tag{24}$$

If we denote
$$D := A_{E:11} - \lambda B_{E:11}$$
then any value λ for which D is an SPSD matrix will provide a lower bound of λ_{min}. We can present the matrix D in the form

$$D = D_0 + \varepsilon D_\varepsilon - \lambda D_\lambda - \varepsilon \lambda D_{\varepsilon \lambda},$$

where the matrices D_0, D_ε, D_λ and $D_{\varepsilon\lambda}$ are SPSD and do not depend on any parameter.

Since $0 < \varepsilon \leq 1$ and $v^T D v$ depends linearly on ε it suffices to consider only the two cases $\varepsilon = 0$ and $\varepsilon = 1$ in order to find the largest possible lower bound $\underline{\lambda} = 1$ for λ_{min}.

In the same way we can determine the smallest possible upper bound $\bar{\lambda} = 4$ for λ_{max}, from where we finally obtain

$$\kappa(B_{11}^{-1} A_{11}) \leq \kappa(B_{E:11}^{-1} A_{E:11}) \leq \frac{\bar{\lambda}}{\underline{\lambda}} \leq 4.$$

∎

Acknowledgment. The partial support of the Bulgarian NSF Grants DCVP 02/01 and DO 02–147/08 and Austrian FWF grant P22989-N18 is highly appreciated.

References

1. Axelsson, O.: Stabilization of algebraic multilevel iteration method; Additive methods. Numerical Algorithms 21(1-4), 23–47 (1999)
2. Axelsson, O., Blaheta, R., Neytcheva, M.: Preconditioning of boundary value problems using elementwise Schur complements. SIAM J. Matrix Anal. Appl. 31, 767–789 (2009)
3. Axelsson, O., Vassilevski, P.: Algebraic multilevel preconditioning methods I. Numer. Math. 56, 157–177 (1989)
4. Axelsson, O., Vassilevski, P.: Algebraic multilevel preconditioning methods II. SIAM J. Numer. Anal. 27, 1569–1590 (1990)
5. Blaheta, R., Margenov, S., Neytcheva, M.: Uniform estimate of the constant in the strengthened CBS inequality for anisotropic non-conforming FEM systems. Numerical Linear Algebra and Applications 11(4), 309–326 (2004)
6. Georgiev, I., Lymbery, M., Margenov, S.: Analysis of the CBS Constant for Quadratic Finite Elements. In: Dimov, I., Dimova, S., Kolkovska, N. (eds.) NMA 2010. LNCS, vol. 6046, pp. 412–419. Springer, Heidelberg (2011)
7. Kraus, J.: Algebraic multilevel preconditioning of finite element matrices using local Schur complements. Numer. Linear Algebra Appl. 13, 49–70 (2006)
8. Kraus, J., Margenov, S.: Robust Algebraic Multilevel Methods and Algorithms. De Gruyter, Germany (2009)
9. Maitre, J.F., Musy, S.: The contraction number of a class of two-level methods; An exact evaluation for Some finite element subspaces and model problems. Lect. Notes Math. 960, 535–544 (1982)

A Computational Approach for the Earthquake Response of Cable-Braced Reinforced Concrete Structures under Environmental Actions

Angelos Liolios[1], Konstantinos Chalioris[1], Asterios Liolios[1],
Stefan Radev[2], and Konstantinos Liolios[3]

[1] Democritus University of Thrace, Dept. Civil Engineering
Division of Structural Engineering, GR-67100 Xanthi, Greece
aliolios@civil.duth.gr
[2] Institute of Mechanics, Bulgarian Academy of Sciences
Acad. G. Bonchev Str. 8, Sofia 1113, Bulgaria
stradev@imbm.bas.bg
[3] Democritus University of Thrace, Dep. Environmental Engineering
Lab. Ecological Mechanics and Technology, Xanthi, Greece
kliolios@env.duth.gr

Dedicated to the Memory of Prof. Panagiotis D. Panagiotopoulos,
(1.1.1950-12.8.1998),
Late Professor of Aristotle-University of Thessaloniki, Greece

Abstract. A numerical treatment for the seismic response of reinforced concrete structures containing cable elements is presented. The cable behaviour is considered as nonconvex and nonmonotone one and is described by generalized subdifferential relations including loosening, elastoplastic - fracturing etc. effects. The problem is treated incrementally by double discretization: in space by finite elements and piece-wise linearization of cable - behaviour, and in time by the Newmark method. Thus, in each time - step an incremental linear complementarity problem is solved with a reduced number of problem unknowns.

Keywords: Computational Earthquake Engineering, Dynamic Unilateral Problems, Cable-braced Structures.

1 Introduction

In Civil Engineering, braces play an important role for the strengthening, repair and earthquake resistant design and construction, see e.g. [1,2]. This holds especially for reinforced concrete structures, exposed to environmental actions and requiring a strengthening procedure to continue to be serviceable. So, the seismic analysis of braced-structures, such as framing systems, suspended roofs and bridges, offshore platforms and braced towers, is an active investigation field [1,2,3].

A special class of braced structures are those containing cable elements. The peculiarity is that these elements can transmit tensile stresses only. The so-caused nonnegativity inequality for cable stresses is the principal condition in

I. Lirkov, S. Margenov, and J. Waśniewski (Eds.): LSSC 2011, LNCS 7116, pp. 590–597, 2012.

the relevant mathematical formulation of the problem. This is nonlinear, not only because of the presence of stress inequalities, but also because the considered cable stress-strain law is nonlinear, non-convex and non-monotone.

Thus, the formulated theory is a large displacement inequality theory, as usual in cable-structures, see e.g. Panagiotopoulos [3,6]. A compact mathematical treatment of the static problem of cable-structures has been also presented by Panagiotopoulos [3,7], on the basis of the variational or hemivariational inequality approach. As well known, the hemivariational inequality concept has been introduced into Mechanics and Applied Mathematics by P.D. Panagiotopoulos for first time in 1983, see [8], and constitutes now the basis of the so-called Non-Smooth Mechanics.

Further, as concerns numerical aspects, a remarkable approach to the above inequality problem has been obtained by piece-wise linearization of the cable constitutive laws and by using mathematical programming [3,4,5].

In general, the solution of the above problem in Civil Engineering praxis usually requires large-scale computations. So, the reduction of the computational procedure is desirable. The aim of this paper is to present such a numerical approach to the seismic problem of cable braced reinforced concrete structures. For this purpose we use a double discretization. First the problem is discretized in space by the Finite Element Method (FEM) and by piece-wise linearization of the constitutive laws of cable-elements. Then, due to large cable deformations, the problem is given an incremental formulation. Further, a time discretization is applied by using the Newmark method [1,2,10]. In each time-step an incremental non-convex linear complementarity problem, with reduced number of unknowns, is formulated and solved. Finally, the developed numerical procedure is applied to a practical Civil Engineering example.

2   Problem Formulation

The structural system is discretized in space by using finite elements of the "natural" type, see e.g. [3,6]. Pin-jointed bar elements are used for the cables. The behaviour of these elements includes loosening, elastoplastic or/and elastoplastic-softening-fracturing and unloading - reloading effects. All these characteristics can be expressed mathematically by the relation:

$$s_i(d_i) \in \hat{\partial} S_i(d_i) \tag{1}$$

where s_i and d_i are the (tensile) force and the deformation (elongation), respectively, of the i-th cable element, $\hat{\partial}$ is the generalized gradient and S_i is the superpotential function, see Panagiotopoulos [7,8]. By definition, relation (1) is equivalent to the following hemivariational inequality, expressing the Virtual Work Principle:

$$S_i^\uparrow(d_i, e_i - d_i) \geq s_i(d_i) \cdot (e_i - d_i) \tag{2}$$

where denotes the subderivative of $S_i^\uparrow$ and e_i, d_i are kinematically admissible (virtual) deformations.

From the mathematical point of view, using (1) and (2), we can formulate the problem as a hemivariational inequality one by following [7] and investigate it. Instead of this, and because here we are interested for the computational treatment of the problem, we proceed directly by piece wise linearizing the constitutive relations (1). So, in a way similar to that in elastoplasticity - see e.g. Maier [5,6] - and in unilateral elastodynamics - see e.g. Liolios [9]-, we have the following constitutive relations (in matrix notation, bold symbols) for the cable elements:

$$\psi = B^T \sigma - A\nu - r, \quad (r \geq 0) \tag{3}$$

$$\psi \leq 0, \quad \nu \geq 0, \quad \psi^T \nu = 0. \tag{4}$$

Here σ is the stress vector of the whole structural system; B is a transformation matrix; A is the current symmetric interaction matrix; and ψ, $(-\nu)$, r are the yield, slackness and ultimate capacity (resistance), respectively, vectors of cable elements. The remaining constitutive relations for the unassembled structural system are

$$e = \varepsilon + B\nu + \theta, \tag{5}$$

$$\sigma = E\varepsilon \quad \text{or} \quad \varepsilon = E^{-1}\sigma, \tag{6}$$

where e, ε, θ are the total, pure elastic and imposed (e.g. thermal or dislocations) strain vectors, respectively, and E is the current elasticity matrix, symmetric and positive definite.

Next, dynamic equilibrium and compatibility for the assembled structural system are expressed, respectively, by the relations

$$G\sigma + K_G u = p - C\dot{u} - M\ddot{u}, \tag{7}$$

$$e = G^T u. \tag{8}$$

Here G is the equilibrium matrix and G^T, its transposed, is the compatibility matrix; u and p are the displacement and the load vectors, respectively; C and M are the damping and mass matrices, respectively, both symmetric and in general positive (semi)- definite. The geometric stiffness matrix K_G depends linearly on preexisting constant load [6]. Through the term $K_G u$ alone the geometry changes affect the equilibrium (second order geometric effects). As usual, dots over symbols denote derivatives with respect to time. On the other hand, for the case of seismic excitation, it is

$$p = -M\ddot{x}_g, \tag{9}$$

where $x_g(T)$ is the ground seismic displacement.

Finally, the initial conditions are

$$u(t = 0) = u_0, \quad \dot{u}(t = 0) = \dot{u}_0, \quad \nu(t = 0) = \nu_0, \tag{10}$$

where $u_0, \dot{u}_0$, and ν_0 are known quantities.

Thus the problem consists in finding the response set $\{\sigma(t), u(t), \varepsilon(t); \psi(t), \nu(t)\}$ which satisfies (3)-(10) for the given excitation set $\{p(t),$ -or $x_g(t)$-, $\theta(t),$ $u_0, \dot{u}_0, \nu_0\}$.

3   The Incremental Linear Complementarity Approach

Due to nonmonotone and nonconvex cable behaviour (large deformations, loosening, elastoplastic-softening-fracturing effects, unloading-reloading etc.), E and A depend on u and ν, respectively. Therefore, an incremental formulation for the problem is more suitable. For this purpose, let E_t, A_t, ψ_t, ν_t etc. denote known quantities at the time t and let Δt be the time increment. Then the linear complementarity conditions (4) at the next time-moment $(t + \Delta t)$ are written as follows:

$$\psi_t + \Delta\psi \le 0, \quad \nu_t + \Delta\nu \ge 0, \quad (\psi_t + \Delta\psi)^T \cdot (\nu_t + \Delta\nu) = 0 \tag{11}$$

Next, the remaining problem conditions (3), (5)-(8) take the incremental form

$$\Delta\psi = B^T \Delta\sigma - A_t \Delta\nu, \tag{12}$$

$$\Delta e = \Delta\varepsilon + B\Delta\nu - \Delta\theta, \tag{13}$$

$$\Delta\sigma = E_t \Delta\varepsilon \text{ or } \Delta\varepsilon = E_t^{-1}\Delta\sigma, \tag{14}$$

$$\Delta\sigma + K_G \Delta u = \Delta p - C\Delta\dot{u} - M\Delta\ddot{u} \tag{15}$$

$$\Delta e = G^T \Delta u. \tag{16}$$

Further, we use a time discretization scheme for the step-by-step solution of problem (11)-(16). As known - see e.g. [10] -, for implicit time integration methods, relations of the following form hold:

$$\Delta\ddot{u} = c_1 \Delta u + a, \quad \Delta\dot{u} = c_2 \Delta u + b \tag{17}$$

where c_1 and c_2 are positive constants in terms of Δt, and a, b known quantities from previous time-steps. So, for the constant-average acceleration method, which is chosen here from the Newmark time-integration schemes, we have

$$c_1 = \frac{4}{\Delta t^2}, \quad c_2 = \frac{2}{\Delta t}. \tag{18}$$

In order to reduce the number of unknowns, we substitute (17)-(18) into (15) and eliminate $\Delta\sigma$, $\Delta\varepsilon$, Δe, Δu from (12)-(16). So we eventually arrive at

$$\Delta\psi = \Lambda\Delta\nu + \lambda \tag{19}$$

where
$$
\begin{aligned}
&\Lambda = H^T K^{-1} H - A - B^T EB, \\
&H = GEB, \\
&K = GEG^T + K_G + c_2 C + c_1 M, \\
&\lambda = H^T K^{-1}(\Delta f + Cb + Ma) - B^T E\Delta\theta, \\
&\Delta f = \Delta p + GE\Delta\theta.
\end{aligned}
\tag{20}
$$

Now, (19) with (11) constitute a Linear Complementarity Problem (LCP). In comparison to problem of (1)-(10), the LCP has a reduced number of unknowns. The LCP is solved in terms of the increments $\Delta\psi$ and $\Delta\nu$ by available computer codes of mathematical programming, see [5,6,7]. Finally, by using Euler relations $u = u_t + \Delta u$, $\nu = \nu_t + \Delta\nu$ etc., we complete the solution at time $(t + \Delta t)$.

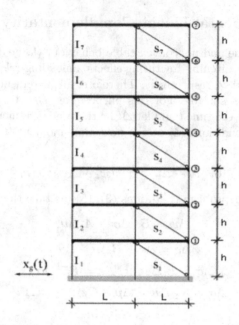

Fig. 1. Numerical example: The cable-braced 7-storey structural system

4   Numerical Example

The presented numerical approach is suitable for Civil Engineering 3-dimensional building structures, where large-scale computations are usually required. Here, the structural case of Fig. 1 is investigated as a simple 2-dimensional representative example. The 7-storey plan framing system of reinforced concrete class C40/45, with L = 7 m and h = 3.5 m, was initially designed and constructed without cable-braces. The beams are of rectangular section 30/75 (width/height, in cm) and have a total vertical distributed load 50 KN/m (each beam). The columns have section dimensions, in cm: 45/45 for the i =1,2 floors, 30/45 for the i = 3,4 floors, and 30/30 for the i = 5,6,7 floors.

Due to environmental actions, corrosion and cracking has been taken place. This had caused a reduction for the section inertia moments, which is estimated [11] to be 10% for the columns and 50% for the beams. So it was necessary for the system to be strengthened. Because of architectural reasons, the cable-braces system shown in Fig. 1 has been applied, and not the usual [1] X-braces. The cable elements, of steel class S500, have a cross-sectional area $F_c = 4cm^2$ and a unilateral behaviour depicted in Fig. 2, with yield strain $\varepsilon_y = 0.2\%$, fracture strain $\varepsilon_f = 2\%$, yield stress $\sigma_y = 40$ kN/cm^2, and elasticity modulus $E_c = 200GPa$. The branch OA is a 2-nd degree parabola. Loading-unloading and fracturing are taken into account.

According to relations (3)-(10), the total number of basic unknowns at every time-step Δt is 70 (three per node, for 3x3x7 =63 nodes, plus the 7 cables).

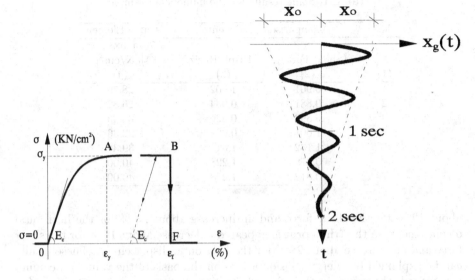

Fig. 2. Cable-elements constitutive law

Fig. 3. Seismic horizontal ground displacement

By using the proposed LCP approach of (11)-(18), only 7 basic unknowns for the 7 cables have to be computed at each time-step. So a significant reduction of the problem unknowns is obtained.

The system is subjected to the horizontal ground seismic excitation:

$$x_g(t) = x_0 e^{-2t} \sin(4\pi t), \tag{21}$$

where $x_0 = 0.025$m. The graphic representation of $x_g(t)$ is shown in Fig. 3. According to usual anti-seismic codes [1,2], the corresponding maximum seismic ground acceleration is about $0.32g$, where g is the gravity acceleration.

Further, for comparison reasons, we introduce the comparison coefficients

$$c = \frac{Q_c}{Q_f}, \tag{22}$$

where Q is the absolutely maximum value which takes a response quantity during the seismic excitation. Index (c) is for the cable-braced system and index (f) for the free (i.e. without cables) system.

Some representative results, obtained by applying the numerical method developed in previous sections, are shown in Table 1. These results concern on the one hand the comparison coefficients c_s for the floor shear forces and c_d for the floor horizontal displacements, and on the other hand the maximum stress s_i for the i-th cable element $(i = 1,\ldots,7)$.

As the table values of column (3) show, the most influenced floors due to cable behaviour are the two higher ones. So, in the 6th floor it is appeared a decrease

Table 1. Some results for the numerical example

Floor	Comparison coefficients		Cable-Element
	c_s	c_d	$\max s_i$
	(Shear Force)	(Displ. Horiz.)	[kN/cm^2]
(1)	(2)	(3)	(4)
1	0.901	1.002	28.79
2	0.881	0.934	26.82
3	0.893	0.920	30.43
4	1.001	0.924	22.36
5	1.132	1.017	39.38
6	0.878	1.228	40.00
7	0.812	1.278	40.00

about 12 % for the shear force and an increase about 23 % for the horizontal displacement. In the 7th floor it is appeared a decrease about 19 % for the shear force and an increase about 28 % for the horizontal displacement. These results can be explained by energy considerations on the basis of the values on column (4). Indeed, the cables in the 6th and 7th floors, with max $s_i = 40$ kN/cm^2), have been plastified. Therefore, the seismic energy, which would been absorbed by them after plastification, has returned partially to the frame system.

Such effects can be of impact type [9] when cable-elements are fractured, and so the change of response values for the frame can be appeared in a sudden and pounding way. On the other hand, we remark from the table values that as long as the cables remain in the elastic or the early elastoplastic range of their behaviour, i.e. without fracturing, the braced structure appears a reduced response in comparison to that one of the free structure. This fact concerns especially the shear forces in the floors $1 \div 4$.

5   Concluding Remarks

Large-scale computational procedures, which are required for the unilateral dynamic problem of the seismic analysis of cable-braced structures in civil engineering praxis, can be reduced significantly by the incremental approach herein presented. This approach takes into account in a realistic way the unilateral behaviour of cable elements and leads to a linear complementarity problem (LCP) in each time increment, with a reduced number of problem unknowns. The numerical realization is obtained by available computer codes of the finite element method, of step-by-step time integration schemes and of mathematical programming (optimization) algorithmes. Moreover, as it has been verified in a simple example, the herein developed approach can treat in a realistic way the seismic problem of cable-braced reinforced concrete structures in Civil Engineering praxis.

Acknowledgement. The author S.R. has been partially supported by the Bulgarian NSF Grants DO-02-115/2008, and DO-02-338/2008.

References

1. Chopra, A.K.: Dynamics of Structures: Theory and Applications to earthquake Engineering. Pearson Prentice Hall, New York (2007)
2. Newmark, N.M., Rosenblueth, E.: Fundamentals of earthquake engineering. PrenticeHall, Inc., Englewood Cliffs (1971)
3. Panagiotopoulos, P.D.: Stress-Unilateral analysis of discretized cable and membrane structure in the presence of large displacements. Ingenieur-Archiv. 44, 291–300 (1975)
4. Maier, G., Contro, R.: Energy approach to inelastic cable-structure analysis. J. Enging Mech. Div., Proc. ASCE, vol. 101, EM5, pp. 531–547 (1975)
5. Contro, R., Maier, G., Zavelani, A.: Inelastic analysis of suspension structures by nonlinear programming. Computer Meth. Appl. Mech. Enging 5, 127–143 (1975)
6. Maier, G.: Incremental plastic analysis in the presence of large displacements and physical instabilizing effects. Int. J. Solids Struct. 7, 345–372 (1971)
7. Panagiotopoulos, P.D.: Hemivariational Inequalities. Applications in Mechanics and Engineering. Springer, Berlin (1993)
8. Panagiotopoulos, P.D.: Non-convex Energy Functions. Hemivariational Inequalities and Substationarity principles. Acta Mechanica 48, 111–130 (1983)
9. Liolios, A.A.: A Linear Complementarity Approach for the Non-convex Seismic Frictional Interaction between Adjacent Structures under Instabilizing Effects. Journal of Global Optimization 17(1-4), 259–266 (2000)
10. Weaver Jr., W., Johnston, P.R.: Structural dynamics by finite elements. Prentice Hall, Inc., Englewood Cliffs (1987)
11. Pauley, T., Priestley, M.J.N.: Seismic design of reinforced concrete and masonry buildings. Wiley, New York (1992)

An Inverse Problem for the Stationary Kirchhoff Equation

Tchavdar T. Marinov[1] and Rossitza Marinova[2,*]

[1] Department of Natural Sciences, Southern University at New Orleans,
6801 Press Drive, New Orleans, LA 70126
tmarinov@suno.edu
[2] Department of Mathematical and Computing Sciences, Concordia University
College of Alberta, Edmonton, AB, Canada T5B 4E4
rossitza.marinova@concordia.ab.ca

Abstract. This work is concerned with the development of numerical methods and algorithms for solving the inverse problem for parameter identification from over-determined data in Kirchhoff plate equations. A technique called Method of Variational Imbedding is used for solving the inverse problem. The original inverse problem is replaced by a minimization problem. The Euler-Lagrange equations comprise a higher-order system of equations for the solution of the original equation and for the coefficients. In the present work, difference scheme and numerical algorithm for solving the Euler-Lagrange system are proposed. Results for different values of the governing parameters and the physical relevance are presented.

1 Introduction

Inverse problems are one of the most active areas of current research in mathematics, science, and engineering. The field of inverse problems for physical systems is concerned with determining system properties, which cannot be measured by direct means. A very common but not well-recognized phenomenon is the inverse problem for identification of the coefficient in partial differential equations when the data are over-posed. The estimation of an unknown coefficient from over-posed boundary data is of significant practical importance when creating non-invasive methods to identify the material properties of a continuum.

Consider the Kirchhoff equation in the form

$$\Delta (D\Delta w) = q, \tag{1}$$

in the unit square Ω. The function w is the deflection of the middle surface of the plate, $\Delta = \frac{\partial^2}{\partial x^2} + \frac{\partial^2}{\partial y^2}$ is the Laplace operator, and D is the bending stiffness of the plate. The function q represents the distributed transverse load applied to the plate.

---

* Adjunct Professor, Department of Computer Science, University of Saskatchewan, SK, Canada.

I. Lirkov, S. Margenov, and J. Waśniewski (Eds.): LSSC 2011, LNCS 7116, pp. 598–605, 2012.
© Springer-Verlag Berlin Heidelberg 2012

In the case of isotropy with a constant bending stiffness D the equation (1) becomes the well-known biharmonic equation of plate flexure

$$D\Delta^2 w = D\left(\frac{\partial^4 w}{\partial x^4} + 2\frac{\partial^4 w}{\partial x^2 \partial y^2} + \frac{\partial^4 w}{\partial y^4}\right) = q. \tag{2}$$

If the coefficient $D(x,y) > 0$ and the right-hand side function $q(x,y) \geq 0$ in (1) are given, under proper boundary conditions, the problem possesses a unique solution, usually referred as a direct solution.

In practice, there exist lots of interesting problems, in which the coefficient D (the bending stiffness of the plate) is not exactly known. In reality, it is expensive, even not possible, to measure the changes of the properties of the materials directly. On the other hand, changes in the physical properties of the materials cause changes in the coefficient D in equation (1) and, respectively, changes in the solution. Thus, a new, so-called inverse, problem appears: to find simultaneously the solution w and the coefficient D of the Kirchhoff equation.

Common approach to the solution of inverse problems is based on regularization methods, which requires finding a minimum with respect to a small regularization parameter of a properly constructed functional (see, for example, [10,4]). However, in general, the numerical solution of the regularization algorithm does not always approach the exact solution when the regularization parameter becomes very small. In the proposed work the inverse problems for coefficient identification from over-determined data is solved by using the Method of Variational Imbedding (MVI), proposed in [1]. MVI gives a solution of the original problem that coincides with the solution of the variational problem.

The idea of MVI is to replace the incorrect problem with a well-posed problem for minimization of a quadratic functional of the original equations, i.e. we "embed" the original incorrect problem in a higher-order boundary value problem which is well-posed (see [1,7,8]). For the latter, a difference scheme and a numerical algorithm for its implementation can easily be constructed. The advantage of MVI comparing to a regularization method is that there are no "boundary layers" at the two ends of the interval as it was observed for a similar problem in [5].

Recently, in [9], the MVI was applied to the problem for identifying the coefficient σ in (1) in the case when the coefficient is a piecewise constant function. In the present work we are considering the case when the coefficient is a piecewise linear function. Although this paper is focused on a fourth order ordinary differential equation, the proposed method can be generalized for identification of coefficient in partial differential equations. Similar to the procedure proposed here, the approach for the identification of a coefficient in parabolic partial differential equation is given in [2]. In [8] MVI is successfully applied to the problem for identification of a coefficient in elliptic partial differential equation.

The paper is organized as follows. In Section 2 the inverse problem for identification of the unknown coefficient is formulated. The application of the MVI to the inverse problem is described in Section 3. The numerical scheme is given in Section 4. Illustration of the constructed numerical scheme is given in Section 5.

2 Inverse Problem Formulation

Consider the Kirchhoff equation (1) where the function $q(x, y)$ is given. We expect that the functions under consideration are as many time differentiable as necessary. If the coefficient D is not given, in order to identify it one needs additional information. Suppose the solution satisfies the conditions

$$w(x, 0) = \alpha_{0,0}(x), \qquad w(x, 1) = \alpha_{1,0}(x), \tag{3}$$
$$w(0, y) = \beta_{0,0}(y), \qquad w(1, y) = \beta_{1,0}(y), \tag{4}$$
$$w_y(x, 0) = \alpha_{0,1}(x), \qquad w_y(x, 1) = \alpha_{1,1}(x), \tag{5}$$
$$w_x(0, y) = \beta_{0,1}(y), \qquad w_y(1, y) = \beta_{1,1}(y), \tag{6}$$
$$w_{yy}(x, 0) = \alpha_{0,2}(x), \qquad w_{yy}(x, 1) = \alpha_{1,2}(x), \tag{7}$$
$$w_{xx}(0, y) = \beta_{0,2}(y), \qquad w_{xx}(1, y) = \beta_{1,2}(y). \tag{8}$$

Obviously, the boundary value problem for the equation (1) with boundary conditions (3–8) is overdetermined. There are three boundary conditions at each boundary point for solving the fourth order equation. Suppose that the coefficient $D = D(x)$ is an unknown function of a single variable, say x. Hence, there are one unknown function of a single variable and an additional function (of a single variable) on the boundary $\partial\Omega$ of the domain Ω.

There may be no solution (w, D), satisfying all of the conditions (3)–(8), for arbitrary $\alpha_{k,l}$, $\beta_{k,l}$, (for $k = 0, 1$ and $l = 0, 1, 2$). For this reason, let us assume that the problem is posed correctly after Tikhonov, [10], i.e., it is known *a-priori* that a solution of the problem exists. In other words, we assume that the data in the boundary conditions (3)–(8) have "physical meaning" and, therefore, a solution exists.

The problem is how to transfer this additional information to the missing information on the coefficients. The solution approach proposed here is a generalization of the implementation of MVI to a similar problem given in [6] for identification of a piecewise coefficient, see also [7,8]. Continuation of the idea is proposed in [9].

3 Variational Imbedding

Following the idea of MVI, we replace the original problem with the problem of minimization of the functional

$$\mathcal{I}(w, D) = \int_0^1 \int_0^1 \mathcal{A}^2(w, D) \, dxdy = \int_0^1 \int_0^1 [\Delta(D\Delta w) - q]^2 \, dxdy \longrightarrow \min, \tag{9}$$

where w satisfies the conditions (3)–(8), $D(x) > 0$ is an unknown function. The functional $\mathcal{I}(w, D)$ is a quadratic and homogeneous function of $\mathcal{A}(w, D)$ and, hence, it attains its absolute minimum if and only if $\mathcal{A}(w, D) \equiv 0$. In this sense there is an one-to-one correspondence between the original equation (1) and the minimization problem (9).

The necessary condition for minimization of the functional $\mathcal{I}$ is expressed by the Euler-Lagrange equations for the functions $w(x, y)$ and $D(x)$.

3.1   Equation for w

The Euler-Lagrange equation with respect to the function w reads

$$\Delta \{D\Delta [\Delta(D\Delta w) - q]\} = 0. \tag{10}$$

The original problem includes the boundary conditions (3)–(8). Since the equation (10) is of the eight order we need one additional boundary condition at each boundary point. We use the original equation (1) as a boundary condition. The original equation is one of the natural boundary conditions for minimization of the functional $\mathcal{I}(w, D)$.

3.2   Equation for $D(x)$

Since D is a function of one variable, we rewrite the functional $\mathcal{I}$ as

$$\mathcal{I}(w, D) = \int_0^1 (C_{22}D''^2 + C_{11}D'^2 + C_{00}D^2 + 4C_{21}D''D' + Q$$
$$+ 2C_{20}D''D - 2C_2D'' + 4C_{10}D'D - 4C_1D' - 2C_0D)\,dx, \tag{11}$$

where

$$C_{22}(x) = \int_0^1 (w_{xx} + w_{yy})^2 dy, \tag{12}$$

$$C_{11}(x) = \int_0^1 (w_{xxx} + w_{xyy})^2 dy, \tag{13}$$

$$C_{00}(x) = \int_0^1 (w_{xxxx} + 2w_{xxyy} + w_{yyyy})^2 dy, \tag{14}$$

$$C_{21}(x) = \int_0^1 (w_{xx} + w_{yy})(w_{xxx} + w_{xyy})dy, \tag{15}$$

$$C_{20}(x) = \int_0^1 (w_{xx} + w_{yy})(w_{xxxx} + 2w_{xxyy} + w_{yyyy})dy, \tag{16}$$

$$C_{10}(x) = \int_0^1 (w_{xxx} + w_{xyy})(w_{xxxx} + 2w_{xxyy} + w_{yyyy})dy, \tag{17}$$

$$C_2(x) = \int_0^1 (w_{xx} + w_{yy})q\,dy, \tag{18}$$

$$C_1(x) = \int_0^1 (w_{xxx} + w_{xyy})q\,dy, \tag{19}$$

$$C_0(x) = \int_0^1 (w_{xxxx} + 2w_{xxyy} + w_{yyyy})q\,dy, \tag{20}$$

$$Q(x) = \int_0^1 q^2\,dy. \tag{21}$$

The Euler-Lagrange equation with respect to the function D reads

$$\frac{d^2}{dx^2}\left(2C_{22}D'' + 4C_{21}D' + 2C_{20}D - 2C_2\right)$$
$$-\frac{d}{dx}\left(4C_{21}D'' + 2C_{11}D' + 4C_{10}D - 4C_1\right)$$
$$+ 2C_{00}D + 2C_{20}D'' + 4C_{10}D' - 2C_0 = 0. \tag{22}$$

The equation (22) is of the fourth order; hence, it requires four boundary conditions. Suppose the values of the function D and the first derivatives of D at $x = 0$ and $x = 1$ are given, i.e.,

$$D(0) = \gamma_{00}, \qquad D(1) = \gamma_{01}, \tag{23}$$

$$D'(0) = \gamma_{10}, \qquad D'(1) = \gamma_{11}. \tag{24}$$

4   Difference Scheme

We solve the formulated eight-order boundary value problem by finite differences. It is convenient for the numerical treatment to rewrite the eight order equation as a system of two fourth order equations:

$$\Delta(D\Delta w) - q = u, \qquad \Delta(D\Delta u) = 0. \tag{25}$$

The additional boundary condition in terms of the new function u is $u = 0$. At each boundary point we have three boundary conditions on the function w and one on u. For this reason we cannot split the system and solve the equations separately.

4.1   Grid Pattern and Approximations

We introduce a regular mesh with steps h in x direction and k in y direction (see Fig. 1) allowing to approximate all operators with standard central differences with second order of approximation.

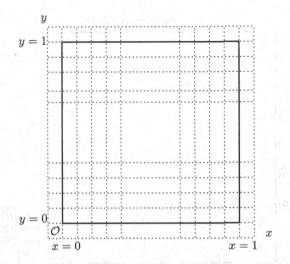

Fig. 1. The mesh used in the numerical experiments

The grid spacing are $h \equiv \frac{1}{N-3}$, where N is the total number of grid points in the x direction and $k \equiv \frac{1}{M-3}$, where M is the total number of grid points in the y direction. Then, the grid points are defined as follows: $x_i = (i-2)h$ for $i = 1, 2, \ldots, N$ and $y_j = (j-2)k$ for $j = 1, 2, \ldots, M$. We employ symmetric central differences for approximating the differential operators. The differential operators in the boundary conditions are approximated with second order formulae using central differences and half sums.

4.2   General Construction of the Algorithm

(I) With the obtained "experimentally observed" values of $\alpha_{i,j}$ and $\beta_{i,j}$, (for $i = 0, 1$ and $j = 0, 1, 2$), the eight-order boundary value problem (10), (3)–(8), is solved for the function w with an initial guess for the function D.

(II) The approximation of the function D for the current iteration is calculated from the equation for D (22) under boundary conditions (23), (24). If the l^1-norm of the difference between the new and the old field for D is less than ε_0 then the calculations are terminated. Otherwise, the algorithm returns to (I) with the newly calculated D.

5   Numerical Experiments

The accuracy of the difference scheme is validated with tests involving various grid spacing h and k. We have run a number of calculations with different values of the mesh parameters and verified the practical convergence and the $O(h^2+k^2)$ approximation of the difference scheme. For all calculations presented here, $\varepsilon_0 = 10^{-12}$ and the initial guess for the coefficient $D(x)$ is $D(x) \equiv 1$.

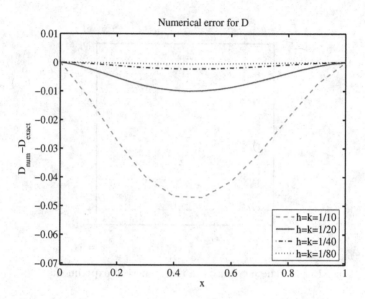

Fig. 2. The difference between numerical and exact values of $D(x)$ for $h = k = 0.1, 0.05, 0.025, 0.0125$

One exact solution can be obtained for $q(x, y) = 10e^{2x+y}$. In this case the exact solution is

$$w(x, y) = e^{x+y}, \quad D(x) = e^x. \tag{26}$$

The above solution has no practical meaning; however, it serves the purpose of verifying the properties of the numerical scheme. This test is used to confirm the second order of approximation of the proposed scheme. Because of the space limit, here we will provide only results for the obtained coefficient D. Since the accuracy for D depends on the accuracy for w the convergence for D implies the convergence for w.

The point-wise numerical errors for the function D, calculated for different grid steps $h = k = 0.1; 0.05; 0.025; 0.0125$, are shown at Fig. 2.

The number of iterations with respect to D and the l^1 norm of the difference between the identified coefficient D and the exact one for four different steps $h = k = 0.1; 0.05; 0.025; 0.0125$ are given in Table 1. The rates of convergence, calculated as

$$\text{rate}_D = \log_2 \frac{\|D_{2h} - D_{\text{exact}}\|}{\|D_h - D_{\text{exact}}\|}, \tag{27}$$

are also shown in Table 1.

As expected, the results confirm the $O(h^2 + k^2)$ of approximation of the numerical solution to the exact solution. The numerical experiments also show that the rate of convergence is two.

Table 1. The number of iterations, l^1 norm of the difference $D - D_{\text{exact}}$, and the rates of convergence for four different values of the mesh spacing

$h = k$	Iterations	$\|D - D_{\text{exact}}\|_{l^1}$	rate_D
0.1	87	0.0272	—
0.05	101	0.0056	2.2801
0.025	118	0.0012	2.2224
0.0125	142	2.9309e-004	2.0336

Acknowledgment. This work was partially supported by MITACS and NSERC.

References

1. Christov, C.I.: A method for identification of homoclinic trajectories. In: Proc. 14-th Spring Conf., Sunny Beach, Union of Bulg. Mathematicians, Sofia, Bulgaria (1985)
2. Christov, C.I., Marinov, T.T.: Identification of heat-conduction coefficient via method of variational imbedding. Mathematics and Computer Modeling 27(3), 109–116 (1998)
3. Hadamard, J.: Le Probleme de Cauchy et les Equations aux Derivatives Partielles Lineares Hyperboliques. Hermann, Paris (1932)
4. Lattès, R., Lions, J.L.: Mèthode de quasi-reversibilite et applications. Dunod, Paris (1967)
5. Lesnic, D., Elliott, L., Ingham, D.B.: Analysis of coefficient identification problems associated to the inverse Euler-Bernoulli beam theory. IMA J. of Applied Math. 62, 101–116 (1999)
6. Marinov, T.T., Christov, C.I.: Identification the unknown coefficient in Ordinary Differential Equations via Method of Variational Imbedding. In: Deville, M., Owens, R. (eds.) 16th IMACS World Congress 2000 Proceedings, paper 134–2. Rutgers University, New Brunswick (2000) ISBN 3-9522075-1-9
7. Marinov, T.T., Christov, C.I., Marinova, R.S.: Novel numerical approach to solitary-wave solutions identification of Boussinesq and Korteweg-de Vries equations. Int. J. of Bifurcation and Chaos 15(2), 557–565 (2005)
8. Marinov, T.T., Marinova, R.S., Christov, C.I.: Coefficient Identification in Elliptic Partial Differential Equation. In: Lirkov, I., Margenov, S., Waśniewski, J. (eds.) LSSC 2005. LNCS, vol. 3743, pp. 372–379. Springer, Heidelberg (2006)
9. Marinov, T.T., Vatsala, A.: Inverse Problem for Coefficient Identification in Euler-Bernoulli Equation. Computers and Mathematics with Applications 56(2), 400–410 (2008)
10. Tikhonov, A.N., Arsenin, V.: Methods for Solving Incorrect Problems. Nauka, Moscow (1974)

An Improved Sparse Matrix-Vector Multiply Based on Recursive Sparse Blocks Layout

Michele Martone[1], Marcin Paprzycki[2], and Salvatore Filippone[1]

[1] University of Rome "Tor Vergata", Via del Politecnico 1, 00133 Rome, Italy
[2] Systems Research Institute, Polish Academy of Sciences, ul. Newelska 6, 01-447
Warsaw, Poland

Abstract. The *Recursive Sparse Blocks* (*RSB*) is a sparse matrix layout designed for coarse grained parallelism and reduced cache misses when operating with matrices, which are larger than a computer's cache. By laying out the matrix in sparse, non overlapping blocks, we allow for the shared memory parallel execution of transposed *SParse Matrix-Vector multiply* (*SpMV*), with higher efficiency than the traditional *Compressed Sparse Rows* (CSR) format. In this note we cover two issues. First, we propose two improvements to our original approach. Second, we look at the performance of standard and transposed shared memory parallel *SpMV* for unsymmetric matrices, using the proposed approach. We find that our implementation's performance is competitive with that of both the highly optimized, proprietary Intel MKL Sparse BLAS library's CSR routines, and the *Compressed Sparse Blocks* (CSB) research prototype.

1 Introduction and Related Work

Many scientific/computational problems require the solution of systems of partial differential equations (PDEs). Often, discretization of these problems result in *sparse* matrices. A common approach for the solution of sparse linear systems is through the use of *iterative methods*, whose computational core requires sparse matrix-vector multiplication. In this document, we focus on the efficient implementation of sparse matrix-vector multiplication, on *cache based, shared memory* computers. In this context, we have recently proposed a sparse matrix format, called *Recursive Sparse Blocks* (*RSB*) [3,4]. The central idea of RSB is a recursive partitioning-based organization of matrices, with either *Compressed Sparse Rows* (CSR) or *Coordinate* (COO) format *leaves* of a *quad-tree* structure over matrices. In this paper, we present some optimizations to our RSB-based *SpMV* implementation, and compare performance of the modified approach to that of the Intel's MKL proprietary Sparse BLAS implementation, and the publicly available CSB (see [1]) prototype. To this end, we briefly recall the way that the *SpMV* /*SpMV_T* computational kernels work and behave on computers of our interest in § 2. Next, we introduce the proposed optimizations in § 3. Finally in § 5, we discuss the efficiency of our prototype, by comparing it to the mentioned highly efficient MKL's CSR and CSB implementations.

I. Lirkov, S. Margenov, and J. Waśniewski (Eds.): LSSC 2011, LNCS 7116, pp. 606–613, 2012.

```
1 for l ← 1 to s.nnz do
2     i ← s.IA(l);
3     j ← s.JA(l);
4     s.y(i + s.roff) ← s.y(i + s.roff) + s.VA(l)s.x(j + s.coff);
5 end
```

Fig. 1. *SpMV* listing for a COO submatrix *s*

```
1 for l ← 1 to s.nnz do
2     i ← s.IA(l);
3     j ← s.JA(l);
4     s.y(j + s.coff) ← s.y(j + s.coff) + s.VA(l)s.x(i + s.roff);
5 end
```

Fig. 2. *SpMV_T* listing for a COO submatrix *s*

2 *SpMV* and Transposed *SpMV*

We define the *sparse matrix-vector multiply* (*SpMV*) operation as "$y \leftarrow A\,x$" and its transposed version (*SpMV_T*) as "$y \leftarrow A^T\,x$" (where A is a sparse matrix, while x, y are vectors). With RSB, A is *recursively partitioned* into submatrices, and then the individual N *leaf submatrices* $s_1 : s_N$ are represented in either COO or CSR format (eventually using 16 bits for the local indices); for details, see [3,4]. The leaf submatrices are all disjoint; each submatrix s covers rows indices $[s.roff : s.roff + s.rows]$ and column indices $[s.coff : s.coff + s.cols]$. For this reason, the *SpMV* operation may be decomposed into the following N steps (for $n = 1, ..., N$) $y_{s_n.roff:s_n.roff+s_n.rows} \leftarrow y_{s_n.roff:s_n.roff+s_n.rows} + a_{s_n.roff:s_n.roff+s_n.rows,s_n.coff:s_n.coff+s_n.cols}x_{s_n.coff:s_n.coff+s_n.cols}$. Note that some steps may be executed in parallel by two or more threads. In the case with two different threads i, j operating on two different submatrices s_p, s_q, updating the two y intervals $s_p.roff : s_p.roff + s_p.rows$ and $s_q.roff : s_q.roff + s_q.rows$ is allowed, as long as the intervals do not intersect. In the case when the two intervals intersect, a *race condition* may occur; that is, concurrent updates of vector y may lead to inconsistent results in the intersecting y subvector. In the same spirit, the *SpMV_T* operation may be decomposed into $y_{s_n.coff:s_n.coff+s_n.cols} \leftarrow y_{s_n.coff:s_n.coff+s_n.cols} + a_{s_n.coff:s_n.coff+s_n.cols,s_n.roff:s_n.roff+s_n.rows}x_{s_n.roff:s_n.roff+s_n.rows}$. Clearly, while in the untransposed case the requirement for avoiding race conditions is on the *rows interval*, in the transposed case the *columns intervals* of the participating submatrices shall be disjoint. Our basic shared memory parallel algorithm for RSB/*SpMV* is outlined in Fig. 5 (see also [4]); in the listing, at line 8, assume for the time being $s.r_h = 0, s.r_t = 0$. Workload is partitioned among threads by means of a parallel section (lines 5-15). Repeatedly, each participating thread picks up a submatrix and updates the y array with its contribution to the product. When picking up a submatrix, a thread locks y array's interval corresponding to the submatrix rows interval. The same listing is suitable for *SpMV_T*, after a slight modification; namely, the pairwise exchange in all occurrences of: $s.roff$ and $s.coff$, $s.rows$ and $s.cols$. The bulk of computation is executed at the leaf level, when either COO or CSR submatrices are multiplied by the corresponding subvector. See Fig. 3 for the CSR submatrices code (again, assume $s.r_h = 0, s.r_t = 0$), and Fig. 1 for the COO version of the $SpMV$. For the $SpMV\_T$, see listings Fig. 4 and Fig. 2. We use the most common variant of CSR storing rows in ascending order, and column indices in ascending order within each row. The COO submatrices of RSB are organized exactly in the same way. A consequence of this layout ordered by rows is that for most real world matrices, for each given nonzero coefficient $a_{i,j}$, it is likely that the next stored nonzero $a_{i,j'}$ is quite near, i.e. with $\Delta = j' - j$ *reasonably small*. If $\Delta < C_l/N_s$, with C_l the *cache line length*, and N_s the floating point number size, both expressed in bytes, then after computing the contribution $y_i \leftarrow y_i + a_{ij}x_j$ (line 4 in Fig. 1, line 4 in Fig. 3), loading of element $x_{j'}$ will, with very high probability, only require a single fetch from cache memory, with no further cache misses. Normally both CSR and COO $SpMV$ algorithms are written such that the compiler uses *registers* for the accumulation of the y_i contribution, while updating the y_i memory location no more than once per submatrix. In the case of CSR and listing Fig. 3, this is straightforward to achieve — it requires referencing a local variable instead y_i in the inner loop, and update of the y_i location right after the inner loop. In the case of COO listing, Fig. 1, one should reorganize in two loops: an outer one cycling on rows, and an inner one cycling on a single row nonzeroes. If the number of nonzeroes of a COO submatrix is likely to be less than the number of rows, as is the case for our leaf matrices because of the criteria for COO in [3], and discussion of *hypersparsity* in [2], we clearly have a problem, since such a double loop would perform $O(s.nnz + s.rows)$ control instructions, which is always more than $O(s.nnz)$ as in Fig. 1. In the case of CSR submatrices, we are constrained to a $O(s.nnz + s.rows)$ loop complexity by the nature of CSR. However, here we have a guarantee that the number of nonzeroes exceeds the number of rows (by the definition of RSB leaves—see [3]), so the double loop is not a concern.

3   Two Optimizations to RSB $SpMV$ /$SpMV\_T$

In this section we introduce two closely related modifications to our $SpMV$ /$SpMV\_T$ algorithms for RSB. First, we note that our implementations for CSR

```
1 for i ← 1 + s.r_h to s.rows − s.r_t do
2     for l ← s.PA(i) to s.PA(i + 1) − 1 do
3         j ← s.JA(l);
4             s.y(i + s.roff) ← s.y(i + s.roff) + s.VA(l)s.x(j + s.coff);
5     end
6 end
```

Fig. 3. $SpMV$ listing for a CSR submatrix s, with head/tail skipping

```
1  for i ← 1 + s.r_h to s.rows − s.r_t do
2      for l ← s.PA(i) to s.PA(i + 1) − 1 do
3          j ← s.JA(l);
4          s.y(j + s.coff) ← s.y(j + s.coff) + s.VA(l)s.x(i + s.roff);
5      end
6  end
```

Fig. 4. *SpMV_T* listing for a CSR submatrix s, with head/tail skipping

(Fig. 3, Fig. 4, considering $s.r_h, s.r_t, s.c_h, s.c_t$ set to zero) visit the whole PA (*rows pointer*) array once, reading exactly $s.rows + 1$ locations. From [3] we have that for a matrix A stored in RSB, any of its submatrices s is stored in CSR only if $s.nnz > s.rows$ (or $\kappa_s \overset{def}{=} \frac{s.nnz}{s.rows} > 1$). In such a situation, storage of s uses exactly $I_{CSR}(s) \overset{def}{=} s.nnz + s.rows + 1$ indices, and these $I_{CSR}(s)$ indices are all read from memory during $SpMV / SpMV\_T$. Even if A has no empty rows (rows with only zeroes), we have no guarantee that the same applies to any given s, especially given the way RSB *recursive subdivision* of matrices works. Moreover, it is very likely that for most matrices of interest to us, there will be some $s.r_h$ *empty heading rows*. Similarly, it is reasonable to assume that there are some $s.r_t$ *empty tail rows* also. Values of $s.r_h, s.r_t$ may be easily computed at matrix build time, and may be used in Fig. 3 and Fig. 4 to work on the non-empty interval of rows only, and simply *skipping* iterations on the empty ones. For a square submatrix s having $r_e(s) \overset{def}{=} s.r_h + s.r_t$, the use of this row skipping technique allows reducing the amount of indices read up to 50% (consider a square s with $s.nnz = s.rows + 1$ nonzeroes distributed on two rows) with 4 byte indices and up to 66% with 2 bytes indices. In a more common situation, say $r_e(s) = s.rows/\nu_s$ (for some ν_s), one would save $s.rows/\nu_s$ accesses out of $s.rows(\kappa_s + 1) + 1$. For $s.rows \gg 1$, the saved fraction is $\frac{1}{\kappa_s \nu_s}$; for realistic cases, e.g.: $\kappa_s = 2, \nu_s = 2$, this amounts to 25%, which is not bad. A good property of this optimization is that in the case of no empty rows, there is no runtime performance loss compared to the base implementation. We also observe that this optimization is valid for both $SpMV$ and $SpMV\_T$. The second optimization also relies on the computation of $s.r_h/s.r_t$, but is applied to the outer parallel algorithm shown in Fig. 5. The base version of this algorithm considers to be zero both $s.r_h, s.r_t$, on all submatrices; thus when a submatrix is picked up by a thread, the entire $s.roff...s.roff + s.rows$ interval of the output vector y may have to be updated, and therefore is locked. But if a submatrix s has $s.r_h$ empty heading rows, and $s.r_t$ empty tail rows, then rows outside the $s.roff + s.r_h...s.roff + s.rows − s.r_t$ range will not be modified; therefore we observe that only the corresponding subvector of y must be really locked. We apply our optimization by locking only the effective rows interval, thus allowing for a reduced degree of resource contention among threads and enhancing potential parallelism. The worst case is when $s.r_h = 0$ and $s.r_t = 0$, and this is no worse than (indeed, identical to) without the optimization. The best case may be when N_s submatrices extending on the same rows range exists, but

```
1  S ← [s₀, s₁, ..., s_{N-1}] /*an array of terminal submatrices, in any order*/ ;
2  B ← [0, 0, .., 0] /*a zero bit for each submatrix*/ ;
3  n ← 0 /*count of visited submatrices so far*/;
4  while n < N do
5      begin parallel ;
6          s ← pick an unvisited submatrix s from S;
7          /*(should have picked up s ← S[i], with B[i] = 0)*/ ;
8          [f, l] ← [s.roff+s.rₕ , s.roff+s.rows−s.rₜ] ;
9          if locked([f ... l]) then cycle ;
10         lock([f ... l]) /*we lock y on s's effective rows interval*/ ;
11         /*perform SpMV on s and x[s.coff:s.coff+s.cols] into y[f : l]*/ ;
12         y[f : l] ← y[f : l] + s · x[s.coff:s.coff+s.cols] ;
13         B[i] ← 1; n ← n + 1 ;
14         unlock([f ... l]) ;
15     end parallel;
16 end
```

Fig. 5. Multithreaded *SpMV* for leaf submatrices of a RSB matrix, with head/tail skipping

each submatrix has its nonzeroes laid out only on a contiguous group of rows, in a way that there is no intersection of non-empty row intervals, for any given pair of submatrices. In this limit case, all of the submatrices may be processed in parallel, with potential N_s-fold parallelism. For a more realistic case, consider a pair of matrices s, s', whose nonzeroes have no common row (think of a case when a large banded matrix is subdivided), but both have the same row offset ($s.roff = s'.roff$) and extension ($s.rows = s'.rows$). In this case, we double the potential parallelism with little effort. This optimization may also be applied to the transposed *SpMV*, if we use the *empty heading columns* $s.c_h$, instead of $s.r_h$, *empty tail columns* $s.c_t$ instead of $s.r_t$, and swap usage of $s.roff$ with $s.coff$ in Fig. 5.

4   Experimental Setup and Methodology

For space reasons we report only a limited set of experimental data. We chose to use a sample of large (exceeding hardware cache), sparse square matrices obtained from the *University of Florida Sparse Matrix Collection* (see [5]). Matrices information is summarized in Table 1. We report results of experiments performed on an Intel Xeon 5670, supporting up to 12 hardware threads, with 3 levels of cache memory (sized respectively 32KB/256KB/12MB). Our codes were implemented in C99, and compiled with Intel's ICC v.12.0.2 compiler, with the optimization -O3 flag (no machine specific optimization flags were used). Our parallel RSB implementation using OpenMP is compared against the CSR implementation present in the proprietary Intel MKL 10.3-2 library, and the publicly available CSB (see Buluç et al. [1]) prototype (compiled with the special purpose CILK++ compiler, version 8503, with -O3 -fno-rtti -fno-exceptions flags).

Table 1. List of matrices used in the experiments, with their row/column dimensions (r/c), nonzeroes count (nnz), average nonzeroes per row (nnz/r) count

matrix	r	c	nnz	nnz/r
cage15	5154859	5154859	99199551	19.24
circuit5M_dc	3523317	3523317	19194193	5.45
fem_hifreq_circuit	491100	491100	20239237	41.21
GL7d18	1955309	1548650	35590540	18.20
patents	3774768	3774768	14970767	3.97
RM07R	381689	381689	37464962	98.16
TSOPF_RS_b2383	38120	38120	16171169	424.22
wikipedia-20070206	3566907	3566907	45030389	12.62

Here we also use explicit *loop unrolling* (four-fold) on the COO and inner CSR loops of our codes. We express the performance of a computation in MFLOPS (that is, *time efficiency*); we count 2 operations for each nonzero of a matrix involved in the multiplication by a vector.

5 Results

Let us now discuss the experimental results. In Fig. 6 we report results obtained using 12 threads; in Fig. 7 results for a single thread. The first thing we note, is that the MKL (CSR) results for *SpMV_T* are consistently lower than for *SpMV*. This performance gap is due to the row-major layout of CSR, which requires the transposed update of the results array to be written at *random* locations (unlike the normal update, which reads random locations of the multiplicand vector, but updates a sequentially accessed array). This gap is almost absent in the case of CSB: recall (see [1]) the unbiased Z-ordering in its *sparse blocks*. Regarding the *SpMV* /*SpMV_T* gap, RSB falls in between MKL and CSB, due to its storage of consecutive rows of *sparse submatrices*. We notice that running in parallel (Fig. 6), the aforementioned efficiency gap is much more pronounced for CSR. The reason for this is the lack, in the CSR format, of immediate information for the serialization of the threads write instructions in updating the result

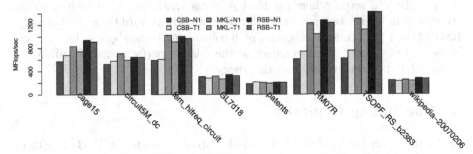

Fig. 6. *SpMV* and *SpMV_T* performance, 12 threads

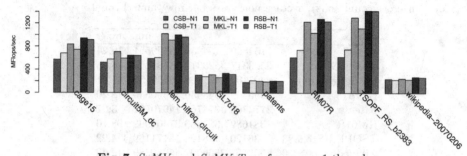

Fig. 7. *SpMV* and *SpMV_T* performance, 1 thread

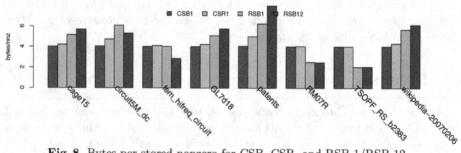

Fig. 8. Bytes per stored nonzero for CSB, CSR, and RSB-1/RSB-12

vector. CSB and RSB are structured in blocks, offering a coarse grained way to parallelization of *SpMV_T*. Summarizing, for the chosen set of matrices, RSB performs always better than CSB and MKL with a single thread, and almost always, in parallel runs. In interpreting performance results for the three formats (especially when running in parallel, when the CPU-memory communications channels are likely to be saturated), we may consider the *bytes per indexing nonzero* metric. Since RSB's index usage depends on the subdivisions/threads count, in Fig. 8 we report this value for both single and 12 threaded runs. For CSB, we report the size of the arrays allocated in the source code, although we do not take into account here the *block pointers* array (see [1]), which also contributes as index-related memory traffic. It is straightforward to see that usually, RSB's higher performance cases in Fig. 6 coincide with the shortest index usage cases, and vice-versa. We also note that the relative performance of CSB/MKL seems related to the average indexing usage.

6   Concluding Remarks

In this paper, we have proposed two simple optimizations to our RSB algorithms for *SpMV* /*SpMV_T*, and performed experiments on large sparse matrices.

Since the two optimizations have no potential negative impact, and since their benefit is difficult to quantify in advance, we skipped the comparison to the past code versions, and compared the code directly to two different, efficient *SpMV* implementations: Intel MKL's CSR and CSB (see Buluç et al. [1]). Our main finding is that the block structure of RSB allows the parallel implementations of both *SpMV* /*SpMV_T* to be efficient without the prominent performance gap which is inherent in a CSR implementation (in this case, MKL's). We also find confirmation that RSB's hybrid structure (a recursive layout on outside, with a row-major layout on the inside) is advantageous when performing *SpMV* /*SpMV_T* on large matrices serially. Furthermore, the technique of enhancing RSB's parallelism by using empty rows information applies to triangular solve and symmetric *SpMV* kernels as well.

We wish to thank Paweł Gepner and Jamie Wilcox at Intel Corporation for giving us access and technical support for the machine used in the experiments.

References

1. Buluç, A., Fineman, J.T., Frigo, M., Gilbert, J.R., Leiserson, C.E.: Parallel sparse matrix-vector and matrix-transpose-vector multiplication using compressed sparse blocks. In: auf der Heide, F.M., Bender, M.A. (eds.) SPAA, pp. 233–244. ACM (2009)
2. Buluç, A., Gilbert, J.R.: On the Representation and Multiplication of Hypersparse Matrices. In: IEEE International Parallel and Distributed Processing Symposium (IPDPS 2008), pp. 1–11 (April 2008)
3. Martone, M., Filippone, S., Paprzycki, M., Tucci, S.: About the assembly of recursive sparse matrices. In: Proceedings of the International Multiconference on Computer Science and Information Technology, Wisła. Poland, pp. 317–325. IEEE Computer Society Press, Los Alamitos (2010)
4. Martone, M., Filippone, S., Paprzycki, M., Tucci, S.: On BLAS operations with recursively stored sparse matrices. In: Proceedings of the International Symposium on Symbolic and Numeric Algorithms for Scientific Computing. Timisoara, Romania, pp. 49–56. IEEE (September 2010)
5. The University of Florida Sparse Matrix Collection,
   http://www.cise.ufl.edu/research/sparse/matrices

On the Differences of the Discrete Weak and Strong Maximum Principles for Elliptic Operators

Miklós E. Mincsovics[1] and Tamás L. Horváth[1,2]

[1] Dep. of Appl. Anal. and Comp. Math., Eötvös Loránd University,
Pázmány Péter sétány I/C, Budapest H-1117, Hungary
m.e.mincsovics@gmail.com
[2] Dep. of Math. and Comp. Sci., Széchenyi University,
Egyetem tér 1, H-9026 Győr, Hungary
thorvath12@gmail.com

Abstract. When choosing a numerical method to approximate the solution of a continuous mathematical problem, we need to consider which method results in an approximation that is not only close to the solution of the original problem, but possesses the important qualitative properties of the original problem, too. For linear elliptic problems the main qualitative properties are the various maximum principles. The preservation of the weak maximum principle was extensively investigated in the last decades, but not the strong maximum principle preservation. In this paper we focus on the latter property by giving its necessary and sufficient conditions, investigating the relation of the preservation of the strong and weak maximum principles and illustrating the differences between them with numerous examples.

1 Introduction

In the early theory of PDE's maximum principles played an important role. They provide an efficient tool to prove uniqueness and stability for the classical solutions of linear elliptic and parabolic problems. Later, when the concept of weak solution had been introduced, they lost a little bit from their importance. Now, in the age of computers and numerical methods, the investigation of maximum principles came into fashion again.

A numerical method is a sequence of simpler problems, whose solutions hopefully tend to the solution of the original problem. When this holds, the numerical method is called convergent. However, convergence is a theoretical question, in the application we must choose some parameter setting, and not an infinite sequence of it. Thus, we need to decide between convergent numerical methods from another point of view. This leads to the investigation of what qualitative properties can be preserved when we apply some numerical method, e.g., we usually prefer one in which the simpler problems possess the same important qualitative properties as the original problem.

I. Lirkov, S. Margenov, and J. Waśniewski (Eds.): LSSC 2011, LNCS 7116, pp. 614–621, 2012.

Maximum principles are essential qualitative properties of linear elliptic problems. When for a simpler problem the maximum principle holds, we say that it possesses the discrete maximum principle, since the simpler problems are usually defined in finite dimensional spaces.

The first paper in which a discrete maximum principle was formulated is probably [9]. The definition of the discrete weak maximum principle which is used today appeared first in [1] (but it was named differently). While the discrete weak maximum principle was extensively investigated in the last decades, the discrete strong maximum principle has not been thoroughly analysed. In [6] and in [7] a sufficient algebraic condition was given, while in [2] the positivity of the discrete Green function was investigated (which is in a close relation with the discrete strong maximum principle) in a special case. However, a sufficient and necessary condition was missing. The authors' intention was to fill this gap with this paper.

The paper is organized as follows. In Section 2 we list important pieces of information on the elliptic maximum principles. In Section 3 we give the definition of the discrete elliptic maximum principles. In Section 4 we present the main results about the discrete strong maximum principles, pointing out the differences between the weak and strong maximum principles. Finally, in Section 5 we illustrate our results with several numerical examples.

2 Maximum Principles

In this section we list the definitions of continuous maximum principles for linear elliptic operators and the important theorems about them, based on [3]. We study elliptic operators, and not elliptic PDEs, since this way is more comfortable, and clearly the qualitative properties of some PDEs depend on the qualitative properties of the corresponding operators.

Let $\Omega \subset \mathbb{R}^d$ be an open and bounded domain with boundary $\partial\Omega$, and $\bar{\Omega} = \Omega \cup \partial\Omega$. We investigate the elliptic operator K, $\operatorname{dom} K = C^2(\Omega) \cap C(\bar{\Omega})$, defined as

$$Ku = -\sum_{i,j=1}^{d} a_{ij} \frac{\partial^2 u}{\partial x_i \partial x_j} + \sum_{i=1}^{d} b_i \frac{\partial u}{\partial x_i} + cu, \tag{1}$$

where $a_{ij}(\mathbf{x}), b_i(\mathbf{x}), c(\mathbf{x}) \in C(\Omega)$.

Definition 1. *We say that the operator K, defined in (1), possesses*

– *the weak maximum principle (wMP) if the following implication holds:*

$$Ku \leq 0 \ in \ \Omega \quad \Rightarrow \quad \max_{\bar{\Omega}} u \leq \max\{0, \max_{\partial\Omega} u\};$$

– *the strong maximum principle (sMP) if it possesses the wMP, moreover, the following implication holds:*

$$Ku \leq 0 \ in \ \Omega \quad and \quad \max_{\Omega} u = \max_{\bar{\Omega}} u = m \geq 0 \quad \Rightarrow \quad u \equiv m \ in \ \bar{\Omega};$$

- *the strictly weak maximum principle (WMP) if the following implication holds:*

$$Ku \leq 0 \text{ in } \Omega \quad \Rightarrow \quad \max_{\partial\Omega} u = \max_{\Omega} u;$$

- *the strictly strong maximum principle (SMP) if it possesses the WMP, moreover, the following implication holds:*

$$Ku \leq 0 \text{ in } \Omega \quad \text{and} \quad \max_{\Omega} u = \max_{\bar{\Omega}} u = m \quad \Rightarrow \quad u \equiv m \text{ in } \bar{\Omega}.$$

Theorem 1. *If operator K, defined in (1), is uniformly elliptic and*

- *$c \geq 0$, then it possesses the wMP;*
- *$c \geq 0$, moreover Ω is connected, then it possesses the sMP;*
- *$c = 0$, then it possesses the WMP;*
- *$c = 0$, moreover Ω is connected, then it possesses the SMP.*

Remark 1. Sometimes the case $c = 0$ is called strong elliptic maximum principle, but we wanted to reserve this name to the other property.

The requirements under which the operator possesses a weak maximum principle can be weakened, see, e.g., [1].

Finally, we mention that it is possible to define minimum principles and to get similar theorems, due to the linearity of operator K. More information about maximum and minimum principles can be found in [3].

3 Discrete Maximum Principles

To obtain a simpler problem (or a sequence of simpler problems) from an elliptic PDE, usually some discretization method is applied, e.g., a finite difference (FDM) or a finite element method (FEM). Almost every discretization method leads to a system of linear algebraic equations, where the discrete operator corresponding to the operator K in (1) is the coefficient matrix.

In the following we define discrete maximum principles for such a discrete operator, i.e., for a matrix. We choose the natural way (independently of the original problem), which results in the adequate definition, corresponding to the definition of the last section, if an FDM or FEM with linear or multilinear elements is applied. However, it should be mentioned that in case of an FEM with higher order elements this approach is not applicable.

We use the following types of typesetting: $\mathbf{A}$ for matrices, $\mathbf{a}$ for vectors. $\mathbf{0}$ denotes the zero matrix (or vector), $\mathbf{e}$ is the vector all coordinates of which are equal to 1. The dimensions of these vectors and matrices should be clear from the context. $\mathbf{A} \geq \mathbf{0}$ ($\mathbf{A} > \mathbf{0}$) or $\mathbf{a} \geq \mathbf{0}$ ($\mathbf{a} > \mathbf{0}$) means that all the elements of $\mathbf{A}$ or $\mathbf{a}$ are non-negative (positive). $\max \mathbf{a}$ denotes the maximal element of the vector $\mathbf{a}$. The symbol $\max\{0, \mathbf{a}\}$ denotes $\max\{0, \max \mathbf{a}\}$.

We will investigate the matrix $\mathbf{K} = [\mathbf{K}_0 | \mathbf{K}_\partial] \in \mathbb{R}^{N \times \bar{N}}$, where $\mathbf{K}_0 \in \mathbb{R}^{N \times N}$, $\mathbf{K}_\partial \in \mathbb{R}^{N \times N_\partial}$, $\bar{N} = N + N_\partial$, acting on the vector $\mathbf{u} = [\mathbf{u}_0 | \mathbf{u}_\partial]^T \in \mathbb{R}^{\bar{N}}$, $\mathbf{u}_0 \in \mathbb{R}^N$, $\mathbf{u}_\partial \in \mathbb{R}^{N_\partial}$. The partitioned forms are constructed by taking into consideration the separation of the interior and boundary points. We assume that $N, N_\partial \geq 2$.

Definition 2. *We say that a matrix* $\mathbf{K}$ *possesses*

- *the discrete weak maximum principle (DwMP) if the following implication holds:*
$$\mathbf{K}\mathbf{u} \leq \mathbf{0} \quad \Rightarrow \quad \max \mathbf{u} \leq \max\{0, \mathbf{u}_\partial\}\,;$$

- *the discrete strong maximum principle (DsMP) if it possesses the DwMP, moreover, the following implication holds:*
$$\mathbf{K}\mathbf{u} \leq \mathbf{0} \quad and \quad \max \mathbf{u} = \max \mathbf{u}_0 = m \geq 0 \quad \Rightarrow \quad \mathbf{u} = m\mathbf{e}\,;$$

- *the discrete strictly weak maximum principle (DWMP) if the following implication holds:*
$$\mathbf{K}\mathbf{u} \leq \mathbf{0} \quad \Rightarrow \quad \max \mathbf{u}_\partial = \max \mathbf{u}\,;$$

- *the discrete strictly strong maximum principle (DSMP) if it possesses the DWMP, moreover, the following implication holds:*
$$\mathbf{K}\mathbf{u} \leq \mathbf{0} \quad and \quad \max \mathbf{u} = \max \mathbf{u}_0 = m \quad \Rightarrow \quad \mathbf{u} = m\mathbf{e}\,.$$

4 Differences between the Discrete Weak and Strong Maximum Principles

The only difference between the conditions in Theorem 1 for the weak and strong maximum principles is the connectedness of the domain Ω. Next, we investigate this question in the discrete case.

The following theorem gives necessary and sufficient conditions for the discrete weak and strong maximum principles. The first part (DwMP) of it is from [1], the third (DWMP) is a joint result of Faragó and Mincsovics, published in [4]. The second (DsMP) and fourth (DSMP) parts of it are new.

Theorem 2. *The matrix* $\mathbf{K}$ *possesses*

- *the DwMP if and only if the following three conditions hold:*

  (w1) $\mathbf{K}_0^{-1} \geq \mathbf{0}$;   (w2) $-\mathbf{K}_0^{-1}\mathbf{K}_\partial \geq \mathbf{0}$;   (w3) $-\mathbf{K}_0^{-1}\mathbf{K}_\partial \mathbf{e} \leq \mathbf{e}$.

- *the DsMP if and only if the following three conditions hold:*

  (s1) $\mathbf{K}_0^{-1} > \mathbf{0}$;   (s2) $-\mathbf{K}_0^{-1}\mathbf{K}_\partial > \mathbf{0}$;
  (s3) $-\mathbf{K}_0^{-1}\mathbf{K}_\partial \mathbf{e} < \mathbf{e}$ *or* $-\mathbf{K}_0^{-1}\mathbf{K}_\partial \mathbf{e} = \mathbf{e}$.

- *the DWMP if and only if the following three conditions hold:*

  (W1) $\mathbf{K}_0^{-1} \geq \mathbf{0}$;   (W2) $-\mathbf{K}_0^{-1}\mathbf{K}_\partial \geq \mathbf{0}$;   (W3) $-\mathbf{K}_0^{-1}\mathbf{K}_\partial \mathbf{e} = \mathbf{e}$.

- *the DSMP if and only if the following three conditions hold:*

  (S1) $\mathbf{K}_0^{-1} > \mathbf{0}$;   (S2) $-\mathbf{K}_0^{-1}\mathbf{K}_\partial > \mathbf{0}$;   (S3) $-\mathbf{K}_0^{-1}\mathbf{K}_\partial \mathbf{e} = \mathbf{e}$.

Proof. – We begin with the DSMP case.

- First, we assume (S1)–(S3), then

$$\mathbf{u}_0 = \mathbf{K}_0^{-1}\mathbf{K}\mathbf{u} - \mathbf{K}_0^{-1}\mathbf{K}_\partial\mathbf{u}_\partial$$

holds. (It follows from (S1) that $\mathbf{K}_0^{-1}$ exists.) Let us assume that $\mathbf{K}\mathbf{u} \leq \mathbf{0}$. We write $\mathbf{u}_0 = m\mathbf{e} - \mathbf{h}_0$, $\mathbf{u}_\partial = m\mathbf{e} - \mathbf{h}_\partial$, where both $\mathbf{h}_0, \mathbf{h}_\partial \geq \mathbf{0}$ have a 0 coordinate (i.e., $\max \mathbf{u} = \max \mathbf{u}_0 = m$). Thus

$$m\mathbf{e} - \mathbf{h}_0 = \mathbf{K}_0^{-1}\mathbf{K}\mathbf{u} - \mathbf{K}_0^{-1}\mathbf{K}_\partial m\mathbf{e} + \mathbf{K}_0^{-1}\mathbf{K}_\partial\mathbf{h}_\partial. \qquad (2)$$

Using (S3) we get

$$\mathbf{h}_0 = \mathbf{K}_0^{-1}(-\mathbf{K}\mathbf{u}) - \mathbf{K}_0^{-1}\mathbf{K}_\partial\mathbf{h}_\partial. \qquad (3)$$

Using (S1), (S2) and the fact that $\mathbf{h}_0$ has a 0 coordinate yields that $-\mathbf{K}\mathbf{u} = \mathbf{0}$ and $\mathbf{h}_\partial = \mathbf{0}$. These imply $\mathbf{h}_0 = \mathbf{0}$.

- Second, we assume the DSMP. Then the DWMP holds, thus (W1)–(W3) hold. We can choose freely $\mathbf{K}\mathbf{u} \leq \mathbf{0}$, $\mathbf{h}_\partial \geq \mathbf{0}$ in (3).

  First, we set $\mathbf{h}_\partial = \mathbf{0}$ and we assume that $\mathbf{K}_0^{-1}$ has a 0 element, let it be the ij-th entry of the matrix. We choose the j-th coordinate of $-\mathbf{K}\mathbf{u}$ as 1, the others as 0, then the i-th coordinate of $\mathbf{h}_0$ is 0. If in the j-th column there is a positive entry, then $\mathbf{h}_0 \neq \mathbf{0}$, which is a contradiction. Otherwise, the matrix $\mathbf{K}_0^{-1}$ has a zero column, which is a contradiction, too, since it is invertible.

  Second, we set $\mathbf{K}\mathbf{u} = \mathbf{0}$, and we assume that $-\mathbf{K}_0^{-1}\mathbf{K}_\partial$ has a 0 element, let it be the ij-th entry of the matrix. We choose the j-th coordinate of $\mathbf{h}_\partial$ as 1, the others as 0, then the i-th coordinate of $\mathbf{h}_0$ is 0, but $\mathbf{h}_\partial \neq \mathbf{0}$, which is a contradiction. □

– We finish with the DsMP case.

- First, we assume (s1)–(s3). If $-\mathbf{K}_0^{-1}\mathbf{K}_\partial\mathbf{e} = \mathbf{e}$ holds, then we can adopt the proof of the DSMP case. If $-\mathbf{K}_0^{-1}\mathbf{K}_\partial\mathbf{e} < \mathbf{e}$ holds, then (3) is modified as $\mathbf{h}_0 > \mathbf{K}_0^{-1}(-\mathbf{K}\mathbf{u}) - \mathbf{K}_0^{-1}\mathbf{K}_\partial\mathbf{h}_\partial$, which excludes the possibility that $\mathbf{h}_0$ has a 0 coordinate.

- Second, we assume the DsMP. We get (s1), (s2) by putting $m = 0$ into (2), then the argumentation of the DSMP case can be repeated. To get (s3), we assume that $-\mathbf{K}_0^{-1}\mathbf{K}_\partial\mathbf{e} \not< \mathbf{e}$ and $-\mathbf{K}_0^{-1}\mathbf{K}_\partial\mathbf{e} \neq \mathbf{e}$, i.e., $\mathbf{e} + \mathbf{K}_0^{-1}\mathbf{K}_\partial\mathbf{e}$ has a 0 and a positive coordinate, too, let them be the i-th one and the j-th one, respectively. Choosing $m = 1$, $\mathbf{K}\mathbf{u} = \mathbf{0}$, $\mathbf{h}_\partial = \mathbf{0}$ yields that the i-th coordinate of $\mathbf{h}_0$ is 0 and the j-th one is positive, which is a contradiction. □

If we compare the continuous and the discrete case, we can conclude that the condition (w3) $-\mathbf{K}_0^{-1}\mathbf{K}_\partial\mathbf{e} \leq \mathbf{e}$ corresponds to $c \geq 0$, (W/S3) $-\mathbf{K}_0^{-1}\mathbf{K}_\partial\mathbf{e} = \mathbf{e}$ corresponds to $c = 0$, moreover, (s3) can be shed more light on if we notice the fact that u constant implies $c = 0$. Note that $-\mathbf{K}_0^{-1}\mathbf{K}_\partial\mathbf{e} = \mathbf{e}$ is equivalent to $\mathbf{K}\mathbf{e} = \mathbf{0}$, and $\mathbf{K}\mathbf{e} \geq \mathbf{0}$ implies $-\mathbf{K}_0^{-1}\mathbf{K}_\partial\mathbf{e} \leq \mathbf{e}$, but here the reverse is not true.

(s/S1) and (s/S2) correspond to the connectedness of the domain Ω. One can easily see that (s/S1) implies the irreducibility of $\mathbf{K}_0$, which means that all the discrete interior points are in contact with each other, which is clearly some discrete connectedness property. [1] gives practical conditions to satisfy the DwMP by introducing the notion of generalized nonnegative type. In [6] the DsMP was proved for this class of matrices and it was observed that if the matrix $\mathbf{K}_0$ is irreducibly diagonally dominant (the generalized nonnegative type contains this property), then (s/S1) is fulfilled, the proof can be found in [8]. However, for discrete weak maximum principles it is not needed, as it was observed in [5]. To ensure (s/S2), one possibility is to require $\mathbf{K}_\partial \leq \mathbf{0}$ and at least one nonzero element in every column (with (s/S1)), which can be interpreted as all of the discrete boundary points are in contact with the discrete interior points.

We can conclude that irreducibility is necessary for DsMP and DSMP (but it is not sufficient). Anyway, this would be the key-concept, if we want something to emphasize.

5   Numerical Examples

In this final section we present numerical examples, constructed with the help of Matlab. In all examples we used linear finite element discretization. We focus on the irreducibility property, i.e., we give examples in which the discrete domain is not connected from some point of view. This can easily happen when the domain consists of two relatively large areas connected in the middle with a thin "path". In this case the program package COMSOL can produce qualitatively incorrect mesh, too.

In the first three examples $K = -\Delta$, in the fourth it is defined as $Ku = -\Delta u + 128u$. In all examples $Ku = 0$. In the first two u is defined as 1 on the boundary of the left square, 0 on the boundary of the right square and linearly decreasing from 1 to 0 on the boundary of the middle square. The boundary condition of the third example differs only on the middle part: on the left part of the boundary of it, i.e. on $\{(x,y) : x \in [3, 3.5], y \in \{1, 2\}\}$, u is 1, then linearly decreasing from 1 to 0 on the right part of the boundary of the middle square i.e. on $\{(x,y) : x \in [3.5, 4], y \in \{1, 2\}\}$. The fourth example is similar to the first two.

The arrangement within the figures is as follows. The top left panel presents the mesh, the top right panel presents the nonzero elements of the matrix $\mathbf{K}_0$, and in the bottom panels $\mathbf{u}$ is plotted from two different angles, the right one shows us better where the function is constant.

The first example shows us how an inadequate mesh can result in a reducible matrix and so losing the DSMP (but the DWMP is fulfilled). The second is the "good" example, here both discrete maximum principles are fulfilled. In [2] a mesh is presented, this is the third example here, which seems to be good at first sight, but the two right angles damage the connection of the two seemingly connected points in the middle, cf. [5], too.

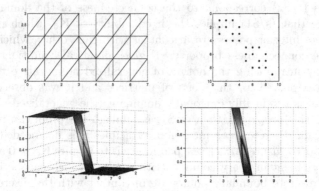

Fig. 1. 1. Example: The mesh results in a reducible matrix. The DsMP failed while the DwMP was fulfilled.

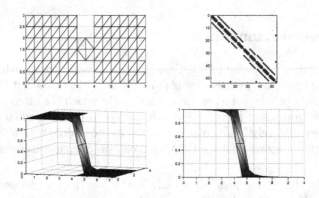

Fig. 2. 2. Example: The mesh results in an irreducible matrix. Both of the DsMP and DwMP were fulfilled.

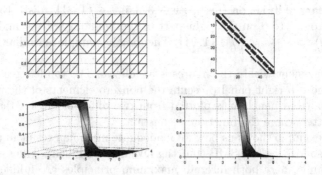

Fig. 3. 3. Example: The mesh results in a reducible matrix. The DsMP failed while the DwMP was fulfilled.

The fourth example presents a mesh, which results in losing the DsMP, in addition to that the DwMP is fulfilled. It is caused surprisingly by the usage of equilateral triangles.

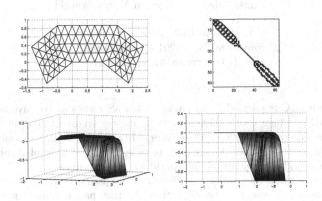

Fig. 4. 4. Example: The mesh which contains equilateral triangles can results in a reducible matrix, too. The DsMP failed while the DwMP fulfilled.

Acknowledgments. The Project is supported by the European Union and co-financed by the European Social Fund (grant agreement no. TAMOP 4.2.1./B-09/1/KMR-2010-0003).

References

1. Ciarlet, P.G.: Discrete maximum principle for finite-difference operators. Aequationes Math. 4, 338–352 (1970)
2. Draganescu, A., Dupont, T.F., Scott, L.R.: Failure of the discrete maximum principle for an elliptic finite element problem. Math. Comp. 74(249), 1–23 (2005)
3. Evans, L.C.: Partial Differential Equations. Graduate Studies in Mathematics, vol. 19. AMS (1997)
4. Faragó, I.: Numerical Treatment of Linear Parabolic Problems. Dissertation for the degree MTA Doktora (2008)
5. Hannukainen, A., Korotov, S., Vejchodský, T.: On Weakening Conditions for Discrete Maximum Principles for Linear Finite Element Schemes. In: Margenov, S., Vulkov, L.G., Waśniewski, J. (eds.) NAA 2008. LNCS, vol. 5434, pp. 297–304. Springer, Heidelberg (2009)
6. Ishihara, K.: Strong and weak discrete maximum principles for matrices associated with elliptic problems. Linear Algebra Appl. 88-89, 431–448 (1987)
7. Knabner, P., Angermann, L.: Numerical Methods for Elliptic and Parabolic Partial Differential Equations. Springer, New York (2003)
8. Varga, R.S.: Matrix Iterative Analysis. Prentice-Hall, Englewood Cliffs (1962)
9. Varga, R.: On discrete maximum principle. J. SIAM Numer. Anal. 3, 355–359 (1966)

Adaptive FEM Package with Decentralized Parallel Adaptation of Tetrahedral Meshes

Tomasz Olas and Roman Wyrzykowski

Czestochowa University of Technology
Dabrowskiego 73, 42-201 Czestochowa, Poland
{olas,roman}@icis.pcz.pl

Abstract. Our parallel FEM package NuscaS allows us to solve adaptive FEM problems with 3D unstructured meshes on distributed-memory parallel computers such as PC-clusters. For solving sparse systems of equations, NuscaS uses the message-passing paradigm to implement the PCG method with geometric multigrid as a preconditioner.

For the mesh adaptation, the 8-tetrahedra longest-edge partition is used as a refinement mesh algorithm. In this paper, a new method for parallelizing this algorithm is presented. It was developed for the message-passing model, and implemented using the MPI standard. The new solution is based on a decentralized approach. So it is more scalable in comparison to previous implementations, where a centralized synchronizing node (coordinator processor or gateway node) is required.

Both the sequential and parallel versions of the mesh adaptation are carefully optimized to maximize performance. One of key solutions is the usage of suitable data structures, such as hash tables. They allow for high performance while preserving modest memory requirements.

1 Introduction

The finite element method (FEM) is a powerful tool for studying different phenomena in various areas. However, many applications of this method have too large computational or memory costs for a sequential implementations, to be useful in practice. Parallel computing allows FEM users to overcome this bottleneck [4]. For this aim, an object-oriented environment for the parallel FEM modeling, called *NuscaS*, was developed at the Czestochowa University of Technology [9]. Being primarily intended for PC-clusters, NuscaS was also ported to other parallel and distributed architectures, like the CLUSTERIX grid.

Numerical modeling of 3D thermomechanical problems described by time-dependent PDEs is a complex and time-consuming issue. Adaptive techniques are powerful tools to perform efficiently such modeling using the FEM analysis [6]. They allow the solution error to be kept under control; therefore computation costs can be minimized. However, during the adaptation computational workloads change unpredictably at the runtime, and dynamic load balancing is required.

In our previous work [5], we focused on the problem how to implement dynamic load balancing efficiently on distributed-memory parallel computers such

I. Lirkov, S. Margenov, and J. Waśniewski (Eds.): LSSC 2011, LNCS 7116, pp. 622–629, 2012.

as PC clusters. This implementation is based on using the proposed performance model. At the same time, the serious performance bottleneck of our previous numerical code was a sequential implementation of the mesh adaptation step. In this paper, we propose a method which eliminates this drawback by implementing the mesh adaptation in parallel.

For the mesh adaptation, the 8-tetrahedra longest-edge partition [7] is used as a refinement mesh algorithm in NuscaS. In this paper, a new method for parallelizing this algorithm is presented. It was developed for the message-passing model, and implemented using the MPI standard. The new solution is based on a decentralized approach. So it is more scalable in comparison to previous implementations [1,7,8], where a centralized synchronizing node (coordinator processor or gateway node) is required.

2   NuscaS Package with Parallel Computing Capabilities

2.1   Basics of NuscaS

The basic part of the NuscaS package is the kernel - a class library which provides an object-oriented framework for developing finite element applications. This library consists of basic classes necessary to implement FEM modeling, irrespective of the type of a problem being solved. The inheritance is used to develop new applications classes for a particular modeling problem.

The previous class library was useful only for 2D problems. So a completely new library with 3D finite element solvers was developed [5]. It was called the Numerical Simulations and Computations (NSC) library. Although the internals of the library are unavoidably complex, the interface is as easy to use as possible. The library is maintainable and extensible using design patterns. The NSC library is usable for large-scale problems, where the number of nodes in a mesh can reach millions.

2.2   Parallel Computing in NuscaS

The parallel version of the library is based on the geometric decomposition applied for uniprocessor nodes of a parallel system; the distributed-memory architecture and message-passing model of parallel programming are assumed. Moreover, to solve linear systems of equations with sparse matrices, which are results of FEM discretization, we use iterative methods based on Krylov subspace methods. The computational kernel of these methods is the matrix-vector multiplication with sparse matrices. When implementing this operation in parallel, the overlapping of computation and communication is exploited to reduce the execution time of the algorithm.

In this case, a FEM mesh is divided into p submeshes (domains), which are assigned to separate processors of a parallel architecture. Every processor (or process) keeps the assigned part of the mesh. As a result, the j-th domain will own a set of N_j nodes selected from all the N nodes of the mesh. For an arbitrary domain with index j, we have three types of nodes:

- N_j^i of *internal nodes* - they are coupled only with nodes belonging to this subdomain;
- N_j^b of *boundary nodes* - they correspond to those nodes in the j-th subdomain that are coupled with nodes of other subdomains; internal and boundary nodes are called *local* ones, so the number of *local* nodes is $N_j^l = N_j^i + N_j^b$;
- N_j^e of *external nodes* - they correspond to those nodes in other domains that are coupled with boundary nodes of the given domain.

When solving a system of equations in parallel using iterative methods, values of unknowns computed in boundary nodes are exchanged between neighbor domains. Hence at the preprocessing stage, it is necessary to generate data structures for each domain, to provide an efficient implementation of communication. Furthermore, values computed in boundary nodes must be agreed between neighbor domains. For this aim, in each domain (process) j for every neighbor process k the following sets are stored:

- S_j^k - set of those indexes of boundary nodes in process j that are external nodes for process k;
- R_j^k - set of those indexes of external nodes in process j that are assigned to process k.

To provide success of a Krylov subspace method by its acceleration, a suitable preconditioner has to be used. In NuscaS, we use the geometric multigrid [2] as a preconditioner for unstructured 3D meshes. Both the V-cycle and full multigrid algorithms were implemented, with the weighted Jacobi smoother. As a solver for the coarsest grid, we adopted the PCG algorithm with the Jacobi preconditioner.

3   Parallel Adaptation of Tetrahedral Meshes

3.1   Mesh Adaptation

In general, mesh adaptation methods can be grouped into three different classes, named r-, p-, and h-methods, depending on the level of mesh modifications. In this paper, we focus only on adaptation by h-methods [7]. The first step in such a mesh adaptation procedure is selection of elements which should be partitioned, based on estimating a discretization error [5]. The next step divides elements using the longest-edge bisection method (see Fig. 1). The presented solution can be easily adapted to thermomechanical problems of different types.

   To implement the mesh adaptation (refinement) in parallel, the Longest-Edge Propagation Path (LEPP) method is used in this work [7]. The partitioning of tetrahedral elements is performed based on the iterative algorithm of 8-tetrahedra longest edge partition (shortly 8T-LT). At this moment, our parallel adaptation algorithm works only for tetrahedral elements (Fig. 1). How to implement this adaptation in parallel for other types of FEM elements will be investigated in future works.

a)   b)   c)

d)   e)

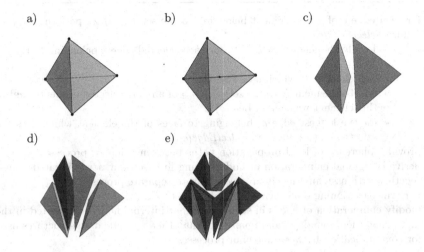

Fig. 1. Refinement of a tetrahedral element using the 8T-LE algorithm: a) original element; b) selection of the longest edge which is then used for partitioning (bisection); c) bisection of the original element; d) and e) results of applying the bisection procedure iteratively

3.2 Parallel Algorithm for Mesh Adaptation

When implementing the mesh adaptation in parallel, we base on the available mesh decomposition and coupling between neighbor domains (processes). Every process stores information about the domain assigned to it, sequence and number of internal, boundary and external nodes, as well as information necessary to organize communication with the neighbor processes. Additionally, the global enumeration of nodes is used, including the external nodes. Each process holds data structures which make possible transition from the global enumeration to the local one, and vice versa.

The proposed parallel algorithm of mesh adaptation is presented in Fig. 2. Communication between neighbor processes takes place in steps 3 and 4, based on information about edges which are exchanged between processes. This information is stored in sets e^i_{send} ($i = 0, \ldots, p - 1$), after performing the procedure *SelectEdge* (Fig. 3). As a result, in step 3 which is responsible for providing coherency of local propagation paths between neighbor processes, pairs of global indexes (n^1_g, n^2_g) describing edges are sent, using the non-blocking MPI routine `MPI_Isend`. After receiving this information, the mapping from the global enumeration to local one is performed, to allow placing edges in local data structures.

When performing FEM computation in parallel, elements which are located on boundaries of domains are duplicated in neighbor processes. This solution allows for avoiding communications at the cost of additional computation. We follow this concept when performing the mesh adaptation. In this case, the partitioning of elements must be realized in the same way in neighbor processes. The only difficulty emerges when during the element partitioning there are two or more

1. for each edge e of every element belonging to the set $E_{selected}$, perform the procedure *SelectEdge(e)*
2. perform locally the algorithm *LEPP*: for each selected edge e belonging to the set $e_{selected}$
   - for each element E which uses the edge e
     - add the element E to the set $E_{selected}$ of already selected elements (unless this element was added before)
     - for the longest edge e_l belonging to faces of the element which uses the selected edge e, perform *SelectEdge(e_l)*
3. provide coherency of local propagation paths between neighbor processes
4. derive the global enumeration of nodes taking into account newly created nodes, together with assigning newly created nodes to separate processes
5. perform partitioning of elements which belong to $E_{selected}$
6. modify enumeration of nodes in such a way that internal nodes are located in the first place, then boundary, and finally external; then upgrade data structures used for communication between neighbor processes

Fig. 2. Parallel algorithm for mesh adaptation

- add the edge e to the list $e_{selected}$ of already selected edges
- check if the edge e is located on the boundary with a neighbor domain (process); if so, add this edge to the list e^i_{send} of edges which are sent to this neighbor
- divide the edge e

Fig. 3. Procedure *SelectEdge(e)*

selected edges with the same length. To avoid this difficulty, we propose a solution based on the global indexes of nodes. So, when determining the longest edge in case of edges with the equal length, we compare additionally the global indexes of edges to choose an edge with the highest value of global indexes. This solution allows us to avoid communication between processes. It is sufficient to derive the global enumeration of nodes.

When deriving the global enumeration, it is necessary to determine to which process newly created nodes should be assigned in case when they are located on boundaries of domains. Possible solutions are assumed to not require any communications, and to be based entirely on global indexes of nodes and the previously performed decomposition of the FEM mesh. In our implementation, we arbitrarily choose a solution when a newly created node n_i which belongs to an edge (n_g^1, n_g^2) is assigned to a process in accordance with the following expression:

$$\mathcal{P}(n_i) = \begin{cases} \mathcal{P}(n_g^1) \text{ if } min\{n_g^1, n_g^2\} \ mod \ 2 = 0 , \\ \mathcal{P}(n_g^2) \text{ otherwise} . \end{cases} \tag{1}$$

The implementation of step 4 starts with the parallel procedure of deriving a new global enumeration of nodes, taking into account newly created nodes. This procedure includes the following three stages:

4.1 Determine Interval of Global Indexes for Each Process, as Well as Assign Global Indexes to the Local Nodes of Processes

For this aim, the information about the number n_l^i of local nodes in each process is spread among all the processes, using the `MPI_Allgather` routine. The global index n_g^i of a node i in process j is determined by adding the local index n^i of this node to the sum of numbers of nodes in all the processes from 0 do $j - 1$:

$$n_g^i = n^i + \sum_{k=0}^{j-1} N_k^l. \tag{2}$$

4.2 Exchange Global Indexes of Nodes Located on Boundaries of Domains

For this purpose, to update indexes of external nodes, the global indexes of these nodes are exchanged between neighbor processes, using sets S_j^k and R_j^k.

4.3 Exchange Global Indexes of Newly Created Nodes

- for each process j, look through the list e_{send}^j of edges: if an edge corresponds to newly created nodes assigned to process j, then add this edge to the list e_{rn}^j, otherwise add this edge to the list e_{sn}^j
- for each process j, send to its neighbor processes global indexes of nodes corresponding to edges from the list e_{sn}^j
- for each process j, receive from its neighbor processes information about global indexes of newly created nodes, which are finally determined using information contained in the list e_{rn}^j

4   Details of Implementation

The proposed parallel algorithm is implemented using the object-oriented programming technique in C++ language, and MPI standard. To perform the mesh decomposition, the ParMetis package [3] is utilized.

To provide the implementation efficiency, one of key is the usage of suitable data structures allowing for achievement of high performance, while preserving modest memory requirements. The demand for RAM is especially crucial in case of processing large FEM meshes on a relatively small number of processors.

During the adaptation, the FEM mesh is stored in three classes: Node , Edge, and Element. Additionally, the class AdaptationCom is introduced for storing information about the parallel implementation, as well as the class Adaptation where all the object related to the adaptation are grouped.

In the algorithm implementation, the frequent operation is searching for an edge based on nodes which describe it. To implement this operation efficiently, the solution based on hash tables is proposed.

Table 1. Parameters of FEM meshes before and after adaptation

mesh	before adaptation		after adaptation	
	number of nodes	number of elements	number of nodes	number of elements
100K	109009	622592	850849	4980736
400K	402017	2293760	3136705	18350080

During the mesh adaptation, there appears the problem of degenerating elements located on boundaries of modeled objects. Our adaptation module allows for adjusting FEM meshes to object geometries (Fig. 4). This is performed after the adaptation through changing positions of nodes located on the boundary. For this aim, we use the information about geometry, available as NURBS surfaces which are imported from the GID package. The described adjustment is performed locally in processors, and does not require communication.

5   Performance Results

Parameters of FEM meshes which were used in our experiments are presented in Table 1. These meshes were generated for the geometry shown in Fig. 4. Since the main purpose of these experiments was to investigate the scalability of the proposed algorithm and its implementation, the adaptation was performed for all the elements of FEM meshes. Each node of the computer cluster which was used for experiments contains two Dual Core Xeon 64-bit processors operating at a core frequency of 2.66 GHz, with 4 GB RAM and Gigabit Ethernet interface.

Fig. 5 presents the execution time and speedup for different numbers of processors (cores). The presented results take into account only the process of mesh adaptation, without auxiliary operations on data structures, as well as without adjustment to object geometry and load balancing. Because of an insufficient amount of RAM memory in a single node of the utilized computer cluster, the speedup for the mesh 400K was normalized with respect to the algorithm execution on four nodes.

The achieved performance results show quite good scalability of the proposed parallel algorithm of mesh adaptation, even for Gigabit Ethernet interconnect. At the same time, there is obviously a strong need for extensive experimental

a)   b)   c)

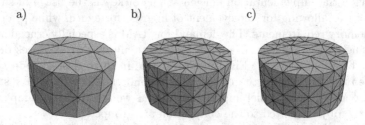

Fig. 4. Visualization of an original FEM mesh (a) after adaptation without (b) and with adjustment to object geometry (c)

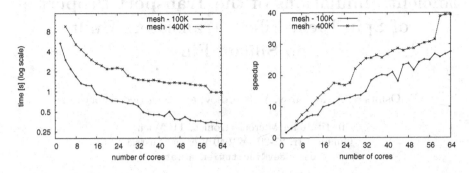

Fig. 5. Execution time and speedup versus number of cores, for different FEM meshes

studies of performance characteristics of this algorithm for different FEM meshes, and different computer architectures, including, for example, more advanced communication networks such as InfiniBand.

References

1. Balman., M.: Tetrahedral Mesh Refinement in Distributed Environments. In: Int. Conf. Parallel Processing Workshops (ICPPW 2006), pp. 497–504. IEEE Computer Soc. (2006)
2. Hulsemann, F., Kowarschik, M., Mohr, M., Rude, U.: Parallel Geometric Multigrid. Lecture Notes in Computational Science and Engineering, vol. 51, pp. 165–208 (2006)
3. Karypis, G., Schloegel, K., Kumar, V.: PARMETIS Parallel Graph Partitioning and Sparse Matrix Ordering Library Version 3.1. Univ. Minnesota, Army HPC Research Center (2003),
   http://glaros.dtc.umn.edu/gkhome/fetch/sw/parmetis/manual.pdf
4. Olas, T., Karczewski, K., Tomas, A., Wyrzykowski, R.: FEM Computations on Clusters Using Different Models of Parallel Programming. In: Wyrzykowski, R., Dongarra, J., Paprzycki, M., Waśniewski, J. (eds.) PPAM 2001. LNCS, vol. 2328, pp. 170–182. Springer, Heidelberg (2002)
5. Olas, T., Leśniak, R., Wyrzykowski, R., Gepner, P.: Parallel Adaptive Finite Element Package with Dynamic Load Balancing for 3D Thermo-Mechanical Problems. In: Wyrzykowski, R., Dongarra, J., Karczewski, K., Wasniewski, J. (eds.) PPAM 2009. LNCS, vol. 6067, pp. 299–311. Springer, Heidelberg (2010)
6. Patzak, B., Rypl, D.: A Framework for Parallel Adaptive Finite Element Computations with Dynamic Load Balancing. In: Proc. First Int. Conf. Parallel, Distributed and Grid Computing for Engineering, Paper 31. Civil-Comp Press (2009)
7. Plaza, A., Rivara M.: Mesh Refinement Based on the 8-Tetrahedra Longest-Edge Partition. In: Proc. 12th Int. Meshing Roundtable, Sandia National Laboratories, pp. 67–78 (2003)
8. Rivara, M., Pizarro, D., Chrisochoides, N.: Parallel Refinement of Tetrahedral Meshes using Terminal-Edge Bisection Algorithm. In: Proc. 13th Int. Meshing Roundtable, Sandia National Labs, pp. 427–436 (2004)
9. Wyrzykowski, R., Olas, T., Sczygiol, N.: Object-Oriented Approach to Finite Element Modeling on Clusters. In: Sørevik, T., Manne, F., Moe, R., Gebremedhin, A.H. (eds.) PARA 2000. LNCS, vol. 1947, pp. 250–257. Springer, Heidelberg (2001)

Efficient Simulations of the Transport Properties of Spin Field-Effect Transistors Built on Silicon Fins

D. Osintsev, A. Makarov, V. Sverdlov, and Siegfried Selberherr

Institute for Microelectronics, TU Wien,
Gußhausstr. 27-29, A-1040 Vienna, Austria
osintsev@iue.tuwien.ac.at

Abstract. Significant progress in integrated circuits performance has been supported by the miniaturization of the transistor feature size. With transistor scalability gradually slowing down new concepts have to be introduced in order to maintain the computational speed increase at reduced power consumption for future micro- and nanoelectronic devices. A promising alternative to the charge degree of freedom currently used in MOSFET switches is to take into account the spin degree of freedom. We computationally investigate transport properties of ballistic spin field-effect transistors (SpinFETs). These simulations require a significant amount of computational resources. To achieve the best performance of calculations we parallelize the code for a shared-memory multi-CPU system. As the result of the optimization of the whole model a significant speed-up in calculations is achieved. We demonstrate that the [100] oriented silicon fins are best suited for practical realizations of a SpinFET.

1   Introduction

The outstanding increase of the computational speed in present integrated circuits is supported by the continuing miniaturization of the semiconductor device feature size. With scaling approaching its physical limits, the semiconductor industry is facing the challenge to introduce new innovative elements to increase integrated circuit performance. Employing spin as an additional degree of freedom is promising for boosting the efficiency of future low-power integrated electronic circuits. Indeed, the spin of an electron can change its orientation to opposite very fast by consuming an amazingly small amount of energy. Thus utilizing spin properties for future microelectronic devices opens a great opportunity to reduce power consumption.

The spin field-effect transistor is a future semiconductor spintronic device promising a performance superior to that achieved in the present transistor technology. SpinFETs are composed of two ferromagnetic contacts (source and drain) connected to the semiconductor channel. The ferromagnetic source (drain) contact injects (detects) spin-polarized electrons to (from) the semiconductor region.

I. Lirkov, S. Margenov, and J. Waśniewski (Eds.): LSSC 2011, LNCS 7116, pp. 630–637, 2012.

Thus ferromagnetic contacts play the role of polarizer and analyzer for the electron spin as described by Datta and Das [1]. Because of the non-zero spin-orbit interaction the electron spin precesses during the propagation through the channel. At the drain contact only the electrons with the spin aligned to the drain magnetization can leave the channel and contribute to the current. Current modulation is achieved by changing the strength of the spin-orbit interaction in the semiconductor region and thus the degree of the spin precession. The strength of the spin-orbit interaction can be controlled by applying the external gate voltage which introduces the structural inversion asymmetry. This is the dominant mechanism of the spin-orbit interaction in confined structures of III-V semiconductors. The corresponding effective Hamiltonian is in the Rashba form [2]

$$H_R = \frac{\alpha_R}{\hbar}(p_x\sigma_y - p_y\sigma_x),\tag{1}$$

where α_R is the effective electric field-dependent parameter of the spin-orbit interaction, $p_{x(y)}$ is the electron momentum projection on the $x(y)$ axis, σ_x and σ_y are the Pauli matrices.

Silicon is characterized by a weak spin-orbit interaction and long spin life time. It is therefore an attractive material for spin-driven applications. However, because of the weak spin-orbit interaction, silicon was not considered as a candidate for the SpinFET channel material. Recently it was shown [3] that thin silicon films in the SiGe/Si/SiGe structures have enhanced values of the spin-orbit interaction. It turns out, however, that the Rashba spin-orbit interaction in confined silicon structures is relatively weak and is approximately ten times smaller than the value of the dominant contribution which is of the Dresselhaus type with a corresponding effective Hamiltonian in the form

$$H_D = \frac{\beta}{\hbar}(p_x\sigma_x - p_y\sigma_y).\tag{2}$$

This term is due to interfacial disorder induced inversion symmetry breaking and depends almost linearly on the effective electric field [4]. For a built-in field of 50kV/cm, the strength of the Dresselhaus spin-orbit interaction is found to be $\beta \approx 2\mu eV$nm, which is in agreement with the value found experimentally [5], while $\alpha_R \approx 0.1\mu eV$nm. This value of the spin-orbit interaction in confined silicon systems is sufficient for application as SpinFET channels.

To calculate properties of the spin field-effect transistor we consider a model similar to [6,7]. Contrary to [6,7] we introduce the spin-orbit interaction in the Dresselhaus form. The effective mass Hamiltonian for the source and the drain contacts is taken in the form

$$\hat{H}_{source} = \frac{\hat{p}_x^2}{2m_f^*} + h_0\hat{\sigma}_z, \quad x < 0,\tag{3}$$

$$\hat{H}_{drain} = \frac{\hat{p}_x^2}{2m_f^*} \pm h_0\hat{\sigma}_z, \quad x > L,\tag{4}$$

where m_f^* is the effective mass in the contacts, h_0 is the exchange splitting energy, and $\hat{\sigma}_z$ is the Pauli matrix; $\pm$ in (4) stands for the parallel and antiparallel configuration of the contact magnetization, respectively. For the silicon region the Hamiltonian reads

$$\hat{H}_S = \frac{\hat{p}_x^2}{2m_s^*} + \delta E_c - \frac{\beta}{\hbar}\hat{\sigma}_x\hat{p}_x + \frac{1}{2}g\mu_B B\hat{\sigma}^*, \tag{5}$$

for [100] oriented fins and

$$\hat{H}_S = \frac{\hat{p}_x^2}{2m_n^*} + \delta E_c - \frac{\beta}{\hbar}\hat{\sigma}_y\hat{p}_x + \frac{1}{2}g\mu_B B\hat{\sigma}^*, \tag{6}$$

for [110] oriented fins. Here m_n^* is the subband effective mass, δE_c is the band mismatch between the ferromagnetic and the silicon region, β is the strength of the spin-orbit interaction, g is the Landé factor, μ_B is the Bohr magneton, B is the magnetic field, and $\hat{\sigma}^* \equiv \hat{\sigma}_x cos\gamma + \hat{\sigma}_y sin\gamma$ with γ defined as the angle between the magnetic field and the transport direction.

To calculate the dependence of the transport properties on the spin-orbit interaction, we need the electron eigenfunctions in the ferromagnetic and the semiconductor regions. We are looking for a wave function in the left contact in the following form

$$\Psi_L(x) = (e^{ik_\uparrow x} + R_\uparrow e^{-ik_\uparrow x})\begin{pmatrix} 1 \\ 0 \end{pmatrix} + R_\downarrow e^{-ik_\downarrow x}\begin{pmatrix} 0 \\ 1 \end{pmatrix}, \tag{7}$$

$$\Psi_L(x) = R_\uparrow e^{-ik_\uparrow x}\begin{pmatrix} 1 \\ 0 \end{pmatrix} + (e^{ik_\downarrow x} + R_\downarrow e^{-ik_\downarrow x})\begin{pmatrix} 0 \\ 1 \end{pmatrix}, \tag{8}$$

where (7) describes a spin-up induced electron and (8) describes a spin-down induced electron, $k_{\uparrow(\downarrow)}$ is the wave vector of the spin-up(spin-down) electron, $R_{\uparrow(\downarrow)}$ are the corresponding reflection coefficients. For the right contact the wave function is given by

$$\Psi_R(x) = C_\uparrow e^{ik_\uparrow x}\begin{pmatrix} 1 \\ 0 \end{pmatrix} + C_\downarrow e^{ik_\downarrow x}\begin{pmatrix} 0 \\ 1 \end{pmatrix}, \tag{9}$$

where $C_{\uparrow(\downarrow)}$ are the transmission amplitudes. For the silicon region the wave function can be written as

$$\psi_S(x) = A_+ e^{ik_{x1}^{(+)}x}\begin{pmatrix} k_1 \\ 1 \end{pmatrix} + B_+ e^{ik_{x2}^{(+)}x}\begin{pmatrix} k_2 \\ 1 \end{pmatrix} \tag{10}$$

$$+ A_- e^{ik_{x1}^{(-)}x}\begin{pmatrix} k_3 \\ -1 \end{pmatrix} + B_- e^{ik_{x2}^{(-)}x}\begin{pmatrix} k_4 \\ -1 \end{pmatrix}, \tag{11}$$

where $k_{x1(2)}^{+(-)}$ are the wave vectors corresponding to the Hamiltonian (5) in case of a [100]-oriented fin or (6) in case of a [110]-oriented fin. The coefficients k_1, k_2, k_3, k_4 are calculated depending on the fin orientation. For a [100]-oriented fin the coefficients are

$$k_1 = -\frac{\frac{iBg\mu_B \sin(\gamma)}{2} - \frac{Bg\mu_B \cos(\gamma)}{2} + \beta k_{x1}^{(+)}}{\sqrt{\left(\frac{Bg\mu_B \cos(\gamma)}{2} - \beta k_{x1}^{(+)}\right)^2 + \left(\frac{Bg\mu_B \sin(\gamma)}{2}\right)^2}}, \tag{12}$$

$$k_2 = -\frac{\frac{iBg\mu_B \sin(\gamma)}{2} - \frac{Bg\mu_B \cos(\gamma)}{2} + \beta k_{x2}^{(+)}}{\sqrt{\left(\frac{Bg\mu_B \cos(\gamma)}{2} - \beta k_{x2}^{(+)}\right)^2 + \left(\frac{Bg\mu_B \sin(\gamma)}{2}\right)^2}}, \tag{13}$$

$$k_3 = \frac{\frac{iBg\mu_B \sin(\gamma)}{2} - \frac{Bg\mu_B \cos(\gamma)}{2} + \beta k_{x1}^{(-)}}{\sqrt{\left(\frac{Bg\mu_B \cos(\gamma)}{2} - \beta k_{x1}^{(-)}\right)^2 + \left(\frac{Bg\mu_B \sin(\gamma)}{2}\right)^2}}, \tag{14}$$

$$k_4 = \frac{\frac{iBg\mu_B \sin(\gamma)}{2} - \frac{Bg\mu_B \cos(\gamma)}{2} + \beta k_{x2}^{(-)}}{\sqrt{\left(\frac{Bg\mu_B \cos(\gamma)}{2} - \beta k_{x2}^{(-)}\right)^2 + \left(\frac{Bg\mu_B \sin(\gamma)}{2}\right)^2}}. \tag{15}$$

We compute the current through the device as

$$I^{P(AP)} = \frac{e^2}{h} \int_{\delta E}^{\infty} \left[T_\uparrow^{P(AP)}(E) + T_\downarrow^{P(AP)}(E) \right]$$
$$\left\{ \frac{1}{1 + e^{\frac{E - E_F}{kT}}} - \frac{1}{1 + e^{\frac{E - E_F + eV}{kT}}} \right\} dE, \tag{16}$$

where k is the Boltzmann constant, T is the temperature, and V is the voltage. The transmission coefficients are determined by applying the standard boundary conditions at the ferromagnetic/silicon interfaces. The spin-up ($T_\uparrow^P$) and spin-down ($T_\downarrow^P$) transmission coefficients for the parallel configuration of the contact magnetization are defined as

$$T_\uparrow^P = |C_\uparrow|^2 + \frac{k_\downarrow}{k_\uparrow}|C_\downarrow|^2, \tag{17}$$

$$T_\downarrow^P = \frac{k_\uparrow}{k_\downarrow}|C_\uparrow|^2 + |C_\downarrow|^2. \tag{18}$$

For the anti-parallel configuration of the contact magnetization the transmission probabilities are given by

$$T_\uparrow^{AP} = \frac{k_\downarrow}{k_\uparrow}|C_\uparrow|^2 + |C_\downarrow|^2, \tag{19}$$

$$T_\downarrow^{AP} = |C_\uparrow|^2 + \frac{k_\uparrow}{k_\downarrow}|C_\downarrow|^2. \tag{20}$$

The conductance is defined as

$$G^{P(AP)} = \lim_{V \to 0} \frac{I^{P(AP)}}{V}. \tag{21}$$

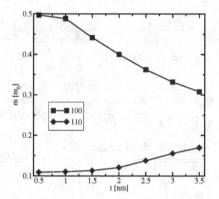

Fig. 1. Ground subband effective mass dependence on t in [100] and [110] fins

Finally, the tunneling magnetoresistance (TMR) is defined as

$$\text{TMR} \equiv \frac{G^P - G^{AP}}{G^{AP}}. \tag{22}$$

2   Simulations

We investigate the dependence of the conductance through the silicon Spin-FET on the physical parameters. In order to calculate the conductance one has to determine the wave vectors, compose the system of equations corresponding to the boundary conditions, and solve the obtained system of linear equations to determine the transmission coefficients for the spin-up and spin-down electrons. These calculations must be performed for the parallel and the anti-parallel configuration of the contact magnetization for each energy point of the half-infinite integrand (16). It follows from (21) that the integral evaluated numerically describes the conductance for a single point of the conduction band mismatch δE_c, at a single value of temperature T. Thus to investigate the transport properties of the silicon SpinFET at various parameter values a huge amount of calculations must be carried out. To reduce the simulation time the code for the model must be heavily optimized and parallelized.

The usual techniques for parallelization are divided into two groups - parallelization in case of shared-memory and parallelization in case of distributed memory systems. The advantage of shared-memory parallelization is that it guarantees uniform access to the memory for each process. This means that the time spent for data manipulation in the memory is approximately the same for each process. The advantage of distributed calculations is that the number of processors used for the calculations is not limited to the number of processors on a single node. In our simulations the code is parallelized with the OpenMP library. Because of the absence of the correlation between the conductances at different energy points the whole calculations are distributed between a large number of

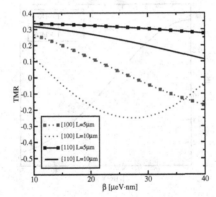

Fig. 2. Tunneling magnetoresistance dependence on the value of the Dresselhaus spin-orbit interaction for a magnetic field orthogonal to the transport direction. The source and the drain contact polarization $P = 0.4$, the delta-function barrier strength $z = 5$.

computing threads. Thus the time spent for the calculations is reduced proportionally to the number of parallel threads. Although threads perform calculations in parallel, the time spent by each thread to fulfill the tasks is not the same. Therefore one has to take proper care of a uniform distribution of the computational load between the threads. This problem is crucial for obtaining the maximum possible speed-up. Since OpenMP provides a possibility to control the amount of calculations for each thread in the run-time mode, the problem of the uniform distribution of the computational load is solved by standard tools.

3   Results

We have improved the performance of the code by utilizing the OpenMP approach. The results of using the parallel implementation are presented in Table 1. The actual speedup of the code is very close to the ideal speedup, in which the increase of the number of processors (cores) twice leads to the decrease in the computing time twice, due to the already mentioned fact that there is no correlation between the calculations of the conductances.

Calculations in the presence of temperature require high accuracy. Therefore, the adaptive methods from the GNU Scientific Library [8] have been used for the numerical integration.

We consider square silicon fins with [100] or [110] orientation, with (001) horizontal faces. The parabolic band approximation is not sufficient in thin and narrow silicon fins. In order to compute the subband structure in silicon fins we employ the two-band **k·p** model, which has been shown to be accurate up to 0.5eV above the conduction band edge. The resulting Schrödinger differential equation, with the confinement potential appropriately added in the Hamiltonian, is discretized using the box integration method and solved for each value of the conserved momentum p_x along the current direction using efficient numerical algorithms available through the Vienna Schrödinger-Poisson framework

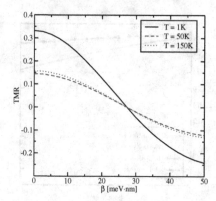

Fig. 3. Tunneling magnetoresistance dependence on the value of the Dresselhaus spin-orbit interaction for a semiconductor channel of length L=5μm; the spin polarization P in the source and the drain contacts is P =0.4; the strength z of the delta-function-like barriers at the contacts between the channel and source/drain z =5, the voltage V =0.001V, and the effective mass in the ferromagnetic region m$_s^*$=0.438m$_0$

Table 1. Calculation time (seconds) depending on the number of processors and the number of points

Number of Processors	N = 24	N = 48	N = 96
1	1036	2057	4129
2	549	1094	2198
4	286	570	1138
8	160	318	633

(VSP) [9]. The dependence of the effective mass of the ground subband on t in [100] and [110] fins is shown in Fig. 1.

Some results obtained with our simulation procedure are shown in Fig. 2 and Fig. 3. For the channel we chose the silicon fins with [100] and [110] orientations and square cross-section with (001) horizontal faces. Fig. 2 shows the dependence of the TMR for the [100] and [110] oriented fins with a thickness t =1.5nm on the value of the spin-orbit interaction β. The fins with [100] orientation show a stronger dependence on β compared to the [110] oriented fins. The reason of the stronger dependence is that the influence of the spin-orbit interaction on the conductance is determined by the wave vector $k_D = m_n^*\beta/\hbar^2$. As the effective mass value for the [110] oriented fins is smaller compared to the [100] oriented fins, the same variation of k_D results in a larger variation of β in case of the [110] oriented fins. Thus the [100] oriented fins are preferred for silicon SpinFETs. Fig. 3 demonstrates the dependence of the TMR on the spin-orbit interaction at various temperatures. One can see that the oscillatory amplitude of the TMR survives even at T=150K. This fact is encouraging for utilizing silicon as the semiconductor channel material in SpinFETs at room temperature.

4 Conclusion

We investigated the transport properties of SpinFETs and show that silicon can be considered as a promising material for spin-driven applications. We sketch our way to speed-up the calculations by employing an efficient parallelization scheme. We show that because of the Dresselhaus form of the spin-orbit interaction the TMR of silicon fins can be modulated by the spin-orbit interaction even at relatively high temperatures. We demonstrate that because of the larger subband effective mass the [100] oriented silicon fins are best suited for the realization of a SpinFET.

Acknowledgments. This work is supported by the European Research Council through the grant #247056 MOSILSPIN.

References

1. Datta, S., Das, B.: Electronic analog of the electro-optic modulator. Applied Physics Letters 56(7), 665–667 (1990)
2. Giglberger, S., Golub, L.E., Bel'kov, V.V., Danilov, S.N., Schuh, D., Gerl, C., Rohlf-ing, F., Stahl, J., Wegscheider, W., Weiss, D., Prettl, W., Ganichev, S.D.: Rashba and Dresselhaus spin splittings in semiconductor quantum wells measured by spin photocurrents. Phys. Rev. B 75(3), 35327 (2007)
3. Nestoklon, M.O., Ivchenko, E.L., Jancu, J.-M., Voisin, P.: Electric field effect on electron spin splitting in SiGe/Si quantum wells. Phys. Rev. B 77(15), 155328 (2008)
4. Prada, M., Klimeck, G., Joynt, R.: Spin-orbit splittings in Si/SiGe quantum wells: from ideal Si membranes to realistic heterostructures. New J. Phys. 13, 13009 (2011)
5. Wilamowski, Z., Jantsch, W.: Suppression of spin relaxation of conduction electrons by cyclotron motion. Phys. Rev. B 69(3), 35328 (2004)
6. Cahay, M., Bandyopadhyay, S.: Phase-coherent quantum mechanical spin transport in a weakly disordered quasi-one-dimensional channel. Phys. Rev. B 69(4), 45303 (2004)
7. Jiang, K.M., Zhang, R., Yang, J., Yue, C.-X., Sun, Z.-Y.: Tunneling magnetoresis-tance properties in ballistic spin field-effect transistors. IEEE T-ED 57, 2005 (2010)
8. GNU Scientific Library, http://www.gnu.org/s/gsl/
9. Karner, M., Gehring, A., Holzer, S., Pourfath, M., Wagner, M., Gös, W., Vasicek, M., Baumgartner, O., Kernstock, C., Schnass, K., Zeiler, G., Grasser, T., Kosina, H., Selberherr, S.: A multi-purpose Schrödinger-Poisson solver for TCAD applications. Journal of Computational Electronics 6, 179–182 (2007)

Large-Scale Simulation of Uniform Load Traffic for Modeling of Throughput on a Crossbar Switch Node

Tasho Tashev and Vladimir Monov

Institute of Information and Communication Technologies,
Bulgarian Academy of Sciences, Acad. G. Bonchev, bl.2, 1113 Sofia, Bulgaria
{ttashev,vmonov}@iit.bas.bg
http://www.iict.bas.bg

Abstract. In the present paper we propose a family of patterns for uniform traffic simulating. The results from computer simulations of the throughput on a switch node with these patterns are presented. The necessary computations have been performed on grid-clusters of IICT-BAS and CERN. Our simulations utilize the PIM-algorithm for non-conflict schedule, specified by apparatus of Generalized Nets. It is shown that the usage of the suggested family of patterns enables us to evaluate the influence of large scale changes of the input buffer on the throughput.

Keywords: large-scale simulation, generalized nets, switch node.

1 Introduction

In modern communication systems, crossbar packet switches route traffic from the input to output where a message packet is transmitted from the source to the destination. The randomly incoming traffic must be controlled and scheduled to eliminate conflict at the crossbar switch. The goal of the traffic-scheduling for the crossbar switches is to maximize the throughput of packet through a switch and to minimize packet blocking probability and packet waiting time [1].

The problem of calculating of non-conflict schedule is NP-complete [2]. Algorithms are suggested which solve the problem partially. A source of a series of parallel algorithms is the PIM-algorithm (Parallel Iterative Matching) [3]. Its development goes through Round-Robin schemes and is universally recognized by the iSLIP-algorithm [1]. The latter is efficient enough for size of the matrix switch up to 32x32 lines. However, progress in optic fibers utilization as well as increasing the number of personal computers require larger size. One part of the investigations represent developing of modifications of iSLIP, still using input buffering with VOQ (Virtual Output Queuing). The obtained results are effective when the size is 64x64 and larger, for example GMM algorithm [4]. The approach of Birkhoff-von Neumann is very interesting, too [5].

Another group of researchers use input and intermediate buffering (CICQ), by applying a buffer associated with commutation field. Of course, more and more investigations are directed to a completely optical commutation [6].

I. Lirkov, S. Margenov, and J. Waśniewski (Eds.): LSSC 2011, LNCS 7116, pp. 638–645, 2012.

As a formal means for description and investigations of the characteristics of switch algorithms many authors used cellular automata, neural networks, queue theory, etc. In our research we apply Generalized Nets (GN) apparatus [7,8]. The efficiency of the switch functioning can be firstly assessed by the obtained throughput. The load (input) packet traffic is obligatory divided into two types - uniform load traffic and non-uniform [6]. These two types additionally are divided into sub-types traffic. When such versatility is available a problem for adequate comparison between them arises.

As far as the effectivity check of the algorithms always begins with throughput modeling of the switch node with uniform load traffic, in the present paper we propose a family of patterns for uniform load traffic simulating. The aim is to use this family as a point of quick reference for variety of algorithms. In order to clarify the correctness of comparison with results obtained with other algorithms, we synthesized a model of the well known PIM-algorithm by means of GN. Then we carried out computer simulations of its throughput by using the proposed patterns for uniform load traffic.

The paper is structured as follows. Section 2 briefly describes the PIM-algorithm. Section 3 presents a GN-model of the PIM-algorithm. Section 4 describes our patterns for uniform load traffic. Results from simulations using grid-resources are presented in section 5, while section 6 outlines the conclusions and some possible lines of future research.

2   PIM-Algorithm of Non-conflicts Schedule for Commutation

The requests for transmission through switching $n \times n$ line switch node is presented by an $n \times n$ matrix T, named traffic matrix (n is integer) [4]. Every element t_{ij}, ($t_{ij} \in \{0, 1, 2, \ldots\}$) of the traffic matrix represents a request for a packet from input i to output j. For example $t_{ij} = 2$ means that two packets from the i−th input line have to be send to j−th output line of the switch node.

It is assumed that a conflict situation is created when in any row of the T matrix the number of requests is more than 1. This corresponds to the case when one source declares connection with more than one receiver. If any column of the T matrix hosts more than one nonzero element, this also indicates a conflict situation. Avoiding conflicts is related to the switch node efficiency.

We will give a succinct description of the PIM-algorithm. It has three phases.

1. **Request:** Every input sends request to every output for which it has a packet for transmission.
2. **Grant:** Every output chooses randomly one of the received requests and grants permission for sending to the corresponding input.
3. **Accept:** Every input received grants chooses randomly one of them. This packet will be accepted for commutation.

Inputs execute in parallel the first phase. Outputs execute in parallel second phase. Inputs are working in parallel in the third phase [3]. This parallelism is suitable for applying of GN apparatus.

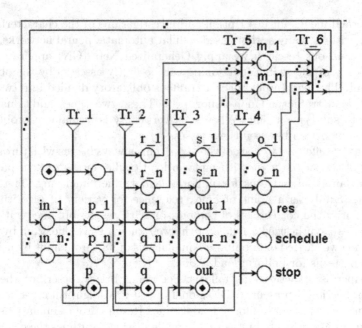

Fig. 1. Graphical form of GN model of the PIM algorithm

3 Generalized Net Model of PIM-Algorithm

In our previous investigations algorithms for computing of non-conflict schedule are modeled by GN based on the principle of sequent-random choice [9]. The PIM-algorithm is based on distributive-random choice where stages of parallel processing of information are clearly defined. By means of GN we can effectively model these processes. For this purpose we have to synthesize a GN model of the PIM-algorithm.

The described three phases of algorithm lead to at least three transitions in the GN model. The model is developed for switch node with n inputs and n outputs. Based on a previous work [10], here, we specify VOQ (places $m_1, \ldots, m_n$) in an explicit form. The graphic form of this model is shown in Figure 1.

The places in the model have the following significance: $in\_1, \ldots, in\_n$ – inputs of the switch node in the initial moment; $p\_1, \ldots, p\_n$ – outputs of the switch node in phase two; $q\_1, \ldots, q\_n$ – inputs of the switch node in phase three; $r\_1, \ldots, r\_n$ – non-chosen requests after phase two; $s\_1, \ldots, s\_n$ – non-chosen requests after phase three; *start* – place for initiating of schedule computing; p – place for starting of phase one; q – place for starting of phase two; *out* – place for starting of phase three; $out\_1, \ldots, out\_n$ – places for current solution; $o\_1, \ldots, o\_n$ – non-accepted requests after phase three; *schedule* – place for solutions; *res* – null VOQ; *stop* – place for the end of schedule computing.

Every token in places $in\_1, in\_2, \ldots, in\_n$ represents a request for sending a packet. It has an initial characteristic: ordered triple of numbers including the number of the input (noted by i), the number of output (noted by j) for which the packet is directed, and the number of packets (noted by l) in VOQ:

$$ch_0 = (pr_1ch_0, pr_2ch_0, pr_3ch_0) = (i, j, l).$$

Each of the transitions has one and the same priority. The same refers to the tokens. The analysis of the model by the means of GN proves that it has non-conflict schedule. The model has the abilities to provide information about the number of switchings of the crossbar matrix, which will be used for calculating of the average throughput.

4   Family of Patterns for Uniform Traffic

The matrix T defines a uniform traffic demand matrix if the total number of packets in each row and each column are equal [1,4]. For our large-scale computer simulation we suggest several types of uniform matrices T, which will be called a family of patterns for T. They posses the following properties:

1. easy generation for any size of the switch (nxn);
2. generation does not depend on the type of hardware used, compiler and operation system;
3. their exact, optimal, non-conflict schedule is known.

The first type matrix is called $Pattern_1$. Its optimal schedule requires $1 \times n$ switchings of crossbar matrix for $n \times n$ switch. In general, this type matrix is denoted by $Pattern_i$. Its optimal schedule requires ($i \times n$) switchings of crossbar matrix for $n \times n$ switch. This type of matrices is shown in Figure 2.

The second type matrix is called $Pattern_N$. Its optimal schedule requires n^2 switchings of crossbar matrix for $n \times n$ switch. The third type matrix is called $Pattern_P$. Its optimal schedule also requires n^2 switchings of crossbar matrix for $n \times n$ switch. These types of matrices are shown in Figure 3. The $Pattern_P$ has the property of triangular Pythagorean numbers: the sum of n consecutive odd numbers is equal to n^2.

The first two types represent balanced traffic, and the third – non-balanced. The second type will be used for checking of uniformity of the generator of pseudo-random sequences during the simulations because

$$T(Pattern_N, k = n) = T(Pattern_i, i = n, k = n), n = 2, 3, \ldots$$

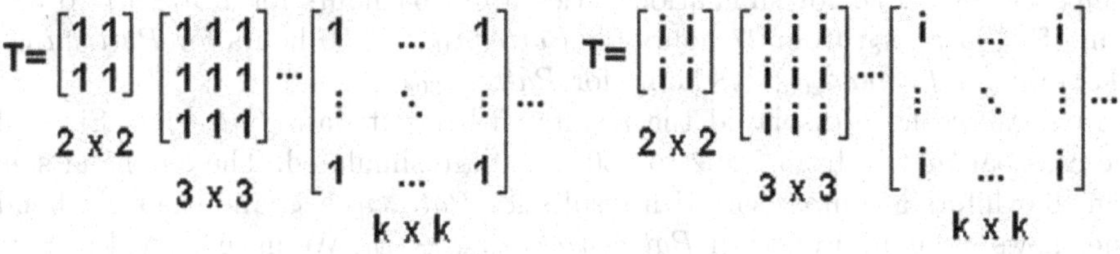

$$T=\begin{bmatrix}1&1\\1&1\end{bmatrix}\begin{bmatrix}1&1&1\\1&1&1\\1&1&1\end{bmatrix}\cdots\begin{bmatrix}1&\cdots&1\\ \vdots&\ddots&\vdots\\1&\cdots&1\end{bmatrix}\cdots \qquad T=\begin{bmatrix}i&i\\i&i\end{bmatrix}\begin{bmatrix}i&i&i\\i&i&i\\i&i&i\end{bmatrix}\cdots\begin{bmatrix}i&\cdots&i\\ \vdots&\ddots&\vdots\\i&\cdots&i\end{bmatrix}\cdots$$

$$2 \times 2 \qquad 3 \times 3 \qquad\qquad k \times k \qquad\qquad 2 \times 2 \qquad 3 \times 3 \qquad\qquad k \times k$$

Fig. 2. Matrices of types $Pattern_1$ and $Pattern_i$

$$T=\begin{bmatrix}2\,2\\2\,2\end{bmatrix}\begin{bmatrix}3\,3\,3\\3\,3\,3\\3\,3\,3\end{bmatrix}\cdots\begin{bmatrix}k&\cdots&k\\\vdots&\ddots&\vdots\\k&\cdots&k\end{bmatrix}\cdots \qquad T=\begin{bmatrix}1\,3\\3\,1\end{bmatrix}\begin{bmatrix}1\,5\,3\\3\,1\,5\\5\,3\,1\end{bmatrix}\cdots\begin{bmatrix}1&\cdots&3\\\vdots&\ddots&\vdots\\2k\text{-}1&\cdots&1\end{bmatrix}\cdots$$

$$\underset{2\times2}{}\quad\underset{3\times3}{}\quad\underset{k\times k}{}\qquad\qquad\underset{2\times2}{}\quad\underset{3\times3}{}\quad\underset{k\times k}{}$$

Fig. 3. Matrices of types $Pattern_N$ and $Pattern_P$

For example :

$$T(Pattern_N, k = 10) = T(Pattern_{10}, i = 10, k = 10), n = 10$$

where $(k \times k) = 10 \times 10$ is the current size of the simulated crossbar field.

5   Result of Grid-Simulations

The transition from a GN-model to executive program is performed as in [9]. The program package Vfort of the Institute of mathematical modeling of Russian Academy of Sciences is used [11]. The source code has been tested on Vfort and then compiled by means of the grid-structure of the IICT-BAS. The resulting executive code is executed in the grid-structure. A main restriction is the time for execution.

In the figures below, $Pattern_i$ is denoted as P-i for $i = 1, 2, \ldots,$ $Pattern_N$ and $Pattern_P$ are denoted as P-N and P-P, respectively. Fig.4 shows the results from computer simulation of the PIM-algorithm with input data $Pattern_{1,\ldots,8}$. Sizes of the crossbar matrix from 2×2 to 260×260 are simulated. The resulting throughput is the average for 10 000 simulations for each size.

Fig.5 (left) presents the results with input data $Pattern_{10,20,30,40,50,100}$. Sizes of the crossbar matrix from 2×2 to 130×130 are simulated. The right hand side of Fig.5 shows the results for $Pattern_{200,300,400,500,1000}$. The size is decreased, but the speed of approaching the known theoretical boundary (1-1/e 0,632) is increasing.

As a conclusion, the family of patterns $Pattern_i$ indicates the relevance of the size of VOQ to the throughput: larger input buffer has larger throughput. The cost is an increased time for calculation of the non-conflict schedule. The computer time used for simulation varies: from 64 hours for $Pattern_1$ to 387 hours for $Pattern_8$; from 27 hours for $Pattern_{10}$ to 260 hours for $Pattern_{100}$; 194 hours for $Pattern_{200}$; 348 hours for $Pattern_{300}$.

In Fig.6 we have displayed the results with input data $Pattern_N$. Sizes of the crossbar matrix from 2×2 to 130×130 are simulated. The left hand side of fig.6 exhibits a comparison with results for $Pattern_{1,\ldots,8}$, and the right hand side shows a comparison with $Pattern_{10,20,30,40,50,100}$. We have a good conformity of the results as it was pointed out in Section 4. For example: throughput $(Pattern_N, n = 8) \approx$ throughput $(Pattern_8, n = 8)$, fig.6(left). The precision of

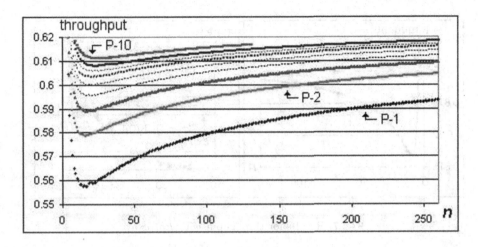

Fig. 4. Results for throughput with $Pattern_1$ to $Pattern_8$

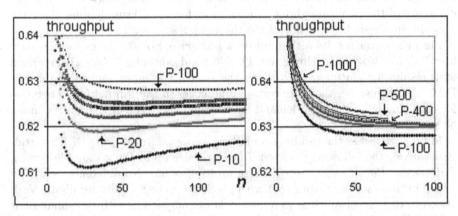

Fig. 5. Results for throughput with $Pattern_{10}$ to $Pattern_{1000}$

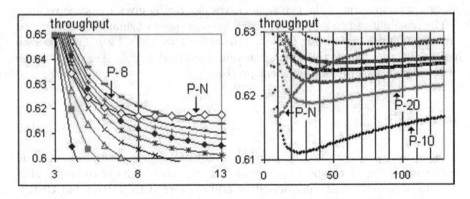

Fig. 6. Results for throughput with $Pattern_N$

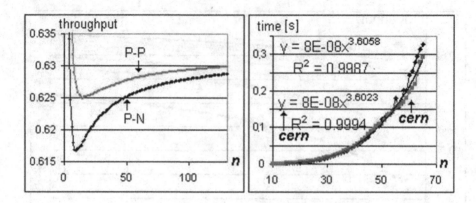

Fig. 7. Results for throughput with $Pattern_P$

the results can be estimated by the position of the first significant digit in the difference of the two throughputs. In our case, for 10 000 simulations, there are 3 significant digits in the value of the throughput.

Fig.7 shows the results with input data $Pattern_P$. Sizes of the crossbar matrix from 2×2 to 130×130 are simulated. The left hand side of fig.7 gives a comparison with results for $Pattern_N$. Although the size of the input buffers of the two Patterns is the same, there are certain differences in the throughput. Therefore for each simulated uniform demand traffic, the exact type of the matrix must be specified.

In order to assess the results, simulations were performed using CERN's grid-structure for the following patterns: $Pattern_1$ (size up to 130×130), $Pattern_{50}$, $Pattern_{100}$, $Pattern_N$, $Pattern_P$ (size up to 65×65). No differences were detected in the resulting throughput with precision up to 8 significant digits. Variations in the execution time were observed in comparison with the simulations in IICT-BAS . Fig.7 (right) shows the calculation time of a non-conflict schedule averaged from 10 000 simulations for $Pattern_P$ of the two grid-structures. A power approximation of the two time curves practically gives the same result.

The possibility to chose between precision and speed of simulation was tested with 100 000 simulations for $Pattern_1$ (from 2×2 to 130×130). The precision increased 3 to 4 times; the simulation time increased tenfold. For our purpose 10 000 simulations provide sufficient precision.

6   Conclusion

In the present paper we have developed a family of patterns for uniform demand traffic simulating. A comparison is made between the results of computer simulations of pattern switch performed on grid-clusters of IICT-BAS and CERN. It is shown that the suggested family of patterns enables us to evaluate the influence of large scale size changing of the input buffer on the throughput. In the future studies we will search for uniform-patterns for fast discovery of instability regions in the work of the algorithms for non-conflict schedule.

Acknowledgments. The authors thank Assoc.Prof. PhD. Leandar Litov, chair of Department of Atomic physics, Faculty of Physics, Sofia University "St. Kliment Ohridski" for providing necessary computer resource for CERN-grid.

References

1. Gupta, P., McKeown, N.: Designing and Implementing a Fast Crossbar Scheduler. IEEE Micro, 20–28 (January-February 1999)
2. Chen, W., Mavor, J., Denyer, P., Renshaw, D.: Traffic routing algorithm for serial superchip system customisation. IEE Proc. 137, [E]1 (1990)
3. Anderson, T., Owicki, S., Saxe, J., Thacker, C.: High speed switch scheduling for local area networks. ACM Trans. Comput. Syst. 11(4), 319–352 (1993)
4. Al Sayeed, C., Matrawy, A.: Guaranteed Maximal Matching for Input Buffered Crossbar Switches. In: Proc. of the 4th Annual Communication Networks and Services Research Conference (CNSR 2006), May 24-25, pp. v–ix (2006)
5. Cheng, C., Chen, W., Huang, H.: Birkhoff-von Neumann input buffered crossbar switches. In: Proc of IEEE INFOCOM 2000, vol. 3, pp. 1624–1633 (2000)
6. Elhanany, I., Hamdi, M.: High-performance packet switching architectures. Springer-Verlag London Limited (2007)
7. Atanassov, K.: Generalized Nets. World Scientific, Singapore (1991)
8. Atanassov, K.: On Generalized Nets Theory. Prof. M. Drinov Publishing House, Sofia (2007)
9. Tashev, T., Vorobiov, V.: Generalized Net Model for Non-Conflict Switch in Communication Node. In: Proc. of Int. Workshop DCCN 2007, September 10-12, pp. 158–163. IPPI Publ., Moscow (2007)
10. Tashev, T., Gochev, V.: One Generalized Net Model for Estimation of Decisions for PIM-Algorithm in Crossbar Switch Node. In: Proc. of the Tenth Int. Workshop on Generalized Nets, Sofia, December 5, pp. 51–58 (2009)
11. http://www.imamod.ru/~vab/vfort/download.html

Two-Phase Porous Media Flow Simulation on a Hybrid Cluster

Marina Trapeznikova[1], Boris Chetverushkin[1],
Natalia Churbanova[1], and Dmitrii Morozov[2]

[1] Keldysh Institute of Applied Mathematics, Russian Academy of Sciences
4 Muisskaya Square, Moscow 125047, Russia
`marina@imamod.ru`
[2] Moscow Institute of Physics and Technology (State University)
9 Institutskij per., Dolgoprudny, Moscow Region 141700, Russia

Abstract. A kinetically-based model is developed to describe flow of slightly compressible two-phase fluid in a porous medium. The continuity equations for phases are modified taking into account the minimal scales of averaging on space and on time, as a result regularizing terms and the second order time derivative with small parameters are present in the equations. They are approximated by the three-level explicit difference scheme with a mild enough stability condition. The proposed algorithm is easily adapted to modern hybrid supercomputers. The problem of contaminant infiltration into the soil is solved on a cluster with graphics accelerators. High speed-up of GPU computations in comparison with CPU is demonstrated.

Keywords: Porous Medium, Two-Phase Fluid, Hyperbolic Equation, Explicit Difference Scheme, Stability Condition, GPU Computing.

1 Introduction

Modern computer systems of the super high performance give unique chances for mathematical modeling processes in the subsurface. Among them, there are industrial problems of hydrocarbon recovery, important ecological problems of oil and groundwater contamination etc. The possibility to use a huge amount of computational grid nodes helps to describe large geological strata of complicated structures and to obtain high-accurate predictions for the reasonable time. Unfortunately realization of these opportunities faces significant difficulties caused by the hybrid computer architecture. As a rule, modern supercomputers combine shared and distributed memory, multicore CPUs, GPUs and EPLDs. For their efficient employment the necessity of general-purpose computational algorithms with logical simplicity arises.

Explicit finite difference schemes for numerical solution of the problems of mathematical physics belong to such simple algorithms. However these schemes posses often a very strong time-step restriction. At small steps of the spatial grid this restriction becomes unacceptable severe. Therefore one has to use implicit

I. Lirkov, S. Margenov, and J. Waśniewski (Eds.): LSSC 2011, LNCS 7116, pp. 646–653, 2012.

schemes what leads to iterative algorithms and consequently to the efficiency decrease at parallel implementation. In the present paper an original mathematical model of two-phase fluid flow in a porous medium is proposed. It allows implementation by explicit schemes with a mild enough stability condition.

The new approach is verified by solving the 2D problem of contaminant infiltration into the water-saturated soil. Computations on a hybrid cluster based on graphics accelerators demonstrate high speed-up of GPU vs. a single core of CPU. Comparison of the proposed approach with some traditional algorithms shows the close agreement of results and significant reduction of computational costs.

2   Hyperbolic Model of Two-Phase Flow in the Subsurface

For many problems of continuum mechanics the so-called "principle of minimal sizes" is valid. It means that some minimal sizes (to be exact some minimal reference scales) exist and act as the lower limits of description details. For example, in gas dynamics the free path length of a molecule is such a scale. The known kinetically-consistent finite difference (KCFD) schemes and the related quasi gas dynamic (QGD) system of equations have been derived on the basis of the given principle [1].

In previous works [2,3] the authors proposed to use this principle for porous media flow simulation and to develop a kinetically-based model by the analogy with the QGD system. In this case the minimal reference scale l is a distance of the order of a hundred rock grain sizes. Thus considering fluids as slightly compressible the next model is obtained:

$$\frac{\partial \rho}{\partial t} + \operatorname{div}\rho\mathbf{u} = \operatorname{div}\frac{lc}{2}\operatorname{grad}\rho \qquad (1)$$

$$\mathbf{u} = -\frac{K}{\mu}\operatorname{grad}p \qquad (2)$$

$$p = p_0 + \beta(\rho - \rho_0). \qquad (3)$$

Here ρ is the density, p is the pressure, $\mathbf{u}$ is the Darcy velocity, K is the absolute permeability, μ is the dynamic viscosity, β is the compressibility factor, p_0 and ρ_0 are constant reference values of the pressure and the density, l is the minimal reference scale, c is a magnitude of the order of the sound speed in fluid.

In contrast to the classic continuity equation, equation (1) can be approximated by the explicit scheme with central differences for the convective term discretization [2,3] what is very convenient for many cases. Unfortunately usage of explicit schemes for equation (1) leads to the strong time-step restriction:

$$\Delta t \sim h^2 \qquad (4)$$

where h is the spatial grid step.

To overcome this restriction equation (1) can be replaced by the hyperbolic one [2,3]:

$$\frac{\partial \rho}{\partial t} + \tau \frac{\partial^2 \rho}{\partial t^2} + \mathrm{div}\,\rho\mathbf{u} = \mathrm{div}\,\frac{lc}{2}\,\mathrm{grad}\,\rho \qquad (5)$$

where parameter τ is evaluated as the minimal reference time with the next order of magnitude:

$$\tau \sim h/c. \qquad (6)$$

Equation (5) is approximated by the following three-level explicit scheme of the second order of approximation on time and on space (the 1D case is presented for the simplicity):

$$\frac{\rho_i^{j+1} - \rho_i^{j-1}}{2\Delta t} + \tau \frac{\rho_i^{j+1} - 2\rho_i^j + \rho_i^{j-1}}{\Delta t^2} + \frac{(\rho u)_{i+1}^j - (\rho u)_{i-1}^j}{2h} = \left(\frac{lc}{2}\rho_{\overline{x}}\right)_x^j. \qquad (7)$$

In order to avoid the uncontrolled loss of the numerical solution accuracy the τ-parameter should be chosen from special considerations [2,3] what leads to the necessity of complying with the next stability condition for the three-level scheme:

$$\Delta t \sim h^{3/2}. \qquad (8)$$

This condition is milder and more acceptable than restriction (4).

Modified models (1)–(3) and (5), (2)–(3) have been verified by solving a large number of test problems. Among them the plane-radial fluid flow to a single producing well has been predicted in the 1D statement. Coincidence of analytical and numerical results for the steady-state flow has been observed [2,3]. Computations of the similar 2D problem concerning fluid inflow to a well placed in the center of the square domain also demonstrate the smoothing effect of the regularizing terms in the right-hand sides of (1) and (5) [2,3]. Corresponding to theoretical estimations a noticeable increase of the stability threshold is obtained for all test problems when employing the three-level scheme (7) for the hyperbolic equation (5) instead of the two-level scheme for the parabolic equation (1).

As a rule engineering applications of the subsurface processes require considering multiphase fluid flows. For example, problems of oil recovery by means of secondary hydrodynamic methods as well as ecological problems aimed at prevention of the soil contamination assume modeling joint flows of water, gas and so-called Non-Aqueous Phase Liquid (NAPL) [4]. Oil, petrol or tetrachlorethylene can be treated as NAPL depending on the application. Besides that in many situations the capillary and gravity forces should be taken into account.

Below the generalization of model (5), (2)–(3) to the two-phase case is given (the subscript $\alpha = w, n$ indicates water or NAPL correspondingly):

$$m\frac{\partial(\rho_\alpha S_\alpha)}{\partial t} + \tau \frac{\partial^2(\rho_\alpha S_\alpha)}{\partial t^2} + \mathrm{div}\,(\rho_\alpha \mathbf{u}_\alpha) = q_\alpha + \mathrm{div}\frac{l_\alpha c_\alpha}{2}\mathrm{grad}\,(\rho_\alpha S_\alpha) \qquad (9)$$

$$\mathbf{u}_\alpha = -K\frac{k_\alpha}{\mu_\alpha}(\mathrm{grad}\,p_\alpha - \rho_\alpha \mathbf{g}) \qquad (10)$$

$$\rho_\alpha = \rho_{0\alpha} \left[1 + \beta_\alpha \left(p_\alpha - p_{0\alpha} \right) \right] \tag{11}$$

$$\sum_\alpha S_\alpha = 1 \tag{12}$$

$$p_\alpha - p_\beta = p_{c\alpha\beta} \left(S_\alpha, S_\beta \right), \qquad \alpha \neq \beta \tag{13}$$

where m is the porosity, S_α is the α-phase saturation, k_α is the relative phase permeability, c_α is the sound speed in the α-phase, $\mathbf{g}$ is the gravity vector, q_α is the source of the fluid, p_c is the capillary pressure.

The regularizing term depends on the phase saturation S_α. The minimal reference scale l_α differs depending on the phase.

The capillary pressure and the relative phase permeability are strongly nonlinear functions of the saturation. In the current research the Brooks&Corey constitutive relationships [4] are chosen to describe them:

$$p_c \left(S \right) = p_d S_e^{-\frac{1}{\lambda}} \tag{14}$$

$$S_e = \frac{S_w - S_{wr}}{1 - S_{wr}} \tag{15}$$

$$k_w(S) = S_e^{\frac{2+3\lambda}{\lambda}}, \qquad k_n(S) = (1 - S_e)^2 \left(1 - S_e^{\frac{2+3\lambda}{\lambda}} \right) \tag{16}$$

where S_e is the effective saturation, S_{wr} is the residual saturation of water, p_d is the entry pressure, λ is the indicator of the pour size distribution for the given medium.

The primary unknown quantities are the NAPL saturation S_n and the water pressure p_w. The computational algorithm of the explicit type for the above model implementation consists of the next stages (for the simplicity the 1D case is presented, $\alpha = w, n$).

1. Calculation of the Darcy velocities for the both phases on the current time level:

$$u_{\alpha i}^j = -K \frac{k_\alpha \left(S_{w i}^j \right)}{\mu_\alpha} \left(\frac{p_{\alpha i+1}^j - p_{\alpha i-1}^j}{2h} - \rho_\alpha g \right). \tag{17}$$

2. Solving for $(\rho_\alpha S_\alpha)_i^{j+1}$ on the new time level with the use of the three-level explicit difference scheme:

$$(\rho_\alpha S_\alpha)_i^{j+1} = \frac{2\Delta t^2}{m\Delta t + 2\tau} \left[q_i + \left(\frac{l_\alpha c_\alpha}{2} (\rho_\alpha S_\alpha)_{\overline{x}} \right)_x^j - \frac{(\rho_\alpha u_\alpha)_{i+1}^j - (\rho_\alpha u_\alpha)_{i-1}^j}{2h} \right] + $$

$$+ \frac{4\tau}{m\Delta t + 2\tau} (\rho_\alpha S_\alpha)_i^j + \frac{m\Delta t - 2\tau}{m\Delta t + 2\tau} (\rho_\alpha S_\alpha)_i^{j-1}. \tag{18}$$

3. Obtaining the NAPL saturation $S_{n i}^{j+1}$ and the water pressure $p_{w i}^{j+1}$ on the new time level by means of the next system solution in each node of the spatial grid:

$$\begin{cases} \rho_{0 w} \left[1 + \beta_w \left(p_{w i}^{j+1} - p_{0 w} \right) \right] \left(1 - S_{n i}^{j+1} \right) = (\rho_w S_w)_i^{j+1} \\ \rho_{0 n} \left[1 + \beta_n \left(p_{w i}^{j+1} + p_c \left(S_{n i}^{j+1} \right) - p_{0 n} \right) \right] S_{n i}^{j+1} = (\rho_n S_n)_i^{j+1} \end{cases} \tag{19}$$

This system can be solved, for example, by the Newton method what takes only a few iterations.

4. Calculation of S_{wi}^{j+1}, p_{ni}^{j+1} and $\rho_{\alpha i}^{j+1}$ from (12), (13) and the state equation (11) correspondingly.

3   Solution of the Infiltration Problem on a Hybrid Cluster

The test problem of the ecological content concerning contaminant infiltration into the subsurface is used to verify the proposed model and algorithm. As a contaminant the organic solvent tetrachloroethylene (TCE) is considered. It penetrates from the earth surface into the fully water-saturated soil under the gravity force. TCE is denser than water therefore it is classified as Dense Non-Aqueous Phase Liquid (DNAPL) [4]. The process is governed by equations (9)-(16). For computations the following values of parameters are taken:

$$m = 0.4, \quad K = 6.64 \cdot 10^{-11} m^2, \quad \lambda = 2.7, \quad p_d = 755\, Pa, \quad S_{wr} = 0.09,$$

$$\rho_{0w} = 1000\frac{kg}{m^3}, \quad \rho_{0n} = 1460\frac{kg}{m^3}, \quad \mu_w = 10^{-3}\frac{kg}{m \cdot s}, \quad \mu_n = 9 \cdot 10^{-4}\frac{kg}{m \cdot s}$$

$$l_w = 10^{-6} m, \quad l_n = 10^{-7} m, \quad c_w = 1407\frac{m}{s}, \quad c_n = 1225\frac{m}{s},$$

$$\beta_w = \beta_n = 10^{-4}\frac{m \cdot s^2}{kg}$$

The problem is solved in the 2D statement (see Fig. 1). At the initial moment the water pressure has the hydrostatic distribution, DNAPL is absent in the domain. On the top boundary there is a spot of TCE which is described by the boundary condition $S_n = 0.4$ in the corresponding points while $q_n = 0$. The side and bottom boundaries are permeable. For this problem there are experimental data and numerical results of other authors [4] to be compared.

The goal of investigations is to compare the classic approaches to porous media flow simulation with the new kinetic one. Three approaches have been

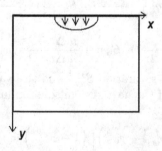

Fig. 1. The infiltration problem statement

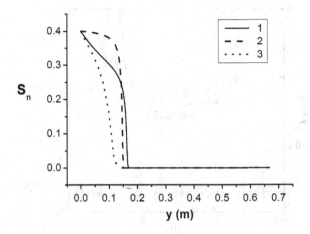

Fig. 2. DNAPL saturation profiles (1 - the modified model, 2 - the classic model for compressible fluid, 3 - the classic model for incompressible fluid)

realized: 1 – the modified model (9)-(16) and algorithm (17)-(19); 2 – the classic model for slightly compressible fluid with the use of the two-level explicit scheme of the first order with upwind differences for convective term approximation (instead of scheme (18)); 3 – the model for incompressible fluid which is employed traditionally to solve a wide range of porous media flow problems, the well-known IMPES (IMplicit Pressure - Explicit Saturation) method [5] is applied for its numerical implementation.

Fig. 2 shows the DNAPL saturation profiles in the middle cross section at some time moment obtained via the above mentioned approaches. One can see that approaches 1 and 2 ensure approximately equal front propagation velocities as they both assume the compressibility of fluid. However the curve 1 agrees better with the curve 3 than the curve 2.

Fig. 3 presents isolines of the DNAPL saturation field at the same time moment, obtained by approach 1. The contamination field has the typical view.

One of the major goals of the research is to prove the effectiveness of GPU computing for solving problems of mathematical physics. Therefore computations of the above infiltration problem have been performed on a cluster having graphics accelerators. The MVS-Express supercomputer is created in the Keldysh Institute of Applied Mathematics (KIAM), Moscow, based on the Non-Uniform Memory Access (NUMA) technology. Each node of the cluster includes 2 four-core CPUs (AMD Opteron 2.6 GHz) and GPU (NVIDIA GeForce 295GTX) supporting the Compute Unified Device Architecture (CUDA) technology. For data exchange between nodes the original library Shmem-Express (a dialect of the communication system Shmem) is developed in KIAM.

For computations on a graphics accelerator the code is organized as a sequence of few CUDA kernels to achieve the maximal parallelization efficiency. A kernel is

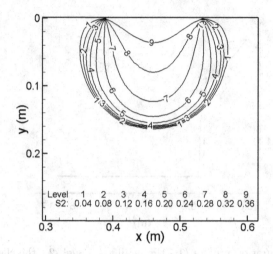

Fig. 3. Isolines of the DNAPL saturation field obtained via the modified model

a function with the data parallelism over a large number of threads. Each thread is used for treating one point of the computational grid. A lot of threads are processed by GPU simultaneously. To gain the highest performance the access to different memory types should be optimized. Invariable parameters of the problem are loaded to the constant memory. The number of the global memory accesses is reduced if possible. Instead the local memory is used.

First of all the code developed has ran on a single node of the cluster. Comparison of the GPU run time with the run time of a single CPU core demonstrates benefits of the GPU usage. Fig. 4 reflects the speed-up for different grid sizes. The highest speed-up is achieved if the computational grid size is about one million points, then it comes approximately to 48 times. Further grid refinement does not lead to the speed-up increase on the given hardware.

For the algorithm implementation on several GPUs belonging to different nodes the data partitioning technique is applied. The computational domain is divided geometrically in one direction into subdomains with equal numbers of grid points such as the number of subdomains equals the number of exploitable nodes. The above CUDA-code is extended by Shmem-Express routines. Due to specific organization of separate kernels there is no need to synchronize parallel processes. While comparing the run time of GPUs from four different nodes with the run time of a single CPU core the speed-up comes to 82.2 times at the grid size of 12288×2048 (see Fig. 4, the right column).

All computations have been performed with double precision.

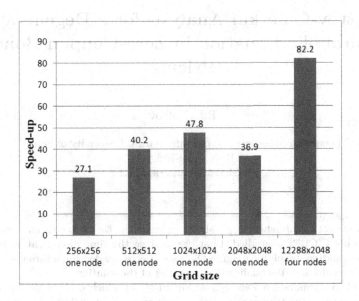

Fig. 4. GPU speed-up vs. CPU, depending on the grid size and the number of nodes

4   Conclusion

Large-scale problems in the 3D statement implying three-phase fluids in hetero-geneous media will be under consideration in the nearest future. The further optimization of the GPU solver and the development of a hybrid CPU/GPU solver are planned by the authors.

Acknowledgments. This work is supported by the Russian Foundation for Basic Research (grants No 09-01-00600, 10-01-90005).

References

1. Chetverushkin, B.N.: Kinetic Schemes and Quasi-Gas Dynamic System of Equations. CIMNE, Barcelona (2008)
2. Chetverushkin, B.N., Morozov, D.N., Trapeznikova, M.A., Churbanova, N.G., Shilnikov, E.V.: An Explicit Scheme for the Solution of the Filtration Problems. Math. Mod. and Comp. Sim. 2(6), 669–677 (2010)
3. Chetverushkin, B.N., Churbanova, N.G., Morozov, D.N., Trapeznikova, M.A.: Kinetic Approach to Simulation of Multiphase Porous Media Flows. In: Pereira, J.C.F. (ed.) CD-Rom Proc. of ECCOMAS CFD 2010 Conf. IDMEC, Lisbon, Portugal (2010)
4. Helmig, R.: Multiphase Flow and Transport Processes in the Subsurface - A Contribution to the Modelling of Hydrosystems. Springer, Berlin (1997)
5. Aziz, K., Settari, A.: Petroleum Reservoir Simulation. Applied Science Publ. Lmt., London (1979)

Petrov-Galerkin Analysis for a Degenerate Parabolic Equation in Zero-Coupon Bond Pricing

R.L. Valkov

Faculty of Mathematics and Informatics, University of Sofia,
1000 Sofia, Bulgaria
rvalkov@fmi.uni-sofia.bg

Abstract. A *degenerate* parabolic equation in the zero-coupon bond pricing (ZCBP) is studied. First, we analyze the time discretization of the equation. Involving weighted Sobolev spaces, we develop a variational analysis to describe qualitative properties of the solution. On each time-level we formulate a Petrov-Galerkin FEM, in which each of the basis functions of the trial space is determined by the finite volume difference scheme in [2, 3]. Using this formulation, we establish the stability of the method with respect to a discrete energy norm and show that the error of the numerical solution in the energy norm is $O(h)$, where h denotes the mesh parameter.

1 Introduction

Degenerate parabolic equations play an important role in creating many models of mathematical finance [1, 5, 10]. We consider the ZCBP problem, [5, 10]:

$$\frac{\partial P}{\partial t} - \frac{w^2(r)}{2}\frac{\partial^2 P}{\partial r^2} - (\theta(r) + \lambda(t)w(r))\frac{\partial P}{\partial r} + rP = 0, \quad (r,t) \in Q_T, \ Q_T = I \times (0,T], \tag{1}$$

$$P(r,0) = P_0(r), r \in I = (0,R). \tag{2}$$

Following the modelling in [5, 10] we assume that $w(r)$ is a non-negative and smooth bounded function and $\theta(r)$ is a smooth function, which satisfy:

$$w(r) = r(R - r)w_0(r), \tag{3}$$

$$\theta(0) \geq 0, \ \theta(R) \leq 0. \tag{4}$$

Being different from the classical parabolic equations in which the principal coefficient is assumed to be *strictly positive*, the parabolic equation (1) belongs to the second order differential equations with *non-negative characteristic form*. The main difficulty in such kind of problems is the *degeneracy*. It can be easily

I. Lirkov, S. Margenov, and J. Waśniewski (Eds.): LSSC 2011, LNCS 7116, pp. 654–661, 2012.

seen that at $r = 0$ and $r = R$, equation (1) degenerates into a hyperbolic equation with positive and negative characteristics respectively

$$\frac{\partial P}{\partial t} - \theta(0)\frac{\partial P}{\partial r} = 0, \tag{5}$$

$$\frac{\partial P}{\partial t} - \theta(R)\frac{\partial P}{\partial r} = -RP. \tag{6}$$

By the Fichera's theory (see [8]) for degenerate parabolic equations, we have that at the degenerate boundaries $r = 0$ and $r = R$ boundary conditions should not be given. Therefore, the maturity data $P(r, T)$ determines the solution $P(r, t)$ of problem (1), (2) *uniquely*. But if one tries to obtain the numerical solution of (1)-(4) the boundary conditions (5), (6) should be considered.

In the recent years the numerical simulations in finance had become increasingly important in the world of finance [6, 9, 11–13]. In [2, 3] was developed a fitted finite volume method for the degenerate partial differential equation (1) and in [4] for a two-dimensional model. The method is based on a finite volume spatial discretization and then an implicit time-stepping (Crank-Nicholson) technique is used. Although the method works well for practical computation, see the experiments in [2–4], its' theoretical justification remains an open question. In this paper we will answer this question by providing convergence analysis on the base of Rothe method. In the *Rothe method (method of horizontal lines)*, the temporal derivative is discretized first and the original parabolic PDE is reduced to elliptic PDEs(which are quite well understood) and therein lies the advantages of the method, [7].

Due to space requirements we will consider in the present paper only the case

$$\theta(r) = r(R - r)\theta_0(r), \ \theta_0(0) \neq 0, \theta_0(R) \neq 0. \tag{7}$$

The rest of the paper is organized as follows. We discuss the variational formulation and implement the Rothe method to problem (1), (2), see Section 2. Also, we derive the existence and uniqueness of a weak solution of (1), (2). Then, in Section 3, using some results obtained in [2, 3], we describe a Petrov-Galerkin formulation of the corresponding elliptic problem on each time level. Error estimates for the Petrov-Galerkin problem approximations are given in Section 4. Finally, in Section 5, we make some concise conclusions.

2   Time Discretization

The space of all weighted square-integrable functions is defined as

$$L_w^2(I) := \left\{ \|v\|_{0,w} := \left(\int_0^R r^2(R - r)^2 v^2 dr \right)^{1/2} < \infty \right\}.$$

Using $L^2(I)$ and $L_w^2(I)$, we define the following weighted Sobolev space $H_w^1(I) :=$ $\{v : v \in L^2(I), \ v' \in L_w^2(I)\}$, where v' denotes the weak derivative of v. Let $\|\cdot\|_{1,w}$ be the functional on $H_w^1(I)$ defined by

$$\|v\|_{1,w} = \left(\|v\|_0^2 + \|v'\|_{0,w}^2\right)^{1/2}.$$

The Rothe method is an efficient tool for solving evolution problems. Replacing the time derivative by the backward derivative, the problem (1), (2) is approximated by a sequence of elliptic problems, which have to be solved successively by increasing the time steps t_j for $j = 1, \ldots, n$. In the weak formulation, we have to find subsequently such functions $P^j \in H_w^1(I)$, for which the integral identity

$$(a(P^j)' + r(R-r)b^j P^j, v') + (c^j P^j, v) = (f, v), \ \forall v \in H_w^1(I), \qquad (8)$$

$$a = \frac{r^2(R-r)^2 w_0^2(r)}{2}, b = \theta_0(r) + (\lambda(t) - w'(r))w_0(r),$$

$$c = r + (r(R-r)b)' + 1/\tau, f = P^{j-1}/\tau.$$

We will establish coercivity of the corresponding bilinear form and use this to derive an upper bound for the approximation error in a proper norm.

Assumption 1. *Assume that the following condition is satisfied:*

$$r + 0.5\left(r(R-r)b\right)' + 1/\tau \geq \beta > 0,$$

for all $r \in I$, where β is a positive constant.

Lemma 1. *Let Assumption 1 be satisfied. Then the bilinear form of (8) is coercive and there exists a unique solution $P^j \in H_w^1(I)$.*

Theorem 2. *Let Assumption 1 holds and $P_0 \in H_w^1(I)$. Then there exists unique solution $P \in C\left([0,T], L^2\right) \cap L_\infty\left([0,T], H_w^1\right)$, obeying $\frac{\partial P}{\partial t} \in L^2\left((0,T), L^2(I)\right)$, of the problem (1)-(2) and*

$$\tilde{P}_n \to \tilde{P} \text{ in } L_2([0,T], H_w^1), \ \tilde{P}_n \to \tilde{P} \text{ in } L_2([0,T], L_2(I)).$$

Moreover

$$\left\|\tilde{P}(r, t_j) - P^j(r)\right\|_{1,w} \leq \frac{\tau M}{2C^2}\left(1 - e^{-C^2 j\tau}\right),$$

where C is the constant of positive definiteness of the bilinear form and M is a constant such that $\|P_0\|_{1,w} \leq M$.

3   Petrov-Galerkin Formulation

In this section we first formulate the finite volume method, developed in [2,3], as a Petrov-Galerkin finite element method for problem (8).

Let the interest rate interval $I = (0, R)$ be divided into N sub-intervals $I_i :=$ (r_i, r_{i+1}), $i = 0, 1, \ldots, N-1$ with the grid $\overline{w}_h = w_h \cup \{r_0\} \cup \{r_N\}$, $\overline{w}_h = \{r_i, \ i =$

$0, 1, \ldots, N$, $0 = r_0 < r_1 < \cdots < r_{N-1} < r_N = R\}$. For each $i = 0, 1, \ldots, N - 1$ we put $h_i = r_{i+1} - r_i$ and $h = \max_{0 \le i \le N-1} h_i$. We also let $r_{i-1/2} = (r_{i-1} + r_i)/2$ and $r_{i+1/2} = (r_i + r_{i+1})/2$ for each $i = 1, 2, \ldots, N - 1$. These mid-points form a second partition grid $\overline{w}_{\tilde{h}} = w_{\tilde{h}} \cup \{r_{-1/2}\} \cup \{r_{N+1/2}\}$ of $[0, R]$ if we define $r_{-1/2} = r_0$ and $r_{N+1/2} = r_N$. Let $\tilde{h}_i = r_{i+1/2} - r_{i-1/2} = 0.5(h_i + h_{i-1})$, $\tilde{h}_0 = h_1/2 = r_{1/2}$, $\tilde{h}_N = R - r_{N-1/2}$.

We now discuss the finite element formulation of the discretization scheme. For any $i = 0, 1, \ldots, N - 1, N$ let ψ_i denote the characteristic function given by $\psi_i(r) = 1$, $r \in (r_{i-\frac{1}{2}}, r_{i+\frac{1}{2}})$; 0, otherwise. We choose the test space to be $T_h = span\{\psi_i\}_0^N$.

Following [2, 3], for the case (7), we introduce the functions

$$\phi_i(r) = \begin{cases} \dfrac{\left(\frac{R}{r_{i-1}} - 1\right)^{\frac{\alpha_{i-1}}{R}} - \left(\frac{R}{r} - 1\right)^{\frac{\alpha_{i-1}}{R}}}{\left(\frac{R}{r_{i-1}} - 1\right)^{\frac{\alpha_{i-1}}{R}} - \left(\frac{R}{r_i} - 1\right)^{\frac{\alpha_{i-1}}{R}}}, & r \in (r_{i-1}, r_i), \\[3ex] \dfrac{\left(\frac{R}{r_{i+1}} - 1\right)^{\frac{\alpha_i}{R}} - \left(\frac{R}{r} - 1\right)^{\frac{\alpha_i}{R}}}{\left(\frac{R}{r_{i+1}} - 1\right)^{\frac{\alpha_i}{R}} - \left(\frac{R}{r_i} - 1\right)^{\frac{\alpha_i}{R}}}, & r \in (r_i, r_{i+1}); \\[3ex] 0, \text{otherwise.} \end{cases}$$

On the intervals $(0, r_1)$ and (r_{N-1}, R) we define the linear functions

$$\phi_0(r) = \begin{cases} 1 - \frac{r}{r_1}, & r \in (0, r_1), \\ 0, \text{otherwise,} \end{cases} \qquad \phi_N(r) = \begin{cases} \frac{(r - r_{N-1})}{(R - r_{N-1})}, & r \in (r_{N-1}, R), \\ 0, \text{otherwise.} \end{cases}$$

Next we define the linear functions $\phi_1(r)$ and $\phi_{N-1}(r)$ on the intervals $(0, r_2)$ and (r_{N-2}, R)

$$\phi_1(r) = \begin{cases} \frac{r}{r_1}, & r \in (0, r_1), \\[2ex] \dfrac{\left(\frac{R}{r_2} - 1\right)^{\frac{\alpha_1}{R}} - \left(\frac{R}{r} - 1\right)^{\frac{\alpha_1}{R}}}{\left(\frac{R}{r_2} - 1\right)^{\frac{\alpha_1}{R}} - \left(\frac{R}{r_1} - 1\right)^{\frac{\alpha_1}{R}}}, & r \in (r_1, r_2); \ 0, \text{otherwise,} \end{cases}$$

$$\phi_{N-1}(r) = \begin{cases} \dfrac{\left(\frac{R}{r_{N-2}} - 1\right)^{\frac{\alpha_{N-2}}{R}} - \left(\frac{R}{r} - 1\right)^{\frac{\alpha_{N-2}}{R}}}{\left(\frac{R}{r_{N-2}} - 1\right)^{\frac{\alpha_{N-2}}{R}} - \left(\frac{R}{r_{N-1}} - 1\right)^{\frac{\alpha_{N-2}}{R}}}, & r \in (r_{N-2}, r_{N-1}), \\[3ex] (R - r)/(R - r_{N-1}), & r \in (r_{N-1}, R); \ 0, \text{otherwise.} \end{cases}$$

Lemma 3. *For each $i = 0, \ldots, N$ the function $\phi_i(r)$ is a monotonically increasing on (r_{i-1}, r_i) and decreasing on (r_i, r_{i+1}). Furthermore, $\phi_i(r)$ and $\phi_{i+1}(r)$ satisfy $\phi_i(r) + \phi_{i+1}(r) = 1$ for all $r \in (r_i, r_{i+1})$ and $i = 0, \ldots, N - 1$.*

Now the finite element trial space is chosen to be $S_h = span\{\phi_i\}_0^N$.

Let Π denotes the mass lumping operator from $C^0(I)$ to T_h such that for any $v \in C^0(I)$ $\Pi(v) = \sum_{i=0}^N v(x_i)\psi_i(x)$, where ψ_i is defined above. Using S_h and

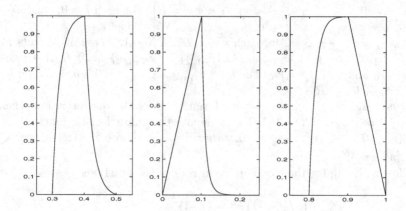

Fig. 1. Hat functions at various nodes for $R = 1$, $\alpha = 10$

T_h, we define the following Petrov-Galerkin problem: find $P_h^j \in S_h$ such that for all $v_h \in T_h$

$$
A(P_h^j, v_h) \equiv \sum_{i=0}^{N} \left[r(R - r) \left(\tilde{a} \frac{\partial P_h^j}{\partial r} + \hat{b} P_h^j \right) \right]_{r_{j-1/2}}^{r_{j+1/2}} v_h \tag{9}
$$
$$
+ (\Pi(c P_h^j), v_h) = (\Pi(f), v_h),
$$

where $\tilde{a}_{i+1/2} = \left(r_{i+1/2}(R - r_{i+1/2}) w_0^2(r_{i+1/2}) \right) /2$ and $\hat{b}$ denotes the piecewise constant approximation of b on I satisfying $\hat{b} = b_{i+1/2}$ if $r \in I_i$.

Note that, when restricted to S_h, the lumping operator Π is surjective from S_h to T_h. Using this Π, we rewrite problem (6) as the following equivalent Galerkin finite element formulation. Find $P_h^{j+1} \in S_h$ such that for all $v_h \in S_h$

$$
B(P_h^{j+1}, v_h) = (\Pi(f), \Pi(v_h)), \text{ where } B(P_h^{j+1}, v_h) = A(P_h^{j+1}, \Pi(v_h)). \tag{10}
$$

Before further discussions, we first define a functional $\| \cdot \|_{1,h}$ on S_h by

$$
\|v_h\|_{1,h}^2 = \sum_{j=1}^{N-2} b_{j+1/2} r_{j+1/2} (R - r_{j+1/2}) \frac{\left(\frac{r_{j+1}}{R - r_{j+1}} \right)^{\alpha_j} + \left(\frac{r_j}{R - r_j} \right)^{\alpha_j}}{\left(\frac{r_{j+1}}{R - r_{j+1}} \right)^{\alpha_j} - \left(\frac{r_j}{R - r_j} \right)^{\alpha_j}} (v_{j+1} - v_j)^2
$$

for any $v_h = \sum_{j=0}^{N} v_j \phi_j \in S_h$. It is easy to show that $\| \cdot \|_{1,h}$ is a norm on S_h, because $\left(\left(\frac{r_{j+1}}{R - r_{j+1}} \right)^{\alpha_j} - \left(\frac{r_j}{R - r_j} \right)^{\alpha_j} \right) / b_{j+1/2} > 0$ for any $\alpha_j = b_{j+1/2}/\tilde{a}_{j+1/2}$. $\| \cdot \|_{1,h}$ is a weighted discrete energy norm on S_h. Using this norm we define the following weighted discrete H^1-norm

$$
\|v_h\|_h^2 = \|v_h\|_{1,h}^2 + \sum_{j=1}^{N-2} v_j^2 l_j.
$$

Theorem 4. *Let Assumption 1 be fulfilled. If τ and h are sufficiently small, then for all $v_h \in S_h$, we have*

$$B(v_h, v_h) \geq C\|v_h\|_h^2, \tag{11}$$

where C denotes a positive constant, independent of h and v_h.

Outline of the proof. By tedious algebraic calculations we obtain

$$B(v_h, v_h) = \tfrac{1}{2}\|v_h\|_{1,h}^2 - \tfrac{1}{2}b_{1/2}r_{1/2}(R - r_{1/2})v_1^2 + \tfrac{1}{2}b_{N-1/2}r_{N-1/2} *$$

$$(R - r_{N-1/2})v_{N-1}^2 - \tfrac{1}{2}r_{1/2}(R - r_{1/2})\left[\left(\tfrac{R-r_{1/2}}{2}w_0^2(r_{1/2}) + b_{1/2}\right)v_1 - \right.$$

$$\left(\tfrac{R-r_{1/2}}{2}w_0^2(r_{1/2}) - b_{1/2}\right)v_0\right](v_0 - v_1) - \tfrac{1}{2}r_{N-1/2}(R - r_{N-1/2}) *$$

$$\left[\left(\tfrac{r_{N-1/2}}{2}w_0^2(r_{N-1/2}) + b_{N-1/2}\right)v_N - \left(\tfrac{r_{N-1/2}}{2}w_0^2(r_{N-1/2}) - b_{N-1/2}\right)v_{N-1}\right] *$$

$$(v_{N-1} - v_N)\sum_{j=1}^{N-1}\left(c_j - \tfrac{1}{2}\tfrac{r_{j+1/2}(R-r_{j+1/2})b_{j+1/2} - r_{j-1/2}(R-r_{j-1/2})b_{j-1/2}}{r_{j+1/2} - r_{j-1/2}}\right) *$$

$$v_j^2(r_{j+1/2} - r_{j-1/2})^2 + c_0 v_0^2 \hbar_0 + c_N v_N^2 \hbar_N,$$

where

$$c_j - \frac{1}{2}\frac{r_{j+1/2}(R - r_{j+1/2})b_{j+1/2} - r_{j-1/2}(R - r_{j-1/2})b_{j-1/2}}{r_{j+1/2} - r_{j-1/2}} =$$

$$r_j + 0.5(r(R - r)b)_j' + 1/\tau \geq \beta$$

The other terms, except the first, are of order $O(h)$.

We can easily see that we get the same condition as in Assumption 1. So, whenever Assumption 1 is fulfilled, $B(v_h, v_h)$ is coercive.

Corrolary 1. *Problem (8) has a unique solution. We remark, as in the continuous case, that Assumption 1 is a sufficient condition for the unique solvability of problem (8), but it is not necessary. As proved in [2, 3] the system matrix of the fully discrete problem is always an M-matrix, and thus the fully discretized problem is uniquely solvable even when Assumption 1 does not hold. In fact, the method works well in practice without Assumption 1, as the numerical results showed in [2, 3].*

4   Error Estimates for Spatial Discretization

Now we derive an upper bound on the difference between the approximate and the exact solution in the norm $\|\cdot\|_h$. The following theorem establishes an estimate for the error in the flux of the S_h-interpolant of a given function.

Lemma 5. *Let $\hat{a}$ and $\hat{b}$ are the flux coefficients, defined in [2, 3]. For a sufficiently smooth function v we denote:*

$$\rho(v) = \hat{a}r(R - r)v' + \hat{b}v, \quad \rho_i(v) := \hat{a}_{i+1/2}r(R - r)v' + \hat{b}_{i+1/2}v,$$

$$\rho_h(v) = \rho_i(v) \; if \, r \in I_i,$$

where v_I is the S_h-interpolant of v. Then

$$\|\rho(v) - \rho_h(v_I)\|_{\infty,I_i} \leq C(\|\rho'(v)\|_{\infty,I_i} + \|b'\|_{\infty,I_i} + \|v\|_{\infty,I_i})h_i$$

for $i = 0, 1, \ldots, N-1$ and C is a positive constant, independent of h_i and v.

Theorem 6. *Let Assumption 1 be fulfilled, and P^{j+1} and P_h^{j+1} be respectively the solution to problems (8) and (10). Then, the following bound holds:*

$$\|P^{j+1} - P_h^{j+1}\| \leq C_1 h(\|P^{j+1}\|_\infty + \|(P_h^{j+1})'\|_\infty + \|b'\|_\infty + \|c'\|_\infty + \|f'\|_\infty) \quad (12)$$

for a positive constant C_1, independent of h and u.

Outline of the proof. Let P_I^{j+1} be the S_h-interpolant of P^{j+1} and C a generic positive constant, independent of h and P^{j+1}. Then after tedious calculations we obtain

$$\left| B\left(P_h^{j+1} - P_I^{j+1}, v_h\right) \right| \leq R_1 + R_2, \quad (13)$$

$$R_1 \leq \sum_{j=0}^{N} |f_i - f| + |(c - c_i)P^{j+1}| + |(P^{j+1} - P_I^{j+1})]v_j \hbar_j \leq$$

$$\leq Ch(\|f'\|_\infty + \|c'\|_\infty + \|(P^{j+1})'\|_\infty)\left(\sum_{j=0}^{N} v_j^2 \hbar_j\right)^{1/2},$$

$$R_2 \leq Ch\left(\|(\rho'(P^{j+1}))\|_\infty + \|b'\|_\infty\|P^{j+1}\|_\infty)\right)\left(\sum_{j=0}^{N} r_{j+1/2}(R - r_{j+1/2})|v_{j+1} - v_j|\right).$$

After estimating the sum and making a few more adjustments we get:

$$R_2 \leq Ch\left(\|\rho'(P^{j+1})'\|_\infty + \|b'\|_\infty\|P^{j+1}\|_\infty\right)(h|v_1| + |v_h|_{1,h}).$$

Now, replacing R_1 and R_2 in (13) and using (11) we obtain (12).

5   Conclusions

In papers [2, 3] was proposed a finite volume difference method for a degenerate parabolic equation in the zero-coupon bond pricing. In the present paper this method is formulated as Rothe-Petrov & Galerkin method. Stability of the discretization has been proved and an $\mathcal{O}(h)$ order upper bound for the approximation error in the numerical solution has been established. Since the main purpose of the current work is to provide a rigorous stability and convergence analysis for the discretization scheme, proposed in [2,3], no numerical experiments were given in the paper. Extensive numerical experiments on the method can be found in [2,3].

The finite volume difference scheme was extended for two-dimensional problems in [4]. We also plan to develop a suitable finite element method for these two-dimensional models.

Acknowledgements. This work is supported by the Bulgarian National Fund of Science under Project DID 02/37-2009.

References

1. Black, F., Sholes, M.: The pricing of options and corporate liabilities. J. Pol. Econ. 81, 637–659 (1973)
2. Chernogorova, T., Valkov, R.: A computational scheme for a problem in the zero-coupon bond pricing. Amer. Inst. of Physics 1301, 370–378 (2010)
3. Chernogorova, T., Valkov, R.: Finite volume difference scheme for a degenerate parabolic equation in the zero-coupon bond pricing. Math. and Comp. Modelling 54(11-12), 2659–2671 (2011)
4. Chernogorova, T., Valkov, R.: Finite-Volume Difference Scheme for the Black-Scholes Equation in Stochastic Volatility Models. In: Dimov, I., Dimova, S., Kolkovska, N. (eds.) NMA 2010. LNCS, vol. 6046, pp. 377–385. Springer, Heidelberg (2011)
5. Deng, Z.C., Yu, J.N., Yang, L.: An inverse problem arisen in the zero bond pricing. Nonl. Anal.: Real World Appls 11(3), 1278–1288 (2010)
6. Huang, C.-S., Hung, C.-H., Wang, S.: A fitted finite volume method for the valuation of options on assets with stochastic volatilities. Computing 77, 297–320 (2006)
7. Kacur, J.: Method of Rothe in Evolution Equations. BSB Teubner Verlagsges, Leipzig (1985)
8. Oleinik, O.A., Radkevic, E.V.: Second Order Differential Equation with Non-negative Characteristic Form. Rhode Island and Plenum Press, American Mathematical Society, New York (1973)
9. Seydel, R.: Tools for Computational Finance, 2nd edn. Springer, Heidelberg (2003)
10. Stamffi, J., Goodman, V.: The Mathematics of Finance: Modeling and Hedging. Thomas Learning (2001)
11. Wang, S.: A novel finite volume method for Black-Sholes equation governing option pricing. IMA J. of Numer. Anal. 24, 699–720 (2004)
12. Wang, I.R., Wan, J.W.I., Forsyth, P.A.: Robust numerical valuation of European and American options under CGMY process. J. Comp. Finance 10(4), 32–69 (2007)
13. Windeliff, H., Forsyth, P.A., Vezzal, K.R.: Analysis of the stability of the linear boundary conditions for the Black-Sholes equation. J. Comp. Finance 8, 65–92 (2004)

Agents in Grid System — Design and Implementation

Katarzyna Wasielewska[1], Michał Drozdowicz[1], Paweł Szmeja[2], Maria Ganzha[1], Marcin Paprzycki[1], Ivan Lirkov[3], Dana Petcu[4], and Costin Badica[5]

[1] Systems Research Institute Polish Academy of Sciences, Warsaw, Poland
(maria.ganzha,marcin.paprzycki)@ibspan.waw.pl
[2] Technical University of Warsaw, Warsaw, Poland
[3] IICT, Bulgarian Academy of Sciences, Sofia, Bulgaria
[4] Dept. of Computer Science, West University of Timisoara, Timisoara, Romania
[5] Dept. of Computer Science, University of Craiova, Craiova, Romania

Abstract. We are developing an agent-based intelligent middleware for the Grid. It is based on agent teams as resource brokers and managers. Our earlier work resulted in a prototype implementation. However, our recent research led to a complete redesign of the system. Here, we discuss the new and main technical issues found during its implementation.

1 Introduction

The *Agents in Grid* (*AiG*) project aims at utilizing teams of software agents as resource brokers in the Grid. The initial overview of the approach can be found in [1], while the two main scenarios: (1) agents seeking teams to execute job(s), and (2) agents attempting to join the team, were discussed in [4,5]. Since one of the main assumptions was that the system will use ontologies and semantic data processing, we have developed an ontology of the Grid (presented in [3]). While the initial work led to a demonstrator system, our recent studies led to a conclusion that redesign of the system is required, to assure more flexibility and use technologies that are rapidly maturing. The aim of this note is to present the new design, and outline main technical issues found during its implementation. To this effect, we start with an overview of the proposed system. In what follows, we discuss issues involved in the implementation of the redesigned system, moving from the front-end to the negotiation module.

2 System Overview

In our work, Grid is considered as an open environment in which *agents* representing *users* interact to either (a) join teams, or (b) find teams to execute job(s). In [1] we have outlined the system based on the following tenets (for more details, see the Use Case diagram and the discussion presented in [8]):

- agents work in teams (groups of agents)
- each team has a single leader—*LMaster agent*

I. Lirkov, S. Margenov, and J. Waśniewski (Eds.): LSSC 2011, LNCS 7116, pp. 662–669, 2012.

- each *LMaster* has a mirror *LMirror agent* that can take over its job
- incoming workers (*worker agents*) join teams based on user-criteria
- teams (represented by *LMasters*) accept workers based on team-criteria
- each *worker agent* can (if needed) play role of the *LMirror* or the *LMaster*
- matchmaking is facilitated by the *CIC* component.

Let us now briefly focus our attention on interactions between components of the system, i.e. the *User* and its representative the *LAgent*, and agent teams represented by *LMaster* agents (more information can be found in [1]). Let us assume that team "advertisements" describing: (1) what resources they offer, and (2) characteristics of workers they would like to "hire," are posted with the *Client Information Center* (*CIC*). Here, we concentrate our attention on two main scenarios in the system: *User* is looking for a team (1) to commission job execution, or (2) to join (to be paid for use of her resources). In both cases, the *User* interacts with her *LAgent*, and formulates conditions for (1) job execution, or (2) team joining. The *LAgent* communicates with the *CIC* to obtain a list of teams that satisfy these criteria. Next, the *LAgent* communicates with *LMasters* of selected teams, and utilizes the *FIPA Iterated Contract-Net Protocol* [8, 12] to negotiate the contract. If the *LAgent* finds an appropriate team, a *Service Level Agreement* (*SLA*) is formed. If no such team is found, the *LAgent* informs the *User* and awaits further instructions. Let us now present the new design of the system, outlining of the initial implementation, and proposed solutions to specific technical problems.

3   Ontologies in the System

As stated above, we assume that all data in the system will be ontologically demarcated. Therefore, we needed a robust ontology, covering concepts ranging from descriptions of hardware and software, through grid structure, to the *SLA* and contract definitions. After a comprehensive investigation of existing grid ontologies (see, [2]) we decided to use the *Core Grid Ontology* (*CGO*, [10,11]). While the *CGO* provided us with excellent base-terms concerning grid resources and structure, we had to extend it to include the remaining concepts. The complete description of the ontology can be found in [2, 3]. The extended *CGO* (the *AiG Ontology*) is structured into three layers (its core classes depicted in Figure 1):

1. *Grid Ontology*—directly extending the CGO concepts.
2. *Conditions Ontology*—includes classes required by the SLA negotiations (e.g. pricing, payment mechanisms, worker availability conditions, etc.); it imports the *Grid Ontology*, to use terms related to the Grid structure and resources.
3. *Messaging Ontology*—contains definitions of messages exchanged by the agents, forming the communication protocols of the system (uses the *Grid Ontology* and the *Conditions Ontology* to specify content of messages).

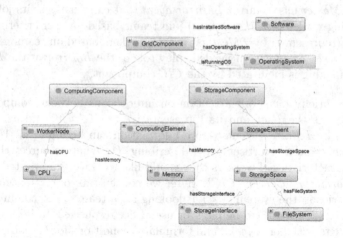

Fig. 1. Ontology diagram for AiG ontologies

The crucial aspect of ontological modeling was the representation of constraints. For example, when a user is looking for a team to have a job executed, she needs to specify the necessary hardware configuration. In this case, the common way of assigning values to class properties is not enough, as we also need to specify minimum, maximum, and range conditions. After considering several approaches, we have settled on class expressions. Here, requirements are defined as a new class that restricts the set of individuals to these satisfying conditions on class properties. We can thus ask a reasoner to infer a list of individuals of the generated class and receive the ones fulfilling the constraints.

4   Negotiations in the System

Obviously, automated negotiations, and *SLA* management, are the key part of system. The *SLA*, is a result of negotiations, and defines agreement reached by the parties. Here, by negotiation we understand flow of messages between parties: *LAgent*s and *LMaster*s. As stated in [8], the negotiation process is based on the *FIPA Iterated Contract-Net Protocol*, and involves both *negotiable* and *static* parameters specified by the *User* through the front-end (described in the next Section). These parameters are passed to the *LAgent*, which forwards them to the *CIC*), which responds with list *LMasters* representing potential partners. Next, the *LAgent* construct a *Call-For-Proposal* message with an OWL instance of required resource description—for the job execution scenario; or of resource that *User* wants to sell—for the team joining scenario. The *CFP* contains also restrictions on contract conditions (for both cases). This message is sent to the selected *LMasters*, and those interested in the proposal reply with OWL instances representing their offers. Both parties shall use multicriterial analysis to evaluate received proposals and make offers that take into consideration their own ability to fulfill required conditions, as well as preferences.

5   Front-End Design and Implementation

In the current implementation, the front-end subsystem is mostly a means to provide ontological data for the negotiations. It consists of three parts, allowing specification of requirements concerning:

1. *Scheduling job execution*—lets *User* specify constraints on conditions of the contract regarding job execution. These include hardware and software requirements that a team has to satisfy in order to be taken into consideration.
2. *Joining a team*—specify information needed for negotiating joining a team, i.e. description of available resources, and restrictions on the contract.
3. *Worker acceptance criteria*—also concerns worker joining a team. Here, the owner of the *LMaster* can specify conditions that must be met by any worker willing to join the team. These may include hardware and software configuration, as well as terms of contract.

In the initial system prototype, the front-end was a desktop application with the *LAgent* running in the background (on the same machine). Advantages of that approach included simple architecture, and ease of interactions between the client application and the agent. However, this also meant that: 1) the *LAgent* could only work while the front-end application was running, and 2) at least part of the *User*'s data was stored on the local machine. Therefore, interacting with the *LAgent* from different machines would be difficult. Since the possibility of accessing an application from any computer becomes a necessity, we decided to develop a web application that can be hosted in a shared environment.

The core of the front-end is a condition builder—a set of condition boxes, each representing a description or constraint on a single class-property relationship (see, Figure 2). Depending on the selected class, the *User* may choose one of properties that the class is in domain of. Next, she can specify an *operator*, from within the applicable ones, to the selected property. For example, for datatype properties these may include: *equal to*, or *greater than*, whereas for object properties these would be *is equal to individual* and *is constrained by*. When an operator is selected, the system generates a fragment of user interface used to specify value of the property. Again, controls depend on the selected operator—be it a simple text box, or a drop down list of applicable individuals.

Interesting is the case of nested constraints. For object properties, when the *User* selects the operator *is constrained by*, for a class to be further specified, a new condition box is created within the existing one. It is used to describe the details, or requirements, regarding the value of the selected property. Front-end supports also setting constraints on multiple properties of the same class, using *and* buttons, which add a new condition box at the same level as the previous.

When the *User* finishes specifying conditions and pushes the *submit* button, the system parses the internal representation of conditions, and transforms it into an *OWL Class Expression* (see, Figure 2). This OWL fragment is passed to a JADE *GatewayAgent*, responsible for passing information between the application server, and the JADE agent container. The *GatewayAgent* forwards the data to the *LAgent*, to handle it within the system.

(a) Condition builder

(b) OWL class

Fig. 2. A condition builder section

In the front-end, all elements from which the *User* builds the ontological conditions and descriptions are generated dynamically, from the structure of the ontology. Therefore, all changes to the ontology can be applied automatically during system runtime. This is extremely important, especially in the case of ontology matching and enriching, based on information received from other agents. It also simplifies maintenance of changes in the ontology.

Furthermore, user interface elements are build dynamically, as responses to user actions. For example, if the *User* wishes to specify a particular CPU architecture, individuals of the *CPUArchitecture* class will only be fetched from the ontology when the *User* selects an *equal to individual* condition on the *hasArchitecture* property. This allows us to process only the needed parts of the ontology. Moreover, it allows to base displayed options on *User*'s previous choices. Such functionality could be a mechanism for providing automated assistance, by suggesting the most useful or common options, or by filtering out inconsistent options.

6   Passing Ontological Information

Communication in the system relies on extracting information from, and manipulating instances of, ontologies. Unfortunately, the default codecs for ontological data, found in the JADE framework, are very limiting in terms of what kind of ontological data can be transferred as a part of the message.

In our case we found it essential to be able to transfer arbitrary fragments of OWL ontologies, including TBox definitions of classes, used for representing constraints and requirements. This problem has been discussed in [7], and resulted in creation of the JADEOWL Codec [6]. Unfortunately, this plug-in was extremely tightly integrated with the commercial RacerPro [17] reasoner, and its development seems to have stopped before the release of the OWL 2.0 specification. Therefore, we have developed our own JADE plug-in, aiming at providing OWL support to the agent message processing.

Direct mapping of OWL 2.0 [15] into any static object-oriented programming language is not possible; i.e. there is no way to represent OWL as Java classes while preserving its dynamic structure and properties (partial solution to this problem can be found in [9]). Therefore, as opposed to the JADEOWL Codec, we have decided that any information instance, such as information about teams or negotiation deals, will be stored and accessed as OWL formatted text files. Thus, the plug-in had to provide interface to files viewed both as raw text, and as OWL ontology; i.e. after passing a raw file, we had to be able to probe the structure of ontology, extract classes and instances, as well as their properties. In this way the plug-in can serve as a high level interface to the structure and content of ontological messages, passed between JADE agents.

In any communication scenario, data is prepared in the form of OWL class or instance. For example, advertising a team by the *LMaster* involves sending an instance of an OWL class (describing the team) to the *CIC*, which recognizes it as a team advertisement and stores it in an OWL file. When asked by the *LAgent*, it filters all stored instances, to satisfy specified constraints.

Although messages contain raw OWL data, their interpretation is done internally by the plug-in. In this way agents can access the information without the need to parse the text. This interpretation requires reasoning about the data, so an instance of a semantic reasoner is bundled with the communication plug-in. Currently, the HermiT [13] reasoner is used. However, OWL API supports also other popular reasoners (e.g. Pellet [16], Fact++ [14]).

To extract data (e.g. an OWL instance) from an ontology, a custom OWL class is created. The exact structure of the class depends on the data that needs to be extracted. For example, if the *CIC* is asked for agent teams with an IBM Linux machine, it sends information received from the *LAgent* to the plug-in. The plug-in creates an OWL class that extends the definition of OWL class describing team advertisements, but also contains an OWL property restrictions that forces any instance of this class to be a team with an IBM Linux computer. Other types of restrictions (like the cardinality restriction) supported by OWL 2.0 [15] are also available.

Here, the reasoner performs consistency and satisfiability tests on the new class in the context of the ontology. If the tests fail, it means that the class cannot have any instances. An exception is thrown and the reasoner output is routed back to the creator of the instance, to inform about the problem and, possibly, how to fix it. After passing the tests, the class prepared in this way is presented to the reasoner that finds all its instances. The prepared OWL instances are sent back (as text) to the *LAgent* that requested the information.

Summarizing, the plug-in aids creation of OWL classes and instances by producing and structuring the actual OWL text, while the reasoner (that is internal to the plug-in) performs validity/consistency checks and filtering. This solution makes full ontological communication available, while preserving constraints set upon ontologies in OWL format.

6.1 Reasoning in Team Selection

After user specifies input data (e.g. worker's description, team members acceptance conditions, etc.), it is passed to the appropriate *LAgent*. The *LAgent*, interprets the received message and locally stores the information. To obtain list of candidate teams, message with the *OWL Class* represented as the raw OWL data describing criteria is constructed, and passed to the *CIC* (see, Section 6). Reasoning in the back-end part of the system is required for both negotiating parties, i.e. the *LAgent* and the *LMasters*. In the proof of concept application, the *LAgent* utilized a linear-additive model for three predefined criteria, in an offer selection process. This model is a simple MCA model, therefore, in the redesigned system implementation we plan to use also other MCA methods to evaluate offers received by the *LAgent*. We are also going to consider additional criteria available in the ontology. On the other hand, *LMasters* use MCA to determine e.g. cost of job execution. Each resource needed for job execution e.g. memory, bandwidth has a pricing property in the ontology that specifies pricing type and price. To evaluate total price of job execution, the *LMaster* combines prices for each required component.

Reasoning is also used by the *LMasters* to verify if they are able to execute a given job i.e. if there is an available member in the team that has resources required to execute a job. So far, team members resource descriptions have been stored in *CIC* component, however, they will be stored also locally so that the *LMaster* can use reasoner on it's local ontological database.

7 Concluding Remarks

The aim of this paper was to discuss an outline of the implementations and solutions applied to selected technical problems within the scope of the *AiG* project. Currently, we are proceeding with testing of the above described components of the system, i.e. the front-end web application, the back-end agent-based application, and a communication bridge between these components—the OWL plug-in for ontology-based interactions.

Acknowledgments. Work of Maria Ganzha and Michał Drozdowicz was supported from the "Funds for Science" of the Polish Ministry for Science and Higher Education for years 2008-2011, as a research project (contract number N516 382434). The Polish-Bulgarian collaboration was partially supported by the

Parallel and Distributed Computing Practices grant. The Polish-Romanian collaboration was partially supported by the *Agent-Based Service Negotiation in Computational Grids*; and the *Agents, Grids and Heterogeneous Computing* grants.

References

1. Dominiak, M., Kuranowski, W., Gawinecki, M., Ganzha, M., Paprzycki, M.: Utilizing agent teams in grid resource management—preliminary considerations. In: Proc. of the IEEE J. V. Atanasoff Conference, pp. 46–51. IEEE CS Press, Los Alamitos (2006)
2. Drozdowicz, M., Ganzha, M., Paprzycki, M., Olejnik, R., Lirkov, I., Telegin, P., Senobari, M.: Ontologies, agents and the grid: An overview. In: Topping, B., Iványi, P. (eds.) Parallel, Distributed and Grid Computing for Engineering, pp. 117–140. Saxe-Coburg Publications, Stirlingshire (2009)
3. Drozdowicz, M., Wasielewska, K., Ganzha, M., Paprzycki, M., Attaui, N., Lirkov, I., Olejnik, R., Petcu, D., Badica, C.: Ontology for contract negotiations in agent-based grid resource management system. In: Iványi, P., Topping, B. (eds.) Trends in Parallel, Distributed, Grid and Cloud Computing for Engineering. Saxe-Coburg Publications, Stirlingshire (2011)
4. Kuranowski, W., Ganzha, M., Gawinecki, M., Paprzycki, M., Lirkov, I., Margenov, S.: Forming and managing agent teams acting as resource brokers in the grid—preliminary considerations. International Journal of Computational Intelligence Research 4(1), 9–16 (2008)
5. Kuranowski, W., Ganzha, M., Paprzycki, M., Lirkov, I.: Supervising agent team an agent-based grid resource brokering system — initial solution. In: Xhafa, F., Barolli, L. (eds.) Proceedings of the Conference on Complex, Intelligent and Software Intensive Systems, pp. 321–326. IEEE CS Press, Los Alamitos (2008)
6. Schiemann, B.: Jadeowl codec,
   http://www8.informatik.uni-erlangen.de/en/demosdownloads.html
7. Schiemann, B., Schreiber, U.: OWL DL as a FIPA ACL content language. In: Ferrario, R., Guarino, N., Vot, L.P. (eds.) Proceedings of the Workshop on Formal Ontology for Communicating Agents, pp. 73–80. University of Malaga (2006)
8. Wasielewska, K., Drozdowicz, M., Ganzha, M., Paprzycki, M., Attaui, N., Petcu, D., Badica, C., Olejnik, R., Lirkov, I.: Negotiations in an agent-based grid resource brokering systems. In: Iványi, P., Topping, B. (eds.) Trends in Parallel, Distributed, Grid and Cloud Computing for Engineering. Saxe-Coburg Publications, Stirlingshire (2011)
9. Xin-yu, Y., Juan-zi, L.: Research on mapping owl ontology to software code model. Computer Engineering 35(3), 36–38 (2009)
10. Xing, W., Dikaiakos, M., Sakellariou, R., Orlando, S., Laforenza, D.: Design and development of a core grid ontology. In: Proc. of the CoreGRID Workshop: Integrated research in Grid Computing, pp. 21–31 (2005)
11. Core grid ontology, http://grid.ucy.ac.cy/grisen/cgo.owl
12. Fipa iterated contract net interaction protocol specification,
    www.fipa.org/specs/fipa00030/PC00030D.pdf
13. Hermit owl reasoner, http://hermit-reasoner.com/
14. Owl: Fact++, http://owl.man.ac.uk/factplusplus/
15. Owl 2 web ontology language, http://www.w3.org/TR/owl2-overview/
16. Pellet: Owl 2 reasoner for java, http://clarkparsia.com/pellet/
17. Racerpro 2.0, http://www.racer-systems.com/

Using Blue Gene/P and GPUs to Accelerate Computations in the EULAG Model

Roman Wyrzykowski, Krzysztof Rojek, and Łukasz Szustak

Czestochowa University of Technology
Dabrowskiego 73, 42-201 Czestochowa, Poland
{roman,krojek,lszustak}@icis.pcz.pl

Abstract. EULAG (Eulerian/semi-Lagrangian fluid solver) is an established computational model developed by the group headed by Piotr K. Smolarkiewicz for simulating thermo-fluid flows across a wide range of scales and physical scenarios. This paper presents perspectives of the EULAG parallelization based on the MPI, OpenMP, and OpenCL standards. We focus on development of computational kernels of the EULAG model. They consist of the most time-consuming calculations of the model, which are: laplacian algorithm (laplc) and multidimensional positive definite advection transport algorithm (MPDATA).

The first challenge of our work was parallelization of the laplc subroutine using MPI across nodes and OpenMP within nodes, on the BlueGene/P supercomputer located in the Bulgarian Supercomputing Center. The second challenge was to accelerate computations of the Eulag model using modern GPUs. We discuss the scalability issue for the OpenCL implementation of the linear part of MPDATA on ATI Radeon HD 5870 GPU with AMD Phenom II X4 CPU, and NVIDIA Tesla C1060 GPU with AMD Phenom II X4 CPU.

1 Introduction

Eulerian/semi-Lagrangian fluid solver (EULAG) [3] is an established computational model for simulating thermo-fluid flows across a wide range of scales and physical scenarios. Its important features are: nonoscillatory integration algorithms, robust elliptic solver, and generalized coordinate formulation enabling grid adaptivity technology. The EULAG model is an ideal tool to perform numerical experiments in a virtual laboratory with time-dependent adaptive meshes and within complex, and even time-dependent model geometries.

In this work, we focus on parallelization of two computational kernels of the EULAG model. They consist of the most time-consuming calculations of the model, which are: laplacian algorithm (laplc) [9] and multidimensional positive definite advection transport algorithm (MPDATA). While the starting point of our development was BlueGene/P - the IBM supercomputer with an innovative massive parallel architecture [4], we focus finally on GPUs which nowadays become [2] extremely promising multi-core architectures for a wide range of general-purpose applications demanding high-intensive numerical computations.

I. Lirkov, S. Margenov, and J. Waśniewski (Eds.): LSSC 2011, LNCS 7116, pp. 670–677, 2012.

In our research we used the BlueGene/P machine, which is located in the Bulgarian Supercomputing Center. This supercomputer consist of two racks, that include 2048 PowerPC 450 processors, which gives 8192 cores. Each node contains 2 GB RAM (4 TB RAM for two racks). This configuration supports the single-precision peak performance of 27.85 Tflops.

Current GPUs are highly efficient, multi-core processors, which have the computing power of several Tflops. GPUs offer a fast, inexpensive solution, but understanding the parallel trade-offs is crucial. These architectures allow for creating many thousands of threads, which has a significant influence on performance of parallel codes. Data transfers between GPU memory and RAM are strongly limited by the PCI Express bandwidth. Available software such as OpenCL and CUDA facilitate the implementation of general-purpose computation on GPUs using high-level programming languages and tools [2,6].

The material of this paper is organized as follows. In Section 2, architectures of BlueGene/P and two graphics cards from leading vendors NVIDIA and AMD are presented. Section 3 introduces computational kernels of the EULAG model, while Section 4 is devoted to OpenCL, an emerging parallel programming standard for multicore architectures. The key issues of parallelizing the computational kernels of the EULAG model on the target architectures are discussed in Section 5, while results of numerical experiments are presented in Section 6. Section 7 gives conclusions, and outlines further work.

2   Architecture Overview

Our research is based on two kind of architectures. The first one is the BlueGene/P supercomputer, while the second one are GPUs. We focus only on features of these architecture which are used in our work.

2.1   Architecture of Blue Gene/P

Each node of BlueGene/P [4] is a single application-specific integrated circuit (ASIC) with four IBM PowerPC 450 (PPC450) embedded 32-bit processor cores, arranged as an SMP. Each core contains L1 cache of a 32 KB instruction cache, and a 32 KB data cache. A dual-pipeline floating-point unit (FPU) is attached to each PPC450 core. It supports two simultaneous double-precision floating-point calculations in SIMD fashion. The dual-pipeline FPUs can simultaneously execute two fused multiply-add instructions per machine cycle, each of which is counted as 2 flops. Thus, each processor unit (PPC450 and FPU) has a peak performance of 4 flops per machine cycle, and the BPC chip with quadruple processor units rates at the peak performance of 16 flops per cycle (850 MHz), or 13.6 Gflops. A BG/P compute chip integrates PPC450 cores with L2 and L3 cache, memory controllers, and external 10 Gb/s network interfaces.

2.2   Architecture of GPUs

Our research is focused on NVIDIA Tesla C1060 and ATI Radeon HD 5870.

Architecture of NVIDIA Tesla C1060. NVIDIA Tesla C1060 [7] includes 10 Thread Processing Clusters (TPC). Every TPC contains 3 compute units. Each compute unit consists of 8 processing elements, and 16KB of local memory. It gives a total number of 240 available processing elements with a clock rate of 1296 MHz. It provides a peak performance of $240 * 1.296 * 2 = 622 \, Gflops$ in single precision. This graphics accelerator card includes 4 GB of global memory with the peak bandwidth of 102.4 GB/s.

Architecture of ATI Radeon HD 5870. ATI Radeon HD 5870 [1] includes 20 compute units. Each compute unit consists of 16 processing elements, and 32KB of local memory. Each of the processing element is built of 5 streaming processors. It gives a total number of 1600 available streaming processors with a clock rate of 850 MHz, and provides the peak performance of $1600*0.850*2 = 2720 \, Gflops$ in single precision. This accelerator card includes only 1 GB of global memory with the peak bandwidth of 153.6 GB/s.

3   The Scope of Our Research on the EULAG Model

The most time-consuming calculations of the EULAG model are two algorithms: laplacian algorithm (laplc) and multidimensional positive definite advection transport algorithm (MPDATA). To identify the most intensive computing we used TAU and gprof profilers.

3.1   Laplacian Algorithm

The laplc routine is summarized in three stages [3,9]. The first one is responsible for computing pressure derivatives:

$$p_i^k = \frac{p_{i+1}^k - p_{i-1}^k}{2} \cdot \delta k, \tag{1}$$

where p_i^k is a pressure in the i-th point of a mesh, k determines the dimension of the mesh, in which i is incremented, and δk is a distance between points in the mesh. The second stage is based on the following equation:

$$f_i^k = -p_i^k \cdot c_i, \tag{2}$$

where f is an interior pressure force, and c_i is a coefficient function. The last stage is responsible for computing the laplacian function, which is finally expressed by the following equation:

$$r_i = -\sum_k \frac{f_{i+1}^k - f_{i-1}^k}{2 \cdot c_i} \cdot \delta k. \tag{3}$$

3.2   Linear Version of MPDATA

The multidimensional positive definite advection transport algorithm [8] is based on the following equation:

$$\Psi_i^{n+1} = \Psi_i^n - \frac{\delta t}{\nu_i} \sum_{j=1}^{l(i)} F_j^\perp S_j, \tag{4}$$

where Ψ is a nondiffusive scalar field, S_j refers both to the face itself and its surface area, ν_i is the volume of the cell containing vertex i, while $F_j^\perp$ is interpreted as the mean normal flux through the cell face S_j averaged over temporal increment δt.

The approximation begins with specifying fluxes $F_j^\perp$:

$$F_j^\perp = 0.5(v_j^\perp + |v_j^\perp|)\Psi_j^n + 0.5(v_j^\perp - |v_j^\perp|)\Psi_j^n, \tag{5}$$

where advective normal velocity $(v_j^\perp)$ is evaluated at the face S_j and assumes the following form:

$$v_j^\perp = S_j \cdot 0.5[v_i + v_j]. \tag{6}$$

4   OpenCL: Emerging Standard for Multicore Architectures

Open Computing Language (OpenCL) is an open, royalty-free standard for parallel programming of heterogeneous computing platforms including CPUs, GPUs, and other processors like Cell/B.E [6,5,10]. It defines the host API for coordinating parallel computation across heterogeneous processors, and a programming language. OpenCL allows for creating portable code across different devices and architectures. The OpenCL allows applications to use OpenCL platform as a single heterogeneous parallel computer system.

The OpenCL platform consists of CPU (host) and one or more graphics cards (compute devices). Compute devices execute functions called kernels. Kernels are instanced as work-items that are grouped in work-groups. Work-items are executed as SIMD or SPMD on processing elements. Work-items executing a kernel have access to distinct memory regions.

5   Accelerating Computations in the EULAG Model

There are different methods of parallelization and parallel programming, which can be used depending on a target computer architecture. In this work, we utilize commonly adopted standards of parallel programing: MPI and OpenMP for BlueGene/P, as well as the emerging standard OpenCL for GPUs.

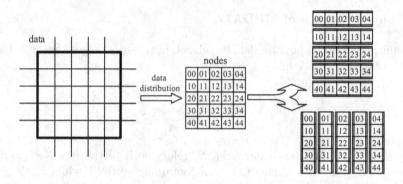

Fig. 1. Data distribution across array of nodes

5.1   Accelerating Computations in Laplc Kernel

One of the most efficient methods of parallelization is to use the message passing (MPI standard) across nodes and multithreading (OpenMP) within each node. In case of the laplc kernel, the first step is parallelization of the algorithm using MPI across a 2D array of nodes, having in mind to provide a load balancing. Fig. 1 shows data distribution across the array of nodes. We can extract two different ways of data distribution due to data dependencies. The first way is based on data dependencies across rows, while the second one is based on data dependencies across columns. To avoid communication between nodes, additional computations are required. Every data chunk is extended by additional rows or columns. Each additional row or column is the same as the row or column in the nearest neighbor chunk, respectively.

The second step is to add the parallelization with OpenMP applied for the laplc routine. This parallelization is implemented using 4 threads within one node. The main challenges for this step are:

1. avoiding reallocation of memory with every call of the routine;
2. possibility of creating OpenMP threads out of this routine;
3. providing shared access to memory by creating threads after memory allocation.

5.2   Accelerating Computations in MPDATA Kernel

The main challenge of parallelization of MPDATA on GPUs is decomposition of a 3D MPDATA mesh (N by M by L) into a 2D grid (N by M) of threads. Fig. 2 shows a 2D grid decomposition into work-groups. Work-groups are mapped into a 2D grid of size RN by CM. A work-group is a collection of work-items of size n=N/RN by m=M/CM. Each work-item computes a single element of the 2D grid (L elements of 3D MPDATA mesh). To avoid communication between work-groups, additional work-items are required. Every work-group is extended by additional rows and columns of work-items. Each additional row and column is the same as the row and column in the nearest neighbor work-group.

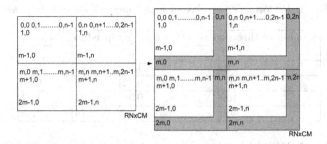

Fig. 2. MPDATA decomposition into work-groups

```
if(j<M && i<N)
    for(k=0; k<L; ++k)
    x(i, j, k)-=
    ( f1(i+1, j, k)-f1(i, j, k)+f2(i, j+1, k)-f2(i, j, k)
    +f3(i, j, k+1)-f3(i, j, k) )/h(i, j, k);
```

Fig. 3. The main part of MPDATA kernel

Fig. 3 presents the main part of the MPDATA kernel. Here $f1, f2, f3$ are computed using donnor-cell (upwind) scheme where $donnor = x - y$ for $x > y$ and 0 otherwise.

6 Performance Results

6.1 Performance Analysis for Laplc

The parallelization of the laplc routine was based on the hybrid model (MPI and OpenMP). The simulations were performed for different data mesh sizes (from 1000x1000 to 9000x9000) using 4, 16, 25, and 100 nodes, that gives 16, 64, 100, and 400 threads, respectively. After introducing some additional computation, the algorithm does not require any communication mechanisms between nodes, so the algorithm is very scalable. The implementation provides a very good load balancing up to 400 threads using the hybrid model. For the mesh of size 9000x9000, the execution time of calculation for 1, 16, 64, 100, and 400 threads are 41.5, 4.57, 0.72, 0.34, and 0.068 seconds, respectively.

Table 1 shows performance results for the laplc routine, for the mesh of size 9000x9000. The speedups for 16, 64, 100, and 400 threads are 9.07, 57.56, 121.91, and 592.14, respectively. In this case, the efficiencies are 0.56, 0.89, 1.21, and 1.48, respectively. The key reason for the super-efficiency is dividing data among nodes into smaller chunks. It improves fitting data-elements into available caches, enhancing cache reuse.

Table 1. Performance results for laplc (mesh of size 9000 by 9000)

Number of threads	Execution time [s]	Speedup	Efficiency
1	41.45	-	-
16	4.57	9.07	0.56
64	0.72	57.56	0.89
100	0.34	121.91	1.21
400	0.07	592.14	1.48

Table 2. MPDATA performance results (mesh of size 90x90x1500)

Hardware	Kernel time [s]	Kernel speedup	Kernel+ data rec. time [s]	Kernel+ data rec. speedup	Kernel+ transfer time [s]	Kernel+ transfer speedup	Host-GPU bandwidth [GB/s]	Memory usage [MB]
CPU	0.75	1	-	-	-	-	-	514.016
NVIDIA Tesla	0.041	18.29	0.06	12.5	0.16	4.68	2.57092	584.543
ATI Radeon	0.039	19.23	0.08	9.38	0.27	2.78	1.35215	584.543

6.2 MPDATA Performance Analysis

The algorithm was tested on the NVIDIA Tesla C1060 card with Linux and ATI Radeon HD 5870 with Windows7. The achieved results are compared with a single-core implementation on the AMD Phenom(tm) II X4 955 processor with Linux.

The MPDATA implementation distinguishes three stages. The first one is sending data, the second one is the kernel responsible for the computations, and the last one is receiving data. Table 2 shows performance results of the MPDATA routine for the mesh of size 90x90x1500. The kernel speedup for NVIDIA Tesla is 18.29, while for ATI Radeon is 19.23. The data transfers have a significant impact on the resulting execution time. Time of data transfers is several times larger than computation time. The achieved bandwidth of data transfer between host and global memory is 2.57 GB/s on NVIDIA Tesla and 1.35 GB/s on ATI Radeon, where the theoretical peak bandwidth is 4 GB/s. As a result, the speedup for NVIDIA Tesla and ATI has decreased to 4.68 and 2.78, respectively.

7   Conclusions and Further Work

The implemented parts of the EULAG model do not require any communication between threads, so the algorithms are very scalable. However, this approach requires some additional calculations for each thread. Our implementation provides a very good load balancing, when using the hybrid model on BlueGene/P, and OpenCL on GPUs.

The first challenge of our work was parallelization of the laplc code, using the hybrid model with MPI across nodes and OpenMP within nodes. This solution allows for a good usage of both shared and distributed-memory system resources, concerning memory capacity, latency, and bandwidth. Also, the hybrid model allows for the adaptation of the EULAG code to other hierarchical architectures such as clusters of multi-core processors.

The second challenge was acceleration of MPDATA using modern GPUs. The code was adopted to two types of GPUs from two leading vendors. NVIDIA TESLA was tested with Linux operating system, while ATI Radeon used Windows7. On ATI we achieved a little better performance of kernel computing, but a worse bandwidth of data transfers than on NVIDIA. In both cases, the performance on GPU is higher than on CPU.

Our parallelization of the EULAG model is still under development. One of leading approaches is using the autotuning technique which allows for algorithm self-adapting to properties of a system architecture. The final result of our work will be adaptation of the EULAG model to heterogeneous clusters with CPUs and GPUs. In this case, the key challenge will be to find an optimal load balance between GPUs and CPUs.

References

1. AMD Corporation: ATI Radeon HD 5870 Feature Summary, http://www.amd.com/
2. Dokken, T., Hagen, T.R., Hjelmervik, J.M.: An Introduction to General-Purpose Computing on Programmable Graphics Hardware. In: Geometric Modelling, Numerical Simulation, and Optimization: Applied Mathematics at SINTEF, pp. 123–161. Springer, Heidelberg (2007)
3. Eulag Research Model for Geophysical Flows, http://www.eulag.com/
4. IBM Blue Gene Team: Overview of the IBM Blue Gene/P project. IBM Journal of Research and Development 52, 199–220 (2008)
5. Khronos OpenCL Working Group: The OpenCL C++ Wrapper API, http://www.khronos.org
6. Khronos OpenCL Working Group: The OpenCL Specification, http://www.khronos.org
7. Lindholm, E., Nickolls, J., Oberman, S., Montrym, J.: NVIDIA Tesla: A Unified Graphics and Computing Architecture. IEEE Micro 28, 39–55 (2008)
8. Smolarkiewicz, P., Szmelter, J.: MPDATA: An edge-based unstructured-grid formulation. Elsevier Journal of Computational Physics 206, 624–649 (2005)
9. Sviercoski, R., Winter, C., Warrick, A.: Analytical approximation for the generalized Laplace equation with step function coefficient. J. Appl. Math. 68, 1268–1281 (2008)
10. Tsuchiyama, R., Nakamura, N., Iizuka, T., Asahara, A., Miki, S.: The OpenCL Programming Book. Fixstars Corporation (2010)

Author Index